Hydrologic Modeling: Statistical Methods and Applications

Hydrologic Modeling: Statistical Methods and Applications

RICHARD H. McCUEN
Department of Civil Engineering
University of Maryland

WILLARD M. SNYDER
Hydrologist
Athens, Georgia

PRENTICE-HALL, Englewood Cliffs, New Jersey 07632

Library of Congress Cataloging-in-Publication Data
McCuen, Richard H.
Hydrologic modeling.

Includes bibliographies and index.
1. Hydrology—Mathematical models. 2. Hydrology—Statistical methods. I. Snyder, Willard M. II. Title.
GB656.2.M33M45 1985 627'.0724 85-25637
ISBN 0-13-448119-4

Editorial/production supervision and interior design: *Theresa A. Soler*
Cover design: *20/20 Services, Inc.*
Manufacturing buyer: *Rhett Conklin*

Printed in the United States of America

10 9 8 7 6 5 4 3 2 1

ISBN 0-13-448119-4 025

Prentice-Hall International (UK) Limited, *London*
Prentice-Hall of Australia Pty. Limited, *Sydney*
Prentice-Hall Canada Inc., *Toronto*
Prentice-Hall Hispanoamericana, S.A., *Mexico*
Prentice-Hall of India Private Limited, *New Delhi*
Prentice-Hall of Japan, Inc., *Tokyo*
Prentice-Hall of Southeast Asia Pte. Ltd., *Singapore*
Editora Prentice-Hall do Brasil, Ltda., *Rio de Janeiro*
Whitehall Books Limited, *Wellington, New Zealand*

To

My brothers, Howard and Bob

R.H.M.

To

My wife, Pat

W.M.S.

Contents

PREFACE xv

PART I FUNDAMENTALS

1 FUNDAMENTALS OF MODELING 1

Introduction 1
Concepts of Statistical Hydrologic Models 3
Development of Models 5
Model Evaluation 6
Problems and Models 8
Modeling and Information Content of Data 10
Summary, 15
Exercises 15
References and Suggested Readings 16

2 **FUNDAMENTALS OF PROBABILITY AND STATISTICS** **18**

Introduction 18

Statistical Frequency Distributions 18

Mathematical Requirements of Frequency Distributions, 21
Samples versus Populations, 24
Moments of a Distribution, 25

Common Probability Functions 31

The Normal Distribution, 31
The Student t *Distribution, 32*
The Chi-Square Distribution, 32
The F *Distribution, 32*
The Binomial Distribution, 33

Confidence Intervals 34

Tolerance Limits 35

Hypothesis Testing 42

Test on a Single Mean, 43
Test of a Distribution Function, 44

Fundamentals of Simulation 49

Random Number Generation, 50
Statistical Tests of Random Numbers, 52
Simulation as a Decision-Making Tool, 53
The Data Analysis Process, 63

Exercises 63

3 **FUNDAMENTALS OF OPTIMIZATION** **68**

Introduction 68

Definitions 69

Elements of Optimization 70

Classification of Optimization Methods 71

Analytical Optimization 72

The Principle of Least Squares, 73
Lagrangian Optimization, 74

Reliability of a Prediction Equation 77

The Correlation Coefficient, 77
The Standard Error of Estimate, 78
The Analysis of Variance, 79
The Standardized Partial Regression Coefficients, 80

Assumptions Underlying the Regression Model, 81

Composite Fitting 83

Applications in Hydrologic Analysis 86

Numerical and Subjective Optimization 94

Exercises 94

PART II UNIVARIATE METHODS

UNIVARIATE ESTIMATION **98**

Introduction 98

The Univariate Modeling Process 99

Elements of Frequency Analysis, 101

Estimation of Model Parameters 101

Types of Parameters, 102
Method of Moment Estimation, 103
Maximum Likelihood Estimation, 105

Construction of Probability Paper 108

Normal Probability Paper, 109
Extreme-Value Plotting Paper, 110

Forms Used as Frequency Distributions 111

Normal Distribution, 111
Poisson Distribution, 114
Gamma Distribution, 116
Pearson Type III Distribution, 119
Extreme-Value Distribution, 121

Moment Ratio Analysis 126

Data Transformations 130

The Log-Normal Distribution, 130
Log Pearson Type III Distribution, 131
The Power Transformation, 131
Change of Variable, 133

Risk Versus Return Period 134

Evaluating the Accuracy of Prediction 135

Tolerance Limits by Simulation 138

Additional Considerations in Univariate Analysis 142

Exercises 144

References and Suggested Readings 148

5

SEMIVARIOGRAM ANALYSIS AND KRIGING 150

Introduction 150

The Estimation Problem 151

Semivariogram Analysis 154

Semivariogram Models, 157
Estimating Semivariogram Parameters, 160
Anisotropy and Boundary Effects, 160
Spatial and Volumetric Estimation, 162

Estimation of Standard Errors 169

Case 1: Point/Point Estimation, 170
Case 2: Point/Vector Estimation, 171
Case 3: Line/Point Estimation, 173
Case 4: Line/Vector Estimation, 176
Case 5: Line/Line Estimation, 176
Case 6: Field/Point Estimation, 177
Case 7: Field/Vector Estimation, 182
Case 8: Field/Line Estimation, 183

Estimation by Kriging 184

Exercises 192

References and Suggested Readings 196

PART III OPTIMIZATION

STATISTICAL OPTIMIZATION 198

Introduction 198

The Calibration and Evaluation of Linear Models 199

The Principle of Least Squares, 201
Goodness of Fit, 204
Matrix Notation, 208
Prediction and Prediction Variance, 212

Stepwise Regression Analysis 218

Model Structure, 218
Forms of Stepwise Regression, 219
Criteria for Model Selection, 223

Regression Analysis of Nonlinear Models 233

Polynomial and Power Model Forms, 234
Transformation and Calibration, 235

Goodness of Fit, 236
The Analysis of Variance for Polynomial Models, 237
Trigonometric Transforms, 243
Circular Variates, 245
Additional Nonlinear Model Forms, 249
Segmented Regression Functions, 249

Exercises 253

References and Suggested Readings 261

7

NUMERICAL OPTIMIZATION **262**

Introduction 262

The Phases of Numerical Optimization 264

Phase 1 Search Procedures 266

The Phase 2 Search 269

Numerics for Nonlinear Least Squares, 269
Variations on Nonlinear Least Squares, 285

A Phase 3 Analysis 286

Applications in Hydrologic Analysis 290

Exercises 300

References and Suggested Readings 302

8

SUBJECTIVE OPTIMIZATION **304**

Introduction 304

An Introductory Example 305

Calibration of a Simple Watershed Model 308

Model Formulation, 309
Model Calibration, 313
Measures of Performance, 318

Application with Hydrologic Data 320

Model Formulation, 320
Assessing Model Performance, 323
Model Calibration, 324
Discussion, 349

Calibration of a Retention Function, 351

Exercises 357

References and Suggested Readings 359

PART IV ADDITIONAL TOPICS

TIME SERIES ANALYSIS AND SYNTHESIS 362

Introduction 362

Components of a Time Series 364

Moving-Average Filtering 365

Autocorrelation Analysis 373

Cross-Correlation Analysis 378

Identification of the Random Component 380

Formulation of Auto- and Cross-Correlation Water Yield Models 380

Autoregression and Cross-Regression 382

Computational Example, 382
Application to Watershed Data, 384

Exercises 398

References and Suggested Readings 401

10 SENSITIVITY ANALYSIS AND PROBABILISTIC MODELING 403

Introduction 403

Mathematical Foundations of Sensitivity Analysis 404

Definition, 404
The Sensitivity Equation, 404
Computational Methods, 405
Parametric and Component Sensitivity, 405
Absolute and Relative Sensitivity, 406
A Correspondence between Sensitivity and Correlation, 407
Time Variation of Sensitivity, 409

Sensitivity in Model Formulation 410

Stability of an Optimum Solution 412

Sensitivity of Initial Value Estimates 414

Sensitivity and Data Error Analysis 416

Limitations of Sensitivity Analysis 431

Probabilistic Modeling 432

Exercises 438

References and Suggested Readings 439

MULTIVARIATE MODELS **442**

Introduction 442

The Effect of Cross Correlation 444

The Basis of Multivariate Analysis 444

The Simple Two-Variable Relationship 445

Multiple-Variable Relationships 448

Eigenvalue–Eigenvector Analysis of a Matrix, 448
Direct Solution of the Characteristic Equation, 449
Iterative Numerical Method, 452
Rotation of Axes, 453
Components Analysis, 456
Illustration of Principal Components Analysis, 458

Components Regression 459

Factor Analysis, 477

Nonlinear Analysis, 492

Summary 504

Exercises 505

References and Suggested Readings 505

APPENDIX **508**

INDEX **565**

Preface

We can only reason from what is;
we can reason on actualities, but not on possibilities.
Henry St. John Bolingbroke (1678–1751)

Art and science have their meeting point in method.
Edward George Bulwer-Lytton (1803–1873)

This book is an attempt to present a few of the computational tools of the modeler. The emphasis has been placed on the computational techniques rather than on the theoretical foundations of the techniques. The tools presented will be of more value to the modeler concerned with analysis and interpretation of real-world data than to the user concerned with an elaboration of the theory.

Recorded hydrologic data are the actualities on which the hydrologist can reason. These data are the known facts that serve as the basis for formulation of new hypotheses. They are also the means by which we verify new hypotheses.

Hydrologic data are, for the most part, obtained by recording events as they happen on watersheds. Watersheds are extremely complex natural systems. Thus the data obtained from them are complex. The role of the hydrologic data analyst is to separate the response of the watershed processes from the inherent noise. This sorting out can be viewed in the perspective of extracting information from recorded data. Rarely are hydrologic data recorded in forms needed for verifying hypotheses. Rarely are they recorded in a form most useful for planning or design. Statistical hydrologic modeling can be defined as the development of forms to extract useful information and the quantification of those forms through optimization to recorded data. This extraction process becomes more important as more hydrologic data are collected, especially when the hydrologic data are just a part of the data base for a physical system. One only needs to think of the large data bases collected either as part of the National Urban Runoff Program or using remote sensing systems to see the need for an understanding of tools for statistical hydrologic modeling.

In addition to emphasizing the computational techniques, the book includes numerous examples that illustrate the application of the tools to actual data. Although the tools are applicable for the analysis of data from any branch of engineering or science, data from the hydrologic sciences were selected because of the authors' interests and experiences in this specialty. In spite of this emphasis, there is no reason that the text could not be of value to someone in other branches of engineering and science. The case studies used were selected for the simplicity of their theoretical basis; thus someone who lacks a background in physical hydrology should not have difficulty in understanding the interpretation of the case studies.

A brief review of the table of contents should suggest that the text covers a broad array of computational tools. An entire book could be devoted to each of the topics, and the reference/reading lists at the end of some of the chapters provide a list of texts and papers devoted to each topic. However, the intent of this book is not to discuss each topic in detail. Instead, a purpose of the book was to provide a survey of the computational tools that are available to the hydrologic modeler and to show how the methods fit together in the decision-making process. This should provide the student of hydrologic modeling with the confidence to study texts that deal with the topics in greater theoretical detail. We have sacrificed the details in order to provide the reader with an overview of the topics and a description of the way that the pieces of the puzzle fit together.

In addition to being separated into 11 chapters, the book was divided into four parts. Part I is intended to be a review of computational tools in statistical analysis and modeling. For those with prior experience or education in the fundamentals of statistical methods, Part I may not be necessary. Part II concentrates on the computational tools used in univariate analyses. Given the importance of calibration in hydrologic modeling, Part III is devoted to three levels of optimization. Three levels are required because of the diverse complexity of hydrologic models, from the simple linear model to the conceptual watershed simulation model. Part IV discusses additional tools with which the hydrologic modeler should be familiar. Hopefully, this separation into both parts and chapters will lead to a better understanding of the way the methods fit together. The flowcharts in Chapter 2 are intended to provide a more descriptive summary of the modeling process.

Richard H. McCuen
Willard M. Snyder

1 Fundamentals of Modeling

INTRODUCTION

Hydrology is essentially a science based on imperfect observation in a complex and sometimes discontinuous domain. The range of hydrologic information is tremendous. The upper ranges include the vast scale of the meteorological phenomena, where information is based on sampling at relatively few points in a great continuum of space and time. The lower ranges include minute details of soil and water interaction. The material of primary interest, water, is nonuniformly diffused in a complicated space. Geometrical description of this space domain must extend from the drainage net of a large river system to the open spaces in a block of soil.

It is a matter of controversy as to how much of hydrology can be regarded as a science based on rigorous statement of cause and effect, and how much is to be relegated to the unexplainable, and called probabilistic, random, or stochastic. It is not the purpose here to probe into the philosophic question of whether any effect can develop without some cause. Much of the controversy in regard to hydrologic determinism stems from two differing viewpoints of the principle of determinism. Hydrologic determinism can be regarded as an abstract, but absolute definition. Any hydrologic event is capable of absolute definition as a rigorously stated consequence of causing processes. Any event, or series of events, not capable of such absolute cause-and-effect definition must be random. A second point of view results, however, when one attempts to use the foregoing definition operationally as a method of classifying events as determinable or random. Classification is done by one who has imperfect, not absolute, understanding. An attempt at classification reflects what hydrologists know, or what they think they know, at the time the classification takes place. A human being is not intellectually static. We learn, we gain understanding, and we transmit knowledge to our own and to subsequent generations. What is imperfectly understood today is better understood tomorrow. If, then, on any day, we have an imperfect explanation of an event, and tomorrow we have a better explanation, to that degree we have moved toward science and away from stochasticism.

In practical hydrologic data analysis it is necessary to consider that a portion of the information in the data is explainable, understandable, and definable. By the very process of formulation and quantification of an informational form the hydrologist factors out that which is explainable. Some residual variability in the data will be unexplainable. Hydrologists can use only that form which represents our present understanding. Poor correspondence between form and data will cause residual unexplained variability. Additionally, lack of precise measurements and human error will cause unexplainable variability. Finally, the inability of the hydrologist to describe exactly the complex real-world watershed and its inputs and outputs can cause nonreproducibility of experiments. Practicality dictates that all unexplainable variabilities in data be treated as randomly generated noise. The noise apparent in hydrologic data may be reduced through increased knowledge of the physical system being modeled, through more detailed models of the system, and through more comprehensive and accurate data collection. Random variation that is considered as noise is not necessarily a reflection that the system is truly stochastic.

Learning with statistical hydrologic models could be defined as progress from concepts of pure stochasticism toward concepts of pure determinism. Hydrologists learn from the information that is extracted from data. As our understanding of watershed processes improves we can formulate better models to apply to data. Eventually, however, the development of form will be limited by the random noise content of the data. Progress can then be made only by design of experiments to generate better data, and by development of means to purge data of as much noise as possible.

The design of hydrologic experiments will normally produce data by definable sets. For example, streamflow may be recorded before and after some physical change in the watershed. Data could be separated by runs of wet and dry years. A pool of data may be obtained from large and small watersheds, from urban watersheds and forested watersheds. To learn the informational makeup of these different data sets, it is necessary that all phases of recording, reduction, and analysis be performed in a systematic, invariant, and unbiased manner. A statistical model, objectively quantified, serves as a control to satisfy these requirements.

CONCEPTS OF STATISTICAL HYDROLOGIC MODELS

This section explores the specific characteristics of models intended for use in hydrologic data analysis. Most of these concepts were introduced previously in general terms. Command of analytical models, however, requires an understanding of their relationship to other model types as well as their function.

Almost everyone has a definition of the word *model.* The hobbyist builds a scale model of a railroad engine. The hydraulic engineer builds a physical model to answer questions about the prototype structure. These are examples of material or physical models. The hydrologic models that we will explore are not material models. They are, instead, formalized ideas, or concepts. Conceptual models are of many kinds and are used in many disciplines. We will study a limited conceptual type, statistically oriented hydrologic models.

Hydrologic models are best defined rigorously in relation to the concept of a system. We will use the following definitions (Diskin, 1970):

> *System*: A system may be considered to be an ordered assembly of interconnected elements that transform, in a given time reference, certain measurable inputs into measurable outputs. Inputs and outputs are usually represented as functions of time. These functions may be continuous or discrete.
>
> *Models*: Models are simplified systems that are used to represent real-life systems and may be substitutes of the real systems for certain purposes. The models express formalized concepts of the real systems.

Statistical models consist of vectors of two elements: input variables and empirical constants. There are two types of variables: the criterion variable and the predictor variables. The criterion variable is the variable for which a predicted value is necessary in design. Predictor variables are variables for which values are measurable and can be used to make estimates of the criterion variable. The drainage area, slope, and 2-year, 2-hour rainfall are examples of predictor variables common to hydrologic models, whereas the peak discharge and monthly water yield are examples of criterion variables. Values for the empirical constants are derived by fitting the model with measured values of the criterion and predictor variables. These constants are called either coefficients or parameters. By definition, parameters and coefficients are:

> *Parameter*: A numerical measure of a property or characteristic that is constant for a system under specified conditions.
>
> *Coefficient*: A variable or constant appearing in a mathematical model, each value of which defines the specific form of the model.

It should be evident from the generality of these definitions that they can be used interchangeably.

The systems we will deal with are real-world watersheds. Such systems receive natural inputs of water and energy, and perhaps human inputs of chemicals, such as fertilizers and pesticides, and release outputs in such forms as quantity and quality of streamflow, evapotranspiration, and other losses of material and energy through advection and radiation. Such input-output processes can be expressed rigorously for simple situations. But watersheds are not simple. Even small watersheds are exceedingly complex. For reasons of practicality we must conceive of simpler, but adequate processes. The formalization of these simplified concepts of processes produces the hydrologic models that we substitute for the real-world watershed.

Whether or not a model is an adequate substitute for the system can be determined only in relation to its purpose. We use models to predict the elements of information we need for reconnaissance, planning, design, operation, and maintenance, of the many facets of human interaction with the natural environment. The model used for such prediction should be the simplest, and probably the cheapest, that will serve the purpose.

Three approaches to modeling are conventionally recognized (DeCoursey, 1971; Snyder, 1971; Woolhiser, 1971). The *stochastic* approach uses the simplest concepts of watershed processes, although not necessarily the simplest mathematics in formalization of those concepts. In the stochastic approach watershed outputs are thought of as a time series of random events. Stochastic elements are not without physical basis. Certainly, some properties of the time series, such as mean values and variabilities, must derive magnitudes from the watershed in which the stochastic generating processes are at work. The essence of stochastic processes is the nonpredictability of exact magnitudes of each element of the series.

At the opposite boundary of predictability are the *deterministic* models. In these models, given values of initial and boundary conditions, a set of input values will always produce exactly the same output values. The generating processes contain no random components. Operationally, however, deterministic models link with the stochastic models, since to predict outputs for the future, the possible future inputs must be stochastically generated.

Parametric models are compromise models in that they contain both stochastic and deterministic component processes. Deterministic models must evolve from rigorous application of physical and chemical principles, to accomplish the structural definition that nothing is left to chance. Such rigor is virtually impossible for real-world watersheds, so simpler watersheds are defined where rigor is attainable. The parametric approach starts from conceptualization of processes on the real watershed, and through rigorous numerical techniques applied to observed inputs and outputs, attempts separation of deterministic and stochastic components. Secondarily, in the parametric approach, the deterministic components derived are associated with the predominant physical characteristics of the watershed.

Statistical hydrologic models evolve from all three approaches to watershed modeling. Stochastic models have mathematical-statistical parameters that express specific properties of probability density functions of the random processes. Such parameters must be evaluated empirically. Deterministic models contain empirically defined parameters such as Manning's *n*, soil permeability, or diffusivity coefficients in turbulent transfer processes. Parametric models usually contain functions of macroscale watershed processes that are expressed by forms intended for empirical quantification. Therefore, any model built or modified to obtain optimum values of any of its elements through rigorous statistical procedures will be considered a statistical hydrologic model.

DEVELOPMENT OF MODELS

Four stages are usually recognized in the development of models. These stages are conceptualization, formulation, programming, and testing. *Conceptualization* covers by far the greatest bulk of effort in development of new models. This stage means the composite of all the thought processes we go through in analyzing a new problem and devising a solution. Probably 80% of the total effort is in conceptualization. It is an indefinable mixture of art and science (James, 1970). Science provides an information base for conceptualization. Included in this base are models developed for previous problems, intellectual abstractions of logical analyses of processes, and mathematical and numerical methodologies, as well as a hard core of observed hydrologic reality. However, the assembly into an "adequate and efficient" problem solution is an art.

Tremendous freedom in practical conceptualization was gained when electronic computers became everyday working tools. Conceptualization with no hope of activation for prediction is meaningless in applied science. The high-speed capacity of the computer opened up entirely new fields of mathematical and numerical analyses. These new techniques, in turn, allowed wider scope in conceptualization. The next stage in model development, formulation, no longer placed so severe a limitation on conceptualization. It was no longer necessary to gear practical conceptualization to such limited formulations as closed integrals, or desk calculator numerics.

The *formulation* stage of model development means the conversion of concepts to forms for calculation. Formulation seems to imply a mathematical equation, or formula. Certainly, equations may be model forms. However, any algorithm, or sequence of calculations converting inputs to outputs, is an acceptable formulation. Such a sequence may contain equations, it may contain graphical relations, it may include "table look-up." It may also include highly sophisticated finite difference solutions of differential equations. Formulation is somewhat like preparation of a detailed flowchart of a problem.

Programming covers the mechanical but highly skilled efforts to translate the computational forms into a computer language. Testing, the last stage in model development, covers all the steps taken to ensure that no errors are present and that computation forms and program structure are satisfactory.

In a wider context *testing* could also mean examination of the model to see how well it performs against recorded data. The goodness of performance is rather simple to quantify for rigorously stated testing criteria. On this quantified scale of goodness, however, it is more difficult to decide how good is good enough.

Obviously, before a model can be tested it must be quantified. The model must be able to accept numerical inputs and convert these to numerical outputs. Some numerical goodness tests can then be devised to show how well the computed numerical outputs match observed numerical outputs. An obvious question arises. How should models be quantified prior to testing? A partial, but again obvious answer is that quantification and testing should go hand in hand. Testing the model on less than some best values of its parameters is meaningless.

Numerical values of parameters can be chosen by a series of assumptions, a trial-and-error process, until some satisfactory level of computed versus observed output test is achieved. Alternatively, the values can be assigned, such as assigning laboratory values of soil permeability to soil in situ. These two methods do not make use of observed outputs. If the model does not test satisfactorily, these methods do not contain features for systematic reevaluation of the parameters to new, and output-conditioned, better values.

An objective of statistical hydrologic modeling is the formulation of means whereby information in the observed outputs is used to quantify the empirical elements of a model. Such procedures are usually called *optimization techniques*, implying optimum values of the parameters for the given set of outputs.

MODEL EVALUATION

In the preceding section we introduced the idea that the last stage of model development, testing, must involve numeric manipulations of the model. Such manipulations mean finding best values of the model parameters as well as use of the model to predict outputs. We will find in later chapters that improvement of prediction is a means to better estimation of parameters.

We previously learned that models contain variables and parameters. Simplistically, parameters can always be evaluated from given or observed sets of variables. For example, we can evaluate the equation for a straight line by passing it through two points. By setting the x and y of the equation equal to the x value and y value of the two given points in turn, we get two simultaneous equations which can be solved for the two parameters of the line. Although this illustration is trivial, it is fundamental to all techniques of numerical evaluation. To generalize, we can state that known values of variables are substituted into the models on a one-to-one basis. One set of variables is needed for each parameter to be evaluated. This basic procedure includes many transformations. If we have observations on rates of change, obviously we can substitute these variables into the differential of our model with respect to time. If we have information on accumulated values, obviously we can substitute these values into appropriate integrals of the models. In all these primary procedures, however, a one-to-one correspondence is maintained. For each parameter to be evaluated, we must have a set of observations.

Practical evaluation of hydrologic models is a much more complicated situation than the condition of one-to-one correspondence above. Whenever we use observed data we must expect some degree of error in these observations. Errors range from simple mistakes, through lack of precision, to inherent variability in data. In the presence of errors we cannot rely on any particular data sets in one-to-one correspondence. We must, instead, find a way to utilize many sets of observations in an attempt to average across the inherent errors. A primary purpose of this text is to present modeling as the development of forms to utilize known averaging techniques. As models of increasing complexity are developed in this text, the reader will find primary emphasis placed on least-squares techniques. Least-squares methods are not the only means of optimization. They are, however, generally simple to apply. They are extendable to complex model forms. In certain situations a least-squares solution has as a by-product a measure of precision.

The tremendous advantage of the least-squares methods is their ability to produce stable reduced data in the presence of errors. Such stability implies an averaging process across a sufficient field of information so that local errors in the field are suppressed. It is well known that the extremely simple process of computing the arithmetic mean by summing and dividing by the number of elements produces such a stable average. It will be found that the principle of least squares provides an extremely powerful averaging and proportioning process.

The need for an averaging process as the means of numeric evaluation is due to an inherent characteristic of hydrology and, in fact, is identical to the need for statistical hydrologic models. Great care has been taken to this point to instill in the reader the idea that the model form is a result of careful and logical effort. Insofar as the model represents physical process, it may be thought of as containing a deterministic core. However, by use of this core one cannot expect to calculate exactly the value of some hydrologic element. Errors will always be present that will prevent such exactitude. The logical structure of the model is, therefore, always surrounded by a cloud of error elements. Numerical evaluation must penetrate this cloud and search for some "best" estimate of the logical core.

The inherent error portion of a statistical model comes from several sources. First, we cannot hope to produce exact hydrologic variables by data processing methods. Take the simple example of rainfall averaged across a drainage area. All we can do is compute the weighted average of rain caught in all gages in the area. We then assume that this is the areal average of rain on the basin. In point of hard fact, we only know what rain was caught in each gage; we do not know exactly how much rain fell on the drainage area.

Another source of error is in the form of the model itself. In the more complex model forms we often need mathematically continuous functions relating variables. In a large proportion of these instances such functions are not known and, indeed, are not possible of exact expression. What exact function, for example, can relate intensity of rainfall to time during a storm? What function can express exactly the elevation of all points in a drainage area using latitudinal and longitudinal coordinates?

Another form of model error results from the great physical scale of logic that the surface water hydrologist is forced to employ. When proceeding according to deterministic principles, one can use exceedingly small scales. For example, a spatial property might be expressed in terms of a differential equation based on an infinitesimal control area. However, the value of the property across finite space must be calculated by a double integration of the equation. This is possible if the property is continuous in the space, or if piecewise integration between discontinuities is permissible. Usually, the discontinuities will be interfaces of two functions of the property. When these two interact through the interface, integration across the discontinuity becomes difficult and complex.

The hydrologist faces the problem of discontinuous space empirically. We can formulate the concept of the effect of the discontinuous space on a hydrologic property. This concept can be incorporated as a component part of a model, and evaluation will give numerical form to the space effect. But to formulate the concept initially, the hydrologist must reason at the macroscale of the space, not at the microscale of the infinitesimal control volume. Needless to say, such a macroscale should reflect such homogeneities as may be found. Total space can sometimes be broken into smaller subspaces with the linkages between subspaces expressed computationally as component parts of a total model.

PROBLEMS AND MODELS

We have discussed models from the viewpoint of definitions, development, and evaluation. We will now look at various types of problems and explore the use of statistical models in each. Dooge (1973) has prepared a classification of systems problems, and we will follow this classification. The words model and system are interchangeable in an operational sense. A model, operationally, is a simple system used in place of a complex system.

Dooge's method of problem classification is based on two major categories, problems of analysis and problems of synthesis. Analysis, which means to break apart, is the objective examination of a given system. The given system has been conceptualized and formalized. In this state it can be used numerically in problems concerned with analysis of form or effect. Three types of problem analysis are possible if we have two of the three elements of input, transfer function, or output. Given inputs and the transfer function we can predict outputs. This problem analyzes the consequences of the given system in operation. Given inputs and outputs we can quantify the given transfer function and analyze magnitudes of the components for the particular physical domain that produced the outputs. Given outputs and a quantified transfer function we can compute inputs. This procedure allows us to analyze a given problem in detection of inputs. We could use this problem type to detect input measurement errors. We might search, for example, for likely pollutant sources in a watershed. Since systems are usually made up of subsystems we could work backward through the last subsystem before output. Computing the input to the last subsystem gives us the output from the next-to-last subsystem.

Synthesis, which means to put together, refers to the design or establishment of a new model. In analytical problems we are given the form of the transfer function and we utilize it. In synthesis problems we search for the needed transfer function. In hydrology we search for a simplified but formalized representation of the complex processes in a real watershed. Our models are the objective of the search for the simplified systems. Analysis is the process of breaking a given totality into essential elements for better understanding. Synthesis is the reverse. It is the assembly of components into a desired totality. Analysis is based on scientific and mathematical method. Synthesis is scientific and mathematical methods combined with generous portions of the art of design.

Three problem types can be recognized in synthesis, just as in analysis. Table 1-1 is an adaptation of Dooge's classification of analysis problems. Given inputs, and a tentative model, we can simulate possible outputs. The intent is to establish a new information base from which to view the adequacy of the design. Given outputs and the tentative model we can generate inputs. This is a procedure that allows us to design an experiment, estimating needed precision of inputs to test the hypothesis of the system. Given inputs and outputs we can add numerical dimension to our tentative conceptualization of the system.

TABLE 1-1 Classification of System Synthesis Problems Adapted from Dooge (1973)

	Element		
Synthesis Problem	Input	System	Output
Simulation	√	√	?
Conceptualization	√	?	√
Experimentation	?	√	√

In summary, we may use the same computational procedures in problems of analysis and synthesis. However, the logical processes are entirely different. In analysis we use deductive logic to proceed from the generality of the given system to the particular of its operation in a specific domain. In synthesis we use inductive logic, attempting to infer the general nature of the system from the particulars of input and output. Statistical hydrologic modeling has found greatest utility in quantification and conceptualization.

System design and system analysis are not substitutes for research. Research must still be regarded as careful addition to the storehouse of human knowledge. Statistical hydrologic modeling is not a substitute for system design and analysis. Statistical hydrologic modeling is a tool to use in system design and analysis. Further, system design and analysis add new dimension and perspective to research. Under the systems concept we conceive of research as inquiry into the total nature of real processes, not as study of artificially isolated parts.

MODELING AND INFORMATION CONTENT OF DATA

Hydrologic modeling can be viewed as a means of getting useful information about watersheds. As stated earlier, we develop models because we need to predict certain elements in the management and utilization of water resources. We need to know how much water is available, when and where it is available, when, during floods, it will be potentially dangerous, and how we, deliberately or necessarily, might threaten its quality.

Such utilitarian predictions are based on watershed processes. Watersheds, for the most part, are natural units of drainage. Only in metropolitan areas have human beings made significant changes in natural drainage patterns. Geomorphological processes over long periods of time tend to establish a balance between the forces of flowing water and the surfaces and channels where it is flowing. In many disciplines models can be constructed that ignore or minimize surface features of the earth. Economic models and population models need have no basis in watersheds. However, when we predict water or water-related phases of our environment, we must rely on nature's water system, the watershed.

The only rational prediction for any watershed is a prediction based on information about that watershed. Prediction is possible when the inputs and the system are known. Obviously, the watershed information leads to the system information to make the prediction. It is also obvious that information on watershed processes comes from watersheds, either the one where we need to predict, or some other, with, hopefully, similar features. In this context we can view modeling as an exercise in organizing the watershed information we need to use in prediction. If we have a formulated model, we can apply it to many sets of input-output data, as a problem in quantification. The quantified parameters are information that we extract from our given data sets. These parameters, modified to the features of our prediction watershed, are a means to transfer information from one watershed to another where we need to use it.

Example 1-1: Monthly Mean Water-Yield Model

Some simple examples of data processing and analysis will serve to convert the abstract processes presented above to more tangible procedures. Consider the problem of finding feasible sites to locate an industrial plant that will need some minimum supply of water. Suppose that the decision is made to consider only streamfront sites where the average streamflow for the month is equal to or above the necessary minimum. A prediction model has now been conceptualized.

Implicitly, we say that future flows, after the plant is built, will average the same as past flows. We need, however, to quantify this model. What are the monthly streamflow values at various sites? Gage heights may have been observed for many years. Our model, however, requires that gage heights be converted to discharges, and that discharge be integrated by calendar months. Averages for January, for February, and for every other month can then be computed, and our elementary model is quantified. Mathematically, the monthly mean model is

$$\hat{R} = \bar{R}_i \qquad \text{for } i = 1, 2, \ldots, 12 \tag{1-1}$$

in which $\bar{R}_i$ is the monthly mean water yield for month i and $\hat{R}$ is the predicted value of water yield for any month.

This quantification of the elementary model of monthly average streamflow is about the simplest processing of data to extract information that can be imagined. The usual requirements in hydrologic data processing involve establishing and numerically evaluating relationships between variables. The variables are the inputs and outputs. The relationships express system operation. The standard error of the mean would be an indicator of the accuracy of the model of Eq. 1-1.

Example 1-2: Stochastic Water-Yield Model

Consider a slightly more complex model to use in our plant-site selection problem. We now decide that simple averages of runoff are not sufficient, and that we want to consider the possible future variability of runoff about this average value. We have now decided on a stochastic model. We conceptualize a runoff process in which mean and variability from the past will repeat in the future. We can evaluate this model, and we can predict with it, as we will see in other chapters of this text. Note that we must extract from the recorded data information on both the means and variabilities. Mathematically, the stochastic model is

$$\hat{R} = \bar{R}_i + \epsilon_j \qquad \begin{array}{l} \text{for } i = 1, 2, \ldots, 12 \\ \text{for } j = 1, 2, \ldots, n \end{array} \tag{1-2}$$

in which ϵ_j is the value of a random variable with a mean of zero and a standard error equal to standard deviation of the monthly mean; the distribution function of ϵ_j would also have to be specified. The model of Eq. 1-2 is more complex than Eq. 1-1 because it is necessary to quantify ϵ_j before the model can be used to predict.

Example 1-3: Conceptual Water-Yield Model

Consider next an extension of our monthly flow prediction problem. We want to consider locating our industrial plant on a stream where we have no flow records (i.e., an ungaged location). Assume that we have usable rainfall records and also have flow records on some adjacent streams. With this limited data base we conceptualize a simple rainfall-runoff model, by considering runoff to be a proportionate part of rainfall that remains constant regardless of the amount of rain. This concept allows us to write

$$\hat{R} = kP \tag{1-3}$$

in which $\hat{R}$ is the predicted runoff for the month, P the precipitation for the month, and k a proportionality constant.

Before proceeding to use the model in Eq. 1-3, we should mentally review the conceptual basis. The model has one very obvious fault. It predicts that runoff is a proportionate part of the total rain that falls. Now, simple observation tells us that this is not so. Many small rains may occur during a month that will not produce measurable streamflow. Rain is caught by trees and buildings. Some may be trapped in street puddles. Only enough may fall to wet the surface of the ground and vegetation. No appreciable runoff results from such small rains. Similarly, no appreciable runoff results from the first rainfall burst during a significant storm.

Our rainfall-runoff concept must be modified to account for an inadequacy that we know it contains by our preset state of hydrologic knowledge. We visualize now that a small amount of rain each month is trapped by the watershed. We do not know how much this will be for a particular month, but we can think of an average amount around which the actual trapped amounts will vary. Our concept now takes the algebraic form

$$\hat{R} = \begin{cases} k(P - P_0) & \text{for } P > P_0 \quad (1\text{-}4a) \\ 0 & \text{for } P \leq P_0 \quad (1\text{-}4b) \end{cases}$$

in which P is the monthly rainfall, P_0 a threshold value equal to the amount of rain needed before runoff begins, and k a proportionality or scaling constant.

The rainfall-runoff model of Eq. 1-4 cannot be used for prediction until values for k and P_0 are set. If measured rainfall and runoff data are not available, values of k and P_0 must be conceptualized from general hydrologic fundamentals. For example, macroscale studies have suggested that 75% of rainfall evaporates and 25% of rainfall appears as runoff. Thus we may assume that the value of k may be on the order of 0.25. Macroscale studies of interception losses suggest that 25% of rainfall is intercepted. Thus we may elect to set P_0 equal to 25% of the mean monthly rainfall. Given these conceptualized values of k and P_0, we could use Eq. 1-4 for predicting water yield $\hat{R}$.

Given a model such as Eq. 1-4, we are not only interested in the predicted value of R; we also want to know how well the model predicts. We can test the model of Eq. 1-4 using the data of Appendix C-1 for the Chattooga River. For the moment, let's assume that P_0 is 1.25; thus Eq. 1-4 becomes

$$\hat{R} = \begin{cases} 0.25\,(P - 1.25) & \text{for } P > 1.25 \text{ in.} \quad (1\text{-}5a) \\ 0 & \text{for } P \leq 1.25 \text{ in.} \quad (1\text{-}5b) \end{cases}$$

We can test the adequacy of the model for 1954. The results are given in Table 1-2.

TABLE 1-2 Prediction for Water-Yield Model (in.) for the Chattooga River, 1954

Month	Measured Rainfall	Measured Runoff, R	$\hat{R}$	$e = \hat{R} - R$	$\hat{R}_1$	$e_1 = \hat{R}_1 - R$
1	11.92	6.36	2.67	−3.69	6.85	0.49
2	4.83	3.63	0.90	−2.73	2.60	−1.03
3	7.40	4.28	1.54	−2.74	4.14	−0.14
4	4.57	4.34	0.83	−3.51	2.44	−1.90
5	2.62	3.10	0.34	−2.76	1.27	−1.83
6	4.24	2.30	0.75	−1.55	2.24	−0.06
7	3.68	1.40	0.61	−0.79	1.91	0.51
8	4.23	1.13	0.75	−0.38	2.24	1.11
9	0.44	0.65	0.00	−0.65	0.00	−0.65
10	0.31	0.55	0.00	−0.55	0.00	−0.55
11	3.54	0.84	0.57	−0.27	1.82	0.98
12	8.41	2.05	1.79	−0.26	4.75	2.70
				−19.88		−0.37

It is evident that the predicted runoff values are considerably less than the measured values, with a mean error of −1.66 in. per month. Thus the model of Eq. 1-5 does not appear to be highly accurate for predicting monthly water yield for the Chattooga River watershed.

Example 1-4: Quantification of the Water-Yield Model

Mathematically, we can manipulate Eq. 1-4a and write it in terms of the equation

$$\hat{R} = kP - kP_0 \tag{1-6}$$

We see now that our equation has the form of a simple straight line, usually written $y = a + bx$. In another chapter we will see how to get numerical values for the a and b of the straight-line equation. This amounts to finding values for k and P_0 in Eq. 1-4, since $k = b$ and $P_0 = a/k$. Now k and P_0 have physical meaning as defined for Eq. 1-4. If we apply our model to streams surrounding our plant location, we extract from these records the numerical physical information of k and P_0. We can then consider using these values of k and P_0 with rainfall for the plant site to predict runoff for the site.

If we plot the rainfall and runoff data of Table 1-2, we can detect a trend in the data (see Fig. 1-1). We will see in later chapters how to systematically evaluate the coefficients k and P_0 of Eq. 1-6; for present purposes we will simply draw a line to represent the "average" relationship between R and P. From this line we get the equation

$$\hat{R}_1 = 0.6(P - 0.5) \tag{1-7}$$

in which the subscript 1 on $\hat{R}$ in Eq. 1-7 is used only to indicate that the model of Eq. 1-7 is different from the model of Eq. 1-5. Equation 1-7 was used to predict values of runoff, with the predicted values and errors (e_1) shown in Table 1-2. It is evident from the graph of Fig. 1-1 that Eq. 1-7 provides a much better method of prediction than does Eq. 1-5. The value for P_0 of Eq. 1-7 indicates that the losses associated with P_0

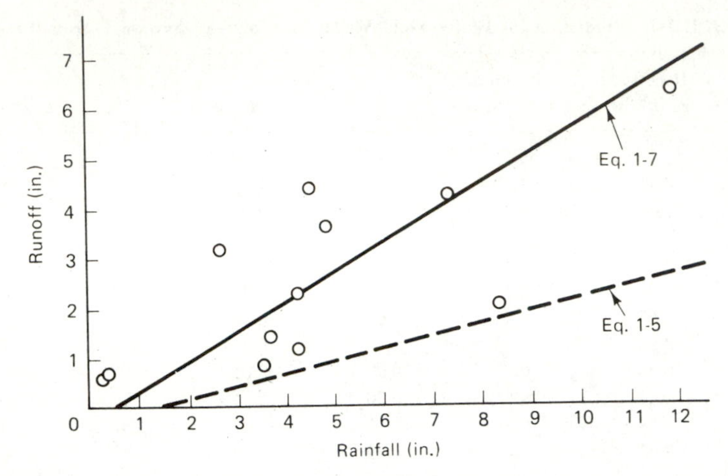

Figure 1-1 Monthly water-yield relationship for the Chattooga River watershed for 1954.

are much less than those estimated using the average interception of 25% of the rainfall. Also, the larger value of k in Eq. 1-7 suggests that a larger portion of the rainfall would appear as runoff. The smaller average error of -0.03 in. per month for Eq. 1-7 would also suggest that Eq. 1-7 is a better model than Eq. 1-5 to use for prediction on the Chattooga River. Thus the quantification of the water-yield model using measured data from the watershed where the model was to be used vastly improved the accuracy of the model.

Example 1-5: Water-Yield Predictions at Ungaged Sites

Example 1-4 showed that a model quantified using on-site data provides more accurate estimates than does a model developed solely from conceptualization. But a model

TABLE 1-3 Water-Yield Predictions (in.) for Wildcat Creek, 1954

Month	Rainfall	Runoff, R	Predicted Runoff, $\hat{R}_1$	$e = \hat{R}_1 - R$
1	6.43	2.99	3.56	0.57
2	2.22	1.20	1.03	−0.17
3	4.22	1.35	2.23	0.88
4	3.06	1.02	1.54	0.52
5	2.50	0.62	1.20	0.58
6	4.27	0.38	2.26	1.88
7	2.63	0.15	1.28	1.13
8	1.62	0.09	0.67	0.58
9	0.51	0.09	0.01	−0.08
10	0.44	0.05	0.00	−0.05
11	3.23	0.24	1.64	1.40
12	3.54	0.38	1.82	1.44
				8.68

user must also be interested in how well the model will predict when the model is transferred to other sites. The model of Eq. 1-7 can be used to predict water yield for the Wildcat Creek watershed (see Appendix C-3). The monthly rainfall for Wildcat Creek for 1954 (the same year for which Eq. 1-7 was quantified) was used to predict the water yield. The results are shown in Table 1-3. In general, Eq. 1-7 tends to overpredict runoff, with an average error of 0.72 in. per month. This suggests that the values of k and/or P_0 from the Chattooga River may not accurately reflect the monthly water yield characteristics of the Wildcat Creek watershed.

Summary

A series of examples have been used to illustrate the types of problems that arise in hydrology and the modeling that is associated with the problem type. Model synthesis and analysis have different purposes. Within these two problem types the modeling effort depends on the information that is known and unknown. Knowledge of this framework for modeling provides a basis for understanding the limits of modeling.

EXERCISES

1-1. Equation 1-4 includes two variables, rainfall P and runoff R. **(a)** Explain how the accuracy of the predicted runoff $\hat{R}$ might differ for seasons when rainfall is generated by widespread cyclonic patterns versus summertime convective showers. **(b)** Would accuracy be expected to vary with topography? If so, how? **(c)** How might runoff accuracy vary between research watersheds with calibrated weirs and watersheds with open-channel rating curves?

1-2. Sets of field measurements of soils are sometimes described as composed of structured and unstructured information. Would stochastic, deterministic, or parametric models be more appropriate for analysis of such sets?

1-3. A watershed is characterized by a constant loss function of 0.07 in./hr. **(a)** If runoff is the rainfall in excess of 0.07 in./hr, what is the total runoff from a 5-hr storm with hourly rainfall increments of 0.08, 0.36, 0.27, 0.05, and 0.18 in.? **(b)** Construct a simple algebraic equation for this conceptual model. **(c)** Suppose that the percolation loss of 0.07 in./hr was unknown, but the volume of runoff was known. How could you find the percolation loss constant?

1-4. If a statistical analysis indicates that the "best" model for the P versus R data of Table 1-2 is $\hat{R} = 0.5686 + 0.4237P$, what would be the values of k and P_0 of Eq. 1-4? Are these physically rational? If not, what does this imply about the model structure?

1-5. Develop a simple, but physically rational model for predicting monthly evaporation (inches) as a function of monthly temperature (°F). **(a)** Evaluate the coefficients of the model using fundamental concepts of heat transfer. **(b)** Show how to evaluate the coefficients of the model using regional climatic information available in most textbooks on general hydrology. **(c)** Graph the 1954 evaporation and temperature data for the Chattooga River watershed (Appendix C-1) and use the graph to evaluate the coefficients of the model. **(d)** Estimate the model accuracy for parts (a), (b), and (c). How could each model be revised to improve the accuracy? **(e)** Use the model to estimate the 1954 evaporation for the Wildcat Creek watershed (Appendix C-3), and assess the accuracy of the model.

REFERENCES AND SUGGESTED READINGS

AMOROCHO, J., Measures of the Linearity of Hydrologic Systems, *Journal of Geophysical Research*, Vol. 68, No. 8, pp. 2237–49, April 1963.

BROWN, J. A. H., Data Needs Aquisition, and Availability for Hydrologic Models, in *Prediction in Catchment Hydrology*, T. G. Chapman, and F. X. Dunin (eds.), Australian Academy of Sciences, 1975.

BURAS, N., Mathematical Modelling of Water Resources Systems, *Hydrological Sciences Bulletin* (International Association of Hydrological Sciences), Vol. 19, No. 4, pp. 393–400, December 1974.

CLARKE, R. T., A Review of Some Mathematical Models Used in Hydrology, with Observations on Their Calibration and Use, *Journal of Hydrology*, Vol. 19, No. 1, pp. 1–20, 1973.

DECOURSEY, D. G., The Stochastic Approach to Watershed Modeling, *Nordic Hydrology*, Vol. 2, No. 3, pp. 186–216, 1971.

DISKIN, M. H., Research Approach to Watershed Modeling, Definition of Terms, *ARS and SCS Watershed Modeling Workshop*, Tucson, Ariz., March 1970.

DOOGE, J. C. I., *Linear Theory of Hydrologic Systems*, Technical Bulletin No. 1468, U.S. Department of Agriculture, U.S. Government Printing Office, Washington, D.C., October 1973.

DORFMAN, R., Formal Models in the Design of Water Resource Systems, *Water Resources Research*, Vol. 1, No. 3, pp. 329–36, 1965.

JACKSON, D. R., and G. ARON, Parameter Estimation in Hydrology: The State of the Art, *Water Resources Bulletin*, Vol. 7, No. 3, pp. 457–72, June 1971.

JAMES, L. D., Watershed Modeling: An Art or a Science? Paper presented at Winter Meeting, ASAE, Chicago, 1970.

LAURENSON, E. M., Modeling of Stochastic-Deterministic Hydrologic Systems, *Water Resources Research*, Vol. 10, No. 5, pp. 955–62, October 1974.

MEYER, C., Using Experimental Models to Guide Data Gathering, *Journal of the Hydraulics Division, Proceedings of the ASCE*, Vol. 97, pp. 1681–97, October 1971.

MEYER, C., Surrogate Modeling, *Water Resources Research*, Vol. 8, No. 1, pp. 212–216, February 1972.

MIHRAM, A., The Modeling Process, *IEEE Transactions on Systems, Man and Cybernetics*, Vol. SMC-2, No. 5, pp. 621–29, November 1972.

SNYDER, W. M., The Parametric Approach to Watershed Modeling, *Nordic Hydrology*, Vol. 2, No. 3, 1971.

SNYDER, W. M., and J. B. STALL, Men, Models, Methods, and Machines in Hydrologic Analysis, *Journal of the Hydraulics Division, ASCE*, Vol. 91, No. HY2, pp. 85–99, 1965.

WITZ, J. A., Integration of Systems Science Methodology and Scientific Research, *Agricultural Science Review*, Vol. 11, No. 2, 1973.

WOODWARD, D. E., Hydrologic and Watershed Modeling for Watershed Planning, *Transactions of the ASAE*, Vol. 16, No. 3, pp. 582–84, 1973.

WOOLHISER, D. A., The Deterministic Approach to Watershed Modeling, *Nordic Hydrology*, Vol. 2, No. 3, 1971.

YEVJEVICH, V., Determinism and Stochasticity in Hydrology, *Journal of Hydrology*, Vol. 22, pp. 225–38, 1971.

2 Fundamentals of Probability and Statistics

INTRODUCTION

It is not the intent of this chapter to provide an in-depth discussion of fundamental methods in probability and statistics. Instead, the material in this chapter is intended as a review of the fundamentals that are necessary to understand the concepts discussed in the other chapters. More detailed discussions of the fundamentals of probability and statistics are provided in other textbooks.

In addition to the review material, the last section of this chapter is presented in the hope that students of hydrologic modeling will recognize that the material presented in the chapters of this text does not represent disjointed topics. Therefore, flowcharts are presented to show how the statistical methods tie together for decision making with hydrologic modeling. These flowcharts should be reviewed periodically as material in the other chapters is studied so that the reader understands just where the new material fits into the decision-making process.

STATISTICAL FREQUENCY DISTRIBUTIONS

We begin with descriptions and definitions of statistical frequency distributions. These distributions are systematic arrangements of measured variates that we presuppose to be basic to analytical hydrology. They are described as systematic because usually our measurements have a limited range of variability and tend to cluster about some central value. For example, we can measure some property, such as the size of a drainage area on a map. If we measure this drainage area several times or have several people measure it once each, we do not get the same value each time it is measured. Hopefully, however, the values would fall to either side and close to the true value of the drainage area. Also, hopefully, more of the measurements would fall close to this central value than would fall at a greater distance from it. The first distribution we will discuss will be built up from a contrived situation but which will nonetheless offer several valuable elementary ideas.

Consider a made-up mixture of a sand to be used in a laboratory permeability experiment. We will imagine that there are 10 different size classes of sand grains in the sample, and an equal number of grains in each size class. Figure 2-1(a) is a bar graph of a small mix of 50,000 grains, 5000 each in 10 different sizes uniformly ranged. In statistics, Fig. 2-1(a) is considered a *population* because it is a complete portrayal of all the sand grains in the mix.

Suppose, now, that we ask the question: What proportion of the total number of grains falls into each size class? Obviously, the answer is 5000/50,000, or $\frac{1}{10}$. Figure 2-1(b) shows a histogram with 10 cells, each having a cell height of $\frac{1}{10}$. The abscissa is still labeled by grain size, but the ordinate has now been called "probability density." This is a subtle change in meaning brought about by an anticipated change in our viewpoint about the ratios we have just computed. As a ratio, $\frac{1}{10}$ describes our sand population. But suppose that we consider drawing a few grains from the sand population for closer inspection. These few grains will be referred to as a *sample.* What are the chances of drawing a sand grain from particular class? The answer, of course is 1 chance in 10, or $\frac{1}{10}$. Thus our viewpoint changed to one of chance or risk inherent in the distribution of grain sizes. The ordinate is labeled "probability" to designate "chance in the draw." It is, in more detail, labeled "probability density" because the height of each cell is actually the density of chances per draw per grain-size class.

The meaning of probability density will be clearer with consideration of the construction that leads to Fig. 2-1(c). Construction proceeds as follows. We ask the question: What are the chances, in the draw of a single sand grain, of getting one in the range 1.0 to 1.1 mm or smaller in size? Since there are none smaller than 1.0 mm, the answer is $\frac{1}{10}$. Now, what are the chances of getting a grain in the range 1.1 to 1.2 mm or smaller? We might draw one in the first class, giving us a $\frac{1}{10}$ chance, or we might draw one in the second class, giving another $\frac{1}{10}$ chance, for a total of $\frac{2}{10}$. This is the height of the cell from 1.1 to 1.2 mm [Fig. 2-1(c)]. We can build up to higher probabilities of "equal or smaller" as we consider larger grain sizes. When we reach range 1.9 to 2.0 mm, the probability of getting one in this range or smaller is 1.0.

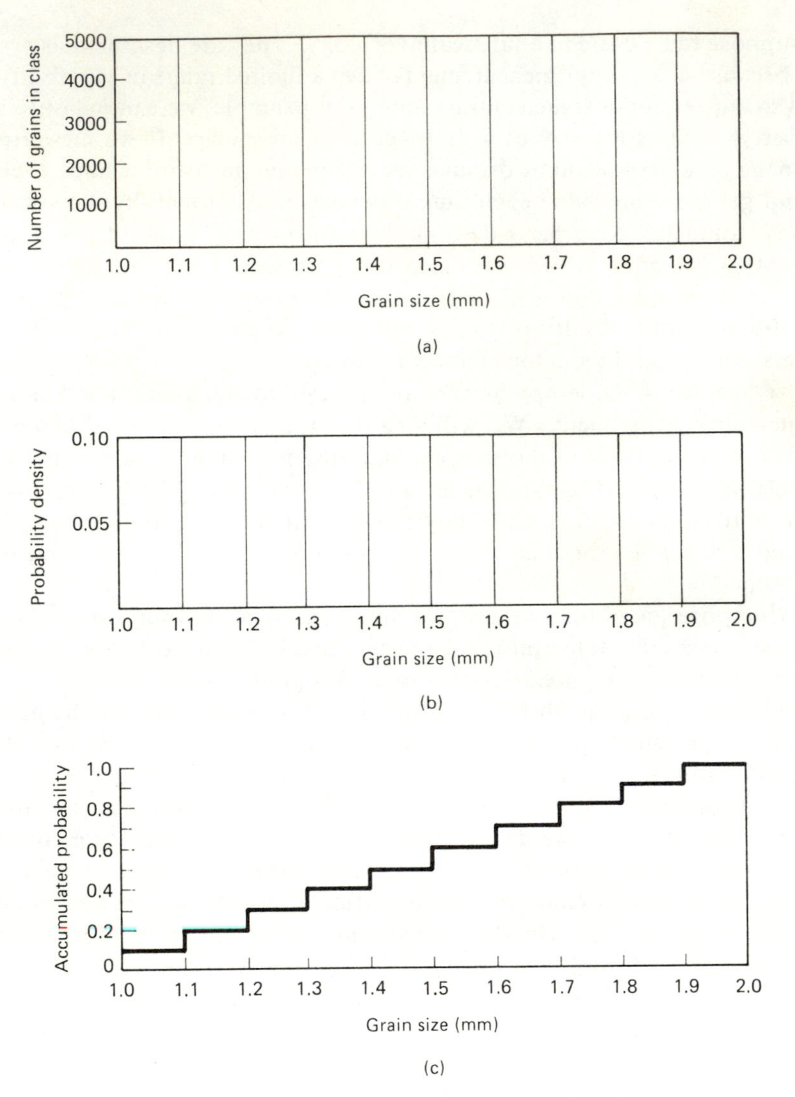

Figure 2-1 Linear distribution of sand grains: (a) distribution by classes; (b) probability distribution by classes; (c) summation of probabilities.

Mathematical unity thus means certainty, with no risk of failure of the draw. While the plot of Fig. 2-1(b) is called a *probability density function*, the plot of Fig. 2-1(c) will be referred to as a *cumulative probability function*, since it is the accumulation of probabilities.

For our purposes, *probability* will mean, as in Fig. 2-1(c), separation of chance into "equal or smaller" and "greater." The probability of "greater" can always be computed by subtracting the probability of "equal or smaller" from 1.0. The probability curve is the integral (or sum or cumulative) of the probability density curve. Conversely, the probability density is the slope of the cumulative probability curve. If we made our grain-size classes narrower and narrower until we reached the infinitesimal concept of the calculus, the probability of "equal" in size would disappear. Then the curve in Fig. 2-1(c) would be a smooth line and the stepped increments would disappear. Probability under this concept of "continuous" distribution of grain sizes then reduces to a concept of "smaller than" or "greater than" for any point on the size scale.

Mathematical Requirements of Frequency Distributions

The distribution of grain sizes in Fig. 2-1(a) is but one of many possible forms which can be used to describe the systematic arrangement of variable measurements. This particular distribution is called *linear* or *uniform* because the frequency (i.e., the number of grains in each size class) is constant across all classes.

A different form of frequency distribution is given in Fig. 2-2(a). For the isosceles triangle, the area enclosed by the triangle is ba. Therefore, the triangle represents a total of ba individual measurements, just as the linear distribution in Fig. 2-1(a) represented 50,000 individual sand grains. Figure 2-1(b) was calculated by dividing each grain-size cell by 50,000. Similarly, we can calculate the probability density function, Fig. 2-2(b), by dividing each ordinate by ab. Because division will not change the triangular form, we need only divide the ordinate a, producing $1/b$.

Figure 2-1(c) was calculated by summing Fig. 2-1(b) from the left side. We can calculate the probability function Fig. 2-2(c) by integrating since Fig. 2-2(b) is a mathematically continuous form. Let us call the scale of measurement x. Considering x zero at the left, we can write

$$P_x = \int_0^x p\,dx = \frac{1}{b^2}\int_0^x x\,dx = \frac{x^2}{2b^2} \qquad \text{for } 0 \le x \le b \tag{2-1a}$$

$$\begin{aligned} P_x &= \frac{1}{2} + \int_b^x p\,dx = \frac{1}{2} + \int_b^x \left(\frac{2}{b} - \frac{x}{b^2}\right) dx \\ &= -1 + \frac{2x}{b} - \frac{x^2}{2b^2} \qquad \text{for } b \le x \le 2b \end{aligned} \tag{2-1b}$$

Here P_x is the probability of an individual lying to the left of x, p the probability density of x, and x the scale of measurement. All the ba measurements fall in the range $0 \le x \le 2b$.

We should now begin to probe a little more deeply into possible hydrologic meaning of distributions such as those in Fig. 2-2. Suppose that the distribution of measurements shown in Fig. 2-2(a) was formed from the record of some meteorological variable. For example, the variable might be the dew-point temperature observed each morning. The lowest dew point (driest

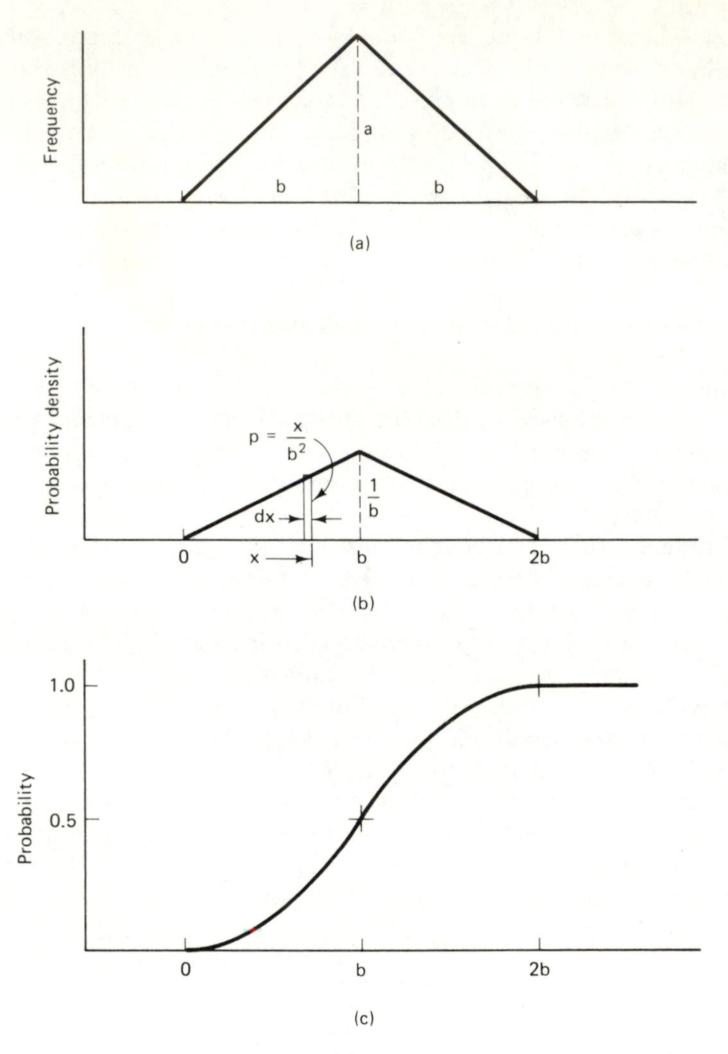

Figure 2-2 Triangular distribution: (a) frequency distribution; (b) probability density function; (c) probability function.

air) ever observed would fall just to the right of the point $x = 0$. The highest dew point would fall just to the left of $x = 2b$. In between these values we could collect, or tally, the number of recordings falling in some number of equal-width subranges, say each 5° interval between $x = 0$ and $x = 2b$. We presuppose a tendency for the dew points to cluster around the central value, and further that we can then "bound" or approximate the actual record by a triangle superimposed on it.

One simple way to superimpose the triangle is to let its area, ab, equal n, the number of observations. Then a will be n/b and will have the units of observations per degree. We will learn more about such superimposed approximations later.

We might be interested in how often, in the recorded dew points, the actual dew points fell in the range $\frac{1}{2}b \leq x \leq \frac{3}{2}b$, or in the central half of the total range. It can readily be seen from Eq. 2-1 that the probability of a dew point lying to the left of $x = \frac{1}{2}b$ is $\frac{1}{8}$. The probability of a dew point lying to left of $x = \frac{3}{2}b$ is $\frac{7}{8}$. The probability of a dew point lying between $x = 1/b$ and $x = \frac{3}{2}b$ is thus $\frac{6}{8}$, or 75%.

We will consider further mathematical properties of frequency curves with one more function. The descending exponential (Eq. 2-2) is a very common form in hydrology as well as in other disciplines:

$$p = C e^{-Ax} \tag{2-2}$$

In this equation C and A are positive parameters and e is the exponential.

To use this curve as a probability density function, we require that it enclose an area equal to unity. The curve will be restricted to positive values of x. The requirement for unit area is expressed by the equation

$$1 = C \int_0^\infty e^{-Ax}\, dx = C\left(-\frac{1}{A} e^{-Ax}\right)_0^\infty = \frac{C}{A} \tag{2-3}$$

Thus unit area requires that $C = A$, and this substitution can be made in Eq. 2-2. A statistical frequency function, enclosing n elements, or recorded values, can be obtained simply by multiplying Eq. 2-2 by n:

$$np = nA\, e^{-Ax} \tag{2-4}$$

An exponential probability function is calculated by integrating Eq. 2-2, such as in

$$P = A \int_0^x e^{-Ax}\, dx = 1 - e^{-Ax} \tag{2-5}$$

The exponential probability function (Eq. 2-5) has an interesting property. This will provide an exercise in probability even if it does not solve, at the moment, any real problems. Consider a point on the probability density function where the density has decreased to one-half the maximum value, A, which occurs at $x = 0$. This half-value point falls at the x value given by

$$\frac{A}{2} = A\, e^{-Ax} \tag{2-6}$$

and x is given by

$$e^{-Ax} = \tfrac{1}{2} \tag{2-7}$$

By putting this value into Eq. 2-5, we get $P = \frac{1}{2}$. The half-value density therefore marks the point where 50% of the records are larger and 50% are smaller.

Samples versus Populations

In the previous sections several elementary properties of statistical distributions were introduced. First, the concept of a graph of the statistical elements was presented. The histogram of the 50,000 sand grains is an example. Because this collection of grains was specified as the complete count, this histogram is a statistical population. Thus the probability density function and the cumulative probability function developed from the histogram also represent populations.

The introduction of the triangular distribution gave two new concepts. First, the idea of the continuous distribution, as opposed to a discrete bar for each class, was introduced. Second, the idea of substituting the continuous triangle for some triangular-shaped bar chart was discussed. No attempt was made to designate these triangles as populations or samples. Neither of the two concepts is pertinent to distinguishing between population and sample. Population distributions may be discrete or continuous, and we may move, within limits, from one to the other by suitable mathematical transforms.

A sample is part of a *population.* It is a partial, or incomplete, count of the statistical items under consideration. Whenever we deal with a total count, we deal with a population. The count may be finite or infinite. It may be by ordinal number units or by measurement so that either discrete or continuous distributions result.

One way to detect a sample is to try to replace it. If a set of elements can be replaced by another set simply by redrawing, the set is a sample. If by redrawing we always get back the same set, the set is a population. The latter situation follows from the condition that if no variability occurs from set to set, we must have the complete totality of possible numbers included in the set, and such completeness has been stated as the fundamental property denoting a population.

Consider the single digits from 0 through 9. If these digits were on 10 cards in a stack, and if we drew 10 cards, we would always get the same set of 10 single digits. These 10 digits form a population because the set is complete. Now draw just three cards. There is no way to tell exactly which three will be drawn. (We eliminate the possibility of a marked deck.) In fact, a standard computation of combinatorial possibilities shows that one of 120 different sets of 3 drawn from 10 will occur.

$$\frac{10!}{(10-3)!\,3!} = 120$$

Obviously, each and every such set of 3 is a sample of the population of the 10 cards.

Moments of Distribution

A method of rigorous mathematical description is needed if we are to compare populations and samples for similarities and differences. We will, from time to time, need to test such things, as whether a sample could come from a particular population, whether two samples indicate two different populations, and also how much variability to expect in a series of samples drawn from the same population. All of these measures, we will see, can be stated with a specified risk of being in error. In engineering terminology, we need measures of the reliability of our predictions or estimates when we use univariate models (i.e., models that include a single random variable). A widely used system for specifying characteristics of univariate models, or distributions, is that based on moments of increasing order.

The concept of statistical moments is similar, but not identical, to mechanical moments. The mechanical first moment is the product of a force times the perpendicular distance from the line of action to the point of reaction. Statistical moments correspond to the perpendicular distance, the moment arm of mechanics. In addition, statistical moments are the result of averaging over many elements in a distribution. In this respect computation of the first two moments will be seen to correspond to the calculation of the center of gravity and the radius of gyration in mechanics.

Since a statistical distribution may be either discrete or continuous, it will be necessary to have two procedures for calculation of moments. For discrete distributions the computation of the ith moment (M_i) is given by

$$M_i = \sum_{j=1}^{n} x_j^i f(x) \tag{2-8}$$

When each observation in the sample is given equal weight, $f(x)$ in Eq. 2-8 equals $1/n$, where n is the number of observations in the sample. For continuous distributions the computation of the ith moment is given by

$$M_i = \frac{\int_a^b x^i p(x)\, dx}{\int_a^b p(x)\, dx} \tag{2-9}$$

The use of Eq. 2-8 is illustrated by the computation in Table 2-1. Ten items that comprise the sample are listed. The random variate x is seen to describe the even numbers from 12 to 30 inclusive. These numbers are raised to the power corresponding to the moment desired, and summed. Division of this sum by 10, the number of items, gives the moment. We can see that the first moment is the simple and familiar arithmetic mean. The second moment could be called the mean square, the third moment the mean cube, and so on. It should be noted in particular that the only unit of measurement that appears in Table 2-1 is the scale of x by which each item is designated or measured. The origin of this scale is the zero of x. Therefore, all four moments in Table 2-1 are moments about the point which is the zero of x.

TABLE 2-1 Example for Calculation of Moments of Discrete Distributions

Item	x	x^2	x^3	x^4
1	12	144	1,728	20,736
2	14	196	2,744	38,416
3	16	256	4,096	65,536
4	18	324	5,832	104,976
5	20	400	8,000	160,000
6	22	484	10,648	234,256
7	24	576	13,824	331,776
8	26	676	17,576	456,976
9	28	784	21,952	614,656
10	30	900	27,000	810,000
Total	210	4740	113,400	2,837,328
Total/n	21	474	11,340	283,732.8

The calculation of statistical moments of continuous distributions will be illustrated first by the triangular distribution in Fig. 2-2 and then by the exponential distribution in Eq. 2-4. The first moment for the triangular distribution is found by placing the equations for the two straight-line segments in Eq. 2-9 as shown in the equation

$$M_1 = \frac{\int_0^b x\left(\frac{x}{b^2}\right) dx + \int_b^{2b} x\left(\frac{2}{b} - \frac{x}{b^2}\right) dx}{\int_0^b \frac{x}{b^2}\, dx + \int_b^{2b} \left(\frac{2}{b} - \frac{x}{b^2}\right) dx} \tag{2-10}$$

Solution of this equation produces the value of b for M_1. This agrees with our intuitive feeling that the average of a symmetrical distribution should be at the midpoint of the range of the variate.

Calculation of the second moment of the triangular distribution is given by

$$M_2 = \frac{\int_0^b x^2\left(\frac{x}{b^2}\right) dx + \int_b^{2b} x^2\left(\frac{2}{b} - \frac{x}{b^2}\right) dx}{\int_0^b \frac{x}{b^2}\, dx + \int_b^{2b} \left(\frac{2}{b} - \frac{x}{b^2}\right) dx} \tag{2-11}$$

Solution of Eq. 2-11 gives a value for M_2 of $\frac{7}{6}b^2$. Both the first and second moments of the triangular distribution were computed using the zero of the scale of x as the center of moments. This computation produces a meaningful value for the first moment, which is a measure of the distance from zero along the x scale to a central value of the distribution. The value of the second moment about the zero of x is not very meaningful. Taking the square root of the second moment will linearize this characteristic. This value is $\sqrt{\frac{7}{6}}b$, which is larger than the distance from the origin of moments to the mean value. This measure is thus not a useful characteristic of the distribution.

The second moment could be computed about any point as an origin of moments. Consider such a point as a on the x scale. Then the moment distance is $(x - a)$. This value can be substituted for the moment distance in Eq. 2-11 as given in

$$M_2' = \frac{\int_0^b (x-a)^2 \frac{x}{b^2}\, dx + \int_b^{2b} (x-a)^2 \left(\frac{2}{b} - \frac{x}{b^2}\right) dx}{\int_0^b \frac{x}{b^2}\, dx + \int_b^{2b} \left(\frac{2}{b} - \frac{x}{b^2}\right) dx} \tag{2-12}$$

We can now ask: What should be the value of a to make M_2' as small as possible? Equation 2-12 can be solved by direct integration. If the resulting expression for M_2' is differentiated and set equal to zero, it is found that $a = b$. In other words, the second moment has a minimum value when the center of moment is placed at the mean value. If we replace a with b in Eq. 2-12 and solve this equation by integration, we find that the second moment about the mean is $\frac{1}{6}b^2$. This is called the *variance.* It is customary to designate this value as m_2 to distinguish it from M_2', which we computed as the moment about the zero of x. (Later we will use the following notation: $\bar{x}$ = the sample mean; μ = population mean; s^2 = sample variance, and σ^2 = population variance.)

This value for the second moment is much more meaningful than the moment about the zero of x. The square root of this quantity is usually called the *standard deviation* and symbolized by s for a sample and σ for the population. It measures the compactness of the distribution. In the triangular distribution, if b is large, the individual items making up the distribution scatter through a wide range equal to $2b$ and the standard deviation, $(1/\sqrt{6})b$, is large. With b small, the peak ordinate $1/b$ is large and many of the individual items are close to the mean. The distribution is compact and $(1/\sqrt{6})b$ is small.

The third and fourth moments are also usually computed about the mean value. Both of these moments are computed by integrating (over the bounds of the domain) the difference between the random variable x and the mean raised to the ith power. The equation for the third moment of the triangular distribution is

$$m_3 = \int_0^b (x-b)^3 \frac{x}{b^2}\, dx + \int_b^{2b} (x-b)^3 \left(\frac{2}{b} - \frac{x}{b^2}\right) dx \tag{2-13}$$

The denominator is not shown because the area is unity. In Eq. 2-13 inspection will show that for every positive value of $(x - b)^3$ in the second term on the right there will be a negative value of $(x - b)^3$ in the first term. Since, in addition, the triangle is symmetrical about the mean value b, the products in the two terms will be equal for equal positive and negative values of $x - b$.

Thus the two integral terms cancel each other, and the third moment is zero. This must be the case for every symmetrical distribution. Since m_3 can only have a nonzero value when the distribution is not symmetrical, this moment measures the skewness.

The fourth moment is given by

$$m_4 = \int_0^b (x-b)^4 \frac{x}{b^2}\,dx + \int_b^{2b} (x-b)^4 \left(\frac{2}{4} - \frac{x}{b^2}\right) dx \tag{2-14}$$

Again, the unit denominator is omitted. Solution of Eq. 2-14 gives a value of $\frac{1}{15}b^4$ for m_4. Since this moment varies as the fourth power of b, it becomes very large when items are an appreciable distance from the mean. Inversely, for the fourth moment to be small, the items must be close to the mean. This denotes a high density of items and consequently a high central ordinate of the distribution. This characteristic measured by the fourth moment is called the *kurtosis*; it is a measure of the peakedness of a distribution.

Calculation of moments of continuous distributions will be further illustrated with the exponential distribution presented earlier. The first moment is given by

$$M_1 = A \int_0^{\infty} x\, e^{-Ax}\,dx \tag{2-15}$$

Integration of this equation gives a value for M_1 of $1/A$.

The next three higher moments may be computed by

$$m_i = A \int_0^{\infty} \left(x - \frac{1}{A}\right)^i e^{-Ax}\,dx \tag{2-16}$$

The index i in this equation takes on, successively, values of 2, 3, and 4. Integration yields the successive values $m_2 = 1/A^2$, $m_3 = 2/A^3$, and $m_4 = 9/A^4$.

It is now necessary to return to calculation of discrete moments. Table 2-1 illustrated calculation of the first four moments about the zero of x. Just as with continuous distributions, the second and higher moments should be computed about the mean value in order to produce meaningful measures of compactness, skewness, and peakedness.

Modification of Eq. 2-8 for moments about the mean gives

$$m_i = \frac{\sum_n (x - \bar{x})^i}{n} \tag{2-17}$$

Although Eq. 2-17 could be used for computation of higher moments above M_1, it is not convenient. Each separate item in a sample would have to be reduced by subtraction of $\bar{x}$. An alternative method is to expand the term in parentheses:

$$\left(x - \frac{\sum x}{n}\right)^2 = x^2 - 2\frac{\sum x}{n} + \left(\frac{\sum x}{n}\right)^2$$

$$\left(x - \frac{\sum x}{n}\right)^3 = x^3 - 3\frac{\sum x}{n}x^2 + 3\left(\frac{\sum x}{n}\right)^2 x - \left(\frac{\sum x}{n}\right)^3$$

$$\left(x - \frac{\sum x}{n}\right)^4 = x^4 - 4\frac{\sum x}{n}x^3 + 6\left(\frac{\sum x}{n}\right)^2 x^2 - 4\left(\frac{\sum x}{n}\right)^3 x + \left(\frac{\sum x}{n}\right)^4$$

Substituting these values into Eq. 2-17 and simplifying yields the set

$$m_2 = \frac{\sum x^2 - \dfrac{(\sum x)^2}{n}}{n} \tag{2-18a}$$

$$m_3 = \frac{\sum x^3 - 3\dfrac{\sum x \sum x^2}{n} + 2\dfrac{(\sum x)^3}{n^2}}{n} \tag{2-18b}$$

$$m_4 = \frac{\sum x^4 - 4\dfrac{\sum x \sum x^3}{n} + 6\dfrac{(\sum x)^2 \sum x^2}{n^2} - 3\dfrac{(\sum x)^4}{n^3}}{n} \tag{2-18c}$$

Equation 2-18a provides a biased estimate of the variance. The following provides an unbiased estimate of the standard deviation:

$$s = \left[\frac{1}{n-1}\left(\sum x^2 - \frac{1}{n}(\sum x)^2\right)\right]^{0.5} \tag{2-18d}$$

The summations in Eq. 2-18 mean summation over the entire sample of n values, although the n was omitted for simplicity of notation. All the terms in Eq. 2-18 can be computed from the sample measurements, as was illustrated by the original computations in Table 2-1. Substitution of the appropriate summations from this table into Eqs. 2-18 produces values of 34, 0, and 1933.8 for m_2, m_3, and m_4, respectively. As was the case for the illustrated continuous distributions, these moments about the mean are much more meaningful. The standard deviation, for example, is $\sqrt{34} = 5.83$, or not quite one-third of the range. The third moment is zero, as required by symmetry.

The first four statistical moments have been presented for three examples. These first four moments are the only ones in common usage. In fact, for short records the values of the third and fourth moments may be highly inaccurate. In summary, the first moment measures from the origin of measurements (i.e., the zero of x) to the center of the distribution. The three higher moments are determined using the first moment, or mean, as the center of moments. Thus these three higher moments are independent of the origin of measurement. They are, however, still dependent on the scale.

The second moment, being a measure of dispersion of the items, is naturally scale dependent. The dispersion, or scatter, of the items must be measured in real scale to be physically meaningful. The third and fourth moments measure more abstract characteristics. It is convenient to convert the second, third, and fourth moments to dimensionless characteristics that will be an aid later in comparison of different samples and distributions. The dimensionless coefficients are the coefficient of variation (C_V), the coefficient of skewness (β_1) and the coefficient of kurtosis (β_2):

$$C_V = \frac{m_2^{1/2}}{m_1} \tag{2-19a}$$

$$\beta_1 = \frac{m_3}{m_2^{3/2}} \tag{2-19b}$$

$$\beta_2 = \frac{m_4}{m_2^2} \tag{2-19c}$$

These dimensionless quantities can be computed for either a population or a sample. Sample values are usually denoted as $\hat{C}_V$, $\hat{\beta}_1$, and $\hat{\beta}_2$, respectively.

TABLE 2-2 Summary of Moments

Distribution	m_2	m_3	m_4	β_1	β_z
Discrete linear	34	0	1933.8	0	1.673
Triangular	$\frac{1}{6}b^2$	0	$\frac{1}{15}b^4$	0	$\frac{12}{5}$
Exponential	$\frac{1}{A^2}$	$\frac{2}{A^3}$	$\frac{9}{A^4}$	2	9

The moments and the dimensionless coefficients for the three distributions used as examples are summarized in Table 2-2. Inspection of this table shows several noteworthy points. The discrete linear distribution, being purely numeric in its definition, has purely numeric values for its moments. The triangular and exponential distributions were defined by shape parameters b and A. Their moments are given by expressions with equivalent powers of the parameters. It may also be noted that comparison of the fourth moment of the three distributions is difficult. The discrete linear distribution apparently has a large fourth moment compared to the other two. However, if we look at the dimensionless coefficient of peakedness, β_2, we see that it has the lowest value of the three. The high value of 9 for β_2 for the exponential distribution shows that it has a wide range of values as measured by its standard deviation, and consequently, a low peak. It is the only one of the three distributions that is not symmetrical.

COMMON PROBABILITY FUNCTIONS

Five probability functions that are used frequently in hydrologic analysis are the normal, the Student t, the chi-square, the F, and the binomial distributions. Whereas the last four are used primarily for support in making statistical tests of hypothesis, the normal distribution is used for prediction as well as support. Since these distributions serve primarily a support function, the discussion here will center on obtaining critical values from tables provided in the Appendices.

The Normal Distribution

The normal probability density function for a random variable x is given by

$$f(x) = \frac{1}{\sigma\sqrt{2\pi}} \exp\left[-\frac{1}{2}\left(\frac{x-\mu}{\sigma}\right)^2\right] \qquad \text{for } -\infty < x < \infty \tag{2-20}$$

in which μ and σ are the population parameters of the distribution. It can be shown that the best estimates of μ and σ are the sample mean ($\bar{x}$) and standard deviation (s), respectively. For computing probabilities based on sample statistics, $\bar{x}$ and (s) can be substituted for μ and σ, respectively. Probabilities could be computed by integrating Eq. 2-20 over a range of values of x. However, since there are an infinite number of values of μ and σ (or $\bar{x}$ and s), numerous such integrations would become very tedious. The problem can be circumvented by making a transformation of the random variable x.

If the random variable x has a normal distribution, the following transformation of x leads to a new random variable z that also has a normal distribution, but with a mean and standard deviation of 0 and 1, respectively:

$$z = \frac{x-\mu}{\sigma} \tag{2-21}$$

Using Eq. 2-21 it can be shown that Eq. 2-20 can be transformed to

$$f(z) = \frac{1}{\sqrt{2\pi}} e^{-z^2/2} \qquad \text{for } -\infty < z < \infty \tag{2.22}$$

Although there are an infinite number of normal distributions defined by Eq. 2-20, there is only one distribution for Eq. 2-22; thus it is called the *standard normal distribution*, and z is called the *standardized variate.* Since there is only one standard normal distribution, probabilities can be computed using Eq. 2-22 and placed in tabular form as a function of z (see Appendix A-1). The table is structured with values of z at increments of 0.1 down the left margin and at increments of 0.01 across the top. The cumulative probability from $-\infty$ to the desired value of z is given within the table. For example, the probability that z is less than 0.23 is 0.5910. Also, the probability that z is between -0.47 and 0.23 equals $(0.5910 - 0.3192)$ or 0.2718. The figure shown at the top of the table provides a good understanding of the relationship between the probability and the value of z. The table can also be used to find the value of z that corresponds to a certain probability. For example, for a probability of 0.05 in the left tail, we enter the table with 0.0500 and find the corresponding value of z, which is -1.645. Since the distribution is symmetric, 5% of the area under the curve $f(z)$ lies to the right of a value of z of 1.645.

The Student *t* Distribution

The *Student t*, or *t*, *distribution* is similar to the normal distribution in that it is symmetric; however, it differs from the normal distribution in that it is a function of a single parameter ν, which is often called the *degrees of freedom* and controls the spread of the t distribution. Since it is a function of the parameter ν, which is a positive integer, there are many t distributions. Thus the table of t values has a slightly different structure than the normal table. In the table of Appendix A-2, the value of ν is given in the left margin, the probability in the right tail of the distribution is given across the top, and the value of the t distribution is given within the table. For example, for $\nu = 7$ and a probability of 0.05, the t value is 1.895. Since the t distribution is symmetric, 5% of the area in the left tail is to the left of a t value of -1.895 for $\nu = 7$. For the case where one is interested in 5% of the area but with $2\frac{1}{2}$% in each tail, the critical t values would be -2.365 and 2.365 for $\nu = 7$. One final point: it is usually acceptable to use the normal distribution in place of the t distribution for $\nu > 30$.

The Chi-Square Distribution

The *chi-square* (χ^2) *distribution* is similar to the t distribution in that it is a function of a single parameter ν, the degrees of freedom; however, it differs from the t distribution in that the χ^2 distribution is not symmetric. The table of chi-square values (Appendix A-3) is identical in structure to the t table, with ν down the left margin, the probability across the top, and the value of the random variable (χ^2) in the table. For 11 degrees of freedom, 5% of the area in the left tail is from 0 to 4.575. For 17 degrees of freedom, there is a probability of 0.025 that χ^2 will be between 30.191 and ∞.

The *F* Distribution

The *F distribution* is a function of two parameters, ν_1 and ν_2. To obtain F values from Appendix A-4, the value of ν_1 is entered along the top of the table and the value of ν_2 along the left margin. The value of F corresponding to the appropriate probability in the right tail of the distribution is obtained from the table. For example, if $\nu_1 = 10$ and $\nu_2 = 20$, there is a 5% chance that F will be greater than 2.35. Note that the table is not symmetric. For $\nu_1 = 20$ and $\nu_2 = 10$, the critical F value for 5% in the right tail is 2.77. Thus the values of ν_1 and ν_2 must not be switched.

The Binomial Distribution

Unlike the previous four distributions, which are to be used with continuous random variables, the *binomial distribution* is intended to be used with a discrete random variable that satisfies the following four assumptions:

1. The experiment consists of n trials.
2. There are only two possible outcomes for each trial.
3. The probability of an outcome remains constant from one trial to the next.
4. The n trials are independent.

It may be instructive to illustrate these four assumptions with an example. First, define a trial as a period of 1 year. Next, let the random variable x be the occurrence of a flood in any year; thus there are only two possible outcomes per trial, either a flood occurs or it does not occur in any one year. Assuming that the probability of a flood of a specified magnitude remains constant from year to year and that floods in successive years are independent, the random variable (flooding) follows a binomial distribution.

For a binomially distributed random variable the probability of *exactly* i outcomes with probability p, in n trials can be determined by

$$b(i; n, p) = \binom{n}{i} p^i (1-p)^{n-1} \tag{2-23a}$$

$$= \frac{n!}{i!(n-i)!} p^i (1-p)^{n-i} \tag{2-23b}$$

in which, in general, $k!$ is read k *factorial* and is computed by $k! = k(k-1)(k-2) \cdots 1$. Values of the cumulative function, which is denoted as $B(i; n, p)$, are given in Appendix A-5. The following identities hold:

$$b(i; n, p) = B(i; n, p) - B(i-1; n, p) \tag{2-24}$$

$$b(i; n, p) = B(n-i; n, 1-p) - B(n-i-1; n, 1-p) \tag{2-25}$$

$$b(i; n, p) = b(n-i; n, 1-p) \tag{2-26}$$

$$B(i; n, p) = 1 - B(n-i-1; n, 1-p) \tag{2-27}$$

To illustrate the use of the tables of Appendix A-5, let's assume that a flood which would cause damage has a probability of occurring in any one year of 0.05. If the project life is considered to be 20 years, the probability of exactly two such floods occurring during the project life is $b(2; 20, 0.05) = B(2; 20, 0.05) - B(1; 20, 0.05)$ or $0.9245 - 0.7358 = 0.1887$. The probability of getting two or more such floods would be $1 - B(1; 20, 0.05)$ or $1 - 0.7358 = 0.2642$.

CONFIDENCE INTERVALS

We have seen that the mean and standard deviation are sufficient to describe a normal distribution. Great care must be exercised, however, in specifying whether values of the mean and standard deviation are computed from samples or from the population. In hydrologic problems, future events, and therefore the total population of our hydrologic experience, are unknown. We can never know our hydrologic statistical population. We only know samples of it from our records and observations in the past.

Since our hydrologic means and variances (standard deviation squared) are thus computed from a past sample, we must expect that in future samples the numerical values will be different. In fact, we must regard the means and variances as being new statistical random variables and having frequency distributions of their own. It is obviously not possible to state exactly the values of $\bar{x}$ and s that would result from a new sample of the same size. It is, however, possible to compute intervals using the sample data and state the probability that these intervals would contain, respectively, the population mean and standard deviation. Also, "on the average," future sample means and variances would be expected to cluster on these population values.

A frequent engineering requirement is an estimate of probable values of future means and variances. The intervals within which we expect the population values to be found with a certain probability are called *confidence intervals.* We make statements such as the following: "The expected mean annual runoff will be greater than A inches and less than B inches with 90% confidence." Actually, we are stating that, judging from one sample, the interval A to B would contain the true mean in 90% of future samples, if we could draw a large number of such samples.

It is important to note that in computing a confidence interval on the mean we must specify the mean, $\bar{x}$, some measure of the dispersion statistic, either σ or s, the level of confidence γ, and the sample size n. From this information we can compute one or two values. If we want a *one-sided confidence interval,* we will compute either the lower limit or the upper limit. If we want a *two-sided confidence interval,* we will compute both an upper and a lower limit. The statement of the problem determines whether we want a one-sided or a two-sided interval.

The calculation of a one-sided confidence interval follows one of the following forms:

$$\text{lower limit:} \quad M_L = \bar{x} - t(n-1, \alpha)\frac{s}{\sqrt{n}} \tag{2-28a}$$

$$\text{upper limit:} \quad M_U = \bar{x} + t(n-1, \alpha)\frac{s}{\sqrt{n}} \tag{2-28b}$$

If we want a two-sided confidnece interval, we use $t(n-1, \alpha/2)$ instead of $t(n-1, \alpha)$, where $\alpha = 1 - \gamma$. If σ is known, we can use σ in place of s and the standard normal variate z in place of t. One could then state that with γ percent confidence, the true mean lies within the confidence interval defined by the limits of Eqs. 2-28. For a one-sided lower limit, the upper limit is defined as ∞, while for the one-sided upper limit, the lower limit is defined as $-\infty$.

TOLERANCE LIMITS

In the preceding section on confidence intervals, we discussed the variability of the mean of a repeated number of samples. In any one sample that we may have, which we consider a record, the mean has a particular value. In future samples, this mean value will not repeat exactly, but we have seen that we could determine limits within which it would be expected to fall with a certain degree of confidence in future samples.

In this section on tolerance limits, we discuss the variability of a proportion of the sample. Again, we start with the concept that we have one sample. It is a record from the past. If we refer to Fig. 2-3(a), we can see that in such a sample some proportion p of the elements in the sample will lie above a value x_p, while a fraction of $1-p$ of the values of the sample will lie below value x_p. Now consider Fig. 2-3(b). Consider that this is some future sample. The mean of this sample and the standard deviation of this future sample must be expected to vary from our past sample of record. Then the value of x_p that measures those items for which a proportion p is larger and $1-p$ is smaller will also be different in this sample. Consider also Fig. 2-3(c). We have a new value x_p which determines the proportion p larger and the proportion $1-p$ smaller. In fact, if we took a large number of samples we would find that this value x_p is the value of a random variable and has a distribution all its own.

If x_p has a known distribution, we are able to state with some confidence where the value of x_p would lie if we had an original sample from which to compute and from which to estimate the range of x_p in future samples. The

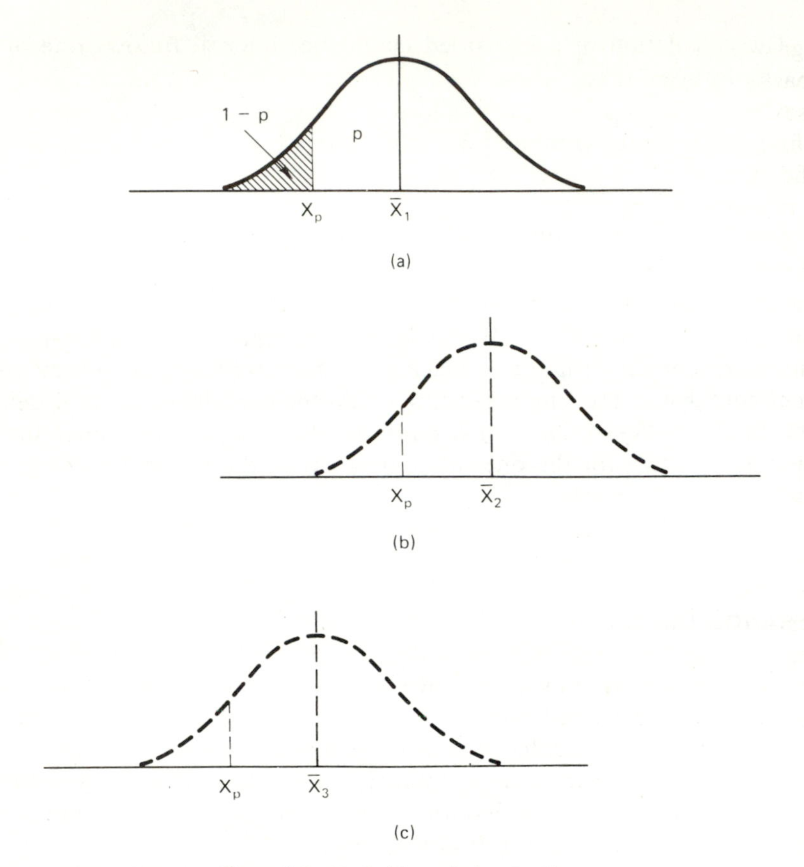

Figure 2-3 Probable variation in X_p.

form for the computation is shown in the equations

$$x_U = \bar{x} + k(n, p, \gamma)s \tag{2-29a}$$

$$x_L = \bar{x} - k(n, p, \gamma)s \tag{2-29b}$$

in which x_L and x_U are the lower and upper limits that can be expected to contain a proportion p of the values of the population. The form of Eqs. 2-29 is very simple. If in our record sample we compute the mean $\bar{x}$ and the standard deviation s, we can compute an upper and a lower limit if we know a value of k. Now this value k is very similar to the value t in Eqs. 2-28. We can determine this value k simply by looking up an appropriate value from a table for given values of the sample size n, the confidence level γ, and the proportion p. Selected values are given in Appendix A-6.

To look up an appropriate value k we must know both n (the sample size) and p (the proportion of the sample we wish to bracket with upper and lower tolerance limits). We must also specify a confidence level, usually designated γ (gamma), with which we wish to make a statement about the probable variation of x_p. Thus there are two probability values that enter into the selection of an appropriate value of k. To repeat, these probability values are first, p, the proportion of the sample that we are specifying, and γ, the confidence level with which we can make a statement about the level of x_p that marks off the proportion p.

Figure 2-3 illustrates a lower tolerance limit. That is, the proportion p of the sample always lies above the x_L of Eq. 2-29b. When we select k for a particular confidence level with this concept of p always being larger, we are in fact considering a one-sided distribution. Equation 2-29a shows that by adding the product ks to the mean value $\bar{x}$, we can compute an upper tolerance limit. This upper limit simply says that the proportion p now is smaller than x_U. The interpretation here is that the distribution is still one-sided. We may compute either an upper limit or a lower limit but we interpret our statement as having a confidence level or γ.

In the preceding section on *confidence intervals*, where we computed a likely range of the mean value of the sample, we showed that we could compute either a one-sided or a two-sided distribution (or a one-sided or two-sided statement about the probable range of the mean value) simply by combining the exceedence probabilities to the left and to the right of our computed range. It is *not* possible to convert the one-sided tolerance limit to a two-sided tolerance limit in the same way. The distribution of the means of a sample is in itself a symmetrical distribution, and therefore we can combine the two parts, the two exceedences, above and below the range. The distribution of a proportion of a sample, however, is not symmetrical. Therefore, we cannot combine the exceedences above and below the range x_U and x_L from a one-sided distribution and obtain a two-sided distribution.

We can compute two-sided tolerance limits. The form of the equation is identical to Eq. 2-29. The difference, however, is in the definition of the factor k in the equation. We must be very careful to use the correct tables of k. It will be found that there is a table of k values for one-sided tolerance limits and a completely different set of k values for two-sided tolerance limits. Therefore, we must specify, first, whether we are interested in whether the proportion of our sample in the future is expected to lie above a certain value x_L, or below a certain value x_U, in which case we are making two separate one-sided statements. If, on the other hand, we are interested in whether the proportion p of the sample will lie between a range of x_L on the left and x_U on the right, we are in fact making a two-sided statement of confidence and we must look up the appropriate k value in the two-sided table.

The exact meaning and application of tolerance limits at this point are probably very confusing and very obscure. Later, we will discuss flood-frequency curves and flow-duration curves. When we take up this practical application of univariate statistical models, the meaning of the proportion of a sample p will become clear. From this clearer understanding of p, there will

TABLE 2-3 Average Air Temperature (°F) by Months for Murray, Kentucky

	June	July	August
1944	77	77	77
1945	72	76	76
1946	76	78	74
1947	74	74	83
1948	78	80	76
1949	75	79	74
1950	75	73	70
1951	73	78	78
1952	82	81	77
1953	79	80	78
1954	78	83	80
1955	69	80	79
1956	74	77	77
1957	75	76	74
1958	72	76	74
1959	72	75	76

Sum: 3667
Sum of squares: 280,597
Mean: 76.4
Variance: 9.65
Standard deviation: 3.11

also develop a clearer understanding of how this proportion p must be expected to vary when we attempt to predict what may happen in the future.

Example 2-1: Confidence Intervals and Tolerance Limits on Mean Monthly Temperature

Table 2–3 shows the average air temperature for the months of June, July, and August for the period from 1944 to 1959 for Murray, Kentucky. Calculations for the first two moments of this distribution are shown at the bottom of Table 2-3. These temperature data are plotted in Fig. 2-4 as a histogram. The width of each cell, called the *class width*, is two degrees. To construct such a histogram it is necessary only to tally the number of temperatures recorded in Table 2-3 that fall in each class.

Before we apply confidence intervals and tolerance limits to a distribution such as shown by the histogram in Fig. 2-4, we should check to see that the distribution is approximately normally distributed. A simple way to do this is to superimpose a normal curve on the histogram of temperatures. We make use of the method of moments; the first moment of the sample is set equal to the first moment of a normal distribution (i.e., $\bar{x} = \mu$), and the second moment of the sample is set equal to the second moment of the distribution (i.e., $s^2 = \sigma^2$). Since the sample moments are taken to be estimates of the population from which the sample was drawn, we use the unbiased estimate of the population variance. We can, of course, compute the variate corresponding to any z value if we know the mean and standard deviation, as shown in Eq. 2-30, which is Eq. 2-21 rearranged:

$$x = \bar{x} + zs \tag{2-30}$$

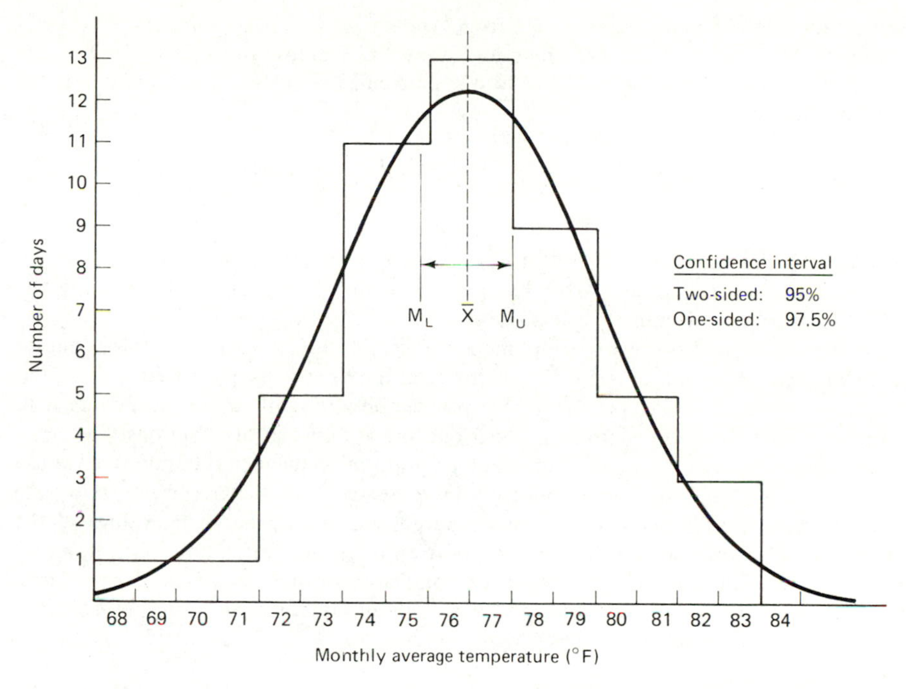

Figure 2-4 Confidence interval on the mean of the monthly average temperature.

We must remember that z can be positive or negative and that a normal curve is symmetrical.

Calculation of the ordinates of a normal curve to superimpose on our sample of temperatures is shown in Table 2-4. A range of z values is selected. Corresponding

TABLE 2-4 Calculation of Ordinates of a Normal Curve

z	zs	$\bar{x} + zs$	$\bar{x} - zs$	Standard Ordinate	Sample Ordinate
0.0	0.0	76.40		0.3989	12.23
0.2	0.62	77.02	75.78	0.3910	12.08
0.4	1.24	77.64	75.16	0.3683	11.38
0.6	1.86	78.26	74.54	0.3332	10.30
0.8	2.48	78.88	73.92	0.2897	8.96
1.0	3.11	79.51	73.29	0.2420	7.48
1.25	3.88	80.28	72.52	0.1826	5.64
1.50	4.66	81.06	71.74	0.1295	4.00
1.75	5.44	81.84	70.96	0.0863	2.67
2.00	6.21	82.61	70.19	0.0540	1.67
2.25	6.99	83.39	69.41	0.0317	0.98
2.50	7.77	84.17	68.63	0.0175	0.54
2.75	8.54	84.94	67.86	0.0091	0.28
3.00	9.32	85.72	67.08	0.0044	0.14

standard ordinate values are obtained from Appendix A-1. Also, temperatures corresponding to these z values are computed above and below the mean. Finally, the standard ordinates must be modified to our particular problem. This modification is shown in the equation

$$\text{sample ordinate} = \frac{nW\,(\text{standard ordinate})}{s} \tag{2-31}$$

where W is the class width of our histogram, n the number of items in our sample, and s the standard deviation. W and s in Eq. 2-31 adjust for scale, and n expands the ordinate from unity to our sample size.

The computed ordinates are plotted in Fig. 2-4. It is evident that this fitting of a normal curve to our data by the method of moments has produced quite good correspondence. Later in this chapter we will see how to apply more stringent tests to correspondence between sample and distribution, or between distributions.

Since we have computed estimates of the population mean and standard deviation we can use Eqs. 2-28 to compute confidence intervals and Eqs. 2-29 to compute tolerance limits. These quantities, it will be remembered, express probabilistic values of the means and of proportions of future samples from the same population. Therefore, we must set a level of probability by which we condition (or inversely, express confidence in) our statements about these future values. Quite frequently, the values of 95% and 99% are used to express moderately high confidence and very high confidence, respectively.

The value of t (47, 97.5) is about 2.01. This value can be found in Appendix A-2. Using the value of 2.01 for t in Eqs. 2-28, we can calculate

$$M_L = 76.4 - \frac{2.01(3.1)}{\sqrt{48}} = 75.3°\text{F}$$

$$M_U = 76.4 + \frac{2.01(3.1)}{\sqrt{48}} = 77.5°\text{F}$$

These values are plotted in Fig. 2-4. Their meaning can be interpreted as follows. In 97.5% of future samples, the monthly means will be less than M_U (i.e., less than 77.5°F). We can also say that in 97.5% of future samples their means will be greater than M_L (i.e., greater than 75.3°F). These are "one-sided" statements. From the above, we could *not* say that the monthly means will be between 75.3 and 77.5°F in 97.5% of future samples because the computations above are for one-sided intervals. We should note that 2.5% of future samples will have means greater than 77.5°F and 2.5% will have means less than 75.3°F. Then a total of 5% of means of future samples will fall outside the interval between M_L and M_U. Or we can say that 95% of future samples will have their means between 75.3 and 77.5°F. This is a "two-sided" statement about the confidence interval of the means of future samples.

In illustrating the calculation of tolerance limits, we could arbitrarily set some value of p in Eqs. 2-29 and compute the probable variation of this proportion of events in future samples. Instead of doing this, however, it will be more instructive to choose some temperature and investigate how the proportion of the sample corresponding to this temperature will vary. Suppose that we choose the temperature 82°F. This corresponds to a z value (standardized deviate) of

$$z = \frac{82 - 76.4}{3.1} = 1.8$$

From Appendix A-1 we can find that this corresponds to a proportion of 96.4% of our sample. If we set a confidence level of 95%, $k(96.4, 95, 48)$ for a one-sided statement can be found from standard tables to equal about 2.36 (double interpolation in the tables will probably be necessary.) Equation 2-29a shows that

$$X_U = 76.4 + (2.36)(3.1) = 83.7°\text{F}$$

In other words, in 95% of future samples the temperature marking 96.4% of the items will lie below 83.7°F. Conversely, in 5% of future samples the temperature marking 96.4% of the items will lie above 83.7°F. This tolerance limit is plotted in Fig. 2-5.

Confidence intervals and tolerance limits are extremely important "predictions" for planning future operations. In Example 2-1 the confidence interval of the mean predicted the general level, or average temperature, we should expect, and therefore plan for. The tolerance limit shows how often an extreme temperature is likely to occur. If such an extreme is critical, we would know how often to expect this critical value, and take precautions. Although this explanation is based on a temperature example, the same ideas apply to means and extremes of rainfall, means and extremes of streamflow,

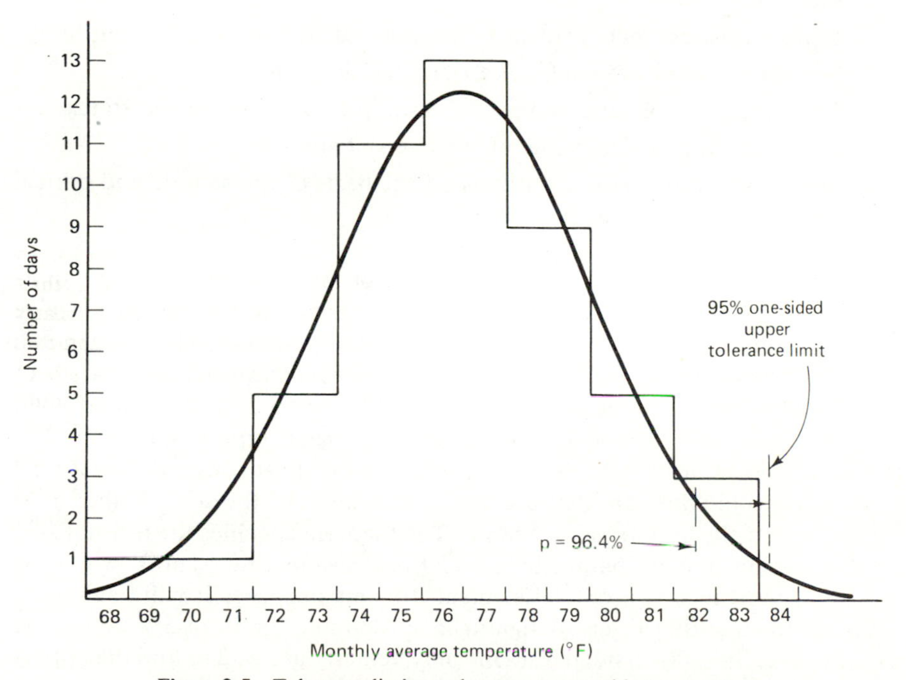

Figure 2-5 Tolerance limit on the average monthly temperature.

means and extremes of wind velocities, or any other information that we cannot forecast exactly.

HYPOTHESIS TESTING

Interval estimation provides a range of values in which one can expect, with a certain degree of confidence, the true value to lie. In some analyses, we are interested in a different type of solution, specifically, a decision requiring a "yes" or "no" response. For example, with the data from Table 2-3 we may ask the question: Recognizing that the mean of 76.4°F is only a sample statistic, can we reasonably conclude that the true mean is greater than 80°F? With respect to Fig. 2-4, we made a somewhat subjective conclusion that the sample points could be represented by a normal distribution. In other cases, the decision may not be obvious and we may want an answer to the question: Can a sample of data be represented by some specific distribution function with specified parameters? Both the questions on the mean and distribution require a "yes–no" response. *Hypothesis testing* is a systematic, statistical method of making a response to such questions.

Hypothesis tests use the following six-step procedure:

1. State the hypotheses.
2. Using statistical theory, identify the test statistic and its distribution.
3. Specify the level of significance (i.e., decision risk).
4. Obtain the sample and compute the sample value of the test statistic.
5. Determine the critical value of the test statistic.
6. Select the appropriate hypothesis on the basis of the sample and critical values of the test statistic.

While other texts provide a detailed description of these steps, a few points are of importance. First, there are two hypotheses, both of which must be stated in terms of population parameters. The *null hypothesis*, which is denoted by H_0, is a statement of equality. The *alternative hypothesis*, H_A, is a statement of inequality and can be either one-sided ($<$ or $>$) or two-sided ($\neq$). Examples will be provided below. Second, for each type of problem (i.e., testing of means, testing of variances, or testing of probability distributions) there is a specific theorem that provides a statement of the statistic that must be used to test the hypotheses of step 1. The theorem specifies the test statistic, its distribution and its parameters, and the decision rule. Third, while the level of significance is a critical element, for our purposes it will have to be sufficient to state that levels of significance (denoted by α) of 5% or 1% are usually used. For a two-sided test, one may actually use $\alpha/2$ to find the critical value of the test statistic. Fourth, critical values of the test statistic are obtained from tables such as the normal, t, χ^2, and F tables of Appendices A-1, A-2, A-3, and A-4, respectively.

Test on a Single Mean

The test for a single mean can be summarized using the six steps listed above:

1. H_0: $\mu = \mu_0$

   H_A: $\left.\begin{matrix}\mu < \mu_0 \\ \mu > \mu_0\end{matrix}\right\}$ one-sided

   $\mu \neq \mu_0$ two-sided
2. Theorem: If $\bar{x}$ is the mean of a random sample of size n taken from a normal population with mean μ and standard deviation σ, the test statistic is

$$t = \frac{\bar{x} - \mu}{s/\sqrt{n}} \tag{2-32}$$

   in which t is the value of a random variable having the Student t distribution with degrees of freedom $\nu = n - 1$.
3. Specify α, the level of significance.
4. Find n, $\bar{x}$, and s from the sample and compute a sample estimate of t using the equation in step 2.
5. For a value of ν, which equals $n - 1$, find the critical t value from Appendix A-2 for the level of significance identified in step 3; this value is denoted as t_α.
6. Depending on the alternative hypothesis selected in step 1, use the following decision criteria to select the most appropriate hypothesis:

If H_A is:	*Then reject H_0 if:*
$\mu < \mu_0$	$t < -t_\alpha$
$\mu > \mu_0$	$t > t_\alpha$
$\mu \neq \mu_0$	$t < -t_{\alpha/2}$ or $t > t_{\alpha/2}$

In some cases, the standard error of the mean $(\sigma/\sqrt{n})$ may be known; in such a case the following test statistic can be used in place of the t statistic of step 2:

$$z = \frac{\bar{x} - \mu}{\sigma/\sqrt{n}} \tag{2-33}$$

in which z is the value of a random variable having a standard normal distribution. If $\sigma/\sqrt{n}$ is known, the assumption of normality is not necessary.

Example 2-2: Hypothesis Test on the Mean Monthly Temperature

Using the data of Table 2-3, one might be interested to know if it is likely that the true mean is greater than 80°F even though the sample mean equaled 76.4°F. The following illustrates a test of this hypothesis:

1. H_0: $\mu = 80°F$
   H_A: $\mu > 80°F$.
2. Use the t statistic.
3. $\alpha = 0.05$.
4. $t = \dfrac{76.4 - 80}{\sqrt{9.4/48}} = 8.135$.
5. For $\nu = 47$, $t_\alpha = 1.68$.
6. Reject H_0 since $t > t_\alpha$.

The decision to reject the null hypothesis is not surprising given the small sample size and the large scatter of the data (69 to 83), with 8 of the 48 values being 80 or above. The one-sided alternative hypothesis was used because we were interested in the possibility of μ being greater than 80°F. If the problem had been stated, for example, as an interest in whether or not μ was different from 75°F, the two-sided alternative would have been used.

Table 2-5 provides a summary of commonly used hypothesis tests.

Test of a Distribution Function

Another common problem in statistical analysis is the identification and verification of the underlying population. For example, is the sample of data in Fig. 2-4 from a normal population? In such cases, we can propose a null hypothesis stating that the random variable has a normal population with specific values of μ and σ. It is important to note that the null hypothesis is a statement of equality and is expressed in terms of population parameters rather than sample statistics. The alternative hypothesis would indicate that the random variable was *not* from the stated population, which includes specific values of the parameters. It is important to note that rejection of the null hypothesis may be because (1) the stated distribution is incorrect, (2) one or more of the parameters are incorrect, or (3) both the distribution function and one or more of the parameters are incorrect. To summarize, we can express the hypotheses as

$$H_0\text{: } x = \text{pdf (parameters)} \tag{2-34a}$$

$$H_A\text{: } x \neq \text{pdf (parameters)} \tag{2-34b}$$

in which pdf indicates a *probability density function.* For example, one of the following null hypotheses may be tested for the data of Fig. 2-4:

$$H_0\text{:}\quad x \sim N(76.4, \sqrt{9.4})$$

TABLE 2-5 Summary of Hypothesis Tests

H_0	Test Statistic	H_A	Region of Rejection
$x \sim f(x, p)$	$\chi^2 = \sum_{i=1}^{k} \frac{(o_i - e_i)^2}{e_i}$ $\nu = k - 1$	$x \neq f(x, p)$	$\chi^2 > \chi^2_\alpha$
$\mu = \mu_0$ (σ known)	$z = \frac{\bar{x} - \mu_0}{\sigma/\sqrt{n}}$	$\mu < \mu_0$ $\mu > \mu_0$ $\mu \neq \mu_0$	$z < -z_\alpha$ $z > z_\alpha$ $z < -z_{\alpha/2}$ and $z > z_{\alpha/2}$
$\mu = \mu_0$ (σ unknown)	$t = \frac{\bar{x} - \mu_0}{s/\sqrt{n}}$ $\nu = n - 1$	$\mu < \mu_0$ $\mu > \mu_0$ $\mu \neq \mu_0$	$t < -t_\alpha$ $t > t_\alpha$ $t < -t_{\alpha/2}$ and $t > t_{\alpha/2}$
$\sigma^2 = \sigma_0^2$	$\chi^2 = \frac{(n-1)s^2}{\sigma_0^2}$ $\nu = n - 1$	$\sigma^2 < \sigma_0^2$ $\sigma^2 > \sigma_0^2$ $\sigma^2 \neq \sigma_0^2$	$\chi^2 < \chi^2_{1-\alpha}$ $\chi^2 > \chi^2_\alpha$ $\chi^2 < \chi^2_{1-\alpha/2}$ and $\chi^2 > \chi^2_{\alpha/2}$
$\mu_1 = \mu_2$ ($\sigma_1^2 = \sigma_2^2$, but unknown)	$t = \frac{\bar{x}_1 - \bar{x}_2}{s\left[\left(\frac{1}{n_1} + \frac{1}{n_2}\right)\right]^{0.5}}$ where $\nu = n_1 + n_2 - 2$ and $s = \left[\frac{(n_1 - 1)s_1^2 + (n_2 - 1)s_2^2}{n_1 + n_2 - 2}\right]^{0.5}$	$\mu_1 < \mu_2$ $\mu_1 > \mu_2$ $\mu_1 \neq \mu_2$	$t < -t_\alpha$ $t > t_\alpha$ $t < -t_{\alpha/2}$ and $t > t_{\alpha/2}$

$$H_0: \quad x \sim N(76.4, 3)$$

$$H_0: \quad x \sim N(80, 3)$$

The latter two indicate that specific values of the population parameters may not equal the sample statistics.

There are two common alternatives for step 2: the chi-square goodness-of-fit test and the Kolmogorov–Smirnov one-sample test. The characteristics of the sample data determine which of the two tests to use.

The chi-square goodness-of-fit test. The theorem for the chi-square test states:

> A test of fit between frequencies observed in a sample (o_i) and frequencies expected (e_i) for a random variable from the population described in the H_0 is

based on the statistic

$$\chi^2 = \sum_{i=1}^{k} \frac{(e_i - o_i)^2}{e_i} \tag{2-35}$$

where χ^2 is the value of a random variable whose sampling distribution is approximated very closely by the chi-square distribution and k is the number of cells into which the range of the random variable is separated.

It is important to note that there are two populations associated with this test, the population stated in H_0 and the chi-square population of the test statistic. The two populations should not be confused.

After selecting a level of significance α (step 3) and collecting a sample, the range of the sample is separated into cells and the expected (population) and observed (sample) frequencies computed for each cell. The value of the test statistic (Eq. 2-35) is then computed.

The critical value of χ^2_α (step 5) is obtained from Appendix A-3 for level of significance α and degrees of freedom $(k - p)$, in which p is the number of pieces of sample information used to derive the test statistic of Eq. 2-35. The calculation of p will be discussed later. The chi-square test is always a one-tailed test, with the value of χ^2 based on a proportion α in the right tail of the χ^2 distribution.

The null hypothesis is rejected if the computed χ^2 value is greater than χ^2_α. Again, recall that if the null hypothesis is rejected, it can be the result of either an incorrect density function or incorrect parameters, or both.

The data of Table 2-3 and Fig. 2-4 can be used to illustrate the chi-square goodness-of-fit test. Using the six steps for a hypothesis test, we get:

1. H_0: temperature $N(\mu = 76.4°\text{F}, \sigma^2 = 9.4)$
   H_A: $T \neq N(76.4, 9.4)$.
2. Theorem for chi-square test.
3. $\alpha = 0.05$.
4. The range of temperatures was separated into seven cells, with each internal cell covering a range of 2°F; the observed frequencies (o_i) were determined for each cell (see Table 2-6). For the distribution of the null

TABLE 2-6 Chi-Square Goodness-of-Fit Test for Temperature Data of Table 2-3

Cell Bounds (°F)	o_i	z_i	$F(z_i)$	p_i	e_i	e_i
0–71.5	2	−1.598	0.0550	0.0550	2.64	8.26
71.5–73.5	5	−0.946	0.1721	0.1171	5.62	
73.5–75.5	10	−0.294	0.3844	0.2123	10.19	10.19
75.5–77.5	14	0.359	0.6340	0.2496	11.98	11.98
77.5–79.5	9	1.011	0.8440	0.2100	10.08	10.08
79.5–81.5	5	1.663	0.9518	0.1078	5.18	7.49
81.5–∞	3	∞	1.0000	0.0482	2.31	
	48			1.0000	48.00	48.00

hypothesis the standard normal deviate z_i for the upper bound of each cell was determined; the corresponding cumulative probability was obtained from Appendix A-1. The probability of a value being in each cell was computed from the cumulative distribution. The expected frequency e_i for each cell was computed by np_i. Since the expected frequencies for the two outer cells were less than 5, both the expected and observed frequencies were combined with the values for the adjacent cells. Thus, instead of seven cells, five cells will be used to compute the value of the test statistic.

$$\chi^2 = \frac{(8.26-7)^2}{8.26} + \frac{(10.19-10)^2}{10.19} + \frac{(11.98-14)^2}{11.98} + \frac{(10.08-9)^2}{10.08} + \frac{(7.49-8)^2}{7.49} = 0.687 \qquad (2\text{-}36)$$

5. The following three values from the sample were used to compute the expected frequencies: $\bar{x}$, s, and n. Thus the degrees of freedom equal $(k - p) = (5 - 3) = 2$. For a 5% level of significance, the critical value χ^2_α from Appendix A-3 is 5.991.
6. Since $\chi^2 < \chi^2_\alpha$, the null hypothesis must be accepted.

Two points of caution. First, frequencies from adjacent cells are combined only when the *expected* frequency of a cell is less than 5; this rule of thumb does not apply when the observed frequency is less than 5, only when the expected frequency is less than 5. Second, two degrees of freedom were lost in step 5 because values of $\bar{x}$ and s^2 were used in the hypotheses of step 1 as the population parameters μ and σ^2; the additional degree of freedom was lost because expected frequencies were computed by np_i, with n being a sample value.

The chi-square test requires a fairly large sample, say $n > 30$. For smaller sample sizes, it is difficult to obtain measurements that will provide a shape that represents the underlying population. Also, the degrees of freedom will be small and thus the reliability of the decision may be questionable.

The Kolmogorov–Smirnov one-sample test. The objective of the Kolmogorov–Smirnov one-sample test is to test the null hypothesis that the cumulative distribution of a variable agrees with the cumulative distribution of some specified probability function; the null hypothesis must specify both the distribution function and its parameters. The alternative hypothesis is accepted if the distribution function is unlikely to be the underlying function; this will result when either the density function or the specified parameters are incorrect.

The test statistic, which is denoted as D, is the maximum absolute difference between the values of the cumulative distribution of a random sample and the cumulative function of a specified probability distribution. Critical values of the test statistic are usually available only for limited values of the level of significance; those for 5% and 1% are given in Appendix A-7.

The Kolmogorov–Smirnov test may be used for small samples; it is generally more efficient than the chi-square test when the sample is small. The test requires data on at least an ordinal scale, but it is applicable for comparisons with continuous distributions. (The chi-square test may also be used with discrete distributions.)

The Kolmogorov–Smirnov test is computationally simple; the computational procedure requires the following six steps:

1. State the null and alternative hypotheses in terms of the proposed probability distribution and its parameters.
2. The test statistic, D, is the maximum absolute difference between the cumulative function of the sample and the cumulative function of the distribution specified in the null hypothesis.
3. The level of significance should be set; values of 0.05 and 0.01 are usually used.
4. A random sample should be obtained and the cumulative probability function derived. After computing the cumulative probability function for the population, the value of the test statistic should be computed.
5. The critical value, D_α, of the test statistic should be obtained from tables of D_α. The value of D_α is a function of α and the sample size, n.
6. If the computed value D is greater than D_α, the null hypothesis should be rejected.

The data of Table 2-3 can be used to illustrate the computations involved in the Kolmogorov–Smirnov one-sample test. The plot of Fig. 2-4 suggests that the data are normally distributed. While the sample mean and standard deviation equal 76.4°F and 3.1°F, respectively, the null and alternative hypotheses to test are

$$H_0\colon \quad X \sim N(\mu = 75; \sigma = 3) \tag{2-37a}$$

$$H_A\colon \quad X \neq N(\mu = 75; \sigma = 3) \tag{2-37b}$$

The test can be made using an interval of 1°F. Table 2-7 shows the cumulative frequency distribution of the sample $F(T)$ and the cumulative probability distribution, which was obtained by dividing $F(T)$ by the sample size n. For the parameters defined in Eqs. 2-37, values of the standardized variate z can be computed using Eq. 2-21. The corresponding probabilities $P(z)$ are obtained from Appendix A-1, and the absolute differences between the sample and expected population cumulative distributions are computed (Table 2-7).

TABLE 2-7 Kolmogorov–Smirnov Test of Temperature Data for a Normal Population with $\mu = 75$ and $\sigma = 3$

T (°F)	$f(T)$	$F(T)$	$F(T)/n$	$P(z)$	$\left\lvert \frac{F(T)}{n} - P(z) \right\rvert$
68	0	0	0.0000	0.0098	0.0098
69	1	1	0.0208	0.0228	0.0020
70	1	2	0.0417	0.0478	0.0061
71	0	2	0.0417	0.0912	0.0495
72	3	5	0.1042	0.1587	0.0545
73	2	7	0.1458	0.2525	0.1067
74	7	14	0.2917	0.3694	0.0777
75	4	18	0.3750	0.5000	0.1250*
76	7	25	0.5208	0.6306	0.1098
77	6	31	0.6458	0.7475	0.1017
78	6	37	0.7708	0.8413	0.0705
79	3	40	0.8333	0.9088	0.0755
80	4	44	0.9167	0.9522	0.0355
81	1	45	0.9375	0.9772	0.0397
82	1	46	0.9583	0.9902	0.0319
83	2	48	1.0000	0.9962	0.0038
84	0	48	1.0000	0.9987	0.0013

*Maximum absolute difference

The sample value of the test statistic equals the largest difference, which is 0.1250. The critical value for a 5% level of significance, which is obtained from Appendix A-7, equals $1.36/\sqrt{n} = 0.1963$. Thus the null hypothesis cannot be rejected at the 5% level of significance.

FUNDAMENTALS OF SIMULATION

Before approaching the concept of hydrologic simulation, let's examine the question: What is simulation? Suppose we did not know that the sample mean had a normal distribution with mean μ and variance σ^2/n. How could we identify its distribution? If we assume that we cannot make a theoretical analysis, we could take an empirical approach. We could locate a population, take many samples, say m, each with sample size of n, compute the mean of each of the m samples, and derive a histogram of the m sample means. From the histogram we could hypothesize a population (both the distribution function and parameters) and use statistical hypothesis tests to test the empirical findings. But what if we do not have a population from which we can sample? Then we must build one. One means of doing this is with a random number generator, which is available as an element of most computerized statistical packages. The generator could be used to generate the m samples from which the analysis above could be conducted. This would be a basic form of simulation, that is, the generation of random numbers for the purpose of identifying the probability distribution of some random variable is one example of simulation.

Simulation should be viewed in a much broader context. Simulation is a process in which the characteristics of a system are analyzed using a model rather than the system itself. In addition to identifying statistical characteristics of a random variable, the simulation process can be used to develop decision rules, evaluate alternative performance criteria, or evaluate the benefit of alternative data sources. In Chapter 1 simulation was introduced as system synthesis, where the input and transfer function are known and the system output is unknown and must be synthesized.

Before demonstrating the use of simulation in hydrologic modeling, concepts of random number generation will be introduced.

Random Number Generation

A central element in simulation is the random number generator. In practice, a computer package will be used to generate the random numbers used in simulation; however, it is important to understand that these random numbers are generated from a deterministic process and thus are more correctly called *pseudo random numbers.* Because the random numbers are derived from a deterministic process, it is important to understand the limitations of the generators and be aware of methods for statistically analyzing the simulated values for random characteristics.

Random number generators generate numbers having specific statistical characteristics. Obviously, if the generated numbers are truly random, there is an underlying population, which can be represented by a known probability function. A single die is the most obvious example of a random number generator. If we rolled a single die many times, we could tabulate a frequency histogram. If the die were a fair die, we could expect the sample histogram to consist of six bars with almost equal heights. Of course, the histogram for the *population* would consist of six bars of equal height. Other random number generators would produce random numbers having different distributions, and when a computerized random number generator is used, it is important to know the underlying population.

In addition to dice-like devices for generating random numbers, random numbers can be generated using either physical devices or arithmetic algorithms. Physical devices, which have not proved to be highly successful, are usually electronic or radioactive based. Machines that generate electronic noise, from which the numbers are obtained, suffer because the numbers are not reproducible. Although arithmetic procedures are reproducible, they have periods. That is, because they are deterministic, once a value is repeated, the same sequence occurs. This makes it important either to know the characteristics of the generator and/or to test the generated sequence for statistical randomness.

Midsquare method. The midsquare method is one of the simplest and least reliable methods. But it is a method that illustrates problems associated with arithmetic procedures. The general procedure is:

1. Select at random a four-digit number.
2. Square the number and write the square as an eight-digit number (use preceding zeros if necessary).
3. Use the four digits in the middle as the new random number.
4. Repeat steps 2 and 3 to generate as many numbers as necessary.

As an example, consider the "seed" number of 2189. This value produces the following sequence of four-digit numbers:

$$04\underline{7917}21$$

$$62\underline{6788}89$$

$$46\underline{0769}44$$

$$00\underline{5913}61$$

$$34\underline{9635}69$$

$$92\underline{8332}25$$

$$69\underline{4222}24$$

Obviously, at some point one of these numbers must recur, which will begin the same sequence that occurred on the first pass. For example, if the four-digit number of 3500 occurred, the following sequence would result:

$$12\underline{2500}00$$

$$06\underline{2500}00$$

$$06\underline{2500}00$$

Obviously, such a sequence would not pass statistical tests for randomness. While the procedure could be used for very small samples or five-digit numbers could be used to produce 10-digit squares, the midsquare method has serious flaws that limit its usefulness.

Congruential methods. Congruential methods are based on the following relation:

$$X_{i+1} = (aX_i + c) \quad \text{mod } m \tag{2-38}$$

in which X_i is the ith value of the random variable and a, c, and m are constants. Equation 2-38 is read "X_{i+1} is congruent to $(aX_i + c)$ modulo m" and is computed by dividing $(aX_i + c)$ by m and setting X_{i+1} equal to the remainder. For example, if $a = 7$, $c = 1$, $m = 25$, and $X_0 = 3$, we would get the following sequence:

$$X_1 = [7(3) + 1] \quad \text{mod } 25 = 22$$

$$X_2 = [7(22) + 1] \quad \text{mod } 25 = 5$$

$$X_3 = [7(5) + 1] \quad \text{mod } 25 = 11$$

$$X_4 = [7(11) + 1] \quad \text{mod } 25 = 3$$

Since X_4 equals the seed, X_0, the sequence of values (22, 5, 11, 3) would be repeated continually. Thus it has a period of 4 when the seed is 3. For very large values of m, the period should also be much larger. But a sequence should be checked for randomness. There are special cases of the mixed congruential relation of Eq. 2-38. For example, if c equals zero, the relation is referred to as a *multiplicative* congruential method. Congruential methods are used frequently because their characteristics can be derived theoretically.

Statistical Tests of Random Numbers

Tests of random numbers are made because the numbers are generated using deterministic processes and there is always the chance that the recurrence period will be less than the number of random numbers generated. In making statistical tests, it is important to keep in mind that the underlying objective of the testing is to ensure that the values can be treated as values of a random variable. If we have a single sequence of numbers, the analysis is basically univariate. We seek to determine whether or not the sample is a reasonable sample from some population.

A first step in analyzing random numbers would be to perform a frequency analysis. This is especially wise if we are dealing with a uniform, extreme value, normal, or log-normal population. The method for making a frequency analysis is discussed in Chapter 4. Of course, a frequency analysis is not a statistical test of a hypothesis; it does not provide a yes or no answer to the question: can we conclude that the sequence of generated numbers is from the population specified in the null hypothesis? Thus the frequency analysis must be followed by a statistical test such as the chi-square goodness-of-fit test or the Kolmogorov–Smirnov one-sample test.

In addition to checking the underlying population, we are also interested in testing the randomness of the data. Several methods exist for this. For example, the serial correlation coefficient could be computed for one or more lags (see Chapter 9). If we assume independence of adjacent values, we would assume that the lag 1 correlation coefficient equals zero. Thus a test of significance on the correlation coefficient could be made. Rejection of the null hypothesis (i.e., H_0: $\rho = 0$) would suggest that the data are not random. However, acceptance of the null hypothesis does not assure randomness. There may be a recurring association related to some other lag.

Other types of tests are available. The poker test is used to study the occurrence of combinations of digits among five nonoverlapping digits. Assuming a uniform distribution, the exact probability of any combination of five digits can be computed theoretically. The sample probability, which is based on the frequency of occurrence in the sample, can be compared to the true probability using a hypothesis test for probabilities (or proportions). Gap tests can test the significance of the number of digits between occurrence of a specific digit. Actually, there are several types of gap tests. *Run tests*, of which there are several available, can be used to test for randomness. The most common run test, which may also be called a *sign test*, defines a single run as the occurrence of one or more digits above (or below) the median (or some other value). For example, if the numbers consist of the digits from 0 to 9, the following sequence would consist of seven runs (the median of the population is 4.5):

68 01 5 49 7 6 3 18
1 2 3 4 5 6 7 8

Another run defines a run as a sequence of increasing (or decreasing) numbers. The sequence above would include eight runs:

68 01 5 49 7 6 3 18
1 2 3 4 5 6 7 8

Details of these tests and others are available in other texts. The discussion has indicated that the test for randomness is important and that the selection of a test method involves considerable thought.

Simulation as a Decision-Making Tool

Many hydrologic systems involve components that act as a random variable or are subjected to an input that occurs randomly. For example, those persons involved in regulating releases of water from reservoirs for power, irrigation, flood control, and recreation must contend with a random input, namely rainfall; while the policy for releasing water may be deterministic (i.e., not a random variable), the output from the reservoir will still be the value of a random variable because the input will be random. For many systems, both the system and the input are random. For example, the arrival of cars at a toll booth and the time required to pay the toll are both random. Other systems involving queues, such as the demand for access to computer systems, have both randomly spaced input and random processing of the demand. Proper functioning of such systems requires planning. For hydrologic systems, such as the reservoir, there may not be a sufficient record of past rainfall to determine the best policy for releasing water from the reservoir so that benefits from power, irrigation, flood control, and recreation would be maximized. Specifically, the rainfall record may not include sequences during either long periods of drought or unusually heavy rainfall. Thus the agency involved in developing the reservoir release policy needs a method for comparing the effects of different release policies on project benefits when there is a significant degree of uncertainty involved. In the case of a toll booth or supermarket checkout counter, the manager might be interested in examining different scenarios for relieving congestion. Since it would be costly to try each possible scenario for a period of time in order to select the best operation policy, a method is needed where different scenarios can be tried without actually using the system.

Simulation provides a means of examining the output of a system when either components of the system or input to the system are random variables. Such an approach to decision making requires a model of the system with the probabilistic characteristics of the random variables. For the case of the reservoir operation policy, one would need a statement of alternative policies and a probabilistic description of the input into the reservoir. For the case of a toll booth one would need a probabilistic description of both the arrival of cars and the time required to make payment. Similarly, for the case of the supermarket checkout counter, one would need a probabilistic description of both the arrival of customers and the time required to determine the bill. The time required to determine the bill would depend on the number of items that the customer is buying, the time required to record the purchase and place the items in a grocery bag, and the time required to pay the bill (i.e., transferral of money), which would vary depending on the mode of payment (cash, check, credit card).

For those probabilistic elements of the system it is necessary to provide a probabilistic description. This includes the probability function (i.e., mass function for discrete and density function for continuous random variables) and the parameters of the function. These may be obtained from either a theoretical analysis or an empirical study. The accuracy of the decision will depend on the accuracy of both the description of the model and the specification of the probabilistic description of the random elements. When either the input to the system or a component of the system is a random variable, the output will be a random variable and should be characterized appropriately.

Once random elements of the probabilistic model have been characterized, the model can be used to simulate the behavior of the system. Simulation requires a random number generator to generate the sequence of random numbers to represent the probabilistic element. The random numbers must have the same probability function as the population of the probabilistic element. For example, if the input is the value of a random variable having a normal density function with mean μ and standard deviation σ, the random number generator must be capable of generating random numbers for this density function.

In some cases, empirical evidence will suggest that the random variable has a population that cannot be characterized by a known probability function. In such cases, the sample data can be used with a random number generator that generates random numbers having a uniform density function with values of the location and scale parameter of 0 and 1. The sample data are used to derive a cumulative frequency histogram and cumulative probability function. The cumulative probability histogram is plotted with the ordinate being the probability and the value of the random variable plotted on the abscissa. In some cases, a smooth curve can be constructed from the sample histogram. An example is shown in Fig. 2-6. The probability, which varies from 0 to 1, is shown as the ordinate, while the value of the random variable, which varies from 0 to 5, is shown on the abscissa. A smooth curve, which is shown as a dashed line in Fig. 2-6, was drawn through the sample histogram. The figure is used by generating a random number having a uniform distribution with parameters of 0 and 1 and entering the value of the random number as the ordinate, moving horizontally to the distribution (either the discrete sample curve or the fitted continuous function), and then downward to the abscissa. The value on the abscissa is the value of the random variable. As an example,

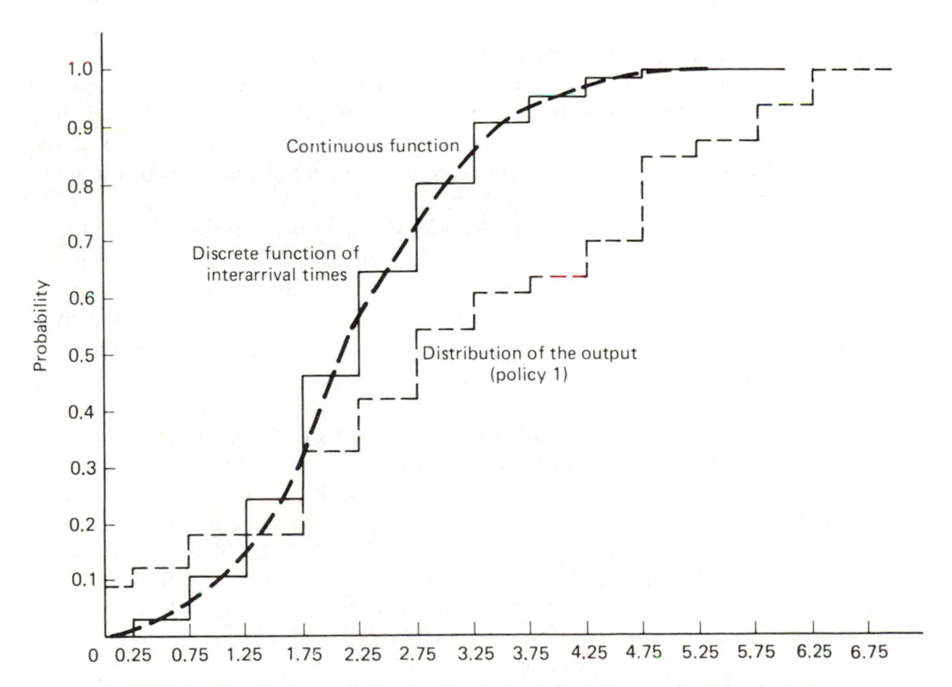

Figure 2-6 Comparison of the distributions of the interarrival times and output from the system.

if the uniform random number generator, which is available on the computer, produced a value of 0.625, a value of 2.43 would result for the random variable. This value was obtained by entering the value of 0.625 on the ordinate and moving horizontally until the dashed line is intersected; the corresponding value of X is 2.43. If the sample histogram was used, the value of the random variable would be 2.25. If a series of these values were generated using a uniform random number generator, the resulting values of the random variable would have the same distribution function as the sample.

Example 2-3: Operation of Queueing Systems

A classical problem in engineering is establishing policies for problems involving queues. The following problem illustrates the use of simulation in the evaluation of policies on the operation of an unloading dock at a waterway transportation center. The manager of the dock is interested in the operation of the station because she is aware that customers whose boats spend excessive time waiting to be unloaded will be less likely to return to the facility. While the total time that a boat spends in the unloading process is of interest, the manager is also interested in both the time spent waiting to enter the unloading facility and the time being unloaded; the total time is the sum of the two. The manager also recognizes that customers are discouraged by long queues; therefore, the number of boats waiting to be served is also a criterion of interest. The manager would be interested in examining alternatives that would reduce both the total time and the length of the queue. For example, the manager could have a second unloading facility installed, which would require increases in costs associated with installation, operation, and maintenance of the added facility. As an alternative, the manager might decide to install a new unloading system that is more efficient than the existing facility; this system would reduce the time required for unloading without the costs associated with the construction of a new facility and the added labor. Of course, there would be a cost associated with the installation of the more efficient unloading system.

Since the manager is most concerned with times of peak demand, she collects data on the current operation of the facility. Using records from previous peak demand periods, she measures the time between arrivals of ships and the time required for unloading. The manager finds that the unloading time can be approximated by a uniform probability function with a mean of 2 days; based on measurements she believes the unloading time can be approximated with probabilities of 0.25 for times of 1.25, 1.75, 2.25, and 2.75 days; these are the center points of the computational interval of 0.5 day. During the period in which data were gathered, a total of 650 ships were unloaded. The time between arrivals is shown in Table 2-8 for 0.5-day intervals. During the time of peak demand all interarrival times were 5 days or less, with an average of 2.2 days. Because the interarrival times did not follow a known probability function, the empirical function was used for simulations. The discrete cumulative function for the interarrival times is shown in Fig. 2-6. The discrete function is used because the data were collected on a 0.5-day interval; the center point of each interval is used to represent the interval. The derivation of the cumulative function for the interarrival times is given in Table 2-8.

To demonstrate the simulation technique a 60-day period was simulated. The interarrival times were simulated by generating a sequence of random numbers from

TABLE 2-8 Computation of Cumulative Probability Function of Interarrival Times

Interarrival Time (days)	Number of Arrivals	Probability Function	Cumulative Function
0.0–0.5	20	0.031	0.031
0.5–1.0	50	0.077	0.108
1.0–1.5	90	0.138	0.246
1.5–2.0	140	0.215	0.462
2.0–2.5	120	0.185	0.646
2.5–3.0	100	0.154	0.800
3.0–3.5	70	0.108	0.908
3.5–4.0	30	0.046	0.954
4.0–4.5	20	0.031	0.985
4.5–5.0	10	0.015	1.000
	650	1.000	

a uniform distribution having location and scale parameters of 0 and 1, respectively. Each of the numbers in the sequence was entered into Fig. 2-6 and the interarrival times determined. The interarrival times are given in Table 2-9 with values ranging between 0.25 and 4.75 days on a 0.5-day interval. The second ship arrived at the unloading area 1.25 days after the first ship. The third ship arrived at a time of 3 days, which is 1.75 days after the second ship. The time of arrival of each ship is also given in Table 2-9.

A second sequence of random numbers having a uniform probability function ($\alpha = 0$, $\beta = 1$) was generated. Since these values are used for simulating the time required for unloading, there are four values (i.e., 1.25, 1.75, 2.25, and 2.75 days) with an equal probability of occurrence. The random numbers were used to generate the unloading times, which are shown in Table 2-9, for the existing situation; this is called policy 1. The time of departure from the unloading facility equals the sum of the unloading time and either the time that the previous ship leaves the facility or the time that the ship arrives at the facility when another ship is not being unloaded. The waiting time for a ship to be unloaded, which is also given in Table 2-9, equals the difference between the time of arrival and the time when the ship begins unloading. The total time equals the sum of the waiting time and unloading time. The number of ships in the queue is determined by comparing the time of departure of a ship and the time of arrival of ships that follow the ship being unloaded. For example, the eleventh ship departs on day 21.75. At that time, ships 12, 13, 14, and 15 have already arrived at the unloading facility (on days 17.75, 19.50, 20.75, and 21.00), so there are four ships waiting in the queue to be unloaded when ship 11 departs.

The manager is interested in determining the effect of a new unloading facility, which would be expected to decrease the unloading time. The simulation based on this time distribution is referred to as policy 2. The probabilities for unloading times of 1.25, 1.75, 2.25, and 2.75 days are 0.35, 0.30, 0.20, and 0.15, respectively. The same sequence of arrivals was used to simulate the effect of policy 2. The unloading times and times of departure were computed and are given in Table 2-9. The total time, waiting time, and number of ships in the queue were computed in the same way as

TABLE 2-9 Simulations of Interarrival and Unloading Times for Alternative Policies

					Policy 1					Policy 2				
Individual	Random Number	Interarrival Time (days)	Time of Arrival (days)	Random Number	Unloading Time (days)	Time of Departure (days)	Waiting Time (days)	Total Time (days)	Number Waiting in Queue	Unloading Time (days)	Time of Departure (days)	Waiting Time (days)	Total Time (days)	Number Waiting in Queue
1	0.19	1.25	0	0.57	2.25	2.25	0	2.25	1	1.75	1.75	0	1.75	1
2	0.37	1.75	1.25	0.84	2.75	5.00	1.00	3.75	2	2.25	4.00	0.50	2.75	2
3	0.09	0.75	3.00	0.54	2.25	7.25	2.00	4.25	2	1.75	5.75	1.00	2.75	2
4	0.30	1.75	3.75	0.25	1.25	8.50	3.50	4.75	2	1.25	7.00	2.00	3.25	1
5	0.76	2.75	5.50	0.95	2.75	11.25	3.00	5.75	2	2.75	9.75	1.50	4.25	1
6	0.53	2.25	8.25	0.14	1.25	12.50	3.00	4.25	2	1.25	11.00	1.50	2.75	1
7	0.09	0.75	10.50	0.21	1.25	13.75	2.00	3.25	2	1.25	12.25	0.50	1.75	1
8	0.26	1.75	11.25	0.51	2.25	16.00	2.50	4.75	2	1.75	14.00	1.00	2.75	1
9	0.43	1.75	13.00	0.16	1.25	17.25	3.00	4.25	2	1.25	15.25	1.00	2.25	1
10	0.13	1.25	14.75	0.39	1.75	19.00	2.50	4.25	2	1.75	17.00	0.50	2.25	1
11	0.25	1.75	16.00	0.85	2.75	21.75	3.00	5.75	4	2.25	19.25	1.00	3.25	1
12	0.42	1.75	17.75	0.82	2.75	24.50	4.00	6.75	4	2.25	21.50	1.50	3.75	3
13	0.31	1.75	19.50	0.16	1.25	25.75	5.00	6.25	4	1.25	22.75	2.00	3.25	2
14	0.15	1.25	20.75	0.14	1.25	27.00	5.00	6.25	3	1.25	24.00	2.00	3.25	2

15	0.01	0.25	21.00	0.80	2.75	29.75	6.00	8.75	4	2.25	26.25	3.00	5.25	2
16	0.54	2.25	23.25	0.24	1.25	31.00	6.50	7.75	3	1.25	27.50	3.00	4.25	2
17	0.28	1.75	25.00	0.12	1.25	32.25	6.00	7.25	3	1.25	28.75	2.50	3.75	2
18	0.50	2.25	27.25	0.87	2.75	35.00	5.00	7.75	4	2.75	31.50	1.50	4.25	2
19	0.12	1.25	28.50	0.06	1.25	36.25	6.50	7.75	3	1.25	32.75	3.00	4.25	1
20	0.67	2.75	31.25	0.33	1.75	38.00	5.00	6.75	3	1.25	34.00	1.50	2.75	1
21	0.22	1.25	32.50	0.44	1.75	39.75	5.50	7.25	3	1.75	35.75	1.50	3.25	1
22	0.54	2.25	34.75	0.49	1.75	41.50	5.00	6.75	3	1.75	37.50	1.00	2.75	1
23	0.58	2.25	37.00	0.22	1.25	42.75	4.50	5.75	2	1.25	38.75	0.50	1.75	1
24	0.18	1.25	38.25	0.13	1.25	44.00	4.50	5.75	2	1.25	40.00	0.50	1.75	0
25	0.53	2.25	40.50	0.03	1.25	45.25	3.50	4.75	1	1.25	41.75	0.00	1.25	0
26	0.75	2.75	43.25	0.79	2.75	48.00	2.00	4.75	2	2.25	45.50	0.00	2.25	0
27	0.76	2.75	46.00	0.01	1.25	49.25	2.00	3.25	2	1.25	47.25	0.00	1.25	1
28	0.05	0.75	46.75	0.27	1.75	51.00	2.50	4.25	1	1.25	48.50	0.50	1.75	0
29	0.57	2.25	49.00	0.18	1.25	52.25	2.00	3.25	1	1.25	50.25	0.00	1.25	0
30	0.69	2.75	51.75	0.28	1.75	54.00	0.50	2.25	0	1.25	53.00	0.00	1.25	0
31	0.61	2.25	54.00	0.14	1.25	55.25	0.00	1.25	0	1.25	55.25	0.00	1.25	0
32	0.83	3.25	57.25	0.06	1.25	58.50	0.00	1.25	1	1.25	58.50	0.00	1.25	1
33	0.00	0.25	57.50	0.65	2.25	60.75	1.00	3.25	0	1.75	60.25	1.00	2.75	0
34	0.81	3.25	60.75	0.34	1.75	62.50	0.00	1.75	—	1.25	62.00	0.00	1.25	—

TABLE 2-10 Computation of Cumulative Probability Function of Waiting Times

Waiting Time (days)	Policy 1			Policy 2		
	Frequency	Probability Function	Cumulative Function	Frequency	Probability Function	Cumulative Function
0.0	3	0.091	0.091	8	0.242	0.242
0.5	1	0.030	0.121	6	0.182	0.424
1.0	2	0.061	0.182	6	0.182	0.606
1.5	0	0.000	0.182	6	0.182	0.788
2.0	5	0.152	0.333	3	0.091	0.879
2.5	3	0.091	0.424	1	0.030	0.909
3.0	4	0.121	0.545	3	0.091	1.000
3.5	2	0.061	0.606			
4.0	1	0.030	0.636			
4.5	2	0.061	0.697			
5.0	5	0.152	0.848			
5.5	1	0.030	0.878			
6.0	2	0.061	0.939			
6.5	2	0.061	1.000			
	33	1.000		33		

TABLE 2-11 Distributions of Total Time in Systems and Unloading Time for Two Policies

Total Time (days)	Frequency for Policy 1	Frequency for Policy 2	Unloading Time (days)	Frequency for Policy 1	Frequency for Policy 2
1.25	2	7	1.25	16	20
1.75	1	5	1.75	7	7
2.25	2	3	2.25	4	5
2.75	0	7	2.75	7	2
3.25	4	5		34	34
3.75	1	2			
4.25	5	4			
4.75	4	0			
5.25	0	1			
5.75	4				
6.25	2				
6.75	3				
7.25	2				
7.75	3				
8.25	0				
8.75	1				
	34	34			

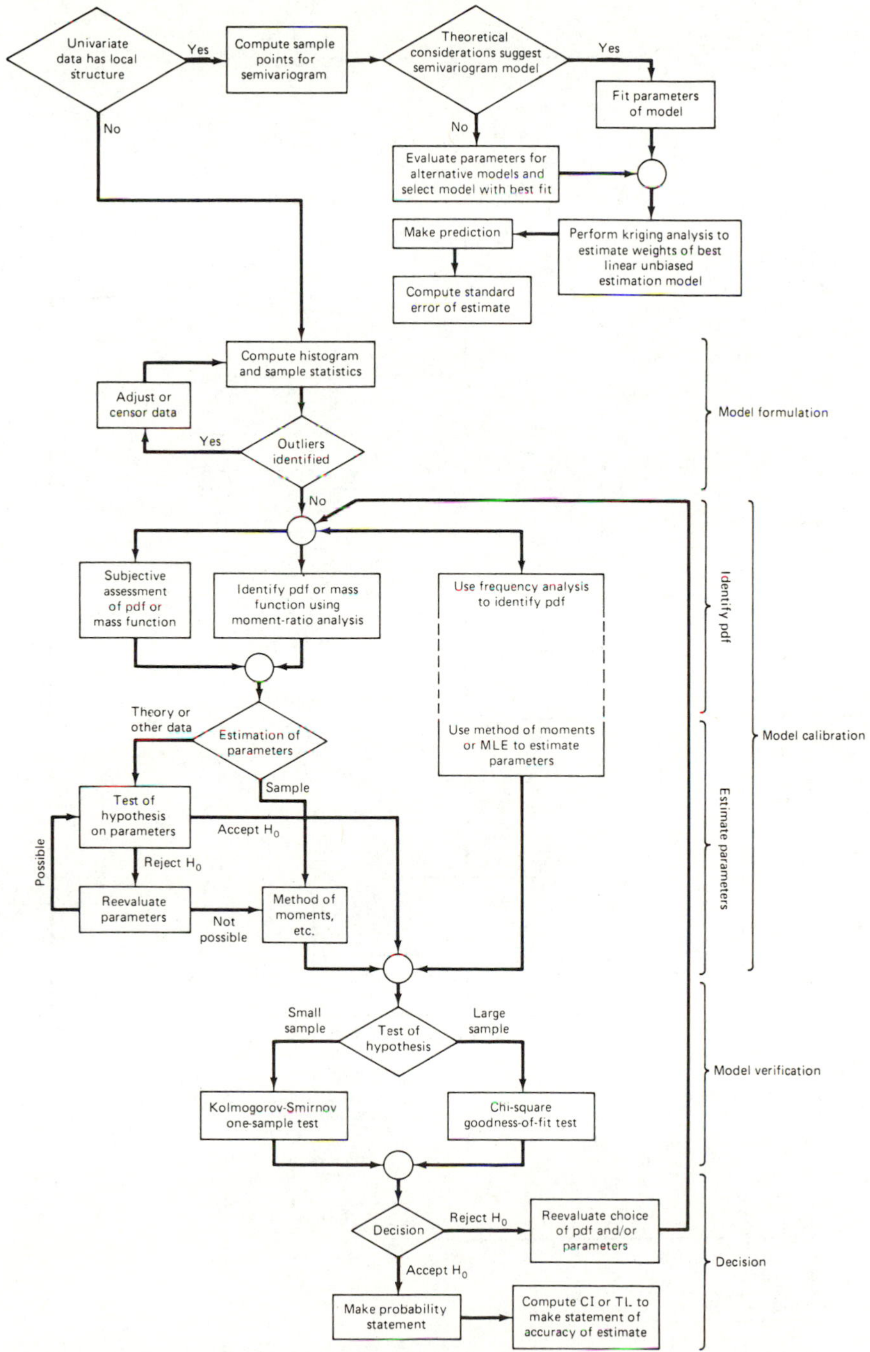

Figure 2-7 Flowchart for statistical decisionmaking of univariate prediction problems.

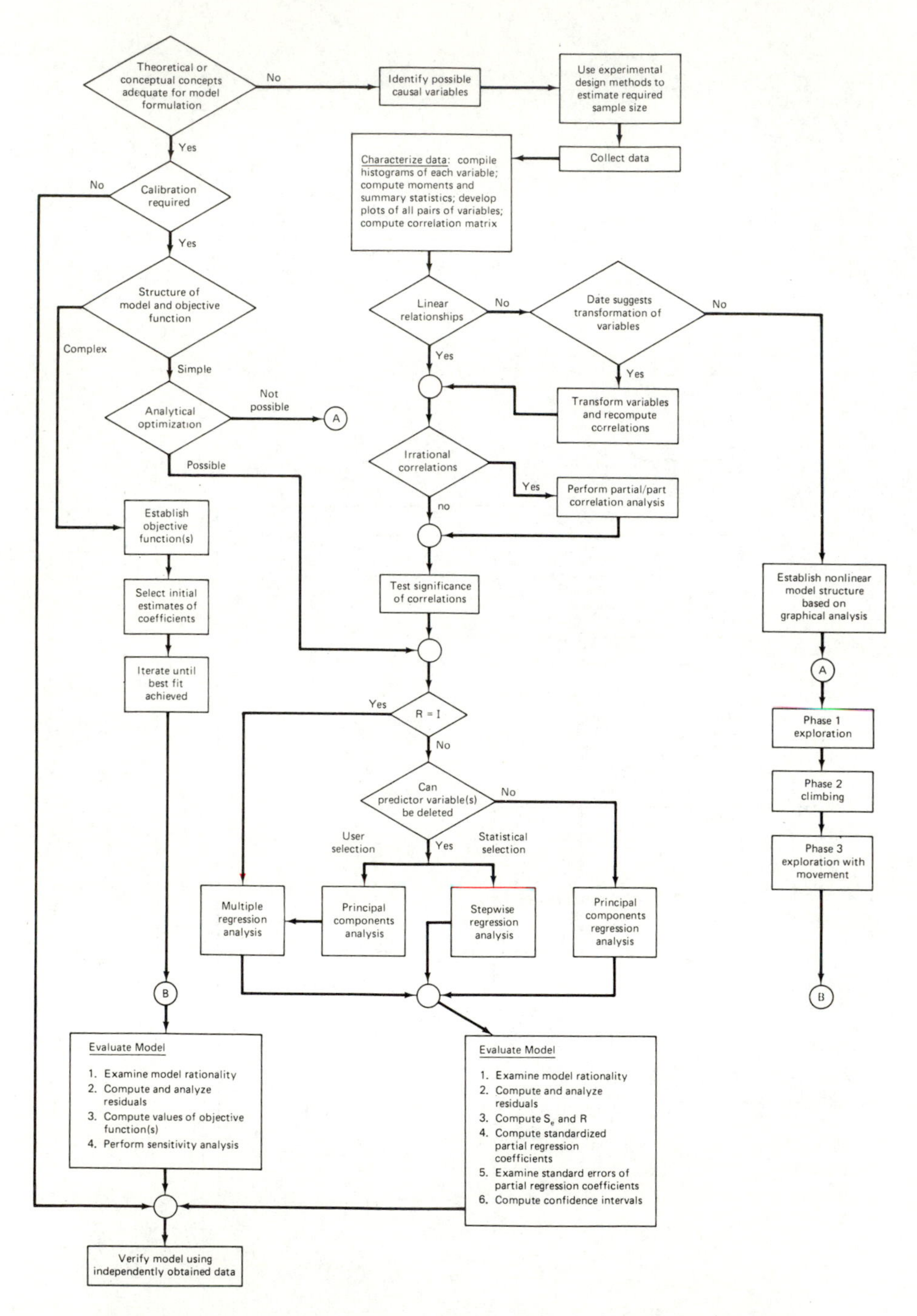

Figure 2-8 Flowchart for statistical decisionmaking of multivariate prediction problems.

for policy 1. The distribution of waiting times is given in Table 2-10. It is quite evident that the waiting time is considerably less with policy 2 compared with policy 1. The mean waiting time for policies 1 and 2 are 3.27 and 1.08 days, respectively. The mean unloading times were 1.78 and 1.59 days, respectively, for policies 1 and 2. The expected mean values were 2.0 and 1.825 days, respectively. It is evident from both the mean values and the distribution of unloading times shown in Table 2-11 that the sample values for the simulated period of 60 days are less than what would be expected. As the length of the period of simulation is increased, one can expect the sample values to be better approximations of the true values.

THE DATA ANALYSIS PROCESS

The greatest concern of students of statistical methods is in knowing which method to apply in solving a specific problem. This concern results from a failure of instructors and textbooks to emphasize the sequential nature of statistical decision making. In most texts different examples are used to illustrate each statistical method, and topics in the various chapters of a text are not tied together. Thus the student fails to see the "big picture." The problem of estimating a population mean can be used to illustrate this. The mean of a population can be estimated using sample data; Eq. 2-8 would be the appropriate "statistical method" to use. But the student of statistics should not just be satisfied with the sample estimate of the mean. The accuracy of the estimate should be assessed using a confidence interval. Equations 2-28 provide confidence intervals on the mean. This sequential approach to estimation is important whether the mean is of interest or the problem involves a more complex estimation problem, such as kriging (Chapter 5) or regression (Chapters 3 and 6). In each case we are interested in a sequential process that includes both estimation and an evaluation of the expected accuracy of the estimate. While the estimation of the mean illustrates the problem, other data analysis problems involve more steps in the sequence. Hydrologic modeling is just one example. The modeling process consists of three phases: formulation, calibration, and verification. An understanding of the number of statistical methods that are necessary for each of these phases is important. The flowcharts of Figs. 2-7 and 2-8 provide a framework for statistical decision making with hydrologic models. The flowcharts show the sequential nature of statistical decision making. Although most of the terms will not be understood at this time, the reader is encouraged to review the flowcharts as new techniques are introduced. Hopefully, this will show how the different methods fit together and make it possible for one to select the appropriate method(s) in solving problems that require a statistical solution.

EXERCISES

2-1. If the experiment is the roll of a fair die, differentiate between the population and a sample.

2-2. If the variable is the mean annual precipitation at a rain gage, differentiate between the population and a sample.

2-3. Find the value of k that makes the following a legitimate probability density function and compute the cumulative probability function.

(a) $f(X) = kX$ for $1 \le X \le 3$

(b) $f(X) = kX$ for $X = 1, 2, 3, 4$

(c) $f(X) = 0.1 + kX$ for $0 \le X \le 2$

(d) $f(X) = \begin{cases} kX & \text{for } 0 \le X \le 1 \\ k & \text{for } 1 \le X \le 2 \end{cases}$

2-4. Calculate the first moments of the functions of Exercise 2-3.

2-5. For the functions of Exercise 2-3, calculate the second moments about the mean.

2-6. A series of flood peaks at the outlet of a watershed are as follows: 210, 270, 340, 230, 280, and 170 cubic feet per second (cfs). Compute the mean and variance of the series.

2-7. Using the flood data of Exercise 2-6, find the mean and variance of the logarithms. Compare these values with the logarithms of the mean and variance computed in Exercise 2-6.

2-8. Obtain a sequence of annual maximum peak discharges from the *U.S.G.S. Water Supply Papers* and derive values for the first four moments and the three dimensionless coefficients of Eqs. 2-19. Use a record of at least 15 years.

2-9. Transform the record of Exercise 2-8 by taking the logarithms of the discharges and recompute both the four moments and the three dimensionless coefficients.

2-10. Assuming that the soil moisture (X) in a field is normally distributed with a mean of 9% and a standard deviation of 2.5%, find the probability that **(a)** $X > 12\%$; **(b)** $X < 8.1\%$; **(c)** $9.2\% < X < 11.3\%$; **(d)** $X > 7.4\%$; **(e)** $x < 5\%$ or $X > 13\%$.

2-11. Assuming that daily evaporation rates (E) from a lake are normally distributed with a mean ($\bar{E}$) of 0.18 in./day and a standard deviation of 0.07 in./day, find E_0 for **(a)** $P(E > E_0) = 0.05$; **(b)** $P(E < E_0) = 0.10$; **(c)** $P(|E - \bar{E}| > 0.05$ in./day$) = 0.2$.

2-12. Using the data of Exercise 2-6, assume a normal distribution and find **(a)** $P(Q > 400 \text{ cfs}) = ?$; **(b)** $P(Q > 150 \text{ cfs}) = ?$; **(c)** $P(Q > Q_0) = 0.01$; **(d)** $P(Q > Q_0) = 0.1$; **(e)** $P(Q < Q_0) = 0.02$.

2-13. Using the data of Exercise 2-8, assume a normal distribution and find the discharges for which the probabilities of exceedence are 0.01, 0.02, 0.04, 0.1, 0.2, and 0.5.

2-14. Using the mean and standard deviation of the logarithms (see Exercise 2-9) and assuming that the logarithms of the discharges are normally distributed, find the values of $\log Q$ for which the probabilities of exceedence are 0.01, 0.02, 0.04, 0.1, 0.2, and 0.5. Convert these to cfs and compare the values with those of Exercise 2-13.

2-15. For the t distribution, find either the value of t, the degrees of freedom, or the probability: **(a)** $P(t > 1.833; \nu = 9) = ?$; **(b)** $P(t < -2.093; \nu = 19) = ?$; **(c)** $P(t >$

$-2.947;\ \nu = 15) = ?$; **(d)** $P(t > 2.056;\ \nu = ?) = 0.025$; **(e)** $P(t > t_0;\ \nu = 2) = 0.005$; **(f)** $P(t < -t_0;\ \nu = 27) = 0.10$; **(g)** $P(t < -t_0 \text{ or } t > t_0;\ \nu = 16) = 0.05$.

2-16. For the chi-square distribution, find either the value of ν, the degrees of freedom, or the probability: **(a)** $P(\chi^2 > 17.275;\ \nu = 11) = ?$; **(b)** $P(\chi^2 < 2.204;\ \nu = 6) = ?$; **(c)** $P(\chi^2 < 12.443 \text{ or } \chi^2 > 28.412;\ \nu = 20) = ?$; **(d)** $P(\chi^2 > 38.968;\ \nu = ?) = 0.02$; **(e)** $P(\chi^2 > \chi_0^2;\ \nu = 5) = 0.05$; **(f)** $P(\chi^2 < \chi_0^2;\ \nu = 18) = 0.10$; **(g)** $P(\chi^2 < \chi_0^2;\ \nu = 13) = 0.90$.

2-17. For the F distribution, find the value of F_0 for each of the following: **(a)** $P(F > F_0;\ \nu_1 = 10,\ \nu_2 = 20) = 0.05$; **(b)** $P(F > F_0;\ \nu_1 = 20,\ \nu_2 = 10) = 0.05$; **(c)** $P(F > F_0;\ \nu_1 = 10,\ \nu_2 = 10) = 0.05$; **(d)** $P(F > F_0;\ \nu_1 = 10,\ \nu_2 = 10) = 0.01$; **(e)** $P(F > F_0;\ \nu_1 = 2,\ \nu_2 = 5) = 0.01$.

2-18. Indicate whether or not the binomial distribution could be used to define probabilities for each of the following random variables (X) and state the reason for your response: **(a)** 10 rolls of a fair die with X being 1, 2, 3, 4, 5, or 6; **(b)** five flips of a fair coin with X being a head; **(c)** six rolls of a fair die with X being the occurrence of an even number of dots; **(d)** 20 weeks of a flood record with the random variable being the overlapping of a coffer dam during any one week.

2-19. Find the following probabilities for a binomially distributed random variable: **(a)** $b(2; 5, 0.1)$; **(b)** $b(4; 6, 0.4)$; **(c)** $B(2; 5, 0.1)$; **(d)** $B(0; 5, 0.2)$; **(e)** $B(0; 5, 0.8)$; **(f)** $B(4; 8, 0.9)$; **(g)** $b(3; 9, 0.7)$.

2-20. What is the probability of a flood with an annual exceedence probability of 0.05 **(a)** not being exceeded in the next 15 years; **(b)** being exceeded exactly twice in the next 20 years; **(c)** not being exceeded more than once in the next 10 years; **(d)** being exceeded in two successive years?

2-21. Compute the one-sided upper confidence interval on the mean of the data of Exercise 2-6 for a confidence level of **(a)** 95% and **(b)** 99%.

2-22. Compute a two-sided 90% confidence interval on the mean for the data of Exercise 2-8.

2-23. Show the effect of sample size on the width of a confidence interval using the data of Exercise 2-6 by computing the width of a two-sided confidence interval assuming that the computed mean and standard deviation are based on sample sizes of 6, 12, 18, 24, and 30.

2-24. Compute a two-sided 95% tolerance limit that places bounds on 99% of the distribution of the population from which the sample of Exercise 2-6 was drawn.

2-25. Compute a one-sided 99% tolerance limit that places an upper bound on 90% of the distribution of the population from which the sample of Exercise 2-8 was drawn.

2-26. Write and execute a computer program to derive an approximation of the sampling distribution of the mean. Use a random number generator that generates standard normal deviates with a mean of zero and a variance of one. Generate 1000 samples of size n and compute the mean of each sample. Develop a histogram using the 1000 values and plot both the histogram and the normal distribution. Determine the number of sample means that lie in each tail for probabilities of 0.01, 0.025, 0.05, and 0.10. Execute the program for sample sizes of 5, 10, and 25. Compare these simulated values with the values estimated using the corresponding confidence interval.

2-27. What four factors influence the critical value of a test statistic? Show pictorially how each factor affects the critical value.

2-28. The frequency histogram of annual maximum discharges for the Piscataquis River is given in the table below. Using a 5% level of significance, test whether or not it can be assumed that this random variable has a normal distribution with **(a)** $\mu = \bar{X}$ and $\sigma = s$, where $\bar{X} = 8634$ cfs and $s = 4095$ cfs; **(b)** $\mu = 8000$ cfs and $\sigma = 4000$ cfs.

Cell Limits (cfs)		
Minimum	Maximum	Frequency
0	4,319	6
4,319	8,137	29
8,137	11,955	12
11,955	15,773	7
15,773	19,591	3
19,591	∞	1

2-29. Daily pan evaporation data were collected for 1 year at a site near Atlanta, Georgia. The frequency histogram of the daily values is given below, with the sample mean and standard deviation equal to 0.14 in. and 0.094 in., respectively. Using a level of significance of 1% and the method of moments for parameter estimation, test the hypothesis that the data are from a population having a **(a)** uniform distribution; **(b)** normal distribution.

Cell Limits (in.)		
Minimum	Maximum	Frequency
0.0100	0.0344	47
0.0344	0.0833	77
0.0833	0.1322	74
0.1322	0.1811	53
0.1811	0.2300	42
0.2300	0.2789	42
0.2789	0.3767	21
0.3767	0.4256	3
0.4256	0.4500	4
0.4500	0.5000	2

2-30. A sample of 20 yields a mean of 32.4. Test the two-sided hypothesis that the sample was drawn from a population with a mean of 35: **(a)** if the variance of the population is 33; **(b)** if the variance of the population is unknown, but the sample variance is 33. Use a level of significance of 5%.

2-31. A random sample of 10 has a mean of 110. Test the null hypothesis that $\mu = 120$ against the alternative hypothesis that $\mu < 120$ at the 5% level of significance. **(a)** Assume that the population standard deviation of 18 is known. **(b)** Assume that the population standard deviation is unknown and the sample value is 18.

2-32. A public water supply official claims that the average household water use is greater than 350 gal/day. To test this assertion, a random sample of 200 households are questioned. If the random sample showed an average use of 308 gal/day and a standard deviation of 35 gal/day, would it be safe to conclude at the 1% level of significance that μ is greater than 350 gal/day?

2-33. The standard deviation of the 58 annual maximum discharges for the Piscataquis River was 4095 cfs. Test the hypothesis that the population standard deviation is 3500 cfs. Use a 5% level of significance.

2-34. The standard deviation of the 58 annual maximum discharges for the Piscataquis River was 4095 cfs. Regional studies indicate that the standard deviation for that watershed should be at least 4500 cfs. Using a 1% level of significance, study whether or not the sample value of σ can reasonably be at least 4500 cfs.

2-35. The downtimes for a university computer system in hours per week for the last 13 weeks are given below. Test the null hypothesis that the downtime can be represented by an exponential distribution $f(t) = Ke^{-Kt}$, where K can be estimated by $K = 1/\bar{X}$. Use $\alpha = 0.20$.

$$\{1.3, 1.0, 1.0, 1.0, 0.0, 1.0, 0.0, 1.3, 0.0, 0.4, 1.0, 0.0, 0.4\}$$

2-36. The data given below are the residuals (e) from a linear bivariate regression analysis and the values of the predictor variable X. Can one reasonably conclude that the residuals are normally distributed? Use $\alpha = 0.01$.

X	e	X	e	X	e	X	e
1	−2	11	1	21	−4	31	−1
2	3	12	3	22	2	32	−2
3	0	13	−6	23	1	33	1
4	−5	14	2	24	−1	34	3
5	2	15	3	25	2	35	2
6	1	16	−2	26	1	36	4
7	3	17	4	27	−3	37	−2
8	−2	18	−1	28	−2	38	−3
9	−1	19	−3	29	4	39	−1
10	−4	20	−2	30	2	40	−1

2-37. The following are 20 soil moisture percentages obtained from field samples. Can one conclude that the soil moisture can be represented by a normal distribution? Use the method of moments and a 5% level of significance.

8.9	10.6	7.3	11.1	12.2	8.3	9.5	7.9	10.3	9.2
12.3	11.7	13.1	10.4	13.2	11.6	10.3	9.5	12.2	8.1

2-38. For the data of Table 2-7, test the null hypothesis that the random variable T is normally distributed with a mean of 80°F and a standard deviation of 3.25°F. Use a 1% level of significance.

3 Fundamentals of Optimization

INTRODUCTION

A useful means of separating data analysis methods is on the basis of the number of variables involved. Part II will be devoted to methods of dealing with a single variable, and thus they are called *univariate methods.* Sets of data that include more than one variable are analyzed using methods (called *multivariate methods* or more correctly, *multiple variable methods*) that are different from those used in univariate analyses. Multivariate methods are discussed in Part III. However, the two cases, univariate and multivariate, have a common thread in that the data are analyzed to derive a model, which, in turn, is used to make predictions. In any empirical model, there are one or more unknowns, which must be assessed before the models can be used to make predictions. The process of deriving values for the unknowns is called *optimization.* It is a necessary step in both univariate and multivariate analysis.

In many engineering projects, estimates of random variables are required. For example, in the design of irrigation projects, it is necessary to provide estimates of evaporation. If we are fortunate to have a past record of daily evaporation rates from a pan, the product of the mean value of these observations and some proportionality constant may be our best estimate. In this case the optimization only involves estimating the mean value. The material in Chapter 2 showed that confidence intervals could be used to indicate the accuracy of the estimated value of the mean. Estimation using the mean is the simplest form of estimation for univariate systems.

If the random variable can be related to other variables, it may be possible to reduce the estimation error compared with errors obtained with univariate analyses. For example, evaporation is a function of the temperature and humidity of the air mass that is over the water body. Thus, if measurements of the air temperature and the relative humidity are also available, a relationship, or model, can be developed. The relationship may provide a more accurate estimate of evaporation for the conditions that may exist in the future. However, since the model will be more complex than a mean value, the optimization process will be more complex. The additional effort required to calibrate or optimize a multivariate model will be worthwhile only if the accuracy of estimation is improved significantly over the accuracy of the univariate model.

DEFINITIONS

A fundamental component of the modeling process is the equation relating two or more variables. The variable for which values must be estimated is called the *criterion* or *dependent variable*; the criterion variable represents the response of the system. For example, in estimating evaporation rates for the design of ponds for irrigation, evaporation would be the criterion variable. It is important to note that the criterion variable is quite often referred to as the dependent variable; this term reflects the dependency of the response variable on other variables in the system.

The objective of regression is to evaluate the coefficients of an equation relating the criterion variable to one or more other variables, which are called the *predictor variables.* Predictor variables are believed to cause variation in the criterion variable. A predictor variable is often called an independent variable; this is a misnomer in that independent variables are usually neither independent of the criterion variable nor independent of the other predictor variables.

The most frequently used linear model relates a criterion variable Y to a single predictor variable X by the equation

$$\hat{Y} = b_0 + b_1 X \tag{3-1}$$

in which b_0 is the intercept coefficient and b_1 is the slope coefficient; b_0 and b_1 are often called *regression coefficients* because they are obtained from a regression analysis. Because Eq. 3-1 involves two variables, Y and X, it is sometimes referred to as the *bivariate model.* The intercept coefficient represents the value of Y when X equals zero. The slope coefficient represents the rate of change in Y with respect to change in X. While b_0 has the same dimensions as Y, the dimensions of b_1 equal the ratio of the dimensions of Y to X.

The *linear multivariate model* relates a criterion variable to two or more predictor variables:

$$\hat{Y} = b_0 + b_1 X_1 + b_2 X_2 + \cdots + b_p X_p \tag{3-2}$$

in which p is the number of predictor variables, X_i the ith predictor variable, b_i the ith slope coefficient, and b_0 the intercept coefficient, where $i = 1, 2, \ldots, p$. The coefficients b_i are often called *partial regression coefficients* and have dimensions equal to the ratio of the dimensions of Y to X_i. Equation 3-1 is a special case of Eq. 3-2.

ELEMENTS OF OPTIMIZATION

There are basically four components of optimization: the objective function, the model, the data, and constraints. All components are not necessary for any one problem. For example, many optimization problems do not involve constraints. Although the following discussion will specifically relate to the optimization of multivariate systems, the discussion is generally applicable to univariate optimization. In fact, univariate optimization should be viewed as a special case of the more general optimization problem.

Each of the four components of the optimization process needs further discussion. First, the function to be optimized is called the *objective function*, which is an explicit mathematical function that describes what is considered the optimal solution. Second, there is a *mathematical model*, such as Eqs. 3-1 or 3-2, that relates the criterion variable to a vector of unknowns (i.e., coefficients) and a vector of predictor variables. The vector of unknowns is the focal point of optimization. The unknowns are the values that are necessary to transform values of the predictor variable(s) into a predicted value of the criterion variable. It is important to note that the objective function and the mathematical model are two separate explicit functions. The third element of optimization is a *data set*. The data set consists of measured values of the criterion variable and the predictor variable(s). In summary, optimization includes (1) an objective function; (2) a mathematical model, which is an explicit function relating a criterion variable to vectors of unknowns and predictor variable(s); and (3) a matrix of measured data. In some cases, one or more *constraints* on the vector of unknowns may be involved.

As an example, one may attempt to relate evaporation (E), the criterion variable, to temperature (T), which is the predictor variable, using the equation

$$\hat{E} = b_0 + b_1 T \tag{3-3}$$

in which b_0 and b_1 are the unknown coefficients, and $\hat{E}$ is the predicted value of E. Because E must be positive and E should increase with increases in T, one may constrain the solution for b_0 and b_1 such that they are nonnegative. To evaluate the unknowns, a set of simultaneous measurements on E and T would be made. For example, if we were interested in daily evaporation rates, we might measure the total evaporation for each day in a year and the corresponding average temperature; this would give us 365 observations from which we could estimate the unknowns. An objective function would have to be selected to evaluate the unknowns; for example, regression minimizes the sum of the squares of the differences between the predicted and measured values of the criterion variable. Other objective functions, however, may be used.

CLASSIFICATION OF OPTIMIZATION METHODS

Optimization methods can be separated into three general categories: (1) analytical, (2) numerical, and (3) subjective. These methods differ by the way in which the objective function is evaluated. *Analytical optimization* uses analytical calculus in deriving the unknowns from the objective function. The coefficients are evaluated numerically in *numerical optimization*; in this case, a finite difference scheme is most often employed. *Subjective optimization* is a trial-and-error process that relies heavily on the users' knowledge of both the model and the physical system being modeled.

Each of the three categories has advantages and disadvantages. Analytical techniques provide a direct solution and will result in an exact solution, if one exists. Analytical methods usually require less time to find a solution. However, this occurs partly because analytical methods are practical only for models that have a simple mathematical structure. The solution procedure becomes considerably more complex when constraints are involved.

Numerical optimization can be used with models with a moderately complex structure, and it is quite easy to include constraints on the unknowns in the solution. However, numerical optimization most often requires a considerable number of iterations (i.e., evaluations of the objective function for different values of the unknowns) in order to approach the optimum solution. Furthermore, the solution usually is not exact, and it is also necessary to provide initial estimates of the unknowns.

While the solution procedures for both analytical and numerical optimization are systematic, subjective optimization is usually nonsystematic. This may be considered as either an advantage or a disadvantage. It is an advantage in that it permits the user to apply his or her knowledge of both the system being modeled and the model in obtaining estimates of the unknowns. On the other hand, it most often results in a nonreproducible solution; that is, two users would probably not obtain the same optimum values for the unknowns. Subjective optimization is most often used with very complex models that involve many unknowns. Because of this, many iterations are required and it is difficult to evaluate the sensitivity of the unknowns.

ANALYTICAL OPTIMIZATION

The derivative of an objective function is used to find the value of an unknown within the function, with the resulting value of the objective function either the minimum or maximum value. For example, if the function $q(b)$ is equal to $8b - b^2$, we may be interested in the value of b that makes $q(b)$ either a minima or maxima. For a function $f(b)$ that is a function of a single unknown b, the following theorem is used:

> If at a point p_0 on a curve $F = f(b)$, $f'(b_0) = 0$ and $f''(b_0) > 0$, where b_0 is the value of the unknown at the point p_0, the point p_0 is a relative minimum; however, if $f'(b_0) = 0$ and $f''(b_0) < 0$, the point p_0 is a relative maximum; if $f'(b_0) = 0$ and $f''(b_0) = 0$, more information is needed to clarify the point p_0.

If both $f'(b_0) = 0$ and $f''(b_0) = 0$, the theorem is applied with the third and fourth derivatives. One can conclude, then, that b_0 is an optimum if, and only if, the lowest-order nonvanishing derivative is nonzero and of even order (i.e., positive for a minima and negative for a maxima). An inflection point occurs when $f'(b_0) \neq 0$ and $f''(b_0) = 0$.

Many engineering systems involve a function that involves two or more unknowns. Therefore, it is also of interest to examine the theorem that applies to a function $f(b_0, b_1)$ that is a function of the two unknowns b_0 and b_1. For the following theorem, let $\bar{b}_0$ and $\bar{b}_1$ be the values of b_0 and b_1 at the optimum. Also, let f'_0 and f'_1 be the first partial derivatives of f with respect to b_0 and b_1, respectively; similarly, we will use f''_{00}, f''_{11}, and f''_{01} to denote the second partial derivatives $\partial^2 f/\partial b_0^2$, $\partial^2 f/\partial b_1^2$, and $\partial^2 f/\partial b_0\,\partial b_1$, respectively.

> If the function f with two unknowns b_0 and b_1 has a relative optimum at the point $Q_0(\bar{b}_0, \bar{b}_1)$, both partial first derivatives must vanish at Q_0:
>
> $$f'_0(\bar{b}_0, \bar{b}_1) = 0 \qquad \text{and} \qquad f'_1(\bar{b}_0, \bar{b}_1) = 0$$
>
> If these equalities hold and if at $(\bar{b}_0, \bar{b}_1)$, $(f''_{01})^2 - f''_{00}f''_{11} < 0$, then for
>
> $$f''_{00}(\bar{b}_0, \bar{b}_1) < 0$$
>
> $f(b_0, b_1)$ has a relative maximum at $(\bar{b}_0, \bar{b}_1)$; and for
>
> $$f''_{00}(\bar{b}_0, \bar{b}_1) > 0$$
>
> $f(b_0, b_1)$ has a relative minimum at $(\bar{b}_0, \bar{b}_1)$; if
>
> $$(f''_{01})^2 - f''_{00}f''_{11} < 0$$
>
> then $f(b_0, b_1)$ has neither a maxima or a minima at $(\bar{b}_0, \bar{b}_1)$; in this case, $(\bar{b}_0, \bar{b}_1)$ is a *saddle point.*

It is of interest to note that the optimization problem for the explicit function $f(b_0, b_1)$ consists of basically two elements, the function and the unknowns.

The Principle of Least Squares

Regression analysis is a frequently used analytical method for developing hydrologic models. The objective function equals the mean square error, and the objective of least squares is to minimize the mean square error. Since the unknowns or coefficients of the model are the only variables whose values are free to vary, the principle of *least squares* is a process of finding the values of the coefficients that achieves the objective. For example, the least-squares solution of the model of Eq. 3-1 would lead to values of b_0 and b_1 that minimize the objective function. In this sense, the coefficients are considered to be the "best" estimates. *Regression* is the tendency for the expected value of one or two jointly correlated random variables to approach more closely the mean value of its set than any other value. The principle of least squares is used to regress Y on either the X or the X_i values of Eqs. 3-1 and 3-2, respectively. To express the principle of least squares, it is important to define the error, e, or residual, as the difference between the predicted and measured value of the criterion variable:

$$e_i = \hat{Y}_i - Y_i \tag{3-4}$$

in which $\hat{Y}_i$ is the ith predicted value of the criterion variable, Y_i is the ith measured value of Y, and e_i is the ith error. It is important to note that the error is defined as the measured value of Y subtracted from the predicted value. Some computer programs use the measured value minus the predicted value; however, this definition indicates that a positive residual implies underprediction. With Eq. 3-4 a positive residual indicates overprediction, while a negative residual indicates underprediction. The objective function (F) for the principle of least squares is to minimize the sum of the squares of the errors:

$$F = \min \sum_{i=1}^{n} (\hat{Y}_i - Y_i)^2 \tag{3-5}$$

in which n is the number of observations on the criterion variable (i.e., the sample size).

The objective function of Eq. 3-5 can be minimized by taking the derivatives with respect to each unknown, setting the derivatives equal to zero, and then solving for the unknowns. The solution requires the model for predicting Y_i to be substituted into the objective function. It is important to note that the derivatives are taken with respect to the unknowns b_i and not the predictor variables X_i.

To illustrate the solution procedure, the model of Eq. 3-1 is substituted into the objective function of Eq. 3-5, which yields

$$F = \sum_{i=1}^{n} (b_0 + b_1 X_i - Y_i)^2 \tag{3-6}$$

The derivatives of Eq. 3-5 with respect to the unknowns are

$$\frac{\partial F}{\partial b_0} = 2 \sum_{i=1}^{n} (b_0 + b_1 X_i - Y_i) = 0 \tag{3-7a}$$

$$\frac{\partial F}{\partial b_1} = 2 \sum_{i=1}^{n} (b_0 + b_1 X_i - Y_i) X_i = 0 \tag{3-7b}$$

Dividing the equations by 2, separating the terms in the summation, and rearranging yields the set of normal equations

$$nb_0 + b_1 \sum X = \sum Y \tag{3-8a}$$

$$b_0 \sum X + b_1 \sum X^2 = \sum XY \tag{3-8b}$$

All the summations in Eq. 3-8 are calculated over all values of the sample; in future equations, the index values of the summation will be omitted when they refer to summation over all elements in a sample of size n. The subscripts for X and Y have been omitted but are inferred. The two unknowns b_0 and b_1 can be evaluated by solving the two simultaneous equations

$$b_1 = \frac{\sum XY - \sum X \sum Y/n}{\sum X^2 - (\sum X)^2/n} \tag{3-9a}$$

$$b_0 = \bar{Y} - b_1 \bar{X} = \frac{\sum Y}{n} - \frac{b_1 \sum X}{n} \tag{3-9b}$$

The values of b_0 and b_1 provided by Eqs. 3-9 provide the minimum value of F (Eq. 3-5).

The principle of least squares is often disassociated from analytical optimization. In fact, it is often identified solely as a statistical method. Although it is a statistical method, the procedure described here uses the concepts of analytical optimization. The only difference is that the objective function involves a discrete set of points, while the theorems provided earlier assumed explicit functions. In spite of this difference, the analytical solution of the unknowns of Eqs. 3-1 and 3-2 represents an example of analytical optimization.

Lagrangian Optimization

As indicated previously, the modeling process involves four components: a model structure, a data set, an objective function, and constraints. For most problems in analytical optimization, the unknowns are not subjected to constraints. However, in many cases involving statistical analysis of data, unconstrained optimization leads to coefficients that are irrational in sign. For example, a model that regresses evaporation on temperature may result in an intercept coefficient that is negative in sign. Physically, this would imply that there is negative evaporation for low temperatures. If one is attempting to place physical significance on the empirical coefficients, they should at least be rational. Thus one may wish to find an optimum set of empirical coefficients subject to one or more constraints, such as constraining the intercept coefficient for the evaporation model to be nonnegative.

The most common method of constrained optimization is Lagrangian optimization of an explicit function. Quite frequently, problems involving constraints on the unknowns are solved using a numerical optimization method rather than attempting an analytical solution. A constrained analytical solution may be preferred to the numerical solution because there is less chance of locating a nonoptimum solution with the analytical approach.

The linear form of Eq. 3-1 is an explicit function involving two unknowns (b_0 and b_1). Values for the unknowns are usually derived using the objective function for a least-squares solution. The usual practice is to find the optimum values of the unknowns without constraining the values of the unknowns. A constrained solution can be obtained by introducing a new objective function. In *Lagrangian optimization* the constraints are written in a form such that the constraint equals zero and are introduced into the objective function by multiplying the zeroed constraint by a constant called a *Lagrange multiplier*, λ. If there is more than one constraint, a multiplier must be introduced for each constraint. Thus the objective function for a least-squares solution would have the form

$$F = \min \sum_{i=1}^{n} e_i^2 - \sum_{j=1}^{m} \lambda_j w_j \tag{3-10}$$

in which e_i is the ith residual (i.e., $\hat{Y}_i - Y_i$), n the sample size, m the number of constraints, λ_j the Lagrange multiplier for the jth constraint, and w_j the jth constraint expressed in a form such that it equals zero. Since all the λ_j values are constants and unknown, it is necessary to differentiate F with respect to each of the unknowns in the model structure and with respect to each of the m values of λ_j. The derivatives are set equal to zero and used to form a set of normal equations.

While Lagrangian optimization is usually used to solve problems in which the objective function is an explicit function, Lagrangian optimization can also be used with statistical objective functions. The method will be demonstrated using the linear bivariate model and the least-squares objective function. For the model $\hat{Y} = b_0 + b_1 X$ in which the intercept coefficient is constrained to be equal to some value c (which is given), we can formulate the objective function as

$$F = \min \sum_{i=1}^{n} (\hat{Y}_i - Y_i)^2 - \lambda(b_0 - c) \tag{3-11}$$

The derivatives of F with respect to the unknowns (b_0, b_1, and λ) are set equal to zero, and expressions for the unknowns are derived. In this case we get

$$b_0 = c \tag{3-12}$$

$$b_1 = \frac{\sum YX - c\sum X}{\sum X^2} \tag{3-13}$$

$$\lambda = cn + b_1 \sum X - \sum Y \tag{3-14}$$

in which each summation is over the n sample values. For example, if we have the data of Table 3-1 and the constraint that $b_0 = 1$, we would get values for b_0, b_1, and λ of 1, $\frac{8}{15}$, and $\frac{1}{3}$, respectively.

TABLE 3-1 Data for Lagrangian Optimization with the Constraint That b_0 Equals a Constant

X	Y	X^2	XY
1	2	1	2
2	1	4	2
3	2	9	6
4	4	16	16
—	—	—	—
10	9	30	26

If we wished to constrain the predicted values to be greater than some constant Y_0, we could formulate the following objective function:

$$F = \min \sum (\hat{Y} - Y)^2 - \lambda(b_0 + b_1 X_0 - Y_0) \tag{3-15}$$

in which X_0 is the value of X when Y equals Y_0. The solution for this problem is

$$b_0 = Y_0 - b_1 X_0 \tag{3-16}$$

$$b_1 = \frac{(Y_0/X_0)\sum X - (1/X_0)\sum XY + \sum Y - Y_0 n}{2\sum X - X_0 n - (1/X_0)\sum X^2} \tag{3-17}$$

$$\lambda = nY_0 - \sum Y + b_1(\sum X - X_0 n) \tag{3-18}$$

If we use the data of Table 3-1 with the constraint that the predicted value must be greater than 2, then, since this occurs for the data of Table 3-1 when $X_0 = 1$, we get values of $\frac{23}{14}$, $\frac{5}{14}$, and $\frac{8}{7}$ for b_0, b_1, and λ, respectively. While the Lagrangian optimization process allows constraints to be applied, it will not necessarily provide either an unbiased model or a minimum mean square error for the least-squares criterion. It will, however, provide the minimum mean square error given the constraint. Also, it should be obvious that if the measured data are used to set X_0 in Eq. 3-15, the constraint will not hold if a smaller value of X_0 is used. Thus it is important that either the data base contain the extreme event or the value of the extreme event be set prior to solving for b_0, b_1, and λ.

RELIABILITY OF A PREDICTION MODEL

In Chapter 2 the mean was shown to be the best estimate of a random variable. The standard error of the mean was used as a measure of the reliability of the estimate. Confidence intervals were used to indicate the likely range of the population value. In summary, the decision making involved making both an estimate of the variable and a statement of the reliability of the estimate. The same applies to decision making involving two or more variables. The models of Eqs. 3-1 and 3-2 can be used to provide "best" estimates, but one should also be interested in the reliability of the estimate.

Having evaluated the coefficients of the prediction equation, it is of interest to evaluate the reliability of the regression equation. The following criteria should be assessed in evaluating the model: (1) the correlation coefficient; (2) the standard error of estimate; (3) the F statistic for the analysis of variance; (4) the rationality of the coefficients and the relative importance of the predictor variables, both of which can be assessed using the standardized partial regression coefficients; and (5) the degree to which the underlying assumptions of the model are met.

The Correlation Coefficient

The correlation coefficient (R) is an index of the degree of linear association between two random variables. The magnitude of R indicates whether or not the regression will provide accurate predictions of the criterion variable. Thus R is often computed before the regression analysis is performed in order to determine whether or not it is worth the effort to perform the regression; however, R is always computed after the regression analysis because it is an index of the goodness of fit. The correlation coefficient measures the degree to which the measured and predicted values agree and is used as a measure of the accuracy of future predictions. It must be recognized that if the measured data are not representative of the population (i.e., data that will be observed in the future), the correlation coefficient will not be indicative of the accuracy of future predictions.

The square of the correlation coefficient (R^2) equals the percentage of the variance in the criterion variable that is explained by the predictor variable; this will be shown below (Eq. 3-23). Because of this physical interpretation, R^2 is a meaningful indicator of the accuracy of predictions.

The Standard Error of Estimate

In the absence of additional information, the mean is the best estimate of the criterion variable; the standard deviation S_y of Y is an indication of the accuracy of prediction. If Y is related to one or more predictor variables, the error of prediction is reduced from S_y to the standard error of estimate, S_e. Mathematically, the standard error of estimate equals the standard deviation of the errors and has the same units as Y:

$$S_e = \left[\frac{1}{\nu} \sum_{i=1}^{n} (\hat{Y}_i - Y_i)^2 \right]^{0.5} \tag{3-19}$$

in which ν is the degrees of freedom, which equals the sample size minus the number of unknowns. For the bivariate model of Eq. 3-1, $p = 1$ and $\nu = n - 2$. For the general linear model with an intercept, Eq. 3-2, there are $(p + 1)$ unknowns; thus $\nu = n - p - 1$. It is important to note that S_e is based on $(n - p - 1)$ degrees of freedom while the error, S_y, is based on $(n - 1)$ degrees of freedom. Thus in some cases S_e may be greater than S_y. To assess the reliability of the regression equation, S_e should be compared with the bounds of zero and S_y. If S_e is near S_y, the regression has not been successful. If S_e is much smaller than S_y and is near zero, the regression analysis has improved the reliability of prediction.

The standard error of estimate is sometimes computed by

$$S_e = S_y \sqrt{1 - R^2} \tag{3-20}$$

Equation 3-20 must be considered only as an approximation to Eq. 3-19 because R is based on n degrees of freedom and S_y is based on $(n - 1)$ degrees of freedom. Using the following separation of variation, the exact relationship between S_e, S_y, and R can be computed:

$$\text{TV} = \text{EV} + \text{UV} \tag{3-21}$$

The total variation (TV) is related to the variance (S_y^2) of Y by

$$S_y^2 = \frac{\text{TV}}{n - 1} \tag{3-22}$$

The correlation coefficient is the ratio of the explained variation (EV) to the total variation (TV):

$$R^2 = \frac{\text{EV}}{\text{TV}} \tag{3-23}$$

and the standard error of estimate is related to the unexplained variation (UV) by

$$S_e^2 = \frac{\text{UV}}{n - p - 1} \tag{3-24}$$

Equation 3-22 can be solved for TV, which can then be substituted into both

Eqs. 3-21 and 3-23; Eq. 3-23 can then be solved for EV, which can be substituted into Eq. 3-21. Equation 3-24 can be solved for UV, which is also substituted into Eq. 3-21. Solving for S_e^2 yields

$$S_e^2 = \frac{n-1}{n-p-1} S_y^2(1 - R^2) \tag{3-25}$$

Thus Eq. 3-25 is a more exact relationship than Eq. 3-20; however, for large sample sizes the difference between the two estimates should be small.

While S_e may actually be greater than S_y, in general, S_e will be within the range from zero to S_y. When the equation fits the data points exactly, S_e equals zero; this corresponds to a correlation coefficient of 1. When the correlation coefficient equals zero, S_e equals S_y; as indicated previously, S_e may actually exceed S_y when the degrees of freedom have a significant effect. The standard error of estimate is often preferred to the correlation coefficient because S_e has the same units as the criterion variable and its magnitude is a physical indicator of the error.

The Analysis of Variance

The regression coefficient (b_1) for the predictor variable X in Eq. 3-1 is a measure of the change in Y that results from a change in X. Recognizing that the regression coefficient is dependent on the units of both Y and X, it is reasonable to ask whether or not a change in X causes a significant change in Y. This is easily placed in the form of a statistical hypothesis test:

$$H_0: \quad \beta_1 = 0 \tag{3-26a}$$

$$H_A: \quad \beta_1 \neq 0 \tag{3-26b}$$

in which β_1 is the population value of the slope coefficient. The hypotheses are designed to test whether or not the relationship between Y and X is significant; this is exactly the same (in concept as well as quantitatively) as testing the hypothesis that the population correlation coefficient equals zero.

The hypothesis test for testing Eq. 3-26 relies on the computations involved in the separation of variation concept. Thus it represents an *analysis of variance* (ANOVA). The total variation is represented by the total sum of squares SS_T:

$$SS_T = \sum Y^2 - \frac{(\sum Y)^2}{n} \tag{3-27}$$

The total sum of squares is separated into the regression (SS_R), or explained variation, and the error (SS_E) sums of squares:

$$SS_R = b_1\left[\sum XY - \frac{\sum X \sum Y}{n}\right] \tag{3-28}$$

$$SS_E = SS_T - SS_R \tag{3-29}$$

in which b_1 is the computed value of the slope coefficient. The mean squares, which correspond to variances, equal the sums of squares divided by the degrees of freedom, which are shown in Table 3-2.

TABLE 3-2 ANOVA Table for a Bivariate Regression

Source of Variation	Sum of Squares	Degrees of Freedom	Mean Squares	F
Regression	SS_R	1	$\frac{SS_R}{1}$	$\frac{MS_R}{MS_E}$
Error	SS_E	$n-2$	$\frac{SS_E}{n-2}$	
Total	SS_T	$n-1$	—	

The null hypothesis is tested for significance using the ratio of the mean square for regression to the mean square for error:

$$F = \frac{MS_R}{MS_E} \tag{3-30}$$

in which F is the computed value of a random variable having an F distribution with 1 and $(n-2)$ degrees of freedom. For a stated level of significance, α, the null hypothesis is accepted if F is less than the critical value, F_α. The region of rejection consists of all values of F greater than F_α.

It is of interest to compare R, S_e, and the F for the ANOVA. The value of R is based on the ratio of EV to TV. The value of S_e is based on UV. The computed value of F is based on EV and UV. Thus the three values that are used to examine the goodness of fit of the data to the linear model are based on the individual elements of the separation of variation concept. The commonality is the reason that the ANOVA test and a test of hypothesis that R is equal to zero must lead to the same conclusion.

The Standardized Partial Regression Coefficient

Because the partial regression coefficient (i.e., b_1 of Eq. 3-1) is dependent on the units of both Y and X, it is often difficult to use its magnitude to measure the rationality of the model. The partial regression coefficient can be standardized by

$$t = \frac{b_1 S_x}{S_y} \tag{3-31}$$

in which t is called the *standardized partial regression coefficient*, and S_x and S_y are the standard deviations of the predictor and criterion variables, respectively. Because the units of b_1 are equal to the ratio of the units of Y to the units of X, t is dimensionless. For the bivariate regression model t equals the correlation coefficient of the regression of the standardized value of Y on the standardized value of X. Thus this suggests that t is a measure of the relative importance of the corresponding predictor variable. The value of t will have the same sign as b_1; therefore, if the sign of t is irrational, one can conclude that the model is not rational. Although it is more difficult to assess the rationality of the magnitude of a regression coefficient than it is to assess the rationality of the sign, the magnitude of the standardized partial regression coefficient can be used to assess the rationality of the model. For rational models t must vary in the range $-1 < t < 1$. Because t equals R for the bivariate model of Eq. 3-1, the value will never fall outside this range; this is not true for multivariate models, such as Eq. 3-2.

Assumptions Underlying the Regression Model

The principle of least squares assumes that the errors (i.e., the differences between the predicted and measured values of the criterion variable) (1) are independent of each other, (2) have zero mean, (3) have a constant variance across all values of the predictor variables, and (4) are normally distributed. If any of these assumptions are violated, one must assume that the model structure is not correct. Violations of these assumptions are easily identified using statistical analyses of the residuals.

When the sum of the residuals does not equal zero, it reflects a bias in the model. The regression approach, when applied analytically, requires that the sum of the residuals equals zero; however, if an inadequate number of significant digits are computed for the regression coefficients, the sum of the residuals may not equal zero. A model may be biased even when the sum of the residuals equals zero. For example, Fig. 3-1(a) shows an $X - Y$ plot in which the trend of the data is noticeably nonlinear, with the linear regression line also shown. If the errors e_i are computed and plotted versus the corresponding values of the predictor variable, a noticeable trend appears [Fig.

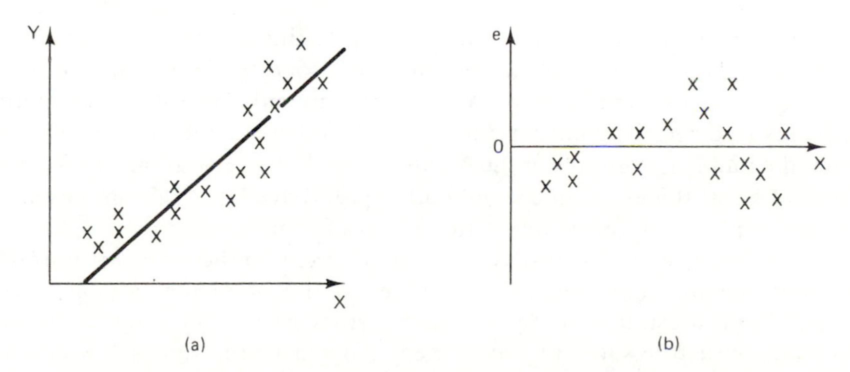

Figure 3-1 Biased linear regression model: (a) $X_i - Y_i$ plot; (b) $e_i - X_i$ plot.

3-1(b)]. The errors are positive for both low and high values of X, while the errors show a negative bias of prediction for intermediate values of X. The trends in the residuals suggest a biased model that should be replaced by a model that has a different structure.

The population model for linear regression has the form

$$Y = \beta_0 + \beta_1 X + \epsilon \tag{3-32}$$

in which β_0 and β_1 are the population values of the intercept and slope coefficients, respectively; and ϵ is a random variable having a normal distribution with zero mean and a constant variance. The variance of ϵ equals the square of the standard error of estimate; therefore, the population of the residuals is defined as

$$\epsilon \sim N(0, \sigma_e^2 = \text{constant}) \tag{3-33}$$

If the error variance (σ_e^2) is not constant across all values of the predictor and criterion variable, the underlying model assumptions are not valid and a more accurate model structure should be identified. The data shown in Fig. 3-2 reveal a trend in the error variance; specifically, as X increases, the error variance increases. Thus the following relationship is apparent:

$$\epsilon \sim N(0, \sigma_e^2 = f(X)) \tag{3-34}$$

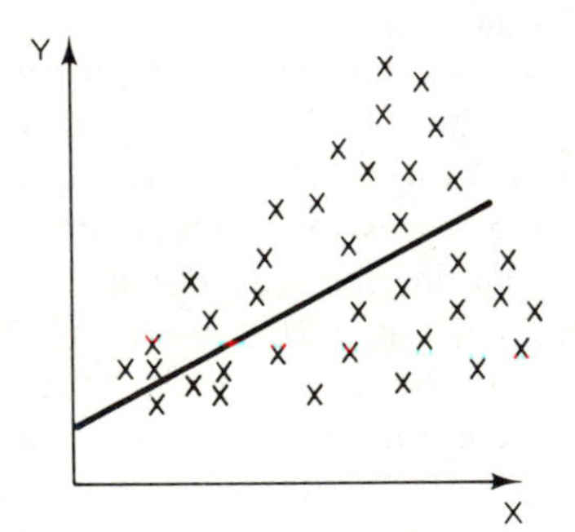

Figure 3-2 Model with a nonconstant error variance.

Although an exact test does not exist for testing a data set for a nonconstant error variance, it is not unreasonable to separate the errors into two or more sets based on the value of X and use a hypothesis test for comparing variances to detect a significant difference. A systematic method for selecting either the number of groups or the points at which the separation will be made does not exist; therefore, one should attempt to have large subsamples and a rational explanation for choosing the points of separation.

The linear regression procedure also assumes that the errors are normally distributed across the range of the values of the predictor variables. This assumption is illustrated in Fig. 3-3. If the errors are not normally distributed, the model structure should be examined. An exact procedure for testing for normality does not exist; however, in a manner similar to the test for constant

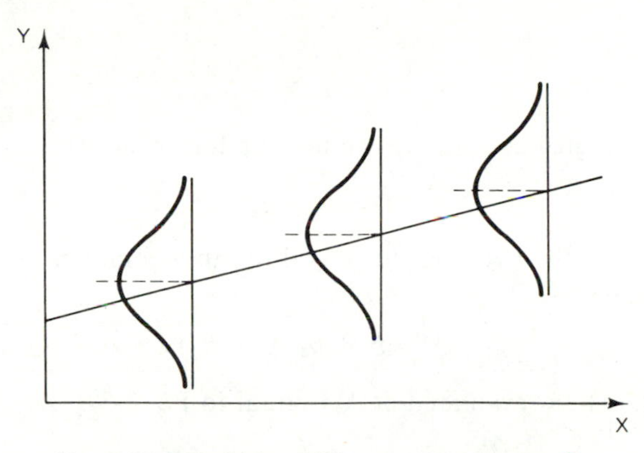

Figure 3-3 Model in which $\epsilon \sim N(0, \text{constant } \sigma_\epsilon^2)$.

variance the residuals could be separated into groups having similar values of X and tests, such as a rank-order frequency analysis, the chi-square goodness-of-fit test, or the Kolmogorov–Smirnov test, could be performed to study the normality question.

The fourth assumption deals with the independence of the observations. This is not usually a problem except when time or space is used as a predictor variable. In such cases, the measurement of Y for one time period may not be completely independent of the measurement for an adjacent time period. A nonparametric test called the run test and/or a serial correlation coefficient may (see Chapter 9) be used to check for statistical independence; statistical independence should not be used in place of physical independence.

COMPOSITE FITTING

For some types of problems, a single equation will not be adequate for accurately fitting a set of data. For example, if the data appear to follow a linear trend over part of the data and a nonlinear trend over another interval, inaccurate solutions will result from attempts to use a single functional form. In other words, a single functional form is not sufficiently flexible to represent the data over the entire range of the data.

One possible solution to this problem is called *composite fitting.* This involves increasing the number of coefficients in the fitting process by using two functional forms, but including constraints so that there are no discontinuities in the resulting model. Consider the case where the data exhibit a linear trend for part of the data and a parabolic trend for another interval. At the same time, we want a model where there is a smooth point of intersection and the two forms have the same slope at that point. If we use a second-order polynomial to represent the data over the range $0 \leq X \leq X_c$, where X_c is the value of the predictor variable at the point of intersection, in this range the model is

$$\hat{Y} = b_0 + b_1 X + b_2 X^2 \tag{3-35}$$

For values of X greater than X_c, we use the linear model

$$\hat{Y} = b_3 + b_4 X \tag{3-36}$$

We can constrain the two lines to pass through a common point (Y_c, X_c) by rewriting Eq. 3-36 as

$$\hat{Y} = b_3 + b_4(X - X_c) \tag{3-37}$$

and setting Eq. 3-35 (evaluated at X_c) equal to Eq. 3-37:

$$b_0 + b_1 X_c + b_2 X_c^2 = b_3 + b_4(X_c - X_c) \tag{3-38}$$

Thus the two lines pass through the point (Y_c, X_c) when

$$b_3 = b_0 + b_1 X_c + b_2 X_c^2 \tag{3-39}$$

For the slopes of the two functional forms to be equal at the point of intersection, we can take the derivatives of Eqs. 3-35 and 3-36 and evaluate them at the point (Y_c, X_c):

$$\left.\frac{dY}{dX}\right|_{X=X_c} = b_1 + 2b_2 X_c \tag{3-40}$$

and

$$\left.\frac{dY}{dX}\right|_{X=X_c} = b_4 \tag{3-41}$$

For the parabola and the straight line to have the same slope at X_c, Eqs. 3-40 and 3-41 can be equated:

$$b_4 = b_1 + 2b_2 X_c \tag{3-42}$$

Thus Eq. 3-35 is used for $X \leq X_c$ and the following is used for $X > X_c$:

$$Y = b_0 + b_1 X_c + b_2 X_c^2 + (b_1 + 2b_2 X_c)(X - X_c) \tag{3-43}$$

which was obtained by substituting Eqs. 3-39 and 3-42 into Eq. 3-37. Equation 3-43 can be simplified to

$$Y = b_0 + b_1 X + b_2(2X_c X - X_c^2) \tag{3-44}$$

The values of b_0, b_1, and b_2 are the same in Eqs. 3-35 and 3-44 for both ranges and must be evaluated across both ranges simultaneously.

This can be illustrated using the data of Table 3-3 and the assumption that the intersection occurs at $X = 3$ (i.e., $X_c = 3$). The values of b_0, b_1, and b_2 can be computed by conventional least squares using the values of X^2 for

TABLE 3-3 Data for Example of Composite Fitting with $X_c = 3$

Y	X	X^2 or $(2X_cX - X_c^2)$
1.5	0.5	0.25
2.0	1.0	1.00
1.0	1.5	2.25
2.0	2.0	4.00
1.5	2.5	6.25
2.5	3.0	9.00
4.5	4.0	15.00
4.0	5.0	21.00
5.5	6.0	27.00
6.5	6.0	27.00

$X \leq X_c$ and $(2X_cX - X_c^2)$ for $X > X_c$. Least squares gives $b_0 = 2.0$, $b_1 = -0.829$, and $b_2 = 0.329$. Equation 3-35 is used for $X \leq X_c$ and Eq. 3-44 for $X > X_c$:

$$Y = \begin{cases} 2.0 - 0.829X + 0.329X^2 & \text{for } X \leq 3 \quad \text{(3-45a)} \\ 2.0 - 0.829X + 0.329(6X - 9) & \text{for } X > 3 \quad \text{(3-45b)} \end{cases}$$

The parabolic equation is plotted in Fig. 3-4 for the range $0 \leq X \leq 3$, and the straight line is plotted for $X > 3$. The intersection at $X = 3$ is smooth, with the two curves held in partial restraint.

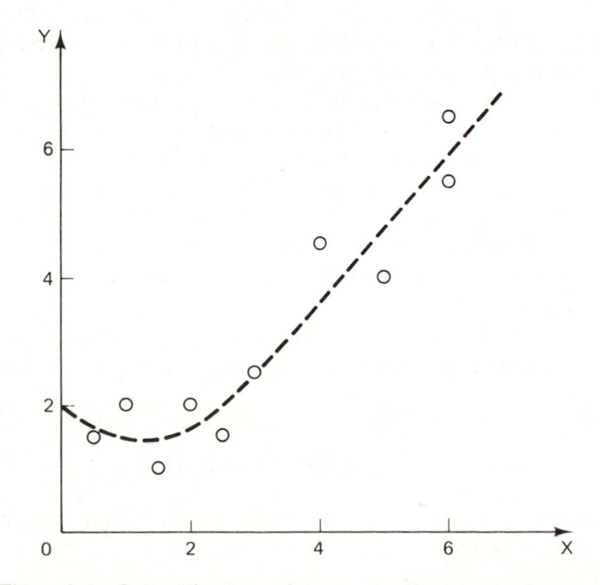

Figure 3-4 Composite regression equation for data of Table 3-3.

APPLICATIONS IN HYDROLOGIC ANALYSIS

The topic of analytical optimization using the principle of least squares will be discussed in detail in Chapter 6. The purpose of the discussion in this chapter is only to introduce the fundamental concepts. Two simple examples will be used to demonstrate the most basic concepts of analytical optimization.

A common problem in hydrologic analysis is the transfer of measured data from one location to another. Quite frequently, measured data must be transferred from a gaged site to an ungaged site. Another transfer problem that requires a model is the transfer of information to a gaged site where part of a record is missing. Example 3-1 will be used to illustrate the use of regression in developing a model that involves data from two sites.

One should not infer from the material in this chapter that the principle of least squares applies only to the linear models of Eqs. 3-1 and 3-2. In some problems, a model should not include an intercept (b_0 of Eqs. 3-1 and 3-2); that is, some physical conditions assume that the relationship between a criterion variable and one or more predictor variables passes through the origin. Example 3-2 will demonstrate how to use the principle of least squares for evaluating data when a zero-intercept model is considered appropriate.

Example 3-1: Spatial Correlation of Temperature Data

Temperature data are an important input to a number of hydrologic analyses, most notably the development of evaporation and snowmelt-runoff models. In most cases, it is necessary to adjust the data that are measured at one site for use at the point of interest. For example, data measured over a long time span can be transferred to a site where only a short data record exists. The period when data were measured at both sites is used to calibrate a model that can be used to transfer the remaining record at the long record site to the short record site.

The mean annual temperatures at Atlanta and Savannah, Georgia, for the period 1946–1966 are given in Table 3-4. The 21-year record for the two stations can be used to demonstrate the use of regression in developing a model for the spatial transfer of data. The temperature at the Atlanta station will be used as the predictor variable, while the temperature at the Savannah station will be used as the criterion variable. The mean temperatures for Atlanta and Savannah are 62.9°F and 67.0°F, respectively. The standard deviations for Atlanta and Savannah are 2.64°F and 2.69°F, respectively. The sums and sums of squares and cross-products are as follows: $\sum t_A = 1{,}321.5$, $\sum t_S = 1{,}407.5$, $\sum t_A^2 = 83{,}300.07$, $\sum t_S^2 = 94{,}480.71$, and $\sum t_A t_S = 88{,}696.24$. Thus we get the following normal equations for the bivariate model of Eq. 3-1:

$$21b_0 + 1321.5b_1 = 1407.5 \tag{3-46a}$$

$$1321.5b_0 + 83{,}300.07b_1 = 88{,}696.24 \tag{3-64b}$$

The two normal equations can be solved for the values of b_0 and b_1, which yields the following prediction equation:

$$\hat{t}_S = 11.1 + 0.888t_A \tag{3-47}$$

TABLE 3-4 Mean Annual Temperature (°F) for Atlanta (t_A) and Savannah (t_S), Georgia

Year	t_A	t_S	$\hat{t}_S$	$\hat{t}_S - t_S$
1946	62.1	68.4	66.3	2.1
1947	66.4	70.8	70.1	0.7
1948	60.4	64.0	64.8	−0.8
1949	67.2	72.6	70.8	1.8
1950	65.9	69.2	69.7	−0.5
1951	65.1	70.1	69.0	1.1
1952	59.0	63.3	63.5	−0.2
1953	64.0	67.1	68.0	−0.9
1954	64.5	66.0	68.4	−2.4
1955	61.9	65.8	66.1	−0.3
1956	64.8	67.8	68.7	−0.9
1957	58.4	63.1	63.0	0.1
1958	62.0	63.9	66.2	−2.3
1959	64.8	70.3	68.7	1.6
1960	65.7	69.3	69.5	−0.2
1961	61.1	65.2	65.4	−0.2
1962	62.7	68.3	66.8	1.5
1963	65.1	66.7	69.0	−2.3
1964	60.5	65.5	64.9	0.6
1965	60.7	65.7	65.0	0.7
1966	59.2	64.4	63.7	0.7
				−0.1

in which t_A is the mean annual temperature at Atlanta and $\hat{t}_S$ is the predicted mean annual temperature at Savannah. The line representing Eq. 3-47 is shown in Fig. 3-5. The measured data show good agreement with the regression line, and the line indicates that the model is unbiased, both for the entire data set and within subranges of the data.

The regression line follows the trend of the data, with little scatter about the line. However, the visual comparison of the sample data and the assumed population (i.e., the regression line) should be supported using statistical tests. The hypotheses of Eq. 3-26 can be tested to determine whether or not the relationship between t_A and t_S is statistically significant. The ANOVA test was made, with the ANOVA table given in Table 3-5. Using the ratio of the mean squares for regression to error (Eq. 3-30), the computed F statistic is 61.03. Although there is no physical basis for selecting a level of significance, values of 0.05 and 0.01 are used frequently in hydrology. The

TABLE 3-5 Analysis-of-Variance Test for Example 3-1

Source of Variation	Sums of Squares	Degrees of Freedom	Mean Squares
Regression	110.35	1	110.35
Error	34.35	19	1.81
Total	144.70	20	

critical F for these two levels of significance are 4.38 and 8.18, respectively. Thus the computed value of 61.03 is highly significant, and it appears safe to reject the null hypothesis. This suggests only that the slope of the line is significantly different from zero, with a value of zero indicating no relationship. The rejection of the null hypothesis does not necessarily mean that the relationship is sufficiently strong that accurate predictions to t_S can be made from measured values of t_A.

The correlation coefficient and standard error of estimate may provide a better assessment of the usefulness of Eq. 3-47. The correlation coefficient can be computed using the square root of the fraction of explained variation (Eq. 3-23) and the values for the separation of variation of Table 3-5:

$$R = \left(\frac{\text{EV}}{\text{TV}}\right)^{0.5} = \left(\frac{110.35}{114.75}\right)^{0.5} = 0.873 \tag{3-48}$$

Thus variation in t_A explains approximately 76% of the variation in t_S. Equation 2-18d can be used to compute the standard deviation of t_S; a value of 2.69°F was computed. The standard error of estimate is obtained from Eq. 3-19 and the values of Table 3-5:

$$S_e = [\tfrac{1}{19}(34.35)]^{0.5} = 1.34°\text{F} \tag{3-49}$$

Thus the error variation decreased from 2.69°F when using the mean value (67.0°F) as the best estimate to 1.34°F when using Eq. 3-47 to estimate t_S. Assuming that the errors are normally distributed about the regression line, the standard error indicates that approximately 68% of the residuals should be within ±1.34°F. Confidence intervals could be computed for estimating the range of errors expected for future conditions. Confidence intervals for a regression equation are discussed elsewhere (pp. 198–202 of *Statistical Methods for Engineers*, Richard H. McCuen, Prentice-Hall, Inc., 1985).

The residuals, which are given in Table 3-4, can be checked to ensure that they meet the four assumptions that underlie the principles of least squares. First, neglecting round-off error the sum of the errors equals zero; thus the model is unbiased. Second, in spite of the somewhat smaller error variation for the points for which t_A is less than 62°F (see Fig. 3-5), there is no reason to believe that the variance of the errors is not constant; thus the assumption of a constant variance does not appear to be violated. Third, there does not appear to be a trend in the residuals due to either time or the temperature. The residuals given in Table 3-4 show a relatively random scatter of positive and negative values with time. The residuals of Fig. 3-5 are apparently uncorrelated with t_A and t_S. Thus the assumption of independence does not appear to be violated. Fourth, regression analysis assumes that the residuals are normally distributed. This assumption can be tested using a Kolmogorov–Smirnov one-sample test (see Chapter 2); the sample size is probably too small to use the chi-square goodness-of-fit test. The range was divided into eight intervals of equal probability, as shown by the equal increment for the cumulative distribution of the expected probabilities. The standard normal deviates were obtained from Appendix A-1 and multiplied by the standard error of estimate (i.e., 1.34°F) to get the bounds on the intervals shown in Table 3-6. The sample probabilities were then determined from the residuals given in Table 3-5; for example, three of the 21 values were less than −1.15, while none were in the interval −1.15 to −0.675. The sample cumulative probability distribution was computed, and the differences in absolute value between the sample and expected cumulative distribution were computed. The computed value of the test

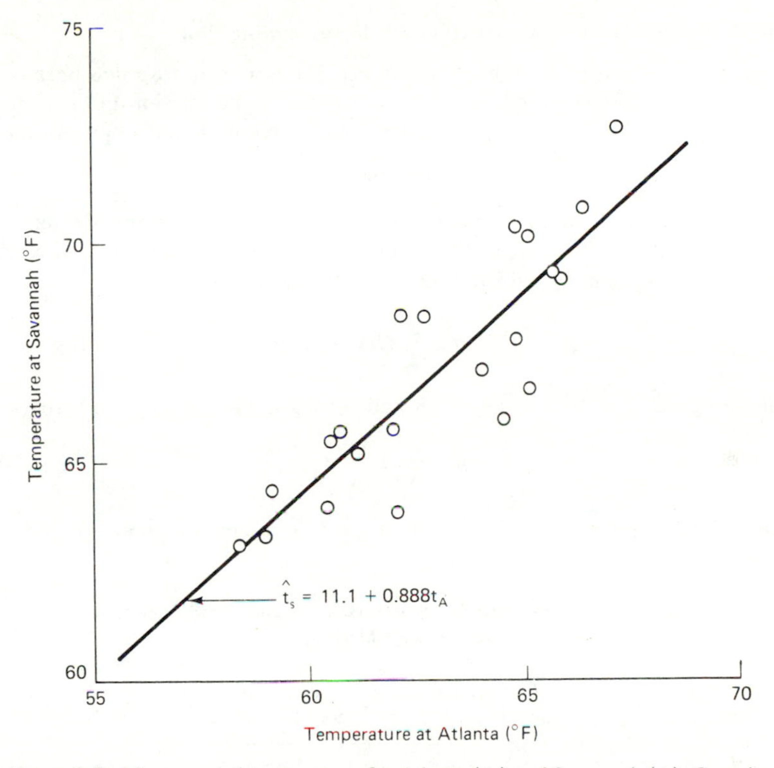

Figure 3-5 Mean annual temperature for Atlanta (t_A) and Savannah (t_S), Georgia, and the linear regression line.

statistic equals 0.107, which is the largest of the computed absolute differences. The critical values of the test statistic for 5% and 1% levels of significance are 0.289 and 0.349, respectively. Thus the null hypothesis of a normal distribution cannot be rejected. In summary, the model of Eq. 3-48 appears to satisfy the four assumptions that underlie the principle of least squares.

TABLE 3-6 Kolmogorov–Smirnov Test for Example 3-1

Interval	Sample Probability	Sample Cumulative	Expected Cumulative	Absolute Difference
$-\infty$ to -1.15	3/21	0.143	0.125	0.018
-1.15 to -0.675	0/21	0.143	0.250	0.107
-0.675 to -0.32	4/21	0.333	0.375	0.042
-0.32 to 0	4/21	0.523	0.500	0.023
0 to 0.32	1/21	0.571	0.625	0.054
0.32 to 0.675	4/21	0.762	0.750	0.012
0.675 to 1.15	2/21	0.857	0.875	0.018
1.15 to ∞	3/21	1.000	1.000	0.000

Example 3-2: Cumulative Soil Loss from a Rill-Susceptible Plot

For some physical processes, the model of Eq. 3-1 is not appropriate because one would expect the intercept coefficient to equal zero. If the relationship between the variables is expected to be linear, the linear, zero-intercept model is appropriate:

$$\hat{Y} = bX \tag{3-50}$$

A zero-intercept model could also be used when there is more than one predictor value (i.e., b_0 of Eq. 3-2 could be zero). The model of Eq. 3-50 can be calibrated using the principle of least squares by substituting Eq. 3-50 into Eq. 3-5:

$$F = \min \sum_{i=1}^{n} (bX_i - Y_i)^2 \tag{3-51}$$

Differentiating Eq. 3-51 with respect to b and setting the result equal to zero yields

$$b = \frac{\sum_{i=1}^{n} Y_i X_i}{\sum_{i=1}^{n} X_i^2} \tag{3-52}$$

The value of b computed using Eq. 3-52 will result in the minimum error variance for any solution of Eq. 3-50.

TABLE 3-7 Cumulative Soil Loss versus Distance Downslope for a Rill-Susceptible Plot

Distance Downslope (ft)	Cumulative Soil Loss (ft^3)	Predicted Soil Loss (ft^3)	Error (ft^3)	
5	1.3	2.3	1.0	
10	2.2	4.6	2.4	
15	4.6	6.9	2.3	
20	6.9	9.2	2.3	
25	8.2	11.5	3.3	
30	11.1	13.8	2.7	
35	17.0	16.1	−0.9	
40	17.9	18.4	0.5	
45	23.4	20.7	−2.7	
50	25.8	23.0	−2.8	
			+8.1	sum

The data of Table 3-7 can be used to illustrate the calibration of the zero-intercept model. The criterion variable is the cumulative soil losses (ft^3) from a rill-susceptible plot 50 ft in length. The distance downslope (ft) will be used as the predictor variable. The data base consists of values taken at 5-ft intervals (i.e., $n = 10$).

Using the data of Table 3-7, the value of the slope coefficient for Eq. 3-50 was computed with Eq. 3-52. The least-squares estimate of b equals 0.46 ft^3/ft. The regression line is plotted in Fig. 3-6. The residuals were computed and are given in Table 3-7. The nonzero sum of the residuals indicates a biased model. In addition to the bias of the total model, it is evident from Fig. 3-6 that there are local biases. Specifically, the model tends to overpredict for locations at the top of the plot and underpredict at the plot outlet. It appears that a zero-intercept nonlinear model would

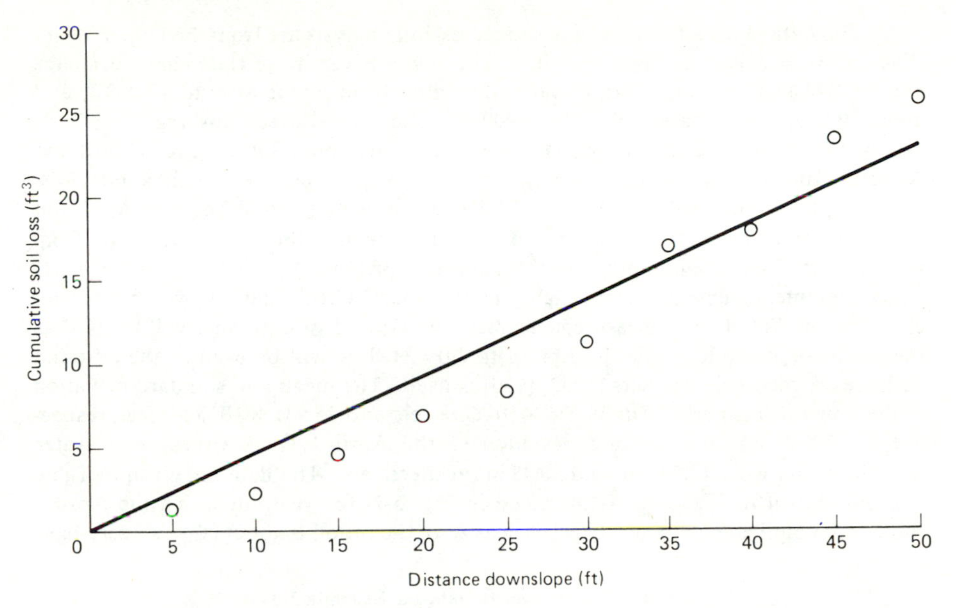

Figure 3-6 Zero-intercept model for predicting cumulative soil loss.

be more appropriate than the model of Eq. 3-50. For example, one might try to fit an equation of the form

$$\hat{Y} = b_0 X^{b_1} \tag{3-53}$$

Thus, while it is rational for the predicted equation to pass through the origin, the linear form of Eq. 3-50 does not provide a rational model for the data of Table 3-7.

Example 3-3: Composite Regression of Snowmelt–Runoff Data

In some regions, snowmelt runoff is a major source of water for irrigation, power, and recreation. Estimates of runoff volumes that are expected to occur during the growing season are needed in the spring for agricultural planning. At the present time, linear regression analysis is the most widely used method for developing forecast models of long-term (i.e., 60 days or longer) runoff volumes. These forecast models use the runoff volume as the criterion variable. Measured snow water equivalents are frequently used as predictor variables, with the measurements often made on the first day of a month such as March or April. Experience indicates that the regression approach provides reasonably accurate forecast models. The location of the station where snow water measurements are made can influence the shape of the relationship between runoff volumes (V) and the snow water equivalents (S). For example, in the case of a watershed that has a large elevation range, the relationship between V and S for snow water data from a station at a relatively low elevation may be characterized by a linear trend for large volumes of V and a parabolic trend that approaches some nonzero volumes as S approaches zero. That is, even in years when measured snow water is near zero at the low elevation station, there may be large amounts of snow at the upper elevations, and this snow can produce reasonably large volumes of runoff. This situation would produce data for which the composite regressions analysis would be appropriate.

The data chosen for the composite regression analysis are from the Upper Sevier River above Hatch in south-central Utah. The Sevier River above Hatch has a drainage area of 340 square miles, with elevations ranging from about 6500 to 11,000-ft and about 70% of the land area being above 9000 ft. The natural streamflow regime consists of low flows from August through March and high snowmelt flows between April and August. The average yearly discharge is about 94,000 acre-feet with about 75% occurring between April and August. While the snowmelt runoff between April and August varies considerably from year to year, the low flow shows little variation from year to year. The mean monthly precipitation in April and May is small. Snow water equivalent measurements are available at the Duck Creek station, which is at an elevation of 8700 ft. The measurements made on April 1 of each year will be used as the predictor variable, while the April 1 to July 31 flow will be used as the criterion variable. A record of 25 years (1952–1976) is used. The mean and standard deviation of the flow for this period was 39.276×10^3 acre-feet and 25.931×10^3 acre-feet, respectively. The mean and standard deviation of the April 1 Duck Creek snow water measurements were 13.324 in. and 8.448 in., respectively. The data are given in Table 3-8 and shown in Fig. 3-7. As indicated in Fig. 3-7, for years in which the April 1 snow water equivalent at Duck Creek is near zero, the runoff is about 20×10^3 acre-feet;

TABLE 3-8 Snowmelt Data for Example 3-3

Year	Snow Water Equivalent at Duck Creek (in.)	4/1 to 7/31 Snowmelt Runoff ($\times 10^3$ acre-feet)
1952	32.6	88.0
1953	6.9	16.7
1954	16.4	35.4
1955	11.6	17.5
1956	9.6	21.0
1957	16.4	40.2
1958	17.9	62.8
1959	7.4	13.8
1960	8.6	17.4
1961	9.0	18.2
1962	24.7	53.8
1963	1.7	18.9
1964	8.2	29.1
1965	13.4	51.9
1966	12.9	39.4
1967	9.7	55.7
1968	18.3	56.4
1969	34.4	107.1
1970	4.3	22.9
1971	8.6	24.3
1972	2.0	20.9
1973	23.8	91.3
1974	6.2	16.6
1975	14.1	37.4
1976	14.4	25.2

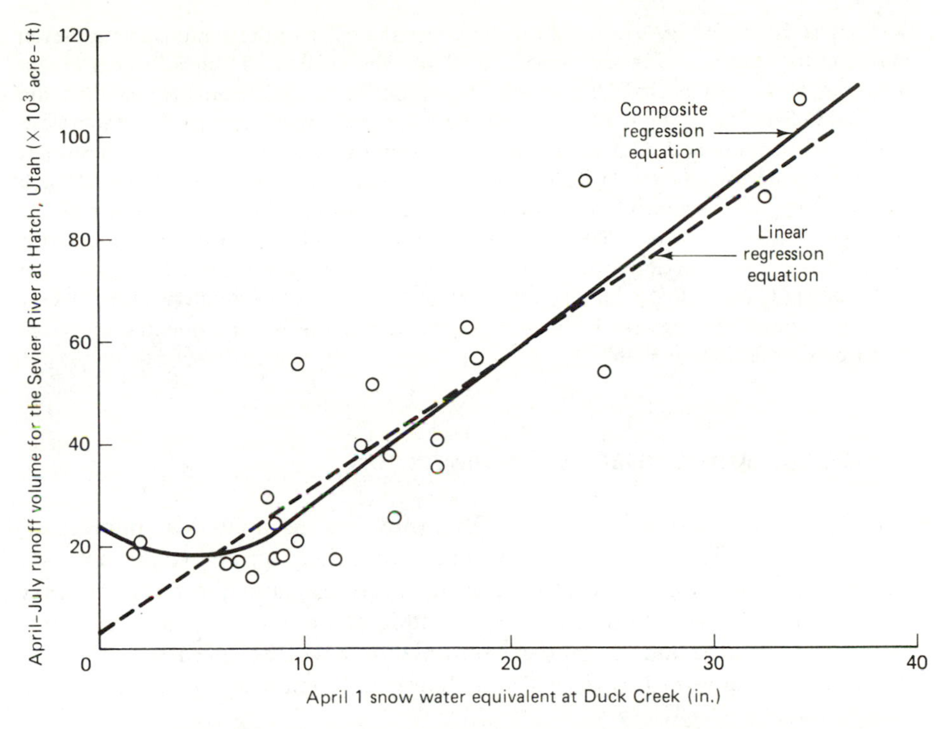

Figure 3-7 Composite regression analysis of Sevier River data.

specifically, the runoff values decrease linearly as the snow water measurements decrease until about 10 in. With one exception (1967) the flow is between 18 and 30 ($\times 10^3$ acre-feet) for snow water values less than 12 in. Thus the data suggest that a composite regression analysis would be appropriate. Using a value of 9.65 in. for X_c, the data were fitted with the following result:

$$\hat{V} = \begin{cases} 24.23 - 2.675S + 0.2961S^2 & \text{for } S \leq 9.65 \text{ in.} \quad (3\text{-}54a) \\ 24.23 - 2.675S + 0.2961(19.3S - 93.1) & \text{for } S > 9.65 \quad (3\text{-}54b) \end{cases}$$

in which $\hat{V}$ is the predicted value of $V(\times 10^3$ acre-feet) and S is the snow water equivalent (inches). The curve is shown in Fig. 3-7. Equations 3-54 appear to provide a reasonable fit to the data for both high and low values of S; it does not appear to result in biased values for years in which the April 1 snow water measurement is low. It is evident from Fig. 3-7 that the model is not entirely rational for values of S below 4.52 in. because the slope becomes negative; however, the slope is not sufficient to cause alarm because the difference is only about 6×10^3 acre-feet (i.e., 24.23×10^3 at $S = 0$ and 18.18×10^3 at $S = 4.52$ in.). Equations 3-54 provide a reasonable fit to the data, with a correlation coefficient of 0.902 and a standard error of estimate of 11.42×10^3 acre-feet. Thus the equations explain over 80% of the variation with no apparent bias.

A bivariate regression model having the form of Eq. 3-1 was computed:

$$\hat{V} = 3.02 + 2.721S \qquad (3\text{-}55)$$

The line is shown in Fig. 3-7. It appears to follow the trend of the composite regression equation for values of S in the range 6 to 30 in. Above 30 in. the linear model of Eq. 3-55 may be slightly biased. Below 6 in. the model of Eq. 3-55 tends to deviate from the trend of the data. While the residuals for measurements where S is less than 5 are not large in comparison to the residuals of the other points, it is especially important to have accurate estimates in years when the snow water equivalent value at Duck Creek is low. The composite regression model provides much greater accuracy in this region of the data. For purposes of comparison, Eq. 3-55 had a correlation coefficient and standard error of estimate of 0.886 and 12.26×10^3 acre-feet, respectively. Although these may not appear to be significantly different from the corresponding values of Eqs. 3-54, the composite regression model appears to provide better estimates for years when S is either low or high.

NUMERICAL AND SUBJECTIVE OPTIMIZATION

Optimization methods were separated into three groups: analytical, numerical, and subjective. This chapter has concentrated largely on analytical methods because they are used most frequently in hydrology and the theorems that underlie analytical optimization are the fundamental basis of numerical and subjective optimization. Analytical optimization will be discussed in more detail in Chapter 6, with Chapters 7 and 8 devoted to numerical and subjective optimization, respectively.

In the introduction to this chapter, the premise was stated that multivariate prediction equations, such as Eqs. 3-1 and 3-2, could be used to improve the accuracy of estimates made using mean value estimation. This is exactly the same reason that numerical and subjective optimization methods are used. When simple models cannot provide highly accurate estimates of the criterion variable, such as with the model of Example 3-2, it is necessary to formulate models having a more complex structure. For such cases, the coefficients cannot be evaluated analytically, and the methods of Chapters 7 and 8 are necessary.

EXERCISES

3-1. Using the method of least squares, derive the normal equations for each of the following models: **(a)** $\hat{y} = b_1 X + b_2 X^2$; **(b)** $\hat{y} = bx^2$; **(c)** $\hat{y} = b_0 X^{b_1}$; **(d)** $\hat{y} = b_0 + b_1\sqrt{x}$.

3-2. Use Lagrangian optimization to find the optimum values of X_1 and X_2 for the objective function $y(X_1, X_2) = X_1^2 + X_2^2 - 3X_1X_2$ subject to the constraint that $X_1^2 + X_2^2 = 4$.

3-3. Use Lagrangian optimization to find the maximum value of X for the objective function $y = X^4$, subject to the constraint that $X \leq 1.5$. Why doesn't $dy/dx = 0$ at the optimum?

3-4. Use Lagrangian optimization to find the optimum value of X_1 and X_2 for the objective function $y(X_1, X_2) = X_1^2 + 3X_1X_2 + 2X_2^2 - 15X_1 - 12X_2$ subject to the following two constraints: **(a)** $2X_1 + 2X_2 \leq 8$ and **(b)** $3X_1 - X_2 \leq 5$. Plot the two constraints using an X_1 versus X_2 graph and show the optimum point.

3-5. Use Lagrangian optimization to find the minimum sum of squares of the errors for the model $\hat{y} = b_1X + b_2X^2$ subject to the constraint that the predicted values be greater than some value greater than y_0.

3-6. **(a)** Find the sample estimates of the regression coefficients for Eq. 3-17 using the following data:

Y	3	2	1	2	2	8
X	2	1	2	3	2	8

(b) Using the first five pairs of Y and X values of part (a) and the point ($Y = 2$, $X = 8$), find the sample estimates of the regression coefficients of Eq. 3-17.

(c) Compute the correlation coefficient for the regressions of parts (a) and (b); discuss the implications of the difference in the correlation coefficients.

3-7. For each of the following data sets, **(a)** graph Y versus X; **(b)** calculate the correlation coefficient; **(c)** determine the slope and intercept coefficients of the linear regression model: $\hat{Y} = b_0 + b_1X$; **(d)** show the regression line on the graph of part (a); **(e)** compute the predicted value of Y for each observed value of X; **(f)** calculate $\sum e^2$ and S_e; **(g)** perform an analysis of variance test of the hypothesis: $\beta_1 = 0$.

Data Set		Data					
1	Y	1	3	5	6	2	1
	X	1	2	4	5	7	8
2	Y	3	2	3	5	8	9
	X	1	4	8	10	11	13
3	Y	9	10	7	3	4	2
	X	2	4	5	7	10	12

3-8. Using the data set below, regress **(a)** Y on X; **(b)** X on Y. For each case compute the correlation coefficient. Transform the regression equation from part (b) to an equation for computing Y and compare the resulting equation with the regression equation of part (a). Explain why the coefficients are different.

Y	3	2	6	8
X	2	5	7	8

3-9. For values of the correlation coefficient from 0 to 1 in increments of 0.1, compute the ratio of S_e/S_y using Eq. 3-20. Discuss the implications of the resulting table with respect to using the correlation coefficient as a measure of goodness of fit.

3-10. For sample sizes of 5, 15, and 25, compute the standard error of estimate (S_e) with Eqs. 3-20 and 3-25 for correlations of 0.05, 0.1, 0.2, 0.8, 0.9, and 0.95. Use

$S_y = 1$. Discuss the implications of the results with respect to using Eq. 3-20 as an approximation of Eq. 3-25.

3-11. For the data of Table 3-1, compute the sum of the squares of the errors for values of b_0 and b_1 shown below. Plot the sum of squares on a graph of b_0 versus b_1 and draw isolines to show the "response surface" (i.e., a plot of the objective function versus the unknowns). How could knowledge of the response surface increase our chances of locating the optimum if the analytical approach to optimization could not be used to find the optimum values of b_0 and b_1?

b_0	−6	−6	−6	−7	−7	−7	−8	−8	−8
b_1	0.6	0.8	1.0	0.6	0.8	1.0	0.6	0.8	1.0

3-12. The equation $q = a\,e^{-bx}$ can be changed to $\ln q = \ln a - bx$ by taking natural logarithms of both sides. Fit this transformed equation to the data set using least squares and Eq. 3-5.

X	q
0	10.1
1	8.0
2	6.9
3	5.0

(*Note*: The least-squares solution of Eq. 3-5 will yield b_0, and the value of a is found from $a = e^{b_0}$.) Using the model form $q = a\,e^{-bx}$ compute the errors $(\hat{q}_i - q_i)$ for $i = 1, 2, 3, 4$. Calculate both the sum of the errors and the sum of the squares of the errors. Explain why the sum of the errors is not equal to zero.

3-13. Write a computer subprogram to calculate the sum of the squares on n differences (n would be 4 for Exercise 3-12) for any values of a and b of the exponential model given in Exercise 3-12. Write a main program that includes nested DO loops to iterate over the coefficients a and b between the ranges $a_L \le a \le a_U$ and $b_L \le b \le b_U$ and uses the subprogram to compute the sum of the squares of the n differences between the predicted and measured values of q.

3-14. Compute the sum of the squares of the errors for each point using the program of Exercise 3-13 with the exponential model, the data of Exercise 3-12, ranges on a and b of $9.5 \le a \le 10.4$ and $-0.26 \le b \le -0.17$, and increments for a and b of $\Delta a = 0.1$ and $\Delta b = 0.01$. Plot the resulting values of the sum of the squares of the errors using the values of a and b as the ordinate and abscissa, respectively. Draw contours for the response surface.

3-15. As in Exercise 3-14, construct a response surface using the ranges for a and b of $10.03 \le a \le 10.12$ and $-0.222 \le b \le -0.212$, with increments for a and b of $\Delta a = 0.01$ and $\Delta b = 0.001$. Plot the data and draw contour lines.

3-16. Read the values of a and b from the graph of the response surface of Exercise 3-14 that yield the minimum sum of squares. Use this pair of values of a and b to compute the four errors and errors squared as in Exercise 3-15. Compare the two sets of errors and errors squares; discuss the reason for the difference.

3-17. The equation $Y = f(X - g)^2$ cannot be converted to a linear form. Modify the program written for Exercise 3-13 to calculate the sums of the squares of the errors for the data

X	3	4	5	6	7
Y	1.55	9.75	26.23	45.31	76.87

Use ranges for f and g of $3.3 \leq f \leq 3.6$ and $2.2 \leq g \leq 2.5$, with an increment of 0.03 for both f and g. Plot the resulting values and draw contour lines. How does this response surface differ from the one drawn for Exercise 3-15?

3-18. Write a simple computer program to iterate on trial values of f and g in the equation of Exercise 3-17. The iteration procedure should be based on the following steps:

1. Read in initial values of the coefficients f and g and the number of rounds to be executed (N).
2. Using the values of f and g compute $\sum (\hat{y}_i - y_i)^2$, which is denoted as F_1.
3. Increment f by df and compute $\sum (\hat{y}_i - y_i)^2$, which is denoted as F_2.
4. If $F_2 < F_1$, compute a new value of f by $f = f + df$.
5. If $F_2 > F_1$, compute a new value of f by $f = f - df$.
6. Repeat steps 2 to 5 for g (instead of f).
7. If df changes from $-$ to $+$ or $+$ to $-$, set df equal to $df/2$; this also applies for g.

Steps 2 to 7 should be repeated for N trials.

3-19. Run the program of Exercise 3-18 for 40 rounds ($N = 40$) using 3.57 and 2.29 for initial values of f and g. Let the initial values of df and dg be 0.005. The data of Exercise 3-17 should be used.

3-20. Run the program as in Exercise 3-19, but use 3.3 and 2.38 for the initial values of f and g, respectively. Compare the results with Exercises 3-18 and 3-19, and compare the results with the contour plot of Exercise 3-17.

4 Univariate Estimation

INTRODUCTION

In many problems in hydrology, the data consist of measurements on a single random variable; hence we must deal with univariate analysis and estimation. The objective of univariate analysis is to analyze measurements on the random variable, which is called *sample information*, and identify the statistical population from which we can reasonably expect the sample measurements to have come. After the underlying population has been identified, one can make probabilistic statements about future occurrences of the random variable; this represents univariate estimation. It is important to remember that univariate estimation is based on the assumed population and not the sample; the sample is used only to identify the population.

The analysis phase of univariate estimation has several steps. The population consists of two elements, the probability density function (pdf) and the parameters of the pdf. Although both elements are necessary to define a population completely, it is important to recognize that there are many density functions and a number of methods for identifying and verifying the parameters of an assumed pdf. After a population (i.e., both a pdf and its parameters) has been identified, the data should be plotted against the population. This step is intended to verify the assumed population, and therefore it is not usually viewed as part of the analysis phase. Statistical hypothesis tests can be used to test the validity of the assumed population; for example, the chi-square and Kolmogorov-Smirnov one-sample tests are used to compare the sample data with an assumed population.

After the population has been identified and substantiated using the measured data and statistical tests, probability statements about the random variable can be made. The output from a univariate analysis can be either a value of the random variable for some level of probability or an expected probability for a specified value of the random variable. In addition to probability statements, one is usually interested in the accuracy of the probability statement. The accuracy can be assessed using confidence intervals and tolerance limits.

THE UNIVARIATE MODELING PROCESS

A model is conceptually a substitution for a system. By *system* we mean a real-world concept which is composed of input elements, a process of conversion, and the generation of output elements. A simple hydrologic system would be rainfall as input, the watershed as a converting device, and the output would be streamflow. This is a real-world system. We model this system with a *mathematical substitution.* The mathematical substitution is a workable simplification of the system. It is used to study the system. We can answer questions about output if we change the input or if we change the form of the process that converts the input to the output without actually experiencing these changes in the real world.

It is, of course, always theoretically possible to change the real-world system. If we had a tiny watershed, we could sprinkle it with water. We could change this sprinkler input and measure directly the effect on the output. Or we could change the shape of our tiny watershed and see what effect this would have on our output. Most of the time, however, it is completely impractical to change the system. We must, therefore, attempt to answer questions about possible changes through our models. It is in this sense that the model represents and substitutes for the system.

There are always three basic steps involved in the use of models to predict hydrologic effect. These steps are as follows:

1. We must specify some conceptual form for our model. Usually, this means that we specify an equation or a set of symbols that defines our model precisely.
2. We must evaluate or quantify this model. That is, we must be able to fit our model to historical, real-world data and see what numerical values the parameters of our model take on for these various past hydrologic records.

3. After our model is evaluated or quantified, we must use our model in prediction. This means that we must specify the state of our system, or the actual process of converting input to output, and we must also specify a series of input values to be converted to output.

When we discuss univariate models, we discuss the simplest possible hydrologic form. A univariate model can be only one of the statistical frequency distributions. Since we have only one variable, we have no relationship to another variable. We, in fact, say that there is no input-output relation but that the system generates a random output. It is uncontrolled in the sense that we cannot predict what any particular output element will be. The output is controlled, however, to that degree which causes it to take a normal, a gamma, or some other distribution. If we know streamflow today, we cannot, with a univariate model, use that streamflow to predict streamflow tomorrow. If we know temperature today, we cannot use that temperature to predict temperature next week. Although the variation in streamflow and the variation in temperature are purely stochastic, we can still predict within the probabilistic constraints of our specified distribution. Although we do not have a relationship form for our model, it is very important to understand that a form exists, and that form is the frequency distribution we have specified as controlling the probable range of our stochastic events.

A simple descriptive example may serve to solidify the concept of a univariate model. Suppose that we have taken the maximum temperature each day during the month of July at some Weather Bureau station. The maximum temperature each day is not constant; it varies between certain limits that cluster fairly closely around the mean daily maximum temperature for July. We may plot a graph of our record of July maximum temperatures for, say, 10 years. Suppose that the graph makes us believe that the distribution of these maximum temperatures is essentially a normal distribution. By choosing the normal distribution we have specified the form of our model. Now, additionally, a normal distribution is completely quantified when we evaluate its mean and its standard deviation. Suppose we say that we will take the mean daily maximum July temperature as the mean of our model and that we will take the standard deviation of our sample of daily maximum July temperatures as the standard deviation of our model. We have then accomplished the second step, which is evaluation or quantification of our model. The third step, prediction, enters when we make some estimate of what the likely July maximum temperature may be. For example, we could look at our quantified model and find that in only 5% of the time was the daily maximum temperature greater than some value, say T. We could then make a predictive statement that on a certain day in July it is unlikely that the maximum temperature would be greater than this value T. We could make a more meaningful prediction if we say that on a certain day, it is only 5% or 10% likely that the temperature will be above our selected temperature T.

Prediction from a statistical model must always have associated with it some statement of the confidence with which we make that prediction. The reason is that the model was quantified from historical data. This is a record (i.e., it is one sample from the past). But our predictions are for the future. We cannot know precisely what the future holds because we can only record statistical samples. We can never record the full population of the events nature could supply us. Therefore, we can only state a likely range of future events based on the probabilities we estimate from the sample provided us.

We have already looked at confidence intervals of the mean and tolerance limits of proportions of the sample. However, the meaning of these specific predictions, which are possible with univariate models, should now be much clearer. They will be taken up again in the following sections of this chapter. The concepts of formulation, evaluation, and prediction will be developed further in future chapters.

Elements of Frequency Analysis

Statistical analysis of univariate data requires several components. First, a model must be selected; for univariate analysis, this is the probability distribution, or pdf. Second, we need a data set; for univariate analysis, this consists of a set of *n* independent observations on or measurements of a random variable. Third, we need a method of fitting; for most applications in univariate analysis, either the method of moments or maximum likelihood estimation is used. Fourth, for purposes of examination we usually use some form of graphical presentation; this requires one of the many types of probability papers and some rule for plotting the data. Before looking at actual methods of univariate analysis, we will examine both the fitting methods and the process of graphical analysis.

The four elements of univariate, or frequency, analysis are very similar to the elements of regression analysis that were introduced in Chapter 3. For example, Eq. 3-1 was used as the model. The data set consisted of *n* simultaneous measurements of two or more variables. Least squares was the fitting method, and the data were graphed and compared to the regression line, with the residuals tested for the four basic assumptions. Thus it should be evident that the elements of frequency analysis given in the preceding paragraph are not entirely new and that univariate modeling has strong similarities with the bivariate modeling of Chapter 3.

ESTIMATION OF MODEL PARAMETERS

One objective of data analysis is to identify the underlying probability density function of a random variable; this includes the specification of both the pdf and the parameters of the distribution. A candidate for the pdf can usually be identified by forming a histogram of the sample data; other methods will be introduced later. It is also necessary to estimate the parameters of the distribution. After discussing the types of parameters, two methods for estimating the parameters will be introduced.

Types of Parameters

A probability mass or density function can be characterized by one or more parameters. Parameters of a distribution control the geometric characteristics of the distribution. Three types of parameters exist: location, scale, and shape parameters.

A *location parameter* identifies the abscissa of a location point of the distribution. All fractiles of the distribution can be located with reference to the location parameter. A measure of central tendency, such as the mean, is used frequently as the location parameter of a distribution; however, the location parameter is not always a measure of central tendency.

For the distributions shown in Fig. 4-1(a), the parameter α is a location parameter; it defines the lower limit of the range of the distribution. In this case, it is neither a mean value nor another measure of central tendency. In

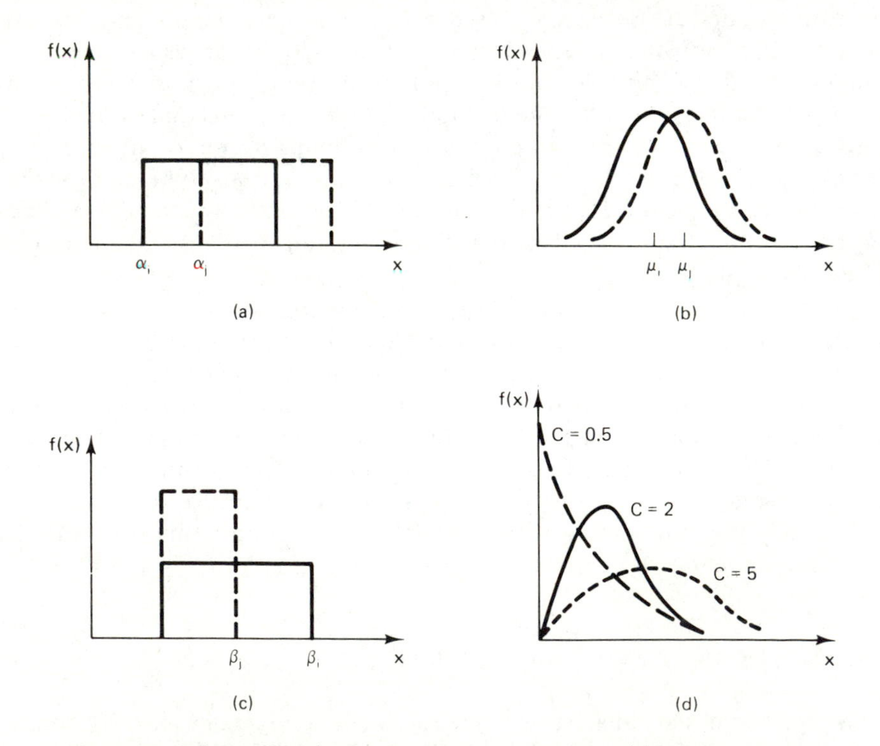

Figure 4-1 Effects of location, scale, and shape parameters: (a) location parameter α; (b) location parameter μ; (c) scale parameter β; (d) shape parameter C.

Fig. 4-1(b) μ is a location parameter. In this case, the location parameter equals the mean of the distribution.

A *scale parameter* is one that determines the location relative to some specified point, often the value of the location parameter, of various fractiles of the distribution. Measures of dispersion, such as the standard deviation and range, are often used as scale parameters. The two distributions shown in Fig. 4-1(c) have the same location parameter and shape but obviously differ in their scale. They are characterized by the scale parameters β_i and β_j.

Shape parameters control the geometric configuration of a distribution; that is, a shape parameter is necessary when the structure, outline, or balance of a distribution changes. Thus a shape parameter is often used to distinguish one density function from other density functions in a family of distributions. A distribution may have more than one shape parameter, or none at all. Figure 4-1(d) shows how a particular distribution function changes as the shape parameter, C, changes. Several of the more common distribution functions do not have a shape parameter.

Method of Moment Estimation

The method of moments is frequently used to provide estimates of the parameters of a distribution, primarily because of its computational simplicity. Equations that relate the moments of a sample to the parameters of a distribution can be derived. Estimates obtained by the method of moments are always consistent but they may not be efficient in a statistical sense.

Because of its structural simplicity, the uniform, or rectangular, distribution will be used to illustrate the method of moments. The uniform distribution is a function of two parameters, α and β. The location parameter α defines the lower limit of the distribution, while the scale parameter β indicates the range of the density function, which is given by

$$f(x) = \begin{cases} \dfrac{1}{\beta - \alpha} & \alpha < X < \beta \\ 0 & \text{otherwise} \end{cases} \tag{4-1}$$

The first moment, or mean, of the distribution can be determined as follows:

$$\begin{aligned} \mu &= \int_{-\infty}^{\infty} Xf(X)\,dX = \int_{\alpha}^{\beta} X\frac{1}{\beta - \alpha}\,dX = \frac{1}{\beta - \alpha}\int_{\alpha}^{\beta} X\,dX \\ &= \frac{1}{\beta - \alpha}\left(\frac{X^2}{2}\right)\Bigg|_{\alpha}^{\beta} = \frac{\beta^2 - \alpha^2}{2(\beta - \alpha)} = \frac{(\beta - \alpha)(\beta + \alpha)}{2(\beta - \alpha)} = \frac{\beta + \alpha}{2} \end{aligned} \tag{4-2}$$

The second moment about the mean, the variance, can be determined using the relationship

$$\mu^2 = \mu_2' - (\mu_1)^2 \tag{4-3}$$

where μ_1 is the mean and μ_2' is the second moment about the origin. Thus Eq. 4-3 becomes

$$\sigma^2 = \int_\alpha^\beta X^2 \frac{1}{\beta - \alpha} \, dX - \left(\frac{\beta + \alpha}{2}\right)^2 = \frac{1}{\beta - \alpha}\left(\frac{X^3}{3}\right)\Bigg|_\alpha^\beta - \left(\frac{\beta + \alpha}{2}\right)^2$$
$$= \frac{\beta^3 - \alpha^3}{3(\beta - \alpha)} - \left(\frac{\beta + \alpha}{2}\right)^2 = \frac{(\beta - \alpha)^2}{12} \tag{4-4}$$

Equations 4-3 and 4-4 provide the basis for evaluating the parameters of the uniform density function. The mean μ and variance σ^2 of the population can be approximated using the sample moments $\bar{X}$ and s^2, respectively. This provides two equations with two unknowns, β and α:

$$\bar{X} = \frac{\beta + \alpha}{2} \tag{4-5}$$

$$s^2 = \frac{(\beta - \alpha)^2}{12} \tag{4-6}$$

Solving Eqs. 4-5 and 4-6 for α and β provides

$$\alpha = \bar{X} - s\sqrt{3} \tag{4-7}$$

$$\beta = \bar{X} + s\sqrt{3} \tag{4-8}$$

To illustrate the method of moments, consider the sample histogram of Fig. 4-2. It is easily shown that the sample mean and variance are 2 and 1.5,

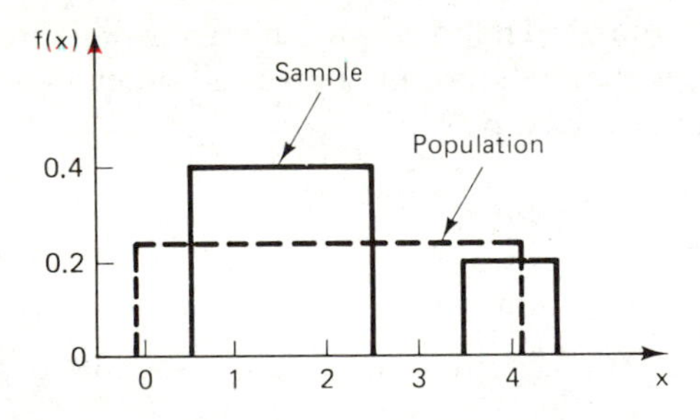

Figure 4-2 Uniform probability function fit using the method of moments.

respectively. Using Eqs. 4-7 and 4-8, the parameters are estimated to be $\alpha = -0.121$ and $\beta = 4.121$. The density function $f(X)$ is thus given by

$$f(X) = \begin{cases} 0.236 & -0.121 < X < 4.121 \\ 0 & \text{otherwise} \end{cases} \tag{4-9}$$

The estimated population density function of Eq. 4-9 is shown in Fig. 4-2. Although the population density function differs from the sample histogram, the population should be used to make probability statements about the random variable X. For example, the probability that a value of X is between 1 and 2 equals $0.236(2 - 1) = 0.236$.

The method of moments is not limited to use with the uniform distribution. It is easy to show that method of moment estimation for the normal distribution leads to the conclusion that the location parameter (μ) equals the sample mean ($\bar{X}$) and the scale parameter (σ) equals the standard deviation of the sample (s). This can be shown by equating the sample mean and standard deviation to the integrals for the mean and standard deviation given in Chapter 2, while using the normal density function of Eq. 2-20 as $f(X)$.

Maximum Likelihood Estimation

The most common statistical method of parameter evaluation is the method of *maximum likelihood.* This method is based on the principle of calculating values of parameters that maximize the probability of obtaining the particular sample.

The *likelihood* of the sample is the total probability of drawing each item of the sample. The total probability is the product of all the individual item probabilities. This product is differentiated with respect to the parameters, and the derivative is set to zero to achieve the maximum.

Maximum likelihood solutions for model parameters are statistically efficient solutions, meaning that parameter values will have minimum variance. This definition of a best method, however, is theoretical. Maximum likelihood solutions do not always produce solvable equations for the parameters. The following examples illustrate easy and moderately difficult solutions. For some distributions, including notably the normal distribution, the method of moments and maximum likelihood estimation (MLE) produce identical solutions for the parameters.

Example 4-1: MLE of a Descending Exponential Function

For this example we will find the maximum likelihood estimate of the parameter A in the density function (Eq. 2-2) with C equal to A as in Eq. 2-3. Consider a sample of N items, $X_1, X_2, X_3, \ldots, X_n$. By definition the likelihood, l, is

$$l = \prod_{i=1}^{N} A\, e^{-Ax_i} \tag{4-10}$$

The product form of the function in Eq. 4-10 is difficult to differentiate. We make use of the fact that the logarithm of a variate must have its maximum at the same place as the maximum of the variate. Taking logarithms of Eq. 4-10 gives

$$L \equiv \ln l = N \ln A\, A - A \sum_{i=1}^{N} x_i \tag{4-11}$$

The differential of L with respect to A, set to zero, is

$$\frac{dL}{dA} = \frac{N}{A} - \sum_{i=1}^{N} x_i = 0 \tag{4-12}$$

Equation 4-12 yields $1/A = (\sum x_i)/N = \bar{X}$. The maximum likelihood value of $1/A$ is thus the mean of the sample of X's.

Example 4-2: MLE of a Double Exponential Function

Consider the problem of finding the maximum likelihood value of parameter A in the function:

$$f(x) = Cx\, e^{-Ax} \tag{4-13}$$

To use this as a probability function we must first find C from

$$C \int_0^\infty x\, e^{-Ax}\, dx = 1 \tag{4-14}$$

Solution of this equation gives $C = A^2$. Thus the likelihood function is

$$l = \prod_{i=1}^{N} A^2 x_i\, e^{-Ax_i} \tag{4-15}$$

The logarithm of this function is

$$L = \ln l = 2N \ln A + \sum_{i=1}^{N} \ln x_i - A \sum_{i=1}^{N} x_i \tag{4-16}$$

and

$$\frac{dL}{dA} = \frac{2N}{A} - \sum_{i=1}^{N} x_i = 0 \tag{4-17}$$

We find that the maximum likelihood value of $1/A$ is one-half the mean of the sample.

Example 4-3: MLE of the Gamma Distribution

The problem here is to find the maximum likelihood expressions for the parameters of the gamma distribution, written as in

$$p(x) = \frac{b^{a+1}}{a!}\, e^{-bx}\, x^a \tag{4-18}$$

$a!$ is the factorial of a and is related to the gamma integral by $a! = \Gamma(a+1)$. Solving for the likelihood expressions:

$$l = \prod_{i=1}^{N} \frac{b^{a+1}}{a!}\, e^{-bx_i} x_i^a \tag{4-19}$$

and

$$L = \ln l = N[(a+1)\ln b - \ln(a!)] - b \sum_{i=1}^{N} x_i + a \sum_{i=1}^{N} \ln x_i \tag{4-20}$$

Therefore,

$$\frac{\partial L}{\partial a} = N \ln b - \frac{N \partial \ln(a!)}{\partial a} + \sum_{i=1}^{N} \ln x_i \tag{4-21}$$

and

$$\frac{\partial L}{\partial b} = \frac{N(a+1)}{b} - \sum_{i=1}^{N} x_i \tag{4-22}$$

Setting the two partial differentials of Eq. 4-21 and 4-22 to zero for their maxima and solving simultaneously for a and b give

$$b = \frac{N(a+1)}{\sum x_i} \tag{4-23a}$$

$$\ln(a+1) - \frac{\partial \ln(a!)}{\partial a} = \ln\left(\frac{\sum x_i}{N}\right) - \frac{\sum \ln x_i}{N} \tag{4-23b}$$

Equations 4-23 illustrate the potential complexity of the maximum likelihood solutions. Equation 4-23b can be solved by some trial-and-error process for a. With a evaluated Eq. 4-23a can be solved for b. The expression $\partial \ln(a!)/\partial a$ in Eq. 4-23b is called the *digamma function* of a. Tables of values of this function can be found in mathematical handbooks.

A sample of 36 synthetic numbers is shown as a histogram in Fig. 4-3. Values for the sample-summation terms in Eq. 4-23b were

$$\ln\left(\frac{\sum x_i}{N}\right) = \ln\left(\frac{116}{36}\right) = 1.17 \quad \text{and} \quad \frac{\sum \ln x_i}{N} = \frac{33.002}{36} = 0.9167 \tag{4-24}$$

Substituting in Eq. 4-23b and rearranging gives

$$\ln(a+1) - \left(\frac{\partial \ln(a!)}{\partial a} + 0.2533\right) = 0 \tag{4.25}$$

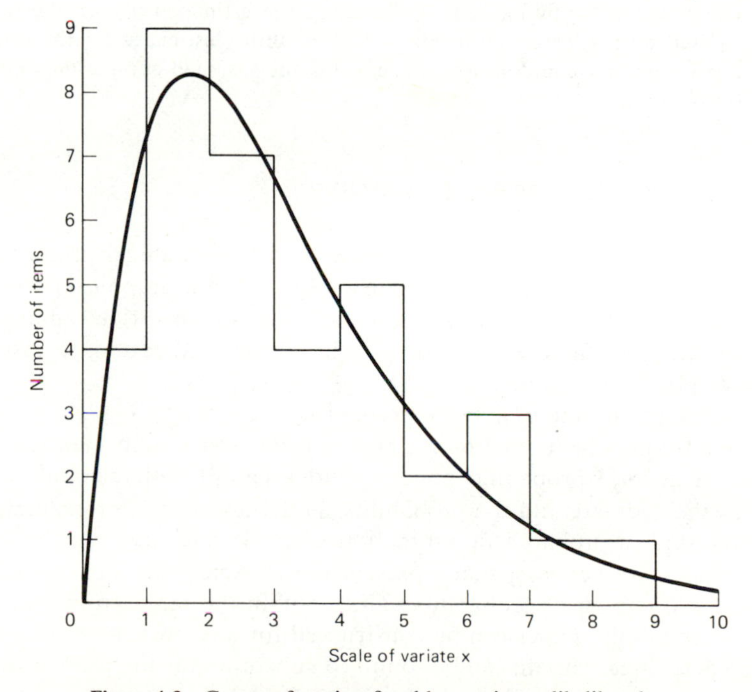

Figure 4-3 Gamma function fitted by maximum likelihood.

We must now search for values of a in Eq. 4-25. One systematic method is to try values of a at the midpoints of intervals of a known to contain a root of the equation. This method is illustrated in Table 4-1. A in this table denotes the first term on the left side of Eq. 4-25, B is the expression in brackets in this equation. Trial

TABLE 4-1 Maximum Likelihood Solution for Gamma Parameter *a*

Trial Value of a	$\psi(a)$	B	A	$A - B$
0	−0.5772	−0.3239	0	0.3239
1.0	0.4228	0.6761	0.6931	0.0170
2.0	0.9228	1.1761	1.0986	−0.0775
1.5	0.7032	0.9565	0.9163	−0.0402
1.25	0.5725	0.8258	0.8109	−0.0149
1.125	0.5034	0.7567	0.7538	−0.0029
	$a \simeq 1.12$	$\frac{1}{b} \simeq \frac{3.222}{2.12}$	$b \simeq 0.658$	

values of a of 0, 1, and 2 were assumed. The quantity $(A - B)$ changed from positive to negative between 1 and 2; therefore, a zero point of the equation falls in this interval. The successive midinterval values of 1.5, 1.25, and 1.125 were assumed, yielding the approximate values of a and b as shown. The digamma function is designated $\psi(a)$.

The gamma distribution (Eq. 4-18), using the estimated values of a and b, is superimposed on the histogram in Fig. 4-3. Probability estimates for the occurrence of various values of x could now be made, using the assumed population, and not the sample histogram.

CONSTRUCTION OF PROBABILITY PAPER

An important element in statistical analysis is the graphical presentation of data; this is especially true in univariate analysis. While graphical presentation (i.e., plotting the data) is usually the last step in univariate analysis, it is important, at least before proceeding to univariate estimation, to discuss the origin of probability paper.

Although a wide variety of probability papers are available, only two are used frequently in hydrology, the normal and Gumbel extreme value papers. In general, probability paper provides a graph with the random variate scale as the ordinate and the probability as the abscissa. For example, if we are interested in probabilistic estimation of peak discharge, q_p, the ordinate will consist of values of q_p that represent the expected range of the data. The abscissa will consist of probability values within the range from 0 to 1.

A probability paper can be constructed for any probability function for which some linear variate can be found to substitute for the probability scale. This is not always possible or practical. For the gamma distribution, for example, there would have to be a different probability paper for each different value of skewness.

Normal Probability Paper

Normal probability paper is based on the standard normal distribution [i.e., $N(0, 1)$]. The first step is to construct a line with values of the standard normal deviate (z) in a linear scale plotted on one side and the area under the standard normal curve (Appendix A-1) to the left of the z value on the other side of the line. The line in Fig. 4-4 shows five values of z and the corresponding areas.

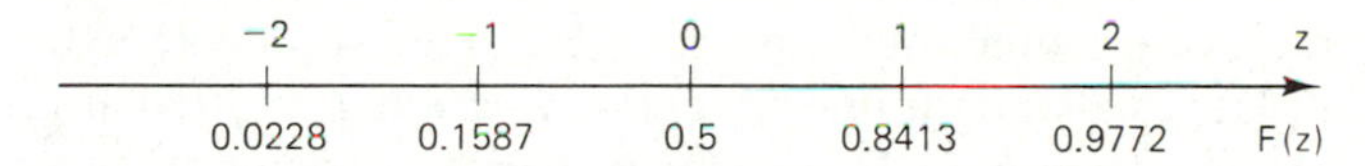

Figure 4-4 Construction of the abscissa of normal probability paper.

The second step is to construct a line perpendicular to the first line. The vertical line should have an arithmetic scale. However, for log-normal probability paper, the vertical scale could be a logarithmic scale.

Additional vertical lines that correspond to convenient probability values (e.g., 0.5 and 0.1) should be constructed by interpolating on the horizontal axis using Appendix A-1. In Fig. 4-5, z values were determined for probabilities of 0.01, 0.10, 0.25, 0.5, 0.75, 0.90, and 0.99 and used in constructing the vertical dashed lines. The location of the lines should be determined using the z axis because it is linear, whereas the $F(z)$ axis is nonlinear.

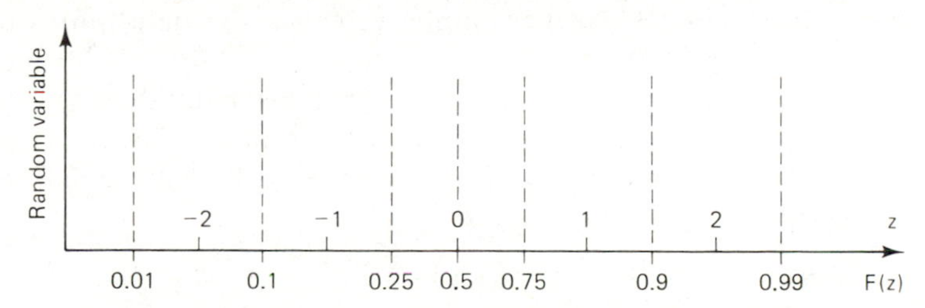

Figure 4-5 Construction of vertical lines for selected probabilities.

On commercially available normal probability paper, probabilities are specified on both the top and bottom of the abscissa. The probabilities given at the top of the paper are presented in decreasing order (i.e., from 99.99% to 0.01%); the probabilities on this axis represent the exceedence probability since a corresponding value of the random variable would have that probability of being exceeded. The probability scale on the bottom of the paper is presented in increasing order (i.e., from 0.01 to 99.99%). These represent the nonexceedence probability (i.e., the probability of not being exceeded in any one period).

Extreme-Value Plotting Paper

The estimation of future floods is so important in the design of hydraulic structures that more effort has probably been expended on statistical flood frequency analysis than on all other hydrologic frequency analyses combined. The importance of flood prediction has led to the development of extreme-value probability paper. The assumption is made that the highest flood each year, being selected from all floods of that year, is an extreme-value variate. The collection of all such variates forms an extreme-value distribution.

Extreme-value probability paper is constructed so that an extreme-value distribution plots as a straight line. A *reduced variate* is plotted at linear scale at the bottom of the graph. This reduced variate is determined as follows. The probabilities in a flood record, or other hydrologic record, are established by the number of times a flood of a certain magnitude has occurred in the record. For example, the largest flood in a 50-year record has a sample probability of $\frac{1}{50}$, or 0.02. We may express this as in Equation 4-26:

$$P = \frac{1}{T} \tag{4-26}$$

where P is the probability that the flood was equaled or exceeded and T is the length of the record. We expect the probabilities given by the sample to repeat in the future. Thus we could expect a flood equal to or larger than the largest flood in the sample to occur *on the average* once in each 50-year record. Therefore, T is defined as the "return period" of the flood. After evaluating a probability distribution from the sample we can associate a return period with each and every probability.

If $P(x)$ gives the probability of an event being smaller than x, $1 - P(x)$ is the probability of an event being equal to or larger than x. The return period T is $1/(1 - P(x))$, or reciprocally, $P(x)$ is $(T - 1)/T$. Now define a new variate, called the *reduced variate*, as in

$$Y = \alpha(X - U) \tag{4-27}$$

Making substitutions of Y and T to eliminate $P(x)$ and X, we get

$$\frac{T-1}{T} = e^{-e^{-Y}} \tag{4-28}$$

We have now established a linear relationship between our flood variate X and the reduced variate Y; this relationship is given by Eq. 4-27. We can also establish a transformed probability scale through Eq. 4-28, since there is a value of T for each Y, and the exceedence probability is $1/T$.

FORMS USED AS FREQUENCY DISTRIBUTIONS

In this section we look at several different forms used as statistical frequency distributions. In particular, the evaluation of the parameters of the forms will be studied. This enables us to fit the forms to simple data and use them for prediction. However, we must note that the normal curve is the one most often used for practical probabilistic interpretation. We can fairly easily compute the moments of distributions other than the normal curve, but confidence intervals and tolerance limits are not readily computed. We will see in a later section that we attempt to solve this problem by transforming our data so that they are normally distributed.

The ultimate objective of frequency analysis is to identify the population that underlies the sample of data. The population, which consists of a probability distribution and the parameters, is used to make probability statements about the occurrence of values of the random variable. In general, the input for a frequency analysis consists of a sample of data and probability paper. The population distribution computed by the frequency analysis is plotted on the probability paper as a cumulative distribution function. After the frequency analysis has been performed, the data are plotted. If the sample data follow the trend of the assumed population, we usually assume that the sample was drawn from that population. If the sample data points depart from the trend of the population line, then either (1) the sample was drawn from a different population, or (2) sampling variation has produced a nonrepresentative sample. In almost all cases, the former reason is assumed to be the cause, especially when the sample size is reasonable.

The following general steps are used to derive a probability frequency curve to represent the population:

1. Hypothesize the underlying density function.
2. Obtain a sample and compute the sample moments.
3. Use either the sample moments or the measured data to estimate the parameters of the density function identified in step 1.
4. Construct the frequency curve that represents the underlying population.

The frequency curve of step 4 can then be used to make probability statements about the likelihood of occurrence of values of the random variable.

Normal Distribution

Quite frequently, one expects the underlying population to be normal. In such cases, normal probability paper, which is readily available, is used.

Following the general procedure outlined above, the following steps are used to fit a normal population using the method of moments:

1. Assume that the random variable has a normal distribution with μ and σ.
2. Compute the sample moments $\bar{X}$ and s.
3. For the normal distribution, the parameters are $\mu = \bar{X}$ and $\sigma = s$.
4. The curve is fitted as a straight line with $(\bar{X} - s)$ plotted at an exceedence probability of 0.8413 and $(\bar{X} + s)$ at an exceedence probability of 0.1587.

Probability statements about any value of the random variable X can then be made using the straight line. The sample points can be plotted to see whether or not the measured values closely approximate the population. If the data provide a reasonable fit to the line, one can assume that the underlying population is the normal distribution and the sample mean and standard deviation are reasonable estimates of the location and scale parameters, respectively. A poor fit would indicate either that the normal distribution is not the population or that the sample statistics are not good estimators of the population parameters, or both.

The temperature data of Table 2-3 can be used to illustrate the calibration of a frequency curve for a normal distribution. The sample mean ($\bar{X}$) and standard deviation (s) were computed as 76.4°F and 3.07°F, respectively. Therefore, the frequency curve can be plotted using two points defined by $(\bar{X} - s, 0.8413)$ and $(\bar{X} + s, 0.1587)$, in which the first value in each set of parentheses is the value of the random variable and the second value is the exceedence probability. The resulting normal population is shown in Fig. 4-6 as a solid straight line. The line also passes through the point $(\bar{X}, 0.5)$, which indicates that 50% of the normal distribution lies below the mean.

Note that we constructed this line without reference to the plotted temperature values. This should always be the case. The individual values are used to estimate the mean and standard deviation of the underlying normal population. The individual values are then plotted only for the purpose of visual comparison of the points and the line. In fact, the points cannot be plotted unless we make some assumptions. Given the 48 monthly average temperatures, what probability can be assigned to each? Briefly, we do not, and we can never, know the answer to this question, because we only have a sample, not the population, and we do not know what future temperatures will be. We do know that we must "spread" our observed temperatures across the probability scale. We can place the 48 temperatures in decreasing order, and assign a rank of 1 to the highest, a rank of 2 to the second highest, and so through the whole sample. Next, we must estimate a probability plotting position from this rank number. The *Weibull plotting position formula* is

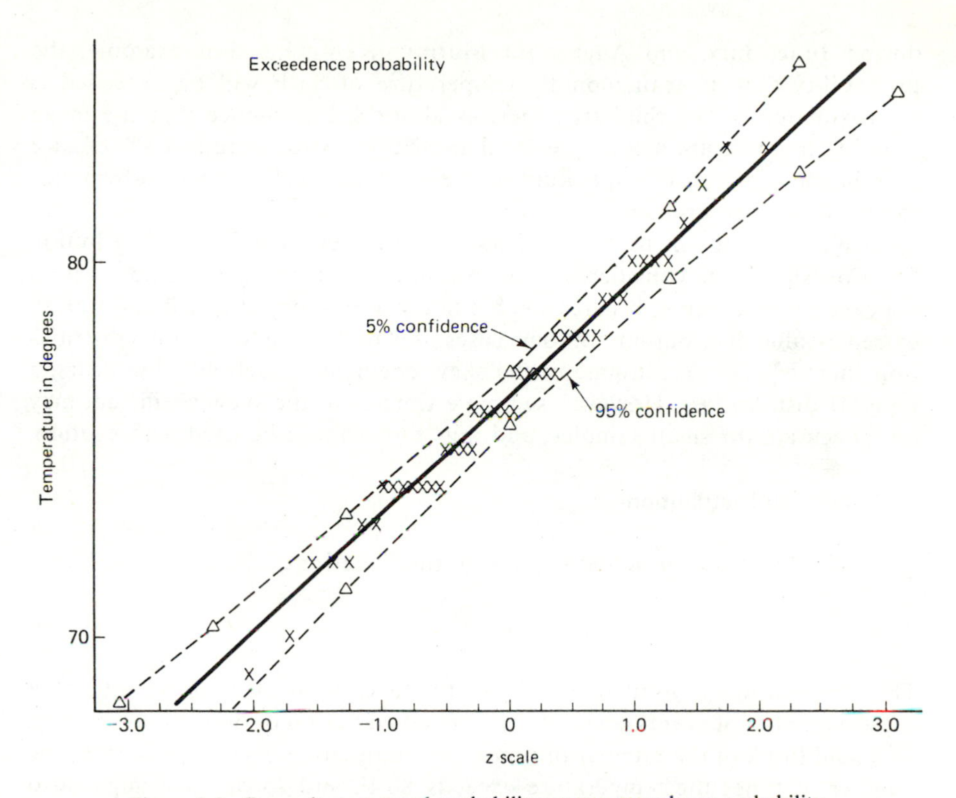

Figure 4-6 Example on normal probability paper: exceedence probability.

commonly used and is given by

$$PP = \frac{i}{n+1} \tag{4-29}$$

in which PP is the estimated plotting position, i the rank number, and n the sample size.

Equation 4-29 was used to compute the plotting positions in Fig. 4-6. Many other plotting position formulas have been proposed and can readily be found in the literature. The entire problem should be kept in proper perspective. The points are plotted for visual comparison only, and they do not determine the line. It is also known from experience that the extreme values will range widely and the interior values will range little no matter what method of estimating plotting positions is used. There seems to be no good way to know what plotting position formula to use, but there also seems to be no good reason to be overly concerned about which one is used so long as our purpose in plotting remains visual comparison.

The normal population line of Fig. 4-6 could be used to make probability statements about the likelihood of occurrence of the mean monthly temperature during June, July, and August for Murray, Kentucky. For example, the probability that a mean monthly temperature of 80°F will be exceeded is approximately 12%. Similarly, there is about a 1% chance that the mean monthly temperature will be greater than 83.6°F. Also, there is a 1% chance that the mean monthly temperature will be less than 69.2°F, which corresponds to an exceedence probability of 99%.

When a poor fit to observed data is obtained, a different distribution function should be considered. For example, when the data demonstrate a concave-upward curve, it is reasonable to try a log-normal distribution or an extreme-value distribution. In some cases, it is preferable to fit with a distribution that requires an estimate of the skew coefficient, such as a log-Pearson type III distribution. However, sample estimates of the skew coefficient may be inaccurate for small samples, and thus they should be used with caution.

Poisson Distribution

The *Poisson distribution* is given in the equation

$$f(x) = \frac{m^x e^{-m}}{x!} \tag{4-30}$$

This distribution is used quite frequently to compute the probability of a certain number of events occurring in a specified period of time. In hydrology, we could think of the number of times the temperature exceeds, say 75°F, the number of times the temperature exceeds 80°F, and so on. We might also think of the number of times a certain level of flooding occurs on a stream.

We can see in Eq. 4-30 that the Poisson distribution contains only one parameter, the m of Eq. 4-30. The moments of the equation are as follows. The mean is equal to m, the standard deviation is equal to m, and the skewness is equal to $1\sqrt{m}$. Since all the moments are functions of the single parameter m, the Poisson distribution is a very rigid curve. In Fig. 4-7, the histogram of a sample of events that are distributed as the Poisson equation is shown. The mean of this sample was computed and was found to be 8.537. If we use the method of moments with the sample value as an estimate of the population value, the m of the Eq. 4-30, we can plot a Poisson curve to enclose our sample. This curve is also shown in Fig. 4-7. To plot the curve, it is necessary to multiply Eq. 4-30 by $2 \times n$, where 2 is the width of the class shown in Fig. 4-7 and n is 82, the number of items in our sample. To use the Poisson distribution we would need to have the integral of the curve. This integral starting at the left-hand side at 0, up to some value, say 8 on the scale on Fig. 4-7, would be the probability that an event would happen at least eight times. The area under the curve from 8 to infinity on the right-hand side would be the probability that an event would happen more than eight times.

We do not need to compute the integrals of Eq. 4-30. The values of the integral have been worked out and are tabulated. It can be shown that the Poisson probabilities are related to the gamma function probabilities. The

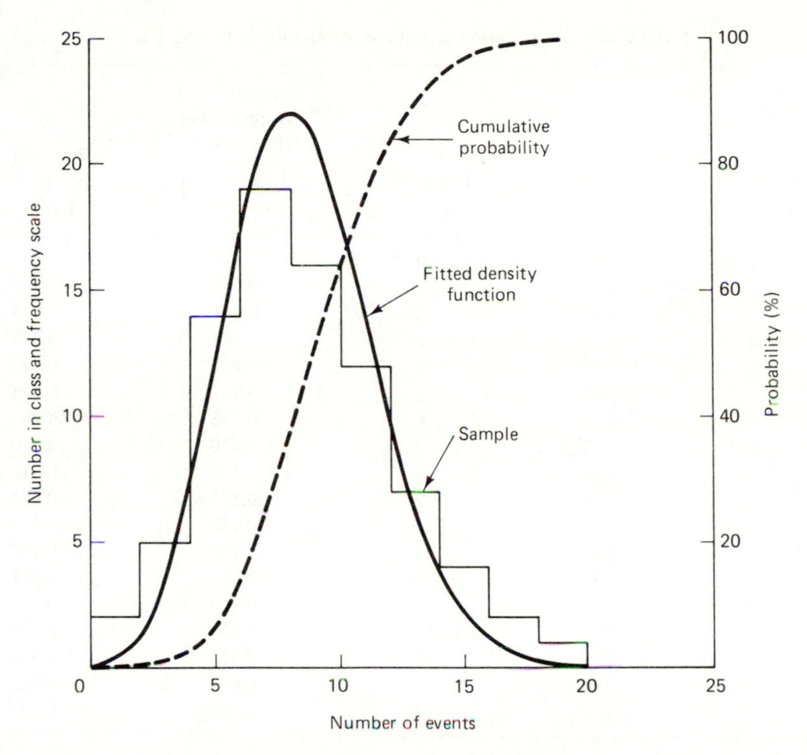

Figure 4-7 Example of Poisson distribution.

gamma function, which was introduced previously as Example 4-3, will be discussed as the next probability distribution. We only need to note here the relationship of the probabilities as given in the equation

$$P(m, i-1) = I\left(\frac{m}{\sqrt{i}}, i-1\right) \tag{4-31}$$

The left-hand side of Eq. 4-31 is a Poisson probability and it expresses the probability that an event whose expected number of occurrences is n will exceed $(i-1)$. The right-hand side of Eq. 4-31 is an incomplete gamma function ratio. For the moment we need only note that values of this probability are tabulated for various values of two parameters. These parameters are expressed in Eq. 4-31 as $m/\sqrt{i}$ and $(i-1)$.

Example 4-4: Analysis of Poisson Probabilities

The Poisson probabilities for the sample shown in Fig. 4-7 are given in Table 4-2. The values are plotted as the shallow S-shaped curve in Fig. 4-7. We can see on this curve that the probability of eight or fewer events happening is about 0.38. The probability of eight or more events occurring is 1 minus 0.38 or 0.62. To the right of the curve we can see that we are almost certain to get 18 or fewer events since the probability is nearly 1. There is very little probability, therefore, that more than 18 events would occur.

TABLE 4-2 Calculation of Poisson Probabilities (Using Eq. 4-31)

Selected Values of i	Compute $8.537/\sqrt{i}$	Compute $i-1$	Probability of More Events Than $(i-1)$ $I\left(\frac{m}{\sqrt{i}}, i-1\right)$	Probability of at Least Events i
1	8.537	0	0.999	0.001
2	6.03	1	0.998	0.002
3	4.93	2	0.991	0.009
4	4.27	3	0.971	0.029
5	3.82	4	0.927	0.073
6	3.48	5	0.852	0.148
7	3.22	6	0.746	0.254
8	3.02	7	0.620	0.380
9	2.85	8	0.484	0.516
10	2.70	9	0.352	0.648
11	2.57	10	0.240	0.760
12	2.47	11	0.156	0.834
13	2.37	12	0.093	0.907
14	2.29	13	0.054	0.946
15	2.21	14	0.029	0.971
16	2.14	15	0.015	0.985
17	2.08	16	0.007	0.993
18	2.02	17	0.003	0.997

Gamma Distribution

The gamma distribution is given in the equation

$$f(x) = C e^{-x/b} x^a \tag{4-32}$$

In Eq. 4-32 C equals $1/b^{a+1}\Gamma(a+1)$ to make the area enclosed by the curve equal to unity. $\Gamma(\cdot)$ is called the gamma function. Values can be found tabulated in mathematical handbooks. The gamma distribution is similar in shape to the Poisson distribution. The curve starts at zero when the variable x is zero, rises to a maximum, and descends to a tail that extends indefinitely to the right. The values that the variable x can take on are thus limited by 0 on the left. Values can extend to infinity on the right.

The gamma distribution differs from the Poisson distribution in that it has two parameters instead of the single parameter of the Poisson. This allows the curve to take on a greater variety of shapes than the Poisson distribution. The parameter a is a shape parameter while b is a scale parameter. The moments of the gamma distribution are given in

$$\text{mean} = b(a+1) \tag{4-33a}$$

$$\text{variance} = b^2(a+1) \tag{4-33b}$$

$$\text{skew} = 2b^2(a+1) \tag{4-33c}$$

Usually, the first two moments of a sample are used to determine the parameters of a gamma distribution. The curve is fitted to a sample by equating the first and second sample moments to the parameter functions given in Eqs. 4-33a and 4-33b. We can see that when we have determined the parameters b and a that we have also specified the third moment of the function, which is Eq. 4-33c. We can summarize this by saying that the gamma function allows us to vary the mean of the distribution and the standard deviation of the distribution according to the sample but the skewness is fixed. It is not an independent value.

Example 4-5: Estimation of Gamma Function Probabilities

In Fig. 4-8 we can consider the histogram as a sample of events that are distributed by the gamma distribution. The mean of this sample was calculated to be 8.717. The second moment of the sample was calculated to be 22.276. From these values we compute the value of b as 2.556 and the value of a as 2.411. With these values and a value for C of Eq. 4-32 we can use Eq. 4-32 to plot a gamma function that represents our sample. It is necessary to multiply Eq. 4-32 by N, the number of observations, which is 113 for our sample, and also to multiply by W, a class width, which is 2 for our sample. The plotted curve is shown superimposed on the histogram in Fig. 4-8.

In order to compute probabilities with a gamma function it is necessary to know the integral of that function. The integral from $x = 0$ to some value, say $x = 10$, as a proportion to the integral of the entire curve, would be the probability of x having

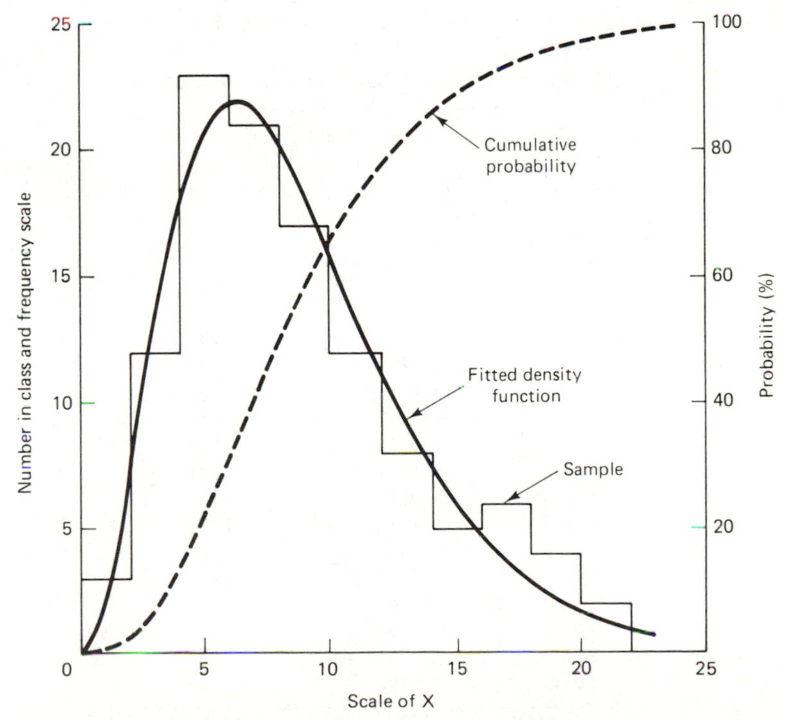

Figure 4-8 Example of gamma function.

values less than 10. The area under the curve above $x = 10$ would also be the probability of x having values greater than 10. We do not need to compute the integrals of Eq. 4-32. They have been computed and tabulated in various texts of mathematical tables. These tabulations are the same as we used to get probabilities for the Poisson distribution. Equation 4-34 defines an incomplete gamma function ratio as a function of x using Eq. 4-32.

$$\mathrm{IGR}(x) = \frac{C\int_0^x e^{-x/b}x^a\,dx}{C\int_0^\infty e^{-x/b}x^a\,dx} \tag{4-34}$$

To use the tabulated values it is necessary to transform the variable x in Eq. 4-34. If we define a new variable, $t = x/b$, we can make necessary substitutions and arrive at the incomplete gamma function ratio as a function of t shown in

$$\mathrm{IGR}(t) = \frac{Cb^{a+1}\int_0^t e^{-t}t^a\,dt}{Cb^{a+1}\int_0^\infty e^{-t}t^a\,dt} = \frac{\int_0^t e^{-t}t^a\,dt}{\Gamma(a+1)} \tag{4-35}$$

Equation 4-35 is identical to Eq. 4-36 with parameter p equal to the shape parameter a:

$$I(u, p) = \frac{1}{\Gamma(p+1)}\int_0^{u\sqrt{p+1}} v^p\,e^{-v}\,dv \tag{4-36}$$

The gamma probabilities for the sample in Fig. 4-8 have been calculated in Table 4-3. One simply selects values of u in Eq. 4-36 and for the specified parameter p looks up the probability, $I(u, p)$. For this particular example, it is necessary to interpolate in the tables between values of p equal 2 and p equal 2.5. For the selected value of u

TABLE 4-3 Calculation of Gamma Function Probabilities (Using Eqs. 4-34 and 4-35)

Selected Values of u	$I(n, 2.41)$	$x = bt = bu\sqrt{p+1}$
0.0	0.00000	0.00
0.2	0.00265	0.94
0.4	0.02040	1.89
0.6	0.06085	2.83
0.8	0.12314	3.77
1.0	0.20184	4.72
1.2	0.29004	5.66
1.4	0.38127	6.60
1.6	0.47039	7.55
1.8	0.55375	8.49
2.0	0.62916	9.44
2.2	0.69556	10.38
2.5	0.77801	11.80
3.0	0.87451	14.15
3.5	0.93209	16.51
4.0	0.96450	18.87
4.5	0.98215	21.23
5.0	0.99102	23.59

the limit of integration is shown in Eq. 4-36 to be $u\sqrt{p+1}$. Relating this to Eq. 4-35, we see that it is a value of t. We must multiply this value by the shape parameter b to get a value of x which is our initial variable. The incomplete gamma function ratio so calculated is the integral of the gamma function. It is plotted in Fig. 4-8.

Pearson Type III Distribution

The equation for the Pearson type III distribution is usually written

$$y = y_0\left(1 + \frac{v}{a}\right)^p e^{-\gamma v} \tag{4-37}$$

In Eq. 4-37, y_0 must equal $p^{p+1}/a\, e^p\Gamma(p+1)$ to make the area under the curve equal to unity.

The variable v can vary from $-a$ to infinity. The origin of v, that is, the place v equals 0, is at the mode, or high point of the distribution. The Pearson type III curve is very similar to a gamma curve. In fact, it is possible to mathematically transform it to a gamma curve.

The first three moments are usually used to fit a Pearson III curve to a sample. This differs from the gamma fitting, where the first and second moments were used. The following equations apply to fitting the Pearson III curve, and in determining probabilities with a Pearson III curve.

$$\gamma = \frac{2m_2}{m_3} \tag{4-38a}$$

$$p = \frac{4m_2^3}{m_3^2} - 1 \tag{4-38b}$$

$$a = \frac{p}{\gamma} \tag{4-38c}$$

$$\beta_1 = \frac{m_3^2}{m_2^3} \tag{4-38d}$$

$$\alpha(\text{skewness}) = \frac{\sqrt{\beta_1}}{2} \tag{4-38e}$$

$$\text{mode} = m_1 - \frac{1}{\gamma} \tag{4-38f}$$

$$x(\text{original variate}) = v + \text{mode} \tag{4-38g}$$

Example 4-6: Fitting a Pearson Type III Distribution

Figure 4-9 illustrates the fitting of a Pearson III distribution to a sample. The sample is the histogram. Calculation of the moments of this sample gave m_1 equal to 10.277, m_2 equal to 27.772, and m_3 equal to 104.092. Using Eq. 4-38 we can calculate γ equal 0.5336, p equal 6.908, and a equal 12.945.

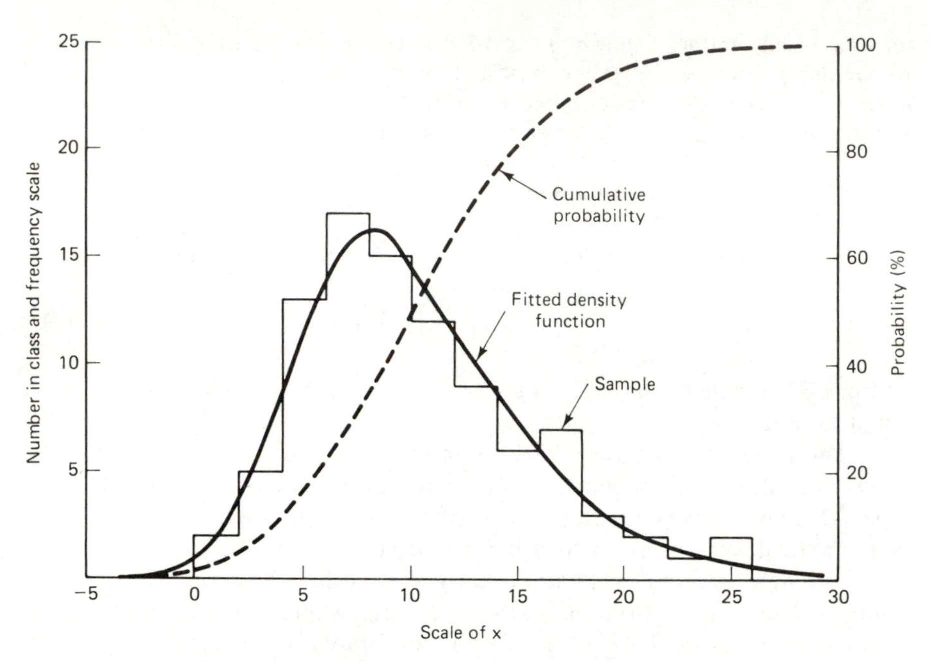

Figure 4-9 Example of Pearson type III distribution.

The mode is at 8.403. The β_1 coefficient was found to be 0.5058, and α is 0.3556. A simple way to plot a Pearson III distribution is to assume values of the variable ν and then transform to the variable x using Eq. 4-38g. To plot the Pearson III ordinates computed with Eq. 4-37, it is first necessary to multiply by the sample size, 94 items for this example, and also to multiply by the class width of 2.

Probabilities for a Pearson III distribution can be computed from the incomplete gamma function ratio. The appropriate equation is

$$P(t \leq x) = I\left(\frac{2}{\alpha} + x, \frac{4}{\alpha^2} - 1\right) \tag{4-39}$$

Equation 4-39 states that if t is a Pearson type III variable, with mean equal to 0 and variance equal to 1, the probability of t being less than some value x is given by the quantity on the right in Eq. 4-39. The two parameters in this quantity are computed from the skewness coefficient α and the critical value x_α. The first parameter, designated u, becomes numerically $5.625 + x$ in our example. This parameter u should not be confused with the moments. The second parameter in the expression in the right of Eq. 4-37 is usually designated p. Its value numerically for this example is 30.63. This parameter p in Eq. 4-39 should not be confused with the p in Eq. 4-37. The simple way to compute the probability curve as shown in Table 4-4 is to select values of x in Eq. 4-39 and look up the probabilities for the appropriate values of the two parameters. The x in Eq. 4-39 is scaled so that it has a mean of 0 and a variance of 1. The last step shown in the column on the right in Table 4-4 is to scale these selected values of x to the scale of the original variable as given in Fig. 4-9. This is done by multiplying

TABLE 4-4 Calculation of Pearson Type III Probabilities (Using Eq. 4-34)

Selected Value of x	$2/\alpha_3 + x = u$	$I(u, p)$	$5.270x + 10.277 = X$
−2.625	3.0	0.0009	−3.557
−2.425	3.2	0.0023	−2.503
−2.025	3.6	0.0115	−0.395
−1.625	4.0	0.0399	1.713
−1.125	4.5	0.1254	4.348
−0.625	5.0	0.2780	6.983
−0.125	5.5	0.4739	9.618
0.375	6.0	0.6652	12.253
0.875	6.5	0.8140	14.888
1.375	7.0	0.9092	17.523
1.875	7.5	0.9606	20.158
2.375	8.0	0.9847	22.793
2.875	8.5	0.9947	25.428
3.375	9.0	0.9983	28.063

by the standard deviation of the original x and adding the mean value of the original x. The probability integral computed in Table 4-4 is plotted in Fig. 4-9.

It was stated earlier that the Pearson type III distribution can be transformed to a gamma distribution. The transformation is given below in Eq. 4-40 and following. Let

$$x = p\left(1 + \frac{\nu}{a}\right) = \frac{p}{a}(a + \nu) = \gamma(a + \nu) \tag{4-40}$$

then

$$\left(1 + \frac{\nu}{a}\right)^p = \frac{x^p}{p^p} \quad \text{and} \quad -\gamma\nu = \gamma a - x$$

Substituting into Eq. 4-37 yields

$$y = y_0\left(\frac{x^p}{p^p}\right) e^{-x+\gamma a}$$

Simplifying results in

$$y = y_0\left(\frac{e^p}{p^p}\right) x^p \, e^{-x} \tag{4-41}$$

It can be seen that this is the same form as the transformed gamma distribution in the numerator of Eq. 4-35.

Extreme-Value Distribution

The extreme-value distribution is usually written as

$$p(x) = e^{-e^{-\alpha(x-u)}} \tag{4-42}$$

In our previous distribution forms we used the distribution function, or probability density function. We also pointed out that the functions could be integrated to produce the probability of events greater or smaller than a particular size within the distribution. Equation 4-42 is the probability form of the extreme-value distribution. It gives the probability that an event is less than the value X of the distribution. The extreme-value distribution has two parameters, the α and u of the equation. These parameters can be evaluated by using the first two moments of a sample. The form to evaluate these moments is

$$\bar{X} = \mu - \frac{\sqrt{6}}{\pi} \gamma\alpha \tag{4-43a}$$

$$\alpha = \frac{\pi}{s\sqrt{6}} \tag{4-43b}$$

In these equations $\bar{X}$ is the sample mean, s is the sample standard deviation, and γ is Euler's constant equal to 0.5772.

The extreme-value distribution is used most frequently in hydrology to give the probability of flood events. A set of flood data normally consists of the largest flood each year in the record of floods. In this manner we select the extreme event each year. The extreme-value distribution is theoretically the distribution of the extremes selected from a normal distribution.

Example 4-7: Extreme Value Analysis of Flood Data

Table 4-5 shows a tabulation of the record of annual maximum floods for the Valley River, at Tomotla, North Carolina. The individual floods are not shown in this table; instead, they have been grouped by 1000-cfs intervals. The number of floods in each interval is shown in the second column. To generate a sample probability to match our probability distribution, we can add the floods sequentially through the intervals starting at the lower end. This summation is shown in the table in the next column,

TABLE 4-5 Annual Peak Discharges Valley River at Tomotla, North Carolina, 1919–1960

Class Range (cfs)	Number of Floods	Summation	Cumulative Probability
0–1000	1	1	0.024
1001–2000	4	5	0.119
2001–3000	7	12	0.286
3001–4000	14	26	0.619
4001–5000	2	28	0.667
5001–6000	7	35	0.833
6001–7000	5	40	0.952
7001–8000	0	40	0.952
8001–9000	2	42	1.000
	42		

headed "summation." A probability distribution must have unit area. We can accomplish this by dividing each of the summation values by the total, 42 for our sample. The results are shown in the column headed "cumulative probability." These values are plotted as the stepped function in Fig. 4-10. Referring again to Table 4-5, if we look at the first summation value, 1, we see that in our record, one flood occurred in the range 0 to 1000 cfs. It could have occurred at any value up to 1000 cfs. We do not know in this table exactly what the flood peak was. We must interpret our probability, therefore, as the probability that the event would not be larger than 1000 cfs. The point in the stepped function where our interpretation of this probability is correct would

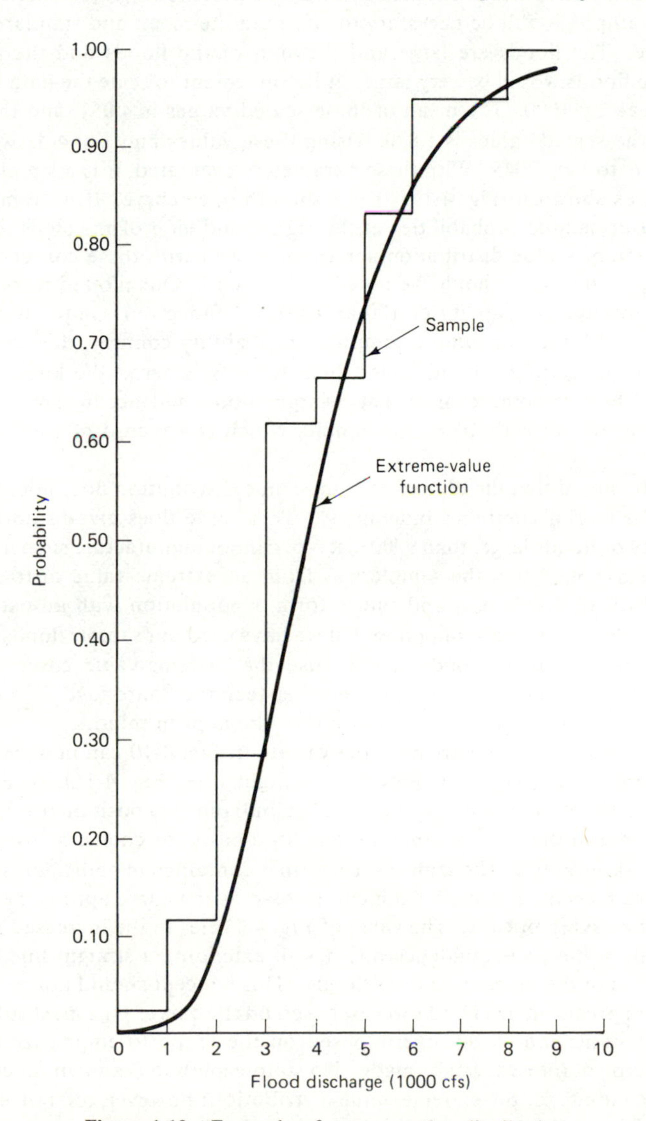

Figure 4-10 Example of extreme-value distribution.

be at the right-hand side of each step: in other words, at 1000 cfs and a probability of 0.024. The probability that a flood would be smaller than 2000 cfs is 0.119, and this is the point at the right-hand edge of our second step.

The step function is only a crude and approximate way to represent our sample. We could plot points where they occurred on the actual scale of peak discharge. However, without making some assumptions, we would not know what probability to assign to these values. We will discuss this point at greater detail in a later section. For this approach to the problem of sample probabilities we will regard a step function as a crude sample representation of probabilities. To plot the extreme-value distribution, fitted to our sample, it will be necessary to compute the mean and standard deviation of the sample. The floods are large and the sum of the floods and the sum of the squares of the floods would be very large. It is convenient to scale the data by dividing each flood peak by 1000. The mean of these scaled values is 4.051, and the standard deviation of the scaled values is 1.820. Using these values and Eq. 4-43 we find u to be 3.232 and α to be 0.7045. With these parameters evaluated, it is a simple matter to plot Eq. 4-42 as shown in Fig. 4-10. It is a smooth ogee curve. If we remember that we interpret our sample probabilities at the right-hand side of the steps, we see that the plotted extreme-value distribution agrees quite well with these corner values.

An important point should be noted in Fig. 4-10. Our plotted record of floods reached a computed probability of 100 at 9000 cfs. Since our sample had no floods beyond this class limit, our sample-computed probability confirms this by saying that the probability of getting a flood larger than 9000 cfs is zero. We know that this is not correct. There is some chance that a larger flood will occur someday. But we cannot develop this directly from our sample, which is a record of what has already occurred.

It can be noted that the plotted extreme-value distribution does not reach 1.00% at 9000 cfs. In fact, it continues indefinitely. This curve does give us an estimate of the probability of floods larger than 9000 cfs. We cannot manufacture such information, however. We assumed that the sample was from an extreme-value distribution (i.e., that the total of all floods past and future form a population with an extreme-value distribution). Only with this assumption that we have used our sample floods to estimate a "true" population of all floods can we use the extreme-value curve to estimate probabilities of future floods. This difference between the "hard fact" of a past record and the "might be" of future floods must always be kept in mind.

The ogee curve of the extreme-value variate in Fig. 4-10 can now be plotted on Gumbel extreme-value paper; it plots as a straight line (Fig. 4-11). The individual annual floods were plotted using Eq. 4-29, the Weibull plotting position formula. Within the range of data in the flood sample there is little basis for choice between Fig. 4-10 or 4-11 as a working graph. The same information is contained in both figures. However, for the more rare events Fig. 4-10 is difficult to read as the curve approaches the 100% probability line asymptotically. The value of Fig. 4-11 lies in the increased readability, and also in the subjective confidence one has in extending a straight line to estimate the probabilities of the higher and rarer floods. This concept should not be interpreted as an increased precision in a flood forecast based on the curve. One must still recognize that tolerance limits should be superimposed on the graph to emphasize that only a probabilistic type of forecast can be made. No simple method is known for constructing exact tolerance limits for the extreme-value distribution; however, tolerance limits may be estimated by simulation, as will be shown later in this chapter.

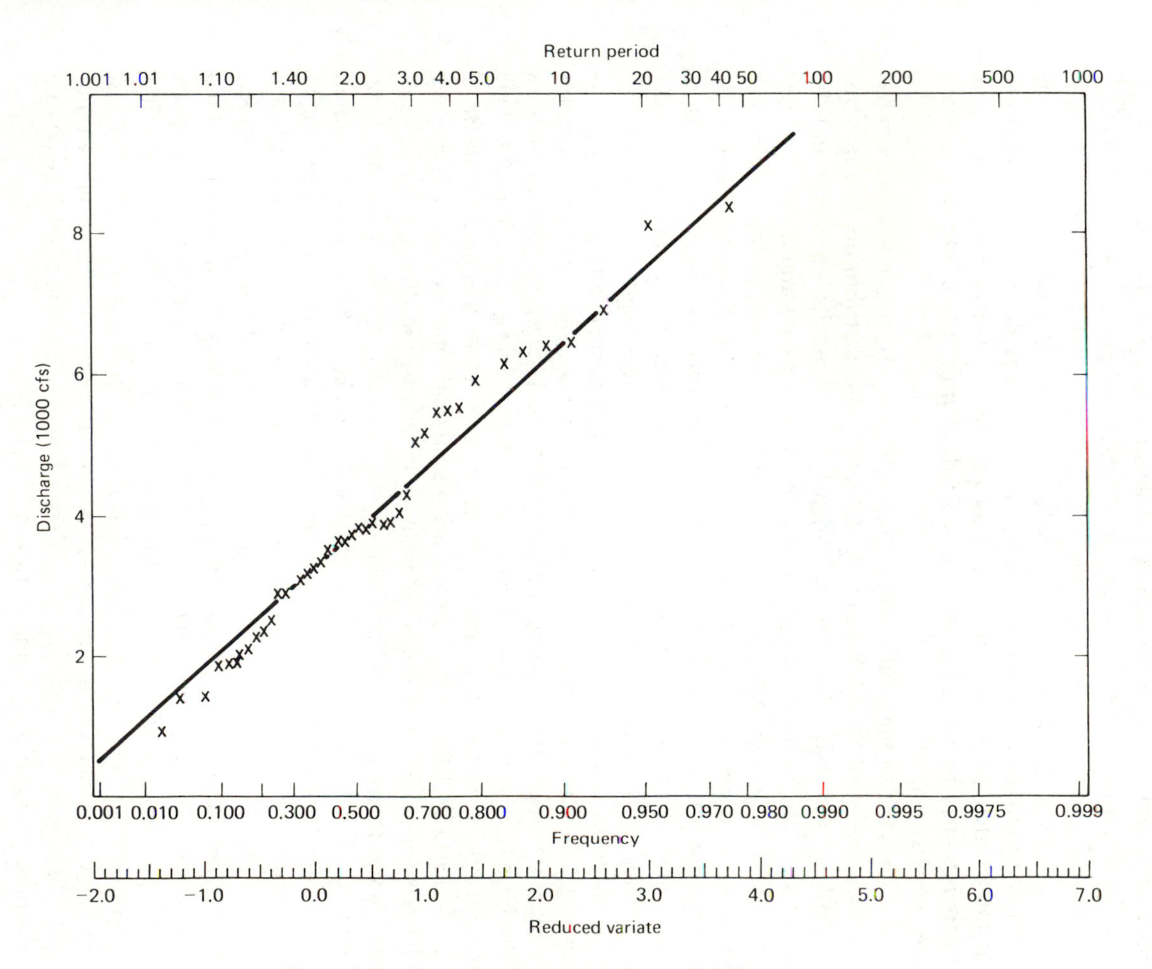

11039 **Figure 4-11** Example of extreme-value probability paper.

MOMENT-RATIO ANALYSIS

As indicated previously, the identification of the distribution function of a random variable is one objective of univariate data analysis. The first step in the identification of the distribution function is histogram analysis. The effectiveness of a graphical analysis in identifying the pdf will depend on the sample size and the interval selected to plot the abscissa. For small samples, it is difficult to separate the data into groups that will provide a reliable indication of the frequency of occurrence. With small samples, the configuration of the histogram may be quite different for a small change in the intervals selected for the abscissa.

Recognizing the limitations of graphical analysis, frequency analysis has become the most popular method of finding the distribution function for hydrologic variables. Its popularity depends on both its simplicity and the fact that it can be used with small samples, which are common in hydrology. However, frequency analysis requires some knowledge of the distribution function, a method for estimating the parameters of the distribution, and probability plotting paper and a plotting position formula. For large samples, the computations and plotting are tedious. However, the method has the advantage of providing a means of assessing the accuracy of estimation.

Moment-ratio analysis is an alternative to graphical and frequency analysis. Identification of a probability distribution from a sample of data requires estimates of the moment ratios of Eqs. 2-19b and 2-19c. Moment-ratio analyses are based on the fact that density functions of known distributions have characteristic values of β_1 and β_2. The moment-ratio diagrams, with β_1 as the abscissa and β_2 as the ordinate, are a systematic means of identifying the distribution function of the population. A moment-ratio diagram for some of the more commonly used probability distributions is shown in Fig. 4-12. Sample estimates of β_1 and β_2 can be used to make a tentative identification of the population underlying the sample. Density functions can be represented by a point, line, or area on a moment-ratio diagram. For example, the uniform density function always has values of 0.0 and 1.8 for β_1 and β_2, respectively. The normal density function is also represented by a point, with values of 0.0 and 3.0 for β_1 and β_2, respectively. The t distribution, which changes scale as its parameter ν changes, is represented by a line on the moment-ratio diagram; since the t distribution is always symmetric, β_1 will equal zero and the line will coincide with the axis for $\beta_1 = 0$. The line extends from the point $(\beta_1 = 0, \beta_2 = 3)$ to $(\beta_1 = 0, \beta_2 = \infty)$ depending on the value of ν. An example of a moment-ratio diagram is shown in Fig. 4-12.

It is important to recognize that sample estimates of β_1 and β_2 that are obtained from a known population will not necessarily equal the population values of β_1 and β_2. For example, if a random sample of size n is obtained from a normal population, the sample estimates of β_1 and β_2 will most probably not equal 0 and 3, respectively; in fact, for small samples the sample estimates

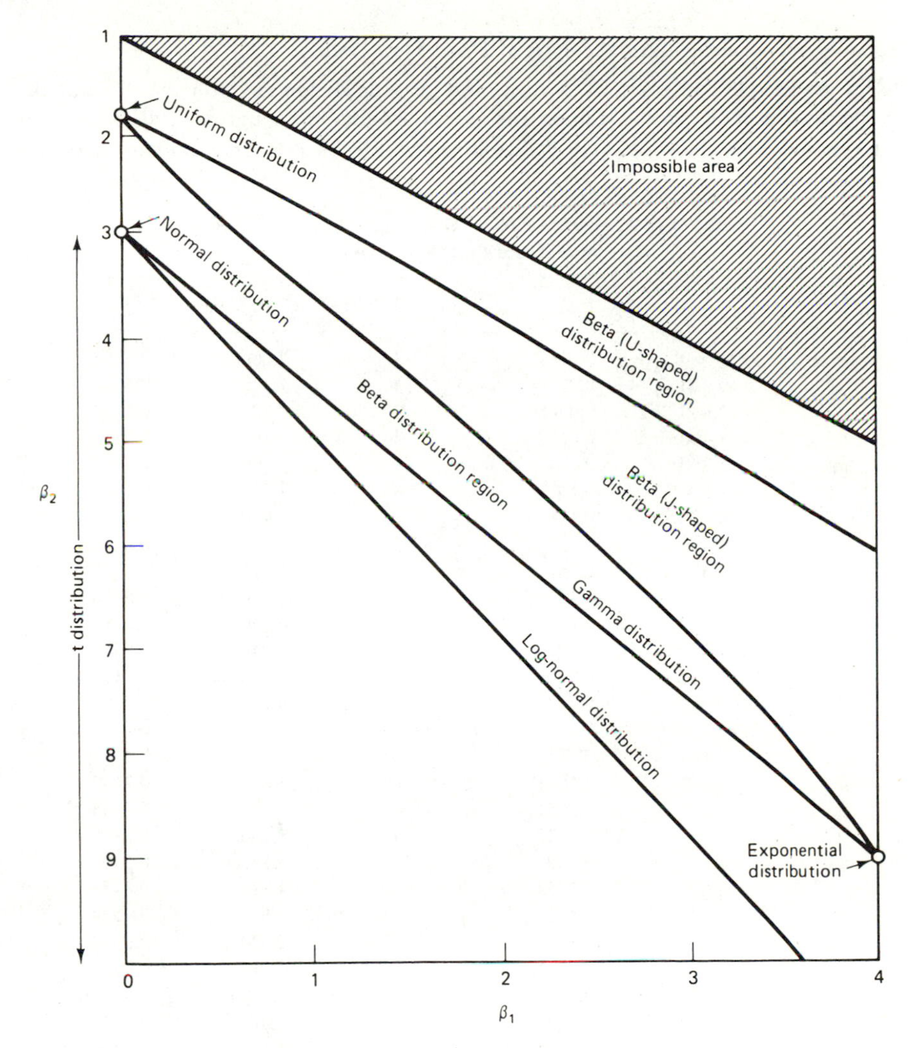

Figure 4-12 Moment-ratio diagram.

may deviate considerably from the population values. As the sample size decreases, the accuracy of a sample estimate decreases; therefore, one must allow a greater sampling variation because the accuracy of the estimators is poorer. It is difficult to obtain accurate estimates of the higher-order moments (i.e., skew and kurtosis) for small samples.

When reporting or summarizing characteristics of a sample of data, it is important to report the moments as well as the moment ratios. Although the moment ratios provide valuable information about the data, they are inadequate by themselves because it would not be known whether an increase in the moment ratio would be due to an increase in the moment in the numerator or a decrease in the moment in the denominator.

Example 4-8: Moment Ratio Analysis of Flood Flow Data

We have a sample consisting of 58 measured peak-discharge rates for the Piscataquis River near Dover-Foxcroft, Maine. The values are the maximum peak discharge for each year over a 58-year span. The values are given in Table 4-6. The summary statistics

TABLE 4-6 Annual Maximum Discharges (Q_p) and Expected Exceedence Probability (pp)

Rank	Q_p (cfs)	pp	Rank	Q_p (cfs)	pp
1	21,500	0.0169	30	7,600	0.5085
2	19,300	0.0339	31	7,420	0.5254
3	17,400	0.0508	32	7,380	0.5424
4	17,400	0.0678	33	7,190	0.5593
5	15,200	0.0847	34	7,190	0.5763
6	14,600	0.1017	35	7,130	0.5932
7	13,700	0.1186	36	6,970	0.6102
8	13,500	0.1356	37	6,930	0.6271
9	13,300	0.1525	38	6,870	0.6441
10	13,200	0.1695	39	6,750	0.6610
11	12,900	0.1864	40	6,350	0.6780
12	11,600	0.2034	41	6,240	0.6949
13	11,100	0.2203	42	6,200	0.7119
14	10,400	0.2373	43	6,100	0.7288
15	10,400	0.2542	44	5,960	0.7458
16	10,100	0.2712	45	5,590	0.7627
17	9,640	0.2881	46	5,300	0.7797
18	9,560	0.3051	47	5,250	0.7966
19	9,310	0.3220	48	5,150	0.8136
20	8,850	0.3390	49	5,140	0.8305
21	8,690	0.3559	50	4,710	0.8475
22	8,600	0.3729	51	4,680	0.8644
23	8,350	0.3898	52	4,570	0.8814
24	8,110	0.4068	53	4,110	0.8983
25	8,040	0.4237	54	4,010	0.9153
26	8,040	0.4407	55	4,010	0.9322
27	8,040	0.4576	56	3,100	0.9492
28	8,040	0.4746	57	2,990	0.9661
29	7,780	0.4915	58	2,410	0.9831

are given in Table 4-7, and the histogram is shown in Fig. 4-13. Since the coefficient of variation was relatively small, it was necessary to use a small number of intervals; otherwise, most of the values would have been assigned to one frequency interval. For the frequency interval of 3818 cfs, one-half of the values were in a single cell. As evident from both the histogram and the skew, the distribution is skewed to the right (i.e., the long tail is above the mean). The width of the first and last frequency intervals is less than that for the other intervals; the width of these intervals was limited by the minimum and maximum values of the random variables. The values of the coefficients of skew ($\beta_1 = 1.2$) and kurtosis ($\beta_2 = 3.9$) suggest a beta distribution; however,

TABLE 4-7 Summary Statistics for Annual Maximum Peak Discharge (cfs) Data for the Piscataquis River near Dover-Foxcroft, Maine

Statistic	Value
Mean	8,634
Variance	1.676×10^7
Standard deviation	4,095
Skew	7.574×10^{10}
Kurtosis	1.109×10^{15}
C_v	0.474
β_1	1.217
β_2	3.945
Minimum	2,410
Maximum	21,500
Range	19,090

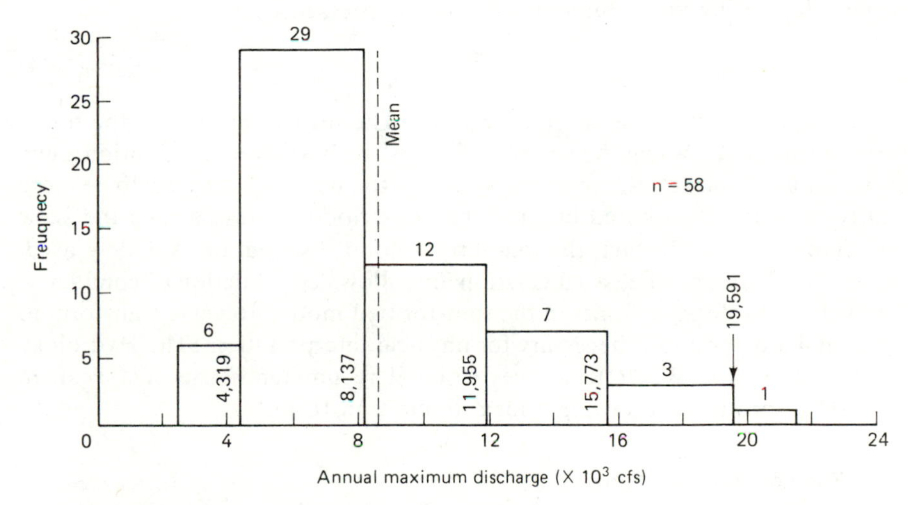

Figure 4-13 Histogram: annual maximum discharge (cfs) for the Piscataquis River at Dover-Foxcroft, Maine.

because of the relatively small sample, it is not possible at this time to eliminate the possibility of using a normal, log-normal, or gamma distribution.

A method for testing the validity of one of these distributions would have to be used to make a final decision. As an example, the data of Table 4-6 and Fig. 4-13 were subjected to a chi-square goodness-of-fit test for testing the null hypothesis that the flows were from a normal distribution with the location and scale parameters equal to the sample mean (8634 cfs) and standard deviation (4095 cfs), respectively. Using the intervals indicated on Fig. 4-13 and combining the cells for $Q > 11{,}955$ cfs to ensure that the expected frequencies were greater than 5, the following value of the χ^2 statistic was computed:

$$\chi^2 = \frac{(8.46 - 6)^2}{8.46} + \frac{(17.75 - 29)^2}{17.75} + \frac{(19.69 - 12)^2}{19.69} + \frac{(12.10 - 11)^2}{12.10} = 10.95$$

Since the sample values of n, $\bar{X}$, and S were used to compute the expected frequencies and there are four cells, there is only one degree of freedom. Even for a level of significance of 0.005% with a critical χ^2 value of 7.88, the null hypothesis would have to be rejected. Thus the data are probably not from a normal distribution with $\mu = 8634$ cfs and $\sigma = 4095$ cfs.

DATA TRANSFORMATIONS

A complete discussion of transformations cannot be given here; however, two particular variate transforms have had wide usage in hydrology and meteorology. The logarithmic transform is expressed as

$$Y = \log X \tag{4-44}$$

In this equation X is the observed variate. The logarithm of X is the transformed variate Y. When the variates of a positively skewed distribution, such as an annual flood series, are so transformed, the new variates tend to be more nearly normally distributed because the large flood values causing the skew are changed most. In fact, the log-normal flood distribution is widely used. It has the advantage of easy calculation and allows construction of confidence intervals and tolerance limits in the transformed mode. Reverse transform to original flood variates is necessary for physical interpretation. The Hydrology Committee of the Water Resources Council recommends use of a Pearson type III distribution following a logarithmic transform.

The Log-Normal Distribution

The same procedure that is used for fitting the normal distribution can be used to fit the log-normal distribution. The underlying population is assumed to be log-normal. The data must first be transformed to logarithms. The mean and standard deviations of the logarithms are computed and used as the parameters of the population; it is important to recognize that the logarithm of the mean does not equal the mean of the logarithms, which is also true for the standard deviation. The population line is defined by plotting the straight line between the points $(\bar{X} - S, 0.8413)$ and $(\bar{X} + S, 0.1587)$, where $\bar{X}$ and S are the mean and standard deviation of the logarithms. In plotting the data, either the logarithms can be plotted on an arithmetic scale or the untransformed data can be plotted on a logarithmic scale.

Log Pearson Type III Distribution

Since the log Pearson type III (LP3) distribution was recommended by the Water Resources Council (WRC) it has been widely used. A basic frequency curve can be developed for the LP3 distribution using the method of moment estimation recommended by the WRC. The general procedure is as follows:

1. For the sample of n observations on the random variable, a new series should be computed by taking the logarithm of each of the n observations.
2. The mean, standard deviation, and skew of the logarithms should be computed.
3. Using the estimate of the skew and selected exceedence probability values, values of the LP3 deviate K can be obtained from Appendix A-8.
4. Using the log mean and log standard deviation of step 1 and the K values of step 3, values of a random variable y can be computed for each exceedence probability of step 3 using

$$y_i = \bar{X} + K_i S \tag{4-45}$$

   in which the subscript i indicates variation of the exceedence probability.
5. Plot y_i versus the exceedence probability on normal probability paper. (*Note*: The ordinate is still expressed in logarithmic terms, so a linear scale should be used for the ordinate.)

To make probability statements about the untransformed random variable, the antilog of the value from the frequency curve should be computed. In using the procedure above for an annual maximum flood series, the WRC recommends using a weighted skew coefficient. The weighted skew is a function of the sample log skew, the sample record length (n), and the value of skew obtained from a map. Recent research has recommended other weighting schemes.

In addition to the computation of the frequency curve for the LP3 distribution, the WRC report provides methods for computing confidence intervals on the log frequency curve. Additionally, methods of testing for outliers and handling historic data are provided. The report provides numerous examples.

The Power Transformation

Stidd (1953) proposed the use of a cube-root transform to normality for precipitation data. This transform is

$$Y = X^c \tag{4-46}$$

In Eq. 4-46 X is the original variate and Y is the transformed variate. For a cube-root transform $c = \frac{1}{3}$. More recently, Stidd (1970) suggested that c varies from $\frac{1}{2}$ to $\frac{1}{3}$. These values were derived from computer-synthesized precipitation data. The change from $\frac{1}{2}$ to $\frac{1}{3}$ is apparently associated with the number of summations of sequential numbers, representing integration of precipitation rates over longer periods of time.

A simple transform that can be very useful in skewed distributions is the reversal of direction of the variate scale. By this means one can change a positively skewed distribution (tail to the right) to a negatively skewed distribution (tail to the left).

Example 4-9: Pearson Type III Analysis with a Linear Transform

Figure 4-14 is a record of monthly average relative humidity for Murray, Kentucky, for the months May through September during the years 1944–1959. The record is plotted as a histogram with a class width of 3%. It is obvious from the plot that the distribution is skewed to the left. Relative humidity cannot normally be greater than 100%, which means air saturated with all the moisture it can hold. Our distribution, therefore, has an upper bound of 100%. Humidity can be 0%, which considering the scale, is almost unbounded to the left in Fig. 4-14.

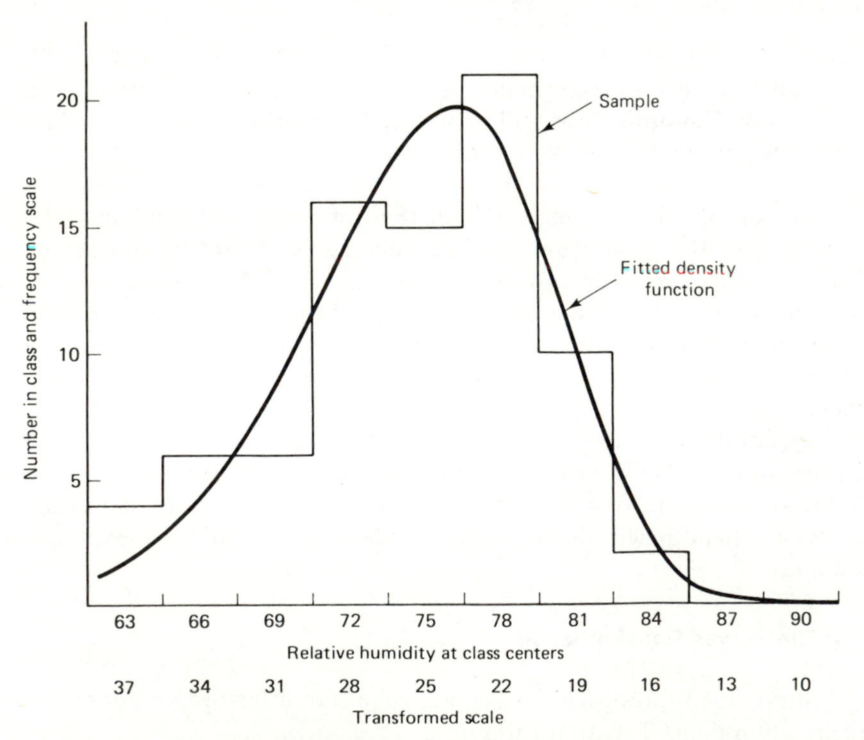

Figure 4-14 Average monthly relative humidity at Murray, Kentucky, for months of April through September 1944–1959.

We should fit a skewed distribution, such as a gamma or Pearson III, to our relative humidity data. But these distributions are skewed to the right. A simple procedure is to transform to a new variate with reversed scale.

Consider a variate defined as

$$W = 100 - T \tag{4-47}$$

in which W is the new variate and T is observed relative humidity. Now if T were to reach its upper limit of 100%, W would be zero. As T gets smaller, W gets larger, becoming 100 when T reaches zero.

A Pearson III distribution was fitted to the relative humidity data measured with transformed variate W. The moments were $\mu_1 = 25.5$, $\mu_2 = 25.19$, and $\mu_3 = 70.06$. These moments gave $\gamma = 0.719$, $p = 12.02$, and $a = 16.72$. The mode is at $25.5 - 1/0.719 = 24.11$ on the W scale. This is $100 - 24.11 = 75.8$ on the T scale. The origin of the curve (the lowest W value) is at $24.11 - 16.72 = 7.39$. Measured on the T scale this corresponds to 92.61%. The full Pearson III distribution is superimposed on the histogram of observations in Fig. 4-14 and shows a good "smoothed" fit.

Change of Variable

When a variable is changed algebraically in a probability density function, a new distribution results. This is a useful transformation in those circumstances where the density of a variable is known and we wish to compute the density function of some other variable that is a function of the original variable. Although the mathematics for a change of variable is straightforward, little use of this device is found in hydrologic literature.

One example of change of variable will be given to illustrate the method. Fry (1965) gives the equation (his number 85.6)

$$p(\xi, \eta, \zeta, \ldots, \omega) = \frac{p(x, y, z, \ldots, w)}{\partial(\xi, \eta, \zeta, \ldots, \omega)/\partial(x, y, z, \ldots, w)} \tag{4-48}$$

for the transformation. In Eq. 4-48, $p(x, y, z, \ldots, w)$ is the density function in multiple dimensions for the variables $x, y, z, \ldots, w$, and $p(\xi, \eta, \zeta, \ldots, \omega)$ is the density function for new variates $\xi, \eta, \zeta, \ldots, \omega$ for which a full set of equations such as Eqs. 4-49 is known:

$$\xi = f_1(x, y, z, \ldots, w) \tag{4-49a}$$

$$\eta = f_2(x, y, z, \ldots, w) \tag{4-49b}$$

$$\zeta = f_3(x, y, z, \ldots, w) \tag{4-49c}$$

$$\vdots$$

$$\omega = f_n(x, y, z, \ldots, w) \tag{4-49d}$$

The denominator of the right-hand side of Eq. 4-48 is the *Jacobian*. It is defined as a determinant in which the elements are all the possible partial derivatives of Eqs. 4-49.

Example 4-10: Exponential Transform for Watershed Length

For a simple hydrologic example, suppose that we have placed a grid overlay on a topographic map. "At random" we choose elements of the grid, and on the map underneath we measure the drainage area of both branches above every stream junction that falls in the chosen grid elements. Now there are many more small drainage areas than there are large ones. If we converted our map sample to a probability density function, we would probably find it to be similar to a descending exponential:

$$p(x) = b\, e^{-bx} \tag{4-50}$$

In Eq. 4-50, x is the drainage area and b is a shape parameter. We may define a "mean length" of a drainage basin as the distance from the outlet (in this example, from stream junctures) through the center of the basin to the divide. For a square basin draining from one corner, the mean length would be a diagonal of length $\sqrt{2x}$. For a circular drainage the mean length would be $2\sqrt{x/\pi}$. For natural drainages we will assume that

$$x = kl^2 \tag{4-51}$$

as the relationship between size and mean length, l, and k is simply some proportionality factor. Equation 4-48 reduces to the simple form

$$p(l) = \frac{p(x)}{\partial l/\partial x} \tag{4-52}$$

From Eq. 4-51,

$$\frac{\partial l}{\partial x} = \frac{1}{2\sqrt{kx}} \tag{4-53}$$

Therefore,

$$p(l) = \frac{b\, e^{-bx}}{2\sqrt{kx}} = \frac{b}{2k}\,\frac{e^{-bkl^2}}{l} \tag{4-54}$$

would be the density function for the mean length.

RISK VERSUS RETURN PERIOD

When we use a flood frequency curve, we set a magnitude of flood that corresponds to a specified design probability. We expect this magnitude of flood to occur in P percent of some long future record of floods. Usually, we think of the return period (T), as in Eq. 4-26, rather than the probability. A 25-year flood would be found to be exceeded, *on the average*, four times in each 100-year period of a large number of 100-year records. By using the return period as a design criterion, we are implying that we expect average conditions to apply over some long future. This is correct procedure when we have long-lived structures, or economic benefits accruing over very long periods. But what risks do we run over short periods in which there is no opportunity for long-time averaging? Such a short period might be the construction period of a project.

Risk estimation is an alternative to the return-period concept. Risk can be introduced using the annual maximum flood as the random variable. Thus the time period is 1 year. We will assume that in any one year, a flood either occurs or it does not occur and that no more than one flood of a certain magnitude will occur in any one year. If we also assume that the probability of that flood occurring remains constant from year to year, we have satisfied the four assumptions underlying the binomial distribution. Thus, if we define the risk as being the probability of one or more floods having probability p occurring in n years, we get the risk as

$$\text{risk} = 1 - \binom{n}{0} p^0 (1-p)^{n-0} = 1 - (1-p)^n \tag{4-55}$$

We can develop the concept of risk in a different way. If p is the probability that a flood will occur in any year, $1 - p$ is the probability that it will not occur. If, further, we need n years for construction, $(1-p)^n$ is the probability that the flood will not occur in these n years. Conversely, $1 - (1-p)^n$ is the probability that the n-year period will not be flood free. In other words, it is the chance of at least one flood equal to or greater than the flood corresponding to p, and represents risk. Table 4-8 shows risks for various values of p and n.

TABLE 4-8 Values of Risk

	$T = 1/p$		
n	25	50	100
1	0.040	0.020	0.010
2	0.078	0.040	0.020
3	0.116	0.059	0.030
4	0.151	0.078	0.039
5	0.184	0.096	0.049
10	0.335	0.183	0.096
25	0.639	—	—
50	—	0.636	—
100	—	—	0.634

EVALUATING THE ACCURACY OF PREDICTION

Earlier in this chapter we studied the use of confidence intervals to make probabilistic predictions about possible future values of the mean of a sample. We also studied tolerance limits to make probabilistic predictions about possible future values of specified proportions of our samples. The reason for

TABLE 4-9 Factors for Tolerance-Limit Envelopes

Confidence Level Percent	Sample Size	Probability of Event Being Exceeded—Percent										
		0.1	1	5	10	25	50	75	90	95	99	99.9
5	40	3.85	2.93	2.12	1.69	0.99	0.26	−0.40	−0.97	−1.29	−1.89	−2.55
	50	3.76	2.85	2.06	1.64	0.95	0.24	−0.43	−1.00	−1.32	−1.93	−2.60
	60	3.69	2.80	2.02	1.60	0.92	0.21	−0.45	−1.02	−1.35	−1.96	−2.64
	70	3.64	2.76	1.98	1.58	0.90	0.20	−0.46	−1.04	−1.37	−1.99	−2.67
	80	3.60	2.73	1.96	1.56	0.89	0.18	−0.48	−1.05	−1.39	−2.01	−2.69
	90	3.56	2.70	1.94	1.54	0.87	0.17	−0.49	−1.06	−1.40	−2.02	−2.71
	100	3.53	2.68	1.92	1.52	0.86	0.16	−0.50	−1.08	−1.41	−2.04	−2.73
10	40	3.66	2.78	2.00	1.59	0.91	0.20	−0.46	−1.03	−1.36	−1.98	−2.66
	50	3.59	2.72	1.96	1.55	0.88	0.18	−0.48	−1.06	−1.39	−2.01	−2.70
	60	3.54	2.68	1.93	1.53	0.86	0.17	−0.50	−1.07	−1.41	−2.04	−2.73
	70	3.50	2.65	1.90	1.51	0.85	0.15	−0.51	−1.09	−1.43	−2.06	−2.75
	80	3.47	2.63	1.88	1.49	0.84	0.14	−0.52	−1.10	−1.44	−2.07	−2.77
	90	3.45	2.61	1.87	1.48	0.83	0.14	−0.53	−1.11	−1.45	−2.08	−2.79
	100	3.43	2.59	1.86	1.47	0.82	0.13	−0.53	−1.12	−1.46	−2.10	−2.80

constructing a graph such as Fig. 4-6 represents the same need to predict future values. Usually, the prediction involves proportions rather than means. Specifically, we try to extrapolate the straight line to higher (or lower) probabilities.

In our example of 48 average monthly temperatures, each element is $\frac{1}{48}$ of the total, and therefore has a sample probability of $\frac{1}{48} = 2.083\%$. If we extended our straight line in Fig. 4-6 to 0.5% (top scale), this corresponds to a sample proportion of $\frac{1}{200}$. The temperature for this point (84.4°F) would be expected once in 200 months if the sample had correctly estimated the population.

Our study of tolerance limits has shown us that we must attach a confidence level to the prediction. The probability of 0.5 is a sample proportion of $\frac{1}{200}$, and we must estimate how this proportion might range. Our correct prediction is that we are *P* percent confident that the temperature which occurs on an average once in 200 months will not be greater than 84.4°F.

Beard (1962) constructed a table for easy calculation of tolerance-limit curves at selected probability values. His table provided coefficients of the standard deviation. Products of the coefficient and standard deviation were added to the values of the variate for a given proportion of the sample. Tolerance-limit curves can also be constructed using values added to the mean of the sample. The coefficients in this calculation are values from one-sided tolerance limit tables. Such coefficients for 5% and 10% confidence levels are shown in Table 4-9 for selected sample sizes and selected proportions of the sample. These values were calculated using the equations on pages 2-5 of Natrella (1963).

TABLE 4-10 Calculation of Confidence Curves

	m = 76.4°					s = 3.11°					
P	0.1	1	5	10	25	50	75	90	95	99	99.9
k(48, .05)	3.78	2.87	2.07	1.65	0.96	0.24	−0.42	−0.99	−1.31	−1.92	−2.59
kS	11.8	8.9	6.4	5.1	3.0	0.8	−1.3	−3.1	−4.1	−6.0	−8.0
m+kS	88.2	85.3	82.8	81.5	79.4	77.2	75.1	73.3	72.3	70.4	68.4
k(48, .95)	2.59	1.92	1.31	0.99	0.42	−0.24	−0.96	−1.65	−2.07	−2.87	−3.78
kS	8.0	6.0	4.1	3.1	1.3	−0.8	−3.0	−5.1	−6.4	−8.9	−11.8
m + kS	84.4	82.4	80.5	79.5	77.7	75.6	73.4	71.3	70.0	67.5	64.6

Tolerance-limit curves for the average monthly temperature problem are given in Table 4-10. These calculations yield curves for 5% and 95% levels of confidence. Values of the coefficient *k* are obtained by interpolation for a sample size of 48 in Table 4-9. The products *ks* are added to the mean temperature of 76.4°F. Coefficients of *k* for 95% confidence are negative and reversed values of the 5% coefficients. The 5% and 95% $m + ks$ values are plotted in Figure 4-6. These tolerance-limit curves are identical to curves obtained using Beard's method of calculation.

Now we can state from Fig. 4-6 that in only 5% of future samples will the 1-in-100 (1%) mean monthly temperature be above 85.3°F. In 95% of future samples it will be above 82.4°F. We expect the average (not the 50% confidence level) of 1 in 100 temperatures to be 83.6°F.

We should note in Fig. 4-6 that the two confidence curves tend to enclose the plotted points. Since, in effect, our future samples will tend to lie (90% of the time) anywhere between these curves, it seems unprofitable to try to determine exact plotting positions.

TOLERANCE LIMITS BY SIMULATION

Simulation was introduced in Table 1-1 as one of the problem types under synthesis. We are given possible inputs and a possible system and we explore the outputs. With a univariate model we have specified that the system is simply one of the frequency distributions. We specify further here that the particular system is modeled by the extreme-value distribution. In simulation we pose the question: If we pulse this univariate model with some number of random numbers, what types of synthetic samples might be generated?

Random numbers are linearly distributed. For example, 10 random numbers, in the scale 0 to 100, can represent 10 equally likely probabilities. Note that we specify that the probabilities are equally likely, not some magnitude of event associated with the probabilities. The form of the extreme-value distribution given by Eq. 4-42 expresses probability, not probability density. Suppose that we have values of α and μ from a real sample. Then using Eq. 4-42, for any random number considered a random probability $p(x)$, we can calculate the value of x. Fifty such random numbers would make a synthetic extreme-value sample of 50 x's. We could repeat this construction 100 times and have 100 synthetic samples of 50 items each. Obviously, the transformation from the random probabilities to the random x's depends on the particular values of α and μ.

If we had performed the simulation of 100 samples, we would be in a position to answer several questions. What are the range and distribution of the largest x in each of the samples? What are the range and distribution of the kth largest value? What are the range and distribution of the middle value of each sample?

Answers to such questions are illustrated by a set of 100 samples of 100 items each. The 100 items can be considered synthetic 100-year flood records and we have 100 of such records. Values of 0.9 for α and 10 for μ were used in the simulation. Two of the samples of 100 events, with the items arranged in order of magnitude, are plotted in Fig. 4-15. The magnitude of the events is not dimensioned. For example, rank number 50 in both samples could be

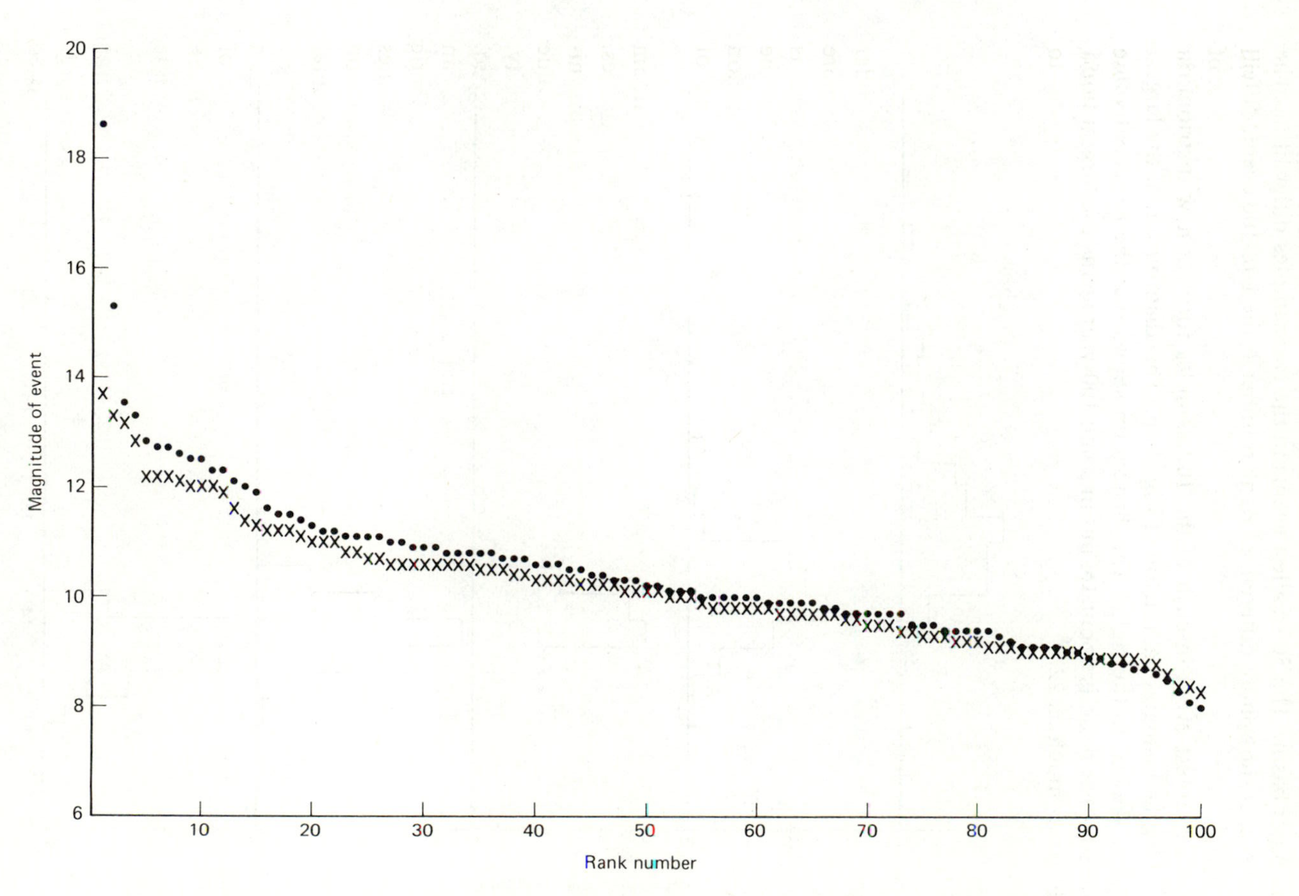

Figure 4-15 Selected simulations of 100-event samples using extreme-value distribution: $\alpha = 0.9$, $\mu = 10$.

considered to have a magnitude of 1000 cfs for a small basin or 100,000 cfs for a major basin. The 50 smallest events in the two samples differ little. The larger events show more difference. Rank number 1, the 1-in-100 event, differs widely.

The largest event in each of the 100 samples forms a new distribution shown by the histogram at the top of Fig. 4-16. This distribution of the largest is highly skewed to the right. This histogram shows that the expected value of the 100-year flood is about 16, but in some 100-year spans the largest flood could be as much as 22.

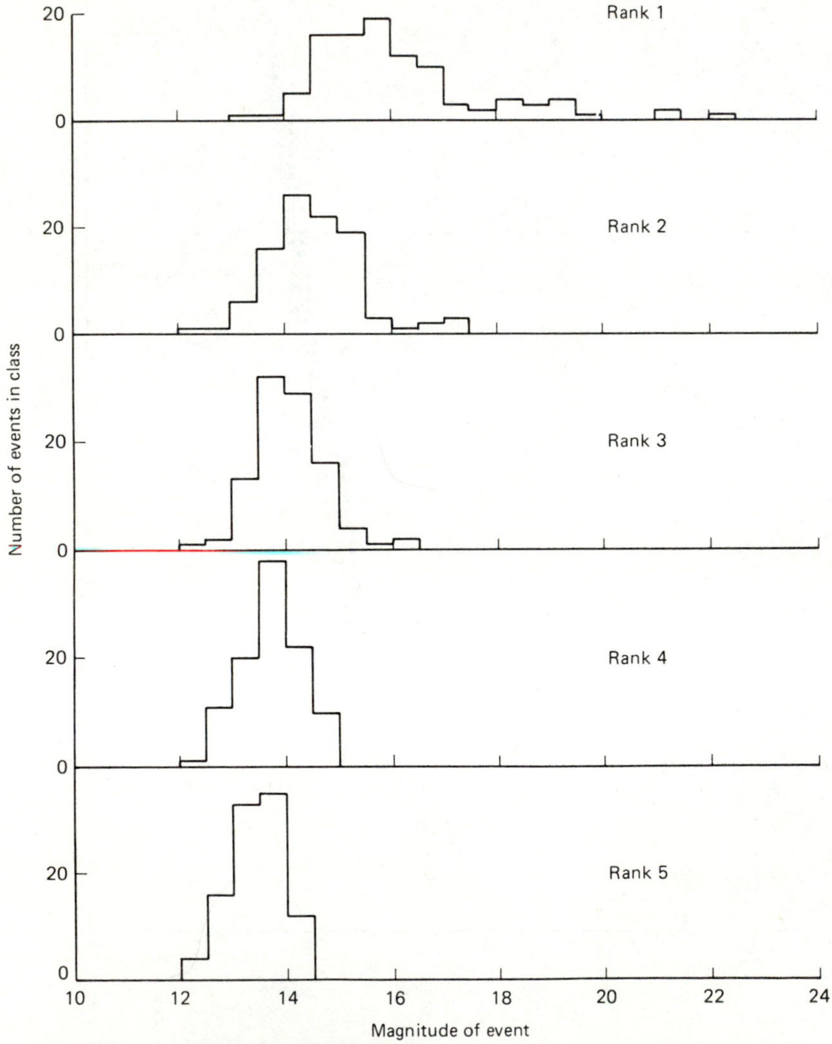

Figure 4-16 Distributions of five largest events for selected simulations of 100-event samples using extreme-value distribution.

Histograms of rank numbers 2 through 5 of each of the 100 samples were also constructed and are shown in Fig. 4-16. The skewness to the right diminishes rapidly, and the histograms for ranks 3 through 5 are nearly symmetrical. Rank 2 indicates the magnitudes that were equaled or exceeded twice in the 100-year synthetic samples. The histogram for this rank number thus estimates the expectation of the 50-year flood. In the same way the histogram for rank 5 estimates the expectation and range of the 20-year flood. These are values that would be equaled or exceeded in various 20-year spans. We must expect that some of the exceeding events in the 20-year spans will be extreme events, such as in rank 1.

All 100 samples have their events arranged in order of magnitude. As a second step, each of these rank numbers may also be arranged in order of magnitude. We thus have ranks 1 to 100 of the individual samples ranked across the 100 samples. Selected ranks numbers of the 10 highest sample ranks are plotted in Fig. 4-17.

Rank 1 of sample rank 1 in Fig. 4-17 is about 22.5. Rank 100 of sample rank 1 is about 13. This is, of course, the range of the histogram at the top of Fig. 4-16. The range of rank number diminishes rapidly with sample rank, and for sample rank 10 ranges only from 13.2 to 11.7. Rank 5 of the sample ranks is also plotted in Fig. 4-17. This value is 18.9 for sample rank 1. We interpret this to mean that 95 times out of 100 the 100-year flood is estimated to be less than 18.9. Similarly, 95 times out of 100 the 50-year flood event is

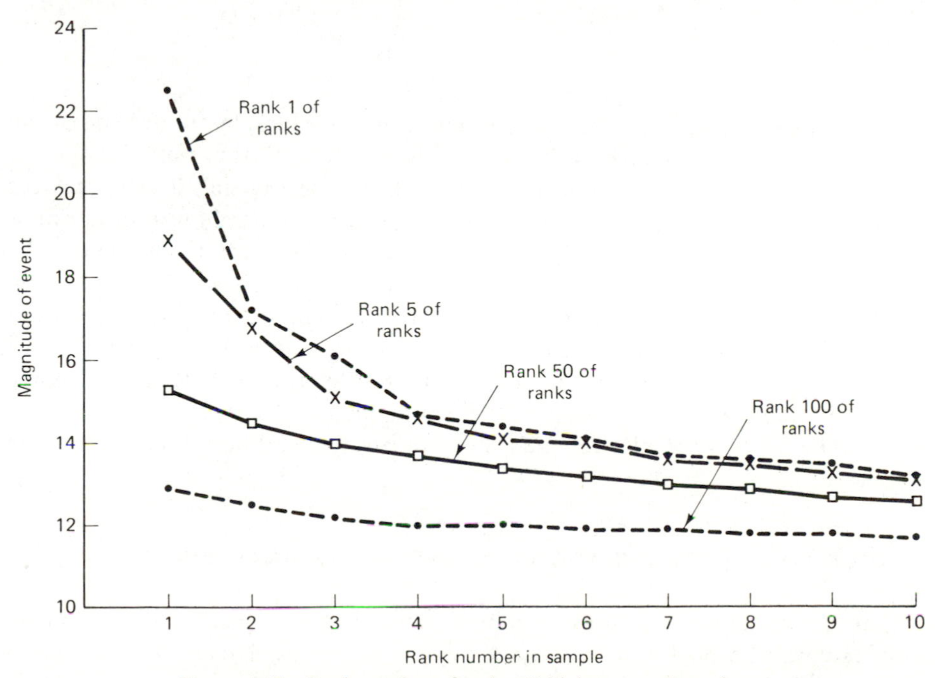

Figure 4-17 Rank numbers for the 10 highest sample ranks.

estimated to be less than 16.8, and 95 times out of 100 the 20-year flood is estimated to be less than 14.1.

Rank 50 of the sample ranks estimates closely the median value. About 50% of the time the 100-year flood will be larger than 15.3, and 50% of the time it will be smaller. The median expectation of the 10-year flood is 12.6.

The median expectation of all floods can be estimated from the distribution of sample rank 50. This distribution is shown by the sample histogram in Table 4-11. The median value has a very narrow range, between 11 and 10. Rank 3 of sample rank 50 was 10.7. Rank 98 was 10.1. Therefore, we estimate that 94% of the time the median flood will lie in this interval.

TABLE 4-11 Histogram of Rank 50 of the Extreme-Value Distribution

Class Limits	Number in Class
10.9–11.0	1
10.8–10.9	0
10.7–10.8	5
10.6–10.7	11
10.5–10.6	13
10.4–10.5	26
10.3–10.4	22
10.2–10.3	11
10.1–10.2	9
10.0–10.1	2
	100

The limits and intervals as estimated in this example are dependent on the sample size simulated, 100 events. We pointed out that rank 5 estimates the 1-in-20 event. We should be more specific and say that it estimates the 1-in-20 event based on 100-event samples. Shorter samples show more variance of the same return period. If we generate 100 samples of 20 events, the average value will be nearly the same as the average of rank 5 in 100-event samples. However, the range of rank 1 in 20-event samples will be greater than that of rank 5 of 100-event samples. For this reason the length of sample simulated should be based on some "planning period" of design or design life. If we choose a 20-year risk level for a small highway bridge, we should simulate 20-year samples to study the expected variabilities within the design period.

ADDITIONAL CONSIDERATIONS IN UNIVARIATE ANALYSIS

The univariate methods presented here should not be viewed solely as a means of estimating probabilities. The methods can also be used to check assumptions that underlie other statistical procedures. For example, regression analysis (see Chapters 3 and 6) assumes that the residuals [i.e., the differences between the predicted and measured values of the dependent (random) variable] are normally distributed with a mean of zero and a variance equal to the square of the standard error of estimate [i.e., $N(0, S_e^2)$]. It should be evident that the residuals are values of a single random variable and that the assumption about the residuals can be tested using the methods outlined in this chapter. A histogram could be plotted to see whether or not the residuals follow the bell shape of a normal distribution. A frequency analysis could be performed. If the residuals fall randomly about the straight line that is defined by the two points ($\bar{y} = 0, p = 0.5$) and ($\bar{y} = S_e, p = 0.1587$), we can assume that the assumption of normality has been met. If the plotted points tend to deviate from the line, especially at high and/or low values of the exceedence probability, an incorrect structure may have been used as the regression model. A moment-ratio analysis could also be used to check the assumption of normality. If the sample values of β_1 and β_2 are near 0 and 3, respectively, we may assume that the assumption of normality is valid. Whether a histogram analysis, a frequency analysis, or a moment-ratio analysis is used, we cannot expect a perfect fit. The histogram will not be perfectly symmetric; the plotted residuals will not fall exactly on the line, and the sample values of β_1 and β_2 will not be 0 and 3. We have to expect sampling variation, and ultimately, we must address the question of whether the observed variation is greater than we would expect from sampling error. Even when statistical tests of hypothesis (see Chapter 2) are used to support the univariate analysis, the level of significance must be selected. Regardless of the path selected, a somewhat arbitrarily selected decision criterion must be selected to make the statistical decision.

In the discussion on the construction of probability paper, we stated that there was a unique probability paper and plotting position formula for each distribution function; however, in the preceding section, we concluded that it was unprofitable to try to determine exact plotting positions. Thus most analyses use either the normal or Gumbel probability paper, with the Weibull plotting position formula used most often. This should not imply that other plotting position formulas are not available. Table 4-12 summarizes seven plotting position formulas; the examples shown in Table 4-12 indicate that there is a noticeable scatter of the values. Although the differences near a probability of 0.5 will have little effect on the goodness of fit, the difference in the formulas will be more significant when trying to estimate the more extreme probabilities.

A primary goal of statistical analysis is estimation. Univariate methods are no different. Estimation with univariate methods can take on one of two forms. First, we might be interested in estimating the probability that some specific value of our random variable will be exceeded in a given time period. For example, if levees on a river are a specified height, we would be interested in the probability in any year of the discharge exceeding the value that would

TABLE 4-12 Alternative Plotting Position Formula

Method	Formula	Probability for: ($i = 1, n = 10$)	($i = 5, n = 10$)
Beard	$1 - (0.5)^{1/n}$	0.067	—
Blom	$\frac{i - 0.375}{n + 0.25}$	0.061	0.451
California	$\frac{i}{n}$	0.10	0.500
Chegodayev	$\frac{i - 0.3}{n - 0.75}$	0.067	0.508
Hazen	$\frac{2i - 1}{2n}$	0.05	0.450
Tukey	$\frac{3i - 1}{3n + 1}$	0.065	0.452
Weibull	$\frac{i}{n + 1}$	0.091	0.455

cause overtopping of the levees. Second, we might be interested in estimating the magnitude of the random variable corresponding to some specified probability. For example, we may want to know the 7-day low-flow magnitude that has an exceedence probability of 95%. In either case, the sample must be used to identify the population (i.e., the pdf and its parameters). The univariate model must be calibrated and verified before it can be used for prediction.

EXERCISES

4-1. A first step in univariate analysis is to plot a frequency histogram and compute the sample moments. What is the value of each of these, and what are the limits to their usefulness?

The following data sample can be used for Exercises 4-2 to 4-4, 4-6, 4-8, and 4-9.

71	46	82	79	60	42	93	88	67	65
77	94	57	68	58	84	63	32	83	70
95	87	61	53	18	74	72	66	55	86
73	73	68	91	75	63	49	71	47	69
52	74	34	90	53	71	47	95	74	51
76	22	84	74	63	50	70	18	66	45
41	81	64	6	98	78	39	72	58	79
68	57	79	88	54	76	62	84	81	25
77	71	13	93	83	73	42	8	68	65
37	96	61	75	66	31	99	59	76	53

4-2. Derive a frequency histogram for the given sample with $n = 100$. Convert the frequency histogram to a probability histogram. Use intervals of 0-10, 11-20, 21-30, . . . , 91-100.

4-3. Using the data sample ($n = 100$), construct a probability histogram. Also, construct two probability histograms for $n = 50$ using the first and second sets of five rows. Then construct five probability histograms for $n = 20$ using the five sets of two rows. Discuss the change in shape with sample size. In each case, use the intervals 0-10, 11-20, 21-30, . . . , 91-100.

4-4. Using the first five rows ($n = 50$) of the data sample, construct frequency histograms for 10-point intervals **(a)** 0-10, 11-20, 21-30, . . . , 90-100 and **(b)** 0-5, 6-15, 16-25, . . . , 76-85, 86-95, 96-100. Do the two histograms differ in shape characteristics? If so, why? Would they lead to identification of the same population?

4-5. Construct a frequency histogram for each of the random variables of Appendix B-1. What characteristics of the sample are evident from the histogram? Use a chi-square goodness-of-fit test or a Kolmogorov-Smirnov test to test the null hypothesis that each of the variables is normally distributed with the location and scale parameters equal to the sample mean and standard deviation, respectively.

4-6. Using the 50 values in the first five rows of the data, compute the moment ratios and indicate the most likely probability distribution. Compare the results with that resulting from the histogram of Exercise 4-3.

4-7. Using the data of Appendix B-1, compute the moment ratios and determine the most likely probability distribution function for **(a)** X_1: the precipitation/temperature ratio; **(b)** X_2: the watershed slope; **(c)** X_3: the large particle index; **(d)** X_4: the small particle index.

4-8. Using the 20 values in the first two rows of the data, perform a frequency analysis for a normal distribution. Fit the frequency curve using the method of moments, and use the Weibull plotting position formula to plot the data.

4-9. Using the 20 values in the first two columns of the data, perform a frequency analysis for a log-normal distribution. Fit the frequency curve using the method of moments, and use the Weibull plotting position formula to plot the data.

4-10. Using the data of Appendix B-1, perform a frequency analysis for a normal distribution for **(a)** X_1; **(b)** X_2; **(c)** X_3; **(d)** X_4. Fit the frequency curve using the method of moments, and use the Weibull plotting position formula to plot the data.

4-11. What is the probability of having two or more events with an annual exceedence probability of 0.1 in a period of 10 years?

4-12. What is the probability of an event with an annual exceedence probability of 0.2 not occurring in a period of **(a)** 1 year? **(b)** 5 years? **(c)** 10 years? **(d)** X years?

4-13. Using the frequency curve of Fig. 4-6, what are the exceedence probabilities for **(a)** 75°F? **(b)** 80°F? **(c)** 85°F?

4-14. Based on the frequency curve of Fig. 4-6, what is the probability of not having the mean monthly temperature exceed 80°F for 2 months?

4-15. Using Fig. 4-6, what is the probability that the mean monthly temperature will not exceed 70°F for **(a)** any one month? **(b)** three consecutive months?

4-16. Using the frequency curve of Fig. 4-11, find the probability that a discharge of 8000 cfs will not be exceeded in **(a)** 1 year; **(b)** a 5-year period; **(c)** a period of *n* years.

4-17. Using the frequency curve of Fig. 4-11, what discharge is likely to occur on the average once every **(a)** 10 years? **(b)** 25 years? **(c)** 50 years?

4-18. Given the probability density function

$$f(X) = \begin{cases} kX & 0 \le X \le X_1 \\ 0 & \text{otherwise} \end{cases}$$

where X_1 can take on any positive value, derive an estimator of X_1 using the method of moments.

4-19. Using the method of moments, derive relationships between the location (μ) and scale (σ) parameters of a normal distribution and sample moments.

4-20. Given the probability function

$$f(X) = \begin{cases} kX^2 + 2 & 0 \le X \le X_1 \\ 0 & \text{otherwise} \end{cases}$$

where X_1 can take on any positive value, derive an estimator of X_1 using the method of moments.

4-21. Using the data of Table 4-6, perform a frequency analysis after making a logarithmic transformation of the data. Using the resulting log-normal frequency curves, find the 2-, 5-, 10-, 25-, 50-, and 100-year discharges.

4-22. Using the data of Table 2-3, perform a frequency analysis for a log-Pearson type III distribution. Use the sample skew in constructing the frequency curve. Find the temperature corresponding to an exceedence probability of 0.01.

4-23. Use package software to simulate 100 samples of 25 events each based on the extreme-value distribution. Estimate the tolerance limits of the largest item.

4-24. Modify the package software to generate log-normal samples. Compare the estimated tolerance limits with those calculated exactly by assuming a normal distribution.

4-25. If the pdf of x is $A\,e^{-Ax}$ and $x = fw^g$, what is the pdf of w?

4-26. If the pdf of x is the normal distribution and $x = \ln v$, what is the pdf of v?

4-27. If the pdf of x is the normal distribution and $x = e^{K(v-\alpha)}$, what is the pdf of v?

4-28. Fit an extreme-value distribution to the following data set ($n = 85$):

1	20.75	18	8.56	35	6.38	52	5.21	69	3.57
2	20.16	19	8.09	36	6.22	53	5.21	70	3.44
3	16.13	20	8.09	37	6.22	54	4.99	71	3.17
4	15.19	21	8.02	38	6.19	55	4.98	72	2.96
5	15.16	22	7.95	39	5.91	56	4.83	73	2.82
6	13.78	23	7.88	40	5.90	57	4.71	74	2.67
7	12.73	24	7.85	41	5.85	58	4.36	75	2.31
8	12.28	25	7.73	42	5.82	59	4.33	76	2.27
9	12.00	26	7.66	43	5.68	60	4.31	77	2.20
10	11.73	27	7.50	44	5.67	61	4.28	78	2.16
11	11.49	28	7.43	45	5.61	62	4.17	79	1.83

12	11.41	29	7.42	46	5.56	63	4.11	80	1.43
13	11.32	30	7.24	47	5.54	64	4.04	81	1.02
14	10.71	31	6.99	48	5.49	65	3.82	82	0.92
15	10.46	32	6.84	49	5.33	66	3.67	83	0.47
16	10.30	33	6.73	50	5.33	67	3.63	84	0.29
17	9.42	34	6.58	51	5.31	68	3.59	85	0.08

4-29. The following data are the 7-day rainfall totals in centimeters. Fit the pdf of v derived from Exercise 4-26 to the data set. (*Note*: Compute $x_i = \ln v_i$ for each data point i; fit the normal distribution to x.)

3.86	0.03	0.00	2.18
1.68	0.15	3.86	11.38
1.22	5.84	9.93	0.71
0.43	1.60	0.91	0.69
14.71	5.08	1.63	7.32
1.37	4.19	5.44	0.71
5.16	3.81	1.75	2.67
0.03	0.33	6.10	4.04
4.95	0.00	0.53	2.49
3.76	0.00	0.36	1.27
12.57	3.10	0.00	2.62
0.15	1.73	2.41	2.13
1.22	3.94	6.93	1.40
2.77	0.08	1.65	7.59
0.66	1.02	5.41	1.73
2.74	4.57	1.42	4.01
6.10	0.71	5.05	2.26
0.00	1.91	4.09	0.84
0.69	1.24	0.08	0.15
1.27	4.34	1.68	6.45
0.08	1.24	5.13	3.05
1.42	1.14	5.21	6.58
2.67	4.70	0.00	0.00
1.37	0.00	1.45	1.78
1.42	7.44	0.00	1.30
2.39	0.00	0.00	2.69
0.15	4.17	11.38	5.21
0.69	3.30	1.65	10.06
0.69	0.00	2.13	6.76
0.10	3.63	0.00	2.95
0.64	1.57	0.71	5.18
6.20	1.78	0.91	6.06
1.68	7.04	2.74	2.06
8.26	0.38	3.10	7.16
4.22	3.73	0.79	7.59
3.07	2.95	4.11	1.30

4-30. Write a computer program to generate psuedorandom normal numbers for the pdf of x in Exercise 4-29. Transform the generated numbers from x to v and arrange them in descending order of magnitude. Generate 50 samples of 50 values of v. Plot the 50 sample values of ranks 1 to 5, size against rank number. Discuss the generated distributions of values for the rank numbers. Compare the tolerance limits.

4-31. Fit the derived pdf of v from Exercise 4-27 to the data set from Exercise 4-29. (*Note*: Compute $x_i = e^{K(v_i - \alpha)}$ for each data point i; fit the normal distribution to x.)

4-32. Repeat the steps of Exercise 4-30 for the transform and data set of Exercise 4-31. Plot ranks 1 to 5 and ranks 46 to 50.

4-33. Linearly distributed random numbers can be considered equally likely probabilities in the scale $0 < r < 1$. Write a computer program to obtain a sample of size n random numbers and from these calculate extreme-value variates from Eq. 4-42. Use sample sizes of 10, 25, and 50.

4-34. On log-normal probability paper plot three generated samples from Exercise 4-6: **(a)** that sample with the largest value of the rank number 1 item; **(b)** that sample with the smallest rank number 1 item; **(c)** that sample with the median value of the rank number 1 item. Discuss the variability.

REFERENCES AND SUGGESTED READINGS

ADAMOWSKI, K., Plotting Formula for Flood Frequency, *Water Resources Bulletin*, Vol. 17, No. 2, pp. 197–202, 1981.

BEARD, L. R., *Statistical Methods in Hydrology*, Corps of Engineers, Sacramento, California, 1962.

BETHLAHMY, N., Flood Analysis by SMEMAX Transformation, *Journal of the Hydraulics Division, ASCE*, Vol. 103, No. HY1, pp. 69–78, 1977.

BODHAINE, G. L., and D. M. THOMAS, *Magnitude and Frequency of Floods in the United States: Part 12. Pacific Slope Basins in Washington and Upper Columbia River Basin*, U.S. Geological Survey Water-Supply Paper 1687, U.S. Government Printing Office, Washington, D.C., 1964.

BUCKETT, J., and F. R. OLIVER, Fitting the Pearson Type 3 Distribution in Practice, *Water Resources Research*, Vol. 13, No. 5, pp. 851–52, 1977.

CARRIGAN, P. H., JR., *A Flood-Frequency Relation Based on Regional Record Maxima*, U.S. Geological Survey Professional Paper 434-F, U.S. Government Printing Office, Washington, D.C., 1971.

CONDIE, R., The Log Pearson Type 3 Distribution: The *T*-year Event and Its Asymptotic Standard Error by Maximum Likelihood Theory, *Water Resources Research*, Vol. 13, No. 6, pp. 987–91, 1977.

CUNNANE, C., A Particular Comparison of Annual Maxima and Partial Duration Series Methods of Flood Frequency Prediction, *Journal of Hydrology*, Vol. 18, pp. 257–71, 1973.

DALRYMPLE, T., Flood-Frequency Analysis, *Manual of Hydrology: Part 3. Flood-Flow Techniques*, U.S. Geological Survey Water-Supply Paper 1543-1, U.S. Government Printing Office, Washington, D.C., 1960, pp. 25–47.

DANUSHKODI, V., Flood Flow Frequency by SCS-TR-20, *Journal of the Hydraulics Division, ASCE*, Vol. 105, No. HY9, pp. 1123-35, September 1979.

FRY, T. C., *Probability and Its Engineering Uses*, D. Van Nostrand Co., New York. 1965.

GLADWELL, J. S., and C. LIN, Confidence Limits Determined Using Order Statistics, *Water Resources Research*, Vol. 5, No. 5, pp. 1120-23, 1969.

HUGHES, W. C., Peak Discharge Frequency from Rainfall Information, *Journal of the Hydraulics Division, ASCE*, Vol. 103, No. HY1, pp. 39-50, 1977.

MCGUINNESS, J. L., *Simplified Techniques for Fitting Frequency Distributions to Hydrologic Data*, Agricultural Handbook 259, U.S. Department of Agriculture, 1964.

NOZDRYN-PLOTNICKI, M. J., and W. E. WATT, Assessment of Fitting Techniques for the Log Pearson Type III Distribution Using Monte Carlo Simulation, *Water Resources Research*, Vol. 15, No. 3, pp. 714-18, June 1979.

RAO, D. V., Log Pearson Type 3 Distribution Evaluation, *Journal of the Hydraulics Division, ASCE*, Vol. 106, No. HY5, Proc. Paper 15391, pp. 853-72, May 1980.

REICH, B. M., and K. G. RENARD, Application of Advances in Flood Frequency Analysis, *Water Resources Bulletin*, Vol. 17, No. 1, pp. 67-74, 1981.

STIDD, C. K., Cube-Root Normal Precipitation Distributions, *Transactions of the AGU*, Vol. 34, No. 1, 1953.

STIDD, C. K., The *n*th-Root Normal Distribution of Precipitation, *Water Resources Research*, Vol. 6, No. 4, 1970.

TANG, W. H., Bayesian Frequency Analysis, *Journal of the Hydraulics Division, ASCE*, Vol. 106, No. HY7, Proc. Paper 15532, pp. 1203-18, July 1980.

TASKER, G. D., Flood Frequency Analysis with a Generalized Skew Coefficient, *Water Resources Research*, Vol. 13, No. 2, pp. 373-75, 1978.

U.S. Army Corps of Engineers, Vol. 3, *Hydrologic Frequency Analysis*, IHD-0300, April 1975.

VENUGOPAL, K., Flood Analysis by SMEMAX Transformation, a Review, *Journal of the Hydraulics Division, ASCE*, Vol. 106, No. HY2, pp. 338-40, February 1980.

WALLACE, J. R., and J. L. GRANT, A Least Squares Method for Computing Statistical Tolerance Limits, *Water Resources Research*, Vol. 13, No. 5, pp. 819-24, 1977.

ZHANG, Y., Plotting Positions of Annual Flood Extremes Considering Extraordinary Values, *Water Resources Research*, Vol. 18, No. 4, pp. 859-64, 1982.

5 Semivariogram Analysis and Kriging

INTRODUCTION

Before discussing the method of estimation known as kriging, it is important to understand its place in statistical estimation. Frequency analysis was introduced in Chapter 4 as a univariate estimation method; however, frequency analysis provides only an estimated value, with some probability of occurrence or nonoccurrence. For example, we can estimate the probability that a flood of X cfs will not occur in any one time period (e.g., 1 year). Or for a given probability, we can estimate the magnitude of the flood that will not occur in any one time period. In addition to making an estimate, we are usually interested in the accuracy of the estimate. In Chapter 4 we showed how to compute the standard error of estimate and the confidence interval. These statistics characterize the accuracy of the estimate. For example, for the discharge of X cfs, we can compute both the standard error of the estimated value and the confidence interval of X. The theoretical basis of frequency analysis assumes that the events, or sample points, are independent measures of the random variable. This assumption limits the estimation to making probabilistic statements about the likelihood of either occurrence or nonoccurrence. A lack of dependence on any other factor may contribute to a large standard error and thus a wide confidence interval. It also limits the potential sample size because sample points must be selected in a way that ensures the points to be independent of each other.

Kriging is also an estimation method for univariate data. But it assumes that the sampling domain has a local structure and that knowledge of the local structure can improve the accuracy of the estimated value. The estimation method assumes that the "best" estimate is a weighted average of one or more sample points. Kriging is the method of analysis by which the optimal values of the weights are determined.

In hydrologic decision making we might be interested in the value of the variable at a point or the average value for either a line, a field, or a volume. The sample may consist of a single point, a vector of points, or the values of a line (one-dimensional), a field (two-dimensional), or a volume (three-dimensional). Measurements of rainfall are the most obvious example of a point measurement in hydrology. Samples of water from a stream that are used in determining concentrations of a water-quality indicator would also be considered as point estimates. A vector sample would consist of n such point values. For example, storm rainfall amounts at all rain gages in a region would be a vector sample. Measurements of water content in a snow pack can be made with a snow tube, the diameter of which is usually much smaller than the length of the tube; therefore, such a measurement can be considered as a linear, or line, estimate. As remotely sensed data are integrated into hydrologic design methods, the problem of spatial averaging will become more important to hydrologists because the value of a *pixel* (picture element) represents the spatial average of the surface reflectance for the area within the pixel. Such data would be considered "field" data (two-dimensional) since they represent information obtained in two dimensions. For areas where there is considerable change in the surface reflectance, such as an urban area, over the spatial extent of a pixel, the size of the pixel becomes a factor in spatial estimation. Volumes of soil are collected to determine the soil moisture. The soil moisture will probably not be uniformly distributed within all elements of the sample volume; thus such measurements can be considered as being volumetric measurements (three-dimensional). In summary, hydrologic decision making involves data measured at a point, along a line, within a field, or on a volumetric basis. Kriging estimation may apply to such data.

THE ESTIMATION PROBLEM

It may be instructive to examine the specific estimation problem. In terms of spatial estimation, we have a set of measurements of a property and the coordinates of each measurement, and we wish to make an estimate of the property at a point where measurements were not taken. Additionally, we will assume that the property is statistically stationary over the spatial area of interest; thus if an attempt were made to fit a trend surface to the sample estimates of the property, a significant trend over the sample space would not exist. However, a local systematic structure may exist in the vicinity of any point within the sample space. The spatial extent of this local structure will be referred to as the *region of influence*. Therefore, the estimated value of the property at any point will be influenced only by the values of sample measurements that are located within the region of influence.

The estimation problem can be viewed as a two-step process. First, the range of influence must be identified, and second, a method of estimation must be adopted. Before attacking the first step, let's take a cursory look at the estimation problem.

One alternative to the estimation problem would be to take the arithmetic average of all points in the region of influence as the estimate at the point where an estimate is required. This assumes that each point in the region of influence has an equal effect on the point of estimation. Intuitively, one would guess that the sample measurements should not be given equal weight, but that the "best" estimate could be obtained by assigning weights to the sample measurements within the range of influence that are inversely proportional to the distance from the point of interest. For either estimation scheme, we can make an estimate, $\hat{Y}$, using the following model:

$$\hat{Y} = \sum_{i=1}^{m} w_i Y_i \tag{5-1}$$

in which w_i is the weight given to sample measurement Y_i and m is the number of sample measurements within the region of influence of the point where the estimate $\hat{Y}$ is needed. For an arithmetic average, the weight w_i would be $1/m$. But such a weighting function does not appear to be rational, at least if we are making use of the information content of the entire sample. If the weights are inversely proportional to the separation distance between the locations of the sample point and the point where an estimate is needed, a criterion must be selected to assign values to the weights.

We can summarize the estimation problem using the following questions:

1. Which sample measurements (Y_i) in a sample of size n should be used to make the estimate $\hat{Y}$ at any point?
2. What weighting function, w_i, would provide the most accurate estimate, $\hat{Y}$?
3. Having answered the first two questions, how reliable is the estimated value?

As an illustration, we may wish to know the level of a pollutant at a point in a groundwater system. Assuming that a well does not exist at that point but levels of the pollutant have been measured at wells located within the region, we wish to know which wells to use (i.e., which wells lie within the region of influence), the importance of the values measured at each of the wells within the range of influence, and the accuracy of our best estimate. If we believe that a statistical solution is the best alternative, a spatial analysis with an estimation by kriging will provide the best solution to the estimation problem addressed by the three questions above.

The first step is to address the question of the range or region of influence. The schematic of Fig. 5-1 shows the location of six points of measurement (i.e., points 1 to 6) and the location of the point (i.e., point A) where a

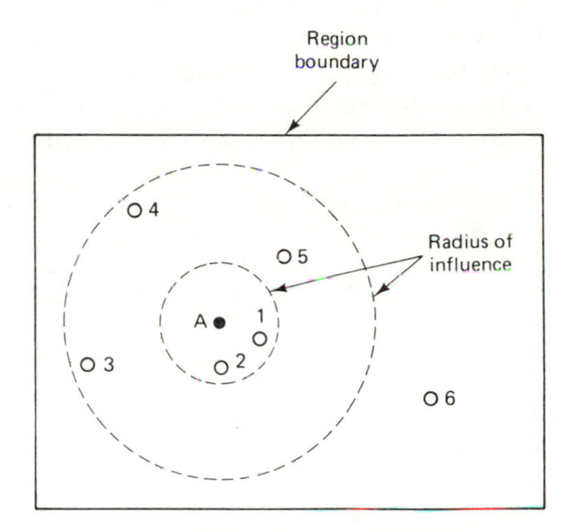

Figure 5-1 Schematic of measurement points.

measurement has not been taken but an estimate is needed. Two regions of influence are shown as dashed lines. If the inner circle represents the true region of influence, which is unknown at this point, only the measurements at points 1 and 2 would be used to estimate the value of the property at point A. If the outer circle represents the true region of influence, points 1 to 5 will be used. Thus, before we can proceed with the estimation of the weights of Eq. 5-1, it is necessary to identify the region of influence so that the sample points to be used in estimation (i.e., the value of m) can be identified.

Once the region of influence is identified, the estimation process can begin. If the mean value concept is used for estimation, the best estimate of $\hat{Y}$ would be

$$\hat{Y} = \frac{1}{m} \sum_{i=1}^{m} Y_i \tag{5-2}$$

in which m would equal either 2 or 5, depending on the region of influence for Fig. 5-1. But as previously stated, there is very little reason to believe that the measurement at either point 4 or point 5 should be given the same weight as the value of the property measured at point 1, since point 1 is much closer to point A than either point 4 or point 5. Therefore, Eq. 5-1 would provide the "best" estimate of the property at point A (Fig. 5-1) and the estimation process would involve finding the optimum set of weights (i.e., the w_i of Eq.

5-1). After the weights are determined, the accuracy of the estimate can be determined.

SEMIVARIOGRAM ANALYSIS

Before defining a *variogram*, let's establish some basic assumptions. Assume that we have a space where measurements are taken on a grid a distance h apart (see Fig. 5-2). Although a regular grid is used for this introduction, it is not a basic requirement of variogram analysis for the data to be measured at grid nodes. The grid points are spaced a distance h apart in both directions.

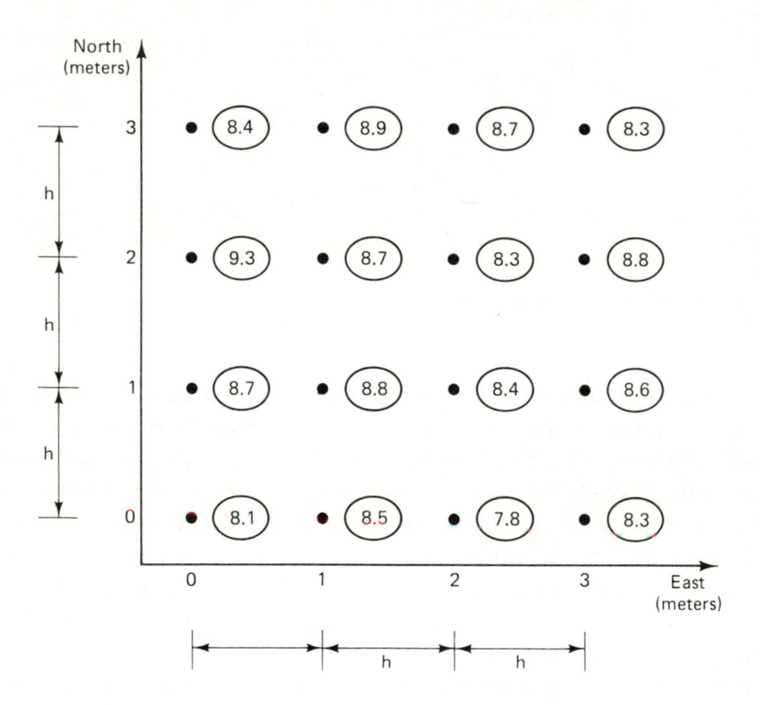

Figure 5-2 Data for computational example.

First, we will assume that the field of measurements represents a statistically stationary field, which means that the difference in the measured values of any two points in the field depends only on the distance between the two points, and possibly their relative orientation within the field but not the location within the field. For example, the magnitude of the difference in the values of Y for points with coordinates $(1, 1)$ and $(1, 2)$ is from the same statistical population as the magnitude of the difference in the values of Y for points $(2, 3)$ and $(3, 3)$, or for points $(2, 3)$ and $(2, 2)$. Again, note that the statistical characteristics of the difference in the values of Y do not depend on the location within the sample space. It is also important to note that the above does not imply that the differences for each pair of points a distance h apart are equal, only that they are from the same statistical population. However, for a different separation distance, say $2h$, the statistical population may be different. Of course, in some cases the populations may be the same.

With this background, let's assume that each point in the field of Fig. 5-2 has a measured value of $Y(\mathbf{X})$, where $\mathbf{X}$ indicates a two-coordinate vector system. For any sample point $Y_i(\mathbf{X})$ there is at least one point separated by distance h from $Y(\mathbf{X})$; this point will be denoted as $Y(\mathbf{X}+h)$. If we take the differences in the magnitude of Y between each pair of points separated by distance h, we define the ith difference as $[Y_i(\mathbf{X}) - Y_i(\mathbf{X}+h)]$. From this we can define the average sum of squares of the differences in the sample as

$$2\hat{\gamma}(h) = \frac{1}{n} \sum_{i=1}^{n} [Y_i(\mathbf{x}) - Y_i(\mathbf{x}+h)]^2 \tag{5-3}$$

The quantity $2\hat{\gamma}(h)$ is the sample variance of the differences; it is the sample estimate of the population value $2\gamma(h)$. The quantity $2\gamma(h)$ is the variogram value for separation distance h, and $\gamma(h)$ is the semivariogram value. Both are a function of the separation distance h. The units of $2\gamma(h)$ and $\gamma(h)$ are the same and equal to the units of the square of Y. If Y is the value of a pollutant measured in ppm, the units of the variogram and semivariogram would be $(\text{ppm})^2$. For the rectangular region of Fig. 5-2, the semivariogram for distance h would be based on 24 values (i.e., $n = 24$), with 12 of the 24 differences measured in the north–south direction. For a separation distance of $2h$, the variogram value $2\hat{\gamma}(2h)$ would be based on 16 values, with eight differences in the north–south direction and eight in the east–west direction.

Before presenting an example to illustrate the calculations, the characteristics of a *semivariogram* need to be specified. A semivariogram is most often plotted as $\gamma(h)$ versus h. Of course, at a separation distance of zero, $Y(\mathbf{X}) = Y(\mathbf{X}+h)$, so $\gamma(h) = 0$. Thus, the semivariogram passes through the origin of the $\gamma(h)$ versus h graph. As h increases from zero, the values of $Y(\mathbf{X})$ and $Y(\mathbf{X}+h)$ will begin to differ by some small amount, so the variable $\gamma(h)$ must be nonzero. Since each difference in Eq. 5-3 is squared, each term of Eq. 5-3, and thus $\hat{\gamma}(h)$, must be positive. The variance will tend to increase with increased separation distance until some point where further increases in the separation distance are not accompanied by increased variance. Thus the semivariogram has the shape indicated in Fig. 5-3. The shape in Fig. 5-3 is typical of the semivariogram of the population. For a sample, the semivariogram consists of a set of points, and in some cases, the semivariogram value may actually show decreases in $\gamma(h)$ as h increases; this is the result of sample variation and is more likely to occur for small samples, especially at large separation distances.

For the general shape of the semivariogram, there are two characteristics of special importance. At some separation distance, r, the semivariogram

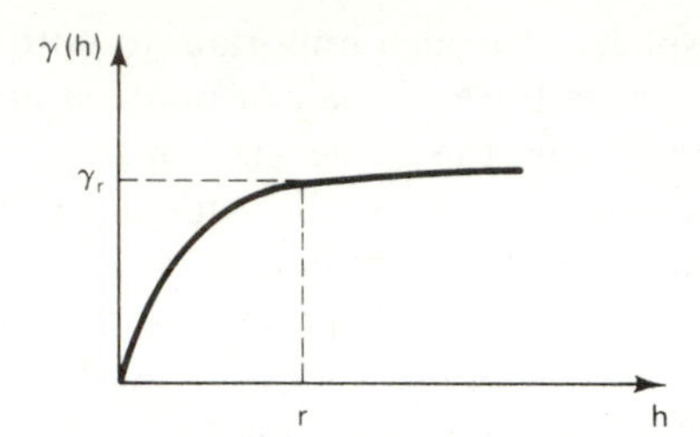

Figure 5-3 Characteristic shape of a semivariogram.

approaches a constant value γ_r; this separation distance is called the *radius of influence* and occurs when the semivariogram approaches the sample variance. The part of the semivariogram where $\hat{\gamma}(h)$ approximates the sample variance is called the *sill* and is denoted as γ_r. These are indicated in Fig. 5-3.

Before providing a case study, it may be worthwhile to provide a simple example to illustrate the computational procedure. A 3 m by 3 m plot is shown in Fig. 5-2, with the soil moisture (in percent) shown within the circle that is adjacent to the sample point location. Since the minimum separation distance is 1 m, the first point on the sample semivariogram will be at a distance of 1 m. The largest separation distance is 3 m and thus there will be only three points on the sample semivariogram. For a separation distance of 1 m, there are 24 pairs of points, with the value of the semivariogram given by

$$\begin{aligned}\hat{\gamma}(1\text{ m}) = \frac{1}{2(24)}[&(8.1-8.5)^2+(8.5-7.8)^2+(7.8-8.3)^2+(8.7-8.8)^2\\ &+(8.8-8.4)^2+(8.4-8.6)^2+(9.3-8.7)^2+(8.7-8.3)^2\\ &+(8.3-8.8)^2+(8.4-8.9)^2+(8.9-8.7)^2+(8.7-8.3)^2\\ &+(8.1-8.7)^2+(8.7-9.3)^2+(9.3-8.4)^2+(8.5-8.8)^2\\ &+(8.8-8.7)^2+(8.7-8.9)^2+(7.8-8.4)^2+(8.4-8.3)^2\\ &+(8.3-8.7)^2+(8.3-8.6)^2+(8.6-8.8)^2+(8.8-8.3)^2]\\ = 0.102(\%)^2 &\end{aligned} \tag{5-4}$$

For a separation distance of 2 m, there are 16 pairs of points, with the value of the semivariogram given by

$$\begin{aligned}\hat{\gamma}(2\text{ m}) = \frac{1}{2(16)}[&(8.1-7.8)^2+(8.5-8.3)^2+(8.7-8.4)^2+(8.8-8.6)^2\\ &+(9.3-8.3)^2+(8.7-8.8)^2+(8.4-8.7)^2+(8.9-8.3)^2\\ &+(8.1-9.3)^2+(8.7-8.4)^2+(8.5-8.7)^2+(8.8-8.9)^2\\ &+(7.8-8.3)^2+(8.4-8.7)^2+(8.3-8.8)^2+(8.6-8.3)^2]\\ = 0.124(\%)^2 &\end{aligned} \tag{5-5}$$

For a separation distance of 3 m, there are eight pairs of points, with the value of the semivariogram given by

$$\begin{aligned}\hat{\gamma}(3\text{ m}) &= \frac{1}{2(8)}[(8.1-8.3)^2+(8.7-8.6)^2+(9.3-8.8)^2+(8.4-8.3)^2 \\ &\quad +(8.1-8.4)^2+(8.5-8.9)^2+(7.8-8.7)^2+(8.3-8.3)^2] \\ &= 0.086(\%)^2\end{aligned} \tag{5-6}$$

The unbiased variance of the 16 measurements is $0.125(\%)^2$, which would be an estimate of the sill. An attempt to assess the radius of influence for such a small sample size will not be made. The sample semivariogram and sill are shown in Fig. 5-4.

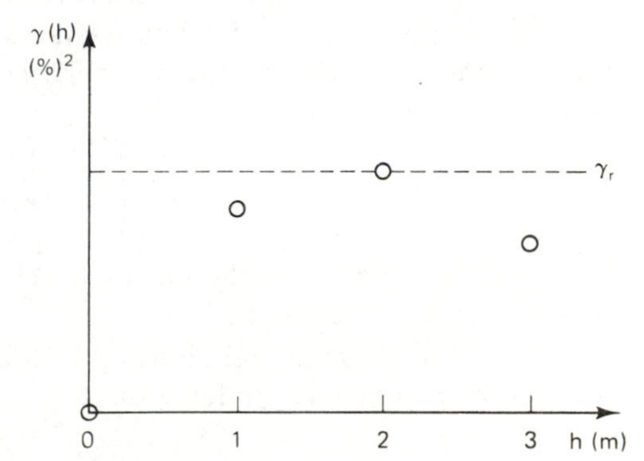

Figure 5-4 Sample semivariogram for separation distances (h) of 0, 1, 2, and 3 m.

Semivariogram Models

It is important to keep in mind the difference between the sample estimate and the population model of the semivariogram. In univariate frequency analysis, probability estimates are made using the assumed population distribution. In variogram analysis, we use the sample data to derive a model to represent the population. The population model is used in kriging estimation. For this reason it is worthwhile examining some of the more common functional forms used to represent the population.

A linear model is the simplest form and easiest to calibrate. The linear model for a semivariogram has the form

$$\gamma(h) = bh \tag{5-7}$$

in which b is the slope of the line and h is the separation distance. In this case, both the sill and the radius of influence have no meaning.

The nonlinear or power model is used quite frequently in hydrologic analysis and it could be used in semivariogram analysis:

$$\gamma(h) = bh^c \tag{5-8}$$

in which c is the power coefficient. Only when c is less than 1 will the power model have a form in which a radius of influence and sill could be inferred. Of course, for c equal to 1, the power model becomes the linear model.

The deWijsian model has the form

$$\gamma(h) = 3b \ln_e h \tag{5-9}$$

For the case where the separation distance scale is logarithmic, the deWijsian semivariogram model will plot as a straight line. But this model, like the others above, does not have a sill. Thus, although it may provide a reasonable fit to some data sets, its structure may limit its usefulness as a descriptive model.

The exponential model follows the general shape of a semivariogram and provides coefficients that can be used as measures of the sill, γ_r, and radius of influence, r:

$$\gamma(h) = \gamma_r\left[1 - \exp\left(-\frac{h}{r}\right)\right] \tag{5-10}$$

The coefficient r controls the rate at which $\gamma(h)$ approaches the sill. Thus it does not have a flexible shape.

The semivariogram relationship of many physical processes can be represented by a spherical model:

$$\gamma(h) = \begin{cases} \gamma_r & \text{when } h > r \quad (5\text{-}11a) \\ \gamma_r\left(\dfrac{3h}{2r} - \dfrac{h^3}{2r^3}\right) & \text{when } h \le r \quad (5\text{-}11b) \end{cases}$$

In general, the spherical model will approach the sill at a smaller separation distance when compared with the exponential model for the same sill. Also, for some values of r, the derivative of $\gamma(h)$ will have a noticeable change at $h = r$. However, this is of little concern for most applications.

Figure 5-5 shows a comparison of the exponential (dashed line) and spherical (solid line) semivariograms for a sill of 1.0 and a radius of influence of 1.0; it is evident that the exponential model approaches the sill much more slowly, with a value of 0.632 at the radius of influence. An exponential semivariogram (- - - -) with a radius of influence of 4 is also shown in Fig. 5-5; while this model approaches the sill near the radius of influence, its rising limb is much steeper than the rising limb of the spherical model. Figure 5-5 illustrates that each model has a distinct shape and that the selection of a model to represent the population must consider the ability of the model to fit the sample values of the semivariogram.

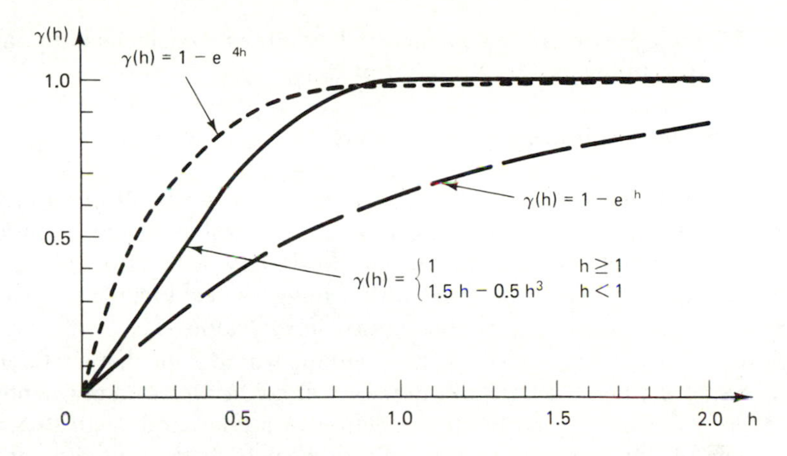

Figure 5-5 Comparison of spherical and exponential semivariograms for $\gamma_r = 1.0$.

In addition to the models above, other functional forms exist. For some physical processes, the variance of the differences increases very rapidly at separation distances near zero. In fact, for some data, the semivariogram value for the smallest separation distance is significantly different from zero and may actually be closer to the sill than to zero. If one tries to fit one of the foregoing model forms to such data, the goodness of fit will be fairly poor. In an attempt to fit the initial rise of the sample points it will produce a nonrepresentative estimate of the radius of influence. To overcome this problem of fitting, combination model forms are sometimes used, with the combination model consisting of the sum of a constant, γ_n, and one of the model structures given above. Because of the origin of many of the concepts in semivariogram analysis, the constant is sometimes termed the *nugget effect*. The following are the combination models using the forms above as the nonconstant component:

$$\gamma(h) = \gamma_n + bh \tag{5-12}$$

$$\gamma(h) = \gamma_n + bh^c \tag{5-13}$$

$$\gamma(h) = \gamma_n + 3b \ln_e h \tag{5-14}$$

$$\gamma(h) = \gamma_n + (\gamma_r - \gamma_n)\left[1 - \exp\left(-\frac{h}{r}\right)\right] \tag{5-15}$$

$$\gamma(h) = \begin{cases} \gamma_r & \text{for } h > r \quad \text{(5-16a)} \\ \gamma_n + (\gamma_r - \gamma_n)\left(\dfrac{3h}{2r} - \dfrac{h^3}{2r^3}\right) & \text{for } h \le r \quad \text{(5-16b)} \end{cases}$$

It should be apparent that the structures of Eqs. 5-12 to 5-16 are more flexible than the corresponding structures of Eqs. 5-7 to 5-11 because there is an additional parameter that can be used to fit the data. For example, the model of Eq. 5-7 includes one fitting coefficient b, while two coefficients (γ_n and b) can be used in fitting Eq. 5-12 to a set of data.

Estimating Semivariogram Parameters

The coefficients of Eqs. 5-7, 5-8, 5-9, and 5-12 can be fit using bivariate regression analysis. The remaining models are best fit using a numerical method; an example of numerical optimization of a semivariogram model is given in Chapter 7. When a model with a nugget effect included is being fit, it may be necessary to include constraints on the values of γ_n; otherwise, an irrational value of γ_n may result. Although one would hope that a model form would be selected on the basis of rationality and consideration of the physical processes, it is quite likely that the selection of a final model will depend on the values of some goodness-of-fit criterion such as the standard error of estimate.

In most actual case studies, the sample size is sufficiently small that the sample yields only a few points on the estimated semivariogram. This makes it difficult to estimate the value of the sill. There is merit in the procedure of setting the value of the sill based on the variance of the differences and then optimizing the remaining coefficient or coefficients to set the range of influence and, if present, nugget effect.

For cases where there are estimated values of $\gamma(h)$ for many separation distances, the fitting process may place too much emphasis on the sill (i.e., points beyond the radius of influence) rather than on obtaining an accurate estimate of the radius of influence and the shape of the rising part of the semivariogram. For example, in using the least-squares principle to fit the coefficients of the spherical or exponential model, too much emphasis may be placed on estimating γ_r if the sample contains many estimates of $\gamma(h)$ beyond the radius of influence. Unless a weighted least-squares procedure is used, the fitting may be dominated by the points beyond r because each sample point is given equal weight when estimating parameters using regression. One procedure that can be used to overcome this limitation of least squares is to use a numerical interpolation method that fits values only on the basis of nearby sample points. Thus the shape of the rising limb of the semivariogram would be influenced only by sample points that are in the rising limb.

Anisotropy and Boundary Effects

In computing the sample semivariogram of Fig. 5-4, no regard was given to direction. That is, in computing the sample values of $\gamma(h)$, differences between pairs of points for both directions (i.e., north–south and east–west) were used in Eqs. 5-4, 5-5, and 5-6. In some cases, one may suspect that the semivariogram for differences in one direction may differ from the semivariogram for differences in the other direction. For example, one might

TABLE 5-1 Semivariogram Values for the Detection of Anisotropic Trends of Figs. 5-2 and 5-6

Separation Distance (m)	Fig. 5-2		Fig. 5-6	
	East-West	North-South	East-West	North-South
1	0.097	0.108	0.027	0.125
2	0.108	0.141	0.037	0.046
3	0.039	0.133	0.028	0.298

reasonably expect the semivariogram for soil moisture across a slope to differ from the semivariogram of soil moisture down a slope. Similarly, semivariograms of a bedload property may be different in the transverse direction of the main channel compared to the longitudinal direction. Rainfall might produce data that have dissimilar directional semivariograms if orographic effects can be defined in one direction but not another. The values in Table 5-1 show the semivariograms for the north-south and east-west directions for the data of Figs. 5-2 and 5-6. Even though the sample sizes are very small,

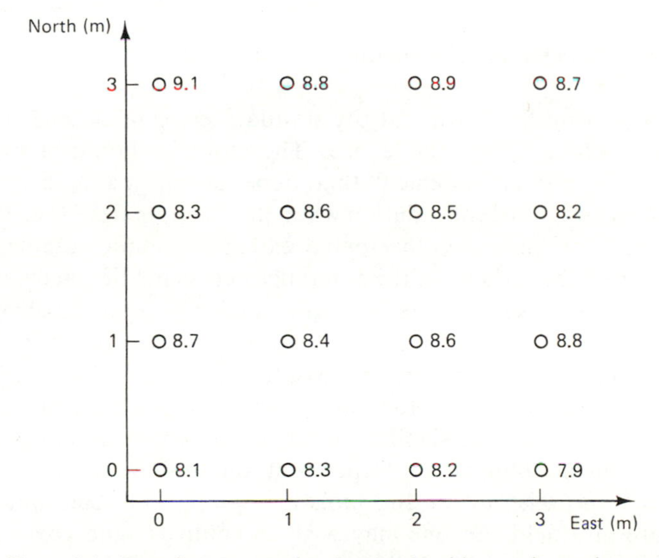

Figure 5-6 Data for computational example with an anisotropic trend.

it should be evident that there is no significant difference in the directional semivariograms for the data of Fig. 5-2, while the semivariogram for the north-south direction in Fig. 5-6 is considerably different from the semivariogram for the east-west direction. When there are directional related differences in the semivariograms, one needs to examine the system for a physical cause.

In addition to directional related trends in semivariograms, actual data can produce other variations in shape and magnitude that are associated with physical characteristics of the system. In a nondirectional semivariogram for an area with nonequal length and width, a discontinuity may result at a separation distance equal to the smaller boundary length. The discontinuity or ridge would suggest that there is a directional effect, and directional semivariograms should be computed.

For a semivariogram that does not have a sill, the data should be checked for stationarity. The lack of the sill may be indicative of a trend in the data, and the trend, which is a deterministic element of the data, should be removed. Most sample semivariograms will be characterized by random scatter, with the amount or scatter depending on the sample size. In some cases it may be difficult to separate the random scatter from the underlying physical shape of the semivariogram. A semivariogram that includes one or more troughs [i.e., significant decreases in $\hat{\gamma}(h)$ in the sill area] may indicate a dominant physical feature of the system that introduces some measure of regularity into the data. For example, for a watershed with a dendretic stream pattern, a directional semivariogram of soil moisture may identify an average length between stream elements.

Spatial and Volumetric Estimation

It was previously shown that the standard error of the mean, $\sigma/\sqrt{n}$ or $S/\sqrt{n}$, is a function of the sample size. Therefore, it should be evident that the value of a field measurement that depends on spatial averaging will demonstrate greater inherent variation when the "sample size" (i.e., the volume or spatial extent) is small. As the spatial extent or sample volume increases, one would expect the variance of the difference between different measurements in a sample to decrease. With respect to a semivariogram, this would correspond to a decrease in the sill. Thus, based on the concept embodied within the standard error of the mean, one would expect the characteristics of a sample semivariogram to be a function of the spatial extent or volume of a sample. This has practical significance in cases where design decisions or computation methodologies are dependent on the size of a sample. For example, one may wish to use soil moisture samples to determine the mean soil moisture of a field. Or one may wish to compare land cover signatures obtained from remotely sensed data based on different pixel sizes. In either case, the sample size must be given consideration.

Before discussing methods for handling data that have been spatially averaged, some clarification of terms is in order. In addition to point measurements, such as rainfall measurements, hydrologic analyses may involve data collected linearly (i.e., one-dimensional), spatially (i.e., two-dimensional), or volumetrically (i.e., three-dimensional). Examples of these three alternatives are snow water estimates obtained with a snow tube from cores in which the length of the core is much greater than the core diameter, spectral signatures for a pixel, and soil moisture values obtained from a volume of soil in which the three linear dimensions are of the same order of magnitude. In some cases, the data base is referred to as the *support* and one refers to the dimensionality of the data as the *type of support* (e.g., volumetric support would denote the three-dimensional case).

Regularization. In addition to defining terms, it is important to state specifically our data analysis objectives. Recognizing that semivariograms depend on the type of support, we need to be sure that the characteristics of the sample semivariogram (i.e., sill and radius of influence) reflect the characteristics of the decision. That is, decisions that depend on estimation using a point semivariogram should not be made with a semivariogram that was computed using data measured on a linear, spatial, or volumetric basis. If the characteristics of the support are different than would be necessary for estimation, we must modify the characteristics of the semivariogram to reflect the decision mode. The terms "deregularization" and "regularization" are used to indicate a change of semivariogram characteristics because of differences in the type of support. *Deregularization* refers to the process of deriving a point sample semivariogram from a core, spatial, or volumetric semivariogram. *Regularization*, which is somewhat the reverse of deregularization, refers to the process of deriving a semivariogram for a core, spatial, or volumentric semivariogram from a point semivariogram. A simple example may illustrate these two opposite processes. Our support may consist of remotely sensed spectral signatures for pixels of size 60 m by 80 m. We may wish to compare the semivariogram for the 60 m by 80 m pixel with a semivariogram from another remote sensing device that has a 30 m by 40 m pixel. In this case we would need to deregularize the semivariogram obtained from the analysis of the data for the 60 m by 80 m pixel and then regularize the computed point semivariogram so that it can be used in comparison with the remote sensing system that uses a 30 m by 40 m pixel. In many cases it is not necessary to perform both deregularization and regularization for the same problem.

Regularization may be viewed as a reduction in the variance due to averaging over a core length, spatial field, or a volume. The concept is not new if one views it in light of the standard error of the mean. Specifically, given a random variable X with mean μ and variance σ^2 we expect individual observations to have a variance about the mean of S^2, the sample variance. If we combine n individual measurements into a sample and compute the sample mean, say $\bar{X}_n$, we know that the variability of $\bar{X}_n$ will be considerably less than the variability of the individual values. Theory indicates that the variance can be estimated by S^2/n, that is, the standard error is $S/\sqrt{n}$. Obviously, the variance is reduced by the averaging process. The same concept applies when comparing the semivariogram for a point with the semivariogram for either a length, spatial field, or a volume. The variation of measurements based on a length, spatial field, or a volume will be less than the variation of point measurements because the nonpoint measurements reflect *volumetric averaging.* That is, the standard error of a nonpoint measurement will be less than the standard error of a point measurement, just as the standard error of the mean is smaller than the standard deviation of point measurements. In semivariogram analysis, this will be reflected by the height of the sill. The sill for point measurements will be higher than the sill for nonpoint measurements. Thus it remains only to show the relationship between the two variations of the point and nonpoint semivariograms. This would correspond to showing the relationship between the standard deviation of point measurements of a random variable and the standard error of the mean (i.e., S versus $S/\sqrt{n}$).

Since the computations for a core length are the simplest of the three types of support, the procedure for regularization will be demonstrated with the line of Fig. 5-7. The semivariogram for points *within* the line can be derived

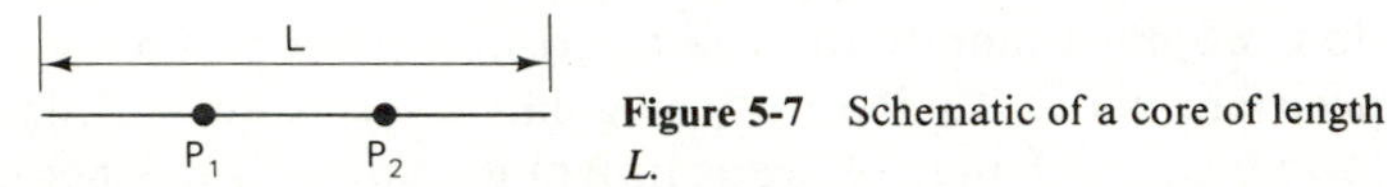

Figure 5-7 Schematic of a core of length *L.*

with Eq. 5-3, which can be denoted as $\gamma_l(h)$. The sill will be denoted as γ_{rl}. By iterating over all possible pairs of points P_1 and P_2 in Fig. 5-7, including the case where $P_1 = P_2$, we can compute the mean of the semivariogram values $\gamma(P_1 - P_2)$ for all pairs. This would give a measure of the variance within the core length, and it would represent the difference between the sills of the point and core length semivariograms [i.e., $(\gamma_r - \gamma_{rl})$]. Alternatively, we could find $(\gamma_r - \gamma_{rl})$ by

$$\gamma_r - \gamma_{rl} = \frac{1}{L^2} \int_0^L \int_0^L \gamma(P_1 - P_2)\, dP_1\, dP_2 \tag{5-17}$$

The difference $\gamma_r - \gamma_{rl}$ is called the *core auxiliary function* because the difference reflects the type of support that exists. The core auxiliary functions for the difference between the point and core sills can be derived for the general models of a semivariogram; Table 5-2 gives the core auxiliary functions for the linear, exponential, and spherical models. The effect on the semivariogram is shown in Fig. 5-8 for the linear, the exponential, and the spherical models. It is important to note that as the sill changes, the radius of influence also changes. The radius of influence for point (r) and core (r_l) semivariograms are related by $r_l = r + L$. It is also important to note that the equations of Table 5-2 apply only when regularizing a semivariogram for a core length; other auxiliary functions must be used for problems in two and three dimensions.

The problem of measuring bedload movement in streams can be used to illustrate the regularization/deregularization process. Bedload movement is highly variable, both laterally and longitudinally. A number of designs for automatic bedload samplers are used for making direct measurements. Since

TABLE 5-2 Core Auxiliary Functions for One-Dimensional Samples with a Length L for Selected Semivariogram Models

Model	Core Auxiliary Function $(\gamma_r - \gamma_{rl})$
Linear	$\frac{bL}{3}$
Exponential	$\gamma_r\left\{1 - \frac{2r}{L} + \frac{2r^2}{L^2}\left[1 - \exp\left(-\frac{L}{r}\right)\right]\right\}$
Spherical	$\frac{\gamma_r L}{20r}\left(10 - \frac{L^2}{r^2}\right)$ for $L \le r$
	$\frac{\gamma_r}{20}\left(20 - \frac{15r}{L} + \frac{4r^2}{L^2}\right)$ for $L > r$

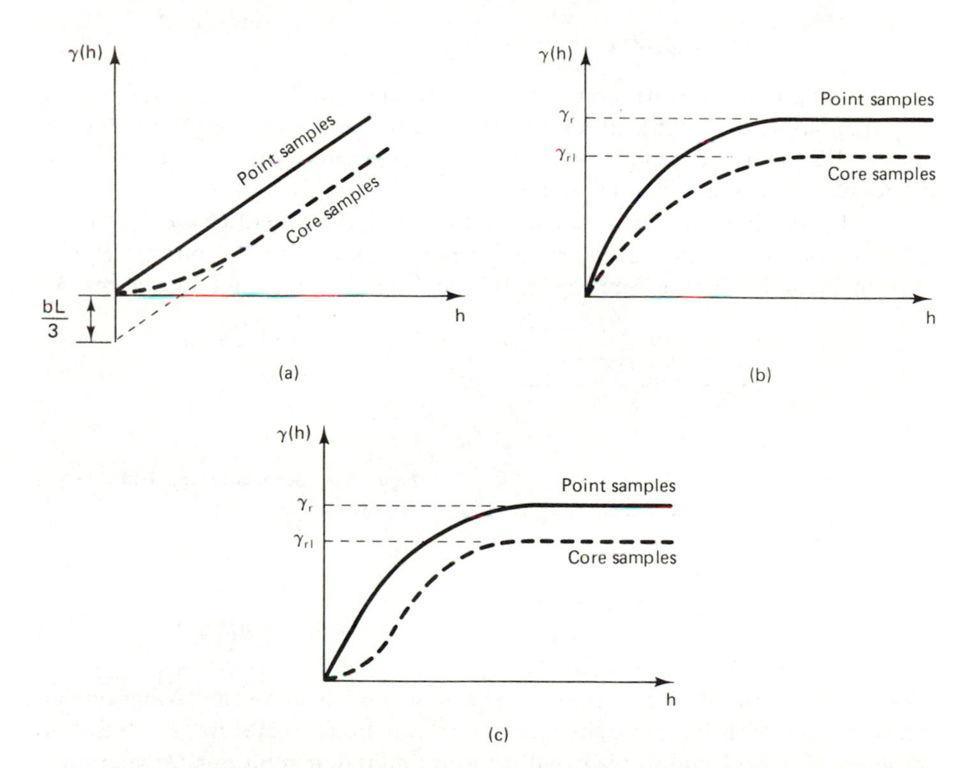

Figure 5-8 Shapes of semivariograms for point and core samples: (a) linear model; (b) exponential model; (c) spherical model.

the design of the sampler can affect the composition of the material collected by the sampler, it is necessary to assess the degree of variation associated with the design of the bedload sampler. Several assumptions will be necessary. First, assume that the apparatus used for a particular channel system has an opening that measures 1 in. by 12 in., with the 12-in. dimension placed vertically. Second, we will assume that the lateral dimension of 1 in. is sufficiently small that the support can be viewed as a one-dimensional case. Third, assume that measurements of the bedload movement with this apparatus can be characterized by an exponential semivariogram with a sill (γ_{rl}) of 20 (tons/year)^2 and a radius of influence of 15 in. To compare the semivariogram for this bedload sampler design with that of other designs, it is necessary to deregularize the semivariogram based on a 12-in. linear dimension to a point semivariogram. For L equal to 12 in. and the radius of influence r_l of 15 in., the radius of influence r of the point semivariogram would be 3 in. Using the 12-in. dimension as the sample length (L), the ratio of L to the radius of influence r is 4.0. Using the expression for the auxiliary function from Table 5-2, the sill for the point semivariogram would be

$$\gamma_r = \frac{\gamma_{rl}}{2r/L - 2(r/L)^2[1 - \exp(-L/r)]} = 53.0\ (\text{tons/day})^2$$

Thus the point semivariogram for this stream can be represented by an exponential model with a sill of 53.0 (tons/day)^2 and a radius of influence of 3 in. This semivariogram model could be compared with the point semivariograms for other channel systems.

The auxiliary function can also be computed for field (panel) samples. In this case, the integration must be for every possible pair of points (P_1 and P_2) on the field surface (see Fig. 5-9), which has the length L and width W.

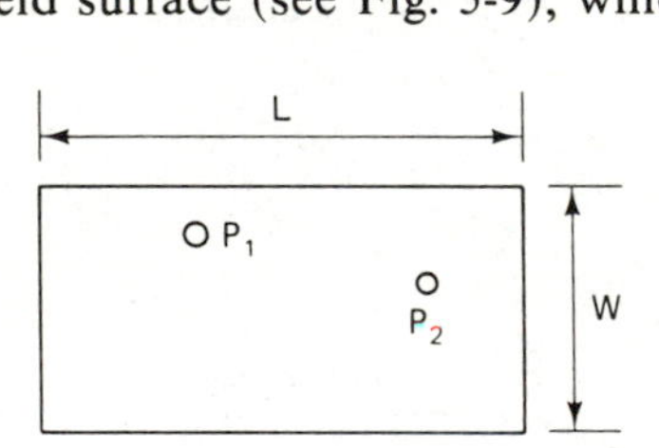

Figure 5-9 Schematic for a field of length L and width W.

The quadruple integral is

$$\gamma_r - \gamma_{rl} = \frac{1}{(LW)^2} \int_0^L \int_0^W \int_0^L \int_0^W \gamma(P_1 - P_2)\, dP_1\, dP_2\, dP_1\, dP_2 \qquad (5\text{-}18)$$

While the solution to Eq. 5-18 for the various models of a semivariogram can be computed with Eq. 5-18, the integration is tedious; therefore, a solution to problems for field samples is usually based on tabular solutions for standardized models. Specifically, the length (L) and width (W) of the panel are standardized by dividing by the radius of influence (i.e., L/r and W/r). The standardized model has a radius of influence of 1.0 and a sill of 1.0. Tables 5-3 and 5-4 give values of the field auxiliary functions for the exponential and spherical models, respectively, which were derived with Eq. 5-18. For a given radius of influence and field of specified length and width, the value of the field auxiliary function of the standardized model can be read from the

TABLE 5-3 Auxiliary Function for Regularization of a Field ($L \times W$) for the *Exponential* Semivariogram with a Radius of Influence (r) of 1.0 and a Sill of 1.0

W/r \ L/r	0.1	0.2	0.3	0.4	0.5	0.6	0.7	0.8	0.9	1.0	1.2	1.4	1.6	1.8	2.0	2.5	3.0	3.5	4.0	5.0	7.5	10.0
0.1	0.051	0.077	0.104	0.130	0.156	0.181	0.204	0.226	0.249	0.270	0.309	0.345	0.378	0.408	0.436	0.497	0.548	0.590	0.626	0.682	0.771	0.822
0.2		0.098	0.122	0.146	0.170	0.193	0.216	0.238	0.259	0.279	0.317	0.352	0.384	0.414	0.442	0.502	0.551	0.593	0.628	0.684	0.773	0.823
0.3			0.143	0.165	0.187	0.209	0.230	0.251	0.271	0.290	0.327	0.361	0.392	0.421	0.448	0.507	0.556	0.598	0.632	0.688	0.775	0.825
0.4				0.185	0.206	0.226	0.246	0.266	0.285	0.303	0.339	0.371	0.402	0.431	0.457	0.514	0.562	0.602	0.637	0.691	0.778	0.827
0.5					0.224	0.243	0.262	0.281	0.299	0.317	0.351	0.383	0.413	0.440	0.465	0.522	0.569	0.608	0.642	0.695	0.780	0.829
0.6						0.261	0.279	0.297	0.315	0.331	0.364	0.395	0.423	0.450	0.475	0.530	0.576	0.615	0.647	0.700	0.783	0.832
0.7							0.296	0.313	0.330	0.346	0.377	0.407	0.434	0.460	0.485	0.538	0.583	0.620	0.653	0.705	0.786	0.834
0.8								0.329	0.345	0.361	0.391	0.419	0.446	0.471	0.494	0.546	0.590	0.627	0.659	0.709	0.790	0.836
0.9									0.360	0.375	0.404	0.431	0.457	0.481	0.504	0.555	0.597	0.634	0.665	0.714	0.794	0.839
1.0										0.389	0.417	0.444	0.468	0.492	0.514	0.563	0.605	0.640	0.671	0.718	0.797	0.841
1.2											0.443	0.467	0.491	0.513	0.533	0.580	0.620	0.653	0.682	0.729	0.803	0.847
1.4												0.490	0.512	0.533	0.552	0.597	0.634	0.666	0.694	0.739	0.810	0.852
1.6													0.533	0.552	0.571	0.613	0.648	0.679	0.706	0.748	0.818	0.857
1.8														0.571	0.588	0.628	0.661	0.691	0.717	0.757	0.824	0.862
2.0															0.605	0.642	0.675	0.703	0.727	0.766	0.830	0.867
2.5																0.676	0.705	0.729	0.751	0.787	0.844	0.879
3.0																	0.730	0.753	0.773	0.805	0.858	0.888
3.5																		0.774	0.792	0.820	0.869	0.897
4.0																			0.808	0.834	0.879	0.905
5.0																				0.858	0.895	0.918
7.5																					0.923	0.939
10.0																						0.952

TABLE 5-4 Auxiliary Function $F(L, W)$ for Spherical Model with Range 1.0 and Sill 1.0

L/r \ W/r	0.1	0.2	0.3	0.4	0.5	0.6	0.7	0.8	0.9	1.0
0.10	0.078	0.120	0.165	0.211	0.256	0.300	0.342	0.383	0.422	0.457
0.20	0.120	0.155	0.196	0.237	0.280	0.321	0.362	0.401	0.438	0.473
0.30	0.165	0.196	0.231	0.270	0.309	0.349	0.387	0.424	0.460	0.493
0.40	0.211	0.237	0.270	0.305	0.342	0.379	0.415	0.451	0.484	0.516
0.50	0.256	0.280	0.309	0.342	0.376	0.411	0.445	0.479	0.511	0.541
0.60	0.300	0.321	0.349	0.379	0.411	0.443	0.476	0.507	0.538	0.566
0.70	0.342	0.362	0.387	0.415	0.445	0.476	0.506	0.536	0.565	0.591
0.80	0.383	0.401	0.424	0.451	0.479	0.507	0.536	0.564	0.591	0.616
0.90	0.422	0.438	0.460	0.484	0.511	0.538	0.565	0.591	0.616	0.640
1.00	0.457	0.473	0.493	0.516	0.541	0.566	0.591	0.616	0.640	0.662
1.20	0.520	0.534	0.551	0.572	0.593	0.616	0.638	0.660	0.682	0.701
1.40	0.572	0.584	0.600	0.618	0.637	0.657	0.677	0.697	0.716	0.733
1.60	0.614	0.625	0.639	0.655	0.673	0.691	0.709	0.727	0.744	0.760
1.80	0.650	0.659	0.672	0.687	0.703	0.719	0.736	0.752	0.767	0.782
2.00	0.679	0.688	0.700	0.713	0.728	0.743	0.758	0.773	0.787	0.800
2.50	0.735	0.743	0.752	0.763	0.775	0.788	0.800	0.813	0.824	0.835
3.00	0.775	0.781	0.789	0.799	0.809	0.820	0.830	0.841	0.851	0.860
3.50	0.804	0.810	0.817	0.825	0.834	0.843	0.852	0.861	0.870	0.878
4.00	0.827	0.832	0.838	0.845	0.853	0.861	0.870	0.878	0.885	0.892
5.00	0.860	0.864	0.869	0.874	0.881	0.887	0.894	0.901	0.907	0.913

L/r \ W/r	1.2	1.4	1.6	1.8	2.0	2.5	3.0	3.5	4.0	5.0
0.10	0.520	0.572	0.614	0.650	0.679	0.735	0.775	0.804	0.827	0.860
0.20	0.534	0.584	0.625	0.659	0.688	0.743	0.781	0.810	0.832	0.864
0.30	0.551	0.600	0.639	0.672	0.700	0.752	0.789	0.817	0.838	0.869
0.40	0.572	0.618	0.655	0.687	0.713	0.763	0.799	0.825	0.845	0.874
0.50	0.593	0.637	0.673	0.703	0.728	0.775	0.809	0.834	0.853	0.881
0.60	0.616	0.657	0.691	0.719	0.743	0.788	0.820	0.843	0.861	0.887
0.70	0.638	0.677	0.709	0.736	0.758	0.800	0.830	0.852	0.870	0.894
0.80	0.660	0.697	0.727	0.752	0.773	0.813	0.841	0.861	0.878	0.901
0.90	0.682	0.716	0.744	0.767	0.787	0.824	0.851	0.870	0.885	0.907
1.00	0.701	0.733	0.760	0.782	0.800	0.835	0.860	0.878	0.892	0.913
1.20	0.736	0.764	0.788	0.807	0.823	0.854	0.876	0.892	0.905	0.923
1.40	0.764	0.790	0.811	0.828	0.842	0.870	0.890	0.904	0.915	0.931
1.60	0.788	0.811	0.829	0.845	0.858	0.883	0.901	0.914	0.924	0.938
1.80	0.807	0.828	0.845	0.859	0.871	0.894	0.910	0.921	0.931	0.944
2.00	0.823	0.842	0.858	0.871	0.882	0.903	0.917	0.928	0.936	0.948
2.50	0.854	0.870	0.883	0.894	0.903	0.920	0.932	0.941	0.948	0.957
3.00	0.876	0.890	0.901	0.910	0.917	0.932	0.942	0.950	0.955	0.964
3.50	0.892	0.904	0.914	0.921	0.928	0.941	0.950	0.956	0.961	0.969
4.00	0.905	0.915	0.924	0.931	0.936	0.948	0.955	0.961	0.966	0.972
5.00	0.923	0.931	0.938	0.944	0.948	0.957	0.964	0.969	0.972	0.977

appropriate table and multiplied by the sill γ_r to get the value of the field auxiliary function, $F(L, W)$. The sill of the regularized model equals the difference between γ_r and $F(L, W)$.

The problem of estimating the mean soil moisture of a field can be used to illustrate the regularization of a field or panel semivariogram. Assume that point soil-moisture measurements in a field yield a semivariogram value that can be best represented by a spherical model with a sill of $23.4(\%)^2$ and a radius of influence of 25 m. Assume that one wants to compare the field measurements with soil-moisture measurements obtained by remote sensing using a pixel of 30 m by 50 m. The sill of the regularized semivariogram can be obtained with Table 5-4. The ratios of the length and width of a 50 m by 30 m pixel to the radius of influence are 2 and 1.2, respectively. From Table 5-4 we get a value for the field auxiliary function of the standardized model of 0.823. This gives a value of $F(L, W)$ of $0.823(23.4) = 19.26(\%)^2$, and the sill of the regularized semivariogram would be $4.14(\%)^2$.

ESTIMATION OF STANDARD ERRORS

Returning to the original problem of estimation, we were interested in estimating a value $\hat{Y}$ using sample information. The value to be estimated will be called the criterion variable and denoted as $\hat{Y}$. The sample information will be denoted as X. It is important to note that the sample information represents the same variable as the criterion. For example, if we want to estimate the average soil moisture in a field, both X and Y represent soil moisture. The different notation serves only to distinguish between the sample and the value to be estimated. In any case, we use sample information to estimate the value of the criterion variable. The estimation problem will depend on the nature of both the sample and the value to be estimated. Table 5-5 shows the possible

TABLE 5-5 Case Number for Type of Estimation Problem and the Nature of the Sample

Value to Be Estimated	Sample		
	Point	Vector	Line
Point	1	2	—
Line	3	4	5
Field	6	7	8

combinations and the case number of those types of problems that will be examined here.

For the present we assume that our best estimate of the criterion will be either the point value for a sample consisting of a single point, the arithmetic mean for a sample consisting of a vector of points, or the average value of the line for a linear sample. Later we will provide a generalization where a weighted average is used as the estimate; this is kriging.

In addition to the estimated value of the sample, we should also be interested in the standard error of estimate. The standard error can be computed for any estimated value, including the mean value of Eq. 5-2, but kriging finds the estimate that has the smallest standard error. The value of the standard error depends on the nature of both the criterion and the sample. From our knowledge of the standard error of the mean, we would assume that the standard error would decrease as the information content of the sample increases. Before using kriging estimation to minimize the standard error, it may be easier to concentrate on deriving an expression for the standard error for the different cases in Table 5-5.

The problem may be easiest to understand if we say that the standard error consists of three components of variation. Two of the components, or sources, measure the variation within the sample and within the criterion to be estimated. Obviously, if either the sample or the criterion consists of a single point, the corresponding component of the "within" variation would be zero. The third component reflects variation associated with the sample and the criterion. Obviously, this source of variation will not be zero unless we know the value of the criterion, in which case we would not really have an estimation problem. Each of these sources of variation will be assessed in the following paragraphs and shown how they are used to compute the standard errors for each of the eight cases.

Case 1: Point/Point Estimation

The simplest estimation problem is where the sample consists of a single point (X) and the value to be estimated ($\hat{Y}$) is a point at some other location. In this case, the best estimate of Y at the unknown point equals the sample value:

$$\hat{Y} = X \tag{5-19}$$

The model of Eq. 5-19 will provide an unbiased estimate if the domain from which the sample is obtained is stationary. Although the expected sum of the errors would be zero, the variance of the errors will not equal zero. For a stationary domain the accuracy of estimation could be estimated by finding the average squared difference between two points separated by distance h. That is, the standard error could be estimated by iterating over all possible points separated from the unknown point by distance h, while squaring and summing the differences. This is, by definition, the value of the variogram for separation distance h. Thus our estimate of the standard error for case 1 is

$$S_e = [2\gamma(h)]^{0.5} \tag{5-20}$$

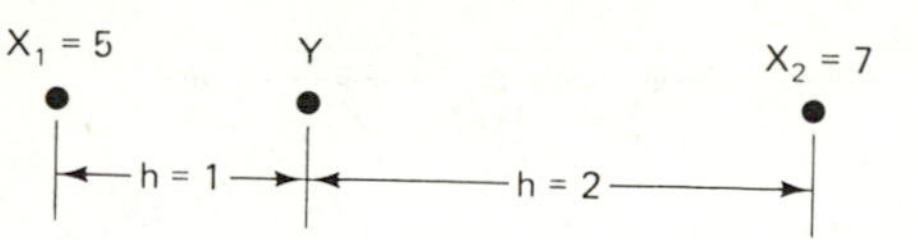

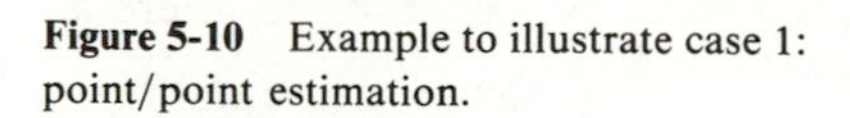
Figure 5-10 Example to illustrate case 1: point/point estimation.

As an example, consider the layout of Fig. 5-10. First, assume that we have the value at a single point X_1 as a sample (at this point assume that X_2 in Fig. 5-10 is not part of the sample), and we need to estimate the value Y at a separation distance of 1 unit from X_1. If the semivariogram is spherical with a sill of 2 and a radius of influence of 3, we can estimate the standard error using Eqs. 5-11 and 5-20:

$$S_e = \left[2\left(h - \frac{h^3}{27}\right)\right]^{0.5} = \left[2\left(1 - \frac{1^3}{27}\right)\right]^{0.5} = 1.388 \quad (5\text{-}21)$$

Thus our estimated value of Y would be 5 and would have a standard error of 1.388. If we wanted to assume either a normal or t distribution, we could derive a confidence interval on the estimated value.

If we only have a sample point X_2 of Fig. 5-10 (assume that X_1 does not exist), our best estimate of Y would be $\hat{Y} = X_2 = 7$. For a separation distance of 2, the standard error of estimate would now be

$$S_e = \left[2\left(2 - \frac{(2)^3}{27}\right)\right]^{0.5} = 1.846 \quad (5\text{-}22)$$

In comparison to the case where X_1 was the sample point, X_2 yields a larger standard error because the separation distance is greater. According to the spherical model of the semivariogram, the expected error increases as the separation distance increases.

Case 2: Point/Vector Estimation

For case 2 we will assume that we wish to estimate the value at a point, with our sample consisting of a vector of n points (i.e., X_i, $i = 1, 2, \ldots, n$). If we assume that each point is given equal weight, the sample mean is our estimator of Y:

$$\hat{Y} = \frac{1}{n} \sum X_i \quad (5\text{-}23)$$

In the case of Fig. 5-10 with both X_1 and X_2 as sample points, our best estimate of Y would be 6.

Equation 5-20 cannot be used to estimate the reliability of values predicted with Eq. 5-23 because it does not account for the variation between the sample points. Variation between sample points influences the reliability of estimates made using the sample points. Consider the case where we need to estimate the average value for the line shown in Fig. 5-11(a), where the sample size is 5. If the range of influence for the underlying semivariogram equals

Figure 5-11 Effect of sample spacing on reliability of estimates: (a) equally spaced sample points; (b) clustered sample points.

the separation distance between the points, h, then each of the sample points represents somewhat independent information. In a sense, our effective sample size is 5. From our knowledge of the standard error of the mean (i.e., $S/\sqrt{n}$) we know that the reliability increases as the sample size increases. Now if the range of influence for the semivariogram that underlies the system shown in Fig. 5-11(a) is greater than $4h$, the sample points do not represent independent information. In a sense, the effective sample size is less than 5. For a very large range of influence, the effective sample size may actually approach 1. Therefore, if Eq. 5-23 is to be used to estimate the criterion Y, the reliability of the estimate is going to depend on the degree of independence of the sample points. If the sample points are relatively far apart, the reliability of an estimate should be greater than the reliability when the sample points are close together. Intuitively, one would expect the reliability of an estimate made from the five points in Fig. 5-11(a) to be greater than the reliability of an estimate made from the five points in Fig. 5-11(b). Thus the degree of variation within a sample must be accounted for when computing the reliability of an estimate.

Before approaching the problem of estimating the reliability of point estimates made from a vector sample, it may be worthwhile formalizing the notation. In developing estimates of reliability, S will be used to indicate the sample, whether or not the sample is a point, vector, or line. While $\hat{Y}$ will be used to indicate the estimated value, Y will represent the criterion being estimated; this may be a point or the average value of the criterion in either a line segment or a field. Where the average semivariogram value is needed, $\bar{\gamma}$ will be used. To indicate the average semivariogram value for points within a sample, the notation $\bar{\gamma}(S, S)$ will be used. For case 1 the reliability was shown to be the value of Eq. 5-20. When both the sample and the criterion are single points, the standard error for case 1 could be represented with the notation above as follows:

$$S_e = [2\bar{\gamma}(S, Y)]^{0.5} \tag{5-24}$$

in which $\bar{\gamma}(S, Y)$ is the average semivariogram value between S and Y; obviously, since the sample and the point to be estimated are both single points, $\bar{\gamma}(S, Y) = \gamma(h)$.

Returning to the problem of case 2, we need to estimate the reliability of estimates made with Eq. 5-23. If Eq. 5-24 represents the error variation for a single point sample, the total variance for a sample of size n will be $2\bar{\gamma}(S, Y)$, which is twice the average semivariogram value for separation distances between each point in the sample and the unknown point. From this, we must subtract the variance that is internal to the sample points (i.e., the "within" sample variation). For the clustered points of Fig. 5-11(b) the internal variation will be quite small and so the error variance of the estimate (Eq. 5-23) will be relatively large. For the equally spaced points of Fig. 5-11(a) the internal variation will be relatively large, so the error variance of the estimate should be reduced significantly compared with a clustered sample [Fig. 5-11(b)]. The internal sample variation depends on the semivariogram. It can be shown that the internal variation will equal the average semivariogram value for the n sample points and will be denoted as $\bar{\gamma}(S, S)$. For a sample of size n, there are $(n \times n)$ pairs of points that must be considered in computing the average semivariogram value. Of course, $\gamma(X_i, X_i)$ will equal zero because the separation distance equals zero, unless there is a nugget effect.

To illustrate the computation of the error variance for case 2, consider the system of Fig. 5-10. In this case the sample consists of X_1 and X_2 (i.e., $n = 2$), and the separation distance for the two sample points equals the radius of influence. The average semivariogram value between the sample points and the criterion point is

$$\bar{\gamma}(S, Y) = \tfrac{1}{2}\left(\tfrac{26}{27} + \tfrac{46}{27}\right) = \tfrac{4}{3} \tag{5-25}$$

The average semivariogram value for the internal sample variance is given by:

$$\bar{\gamma}(S, S) = \tfrac{1}{4}[\gamma(S_1, S_1) + \gamma(S_1, S_2) + \gamma(S_2, S_1) + \gamma(S_2, S_2)] \tag{5-26a}$$

$$= \tfrac{1}{4}(0 + 2 + 2 + 0) = 1 \tag{5-26b}$$

Therefore, we can compute the standard error by

$$S_e = [2\bar{\gamma}(S, Y) - \bar{\gamma}(S, S)]^{0.5} \tag{5-27a}$$

$$= [2(\tfrac{4}{3}) - 1]^{0.5} = 1.291 \tag{5-27b}$$

Thus the reliability of the estimate made with Eq. 5-23 is only slightly better than the reliability of the estimate made using X_1 alone. This is rational since the standard error for X_2 (i.e., 1.846), was not much different from the sill of the semivariogram.

Case 3: Line/Point Estimation

For case 3 we are interested in estimating the mean value of a criterion represented by a line when the sample consists of the value at a single point. Additionally, we need an estimate of the reliability. Our best estimate of the mean value for the line is the value of the single sample point. However, the reliability depends on the location of the point with respect to the line. Intuitively, one would guess that the reliability would be greater if the point was located on the line, as opposed to being separated from the line. Furthermore, it would be rational to believe that the standard error would be smaller when the sample point is at the center of the line rather than at one end of the line. Thus we need a general methodology for predicting the reliability since there are an infinite number of places where the sample point could be taken.

The criterion variable is now the average value of the line. Thus the standard error must account for both the variation between the sample point and each point on the line and the variation within the line. Obviously, $\bar{\gamma}(S, S)$ will be zero, since the sample consists of a single point. Whereas $\bar{\gamma}(S, S)$ accounts for variation within the sample (i.e., between sample points), variation within the criterion (i.e., the line) must be accounted for in calculating the reliability of an estimate for case 3. In general, we will denote this as $\bar{\gamma}(Y, Y)$; however, this source of variation is a function of the length of the line, so we will also denote it as $S_Y^2(L)$, where the subscript Y is used to indicate that the variation is associated with the criterion and the argument L indicates the length of the line. The variation between the sample point and each point on the line will be denoted by $\bar{\gamma}(S, Y)$; thus the standard error will be given by

$$S_e = [2\bar{\gamma}(S, Y) - \bar{\gamma}(Y, Y)]^{0.5} \tag{5-28}$$

In our discussion of regularization it was shown that the variation within the line was defined by the core auxiliary function $F(L)$; thus $\bar{\gamma}(Y, Y)$ equals $F(L)$ for any line of length L, where $F(L)$ is obtained from Table 5-2.

The determination of $\bar{\gamma}(S, Y)$ is more complicated because the line is continuous, and thus integration is involved. For a line of length L where the sample point is at one end of the line, it can be shown that the variance component $\bar{\gamma}(S, Y)$ in case 3 is given by the following equations for the linear, exponential, and spherical semivariogram models, respectively:

$$\bar{\gamma}(S, Y) = \frac{bL}{2} \tag{5-29}$$

$$\bar{\gamma}(S, Y) = \gamma_r \left[1 - \frac{r}{L}(1 - e^{-L/r})\right] \tag{5-30}$$

$$\bar{\gamma}(S, Y) = \begin{cases} \dfrac{0.125\gamma_r L(6 - L^2/r^2)}{r} & \text{for } L \le r \quad (5\text{-}31a) \\[2ex] 0.125\gamma_r \left(8 - \dfrac{3r}{L}\right) & \text{for } L > r \quad (5\text{-}31b) \end{cases}$$

To simplify the notation, the auxiliary functions of Eqs. 5-29 to 5-31 will also be denoted as $\chi(L)$, which indicates that they are a function of the length of the line L. It is important to note that the auxiliary functions $\chi(L)$ of Eqs. 5-29 to 5-31 are different from the auxiliary functions $F(L)$ given in Table 5-2; those in Table 5-2 can be used to compute $\bar{\gamma}(Y, Y)$.

In addition to the cases where the sample point is located at the end of the line, we may be interested in the case where the sample point is located within the line a distance d from one end of the line. It can be shown that $\bar{\gamma}(S, Y)$ can be computed by separating the line into two parts at the location of the sample point and weighting the variance component according to the length of the line segment. For a point located a distance d from one end of the line, there are two line segments, with one having a length d and the other a length of $L - d$. Thus the variance associated with the entire line is the weighted sum of the auxiliary functions for both line segments:

$$\bar{\gamma}(S, Y) = \frac{1}{L}[d\chi(d) + (L - d)\chi(L - d)] \tag{5-32}$$

where the values of $\chi(d)$ and $\chi(L - d)$ are computed from Eqs. 5-29 to 5-31 using either d or $L - d$ as the argument. Equation 5-32 can be used as $\bar{\gamma}(S, Y)$, and Eq. 5-28 can be used to estimate the reliability. For the special case of Eq. 5-32 where the sample point is located at the center of the line, Eq. 5-32 reduces to

$$\bar{\gamma}(S, Y) = \chi\left(\frac{L}{2}\right) \tag{5-33}$$

If the sample point is not on the line but located a distance d from one end of the line, we can compute $\bar{\gamma}(S, Y)$ by considering an imaginary line of length $(L + d)$; the imaginary line is considered to consist of two segments, with one having a length of $(L + d)$ and the other a length of d. The variance associated with each segment is weighted by the length of the segment. The variance associated with the imaginary segment of length d must be subtracted from the variance associated with total length $(L + d)$ in order to compute the within-line variation:

$$\bar{\gamma}(S, Y) = \frac{1}{L}[(L + d)\chi(L + d) - d\chi(d)] \tag{5-34}$$

The line/point estimation problem can be illustrated using a single measurement of the dissolved oxygen (DO) concentration in a stream reach as the sample and the mean value within the stream reach as the criterion. For purposes of illustration we will assume that the process can be represented by a spherical semivariogram with a radius of influence of 0.5 mile (2640 ft) and a sill of 0.25 $(\text{ppm})^2$. A sample is taken at one end of a 2000-ft reach, with the measurement having a DO concentration of 4.2 ppm. Using the sample measurement as the best estimate of the mean value for the 2000-ft reach, the standard error of estimate is obtained using Eq. 5-28:

$$S_e = [2\bar{\gamma}(S, Y) - \bar{\gamma}(Y, Y)]^{0.5} = [2\chi(L) - F(L)]^{0.5}$$

$$= \left[\frac{2\gamma_r L(6 - L^2/r^2)}{8r} - \frac{\gamma_r L}{20r}\left(10 - \frac{L^2}{r^2}\right)\right]^{0.5} = 0.41 \text{ ppm}$$

This is approximately 80% of the square root of the sill.

Case 4: Line/Vector Estimation

For case 4 the sample consists of a vector of n points and the criterion is the average value of a one-dimensional element of length L. This problem is a combination of cases 2 and 3 because both the variation within the line, as in case 3, and the variation within the sample, as in case 2, must be considered. Thus the standard error can be computed by

$$S_e^2 = 2\bar{\gamma}(S, Y) - \bar{\gamma}(Y, Y) - \bar{\gamma}(S, S) \tag{5-35}$$

The variance component $\bar{\gamma}(S, Y)$ is the average semivariogram value between each of the n sample points and the line:

$$\bar{\gamma}(S, Y) = \frac{1}{n} \sum_{i=1}^{n} \bar{\gamma}(S_i, Y) \tag{5-36}$$

in which each element of the summation $\bar{\gamma}(S_i, Y)$ can be computed using the appropriate auxiliary function of Eqs. 5-29 to 5-31. As in case 3, the value of $\bar{\gamma}(Y, Y)$ of Eq. 5-35 is computed using the core auxiliary functions of Table 5-2. The value of $\bar{\gamma}(S, S)$ in Eq. 5-35 is computed using the average semivariogram value for every pair of points in the sample; this was introduced in case 2. In computing $\bar{\gamma}(S, S)$, the averaging should include the case where $S_i = S_j$; unless there is a nugget effect, $\bar{\gamma}(S_i, S_j)$ will equal zero when $i = j$.

The dissolved oxygen problem of case 3 will be used to illustrate case 4. Assume that a second DO measurement is taken at the other end of the 2000-ft reach, with a resulting value of 4.5 ppm. The best estimate would be 4.35 ppm, which is the arithmetic mean of the two measurements. The standard error is computed using Eq. 5-35. The term $\bar{\gamma}(S, Y)$ equals the mean of the auxiliary functions for the two end points; since the two values of $\chi(L)$ are equal, $\bar{\gamma}(S, Y) = \chi(L) = 0.12845$. The term $\bar{\gamma}(Y, Y)$ equals the value computed in case 3 [i.e., $\bar{\gamma}(Y, Y) = 0.08926$]. The term $\bar{\gamma}(S, Y)$ is computed as the average semivariogram between each pair of points:

$$\bar{\gamma}(S, S) = \tfrac{1}{4}[\gamma(S_1, S_1) + \gamma(S_1, S_2) + \gamma(S_2, S_1) + \gamma(S_2, S_2)]$$

where $\gamma(S_1, S_1) = \gamma(S_2, S_2) = 0$, and $\gamma(S_1, S_2) = \gamma(S_2, S_1)$ and are given by

$$\gamma_r\left[\frac{3(2000)}{2(2640)} - \frac{1}{2}\left(\frac{2000}{2640}\right)^3\right] = 0.22974$$

Therefore, $\bar{\gamma}(S, S)$ equals 0.11487, and the standard error equals 0.2297 ppm, which is approximately one-half of the standard error of the single-point sample of case 3.

Case 5: Line/Line Estimation

For case 5, both the sample and criterion are the mean values of one-dimensional elements (i.e., linear segments in which one dimension is significantly greater than the other two dimensions). While the average of the linear sample is used as the estimate of the criterion value, the reliability of the estimate must account for the variation within both the sample and criterion, as well as the variation between the two elements. Equation 5-35 can be used as a general statement for computing the standard error. The values of $\bar{\gamma}(S, S)$ and $\bar{\gamma}(Y, Y)$ are computed using one of the auxiliary functions of Eq. 5-29 to 5-31. The term $\bar{\gamma}(S, Y)$ can be computed using the average semivariogram value between each point in the line representing the sample and each point in the line representing the criterion; this variance component is another auxiliary function that is a function of the semivariogram model, the orientation and distance between the two linear elements, and the lengths of the two elements. For the case of a spherical semivariogram model where the two elements are parallel to each other and of the same length, the auxiliary function is given in Table 5-6; this auxiliary function is standardized for a sill of 1 and a radius of influence of 1, with the arguments being the distance between the two parallel elements (L) and the lengths of the two elements (B). For a given separation distance L and linear length B, the standardized value of the auxiliary function is obtained from Table 5-6. Table 5-6 is entered with the ratios of L/r and B/r. The value is dimensionalized by multiplying by the value of the sill of the underlying semivariogram.

As indicated previously, the value of $\bar{\gamma}(S, Y)$ is a function of the orientation and distance between the two linear elements. The auxiliary function of Table 5-6 is used for the case where the two linear elements are parallel, of equal length, and separated by a distance L; this will be denoted as case 5a. Case 5b will be used to indicate the situation where the two linear elements are perpendicular to one another and share a common end point. For this case Eq. 5-35 can be used to compute the standard error. The values of $\bar{\gamma}(Y, Y)$ and $\bar{\gamma}(S, S)$ will be computed in the same way as for case 5a. The value of $\bar{\gamma}(S, Y)$ is obtained using the auxiliary function of Table 5-7, where L and B are the lengths of the two linear elements. The auxiliary function of Table 5-7 will also be used for case 6. The values of Table 5-7 are for the standardized spherical semivariogram model, with $\gamma_r = r = 1$. The value $\bar{\gamma}(S, Y)$ equals the product of the value of Table 5-7 and γ_r.

Case 6: Field/Point Estimation

For case 6 the sample consists of a single point and the criterion is the average value within a field that has dimensions of L by W. The value of the point sample would be used as the best estimate of the average field value. Since the sample consists of a single point, there would be no within sample variation [i.e., $\bar{\gamma}(S, S) = 0$]; therefore, it would only be necessary to determine both the within field variation [i.e., $\bar{\gamma}(Y, Y)$] and the average variation between the sample point and every point within the field [i.e., $\bar{\gamma}(S, Y)$]. To determine the variation for the average value within the field (not the standard error of

TABLE 5-6 Auxiliary Function for Spherical Model with Range 1.0 and Sill 1.0

L/r \ B/r	0.1	0.2	0.3	0.4	0.5	0.6	0.7	0.8	0.9	1.0
0.05	0.094	0.132	0.175	0.219	0.263	0.306	0.348	0.388	0.426	0.461
0.10	0.161	0.188	0.223	0.261	0.300	0.340	0.379	0.416	0.452	0.486
0.15	0.231	0.252	0.280	0.312	0.347	0.383	0.419	0.453	0.486	0.518
0.20	0.302	0.318	0.341	0.369	0.400	0.432	0.464	0.495	0.526	0.555
0.25	0.372	0.385	0.404	0.428	0.455	0.483	0.512	0.541	0.568	0.594
0.30	0.440	0.451	0.467	0.488	0.511	0.536	0.562	0.588	0.613	0.636
0.35	0.507	0.516	0.529	0.547	0.568	0.590	0.612	0.635	0.657	0.678
0.40	0.571	0.578	0.590	0.605	0.623	0.642	0.662	0.683	0.702	0.721
0.45	0.632	0.638	0.648	0.661	0.677	0.693	0.711	0.729	0.746	0.762
0.50	0.689	0.695	0.703	0.715	0.728	0.742	0.758	0.773	0.787	0.801
0.55	0.743	0.748	0.755	0.765	0.776	0.789	0.802	0.814	0.827	0.838
0.60	0.793	0.797	0.803	0.811	0.821	0.831	0.842	0.853	0.863	0.872
0.65	0.839	0.842	0.847	0.854	0.862	0.870	0.879	0.888	0.896	0.903
0.70	0.879	0.882	0.886	0.892	0.898	0.905	0.912	0.919	0.925	0.930
0.75	0.915	0.917	0.920	0.925	0.930	0.935	0.940	0.945	0.949	0.953
0.80	0.945	0.946	0.949	0.952	0.956	0.960	0.963	0.966	0.969	0.971
0.85	0.968	0.970	0.971	0.974	0.976	0.978	0.981	0.982	0.984	0.985
0.90	0.986	0.987	0.988	0.989	0.990	0.991	0.992	0.993	0.994	0.994
0.95	0.996	0.997	0.997	0.998	0.998	0.998	0.998	0.999	0.999	0.999
1.00	1.000	1.000	1.000	1.000	1.000	1.000	1.000	1.000	1.000	1.000

L/r \ B/r	1.2	1.4	1.6	1.8	2.0	2.5	3.0	3.5	4.0	5.0
0.05	0.524	0.575	0.617	0.652	0.681	0.737	0.777	0.806	0.828	0.861
0.10	0.545	0.594	0.634	0.667	0.695	0.748	0.786	0.814	0.836	0.867
0.15	0.573	0.619	0.656	0.687	0.714	0.764	0.799	0.825	0.846	0.875
0.20	0.605	0.648	0.682	0.711	0.735	0.782	0.814	0.838	0.857	0.884
0.25	0.641	0.679	0.711	0.737	0.759	0.801	0.831	0.853	0.870	0.894
0.30	0.678	0.712	0.741	0.764	0.784	0.822	0.848	0.868	0.883	0.905
0.35	0.715	0.746	0.771	0.792	0.809	0.843	0.866	0.884	0.897	0.917
0.40	0.753	0.780	0.801	0.820	0.835	0.864	0.884	0.899	0.911	0.928
0.45	0.790	0.812	0.831	0.847	0.860	0.884	0.902	0.915	0.924	0.939
0.50	0.825	0.844	0.860	0.872	0.883	0.904	0.918	0.929	0.937	0.949
0.55	0.858	0.873	0.886	0.897	0.906	0.922	0.934	0.943	0.949	0.959
0.60	0.888	0.901	0.911	0.919	0.926	0.939	0.948	0.955	0.960	0.968
0.65	0.915	0.925	0.933	0.939	0.944	0.954	0.961	0.966	0.970	0.976
0.70	0.939	0.946	0.952	0.956	0.960	0.967	0.972	0.976	0.979	0.983
0.75	0.959	0.964	0.968	0.971	0.974	0.978	0.982	0.984	0.986	0.989
0.80	0.975	0.978	0.981	0.983	0.984	0.987	0.989	0.991	0.992	0.993
0.85	0.987	0.989	0.990	0.991	0.992	0.993	0.994	0.995	0.996	0.997
0.90	0.995	0.996	0.996	0.997	0.997	0.997	0.998	0.998	0.998	0.999
0.95	0.999	0.999	0.999	0.999	0.999	1.000	1.000	1.000	1.000	1.000
1.00	1.000	1.000	1.000	1.000	1.000	1.000	1.000	1.000	1.000	1.000

TABLE 5-7 Auxiliary Function $g(L, B)$ for Spherical Model with Range 1.0 and Sill 1.0

L/r \ B/r	0.1	0.2	0.3	0.4	0.5	0.6	0.7	0.8	0.9	1.0
0.10	0.114	0.177	0.243	0.310	0.374	0.436	0.494	0.546	0.593	0.633
0.20	0.177	0.227	0.285	0.346	0.406	0.464	0.518	0.568	0.613	0.651
0.30	0.243	0.285	0.336	0.390	0.445	0.499	0.550	0.597	0.639	0.674
0.40	0.310	0.346	0.390	0.439	0.489	0.539	0.586	0.629	0.668	0.701
0.50	0.374	0.406	0.445	0.489	0.535	0.580	0.623	0.663	0.698	0.728
0.60	0.436	0.464	0.499	0.539	0.580	0.621	0.660	0.697	0.728	0.755
0.70	0.494	0.518	0.550	0.586	0.623	0.660	0.696	0.729	0.757	0.781
0.80	0.546	0.568	0.597	0.629	0.663	0.697	0.729	0.758	0.783	0.805
0.90	0.593	0.613	0.639	0.668	0.698	0.728	0.757	0.783	0.806	0.826
1.00	0.633	0.651	0.674	0.701	0.728	0.755	0.781	0.805	0.826	0.843
1.20	0.694	0.709	0.729	0.751	0.774	0.796	0.818	0.837	0.855	0.869
1.40	0.738	0.751	0.767	0.786	0.806	0.825	0.844	0.861	0.875	0.888
1.60	0.771	0.782	0.797	0.813	0.830	0.847	0.863	0.878	0.891	0.902
1.80	0.796	0.806	0.819	0.834	0.849	0.864	0.879	0.892	0.903	0.913
2.00	0.817	0.826	0.837	0.850	0.864	0.878	0.891	0.902	0.913	0.921
2.50	0.853	0.860	0.870	0.880	0.891	0.902	0.913	0.922	0.930	0.937
3.00	0.878	0.884	0.891	0.900	0.909	0.918	0.927	0.935	0.942	0.948
3.50	0.895	0.900	0.907	0.914	0.922	0.930	0.938	0.944	0.950	0.955
4.00	0.908	0.913	0.919	0.925	0.932	0.939	0.945	0.951	0.956	0.961
5.00	0.927	0.930	0.935	0.940	0.946	0.951	0.956	0.961	0.965	0.969

L/r \ B/r	1.2	1.4	1.6	1.8	2.0	2.5	3.0	3.5	4.0	5.0
0.10	0.694	0.738	0.771	0.796	0.817	0.853	0.878	0.895	0.908	0.927
0.20	0.709	0.751	0.782	0.806	0.826	0.860	0.884	0.900	0.913	0.930
0.30	0.729	0.767	0.797	0.819	0.837	0.870	0.891	0.907	0.919	0.935
0.40	0.751	0.786	0.813	0.834	0.850	0.880	0.900	0.914	0.925	0.940
0.50	0.774	0.806	0.830	0.849	0.864	0.891	0.909	0.922	0.932	0.946
0.60	0.796	0.825	0.847	0.864	0.878	0.902	0.918	0.930	0.939	0.951
0.70	0.818	0.844	0.863	0.879	0.891	0.913	0.927	0.938	0.945	0.956
0.80	0.837	0.861	0.878	0.892	0.902	0.922	0.935	0.944	0.951	0.961
0.90	0.855	0.875	0.891	0.903	0.913	0.930	0.942	0.950	0.956	0.965
1.00	0.869	0.888	0.902	0.913	0.921	0.937	0.948	0.955	0.961	0.969
1.20	0.891	0.907	0.918	0.927	0.935	0.948	0.956	0.963	0.967	0.974
1.40	0.907	0.920	0.930	0.938	0.944	0.955	0.963	0.968	0.972	0.978
1.60	0.918	0.930	0.939	0.945	0.951	0.961	0.967	0.972	0.975	0.980
1.80	0.927	0.938	0.945	0.952	0.956	0.965	0.971	0.975	0.978	0.983
2.00	0.935	0.944	0.951	0.956	0.961	0.969	0.974	0.978	0.980	0.984
2.50	0.948	0.955	0.961	0.965	0.969	0.975	0.979	0.982	0.984	0.987
3.00	0.956	0.963	0.967	0.971	0.974	0.979	0.983	0.985	0.987	0.990
3.50	0.963	0.968	0.972	0.975	0.978	0.982	0.985	0.987	0.989	0.991
4.00	0.967	0.972	0.975	0.978	0.980	0.984	0.987	0.989	0.990	0.992
5.00	0.974	0.978	0.980	0.983	0.984	0.987	0.990	0.991	0.992	0.994

estimate for all points within the field), we must subtract $\bar{\gamma}(Y, Y)$ from $2\bar{\gamma}(S, Y)$; that is, the standard error for case 6 is given by

$$S_e = [2\bar{\gamma}(S, Y) - \bar{\gamma}(Y, Y)]^{0.5} \tag{5-37}$$

The value of $\bar{\gamma}(Y, Y)$ is nothing more than the auxiliary function F; values of the standardized model are given in Tables 5-3 and 5-4 for the exponential and spherical models, respectively. The value of $\bar{\gamma}(S, Y)$ is a variance component that is a function of the underlying semivariogram model, the length and width of the field, and the location of the sample point with respect to the field. A standardized auxiliary function $g(L, W)$ is available for the case of a spherical semivariogram model for a field of length L and width W where the sample point is located at one corner of the field; values of $g(L, W)$ are given in Table 5-7. The auxiliary function measures the average semivariogram value between a point at one corner of the field and all points in the field for a spherical semivariogram model when both the sill and radius of influence equal 1.0. For a field where the sample point is not located at one corner of the field, the value of the variance component $\bar{\gamma}(S, Y)$ can be determined by separating the field into four rectangles with dimensions L_i and W_i and finding the weighted average of the values of $g(L_i, W_i)$ by

$$\bar{\gamma}(S, Y) = \frac{1}{LW} \sum_{i=1}^{4} L_i W_i g(L_i, W_i) \tag{5-38}$$

To illustrate case 6, we can consider any number of hydrologic problems where a single point measurement is used to estimate a spatial average. The criterion variable could be a point rainfall catch, a point snow water equivalent measurement, or a point soil moisture measurement. Specifically, we may wish to use the catch at a rain gage as an estimate of a watershed average, or a point measurement of soil moisture may be used as the best estimate of the average of a field. For purposes of illustration, let's assume that the sample consists of a single measurement of the rainfall depth for a storm event and that the value to be estimated is the mean depth of rainfall over the watershed. The sample value at the point would be the best estimate of the watershed mean. Assuming that the rain gage is located at one corner of the watershed and that the watershed is approximately rectangular, the standard error of estimate can be computed using Eq. 5-37. We will also assume that the process can be represented by a spherical semivariogram with a sill of 0.6 in.^{2} and a radius of influence of 5 miles. Assuming that the watershed is approximately rectangular with a length of 5 miles and a width of 3 miles, Tables 5-4 and 5-7 can be standardized by dividing the length and width dimensions by the radius of influence. The value of $\bar{\gamma}(S, Y)$ equals the product of the sill and the value of $g(L, W)$ obtained from Table 5-8:

$$\bar{\gamma}(S, Y) = \gamma_r g\left(\frac{L}{r}, \frac{W}{r}\right) = 0.6g(\tfrac{5}{5}, \tfrac{3}{5}) = 0.6(0.755) = 0.453 \text{ in.}^2$$

TABLE 5-8 Auxiliary Function for Spherical Model with Range 1.0 and Sill 1.0

L/r \ B/r	0.1	0.2	0.3	0.4	0.5	0.6	0.7	0.8	0.9	1.0
0.10	0.098	0.136	0.178	0.222	0.266	0.309	0.350	0.390	0.428	0.464
0.20	0.164	0.194	0.229	0.268	0.307	0.346	0.385	0.422	0.458	0.491
0.30	0.233	0.257	0.288	0.321	0.356	0.392	0.427	0.462	0.495	0.526
0.40	0.302	0.322	0.348	0.378	0.409	0.441	0.474	0.505	0.535	0.564
0.50	0.368	0.385	0.408	0.434	0.462	0.492	0.521	0.550	0.577	0.603
0.60	0.430	0.445	0.466	0.489	0.515	0.541	0.568	0.594	0.619	0.642
0.70	0.488	0.502	0.520	0.541	0.564	0.588	0.612	0.636	0.658	0.680
0.80	0.542	0.554	0.570	0.589	0.610	0.631	0.653	0.674	0.695	0.714
0.90	0.589	0.600	0.614	0.632	0.650	0.670	0.689	0.708	0.727	0.744
1.00	0.629	0.639	0.653	0.668	0.685	0.703	0.720	0.737	0.754	0.769
1.20	0.691	0.699	0.711	0.723	0.737	0.752	0.767	0.781	0.795	0.808
1.40	0.735	0.742	0.752	0.763	0.775	0.788	0.800	0.812	0.824	0.835
1.60	0.768	0.775	0.783	0.793	0.803	0.814	0.825	0.836	0.846	0.856
1.80	0.794	0.800	0.807	0.816	0.825	0.835	0.845	0.854	0.863	0.872
2.00	0.815	0.820	0.826	0.834	0.842	0.851	0.860	0.869	0.877	0.885
2.50	0.852	0.856	0.861	0.867	0.874	0.881	0.888	0.895	0.902	0.908
3.00	0.876	0.880	0.884	0.889	0.895	0.901	0.907	0.912	0.918	0.923
3.50	0.894	0.897	0.901	0.905	0.910	0.915	0.920	0.925	0.930	0.934
4.00	0.907	0.910	0.913	0.917	0.921	0.926	0.930	0.934	0.938	0.942
5.00	0.926	0.928	0.931	0.934	0.937	0.941	0.944	0.947	0.951	0.954

L/r \ B/r	1.2	1.4	1.6	1.8	2.0	2.5	3.0	3.5	4.0	5.0
0.10	0.526	0.577	0.619	0.653	0.683	0.738	0.777	0.807	0.829	0.861
0.20	0.550	0.598	0.638	0.671	0.698	0.751	0.788	0.816	0.837	0.868
0.30	0.580	0.625	0.662	0.693	0.719	0.768	0.803	0.828	0.848	0.877
0.40	0.614	0.655	0.689	0.718	0.741	0.787	0.819	0.842	0.861	0.887
0.50	0.649	0.687	0.718	0.743	0.765	0.806	0.835	0.857	0.873	0.897
0.60	0.684	0.718	0.746	0.769	0.788	0.825	0.852	0.871	0.886	0.907
0.70	0.717	0.747	0.772	0.793	0.811	0.844	0.867	0.885	0.898	0.917
0.80	0.747	0.774	0.797	0.815	0.831	0.861	0.881	0.897	0.909	0.926
0.90	0.774	0.798	0.818	0.835	0.849	0.875	0.894	0.908	0.919	0.934
1.00	0.796	0.818	0.836	0.851	0.864	0.888	0.905	0.917	0.927	0.941
1.20	0.830	0.848	0.864	0.876	0.886	0.906	0.920	0.931	0.939	0.950
1.40	0.854	0.870	0.883	0.894	0.903	0.920	0.932	0.941	0.948	0.958
1.60	0.873	0.886	0.898	0.907	0.915	0.930	0.940	0.948	0.954	0.963
1.80	0.887	0.899	0.909	0.917	0.924	0.938	0.947	0.954	0.959	0.967
2.00	0.898	0.909	0.918	0.926	0.932	0.944	0.952	0.959	0.963	0.970
2.50	0.918	0.927	0.934	0.940	0.946	0.955	0.962	0.967	0.971	0.976
3.00	0.932	0.939	0.945	0.950	0.955	0.963	0.968	0.972	0.976	0.980
3.50	0.942	0.948	0.953	0.957	0.961	0.968	0.973	0.976	0.979	0.983
4.00	0.949	0.955	0.959	0.963	0.966	0.972	0.976	0.979	0.982	0.985
5.00	0.959	0.964	0.967	0.970	0.973	0.978	0.981	0.983	0.985	0.988

The value of $\bar{\gamma}(Y, Y)$ is also obtained by standardizing the length and width dimensions using the radius of influence. Using Table 5-4, we get

$$\bar{\gamma}(Y, Y) = \gamma_r F\left(\frac{L}{r}, \frac{W}{r}\right) = 0.6F(\tfrac{5}{5}, \tfrac{3}{5}) = 0.6(0.566) = 0.340 \text{ in.}^2$$

Thus, using Eq. 5-37, the standard error is

$$S_e = [2(0.453) - 0.340]^{0.5} = 0.752 \text{ in.}$$

The square root of the sill is 0.775 (in.), which suggests that for the given watershed and semivariogram model, the sample consisting of a single point at one corner of the watershed does not provide a highly accurate estimate of the mean value for the entire watershed.

The standard error would be smaller if the rain gage were located at the center of the watershed. In this case the value of $\bar{\gamma}(Y, Y)$ is the same as the previous case, but the value of $\bar{\gamma}(S, Y)$ will decrease. For this problem the auxiliary function of Table 5-7 can be used by dividing the watershed into four equal rectangles with a length of 2.5 miles and a width of 1.5 miles. Thus the value of $\bar{\gamma}(S, Y)$ is

$$\bar{\gamma}(S, Y) = \frac{\gamma_r}{4}\left[\sum_{i=1}^{4} g\left(\frac{L_i}{r}, \frac{W_i}{r}\right)\right] = \gamma_r g\left(\frac{2.5}{5}, \frac{1.5}{5}\right) = 0.6(0.445) = 0.267 \text{ in.}^2$$

Therefore, the standard error is

$$S_e = [2\bar{\gamma}(S, Y) - \bar{\gamma}(Y, Y)]^{0.5} = [2(0.267) - 0.340]^{0.5} = 0.440 \text{ in.}$$

Thus the standard error decreased by about 41% when the rain gage was located at the center of the watershed, compared with the location at one corner of the watershed.

Case 7: Field/Vector Estimation

If the sample consists of n points rather than just the single point of case 6, the mean of the n sample values can be used as an estimate of the average value of a field of length L and width W. The estimate of the reliability is obtained by extending the concepts of cases 2 and 6. The standard error is given by

$$S_e = [2\bar{\gamma}(S, Y) - \bar{\gamma}(Y, Y) - \bar{\gamma}(S, S)]^{0.5} \tag{5-39}$$

The first two terms are computed in the same way as with case 6. Computation of the third term $\bar{\gamma}(S, S)$ follows the method outlined in case 2 for a vector of points.

Using the rainfall problem of case 6, let's assume that there is a rain gage located at each corner of the rectangular watershed, with storm event catches of 2.6, 3.1, 2.8, and 3.7 in. Using the mean value weighted according to the area it represents (i.e., the Theissen polygon method), the estimate of the average watershed depth is 3.05 in.

Equation 5-39 is used to estimate the reliability. The value of $\bar{\gamma}(S, Y)$ equals the average semivariogram value for the points and the spatial mean. This equals one-fourth of the sum of the auxiliary function value for each sample point and the watershed mean:

$$\bar{\gamma}(S, Y) = \tfrac{1}{4}[\bar{\gamma}(S_1, Y) + \bar{\gamma}(S_2, Y) + \bar{\gamma}(S_3, Y) + \bar{\gamma}(S_4, Y)]$$

$$= \frac{1}{4}\sum_{i=1}^{4} \gamma_r F_i(L, W)$$

Since each point is at a corner of the watershed,

$$\bar{\gamma}(S, Y) = \gamma_r F(L, W) = 0.6F(\tfrac{5}{5}, \tfrac{3}{5}) = 0.6(0.755) = 0.453(\%)^2$$

Since the criterion has not changed, $\bar{\gamma}(Y, Y)$ still equals 0.340. The term $\bar{\gamma}(S, S)$ equals the average semivariogram value between each of the sample points; there would be 16 such pairs:

$$\gamma(S_1, S_1) = \gamma(S_2, S_2) = \gamma(S_3, S_3) = \gamma(S_4, S_4) = 0$$

$$\gamma(S_1, S_2) = \gamma(S_2, S_1) = \gamma(S_3, S_4) = \gamma(S_4, S_3) = \gamma_r\left[\frac{3(3)}{2(5)} - \frac{(3)^3}{2(5)^3}\right] = 0.475$$

$$\gamma(S_1, S_3) = \gamma(S_3, S_1) = \gamma(S_2, S_4) = \gamma(S_4, S_2) = \gamma_r = 0.6$$

$$\gamma(S_1, S_4) = \gamma(S_4, S_1) = \gamma(S_2, S_3) = \gamma(S_3, S_2) = \gamma_r = 0.6$$

Therefore, the term $\bar{\gamma}(S, S)$ is given by

$$\bar{\gamma}(S, S) = \tfrac{1}{16}[4(0) + 4(0.475) + 4(0.6) + 4(0.6)] = 0.419 \text{ in.}^2$$

and the standard error is

$$S_e = [2(0.453) - 0.340 + 0.419]^{0.5} = 0.362 \text{ in.}$$

In comparison to the standard error of estimate for the single sample point located at the center of the watershed, locating four gages at the corners of the watershed has not caused a significant decrease in the standard error. This example shows the importance of location in siting the sample points.

Case 8: Field/Line Estimation

For case 8 the sample consists of the values along a line, while the criterion is the average values in a field. Again, the average value of the points within the line is the linear sample estimate and can be used as an estimate of the field average. The reliability can be estimated using Eq. 5-39. The value of $\bar{\gamma}(Y, Y)$ can be obtained using the standardized auxiliary functions of Tables 5-3 and 5-4. The value of $\bar{\gamma}(S, S)$ is computed using the auxiliary functions of Eqs. 5-29 to 5-31. The value of $\bar{\gamma}(S, Y)$ equals the average semivariogram value for each point on the line and each point within the field. This can be evaluated using the auxiliary function of Table 5-8. These components can be used with Eq. 5-39 to estimate the reliability.

ESTIMATION BY KRIGING

Semivariogram analysis is not an end in itself; it is intended to be used as part of the estimation or prediction process. Two questions about the estimation problem were posed previously. Specifically, we needed to know which sample points should be used in estimating the value at a point where the criterion was not known and what weight should be given to each sample point. The concepts introduced with respect to the semivariogram should provide an obvious answer. Given that we know the radius of influence that describes the process, it should seem reasonable that only sample points located within the radius of influence of the unknown point should be used, and the weight given to each sample point should be inversely proportional to the ordinate of the semivariogram corresponding to the distance separating the sample and unknown point.

To formulate a solution to the problem, it is important to recall the four requirements in statistical modeling: (1) an estimation model; (2) an objective function that defines "best" fit; (3) constraints, when necessary, that place limitations on the solution; and (4) a data base. From Eq. 5-1 it is evident that the problem is being formulated with a linear estimation model in which the w_i serve as unknowns and the Y_i are the sample measurements (i.e., the data base). As discussed previously, in statistical modeling "best" is often taken to imply that the error, or estimation, variance is a minimum. Thus we have as the objective to minimize the error variance. But for the kriging solution, if we want an unbiased model we must impose the constraint that the sum of the weights, w_i, equals 1:

$$\sum_{i=1}^{n} w_i = 1 \tag{5-40}$$

in which n is the number of sample points used to make an estimate. The resulting values of w_i will thus be classed as *best linear unbiased estimators* (BLUE).

It can be shown that the estimation variance, which will be denoted as σ_e^2 or S_e^2 for the population and sample, respectively, depends on the values of the unknown weights, the structure and magnitude of the semivariogram, the location and magnitude of the sample points, and the type of estimation to be made (i.e., point, core length, field, or volumetric). We can minimize the error variance by taking derivatives of the objective function with respect to each unknown and setting the derivatives equal to zero; this provides a set of n equations with n unknowns. While the solution of these n "normal" equations would produce a minimum error variance, the resulting model would not be unbiased. For this we must include the constraint of Eq. 5-40 in the solution procedure. Thus our objective function is to minimize

$$S_e^2 - \lambda\left[\left(\sum_{i=1}^{n} w_i\right) - 1\right] \tag{5-41}$$

in which λ is an unknown. It should be apparent that the solution procedure for kriging is an example of Lagrangian optimization, with λ being the Lagrangian multiplier. There are $(n + 1)$ unknowns (i.e., the n values of w_i and λ), and there are $(n + 1)$ equations (i.e., the n derivatives of Eq. 5-41 with respect to each w_i and Eq. 5-40). Thus we only need an expression for estimating σ_e^2 to find the solution.

If we seek a solution to estimating the value at a point, whether the sample points are distributed linearly in space or time or spatially as in Fig. 5-1, an estimate of the error variance can be made by

$$S_e^2 = 2\sum_{i=1}^{m} w_i\bar{\gamma}(S_i, Y) - \sum_{i=1}^{m}\sum_{j=1}^{m} w_i w_j \bar{\gamma}(S_i, S_j) - \bar{\gamma}(Y, Y) \tag{5-42}$$

in which S_i is the ith sample element, Y indicates the value of the criterion where the estimate is needed, and $\bar{\gamma}(C_i, C_j)$ is the average semivariogram value between all combinations of C_i and C_j, where C_i and C_j are dummy variables and may be either S_i or Y. Equation 5-42 indicates that the error variance consists of three parts. The first term represents the variation associated with differences between the sample S_i and the criterion Y for which a value is needed. The second term reflects the variation within the sample; that is, the average semivariogram value for all elements of the sample $(\bar{\gamma}(S_i, S_j))$ reflects variation that is not error variation, so it must be subtracted from the total expected variation between the sample and the unknown value of the criterion. The third term is similar to the second term in that it represents variation that is not error variation, yet it contributes to the total variation between the sample elements and the unknown value of the criterion. For a system in which there is no nugget effect and a single point of interest, the average semivariogram value for a separation distance of zero must also be zero. This will not be true when dealing with estimation within either a linear, field, or volume. The subtraction of the two terms indicates that we must reduce the error variation because we are interested in a mean value (i.e., the mean of all future estimates).

It is important to compare Eq. 5-42 with Eq. 5-39. Equation 5-39 is valid for estimates when the arithmetic mean is used (i.e., Eq. 5-2). When Eq. 5-1 is used to estimate the value of the criterion, Eq. 5-42 is the appropriate form to use to estimate the standard error. The structures are identical, but Eq. 5-42 includes the weights applied to the sample points. Weights are not applied to $\bar{\gamma}(Y, Y)$ because the value of this source of variation is not influenced by sample elements. For the term of Eq. 5-42 involving $\bar{\gamma}(S_i, S_j)$, the weighting depends on both i and j, so two weights are included. Since the variance $\bar{\gamma}(S_i, Y)$ for any i depends only on S_i, only the weight corresponding to S_i is included.

Having formulated the objective function (Eq. 5-42) the optimal values of the w_i and λ can be obtained by Lagrangian optimization. Specifically, Eq. 5-42 can be differentiated with respect to each of the unknowns and each derivative set equal to zero. From these derivatives we can derive the "normal" equations by algebraic manipulation:

$$\lambda + \sum_{j=1}^{n} w_j \bar{\gamma}(S_1, S_j) = \bar{\gamma}(S_1, Y) \tag{5-43a}$$

$$\lambda + \sum_{j=1}^{n} w_j \bar{\gamma}(S_2, S_j) = \bar{\gamma}(S_2, Y) \tag{5-43b}$$

$$\vdots$$

$$\lambda + \sum_{j=1}^{n} w_j \bar{\gamma}(S_n, S_j) = \bar{\gamma}(S_n, Y) \tag{5-43c}$$

$$\sum_{j=1}^{n} w_j = 1 \tag{5-43d}$$

As an example, if the sample consists of three points, Eqs. 5-43 are

$$\lambda + w_1 \bar{\gamma}(S_1, S_1) + w_2 \bar{\gamma}(S_1, S_2) + w_3 \bar{\gamma}(S_1, S_3) = \bar{\gamma}(S_1, Y) \tag{5-44a}$$

$$\lambda + w_1 \bar{\gamma}(S_2, S_1) + w_2 \bar{\gamma}(S_2, S_2) + w_3 \bar{\gamma}(S_2, S_3) = \bar{\gamma}(S_2, Y) \tag{5-44b}$$

$$\lambda + w_1 \bar{\gamma}(S_3, S_1) + w_2 \bar{\gamma}(S_3, S_2) + w_3 \bar{\gamma}(S_3, S_3) = \bar{\gamma}(S_3, Y) \tag{5-44c}$$

$$w_1 + w_2 + w_3 = 1 \tag{5-44d}$$

Equations 5-43 represent a set of $(n + 1)$ simultaneous equations with $(n + 1)$ unknowns, which can be solved either analytically or numerically. The solution provides the weights that yield the minimum error variance as defined by Eq. 5-42.

Example 5-1: Estimation of a Groundwater Quality Indicator

Assume that data from previous investigations indicate that the semivariogram for a certain groundwater quality indicator has a sill of 4 (ppm)^2, a radius of influence of 2.5 km, and can be represented by a spherical model. Thus the model is

$$\gamma(h) = \begin{cases} 2.4h - 0.128h^3 & \text{for } h \le r \quad (5\text{-}45a) \\ 4 & \text{for } h > r \quad (5\text{-}45b) \end{cases}$$

in which h is the separation distance in km.

At a specific location, an engineer is interested in the value of the water quality indicator at a point. Samples from three wells in the region yield values as shown in Table 5-9. The location of the wells and the unknown point A are shown in Fig. 5-12. Table 5-9 shows the separation distances, and the values of the semivariogram are

TABLE 5-9 Data for Example 5-1

Point	North	East	Y_i	Separation Distance (km)			
				A	1	2	3
A	0	0	—	0	—	—	—
1	0	−1	16.3	1	0	—	—
2	2	2	13.7	$2\sqrt{2}$	$\sqrt{13}$	0	—
3	0	2	12.4	2	3	2	0

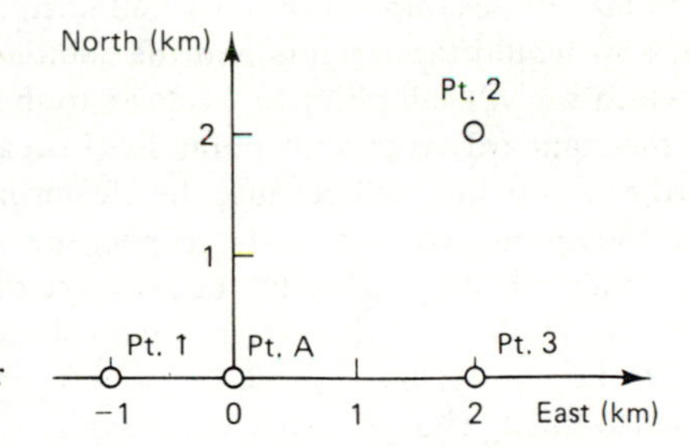

Figure 5-12 Location of points for Example 5-1.

TABLE 5-10 Semivariogram Values for Example 5-1

Point	A	1	2	3
A	0	—	—	—
1	2.272	0	—	—
2	4.000	4	0	—
3	3.776	4	3.776	0

given in Table 5-10. The values of Table 5-10 can be used to set up the normal equations (Eqs. 5-44) for the kriging solution:

$$\lambda + 0w_1 + 4w_2 + \quad 4w_3 = 2.272 \tag{5-46a}$$

$$\lambda + 4w_1 + 0w_2 + 3.766w_3 = 4 \tag{5-46b}$$

$$\lambda + 4w_1 + 3.776w_2 + 0w_3 = 3.776 \tag{5-46c}$$

$$w_1 + \quad w_2 + \ w_3 = 1 \tag{5-46d}$$

Equations 5-46 contain four unknowns, w_1, w_2, w_3, and λ. Solving Eqs. 5-46 yields the following: $w_1 = 0.610$, $w_2 = 0.165$, $w_3 = 0.225$, and $\lambda = 0.712$. Thus the estimate at point A is

$$\hat{Y}_A = 0.61(16.3) + 0.165(13.7) + 0.225(12.4) = 15.0 \text{ ppm} \tag{5-47}$$

The standard error of estimate can be computed from Eq. 5-35:

$$\begin{aligned} S_e^2 &= 2\sum w_i\bar{\gamma}(S_i, Y) - \sum\sum w_i w_j\bar{\gamma}(S_i, S_j) - \bar{\gamma}(Y, Y) \\ &= 2[0.61(2.272) + 0.165(4) + 0.225(3.776)] \\ &\quad - [2(0.61)(0.165)(4) + 2(0.61)(0.225)(4) \\ &\quad + 2(0.165)(0.225)(3.776)] - 0 \\ &= 5.791 - 2.184 - 0 = 3.607\ (\text{ppm})^2 \end{aligned} \tag{5-48}$$

Therefore, the standard error is 1.90 ppm.

It should be of interest to consider the sensitivity of the solution to the location of the points. For example, if point 1 had north/east coordinates of (0, 1), rather than (0, −1), how would the weights and the standard error change? Given that the new point, which we will call point 1a, is closer to the other points in the sample, we would expect the standard error with point 1a (instead of point 1) to be greater than the standard error for Eq. 5-48 because the clustering of points reduces the independence of the information. We would also expect the weight of point 3 to decrease because its importance is being masked by the presence of point 1a between it and the unknown point A. An analysis of the system involving sample points 1a, 2, and 3 shows the weights to be $w_1 = 0.704$, $w_2 = 0.273$, and $w_3 = 0.023$. However, the standard error decreased to 1.830. Thus, as expected, the value of w_3 decreased, with a large proportion of the change in weight going to w_2 rather than w_1. Thus the increase in the variation within the sample was not sufficient to offset the decrease in variation $\bar{\gamma}(S, Y)$ associated with the change in weights. This example demonstrates the importance of accounting for interaction between sample points when assessing the location of samples.

Returning to the original point location of Table 5-9, is the kriging estimate better than the mean of the three points? For an unweighted mean, each point would be given a weight of $\frac{1}{3}$. Thus the standard error would be

$$\begin{aligned} S_e^2 &= 2[\tfrac{1}{3}(2.272 + 4 + 3.776)] - 2(\tfrac{1}{3})^2(4 + 4 + 3.776) \\ &= 4.082\ (\text{ppm})^2 \end{aligned} \tag{5-49}$$

Thus the standard error is 2.02 ppm, which is greater than the kriging error of 1.90 ppm.

Example 5-2: Estimation of Manning's Roughness Coefficient

Estimating the value along a line has important applications in hydrology. Let's illustrate the line/vector estimation problem by estimating the average value of Manning's roughness coefficient for a channel reach using a sample consisting of three point estimates. The 6000-ft reach is shown in Fig. 5-13. The separation distances are shown in Table 5-11. Assuming that past data analyses have shown that the system

TABLE 5-11 Separation Distance (ft) Table for Example 5-2

Point	Y_1	Y_2	Y_3
Y_1	0	—	—
Y_2	1500	0	—
Y_3	4500	3000	0

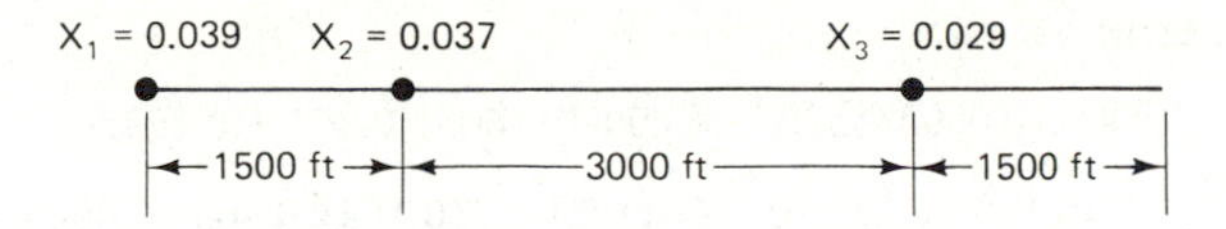

Figure 5-13 Sample-point configuration for Example 5-2.

could be represented by an exponential semivariogram with a sill of 0.000025 and a radius of influence of 2500 ft, we have

$$\gamma(h) = 0.000025\left[1 - \exp\left(\frac{-h}{2500}\right)\right] \tag{5-50}$$

Using the separation distances of Table 5-11, we get the semivariogram values of Table 5-12. The values of $\bar{\gamma}(S_i, Y)$ are computed by Eq. 5-30:

$$\bar{\gamma}(S_1, Y) = 0.000025[1 - \tfrac{2500}{6500}(1 - \exp(-\tfrac{6000}{2500}))] = 0.000015528 \tag{5-51}$$

$$\bar{\gamma}(S_2, Y) = \tfrac{1}{4}(0.000025)[1 - \tfrac{2500}{1500}(1 - \exp(-\tfrac{1500}{2500}))]$$
$$+\tfrac{3}{4}(0.000025)[1 - (\tfrac{2500}{4500})[1 - \exp(-\tfrac{4500}{2500})]\} = 0.000011605 \tag{5-52}$$

$$\bar{\gamma}(S_3, Y) = \gamma(S_3, Y) = 0.000011605 \tag{5-53}$$

TABLE 5-12 Semivariogram Table for Example 5-2

Point	Y_1	Y_2	Y_3
Y_1	0	—	—
Y_2	0.00001128	0	—
Y_3	0.00002087	0.00001747	0

This results in the following set of normal equations:

$$\lambda + 0w_1 + 0.00001128w_2 + 0.00002087w_3 = 0.00001553 \tag{5-54a}$$

$$\lambda + 0.00001128w_1 + 0w_2 + 0.00001747w_3 = 0.000011605 \tag{5-54b}$$

$$\lambda + 0.00002087w_1 + 0.00001747w_2 + 0w_3 = 0.000011605 \tag{5-54c}$$

$$w_1 + w_2 + w_3 = 1 \tag{5-54d}$$

Solving the set of four simultaneous equations for the four unknowns (w_1, w_2, w_3, and λ) yields the following values: $w_1 = 0.164$, $w_2 = 0.373$, $w_3 = 0.463$, and $\lambda = 0.0000016634$. The "best" estimate of the channel average would be

$$\bar{n} = 0.164(0.039) + 0.373(0.037) + 0.463(0.029) = 0.0336 \tag{5-55}$$

To compute the standard error, we obtain the value of $\bar{\gamma}(Y, Y)$ using the function $F(L)$ of Table 5-2:

$$\bar{\gamma}(Y, Y) = 0.000025\{1 - 2(\tfrac{2500}{6000}) + 2(\tfrac{2500}{6000})^2[1 - \exp(-\tfrac{6000}{2500})]\}$$
$$= 0.00001206 \tag{5-56}$$

Therefore, the error variance is given by

$$S_e^2 = 2[0.164(0.000015528) + (0.000011605)(0.373 + 0.463)]$$
$$- [2(0.164)(0.373)(0.00001128) + 2(0.164)(0.463)(0.00002087)$$
$$+ 2(0.373)(0.463)(0.00001747)]$$
$$- 0.00001206 = 0.000024497 - 0.00001058 - 0.00001206$$
$$= 0.000001857 \tag{5-57}$$

Thus the standard error for the estimate given by Eq. 5-55 is 0.00136.

Example 5-3: Estimation of Field Soil Moisture

The sample-point configuration of Fig. 5-14 represents three point soil-moisture samples in a plot measuring 120 ft by 120 ft. The objective is to estimate the average soil moisture for the plot and a measure of the accuracy of the estimate. Table 5-13 gives the

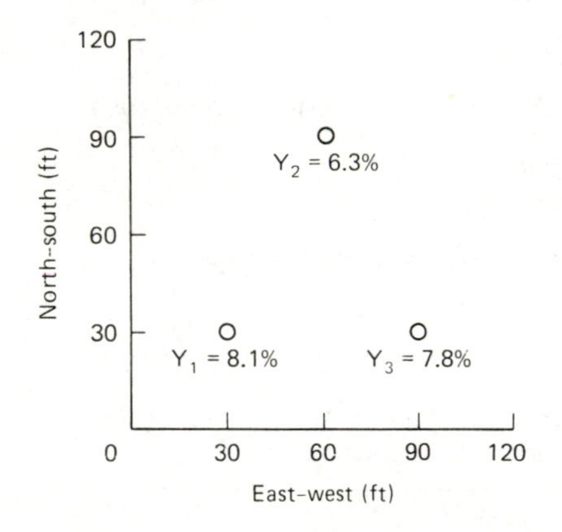

Figure 5-14 Sample-point configuration for Example 5-3.

TABLE 5-13 Separation Distances (ft) for Sample Points of Example 5-3

Point	1	2	3
1	0	—	—
2	50	0	—
3	60	50	0

separation distances for the sample points. Assume that past experience has indicated that soil moisture can be represented by a spherical semivariogram with a sill of 6.8(%)^2 and a radius of influence of 150 ft:

$$\gamma(h) = \begin{cases} 0.068h - \dfrac{6.8h^3}{6,750,000} & \text{for } h \leq 150 \text{ ft} \quad (5\text{-}58a) \\ 6.8 & \text{for } h > 150 \text{ ft} \quad (5\text{-}58b) \end{cases}$$

The semivariogram values for the separation distances of Table 5-13 are given in Table 5-14.

TABLE 5-14 Semivariogram Values (%)^2 for Sample Points of Example 5-3

Point	1	2	3
1	0	—	—
2	3.274	0	—
3	3.8624	3.274	0

To compute the weights, the values of $\bar{\gamma}(S, Y)$ must be computed using the values for the standardized spherical model in Table 5-7. For each of the three sample points, it is necessary to divide the field into four rectangles and compute weighted values of the sample-criterion variances:

$$\bar{\gamma}(S_1, Y) = 6.8\left[\frac{30(40)}{(120)^2} g(40, 30) + \frac{30(80)}{(120)^2} g(80, 30) + \frac{40(90)}{(120)^2} g(90, 40) + \frac{90(80)}{(120)^2} g(90, 80)\right] = 6.8[0.0833g(0.267, 0.2) + 0.167g(0.533, 0.2) + 0.25g(0.6, 0.267) + 0.5g(0.6, 0.533)] = 6.8[0.0833(0.266) + 0.167(0.425) + 0.25(0.488) + 0.5(0.594)] = 6.8(0.512) = 3.4816 \quad (5\text{-}59)$$

$$\bar{\gamma}(S_2, Y) = 6.8\left[\frac{2(40)(60)}{(120)^2} g(60, 40) + \frac{2(80)(60)}{(120)^2} g(80, 60)\right]$$

$$= 6.8[\tfrac{1}{3} g(0.4, 0.267) + \tfrac{2}{3} g(0.533, 0.4)]$$

$$= 6.8[\tfrac{1}{3}(0.376) + \tfrac{2}{3}(0.506)] = 3.1484 \quad (5\text{-}60)$$

$$\bar{\gamma}(S_3, Y) = \bar{\gamma}(S_1, Y) = 3.4816 \quad (5\text{-}61)$$

The values of Eqs. 5-59 to 5-61 and Table 5-14 provide the basis for the following set of normal equations:

$$\lambda + 0w_1 + 3.274w_2 + 3.8624w_3 = 3.4816 \quad (5\text{-}62a)$$

$$\lambda + 3.274\, w_1 + 0w_2 + 3.274\, w_3 = 3.1484 \quad (5\text{-}62b)$$

$$\lambda + 3.8624w_1 + 3.274w_2 + 0w_3 = 3.4816 \quad (5\text{-}62c)$$

$$w_1 + w_2 + w_3 = 1 \quad (5\text{-}62d)$$

Solving Eqs. 5-62 yields the following values for the unknowns: $w_1 = w_3 = 0.3185$, $w_2 = 0.363$, and $\lambda = 1.063$. Thus the weighted mean soil moisture for the field is

$$\bar{Y} = 0.3185(8.1 + 7.8) + 0.363(6.3) = 7.35\% \quad (5\text{-}63)$$

In order to use Eq. 5-35 to estimate variance, the value of $\bar{\gamma}(Y, Y)$ must be computed using the values of Table 5-4:

$$\bar{\gamma}(Y, Y) = 6.8F(120, 120) = 6.8F(\tfrac{120}{150}, \tfrac{120}{150}) = 6.8(0.564) = 3.835 \quad (5\text{-}64)$$

Thus the error variance is

$$\begin{aligned} S_e^2 &= 2[2(0.3185)(3.4816) + 0.363(3.148)] \\ &\quad - [4(0.3185)(0.363)(3.274) + 2(0.3185)^2(3.8624)] - 3.835 \\ &= 6.721 - 2.298 - 3.835 \\ &= 0.5883 \end{aligned} \tag{5-65}$$

Thus the estimate of Eq. 5-63 has a standard error of 0.767%.

EXERCISES

5-1. Samples of the suspended sediment load (L) are taken at 1000-ft intervals along a stream. Compute the semivariogram for the given data for separation distances of up to and including 5000 ft. Assuming a spherical model, find the values of the sill and radius of influence that provide the best fit to the data.

L (ppm): 562 573 546 552 538 537 547 559 543

5-2. Estimates of the saturated hydraulic conductivity (K_s) are made at 100-ft intervals on a hillslope. Compute the semivariogram for the given data for separation distances from 100 to 600 ft. Assuming an exponential model, find the best estimates of the sill and radius of influence.

K_s (cm/hr):
0.211 0.234 0.226 0.251 0.257 0.239 0.246 0.228 0.236 0.225

5-3. Using the data of Exercise 5-2, fit a spherical model that includes a nugget effect. Find γ_r, γ_n, and r that yield the best fit for the sample values of $\gamma(h)$, with separation distances up to 600 ft.

5-4. Snowpack measurements are taken on a 50-ft grid, with the snow water equivalent values shown below. Assuming isotropic conditions, determine the semivariogram.

$$\begin{array}{cc} 20.4 & 21.2 \\ 19.9 & 19.5 \\ 20.7 & 21.3 \end{array}\;\begin{array}{l} 19.7 \\ \left.\begin{array}{l} 20.6 \\ 20.1 \end{array}\right\} 50\text{ ft} \end{array}$$

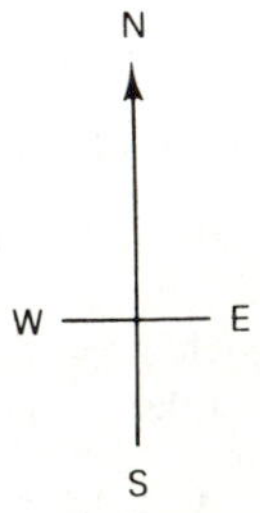

5-5. Using the snow water equivalent values of Exercise 5-4, find the semivariogram for the **(a)** north–south and **(b)** east–west directions.

5-6. Using the soil-moisture data shown on a square grid, find the semivariogram assuming that isotropic conditions exist.

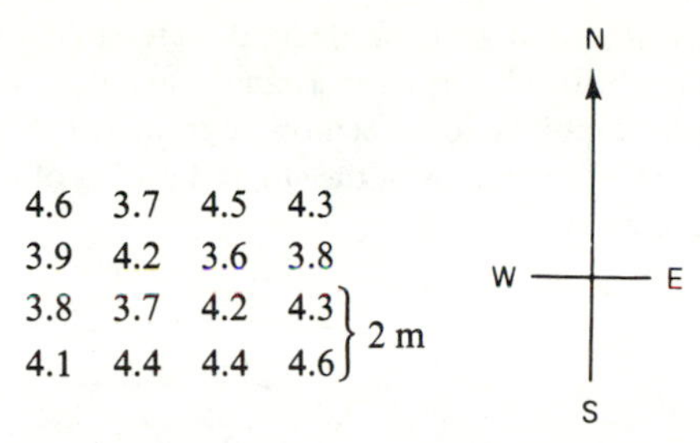

5-7. Using the soil-moisture data given in Exercise 5-6, find the semivariograms for the **(a)** NW–SE and **(b)** SW–NE directions, assuming anisotropic conditions.

5-8. Macroinvertebrate samplings were made along the bottom of a streambed, with samples taken on a rectangular grid. The Shannon–Weaver index of species diversity was determined for each sample. Assuming that each value is representative of a point, derive the semivariogram for the diversity index. Comment on whether or not isotropic conditions exist. For the data shown, three point samples are missing due to mishandling of the data; these points are denoted by X.

6 ft (spacing between columns); 4 ft (spacing between rows)

1.6	2.5	3.1	X	2.6	2.2
X	1.3	2.4	3.2	1.9	2.4
2.4	3.2	2.2	2.5	2.5	X

5-9. For the sample points computed for Exercise 5-4, fit a linear semivariogram using regression analysis.

5-10. For the sample points computed for Exercise 5-6, fit a spherical semivariogram by assuming the sill equals one-half the sample variance and the radius of influence equals 1.5 times the point where a line drawn through the first two points of the sample intersects the sill. Comment on the accuracy of fit and the appropriateness of the spherical model.

5-11. For the sample points computed for Exercise 5-7(a), fit an exponential semivariogram by eye. Comment on the appropriateness of the model and the effect of sample size.

5-12. Field tests on a 300-m square grid at the site of a landfill indicate the levels of chromium (ppm) contamination shown below. Compute semivariograms for both the north–south and east–west directions. Fit semivariograms using a spherical model for both directions.

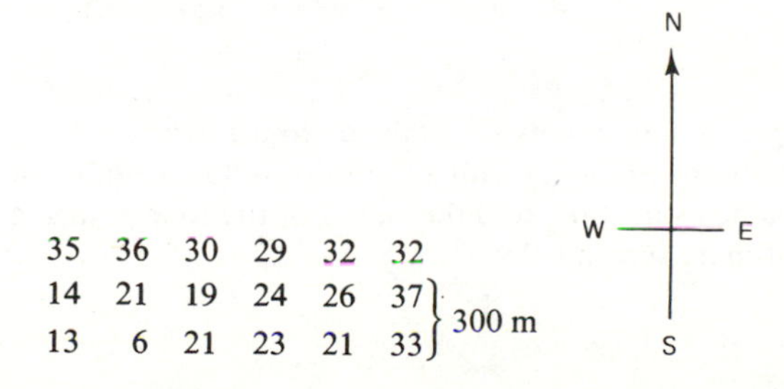

5-13. Raingages are set up at an experimental watershed with a grid spacing of 1 km. The total storm rainfall (in.) for a single event at each gage is shown below. Compute sample points for a semivariogram using separation distances up to 3 km. Find the best estimates of the sill and radius of influence for an exponential semivariogram model.

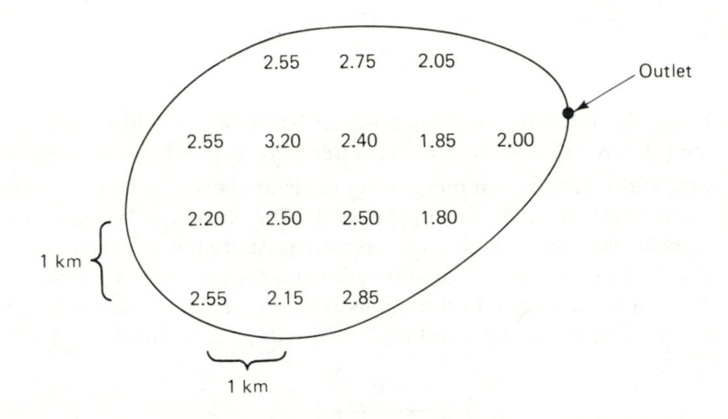

5-14. Using the data of Exercise 5-13, compute semivariograms to check for anisotropic conditions.

5-15. A sample consists of a single point, which has a value of 2.7. The sample is obtained from a process whose population is characterized by the following semivariogram:

$$\gamma(h) = 0.8\left[1 - \exp\left(\frac{h}{4}\right)\right]$$

Determine both the best estimate of the variable and the accuracy of the estimate for a point separated from the sample point by a distance of 1.6.

5-16. A sample of three points is obtained from the process of Exercise 5-15. The spatial distribution of the points is shown below. Using the arithmetic mean of the sample points as the best estimate at point A, find the accuracy of the estimated value.

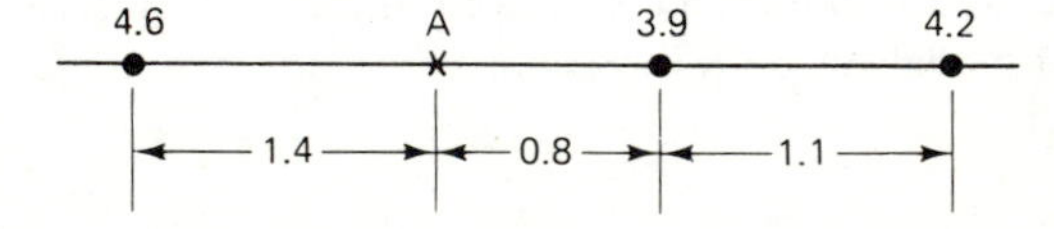

5-17. A sample of four points is obtained from a process that is characterized by a spherical semivariogram with a radius of influence of 6 and a sill of 0.04. Using mean value estimation, find the values of the four points A_i and the accuracy of each estimate shown below.

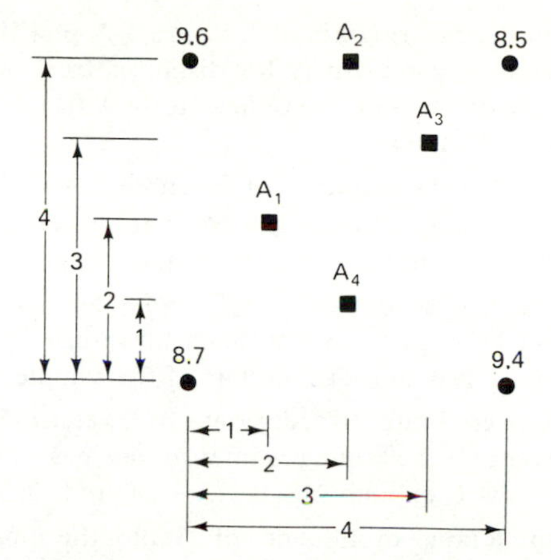

5-18. Using the data of Exercise 5-17, draw isolines within the grid defined by the four sample points of equal levels of accuracy as defined by the standard error. Assume that equal weight is given to each sample point.

5-19. When estimating the mean value for a line of length L using the value of a single sample point, the accuracy of the estimate depends on the underlying semivariogram and the location of the sample point on the line. Compute and plot the accuracy of the estimate (S_e) as the sample point moves from the left end of the line to the right end.

5-20. Determine the standard error of an estimated mean value of a line of length L when the sample consists of a single point located a distance L above the center of the line. Assume an exponential semivariogram with unit sill and a radius of influence of **(a)** L; **(b)** $2L$.

5-21. Assume that the sample consists of a single point and the process can be represented by a spherical semivariogram with a unit sill and a radius of influence of L. If the value to be estimated is the mean value of a line, compute and plot the standard error of estimate as a function of the distance between the sample point and the left end of the line, as the sample moves away from and parallel to the line. Show the curve for the separation distance between the left end of the line and the sample point from 0 to $2L$.

5-22. Compute the standard error for an estimated mean value of a line when the sample consists of two points, which are located at the one-third and two-thirds points on the line, which has length L. Assume that the process can be represented by an exponential semivariogram with a unit sill and a radius of influence of L.

5-23. Using the data and computed semivariogram of Exercise 5-1, find the kriging estimate of the suspended sediment load at a distance downstream of 3500 ft (assume that the sample point on the left is at river location 0 ft). Also, determine the standard error of the estimate.

5-24. Using the data and semivariogram of Exercise 5-2, compute the kriging estimate of the saturated hydraulic conductivity at a distance of 450 ft from the point on the left. Also, determine the standard error of the estimate.

5-25. Using the data and semivariogram of Exercise 5-2, plot the kriging estimate of the saturated hydraulic conductivity for distances from 300 to 400 ft along the hillslope (the point on the left is assumed to be 0 ft). Also, plot the standard error of the estimated values.

5-26. Using the snowpack measurements of Exercise 5-4 and the computed semivariogram, find the kriging estimate of a point at the center of the square formed by the four points in the upper left of the sample field.

5-27. Using the snowpack measurements of Exercise 5-4 and the computed semivariogram, find the kriging estimate of the mean value of a 100-ft line connecting the two points in the NW and NE corners of the sample field.

5-28. Using the macroinvertebrate diversity data of Exercise 5-8 and the computed semivariogram, compute the kriging estimate of the mean value of the area formed by the points of (row 1, column 2), (1, 3), (2, 2), and (2, 3).

5-29. For a sill of 10 and a range of influence of 25, plot the spherical and exponential semivariograms. Find a range of influence for the exponential model that will provide a semivariogram that has approximately the same location and scale as the spherical model with a sill of 10 and a range of influence of 25.

5-30. For the example given with case 3, compute and plot the standard error as a function of the location of the sample point, with the point varying from the center of the line to the end of the line and beyond the end of the line a distance equal to the length of the line.

5-31. Assuming a spherical semivariogram with a sill of 5 and a range of influence of 10, compute and plot the standard error for case 5 for values of the separation distance (L) from $B/2$ to $2B$, where B is the lengths of the two parallel one-dimensional element.

5-32. Assuming a standardized spherical semivariogram ($\gamma_r = r = 1$) plot lines of equal value of the standard error as the location of a single sample point in a field varies over a square field (case 6).

REFERENCES AND SUGGESTED READINGS

CHIRLIN, G. R., and G. DAGAN, Theoretical Head Variograms for Steady Flow in Statistically Homogeneous Aquifers, *Water Resources Research*, Vol. 16, No. 6, pp. 1001–15, 1980.

CHIRLIN, G. R., and E. F. WOOD, On the Relationship between Kriging and State Estimation, *Water Resources Research*, Vol. 18, No. 2, pp. 432–38, 1982.

CLIFTON, P. M., and S. P. NEWMAN, Effects of Kriging and Inverse Modeling on Conditional Simulation of the Avra Valley Aquifer in Southern Arizona, *Water Resources Research*, Vol. 18, No. 4, pp. 1215–34, 1982.

DELHOMME, J. P., Spatial Variability and Uncertainty in Groundwater Flow Parameters: A Geostatistical Approach, *Water Resources Research*, Vol. 15, No. 2, pp. 269–80, 1979.

DUNLAP, L. E., and J. M. SPINAZOLA, *Interpolating Water-Table Altitudes in West-Central Kansas Using Kriging Techniques*, U.S. Geological Survey Open-File Report, OF 81-1062, 1981.

HUGHES, J. P., and D. P. LETTENMAIER, Data Requirements for Kriging: Estimation and Network Design, *Water Resources Research*, Vol. 17, No. 6, pp. 1641–50, 1981.

JOURNAL, A., and C. HUIJBREGTS, *Mining Geostatistics*, Academic Press, New York, 1978.

MIZWELL, S. A., A. L. GUTJAHR, and L. W. GELHAR, Stochastic Analysis of Spatial Variability in Two-Dimensional Steady Groundwater Flow Assuming Stationary and Nonstationary Heads, *Water Resources Research*, Vol. 18, No. 4, pp. 1053–68, 1982.

PALUMBO, M. R., and R. KHOLECL, Kriged Estimates of Transmissivity in the Mesilla Bolson, New Mexico, *Water Resources Research*, Vol. 19, No. 6, pp. 929–36, 1983.

ROGOWSKI, A. S., Use of Geostatistics to Evaluate Water Quality Changes due to Coal Mining, *Water Resources Research*, Vol. 19, No. 6, pp. 983–92, 1983.

RUSSO, D., A Geostatistical Approach to the Trickle Irrigation Design in Heterogeneous Soil, *Water Resources Research*, Vol. 19, No. 2, pp. 632–42, 1983.

6 Statistical Optimization

INTRODUCTION

The methods of Part II dealt with only one variable; thus the methods were called univariate. These models involved statistical frequency distributions. After the parameters of univariate models are evaluated by some method, such as moments of maximum likelihood, they can be used for prediction. However, such prediction has to be limited to likely fluctuations around the mean value. In this chapter we begin a study of models having more than one variable. Multiple-variable models begin the study of functional relationships, which form a deterministic core of statistical models.

Functional relationships in statistical hydrologic models can be either of two kinds. The functions may express an explicit causal relation, or they may express an associative relation. In both kinds of functions, any empirical coefficients must be evaluated by identical statistical procedures. *Causal relations* are always the preferred basis for modeling. In these relations a dependent or criterion variable, denoted as Y, takes on values caused by variations in one or more independent or predictor variables, denoted by X. A simple rainfall–runoff relationship expresses causal relation. Runoff from a watershed is caused by rainfall input to the watershed. The resulting model is an example of macroscale cause and effect. Other causal models may result from microscale concepts, expressing very detailed interactions between soil, water, and vegetation. Circumstances external to the pure cause and effect usually control the choice between macroscale and microscale concepts. The lack of detailed data and the complexity of real watersheds are the usual reasons for choosing macroscale models.

In *associative models* the variables respond to some external cause. The variable Y is now not dependent on X, but both Y and X vary in a similar manner to the external cause. For example, consider two meteorological stations where air temperature is observed. If these stations are reasonably close, the temperatures at both should be similar. Low temperatures at one will be associated with low temperatures at the other, and high temperatures at one will associate with high temperatures at the other. A functional relation exists between temperatures at station A and temperatures at station B. Probably, it is a linear, or straight-line, relationship. If we knew the temperature at B and for some reason had missed recording the temperature at A, we could predict A from B. The first step, of course, would be to look through past records of simultaneous observations of A and B. We would attempt to confirm our linear relationship empirically. After an A versus B temperature relation is formulated and quantified, it could be used for prediction.

In the temperature example above, we must avoid thinking of temperature B as causing temperature A. We must think instead that both temperatures are subject to the same patterns of weather. Both temperatures move up and down with cold- and warm-air movements. Weather patterns are the causative elements in temperature variation. The temperature at B could be regarded as an index of the causative elements, with temperature A related to the causative elements through the index B.

The causal relation and the associative relation described above are both based on rational physical concepts. It is important to realize, however, that the physical concepts are first identified. From this rational base the relation may be formalized. Many associative relations could be constructed without first identifying the underlying implicit causal factor. When such relations are constructed, we may not reason that we have established cause and effect. Cause and effect are established by logic and by understanding of the physical processes.

THE CALIBRATION AND EVALUATION OF LINEAR MODELS

The simplest multiple variable model is the straight-line relationship between two variables. A monthly rainfall–runoff relationship would be such a model. The relationship between evaporation and temperature in Eq. 3-1 is a linear model. Figure 6-1 is a sketch giving some of the definitions of the statistical straight-line model. The dependent or criterion variable, Y, is scaled along the vertical axis. The independent or predictor variable, X, is scaled along the horizontal axis. Adjacent to each scale is sketched a distribution of the variable.

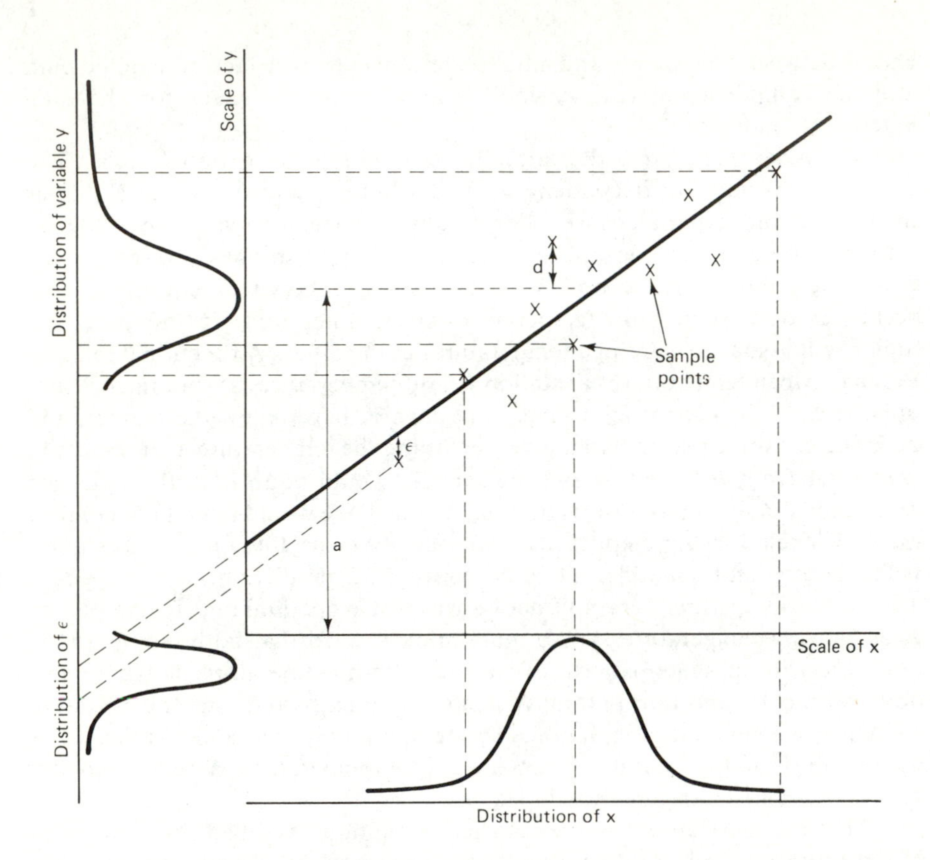

Figure 6-1 Schematic for definitions of statistical straight line.

In order to establish the empirically best values of the parameters in the straight line, we must have observations of (X, Y) pairs. For example, an (X, Y) pair could be rainfall (the X) on a watershed for some month and runoff (the Y) from the watershed for the same month. If we formed a sample histogram of all the X's of our sample pairs, we would be forming the distribution of X, which is shown in Fig. 6-1. If we formed a sample histogram of all the Y's of our sample pairs, we would be forming the distribution of Y, which is also shown in Fig. 6-1.

Before discussing the calibration of the linear multiple variable model, it may be instructive to discuss the estimation problem for the single variable Y. Without knowledge of the relationship between X and Y in Fig. 6-1, our best estimate of Y would be the mean, $\bar{Y}$. It is best because as will be shown below, it is the value that results in the smallest error variation. The error resulting from the use of $\bar{Y}$ as an estimate would be the standard deviation of the dependent variable (i.e., S_y). We seek to develop a relationship between Y and X so that we can reduce the estimation error. Therefore, the value of

S_y can be used as an index of the quality of the estimation. If the linear line of Fig. 6-1 does not produce an expected error, which we will denote here as S_e, that is significantly smaller than S_y, we must conclude that the estimated value from the regression line is not much better than the mean value estimate.

In addition to the distribution of X and Y, the individual (X, Y) pairs can be plotted, as illustrated by sample points in Fig. 6-1. If our linear model is appropriate to the data, the sample points will plot roughly along some line as shown. The (X, Y) pairs cannot be expected to fall exactly along the line because of random errors in X, in Y, or in both. For example, in a rainfall-runoff model we do not know the true amount of rain falling on a watershed. We only know what falls into one or more rain gages.

Each of the sample points in Fig. 6-1 will lie some variable distance from the line. Consider this distance to be measured parallel to the ordinate, such as the ϵ illustrated in Fig. 6-1. The objective is to establish the line in order to predict Y from X. To predict Y as accurately as possible, the line should be located so that the total of the ϵ value of all the points is a minimum. After establishing the line the ϵ can be called the random residuals of Y. The ϵ distribution, which is also shown in Fig. 6-1, can be visualized as being constructed by sweeping up all the sample points parallel to the line.

The establishment of the line and the distribution of the residuals, ϵ, can be considered a partitioning of the information contained in the sample of (X, Y) pairs. Part of this information is used to quantify the parameters of the straight line, while part of the information remains in the residuals. In effect, we estimate a deterministic component, the line, and a random component, the distribution of ϵ. This extraction of information causes a collapse of the standard deviation of the dependent variable Y to a smaller standard deviation of ϵ. One method of locating the line, or partitioning the information, is least squares, which is discussed in the next section; it was discussed briefly in Chapter 3.

Before discussing the calibration of the coefficients that define the line of Fig. 6-1, we now have sufficient background to discuss the quality of the estimation line. Before developing the linear model, our best estimate of Y was $\bar{Y}$, with an expected error of S_y. After we calibrated the line of Fig. 6-1, our best estimate is the value computed from the regression line, with an expected error computed as the standard deviation of the ϵ values. This is called the *standard error of estimate* and denoted as S_e. The regression has led to an increase in estimation accuracy only when S_e is significantly less than S_y.

The Principle of Least Squares

The method of least squares is one method of establishing "best" values of the parameters of a model. This definition of best refers to the line in Fig. 6-1 that yields the lowest possible sum of squares of the residuals for the sample points. Procedures by which this principle can be applied are developed below.

In Chapter 4 the method of moments and maximum likelihood estimation were used to evaluate parameters of univariate models. In almost all cases, the mean of the variates is one parameter. A first, simple application of least squares will demonstrate an additional property of the mean of a set of variates. If we specify some arbitrary point, a, on the ordinate in Fig. 6-1, we can compute the deviates between the observed values of Y and the arbitrary point a. Denote these deviates as d_i, as in

$$d_i = Y_i - a \tag{6-1}$$

The subscript i will take on values from 1 to n, which is the number of values of Y. Now square this equation and sum over the n variates, giving

$$\sum_{i=1}^{n} d_i^2 = \sum_{i=1}^{n} y_i^2 - \sum_{i=1}^{n} (2ay_i) + \sum_{i=1}^{n} a^2 = \sum_{i=1}^{n} y_i^2 - 2a \sum_{i=1}^{n} y_i + na^2 \tag{6-2}$$

We now make use of the first differential in calculus, noting that a function is at a maximum, or minimum, when its first differential is zero. We are interested in $\sum d_i^2$ as a function of our arbitrary point a. Where should we place a to make $\sum d_i^2$ a minimum? This minimum is expressed

$$\frac{d(\sum d_i^2)}{da} = -2 \sum y_i + 2na = 0 \tag{6-3}$$

This yields

$$a = \frac{1}{n} \sum y_i \tag{6-4}$$

In other words, if we want to select a particular point where the sum of squares of the deviations of the Y's from the point is least, we choose the mean of the Y's.

The same simple procedure can be followed to get a least-squares straight line. Write the straight line in the form

$$\hat{Y} = a + bX \tag{6-5}$$

where $\hat{Y}$ means a value of Y computed as a point on the line; X is the independent variate, and a and b are the coefficients that locate the line. The least-squares values of the coefficients a and b are found as follows. The differences between the observed variate Y and the computed values $\hat{Y}$, the residual errors, are specified in the equation:

$$\epsilon_i = \hat{Y}_i - Y_i = a + bX_i - Y_i \tag{6-6}$$

There is an X, and therefore a $\hat{Y}$, for each Y in our sample because our observations must consist of (X, Y) pairs. Now calculate the squares of ϵ_i

and sum them over the sample of size n:

$$\sum \epsilon_i^2 = \sum (a + bX_i - Y_i)^2 \tag{6-7a}$$

$$= \sum Y_i^2 - 2a \sum Y_i - 2b \sum X_i Y_i + na^2 + 2ab \sum X_i + b^2 \sum X_i^2 \tag{6-7b}$$

The summation $\sum \epsilon_i^2$ is a function of both a and b and the expressions for the minimum of both must be found. These minima are found by taking partial derivatives of $\sum \epsilon_i^2$ with respect to a and to b. Both of these two equations are set to zero and solved simultaneously for a and b. The partial derivatives set to zero are (the subscript i was dropped to simplify the notation)

$$\frac{\partial(\sum \epsilon^2)}{\partial a} = -2 \sum Y + 2na + 2b \sum X = 0 \tag{6-8a}$$

$$\frac{\partial(\sum \epsilon^2)}{\partial b} = -2 \sum YX + 2a \sum X + 2b \sum X^2 = 0 \tag{6-8b}$$

The resulting simultaneous equations in a and b, which are usually called the *normal equations*, are shown in

$$na + b \sum X = \sum Y \tag{6-9a}$$

$$a \sum X + b \sum X^2 = \sum XY \tag{6-9b}$$

The quantities $\sum X, \sum Y, \sum Y^2$, and $\sum YX$ are computed from the sample. The sample size, n, of course, is known. Values for a and b can then be found by solving the two simultaneous equations (Eqs. 6-9).

A word of caution is necessary here. The coefficients a and b are computed from the sample quantities $\sum X$, $\sum Y$, $\sum X^2$, and $\sum XY$. These quantities will vary from sample to sample; therefore, a and b are not constants. They are statistical variates just as the mean and standard deviation are variates in univariate models. The line expressed by a and b, after these parameters are quantified by least squares, is not absolute. The line will vary from sample to sample. The hydrologist should be concerned about whether the model has been correctly conceptualized as a straight line, and one should not be lulled into a false sense of mathematical security of particular values just because least squares has been used. The user has the assurance that least squares has given the best line the sample can provide.

The normal equations (Eqs. 6-9) are easily extendable to linear models beyond the simple straight line. For example, we can write

$$\hat{Y} = b_1 X_1 + b_2 X_2 + \cdots + b_p X_p \tag{6-10}$$

to express the concept that the dependent variable Y is the sum of effects of p separate, independent variates. This equation has a form similar to Eq. 6-5 if X_1 takes only values of unity for each observation and b_1 takes the place of a. The normal equations for the linear model of Eq. 6-10 are the simultaneous equations:

$$b_1 \sum X_1 + b_2 \sum X_2 + b_3 \sum X_3 + \cdots + b_p \sum X_p = \sum Y \tag{6-11a}$$

$$b_1 \sum X_2 + b_2 \sum X_2^2 + b_3 \sum X_2X_3 + \cdots + b_p \sum X_2X_p = \sum X_2 Y \tag{6-11b}$$

$$b_1 \sum X_3 + b_2 \sum X_3X_2 + b_3 \sum X_3^2 + \cdots + b_p \sum X_3X_p = \sum X_3 Y \tag{6-11c}$$

$$\vdots$$

$$b_1 \sum X_p + b_2 \sum X_pX_2 + b_3 \sum X_pX_3 + \cdots + b_p \sum X_p^2 = \sum X_p Y \tag{6-11d}$$

There will be as many simultaneous equations as there are unknown coefficients in Eq. 6-10. Note that the product-summation terms to the left of the equal sign in Eqs. 6-11 form a symmetrical arrangement. If we drop the b's these product-summation terms form a matrix in whch the rows and columns can be interchanged.

Goodness of Fit

Once the optimum coefficients of a model have been determined, it is important to determine the accuracy that can be expected when the model is used for prediction. Several measures of how well a linear model fits a data set are in common usage. These are the correlation coefficient, the standard error of estimate (which equals the standard deviation of the residuals), and the efficiency of fit. The definitions of these are given by their computational forms immediately following.

These measures of goodness of fit are based on the partitioning of the sum of squares of the dependent variable. This *sum of squares*, which we can denote as $\sum Y^2$, is a measure of the dispersion of the data that we are going to try to represent with the straight line. This sum of squares breaks into two parts: that portion remaining in the residual distribution at the lower left in Fig. 6-1, and that portion absorbed by the fitting. The sum of the squares of Y is termed SSY. It is identical to $\sum Y^2$. The sum of squares of the residual errors is denoted as SSE. The sum of squares absorbed by the line of regression (our linear model) is denoted as SSR. This partitioning is expressed in the equation

$$\text{SSY} = \text{SSR} + \text{SSE} \tag{6-12}$$

We can compute SSR as

$$\text{SSR} = b_i \sum X_1 Y + b_2 \sum X_2 Y + b_3 \sum X_3 Y + \cdots + b_p \sum X_p Y \tag{6-13}$$

It is necessary to digress slightly here for further discussion of SSY. These values of Y that are squared and summed across the sample are measured from the zero point, or the origin, of Y. However, the set of Y's can always be represented by choosing the mean of the Y's. In fact, if the set is represented by its mean, Eq. 6-4 shows that the least-squares value of Y was chosen. So now we must ask, how much good beyond simply choosing the mean did we accomplish by fitting the line? This question can be answered with equation

$$SS_y = SS_r + SS_e \tag{6-14}$$

In this equation SS_y is the sum of squares of Y about $\bar{Y}$, the mean of Y; it is computed as $SS_y = \sum Y^2 - (\sum Y)^2/n = \text{SSY} - n(\bar{Y})^2$. The value of SS_y is the regression sum of squares adjusted for the mean, and SS_e is the residual sum of squares adjusted for the mean. We should now note that the mean error is zero since a balance of positive and negative deviations is expected. Therefore, there is no adjustment of the sum of squares of the residual error for the mean error, and SSE $= SS_e$. SSY and SS_y are always calculated from the sample. Usually, $\sum Y^2$ is calculated at the same time the other product-summation terms are calculated. Equations 6-12 and 6-13 are used to calculate SSE $= SS_e$, and then Eq. 6-14 yields SS_r. The measures of fit can now be defined.

The correlation coefficient is the square root of the ratio of adjustment sum of squares to the total sum of squares, as in

$$R = \sqrt{\frac{SS_r}{SS_y}} \tag{6-15}$$

The square of the correlation coefficient is also frequently used as a goodness-of-fit statistic because it (R^2) represents the fraction of the variation in the dependent variable that is explained by the variation in the independent variable(s).

The standard error of estimate is given by

$$S_e = \sqrt{\frac{SS_e}{N - p}} \tag{6-16}$$

in which $(N - p)$ measures the degrees of freedom remaining in the residuals. If we consider that the sample contained N observations, or bits of information, we absorbed p of these in evaluating the p regression coefficients, and only $(N - p)$ bits of information remain. The efficiency of fitting is given by

$$\text{Eff} = \frac{SS_y - SS_e}{SS_y} \tag{6-17}$$

The numerator of Eq. 6-17 represents the linear-scale reduction in the dispersion of the Y observations. The efficiency expresses this adjustment of dispersion as a ratio to the initial dispersion.

Example 6-1: Linear Rainfall–Runoff Model

Table 6-1 lists the precipitation for the months of February and March and runoff for the month of March measured on a forest research watershed. We consider that some runoff measured in March is the result of rainfall in March, but some of this runoff also is the result of delayed response of the stream to rain that fell earlier in February.

TABLE 6-1 **Selected Monthly Rainfall and Runoff (in.), White Hollow Watershed**

Year	P_M	P_F	RP_M
1935	9.74	4.11	6.15
1936	6.01	3.33	4.93
1937	1.30	5.08	1.42
1938	4.80	2.41	3.60
1939	4.15	9.64	3.54
1940	5.94	4.04	2.26
1941	2.99	0.73	0.81
1942	5.11	3.41	2.68
1943	7.06	3.89	4.68
1944	6.38	8.68	5.18
1945	1.92	6.83	2.91
1946	2.82	5.21	2.84
1947	2.51	1.78	2.02
1948	5.07	8.39	3.27
1949	4.63	3.25	3.05
1950	4.24	5.62	2.59
1951	6.38	8.56	4.66
1952	7.01	1.96	5.40
1953	4.15	5.57	2.60
1954	4.91	2.48	2.52
1955	8.18	5.72	6.09
1956	5.85	10.19	4.58
1957	2.14	5.66	2.02
1958	3.06	3.04	2.59
$\sum X$	116.35	119.58	82.39
$\sum X_1X_j$	663.1355	589.8177	458.9312
$\sum X_2X_j$		753.1048	435.9246
$\sum Y^2$			331.4729

We may express this concept as the simple linear model,

$$\mathrm{RO}_m = b_1(P_m - I_{\mathrm{am}}) + b_2(p_f - I_{\mathrm{af}}) \tag{6-18}$$

In this equation, RO_m is March runoff; P_m and P_f are March and February precipitation, respectively; I_{am} represents the initial abstraction of March rainfall which does not become runoff; and I_{af} represents the similar abstraction of February rainfall. The coefficients b_1 and b_2, of course, are the regression coefficients that scale the rainfall to the runoff. Equation 6-18 may be expanded to

$$\mathrm{RO}_m = b_1P_m + b_2P_f - b_1I_{\mathrm{am}} - b_2I_{\mathrm{af}} \tag{6-19}$$

If a is substituted for the term $-(b_1I_{\mathrm{am}} + b_2I_{\mathrm{af}})$, Eq. 6-19 converts to

$$\mathrm{RO}_m = a + b_1P_m + b_2P_f \tag{6-20}$$

The sums of variates and sums of squares and products necessary to apply the model, expressed as Eq. 6-20, to the data are given at the bottom of the respective columns in Table 6-1. Thus the simultaneous normal equations are

$$24\ a + \quad 116.35\ b_1 + \quad 119.58\ b_2 = 82.39 \tag{6-21a}$$

$$116.35\ a + 663.1355\ b_1 + 589.8177\ b_2 = 458.9312 \tag{6-21b}$$

$$119.58\ a + 589.8177\ b_1 + 753.1048\ b_2 = 435.9246 \tag{6-21c}$$

which upon solution yields

$$\text{RO}_m = -0.0346 + 0.5880P_m + 0.1238P_f \tag{6-22}$$

Calculation of measures of fit is as follows:

$$\text{SSR} = -0.0346(82.39) + 0.5880(458.9312) + 0.1238(435.9246)$$

$$= 320.9436$$

$$\text{SSE} = 331.4729 - 320.9436 = 10.5293 = \text{SS}_e$$

$$\text{SS}_y = 331.4729 - \frac{(82.39)^2}{24} = 48.6349$$

$$\text{SS}_r = 48.6349 - 10.5293 = 38.1056$$

$$r = \sqrt{\frac{38.1056}{48.6349}} = 0.8852 \tag{6-23a}$$

$$\text{S}_e = \sqrt{\frac{10.5293}{24 - 3}} = 0.7081 \text{ in.} \tag{6-23b}$$

$$\text{S}_y = \sqrt{\frac{48.6349}{23}} = 1.4542 \text{ in.}$$

$$\text{Eff} = \frac{1.4542 - 0.7081}{1.4542} = 0.5131 \tag{6-23c}$$

The runoff and precipitation for March are plotted in Fig. 6-2. The points are labeled with truncated values of February precipitation. The quantified model plots as a family of parallel lines. The slope of these lines is the coefficient b_1, 0.5880. The coefficient b_2, 0.1238, sets the spacing of the lines.

Correlation coefficients can vary between −1 and +1, showing perfect inverse and perfect direct relationships. The correlation coefficient of 0.89 for Eq. 6-20 indicates reasonably good fit. However, the efficiency, 0.51, shows that the original dispersion has been reduced by only a little over one-half.

The threshold values of the model, I_{am} and I_{af}, cannot be determined exactly. However, these values should be nearly the same for the months of February and March during the dormant season of the year. Assuming I_{af} equal to I_{am}, then a equals $0.5880I_{am} + 0.1238I_{am}$. From the value for a of −0.0348, I_{am} is found to be 0.049 in. The quantified physical interpretation of the wintertime monthly runoff process is, therefore, as follows. A potential for runoff is generated as soon as monthly rainfall exceeds about 0.05 in. About 59% of this potential becomes runoff in the month during which the rain fell. An additional 12% becomes runoff the following month. In Fig. 6-2 it should be noted that the proportions 59% and 12% are set reasonably well by the data. The threshold value of 0.05 in. seems small. Figure 6-2 shows that this is

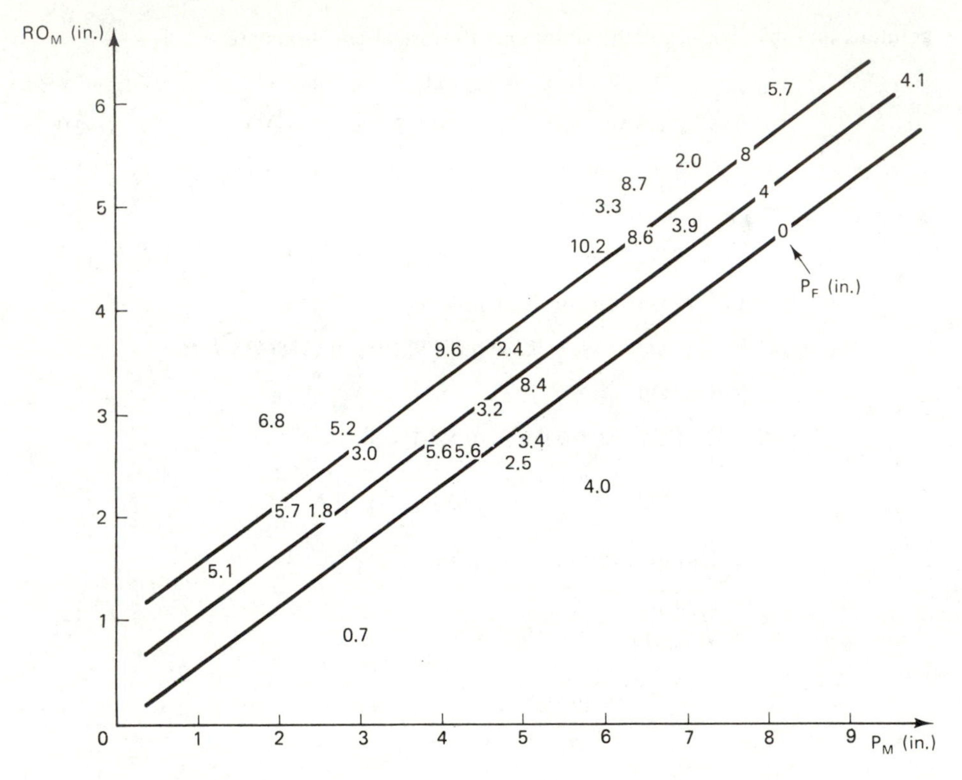

Figure 6-2 Rainfall versus runoff for March, White Hollow watershed.

literally an extrapolation of the lines back to zero runoff. Such an extrapolation may be invalid. However, this structure is incorporated in the conceptualization of the model, and the model can only do what it was designed to do. If the model is unsatisfactory, the hydrologist must design a new model structure that is more applicable to the problem. More advanced models will be introduced throughout the text.

Matrix Notation

The computations needed to perform the quantification of linear models by least squares can be indicated very efficiently and concisely in matrix and vector notation. Several matrix operations must be defined. In the following sections matrices are indicated by capital bold italic letters; vectors are shown by capital bold roman letters. Thus $\boldsymbol{A}$ indicates a matrix, and $\mathbf{B}$ indicates a vector. For our purposes matrices and vectors will be considered defined by the elements of which they are composed.

While matrix addition, subtraction, and multiplication are widely used and understood, matrix inversion and transposing are especially important in regression analysis. Therefore, these two topics will be discussed in the following sections.

Inversion of matrices. There is no matrix operation equivalent to algebraic division. However, if we substitute multiplication by reciprocals for division, we have a matrix analogy.

In simple algebra, if W is the reciprocal of Z, the product WZ is unity. A special matrix takes the place of unity in matrix operations. It is the *identity* or *unit matrix*, (I), which has ones on the main diagonal and zeros elsewhere; for example, a third-order unit matrix would be

$$\boldsymbol{I} = \begin{pmatrix} 1 & 0 & 0 \\ 0 & 1 & 0 \\ 0 & 0 & 1 \end{pmatrix} \tag{6-24}$$

An identity matrix can be of any order, but it must be a square matrix.

Now we can define a matrix $\boldsymbol{B}$ so that

$$\boldsymbol{B} \times \boldsymbol{A} = \boldsymbol{I} \tag{6-25}$$

If $\boldsymbol{B}$ satisfies Eq. 6-25, it is said to be the *inverse* of $\boldsymbol{A}$, and it is usually written $\boldsymbol{A}^{-1}$.

Whereas addition, subtraction, and multiplication of matrices are straightforward operations, inversion of a matrix is not. The most direct means of inversion is to solve n sets of n simultaneous linear equations, where n is the order of the matrix. The equations of the inversion are implicit in Eq. 6-25. We can expand the matrices of this equation in terms of their elements. Suppose that we assign a matrix order of 4. Then

$$\begin{pmatrix} b_{11} & b_{12} & b_{13} & b_{14} \\ b_{21} & b_{22} & b_{23} & b_{24} \\ b_{31} & b_{32} & b_{33} & b_{34} \\ b_{41} & b_{42} & b_{43} & b_{44} \end{pmatrix} \times \begin{pmatrix} a_{11} & a_{12} & a_{13} & a_{14} \\ a_{21} & a_{22} & a_{23} & a_{24} \\ a_{31} & a_{32} & a_{33} & a_{34} \\ a_{41} & a_{42} & a_{43} & a_{44} \end{pmatrix} = \begin{pmatrix} 1 & 0 & 0 & 0 \\ 0 & 1 & 0 & 0 \\ 0 & 0 & 1 & 0 \\ 0 & 0 & 0 & 1 \end{pmatrix} \tag{6-26}$$

Now from the rules for multiplication we can see that

$$\begin{aligned} b_{11}a_{11} + b_{12}a_{21} + b_{13}a_{31} + b_{14}a_{41} &= 1 \\ b_{11}a_{12} + b_{12}a_{22} + b_{13}a_{32} + b_{14}a_{42} &= 0 \\ b_{11}a_{13} + b_{12}a_{23} + b_{13}a_{33} + b_{14}a_{43} &= 0 \\ b_{11}a_{14} + b_{12}a_{24} + b_{13}a_{34} + b_{14}a_{44} &= 0 \end{aligned} \tag{6-27}$$

This set of simultaneous equations can be solved for the b_{ij}, which, by definition, form the first row of the inverse matrix. A similar set of equations is formed using the second row of the inverse matrix and the second column of the identity matrix as the right-hand sides. Solution of this set gives the second row of the inverse. The steps are repeated for the third and fourth rows.

Other methods of inversion are possible. Elaborate numerical methods are necessary for inverting matrices of large order to avoid excessive round-off errors in solving the large number of simultaneous equations. In practical work matrix inversion is normally performed on computers. Most computer libraries will contain programs for matrix inversion, so that little effort is involved.

Transposition of matrices. A second matrix operation is important in the solution of least-squares regression. This is *matrix transposition,* which simply means forming a new matrix by using for its columns, the rows of some other matrix. Thus if

$$\boldsymbol{C} = \begin{pmatrix} C_{11} & C_{12} & C_{13} \\ C_{21} & C_{22} & C_{23} \end{pmatrix}$$

the transpose of $\boldsymbol{C}$, usually written $\boldsymbol{C}^{\mathrm{T}}$, is

$$\begin{pmatrix} C_{11} & C_{21} \\ C_{12} & C_{22} \\ C_{13} & C_{23} \end{pmatrix}$$

Regression operations in matrix notation. Many of the computations performed in fitting multiple-variable models to sets of data can be indicated very efficiently using the matrix notations we have just reviewed. For example, if we had a matrix of observed values of some independent variables, which we can call $\boldsymbol{X}$, the operation $\boldsymbol{X}^{\mathrm{T}}\boldsymbol{X}$ will produce all the product–summation terms on the left-hand side of Eqs. 6-11. If the data matrix consisted of 100 observations on four variables, then X would be a (100, 4) matrix. The transpose of this would be a matrix of (4, 100). Multiplication of the matrix X by its transpose would give a matrix of (4, 4). Similarly, the multiplication of the 100 observations on Y, the dependent variable, which is a matrix of (100, 1), by $\mathbf{X}^{\mathrm{T}}$ will give a matrix of (4, 1). This single-column matrix, or vector, is the right-hand side of Eqs. 6-11. We may now regard Eqs. 6-11 as a matrix–vector operation, written

$$\boldsymbol{X}^{\mathrm{T}}\boldsymbol{X} \times \mathbf{B} = \boldsymbol{X}^{\mathrm{T}}\boldsymbol{Y} \tag{6-28}$$

In this equation the vector $\mathbf{B}$ represents the b coefficients in Eqs. 6-11.

We are interested in finding the coefficients b from our matrix $\boldsymbol{X}^{\mathrm{T}}\boldsymbol{X}$ and our vector $\boldsymbol{X}^{\mathrm{T}}\boldsymbol{Y}$, which we have generated from observed values of the variates. The solution depends on multiplying each side of Eq. 6-28 by the inverse matrix $(\boldsymbol{X}^{\mathrm{T}}\boldsymbol{X})^{-1}$:

$$(\boldsymbol{X}^{\mathrm{T}}\boldsymbol{X})^{-1} \times (\boldsymbol{X}^{\mathrm{T}}\boldsymbol{X}) \times \mathbf{B} = (\boldsymbol{X}^{\mathrm{T}}\boldsymbol{X})^{-1} \times (\boldsymbol{X}^{\mathrm{T}}\boldsymbol{Y}) \tag{6-29}$$

Since the matrix product on the left produces, by definition, the unit matrix, we get

$$\mathbf{B} = (\boldsymbol{X}^{\mathrm{T}}\boldsymbol{X})^{-1}(\boldsymbol{X}^{\mathrm{T}}\boldsymbol{Y}) \tag{6-30}$$

Equation 6-30 expresses in compact form a least-squares solution used in quantifying a linear model by fitting it to data. It is the form frequently used to solve regression problems on a computer.

It may be instructive to solve the regression problem using standardized variables. If the dependent variable and each of the p independent variables are standardized using the standard normal transform of Eq. 2-21, the linear multiple regression model becomes

$$z_y = t_1 z_1 + t_2 z_2 + \cdots + t_p z_p$$

in which the t_i values are called the standardized partial regression coefficients and the z_i values are the standardized values of the variables. It can be shown that the least-squares values of the unknowns (i.e., the t_i values) are related to the correlation matrix $\boldsymbol{R}$ by

$$\boldsymbol{R}_{11}\mathbf{t} = \mathbf{R}_{12}$$

in which $\boldsymbol{R}_{11}$ is the $p \times p$ correlation matrix for the p independent variables and $\mathbf{R}_{12}$ is the vector $p \times 1$ of correlations between the dependent variable and each of the p independent variables. We can solve for the vector of t values by premultiplying both sides of the equation above by $\boldsymbol{R}_{11}^{-1}$ and recognizing that $\boldsymbol{R}_{11}^{-1}\boldsymbol{R}_{11} = I$:

$$\begin{aligned} \boldsymbol{R}_{11}\mathbf{t} &= \boldsymbol{R}_{12} \\ \boldsymbol{R}_{11}^{-1}\boldsymbol{R}_{11}\mathbf{t} &= \boldsymbol{R}_{11}^{-1}\mathbf{R}_{12} \\ \mathbf{It} &= \boldsymbol{R}_{11}^{-1}\mathbf{R}_{12} \\ \mathbf{t} &= \boldsymbol{R}_{11}^{-1}\mathbf{R}_{12} \end{aligned}$$

This equation shows that the standardized partial regression coefficients are a function of both the intercorrelations between the independent variables and the correlations between the dependent and independent variables. Values for the partial regression coefficients (**b**) can be computed from the values of the standardized partial regression coefficients:

$$b_i = \frac{t_i S_y}{S_i}$$

in which S_y is the standard deviation of the dependent variable and S_i is the standard deviation of the ith independent variable.

To compute the **t** vector, it is necessary to compute the inverse $\boldsymbol{R}_{11}^{-1}$. When two or more independent variables are highly correlated, the set of equations that are used to compute $\boldsymbol{R}_{11}^{-1}$ from $\boldsymbol{R}_{11}$ approach linear dependency. $\boldsymbol{R}_{11}$ is said to be a *singular* matrix when linear dependency exists or *near-singular* when the rows or columns are nearly linearly dependent. In the case of a singular matrix, a solution for $\boldsymbol{R}_{11}^{-1}$ does not exist. For a near-singular matrix, the solution may be poor (i.e., the t values may be irrational). The degree of linear dependency can be assessed by computing the determinant of the $\boldsymbol{R}_{11}$

matrix. For the case where there is no intercorrelation, $|\boldsymbol{R}_{11}|$ will equal 1.0. For a singular matrix, $|\boldsymbol{R}_{11}|$ will equal zero. Therefore, when $|\boldsymbol{R}_{11}|$ is near zero, the t_i values may be irrational. As a rule of thumb for large samples, irrational regression coefficients are very likely when $|\boldsymbol{R}_{11}|$ is less than 0.2. Irrational coefficients sometimes occur when $|R_{11}|$ is between 0.2 and 0.5. Thus the value of the determinant $|R_{11}|$ is a useful index to indicate the likelihood that one or more of the independent variables may not be necessary.

The standardized partial regression coefficients, t_i, are useful indicators of the relative importance of the predictor variables; therefore, they should be rational. When the predictor variables are moderately or highly correlated, then the regression coefficients may be irrational. In addition to providing an indication of the relative importance of predictor variables, the standardized partial regression coefficients can also be used to assess the rationality of the regression coefficients. For the linear bivariate model in standardized form $z_y = tz_x$ it can be shown that t equals the correlation coefficient. For the linear multiple regression equation in standardized form, the standardized partial regression coefficient t_i for the ith independent variable will equal the correlation coefficient between the dependent variable and the ith independent variable when $\boldsymbol{R}_{11}$ equals the unit matrix (Eq. 6-24). These two cases suggest that the t_i values are comparable to correlation coefficients; thus, they should be within the range from -1 to $+1$. Values of t_i that are not in this range may be considered irrational. They occur only when the $\boldsymbol{R}_{11}$ matrix has a considerable degree of linear dependency. If a t_i value is irrational, then the b_i value is irrational, and since all of the t_i values are computed from the equation $\mathbf{t} = \boldsymbol{R}_{11}^{-1}\mathbf{R}_{12}$, then, when one is irrational, all should be considered irrational. Irrationality may occur even when all t_i values are within the range from -1 to $+1$. The irrationality is most obvious when a t_i value suggests an indirect (direct) relationship between X_i and Y when one expects a direct (indirect) relationship; one should be cautious about using a model that includes one or more irrational regression coefficients.

Prediction and Prediction Variance

We have seen by Eqs. 6-9 how to represent a bivariate set of observations by a straight line. The underlying principle of fit was that the sum of the squares of the deviations of the dependent variable from the line was minimized. Equations 6-11 provide a solution in the more general case of several independent variables. The solutions of these equations are in the form of numerical values of the regression coefficients, the b's of Eq. 6-10. Now given such a set of values, it is a very easy matter to choose some values of the independent variables, perform the sum of products indicated by Eq. 6-10, and hence arrive at a predicted value of the dependent variable Y. Such a prediction is simple. However, it should never be made without full awareness of the underlying statistical properties. Such properties are brought into clear focus when we think of the ways in which such a predicted value may vary in future samples.

In Chapter 4 we discussed simple univariate models and prediction of future possible events by considering the probable variation of the mean in future samples. In particular, it was pointed out that future samples would not exactly duplicate our sample of record, and the probable variation of the mean was one expression of this future indefiniteness. A little thought will show that exactly the same concepts govern our models in multiple-variable problems. A simple straight line, for example, can move upward or downward, or can rotate, with respect to its position in the sample of record. It may be recalled that our predictions were made by stating a range of variation, defined by confidence intervals, and by further stating the probability that this range would not be exceeded in future samples.

In multiple-variable models our consideration of future variability can take two forms. We can inquire into the likely variations in the structure of our model, or we can inquire into the likely variations in any predictions we make with our model.

Structural variation. Changes in the structure of a least-squares model are expressed as changes in the regression coefficients. The b's of Eqs. 6-11 express the mean slopes of the lines of the relationship between the independent and dependent variables. They are statistical variates just as the means of samples are statistical variates.

The variance of a regression coefficient has the form

$$\sigma^2(b_{kk}) = c_{kk}S_e^2 \tag{6-31}$$

In this equation S_e^2 is estimated from the sample by the squared values of the standard error given in Eq. 6-16, and the c_{kk} are the elements on the diagonal of the inverse matrix $(X^T \cdot X)^{-1}$ in Eq. 6-30. We can then compute two-sided confidence intervals on the regression coefficients:

$$b_u = b + t(n - p, \alpha/2)\sqrt{c_{kk}}S_e \tag{6-32a}$$

$$b_l = b - t(n - p, \alpha/2)\sqrt{c_{kk}}S_e \tag{6-32b}$$

in which $t(n - p, \alpha/2)$ is the value of a random variable having a t distribution with degrees of freedom $(n - p)$ and a probability of $\alpha/2$ in each tail of the distribution. We should remember that possible variations in our structural coefficients as computed by Eqs. 6-32 include the a of Eq. 6-5, as discussed in development of Eqs. 6-11.

Prediction variation. Prediction variation is more complex than structural variation. We must expect that both structure and the scatter of observed values about the structure will vary in future samples. Prediction variance must be an expression of the composite result of both such variances.

Equation 6-33 gives the variance of the average of many future predictions.

$$S^2(\hat{Y}') = S_e^2\left(\frac{1}{n} + \sum_{l=1}^{r}\sum_{j=1}^{r} c_{ij}x_i'x_j'\right) \tag{6-33}$$

In this equation, S_e is from Eq. 6-16, n is the sample size, the c_{ij} are the elements of the inverse matrix $(\boldsymbol{X}^\mathrm{T}\boldsymbol{X})^{-1}$, and the x_i' and x_j' are defined as $(X_i' - \bar{X}_i)$, where x_i' is some selected value of each independent variable for which $S^2(\hat{Y}')$ is desired.

The variance of a single future prediction is given by

$$S_s^2(\hat{Y}') = S^2(\hat{Y}') + S^2 \tag{6-34}$$

For the case of a simple straight line with only one independent variate, Eq. 6-33 reduces to

$$S^2(\hat{Y}') = S^2\left[\frac{1}{n} + \frac{(x')^2}{SS_x}\right] \tag{6-35}$$

Equation 6-34 for this simple case can be written

$$S_s^2(\hat{Y}') = S_e^2\left[1 + \frac{1}{n} + \frac{(x')^2}{SS_x}\right] \tag{6-36}$$

Just as we computed confidence intervals for the structural coefficients we can compute confidence intervals for predicted values. The form is indicated in Eqs. 6-37 for the average of many predictions:

$$\hat{Y}'_U = \hat{Y}' + t(n - p, \alpha/2)S(\hat{Y}') \tag{6-37a}$$

$$\hat{Y}'_L = \hat{Y}' - t(n - p, \alpha/2)S(\hat{Y}') \tag{6-37b}$$

Prediction variance has been discussed primarily in terms of confidence intervals, giving the probabilistic range of variance of the mean value. It should be pointed out, however, that tolerance limits on predictions can also readily be calculated. We might be interested, for example, in what percentage of future expectancies might lie above or below some critical value. We only need note that for a selected value of our independent variate X' the average of predictions is $\hat{Y}'$ and the standard deviation is $S_s(\hat{Y}')$. Now $\hat{Y}'$ and $S_s(\hat{Y}')$ define a particular normal distribution. For this distribution we can compute tolerance limits.

Example 6-2: Calibration of a Peak-Discharge Model

Table 6-2 shows annual flood peak discharges and some associated data for the years 1941 through 1960 for the Watauga River in North Carolina. We will evaluate the model

$$Y = b_1X_1 + b_2X_2 + b_3X_3 + b_4X_4$$

TABLE 6-2 Watauga River Peak-Discharge Data

	Dummy	Precipitation (in.)	Precipitation Duration (in./day)	5-Day Antecedent Precipitation (in.)	Peak Discharge (1000 cfs)
	1	5.71	1.43	1.79	3.02
	1	2.84	1.42	0.38	6.90
	1	2.83	0.94	0.30	5.42
	1	1.91	0.96	1.44	2.32
	1	3.77	1.88	2.85	7.02
	1	3.48	1.16	0.63	3.59
	1	3.50	0.58	0.63	1.84
	1	3.25	3.25	1.11	3.91
	1	3.49	1.74	0.93	7.07
	1	2.60	1.30	1.38	2.88
	1	4.13	2.06	2.03	15.50
	1	2.21	2.21	0.00	4.06
	1	2.04	2.04	0.94	3.49
	1	2.36	2.36	0.63	4.74
	1	3.85	1.28	0.86	7.82
	1	2.30	2.30	1.00	3.88
	1	6.12	1.53	0.14	9.49
	1	2.30	1.15	0.04	7.27
	1	5.02	5.02	0.00	6.42
	1	1.23	0.62	0.28	4.34
$\sum X_1X_j$	20	64.94	35.23	17.36	110.98
$\sum X_2X_j$		241.6450	121.1741	59.1751	389.5694
$\sum X_3X_j$			81.3985	29.6584	204.5627
$\sum X_4X_j$				25.6704	102.4837
$\sum Y^2$					801.1638

where Y is peak discharge in 1000 cfs, X_1 is a dummy variable always equal to 1, X_2 is the associated precipitation, X_3 is the associated precipitation divided by the duration of precipitation, and X_4 is the total precipitation for 5 days antecedent to the peak-producing storm. After evaluation the model will be tested for both the variance of its coefficients and the prediction variance.

The values of sums of squares and products are shown at the bottom of Table 6-2. Since the variance of the coefficients and predictions are required, it is necessary to solve for the inverse matrix. This matrix was calculated using the forms in Eqs. 6-26 and 6-27.

The regression coefficients were next calculated using Eq. 6-30. These coefficients are readily calculated from the data at the top of Table 6-3. Each $\sum X_iY$ term is multiplied by the element in the proper row of the inverse matrix. The sum of these products is the regression coefficient. Calculation of the goodness-of-fit measures follows directly. The value of 0.398 for the correlation coefficient indicates that the model does not represent the data very well.

The variances of the regression coefficients are shown next in Table 6-3. Calculations are based on Eq. 6-31. Additionally, confidence intervals were calculated as

TABLE 6-3 Watauga River Peak-Discharge Auxiliary Calculations

Inverse Matrix c_{ij}

i \ j	1	2	3	4
1	0.48619	−0.08759	−0.05490	−0.06344
2	−0.08759	0.03594	−0.01135	−0.01050
3	−0.05490	−0.01135	0.04853	0.00722
4	−0.06344	−0.01050	0.00722	0.09773

Regression coefficient and correlation

$\sum X_i Y$	110.98	389.5694	204.5627	102.4837
b_i	2.1012	0.8824	0.1531	0.3600

SSR = 645.1784 SSY = 801.1638 SS_e = 155.9854
SSY = 185.3357 SSY = 29.3503
R = 0.398 S^2 = 8.210 S = 2.865

Variance of coefficients

b_i	c_{ii}	$c_{ii}S^2$	$\sqrt{c_{ii}S^2}$	$t\sqrt{C_{ii}S^2}$	$b_i + t\sqrt{c_{ii}S^2}$	$b_i - t\sqrt{c_{ii}S^2}$
1	0.48619	3.9915	1.9988	4.235	6.336	−2.134
2	0.03594	0.2951	0.5432	1.152	2.034	−0.270
3	0.04853	0.3984	0.6312	1.338	1.491	−1.185
4	0.09773	0.8023	0.8957	1.899	2.259	−1.539

Prediction variance: $\bar{X}_2 = 64.94/20 = 3.247$

X_2	$x_2 = X_2 - \bar{X}_2$	X_2^2	$c_{22}X_2^2$	$A = c_{22}X_2^2/n$	$ts\sqrt{A}$	$\hat{Y}$	$\hat{Y} + ts\sqrt{A}$	$\hat{Y} - ts\sqrt{A}$
6	2.753	7.579	0.2724	0.3224	2.038	7.975	10.013	5.937
5	1.753	3.073	0.1104	0.1604	1.437	7.093	8.530	5.656
4	0.753	0.567	0.0204	0.0704	0.952	6.211	7.163	5.259
3	−0.247	0.061	0.0022	0.0522	0.820	5.329	6.149	4.509
2	−1.247	1.555	0.0559	0.1059	1.168	4.447	5.615	3.279
1	−2.247	5.049	0.1815	0.2315	1.726	3.565	5.291	1.839
0	−3.247	10.543	0.3789	0.4289	2.350	2.683	5.033	0.333

indicated by Eqs. 6-32. The value of the t statistic for (20 − 4) = 16 degrees of freedom and a level of significance of 5% is 2.12. The lower confidence limit is negative for all the regression coefficients. The regression coefficients could reasonably be zero, again illustrating the inadequacy of the model.

The prediction variances are calculated at the bottom of Table 6-3. These are values for the average of many future predictions given by Eq. 6-33. The same value of t was used as was used for the calculation of the variance of the regression coefficients. Predicted values of the flood peaks, $\hat{Y}$, were computed with the dummy variable at its value of 1, the variable X_3 at its average value of 1.7615, and the variable X_4 at its average value of 0.868.

Expansion of the term $\sum_{i=1}^{r} \sum_{j=1}^{r} c_{ij} X_i X_j$ in Eq. 6-33 with $r = 4$ gives the following:

$$\begin{aligned} &c_{11}X_1^2 + c_{12}X_1X_2 + c_{13}X_1X_3 + c_{14}X_1X_4 \\ &\quad + c_{21}X_2X_1 + c_{22}X_2^2 + c_{23}X_2X_3 + c_{24}X_2X_4 \\ &\quad + c_{31}X_3X_1 + c_{32}X_3X_2 + c_{33}X_3^2 + c_{34}X_3X_4 \\ &\quad + c_{41}X_4X_1 + c_{42}X_4X_2 + c_{43}X_4X_3 + c_{44}X_4^2 \end{aligned}$$

Now with X_1, X_3, and X_4 set to their respective means, x_1, x_3, and x_4 are zero. The only nonzero term in the expansion above is $c_{22}X_2^2$. This is the term used for calculation of the prediction variance in Table 6-3.

The associated precipitation and the peak discharges are plotted in Fig. 6-3. The straight line represents $\hat{Y}$ from Table 6-3. The curved envelopes are the confidence intervals following Eqs. 6-37. This figure shows that near the mean value of associated

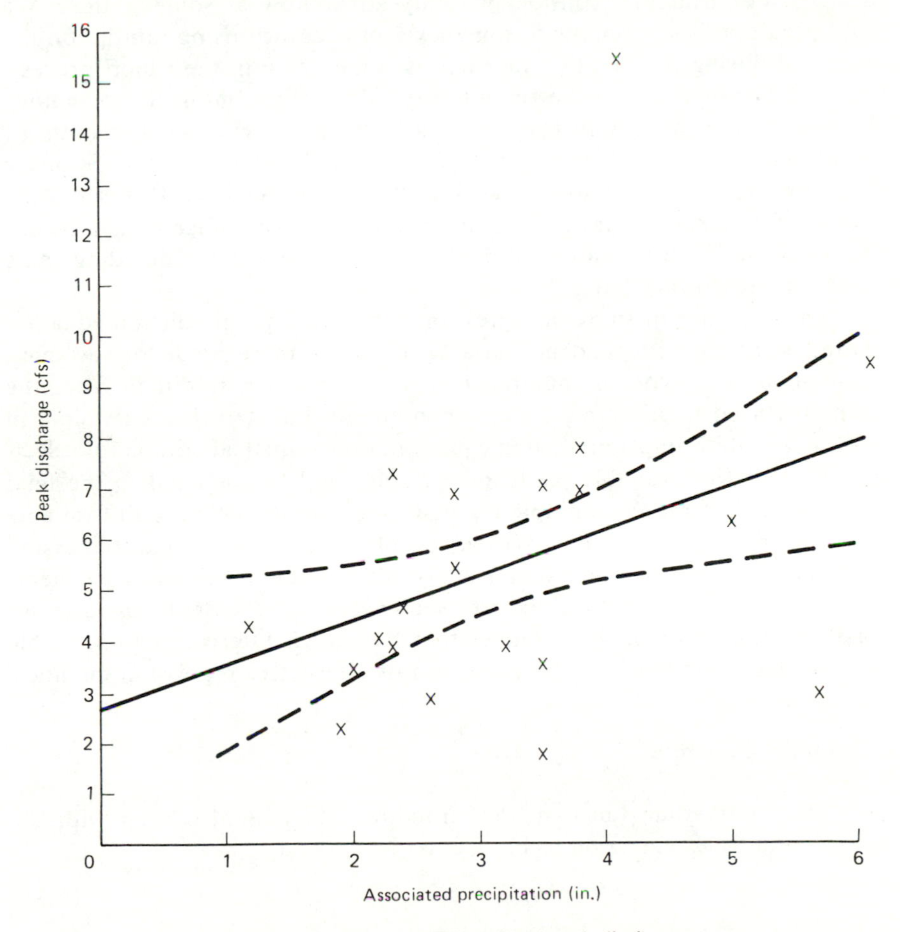

Figure 6-3 Watauga River flood peak discharges.

precipitation we could predict flood peak discharge, on the average, within about 1500 cfs with 90% confidence. However, we would do almost as well by simply predicting the average flood peak of 5550 cfs. For any values of precipitation, any distance from the mean values the confidence envelopes spread too widely to allow the prediction of flood peak with any reasonable confidence.

STEPWISE REGRESSION ANALYSIS

It frequently happens in exploratory studies in hydrologic modeling that we wish to use the basic linear form of Eq. 6-10, but are uncertain as to which hydrologic variables are important. Obviously, in a nice cause-and-effect model such questions do not arise. But in practical hydrologic modeling they do arise. For example, consider monthly streamflow at some station. We could predict this streamflow to some level of accuracy using rainfall on the watershed during the month. But there is some delay in the runoff process. The rainfall from the preceding month may still be contributing to streamflow. The rainfall in prior months may also still be effective. How many months of rainfall should we include in our model? Consider another situation. Suppose that we have two sets of hydrologic data from a watershed. One set is the record before some change in land use in the area. The other set is a record after land use. Can we show a difference in the two sets by including some variable representing change?

The objective of stepwise regression is to develop a prediction equation relating a criterion (dependent) variable to one or more predictor variables. Although it is a type of multiple regression analysis, it differs from the commonly used multiple regression technique in that stepwise regression, in addition to calibrating a prediction equation, uses statistical criteria for selecting which of the available predictor variables will be included in the final regression equation; the multiple regression technique includes all available variables in the equation and often is plagued by irrational regression coefficients. Stepwise regression usually avoids the irrational coefficients because the statistical criteria that are used in selecting the predictor variables usually eliminate predictor variables that have high intercorrelation. Table 6-4 provides a summary of characteristics of alternative regression methods.

Model Structure

The multivariate linear model structure, which is also used with the standard multiple regression technique, is

$$Y = b_0 + b_1X_1 + b_2X_2 + \cdots + b_qX_q \tag{6-38}$$

in which Y is the criterion variable, X_i $(i = 1, 2, \ldots, q)$ are the predictor variables, q is the number of predictor variables, and b_i $(i = 0, 1, \ldots, q)$ are

TABLE 6-4 Summary Characteristics of Alternative Regression Methods

Method	Model	Variable Section	Correlation Analysis	Predictor Intercorrelation
Bivariate	$\hat{Y} = b_0 + b_1 X$	User	Pearson R	Not present
Multiple	$\hat{Y} = b_0 + \sum_{i=1}^{p} b_i X_i$	User	Coefficient of multiple determination	Often significant
Stepwise	$\hat{Y} = b_0 + \sum_{i=1}^{q} b_i X_i$	Statistical	Coefficient of multiple determination	Controlled statistically
Polynomial	$\hat{Y} = b_0 + \sum_{j=1}^{m} b_j X^j$	User or statistical	Coefficient of multiple determination	Extremely significant
Multiple power	$\hat{Y} = b_0 \prod_{j=1}^{p} X_j^{b_j}$	User or statistical	Not applicable	Often significant

the regression coefficients. The value of q can vary from zero to p, where p is the total number of predictor variables in the data set. If $q = 0$, $\hat{Y} = \bar{Y}$ and none of the predictor variables, either alone or in combination with others, can make a significant improvement in the accuracy of prediction. If $q = p$, all the predictor variables are important. In practice, q is less than p.

Forms of Stepwise Regression

Stepwise regression is a term that is applied to several algorithms for simultaneously selecting variables and calibrating the regression equation. The algorithms differ in the sequence for selecting variables. *Forward regression* starts with the predictor variable having the highest correlation with the criterion variable and continues adding variables so that the explained variance is maximized at each step; a test of hypothesis is performed at each step and computation ends when all statistically significant variables have been included. *Backward regression,* a second form of stepwise regression, begins with an equation that includes all the predictor variables in the analysis and sequentially deletes variables, with the variable contributing the least explained variance being deleted first. The third and fourth forms of stepwise regression are variations of the first two forms. *Forward stepwise regression with deletion,* which is the method discussed in this chapter, adds variables to maximize the total variation at each step; however, at each step those variables that are already included in the equation are checked to ensure that they are still statistically significant. A variable found to be no longer significant is deleted. The fourth form of stepwise regression is *backward regression with addition.* This method is similar to the second form except at each step all variables that have been deleted are checked again for statistical significance; if any variable that was previously deleted is found to be significant, then it is added back into the equation. All stepwise regression algorithms are based on the same statistical concepts.

The total *F* test. The objective of a total F test is to determine whether or not the criterion variable is significantly related to the predictor variables that have been included in the equation. It is a test of significance of the following null (H_0) and alternative hypotheses (H_A):

H_0: $\beta_1 = \beta_2 = \cdots = \beta_q = 0$

H_1: at least one regression coefficient is significantly different from zero

in which q is the number of predictor variables included in the equation, and β_i $(i = 1, 2, \ldots, q)$ are the population regression coefficients. The null hypothesis is tested using the following test statistic F:

$$F = \frac{R_q^2/q}{(1 - R_q^2)/(n - q - 1)} \tag{6-39}$$

in which R_q is the multiple correlation coefficient for the equation containing q predictor variables, and n is the number of observations on the criterion variable (i.e., the sample size). The null hypothesis is accepted if F is less than or equal to the critical F value, F_α, which is defined by the selected level of significance α and the degrees of freedom $(q, n - q - 1)$ for the numerator and denominator, respectively. If the null hypothesis is accepted, one must conclude that the criterion variable is not related to any of the predictor variables that are included in the equation. If the null hypothesis is rejected, one or more of the q predictor variables is statistically related to the criterion variable; it does not imply that all predictor variables are necessary.

The partial *F* test. The partial F test is used to test the significance of one predictor variable in the equation. It can be used to test the significance of either the last variable added to the equation or for deleting any one of the variables that are already in the equation; the second case is required to check whether or not the reliability of prediction will be improved if a variable is deleted from the equation. The null and alternative hypotheses are:

$$H_0\colon \quad \beta_k = 0$$

$$H_A\colon \quad \beta_k \neq 0$$

where β_k is the regression coefficient for the predictor variable under consideration. The hypothesis is tested using the following test statistic:

$$F = \frac{\text{fraction increase in explained variation due to subject variable}/\nu_1}{\text{fraction of unexplained variation of the prediction equation}/\nu_2}$$

in which ν_1 and ν_2 are the degrees of freedom associated with the quantities in the numerator and denominator, respectively. In general, $\nu_1 = 1$ and $\nu_2 = n - q - 1$. For example, when selecting the first predictor variable, the test statistic is

$$F = \frac{(R_1^2 - R_0^2)/1}{(1 - R_1^2)/(n-2)} = \frac{R_1^2}{(1 - R_1^2)/(n-2)} \tag{6-40}$$

in which R_1 and R_0 are the correlation coefficients between the criterion variable and the first predictor variable and no predictor variables, respectively; and n is the sample size. The test statistic, which can be used for either the case when a second predictor variable is being added to the equation or the case when the two variables already in an equation are being tested for deletion, is

$$F = \frac{(R_2^2 - R_1^2)/1}{(1 - R_2^2)/(n-3)} \tag{6-41}$$

in which R_1 and R_2 are the correlation coefficients between the criterion variable and a prediction equation having one and two predictor variables, respectively. The null hypothesis is accepted when the test statistic F is less than or equal to the critical F value, F_α. If $F < F_\alpha$, the predictor variable that corresponds to $\beta_k(X_k)$ is not significantly related to the criterion variable when the other $(k-1)$ predictor variables are in the equation; in other words, adding X_k to the prediction equation does not result in a significant increase in the explained variation. If the null hypothesis is rejected (i.e., $F > F_\alpha$), the variable X_k makes a contribution toward the prediction accuracy that is significant even beyond the contribution made by the other $(k-1)$ predictor variables. When the partial F test is used to check for addition of a new predictor variable, a significant F value (i.e., $F > F_\alpha$), indicates that the variable should be added. When the partial F test is used to check for deletion of a predictor variable, a significant F value (i.e., $F > F_\alpha$) indicates that the variable should *not* be dropped from the equation. Partial F statistics are computed for every variable at each step of a stepwise regression.

Procedure. The forward stepwise regression with deletion procedure to be illustrated herein calibrates an equation by stepforward insertion, with predictor variables deleted when appropriate. There are three basic elements: (1) partial F tests for insertion, (2) partial F tests for deletion, and (3) total F tests. The general procedure is outlined in Fig. 6-4 and discussed in the following paragraphs.

1. *Partial F test for insertion*: The partial F values for all predictor variables that are not included in the equation should be computed. The variable with the largest partial F should be selected to enter the equation. On the first step, the partial correlations equal the predictor–criterion correlations, so the predictor variable that has the largest predictor–criterion correlation is added first.

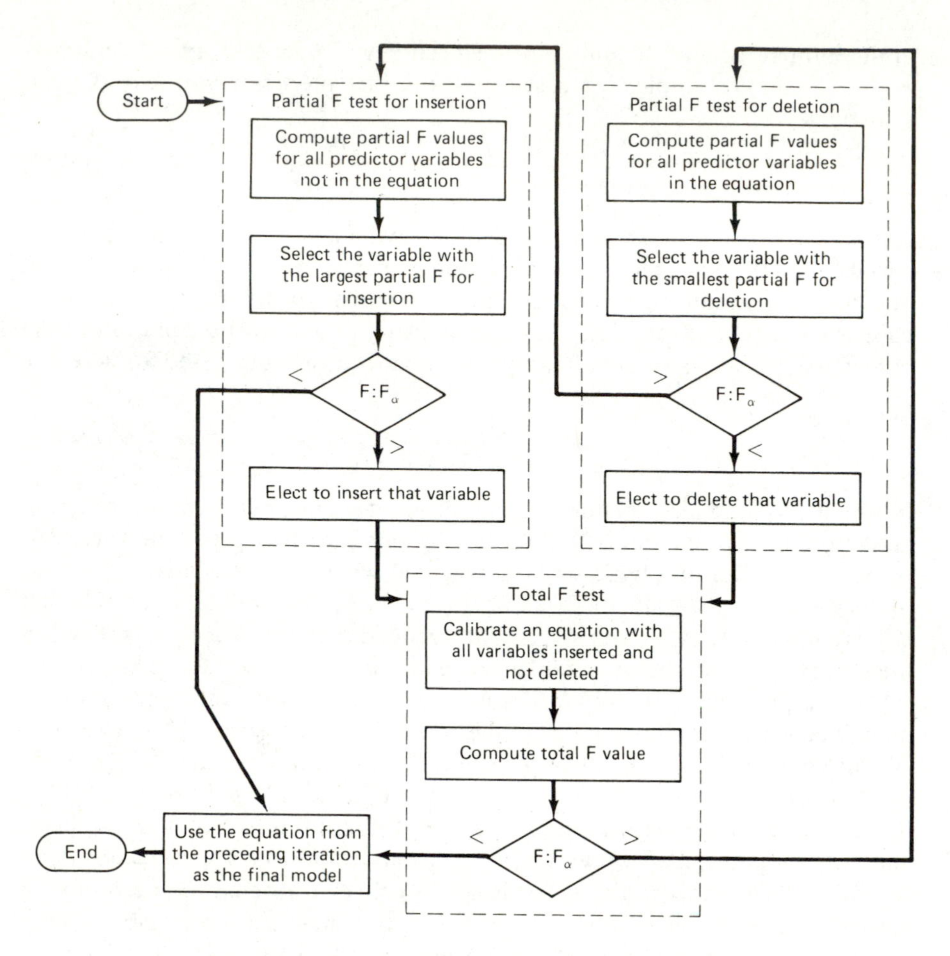

Figure 6-4 Flowchart for stepwise regression procedure.

a. If $F < F_\alpha$, the variable is not significant, and the equation from the previous iteration is the final model (if this option occurs with the first predictor variable tested, the criterion variable is not significantly related to any of the predictor variables).
b. If $F > F_\alpha$, the variable is statistically significant, and the variable should be included in the equation; the process should proceed to the total F test (step 2).

2. *Total F test*: An equation that includes all predictor variables that have been inserted and not subsequently deleted is calibrated. The total F value is calculated and compared with the critical value F_α.
   a. If $F < F_\alpha$, the equation calibrated for the previous iteration should be used as the final model.

   b. If $F > F_\alpha$, the entire model is significant, and all the predictor variables that are in the equation should be tested for significance; control should then proceed to step 3.

3. *Partial F test for deletion*: Partial F values for all predictor variables that are included in the equation are computed. The variable that has the smallest F value is compared with the critical F value, F_α.
   a. If $F < F_\alpha$, the predictor variable is not significant and it is deleted from the equation; control then passes to the total F test (step 2).
   b. If $F > F_\alpha$, all predictor variables included in the equation are significant; control passes to the partial F tests for insertion (step 1).

Criteria for Model Selection

Stepwise regression analysis uses the F statistic as the criterion for selecting the "best" model; however, given the problem of identifying the appropriate level of significance, the statistical criterion should be only one of the criteria for selecting the appropriate model. Other criteria for selecting the best model include (1) the correlation coefficient; (2) the standard error of estimate; (3) the rationality of the intercept coefficient; (4) the rationality of the slope coefficients, which is best evaluated using the standardized partial regression coefficients; (5) the accuracy of the regression coefficients, which is indicated by the standard error of the coefficient, $S_e(b_i)$; and (6) an examination of the residuals to determine whether or not the underlying assumptions have been violated. The square of the correlation coefficient represents the fraction of the total variation that is related to the predictor variables in the regression equation. The change in R^2 from one step to the next is an indication of the change in accuracy that results from adding the predictor variable. When a step involves deleting a predictor variable from the regression equation, the change in R^2 represents the loss of accuracy that results from deleting the predictor variable.

The standard error of estimate S_e is another index that can be used to select a model. The S_e value for any step should be compared with both the standard deviation of the criterion variable S_y and the standard error of estimate from the previous step. In comparing S_e and S_y, one would hope that S_e would be much less than S_y, with a value relatively close to zero being ideal. A value of S_e that is relatively close to S_y indicates a model with relatively poor accuracy. In a few cases, S_e may actually be greater than S_y; this occurs when the sample size is small, the number of predictor variables is large, and the correlation is near zero. This occurs because S_e is based on $(n - q - 1)$ degrees of freedom while S_y is based on $(n - 1)$ degrees of freedom. The relationship between S_e and S_y is

$$S_e = S_y \sqrt{\left(\frac{n - 1}{n - q - 1}\right)(1 - R^2)} \qquad (6\text{-}42)$$

When R^2 is small, the quantity under the square root radical may actually be greater than 1.0; thus S_e would be greater than S_y. The change in S_e from one step to the next is of interest. It is evident from Eq. 6-42 that as R increases, S_e usually decreases. Thus the change in S_e from one step to the next is another indication of the increase or decrease in model accuracy.

The intercept coefficient (i.e., b_0 of Eq. 6-38) should be rational. As the values of the predictor variables go toward specific target values, the intercept coefficient should approach a physically rational value. For example, if the evaporation (E) is related to the air temperature (T), air speed (W), and relative humidity (H), the following model form would result:

$$E = b_0 + b_1 T + b_2 W + b_3 H \tag{6-43}$$

As the air temperature approaches 32°F, E should approach zero because the water will freeze. As W approaches zero, the air mass overlying the water body will remain stagnant and, eventually, saturated. Thus E will approach zero. As H approaches 100%, the air mass overlying the water body will be saturated and the evaporation will go to zero. The intercept coefficient should be such that these conditions should hold. If the resulting predictions at these target conditions are not rational, the intercept coefficient may be irrational.

The slope coefficients should also be rational. Since the slope coefficients represent the change in Y for a given change in the value of the predictor variable X_i, it is easy to check for rationality in sign. In Eq. 6-43, b_1 and b_2 should be positive, while b_3 should be negative; as the relative humidity increases, the evaporation rate should decrease, and thus b_3 will be negative. It is more difficult to check the rationality of the magnitude of a slope coefficient. Since the slope coefficients are a function of the units of both Y and X_i, it may be easier to interpret the rationality of the coefficient by converting a b_i value to the corresponding value of the standardized partial regression coefficient, t_i:

$$t_i = \frac{b_i S_{xi}}{S_y} \tag{6-44}$$

The t_i values are dimensionless and independent of the units of Y and X_i; thus, they are easier to interpret. A t_i value should be between 0 and 1 in absolute value. A value near 1 indicates an important predictor; a value near 0 indicates that the corresponding predictor variable is not important. Thus the larger the value, the more important a predictor variable is considered. Where the intercorrelations between the predictor variables are significant, the t_i values may exceed 1. This should be considered irrational because the t_i values are essentially correlations in the domain of standardized variables. If a t_i value is irrational, the corresponding b_i value should be considered irrational. Irrational models should be used with caution, if at all; this is especially true when used outside the range of the data that was used for calibrating the model. After all, if the b_i value is not rational, it implies that the model does not give a true indication of the rate of change of Y with respect to change in X_i.

The sample regression coefficients b_i are estimates of the population coefficients β_i; therefore, it is of interest to examine the accuracy of the sample estimate. The standard error of the regression coefficient, which is denoted as $S_e(b_i)$, defines the expected dispersion of sample estimate of b_i about the true value. The sample estimate is still the best estimate of the mean of the sampling distribution. While the value of $S_e(b_i)$ is an indication of the accuracy and could be used to develop a confidence interval on β_i, the ratio of $S_e(b_i)$ to b_i provides an index of the relative accuracy and is useful for comparison. Since $S_e(b_i)$ is a dispersion measure and b_i is a measure of central tendency, the ratio $S_e(b_i)/b_i$ has the form of a coefficient of variation, which is a dimensionless quantity. The dimensionless ratios should be used for comparisons. Comparisons can be made both among the variables on any given step of the regression analysis and between values of the same variable at different steps of the analysis. The variable having the smallest error ratio has the greatest relative accuracy. If the ratio increases from one step to the next, the estimate of the regression is considered less accurate; this would suggest that the addition of the new variable at that step of the analysis has not improved the quality of the fit. In such a case, one may choose to stop the stepwise process with the model having the greatest relative accuracy. Of course, the ratio is only one of the criteria that can be used to assess the quality of the regression equation at any step.

The regression line represents the effect of the predictor variables on the criterion variable; therefore, the residuals represent the variation in the criterion variable that is not explained by the predictor variables. There is often more information contained in the residuals than there is explained by the regression equation. Therefore, the residuals should be examined very closely. The residuals will often be helpful in identifying biases of the model. For example, if the residuals are correlated with any of the predictor variables, the linear model does not adequately describe the relationship between the criterion and predictor variables. The residuals should be plotted against each of the predictor variables and the criterion variable because nonlinear relationships in the plots of the residuals will not be detected by computing the correlation coefficients between the residuals and the $(p + 1)$ variables. In addition to detecting bias in a model structure, the residuals can be used to identify variables that were not in the data set but should have been. This requires that the user have a complete understanding of the physical system being modeled. The residuals should be examined to ensure that the four assumptions are not violated: (1) zero mean, (2) constant variance, (3) normal distribution, and (4) independence.

TABLE 6-5 Correlation Matrix

	Variable				
	P	*S*	*C*	*F*	*Y*
P: Precipitation/Temperature Ratio	1.000	0.340	−0.167	−0.445	−0.297
S: Slope		1.000	−0.051	−0.185	0.443
C: Coarse Particle Index			1.000	0.069	−0.253
F: Fine Particle Index				1.000	0.570
Y: Sediment Yield					1.000

Example 6-3: Sediment-Yield Data

The data set of Appendix B-1 was used to perform a stepwise regression analysis. The correlation matrix, which was derived from 37 observations, is shown in Table 6-5; the matrix is characterized by low intercorrelation and moderate predictor–criterion correlations. The largest predictor–criterion correlation is the fine particle index value of 0.570; thus variable 4 enters first. The computed total F statistic is 16.83. For degrees of freedom of 1 and 35 (q, $n - q - 1$) the critical F value for a 5% level of significance is 4.14. Therefore, the equation is statistically significant. The regression equation based solely on X_4 explains 32% of the variation in the sediment-yield data and reduces the standard error of estimate from the standard deviation of 0.829 to 0.691. The intercept (b_0) and slope (b_4) coefficients are given in Table 6-6 with the standard error of the slope coefficient being approximately 25% of the value. The computed partial F to remove is 16.83, which is the same as the total F. Because the value of the partial F to remove is statistically significant, the variable remains in the equation. Thus the first step of the analysis is complete.

In step 2 the largest partial F to enter is 29.07 for the slope (X_2). For degrees of freedom of 1 and 34 and a 5% level of significance, the critical F value is 4.13. Thus the increase in explained variation that will result from X_2 entering the equation is statistically significant; thus X_2 enters the equation. The explained variance doubles to 64%, with a significant decrease in the standard error of estimate to 0.515. The total F statistic for the two-predictor model is significant at the 5% level, with the critical value being 3.28 for 2 and 34 degrees of freedom. The partial F statistics to delete are 29.07 for X_2 and 41.09 for X_4. The critical partial F value for 1 and 34 degrees of freedom is 4.13; thus both F values are significant, and the variables should *not* be deleted. The standard error of b_2 is 0.006, which is about 18% of the coefficient; the standard error of b_4 is 0.008, which is about 16% of b_4. Thus the inclusion of X_2 improved the accuracy of the slope coefficient for X_4.

For step 3, the choice for entering variables is between X_1 and X_3. Since the partial F for X_3 is largest, it is the candidate to enter. For degrees of freedom of 1 and 33, the critical F statistic is 4.13; therefore, X_3 enters the regression equation. The correlation coefficient increases from 0.798 to 0.842, which is an increase in the explained variance (R^2) of 7.2%. The standard error of estimate decreased to 0.467; the change is somewhat marginal. The total F statistic is 26.87; for degrees of freedom of 3 and 33 the critical value is 2.90. Thus the total F statistic is significant. The smallest partial F to remove is 8.35 for X_3. For 1 and 33 degrees of freedom this is significant; thus X_3 should *not* be removed from the equation. Step 3 is complete.

TABLE 6-6 Statistics for Analysis of Sediment Data

Variable Entered	Multiple R	S_e	Total F	Partial Regression Coefficients					Standard Error of Coefficients			
				b_0	b_1	b_2	b_3	b_4	S_1	S_2	S_3	S_4
4	0.570	0.691	16.83	0.662	—	—	—	0.0415	—	—	—	0.010
2	0.798	0.515	29.70	0.102	—	0.0337	—	0.0492	—	0.006	—	0.008
3	0.842	0.467	26.87	0.391	—	0.0331	−0.0113	0.0504	—	0.006	0.004	0.007
1	0.883	0.413	28.23	0.641	−0.797	0.0379	−0.0130	0.0416	0.251	0.005	0.004	0.007

Variable Entered	Standardized Partial Regression Coefficients				Partial F to Enter				Partial F to Remove			
	t_1	t_2	t_3	t_4	1	2	3	4	1	2	3	4
4	—	—	—	0.570	0.11	29.07	4.95	—	—	—	—	16.83
2	—	0.568	—	0.675	5.12	—	8.35	—	—	29.07	—	41.09
3	—	0.557	−0.272	0.692	10.10	—	—	—	—	33.96	8.35	52.30
1	−0.312	0.638	−0.312	0.571	—	—	—	—	10.10	52.15	13.67	37.79

Predictor variable X_1 is the only variable not in the equation. The partial F to enter is 10.10. For degrees of freedom of 1 and 32 the critical F statistic is 4.12. Thus X_1 is a significant variable, and it enters the equation. The explained variance increases by 7.1%, and the standard error of estimate decreases to 0.413. The total F statistic is 28.23, which is significant for 5 and 32 degrees of freedom. For degrees of freedom of 1 and 32, the critical partial F value is 4.11; because the partial F to remove for X_1 is the smallest but still significant, it remains in the equation. Thus at the 5% level of significance, all four predictor variables are considered necessary. The standard errors of the regression coefficients are 0.251, 0.005, 0.004, and 0.007, respectively; these are 31.5, 13.2, 30.8, and 16.8% of the coefficient values. As a percentage, the standard errors of X_3 and X_4 increased from step 3 to step 4; this is not desirable.

The standardized partial regression coefficients, t, are useful in assessing the relative importance of the predictors and the accuracy of the coefficients. The t values indicate that X_2 and X_4 are of similar importance, while X_1 and X_3, which are also similar in importance to each other, are considerably less important in the model. It is important to note that X_2 is considered slightly more important than X_4 in the final model even though X_4 entered before X_2; this occurs because of the higher intercorrelations between X_4 and both X_1 and X_3. Thus, when X_1 and X_3 are included, the importance of X_4 decreases.

Example 6-4: Evaporation Data

A stepwise regression analysis of the evaporation data of Appendix B-2 was performed. The correlation matrix is given in Table 6-7. The partial F statistics to enter for step

TABLE 6-7 Correlation Matrix for the Evaporation Data

	X_1	X_2	X_3	X_4	Y
X_1: Temperature (°F)	1.000	−0.219	0.578	0.821	0.581
X_2: Wind Speed (miles/day)		1.000	−0.261	−0.304	−0.140
X_3: Radiation (equivalent inches)			1.000	0.754	0.578
X_4: Vapor Pressure Deficit				1.000	0.635
Y: Pan Evaporation (in.)					1.000

1 were computed. Since X_4 has the largest partial F statistic, and the largest predictor–criterion correlation coefficient, it is the first variable to enter. At each step the variable with the largest partial F statistic enters the equation (see Table 6-8).

For the first three steps the partial F statistics were significant even at the 1% level of significance; however, for step 4 the partial F statistic corresponding to the wind speed (X_4) is not significant at the 5% level. Thus for a level of significance of 5%, the regression equation containing the temperature (X_1), radiation (X_3), and vapor pressure deficit (X_4) would be selected for use. It has a multiple correlation coefficient of 0.6643 (i.e., $R^2 = 0.441$) and a standard error of estimate of 0.0703; the S_e can be compared with an S_y of 0.0936 in./day.

While it is important to understand the statistical tests of significance in the stepwise regression analysis, it is also important to recognize that other criteria should be considered in selecting a final model. After all, the use of a level of significance of

TABLE 6-8 Stepwise Regression Analysis of Evaporation Data

	Step			
	1	2	3	4
Partial F to enter				
X_1	179.37	6.51	9.23	—
X_2	7.04	1.86	2.51	2.06
X_3	176.60	14.01	—	—
X_4	237.83	—	—	—
Variable entered	X_4	X_3	X_1	X_2
Total F	238.22	130.52	92.13	69.82
Multiple R	0.6353	0.6531	0.6643	0.6667
Standard error of estimate	0.0724	0.0711	0.0703	0.0702
R^2	0.404	0.023	0.015	0.003
Regression coefficients				
b_0	0.01333	−0.00211	−0.06181	−0.07466
b_1	—	—	0.00141	0.00137
b_2	—	—	—	0.00028
b_3	—	0.18207	0.19803	0.20115
b_4	0.35093	0.25508	0.14963	0.16064
Standardized coefficients				
t_1	—	—	0.214	0.208
t_2	—	—	—	0.060
t_3	—	0.230	0.250	0.254
t_4	0.635	0.462	0.271	0.291
Partial F to delete				
X_1	—	—	9.23	8.76
X_2	—	—	—	2.06
X_3	—	14.01	16.77	17.32
X_4	283.22	56.37	9.60	10.82
Standard error: $S_e(b_i)$				
b_1	—	—	0.00046	0.00046
b_2	—	—	—	0.00020
b_3	—	0.04864	0.04836	0.04834
b_4	0.02274	0.03397	0.04830	0.04883
$S_e(b_i)/b_i$				
X_1	—	—	0.326	0.336
X_2	—	—	—	0.714
X_3	—	0.267	0.244	0.240
X_4	0.065	0.133	0.323	0.304

either 5% or 1% is an arbitrary decision. If the value of β (i.e., the probability of a type II error) were computed, a level of significance that is different from the conventional 5% or 1% may be used. Other criteria that are important include the correlation coefficient and the change in the explained variance, the standard error of estimate, the standardized partial regression coefficients, the rationality of the regression coefficients, the standard errors of the regression coefficients, and the magnitude and distribution of the residuals.

The first variable added (X_4) increases the explained variance by 40.4%; the standard error of estimate is reduced from 0.0936 to 0.0724. Both of these changes are physically significant. When the second variable is added to the equation, the explained variation increases by only 2.3% and the standard error of estimate drops less than 2%. It is doubtful that these represent physically significant differences, in spite of the statistical significance. When X_1 and X_2 are added in steps 3 and 4, respectively, the changes are even less. Thus from the standpoint of R and S_e, adding variables beyond the regression of evaporation on the vapor pressure deficit (X_4) does little to improve the prediction accuracy.

The ratio of the standard error of the regression coefficient to the regression coefficient provides an additional indication of model quality. When the ratio is large, one must question the accuracy of the model. For the coefficient b_4 the ratio doubles between steps 1 and 2; an even greater change occurs between steps 2 and 3. The ratio $S_e(b_3)/b_3$ does not change much between steps 2 and 3 and steps 3 and 4; however, in all cases, the ratio is about 25%. For the variable X_2 the ratio is quite high (i.e., 71.4%). This suggests that b_2 is highly inaccurate. In summary, the standard error ratios indicate that the model coefficients decrease in accuracy as more variables are added.

The standardized partial regression coefficients, t_i, provide a means of comparing the relative importance of the predictor variables; they can also be used to judge the rationality of the model. The t_i values for the four regression models are listed in Table 6.8. The magnitude of t_4 decreases significantly from step 1 to 2 and from step 2 to 3; the change from step 3 to 4 is very slight. The reason t_4 is relatively large at step 1 is because X_4 tries to represent the variation of X_1, X_3, and X_4 with the evaporation. As the other predictor variables are added to the equation, the regression coefficient for X_4 approaches a value that represents its effect only. For the model at step 2, X_4 is twice as important as X_3. At step 3, X_3 and X_4 are only slightly more important than X_1. At step 4, X_2 is very small while the other three coefficients are similar, with t_4 being larger than t_3, both of which are slightly larger than t_1.

In model 4 the regression coefficient for X_2 (wind speed) is positive, even though the predictor–criterion correlation is negative. Although the negative sign is irrational because evaporation should increase as wind speed increases, there is some question about whether or not a coefficient that has such a sign change is rational. Sign changes only occur when the intercorrelations are significant. When the signs change, one usually assumes that the coefficients are not rational. Thus it is doubtful that the model with all four predictor variables is rational.

In summary, criteria other than the F statistics should be considered in the selection of a final model. Given that the explained variance does not change much beyond step 1 and that the standard errors of the regression coefficient ratios increase as more than one predictor variable is included, a strong case can be made that the one-predictor variable model (i.e., Y versus X_4) is the best model. The stepwise regression approach is often favored because for a given data set everyone will get the same model for a given level of significance. While there is merit in consistency, rationality is also very important.

Example 6-5: Stepwise Regression of Rainfall–Runoff Data

Table 6-9 gives two sets of rainfall–runoff data for some hypothetical watershed. We consider set 1 of these data as representing the watershed before some experimental land-use treatment and set 2 after the treatment. The objective is to test for significant difference between the two sets, such difference implying that the treatment caused the difference in the relationship between rainfall and runoff.

TABLE 6-9 Data for Dummy Separation of Sets

Set 1		Set 2	
Rainfall	Runoff	Rainfall	Runoff
0.4	0.6	0.6	0.6
1.6	0.6	1.8	1.2
2.5	1.3	2.2	1.0
3.6	1.8	2.6	1.6
5.2	3.0	3.8	2.2
5.6	3.8	6.0	5.0
6.0	4.6		

The data are plotted in Fig. 6-5. It should be noted that set 2 is deficient in high rainfall experiences, there being three events above 4 in. in set 1 but only one in set 2. This is a frequent occurrence in watershed research, caused, obviously, by random differences, or short-term fluctuations, in rainfall. Because of the imbalance of high events, it would seem good analytical strategy to pool the higher events and test for differences between the two sets in the lesser events.

Table 6-10 shows the tabulation of data according to the strategy above. T_1 and T_2 are dummy variables, T_1 being equal to 1 for all events, T_2 being equal to zero for set 1 and 1 for set 2. The variable X is rainfall up to 4 in., but remains constant at 4 in. for all rainfalls larger than 4 in. X_e is a variable expressing rainfall in excess of 4 in. The net consequence of the use of variable X_e is to allow the rainfall-runoff relationship to break into two connected linear segments with the angle point at 4 in. The location (but not the height) must be specified, and any position consistent with the natural alignment of the data can be used. The test model is the regression equation:

$$RO = b_1 T_1 + b_2 T_2 + b_3 X + b_4 T_2 X + b_5 X_e \tag{6-45}$$

The term $b_2 T_2$ allows for a general vertical distinction between the two sets. The term $b_4 T_2 X$ allows for a difference in slope of the linear segment below 4 in. of rainfall. No distinction is made between the sets above 4 in. rainfall.

Since T_2 is zero for set 1, we can write

$$RO = b_1 + b_3 X + b_5 X_e \tag{6-46}$$

and for set 2:

$$RO = (b_1 + b_2) + (b_3 + b_4)X + b_5 X_e \tag{6-47}$$

Also, if we drop the dummy variable T_2 the equation

$$RO = b_1 + b_3 X + b_5 X_e \tag{6-48}$$

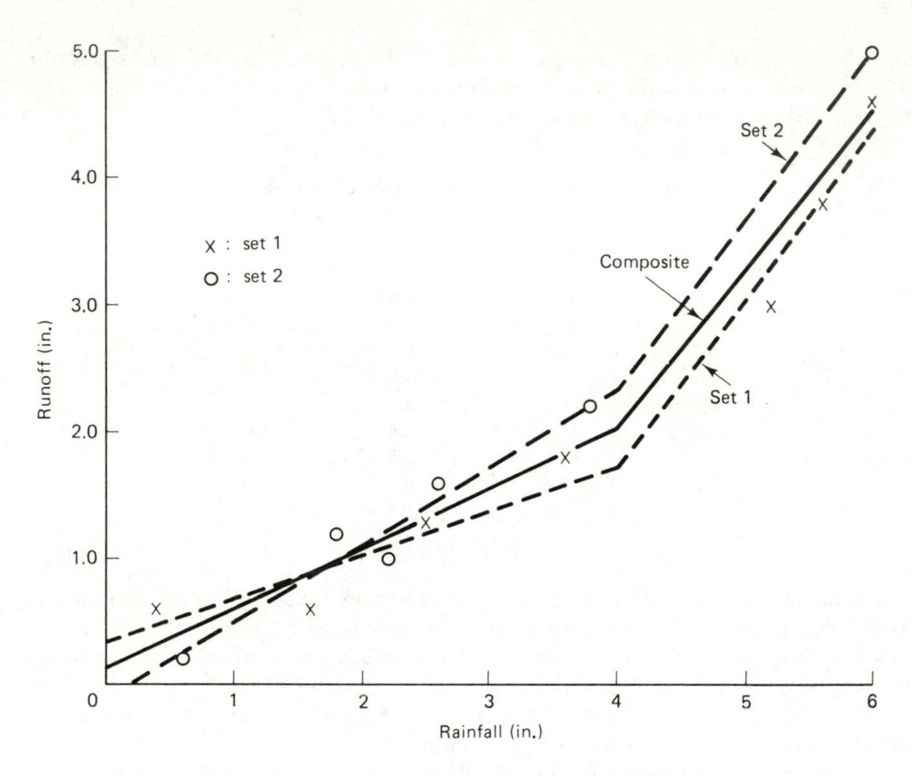

Figure 6-5 Separation of hydrologic data sets using dummy variables: ×, set 1; ○, set 2.

TABLE 6-10 Arrangement of Data for Least-Squares Solution

T_1	T_2	X	T_2X	X_e	RO
1	0	0.4	0	0	0.6
1	0	1.6	0	0	0.6
1	0	2.5	0	0	1.3
1	0	3.6	0	0	1.8
1	0	4.0	0	1.2	3.0
1	0	4.0	0	1.6	3.8
1	0	4.0	0	2.0	4.6
1	1	0.6	0.6	0	0.2
1	1	1.8	1.8	0	1.2
1	1	2.2	2.2	0	1.0
1	1	2.6	2.6	0	1.6
1	1	3.8	3.8	0	2.2
1	1	4.0	4.0	2.0	5.0

applies as a composite of both sets. It is to be understood that the values of b_1, b_3, and b_5 will differ from the similar equation above. Both the full test model and the reduced model with T_2 deleted are fitted to the data. The difference in the sum of squares of the regression between the two models forms the basis of the test of significance.

The calculations are shown in Table 6-11. The distinction between sets is significant at the 5% level. The composited and separated solutions have been plotted in Fig. 6-5.

TABLE 6-11 Dummy Separation of Sets: Computations for Significance

Sums of squares

	X_1	X_2	X_3	X_4	X_5	Y
X_1	13	6	35.1	15	6.8	26.9
X_2		6	15	15	2	11.2
X_3			115.57	45.64	27.2	93.53
X_4				45.64	8	37
X_5					12	28.88
Y						85.13

Solutions

With T_2		Without T_2	
b_1:	0.32758	b_1:	0.14599
b_2:	−0.44857	b_3:	0.46729
b_3:	0.35169	b_5	1.26475
b_4:	0.26732		
b_5:	1.32044		

Significance

$SSR_5 = 84.7066 \qquad SSR_3 = 84.1587$

$SSY = 85.13$

$SS_e = 0.4234 \qquad SSR_5 - SSR_3 = 0.5479$

$$S_e^2 = \frac{0.4234}{13-5} = 0.0529 \qquad S_d^2 = \frac{0.5479}{2} = 0.2740$$

$$F = \frac{0.2740}{0.0529} = 5.18$$

$F(2, 8; 05) = 4.46$

REGRESSION ANALYSIS OF NONLINEAR MODELS

In most empirical analyses linear models are attempted first because of the relative simplicity of linear analysis. Also, linear models are easily applied, and the statistical reliability is easily assessed.

Linear models may be rejected because of either theoretical considerations or empirical evidence. Specifically, theory may suggest a nonlinear relationship between a criterion variable and one or more predictor variables; for example, many biological growth curves are characterized by nonlinear forms. Where a model structure cannot be identified by theoretical considerations, empirical evidence can be used in model formulation and may suggest a nonlinear form. For example, the hydrologic relationship between peak discharge and the drainage area of a watershed has been found to be best represented by a log-log equation, which is frequently referred to as a *power model.* A nonlinear form may also be suggested by the residuals that result from a linear analysis; that is, if a linear model produces nonrandomly distributed residuals (i.e., a bias is apparent), the underlying assumptions are violated; a nonlinear functional form may produce residuals that satisfy the assumptions that underlie the principle of least squares.

Polynomial and Power Model Forms

Linear models were separated into bivariate and multivariate; the same separation is applicable to nonlinear models. It is also necessary to separate nonlinear models on the basis of the functional form. Although polynomial and power models are the most frequently used nonlinear forms, it is important to recognize that other model structures are available and may actually be the correct structure. In addition to the power and polynomial structures, forms such as a square root, exponential, and logarithmic may provide the best fit to a set of data. Since the polynomial and power forms are so widely used, it is of importance to identify their structure:

1. Bivariate
   a. Polynomial

$$Y = b_0 + b_1X + b_2X^2 + \cdots + b_pX^p \tag{6-49}$$

   b. Power

$$Y = b_0X^{b_1} \tag{6-50}$$

2. Multivariate
   a. Polynomial

$$Y = b_0 + b_1X_1 + b_2X_2 + b_3X_1^2 + b_4X_2^2 + b_5X_1X_2 \tag{6-51}$$

   b. Power

$$Y = b_0X_1^{b_1}X_2^{b_2} \cdots X_p^{b_p} \tag{6-52}$$

In these relationships, Y is the criterion variable, X is the predictor variable in the bivariate case, and X_i is the ith predictor variable in the multivariate case, $b_j (j = 0, 1, \ldots, p)$ is the jth regression coefficient, and p is either the number of predictor variables or the order of the polynomial. The bivariate forms are just special cases of the multivariate form.

Transformation and Calibration

The power and polynomial models are widely used nonlinear forms because they can be transformed in a way that makes it possible to use the principle of least squares. Although the transformation to a linear structure is desirable from the standpoint of calibration, it has important consequences in terms of assessing the goodness-of-fit statistics.

The bivariate polynomial of Eq. 6-49 can be calibrated by forming a new set of predictor variables:

$$w_i = X^i \qquad (i = 1, 2, \ldots, p) \tag{6-53}$$

This results in the model

$$Y = b_0 + b_1 w_1 + b_2 w_2 + \cdots + b_p w_p \tag{6-54}$$

The coefficients b_j can be estimated using a standard multiple regression analysis.

The bivariate power model of Eq. 6-50 can be calibrated by forming the following set of variables:

$$z = \ln Y \tag{6-55}$$

$$c = \ln b_0 \tag{6-56}$$

$$w = \ln X \tag{6-57}$$

These transformed variables form the following linear equation, which can be calibrated using bivariate linear regression analysis:

$$z = c + b_1 w \tag{6-58}$$

After values of c and b_1 are obtained, the coefficient b_0 can be determined using

$$b_0 = e^c \tag{6-59}$$

Base 10 logarithms can also be used; in such a case, Eqs. 6-55 to 6-57 will use the base 10 logarithm and Eq. 6-59 will use a base 10 rather than e.

The multivariate polynomial of Eq. 6-51 can also be solved using a multiple regression analysis for a set of transformed variables. For the model given by Eq. 6-51 the predictor variables are transformed as follows:

$$w_1 = X_1 \tag{6-60}$$

$$w_2 = X_2 \tag{6-61}$$

$$w_3 = X_1^2 \tag{6-62}$$

$$w_4 = X_2^2 \tag{6-63}$$

$$w_5 = X_1 X_2 \tag{6-64}$$

The revised model has the form

$$Y = b_0 + \sum_{i=1}^{5} b_i w_i \tag{6-65}$$

It is important to note that the polynomial models do not require a transformation of the criterion variable.

The model given by Eq. 6-51 has only two predictor variables (i.e., X_1 and X_2) and is a second-order equation. In practice, a model may have more predictor variables and may be of higher order. In some cases the interaction terms (i.e., X_iX_j) may be omitted to decrease the number of coefficients that must be calibrated. However, if the interaction terms are omitted when they are actually significant, the goodness-of-fit statistics may suffer unless the variation is explained by the other terms in the model; when this occurs, the coefficients may lack physical significance.

The multivariate power model of Eq. 6-52 can be evaluated by making a logarithmic transformation of both the criterion and the predictor variables:

$$z = \ln Y \tag{6-66}$$

$$c = \ln b_0 \tag{6-67}$$

$$w_i = \ln X_i \qquad (\text{for } i = 1, 2, \ldots, p) \tag{6-68}$$

The resulting model has the form

$$z = c + \sum_{i=1}^{p} b_i w_i \tag{6-69}$$

The coefficients of Eq. 6-69 can be evaluated using a multiple regression analysis. The value of b_0 can be determined by making the transformation of Eq. 6-59. Again, it is possible to use a base 10 transformation rather than a natural log transform.

Goodness of Fit

Since most computer programs that perform multiple regression analyses include goodness-of-fit statistics as part of the output, it is important to recognize the meaning of these statistics. For those nonlinear models in which the criterion variable Y is not transformed, the goodness-of-fit statistics are valid indicators of the reliability of the model; however, when the criterion variable is transformed, such as is necessary for the power model form, the principle of least squares is applied in the log-log space. As a result, the residuals that are used to compute the standard error of estimate, and therefore the correlation coefficient, are measured in the domain of the logarithm of Y and not the Y domain. Therefore, the goodness-of-fit statistics are not necessarily a reliable indicator of model reliability, especially since decisions are made in the Y space and not the log Y space. Therefore, when a model requires a transformation of the criterion variable Y in order to calibrate the coefficients, the goodness-of-fit statistics that are included with the multiple regression output should not be used as measures of reliability. Instead, values for the goodness-of-fit statistics should be computed using the definitions provided previously. The standard error of estimate should be computed using Eq. 6-16. In summary, when an equation is calibrated with the criterion variable transformed such as the log-log space, the least-squares concepts apply only to the transformed space and not the measurement (i.e., $Y - X$) untransformed space. The sum of the residuals will not equal zero and the sum of the squares of the errors may not be a minimum in the Y domain even though they are in the log Y space; furthermore, the assumption of a constant variance may also not be valid. Because these basic assumptions of regression are not valid in the $X - Y$ space, many practitioners object to data transformations. However, in spite of these theoretical considerations, the transformations may provide reasonable estimates in the $X - Y$ space.

The Analysis of Variance for Polynomial Models

Just as the analysis of variance (ANOVA) is used to examine the statistical adequacy of linear models, it can be used to test whether or not the nonlinear polynomial models are adequate. Both the total and partial F statistics can be used. Table 6-12 provides a summary of the ANOVA test for first-, second-, and third-order polynomials that include a single predictor variable. However, the formulas are easily adapted for polynomial models involving more than one predictor variable, including analyses involving interaction terms. While the partial F test can be used to perform a stepwise regression analysis of the polynomial, it is always important to remember that the selection of a 5% level of significance is somewhat an arbitrary decision.

Example 6-6: One-Predictor Polynomial of Sediment Yield versus Slope

A computer program was used to derive first-, second-, and third-order polynomials that regress the sediment yield data of Appendix B-1 on watershed slope. The correlation matrix is given in Table 6-13. It is evident that the intercorrelation coefficients are very significant and should lead to irrational regression coefficients; furthermore, because of the high intercorrelation it is unlikely that the explained variance will improve significantly as the order of the equation increases. Table 6-14 provides a summary of the correlation coefficients, the standardized partial regression coefficients, and the partial regression coefficients. The nonlinear models do not provide a significant increase in the explained variation when compared with the linear model. While the standardized partial regression coefficients for the second-order model are not irrational, those of the third-order model are highly irrational.

The ANOVA results are given in Tables 6-15 and 6-16. The total F tests indicate that all three models are statistically significant at the 5% level of significance. However, the partial F tests indicate that including the quadratic term in the equation does not improve the statistical goodness-to-fit since the computed F value is less than the

TABLE 6-12 ANOVA Table for Polynomial Analysis

Model	Correlation	Test	H_0	Sum of Squares		df	Test Statistic, F
First-order (linear) $Y = a + b_1X$	R_1	Total and partial	$\beta_1 = 0$	Regression	R_1^2	1	$F = \frac{R_1^2}{(1 - R_1^2)/(N - 2)}$
				Error	$1 - R_1^2$	$N - 2$	
Second-order (quadratic) $Y = a + b_1X + b_2X^2$	R_2	Total	$\beta_1 = \beta_2 = 0$	Regression	R_2^2	2	$F = \frac{R_2^2/2}{(1 - R_2^2)/(N - 3)}$
				Error	$1 - R_2^2$	$N - 3$	
		Partial	$\beta_2 = 0$	Regression	$R_2^2 - R_1^2$	1	$F = \frac{R_2^2 - R_1^2}{(1 - R_2^2)/(N - 3)}$
				Error	$1 - R_2^2$	$N - 3$	
Third-order (cubic) $Y = a + b_1X + b_2X^2 + b_3X^3$	R_3	Total	$\beta_1 = \beta_2 = \beta_3 = 0$	Regression	R_3^2	3	$F = \frac{R_3^2/3}{(1 - R_3^2)/(N - 4)}$
				Error	$1 - R_3^2$	$N - 4$	
		Partial	$\beta_3 = 0$	Regression	$R_3^2 - R_2^2$	1	$F = \frac{R_3^2 - R_2^2}{(1 - R_3^2)/(N - 4)}$
				Error	$1 - R_3^2$	$N - 4$	

TABLE 6-13 Correlation Matrix: Sediment Yield (Y) versus Slope (S)

	S	S^2	S^3	Y
S	1.000	0.908	0.791	0.443
S^2		1.000	0.972	0.443
S^3			1.000	0.428
Y				1.000

TABLE 6-14 Summary of Regression Analyses

		Partial Regression Coefficients				Standardized Partial Regression Coefficients		
Model	R	b_0	b_1	b_2	b_3	t_1	t_2	t_3
First-order	0.443	0.220	0.0263	—	—	0.443	—	—
Second-order	0.454	0.323	0.0136	2.31×10^{-4}	—	0.229	0.235	—
Third-order	0.499	0.042	0.0865	−0.0031	3.43×10^{-5}	1.456	−3.164	2.35

TABLE 6-15 ANOVA: Total F Test

Model	R^2	$1 - R^2$	df$_1$	df$_2$	Total F	$F_{0.05}$
First-order	0.196	0.804	1	35	$\frac{0.196/1}{0.804/35} = 8.53$	4.12
Second-order	0.206	0.794	2	34	$\frac{0.206/2}{0.794/34} = 4.41$	3.28
Third-order	0.249	0.751	3	33	$\frac{0.249/3}{0.751/33} = 3.65$	2.90

TABLE 6-16 ANOVA: Partial Tests

Model	ΔR^2	$1 - R^2$	df$_1$	df$_2$	Partial F	$F_{0.05}$
First-order	0.196	0.804	1	35	$\frac{0.196/1}{0.804/35} = 8.53$	4.12
Second-order	0.010	0.794	1	34	$\frac{0.010/1}{0.794/34} = 0.43$	4.13
Third-order	0.043	0.751	1	33	$\frac{0.043/1}{0.751/33} = 1.89$	4.14

critical value. If a stepwise regression analysis procedure were followed, the computations would end with the linear model (i.e., the regression coefficients for the second-order model would not be computed).

Example 6-7: One-Predictor Polynomial of Evaporation versus Temperature

The evaporation data (Appendix B-2) were used to derive polynomials for the regression of evaporation on temperature. The correlation matrix, which is shown in Table 6-17,

TABLE 6-17 Correlation Matrix: Evaporation (E) versus Temperature (T)

	T	T^2	T^3	E
T	1.000	0.995	0.981	0.581
T^2		1.000	0.996	0.602
T^3			1.000	0.627
E				1.000

shows very high intercorrelation. Table 6-18 provides a summary of the regression analyses. While the linear model explains approximately 34% of the variation in evaporation, the nonlinear terms provide only a marginal increase in the explained variation. The quadratic model increases the explained variance by about 8%. The marginal improvement is also evident from the small decrease in the standard error of estimate. The standardized partial regression coefficients indicate that both the second- and third-order models are irrational. In judging the rationality of a model, it is also

TABLE 6-18 Summary of Regression Analysis

	R	S_e[a]	t_1	t_2	t_3	b_0	b_1	b_2	b_3
First-order	0.581	0.0763	0.581	—	—	−0.114	0.00383	—	—
Second-order	0.648	0.0715	2.16	2.76	—	0.425	−0.0142	0.000143	—
Third-order	0.654	0.0711	3.11	−8.44	5.98	−0.239	0.0205	−0.000439	3.14×10^{-6}

[a]$S_y = 0.0936$.

of value to examine the values that would be predicted for values of the predictor variable that are likely to be observed. The three models were used to estimate the daily evaporation depth for selected temperatures (Table 6-19). Because water freezes

TABLE 6-19 Predicted Evaporation Rates (in./day)

	Temperature (°F)			
Model	32	50	75	100
First-order	0.009	0.078	0.173	0.269
Second-order	0.117	0.073	0.164	0.435
Third-order	0.070	0.081	0.154	0.561

at 32°F, evaporation rates near zero are expected; only the linear model gave a rational estimate at 32°F. Estimates made using the second-order model decreases with increases in temperature for temperatures up to about 50°F. Both the second- and third-order models provide especially high rates at 100°F. In summary, the irrational coefficients for this regression analysis, which result from the high intercorrelations, lead to irrational estimates when the nonlinear forms are used.

TABLE 6-20 ANOVA: Total *F* Tests

Model	R^2	$1-R^2$	df_1	df_2	Total F	$F_{0.05}$
First-order	0.338	0.662	1	352	$\frac{0.338/1}{0.662/352} = 179.5$	3.86
Second-order	0.420	0.580	2	351	$\frac{0.420/2}{0.580/351} = 127.1$	3.02
Third-order	0.428	0.572	3	350	$\frac{0.428/3}{0.572/350} = 87.1$	2.62

Tables 6-20 and 6-21 summarize the ANOVA. The total *F* test (Table 6-20) indicates that all models are statistically significant. Similarly, the partial *F* tests (Table 6-21) indicate that each term is significant, and thus if a stepwise regression analysis criteria were used, the third-order equation would be selected.

TABLE 6-21 ANOVA: Partial *F* Tests

Model	ΔR^2	$1-R^2$	df_1	df_2	Partial F	$F_{0.05}$
First-order	0.338	0.662	1	352	$\frac{0.338/1}{0.662/352} = 179.5$	3.86
Second-order	0.082	0.580	1	351	$\frac{0.082/1}{0.580/351} = 49.8$	3.86
Third-order	0.008	0.572	1	350	$\frac{0.008/1}{0.572/350} = 4.58$	3.86

Example 6-7 illustrates that different criteria for model selection (e.g., the partial *F* tests and the standardized partial regression coefficients) may conflict. For this example it appears that the ANOVA is an inadequate criterion for model selection. Quite possibly, the 5% level of significance is not an adequate criterion; after all, use of the 5% level is based solely on convention and not on a rational analysis of the relationship between the level of significance, the power of the test, the sample size, and the physical separation criterion.

Example 6-8: Single-Predictor Power Model

A data set that consists of five observations will be used to illustrate the fitting of a power model that includes a single-predictor variable. The data set is given in Table 6-22. The values were converted to natural logarithms and the coefficients of Eq. 6-58

TABLE 6-22 Data Base for Single-Predictor Power Model

Y	X	$\ln_e Y$	$\ln_e X$	$\hat{Y}$	e	$\hat{Y}_e$	e_e
7.389	2.718	2	1	−214.54	−221.929	6.048	−1.341
20.086	7.389	3	2	53.18	33.095	40.445	20.359
403.429	20.086	6	3	780.93	377.499	270.427	−133.002
2980.96	54.598	8	4	2759.03	−221.927	1,807.96	−1173.00
8103.08	148.41	9	5	8136.00	32.923	12,087.4	3984.31
					$\sum e = 0$		$\sum e \neq 0$

were calibrated using least squares:

$$\ln_e Y = 0.1 + 1.9 \ln_e X \tag{6-70}$$

A correlation coefficient of 0.9851 was computed for Eq. 6-70. Equation 6-70 expressed in the $X - Y$ coordinate space is

$$Y = 0.9048X^{1.9} \tag{6-71}$$

A linear analysis of the values of X and Y provides a correlation coefficient of 0.9975 and the following regression equation:

$$Y = -370.3301 + 57.3164X \tag{6-72}$$

The residuals for Eqs. 6-71 and 6-72 are given in Table 6-19. While the residuals for Eq. 6-72 have a sum of zero, the sum of the residuals for Eq. 6-71 is not equal to zero; however, the sum of the residuals for Eq. 6-70 will equal zero in the natural log space.

The correlation coefficient and standard error of estimate that are computed for the model in the natural log space are not valid indicators of the accuracy of the estimates in the domain of the untransformed criterion variable. For example, the correlation coefficient of 0.9851 computed with Eq. 6-70 is not a measure of the accuracy of Eq. 6-71. Values of Y that were estimated with Eq. 6-71 are given in Table 6-19 and are indicated as $\hat{Y}_e$; the resulting residuals are denoted as e_e. If the separation of variation concept is applied to Eq. 6-71, the components are:

$$\sum (Y - \bar{Y})^2 = \sum (Y - \hat{Y})^2 + \sum (\hat{Y} - \bar{Y})^2 \tag{6-73}$$

$$48{,}190{,}499.4 \neq 17{,}276{,}427.6 + 100{,}525{,}199.1 \tag{6-74}$$

The separation of variation is not valid for Eq. 6-71 because the model was calibrated in the natural log space. The separation of variation does, however, apply in the natural log space. It is evident that if the correlation coefficient were computed as the ratio of the explained variation to the total variation, then from the values of Eq. 6-74, the correlation coefficient would be greater than 1.0, which is not rational. Thus the standard error of estimate should be used as a measure of accuracy:

$$S_e = \left[\frac{17{,}276{,}427.6}{5 - 2}\right]^{0.5} = 2399.75 \tag{6-75}$$

The standard deviation of the criterion variable is 3470.97; therefore, the power model is not highly accurate, and the computed correlation for the log model (i.e., 0.9851) certainly does not reflect the accuracy of the model.

Example 6-9: Multivariate Power Model

The peak discharge data of Appendix B-4 can be used to illustrate a power model that includes more than one predictor variable. The natural logarithms of the seven predictor variables and the criterion were used as input to a linear multiple regression program, with the resulting linear model:

$$\ln Y = 1.6931 + 0.6795 \ln X_1 + 0.5002 \ln X_2 + 0.1683 \ln X_3 + 0.7009 \ln X_4 + 0.1497 \ln X_5 + 0.1652 \ln X_6 + 1.2606 \ln X_7 \tag{6-76}$$

Equation 6-76 results in a correlation coefficient of 0.955 and a standard error of 0.3429, with the units of S_e as the natural logarithm of cfs. This standard error can obviously not be compared with the standard deviation of Y of 5584 cfs. The sum of the errors for Eq. 6-76 equaled zero, which indicates that the model is unbiased in the natural log space.

Equation 6-76 was transformed to the measured (i.e., $X - Y$) space:

$$\hat{Y} = 5.4365 X_1^{0.6795} X_2^{0.5002} X_3^{0.1683} X_4^{0.7009} X_5^{0.1497} X_6^{0.1652} X_7^{1.2606} \tag{6-77}$$

The values of the predictor variables were used to compute values of $\hat{Y}$, from which the errors e_i were computed. The average error [i.e., $(\sum e)/n$] equaled -215 cfs, which indicates that Eq. 6-77 tends to underestimate the peak discharge. The sum of the squares of the errors was used to compute a standard error of estimate, which was 2121 cfs. This is significantly less than the standard deviation of Y of 5584 cfs; however, the model shows a slight bias. Although a correlation coefficient could be computed for Eq. 6-77, it would not be proper to do so because the Pearson correlation coefficient assumes a linear separation of the total variation, which is not possible with the nonlinear model of Eq. 6-77. It is important to emphasize that the correlation coefficient for Eq. 6-76 cannot be used as an indicator of the accuracy of Eq. 6-77 because the correlation is based on a linear separation of variation in the logarithmic space (i.e., $\ln Y$ versus $\ln X_i$).

A linear model was calibrated using the data of Appendix B-4:

$$\hat{Y} = -11{,}011 + 35.93X_1 + 257.39X_2 + 81.642X_3 + 82.379X_4 + 14.209X_5 - 209.85X_6 + 10{,}336X_7 \tag{6-78}$$

Equation 6-78 resulted in a standard error of estimate of 2983 cfs and a correlation coefficient of 0.863. As with all linear multiple regression equations, the sum of the errors equaled zero, which indicates an unbiased model. In comparing the linear and power models, the power model had a smaller standard error of estimate but was biased. Depending on the weight one places on the standard error and the sum of the residuals, selection of a model from these two is not clear cut. Fortunately, there is an alternative that is superior to both; the alternative will be discussed in Example 7-6.

Trigonometric Transforms

Many hydrologic and meteorological variables follow cyclic patterns. The variation between the daily highs and lows of temperature and relative humidity is an example of fairly rapid cyclic variation. The annual cycle of temperature from winter low to summer high is another. Water temperatures in lakes vary somewhat the same way. Measured evaporation from a pan shows an annual cycle reflecting the cycling of solar energy. Many other examples can be mentioned. Most are linked directly to the solar energy cycle, but some, such as seasonal patterns of rainfall, are indirectly linked through changing patterns of hemispherical atmospheric circulation.

A convenient means of representing cyclical patterns is by use of sine functions. The basic expression is

$$Y = A \sin(\omega t + \alpha) \tag{6-79}$$

In this equation A gives the amplitude of the sine wave, ω governs the cycle rate, and α is a phase-shift angle, so that the sine wave is initially zero when $\omega t = -\alpha$. It is, of course, a linear time scale, and Y is the cyclic phenomenon under study. The value of ω is usually dictated by the problem. For example, if we are studying annual variations in monthly data, and if we wish to evaluate the angle in degrees, ωt is 360° when t is 12 months. Thus, ω is $360/12 = 30°$ per month.

The values of A and α in Eq. 6-79 must usually be evaluated from data. While A can be computed directly by least squares, α cannot. However, we can make use of a basic trigonometric identity to produce a transformed linear version of α. The trigonometric identity is

$$\sin(\omega t + \alpha) = \sin \omega t \cos \alpha + \cos \omega t \sin \alpha \tag{6-80}$$

With this substitution, Eq. 6-79 becomes

$$Y = a \cos \alpha \sin \omega t + a \sin \alpha \cos \omega t \tag{6-81}$$

Now substituting $b_2 = a \cos \alpha$ and $b_3 = a \sin \alpha$, there results

$$Y = b_2 \sin \omega t + b_3 \cos \omega t \tag{6-82}$$

With $\sin \omega t$ as one independent variate and $\cos \omega t$ as a second independent variate, the equation is in linear form for solution. Values of A and α can be obtained from b_2 and b_3 since $\alpha = \tan^{-1}(b_3/b_2)$ and $A = b_2/\cos \alpha$.

The values of Y in Eq. 6-82 will swing from positive to negative across zero. The equation can be modified by adding a vertical shifting coefficient, b_1:

$$Y = b_1 + b_2 \sin \omega t + b_3 \cos \omega t \tag{6-83}$$

A sine wave, including the phase-shifted form in Eq. 6-82, is symmetrical. This form can be modified by using a series of sine terms (or sine–cosine) terms for phase shifting:

$$\begin{aligned} Y = b_1 &+ b_2 \sin \omega t + b_3 \cos \omega t + b_4 \sin \omega 2t + b_3 \cos \omega 2t \\ &+ b_6 \sin \omega 3t + b_7 \cos \omega 3t + \cdots \end{aligned} \tag{6-84}$$

Theoretically, any number of terms can be added by extending the series in Eq. 6-84. Practically, the number of coefficients quickly becomes too large for convenient usage. This is especially true if the cyclic model is intended to be incorporated as one component of a larger and more comprehensive model.

Circular Variates

Some few hydrologic and hydrologically related variates are expressed in circular measurements, normally as direction points of the compass. Complete description of slope of land surface must be expressed both as a tangent from the horizontal and as a direction. The vector average of many slope directions gives the aspect, or the direction of the mean slope of a drainage area. Aspect can be an important differential variate between drainage basins because less solar energy is received on drainage basins with a north aspect than on drainages with a south aspect. Wind direction can be an important variate since, for example, moisture could be advected from a lake to a point of humidity measurement, and heat atmospheric pollutants could be advected from a large urban area to an experimental site.

It is bad practice to associate hydrologic or other phenomena directly with the angular measurement of direction. Although it is numerically simple to write $Y = b_1 A$ where Y (the measurement of the hydrologic phenomenon) is made a function of A (the angular measurement) and a regression coefficient b_1. Figure 6-6 shows a plot of this relationship with A expressed in degrees and b_1 as 0.02 unit per degree. Such a function is discontinuous at 360°. Such discontinuity is unrealistic, since a directional effect at, say, 355° should not be greatly different from the effect at 5° of angle.

A trigonometric transform can be used to generate a directional variate that is circularly continuous. It is the same transformation as used for cyclic phenomena in the preceding section. The transformation is

$$Y = b_1 \sin A + b_2 \cos A \tag{6-85}$$

It is necessary to allow two mathematical degrees of freedom to the optimization process since both magnitude and direction of the directional variate should be derived from the data. Figure 6-6(b) shows three evaluations of Eq. 6-85. Values assigned to the coefficients b_1 and b_2 were sequentially X: +1, +1; +1, −1; and −2.5, 2. The curves numbered 1 and 2, with coefficients equal in magnitude, differ only in direction of effect. The curve numbered 3 illustrates a directional effect oriented nearly along the SE–NW axis, with magnitude appreciably greater than the curves numbered 1 and 2.

Example 6-10: Annual Temperature Variation

Figure 6-7 shows monthly average air temperature for Jefferson City, Tennessee, for the years 1955–1959. These cyclic data can be represented by the sine-cosine series

$$T = b_0 + b_1 \sin \omega M + b_2 \cos \omega M + b_3 \sin 2\omega M + b_4 \cos 2\omega M \tag{6-86}$$

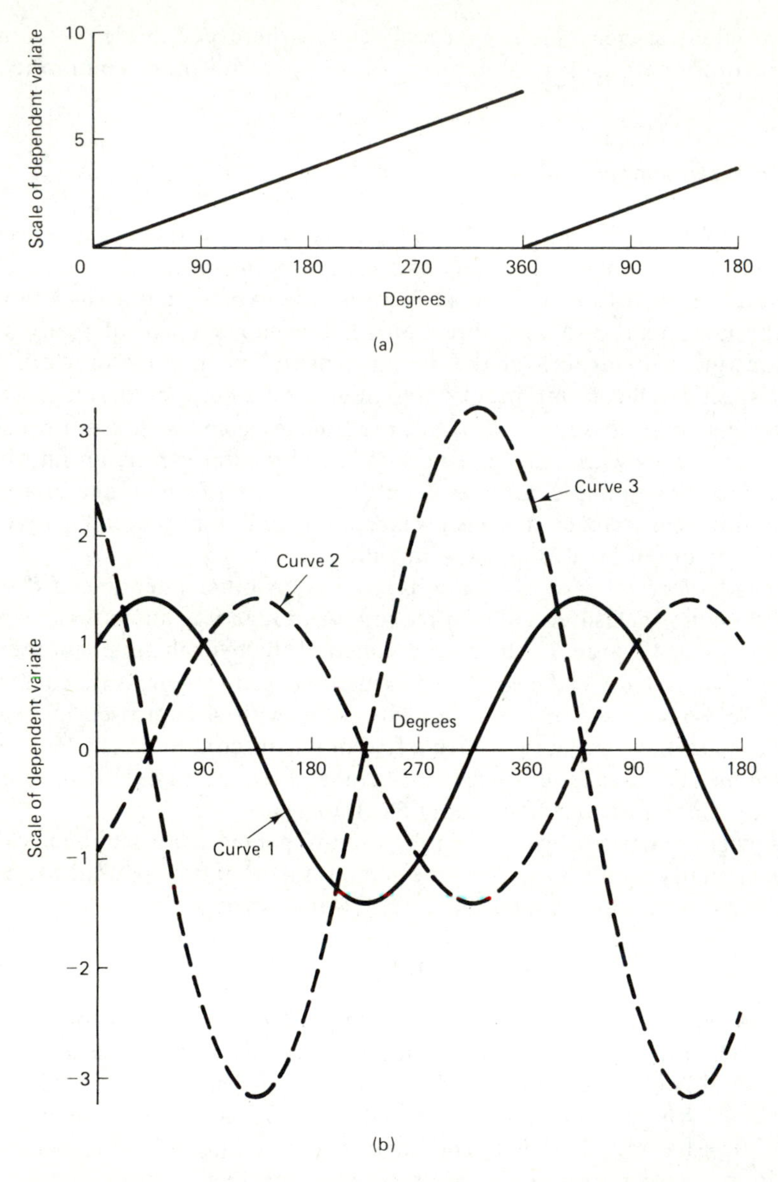

Figure 6-6 Forms of circular functions: (a) discontinuous form; (b) continuous cyclic form.

Given the small sample size it should be of interest to test whether the last two terms are statistically significant. In Eq. 6-86, T is the average monthly air temperature, and M is time measured in months (January = 1, February = 2, etc.). The parameter ω is set by noting that ωM must equal 360 degrees (or a full cycle) when $M = 12$. Therefore, $\omega = 30°$ per month.

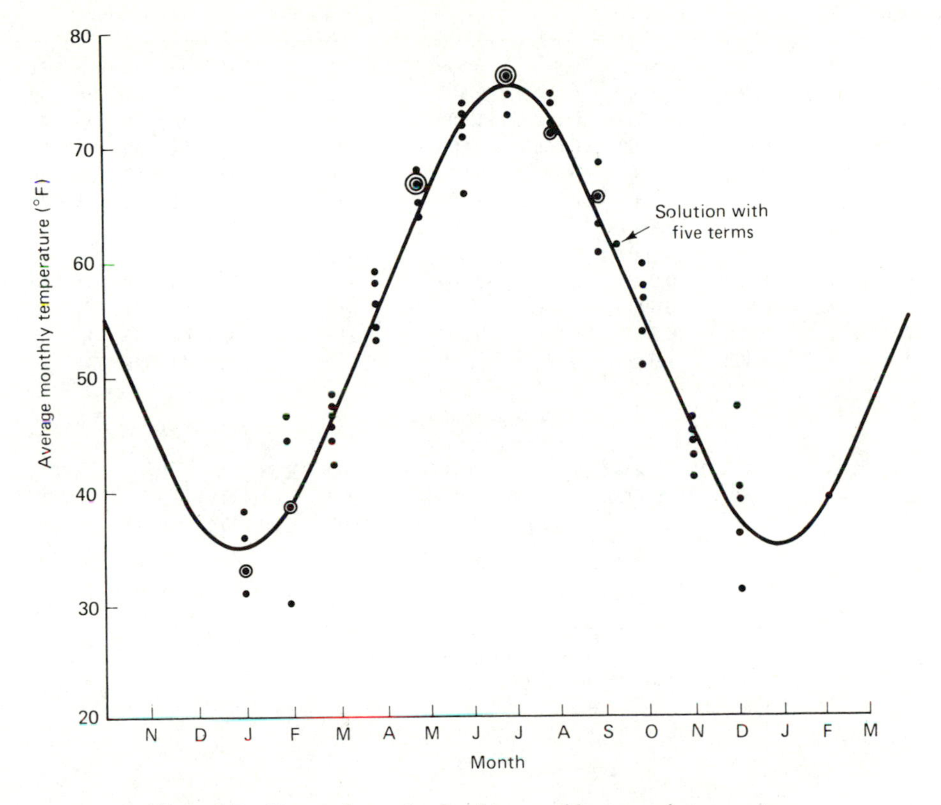

Figure 6-7 Sine-cosine series fitted to monthly mean air temperature.

Equation 6-86 can be written in the more general form

$$T = b_0X_0 + b_1X_1 + b_2X_2 + b_3X_3 + b_4X_4 \tag{6-87}$$

where X_0 is a dummy variable of unity, and the remaining X_i are the trigonometric transforms of the cyclic variable M. Data arranged for least-squares solution of Eq. 6-87 are shown in Table 6-23. Several important properties should be noted. The values of the independent variates are tabulated for only one year by summing the five values per month. This is possible because the curve for all other years will be identical, and thus the sums of products need be computed for only one year. Totals for several years are computed simply by multiplying the one-year totals by the years of record. Also the cross-products with the dependent variable can be computed by first summing the values of the dependent by respective months.

The sums of products are peculiar in that all cross-products of the independent variables add to zero. These elements are all orthogonal. Statistically, we can say that these elements are uncorrelated. The solution of the simultaneous normal equations is therefore trivial since there is no interactive effect of the elements. The solution of the equations without the last two terms is also trivial. Table 6-23 shows only a slight change in correlation caused by deleting these terms. An F test based on Eq. 6-41 shows that the two terms are not very significant.

TABLE 6-23 Least-Squares Solution of Example 6-10

Month	Values of Variables					Sum of Monthly Temperatures	Recorded Temperatures				
	X_0	X_1	X_2	X_3	X_4						
J	1	0.500	0.866	0.866	0.500	171	36	33	38	33	31
F	1	0.866	0.500	0.866	0.500	200	39	45	47	30	39
M	1	1.000	0.000	0.000	−1.000	230	49	48	46	45	42
A	1	0.866	−0.500	−0.866	−0.500	285	59	54	60	57	55
M	1	0.500	−0.866	−0.866	0.500	334	67	67	67	65	68
J	1	0.000	−1.000	0.000	1.000	356	66	71	74	73	72
J	1	−0.500	−0.866	0.866	0.500	376	76	73	76	75	76
A	1	−0.866	−0.500	0.866	−0.500	373	77	73	74	73	76
S	1	−1.000	0.000	0.000	−1.000	341	72	64	69	67	69
O	1	−0.866	0.500	−0.866	−0.500	280	57	60	51	54	58
N	1	−0.500	0.866	−0.866	0.500	219	43	44	45	46	41
D	1	0.000	1.000	0.000	1.000	193	36	47	40	31	39

Sums of products

12	0.000	0.000	0.000	0.000	3358
	6.000	0.000	0.000	0.000	−331.488
		6.000	0.000	0.000	−529.120
			6.000	0.000	1.732
				6.000	−41.000

Coefficients for normal equations

60	0	0	0	0	3358
	30	0	0	0	−331.488
		30	0	0	−529.120
			30	0	1.732
				30	−41.000

Solution for five terms

$b_0 = 55.967, \quad b_1 = -10.050, \quad b_2 = -17.637, \quad b_3 = 0.058, \quad b_4 = -1.366$

Solution for three terms

$b_0 = 55.967, \quad b_1 = -10.050, \quad b_2 = -17.637$

Evaluation of fit

Five terms	Three terms
$SSy = 200{,}974$	
$Sy = 3{,}358$	
$SSy = 13{,}037.93$	
$SSR = 200{,}354.31$	$SSR = 200{,}298.18$
$SS_c = 619.69$	$SS_e = 675.82$
$SS_t = 12{,}418.24$	$SS_t = 12{,}362.11$
$r = 0.976$	$r = 0.974$

Significance

$$F = \frac{(12418.24 - 12362.11)/2}{619.69/(60-5)} = \frac{28.06}{11.27} = 2.49$$

$$F(2, 60, 0.95) = 3.15$$

Additional Nonlinear Model Forms

In addition to the polynomial and power models, other forms can provide good approximations to the underlying population relationship between two or more variables. An exponential model has the form

$$\hat{Y} = b_0 \, e^{b_1 X} \tag{6-88}$$

in which b_0 and b_1 are coefficients requiring values by fitting to data. Values for b_0 and b_1 can be obtained by linear bivariate regression after taking the logarithms of both sides of Eq. 6-88:

$$\ln_e \hat{Y} = \ln b_0 + b_1 X \tag{6-89}$$

In this case Y is transformed; thus the correlation coefficient and standard error of estimate apply to the $\ln_e Y$ versus X space and not the Y versus X space. Also, the intercept coefficient obtained from the regression analysis for Eq. 6-89 must be transformed for use with Eq. 6-88.

A logarithmic curve can also be used to fit values of Y and X:

$$\hat{Y} = b_0 + b_1 \ln X \tag{6-90}$$

In this case Y is not transformed, so the values of the correlation coefficient and the standard error of estimate are valid indicators of the goodness of fit in the Y space.

Segmented Regression Functions

Segmented regression functions are composed of a series of connected linear segments, as shown in Fig. 6-8. The angle points where the segments join are called nodes. At each node an ordinate value Y_i can be specified. Any linear segment, j, can be expressed as

$$y = a_j + b_j x_j \tag{6-91}$$

where x_j is the local scale from $-\frac{1}{2}$ to $\frac{1}{2}$ for the segment. The parameters a_j and b_j can be evaluated by making the line segment pass through the nodal ordinates Y_i and Y_{i+1} at the left and right ends of the segment. Such a solution will yield an alternative equation for the line:

$$y = (\tfrac{1}{2} - x_j)\, Y_i + (\tfrac{1}{2} + x_j)\, Y_{i+1} \tag{6-92}$$

The nodes may be placed at any point in the global hydrologic scale, and when their locations are known, any point in the h scale can be converted to a value in an x_j scale. The y value of the point does not change. In Eq. 6-92 let $w_i = \frac{1}{2} - x_j$ and $w_{i+1} = \frac{1}{2} + x_j$. Then the w's may be regarded as weighting coefficients of the nodal ordinates as

$$y = w_i Y_i + w_{i+1} Y_{i+1} \tag{6-93}$$

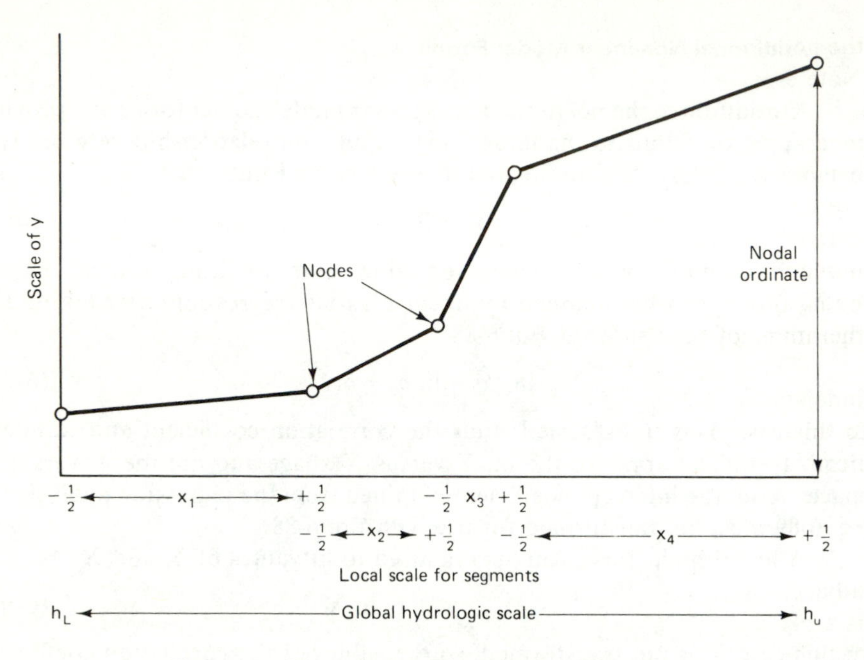

Figure 6-8 Schematic for a piecewise linear function.

Now note that if we had a series of data points in scale (h, y), we could first identify which segment each point falls in, and then we could calculate the appropriate w_i and w_{i+1}. Since we now have values for Y and the w's in Eq. 6-93, we may treat it as a regression equation and calculate "best" values of the nodal ordinates Y_i as regression coefficients. To illustrate, Table 6-24 shows weights for the three nodal values for two connected linear segments. The data points are at the quarter- and half-points of the two segments. Point 4 is located at the middle node. Given values of the dependent variable, y,

TABLE 6-24 Example of Weights for a Segmented Regression Function

Data Point	Weights for Node Y_1	Y_2	Y_3	Data
1	0.75	0.25		y_1
2	0.50	0.50		y_2
3	0.25	0.75		y_3
4		1.0		y_4
5		0.75	0.25	y_5
6		0.50	0.50	y_6
7		0.25	0.75	y_7

the nodal ordinates Y_1, Y_2, and Y_3 can be evaluated directly by least squares. Note especially that Y_2 is required common to the two segments.

One advantage of the piecewise linear structures is that it is not necessary to assume the mathematical form for an unknown functional relationship between variables. Because the nodes express the shape, and because these are derived by least squares, an approximation to the unknown function is derived from the data. This method of analysis is somewhat analogous to numerical solution of differential and integral equations. In such solutions, nodal ordinates are derived analytically, usually under the assumption that the unknown function is linear between nodes.

The segmented approach to data analysis can be readily extended to two independent variables and one dependent variable. This method can be useful in analyzing mapped data, in which case the two independent variables are north–south and east–west coordinates in space, and the dependent variable is the value of some property at the points. We can visualize weather maps of isolines of pressure or temperature, or contours of elevation in a watershed.

The unit structure of Fig. 6-8 was a straight line through two nodes. Two adjacent lines have a common node. The unit structure of a piecewise surface is a plane defined by four nodes as in Fig. 6-9(a). Two adjacent planes have the line between two nodes in common. A total problem space in independent variables h_1 and h_2 is made up of an ensemble of unit planes as illustrated in Fig. 6-9(b). The property under study, the dependent variable y, shows as segmented contours in the $h_1 - h_2$ space. The $h_1 - h_2$ space may be physical space, but obviously, problems can be visualized in mathematical space of independent variables.

To extend from one to two independent variables, the parameters a and b of Eq. 6-91 are made linear functions of a second variable z. This substitution yields

$$y = (a_1 + a_2 z) + (b_1 + b_2 z)x = a_1 + a_2 z + b_1 x + b_2 xz \tag{6-94}$$

Equation 6-94 is the expression for the plane of Fig. 6-9(a). By requiring the plane to pass through the four corner nodes we can derive nodal weights as functions of x and z:

$$\begin{aligned} w_{11} &= \tfrac{1}{4} - \tfrac{1}{2}z - \tfrac{1}{2}x + xz \\ w_{12} &= \tfrac{1}{4} - \tfrac{1}{2}z + \tfrac{1}{2}x - xz \\ w_{21} &= \tfrac{1}{4} + \tfrac{1}{2}z - \tfrac{1}{2}x - xz \\ w_{22} &= \tfrac{1}{4} + \tfrac{1}{2}z + \tfrac{1}{2}x + xz \end{aligned} \tag{6-95}$$

The prediction of y at any point in the unit plane is then given by

$$\hat{y} = w_{11} Y_{11} + w_{12} Y_{12} + w_{21} Y_{21} + w_{22} Y_{22} \tag{6-96}$$

Equation 6-96 is the two-dimensional analogy to Eq. 6-93. Note that Eq. 6-96

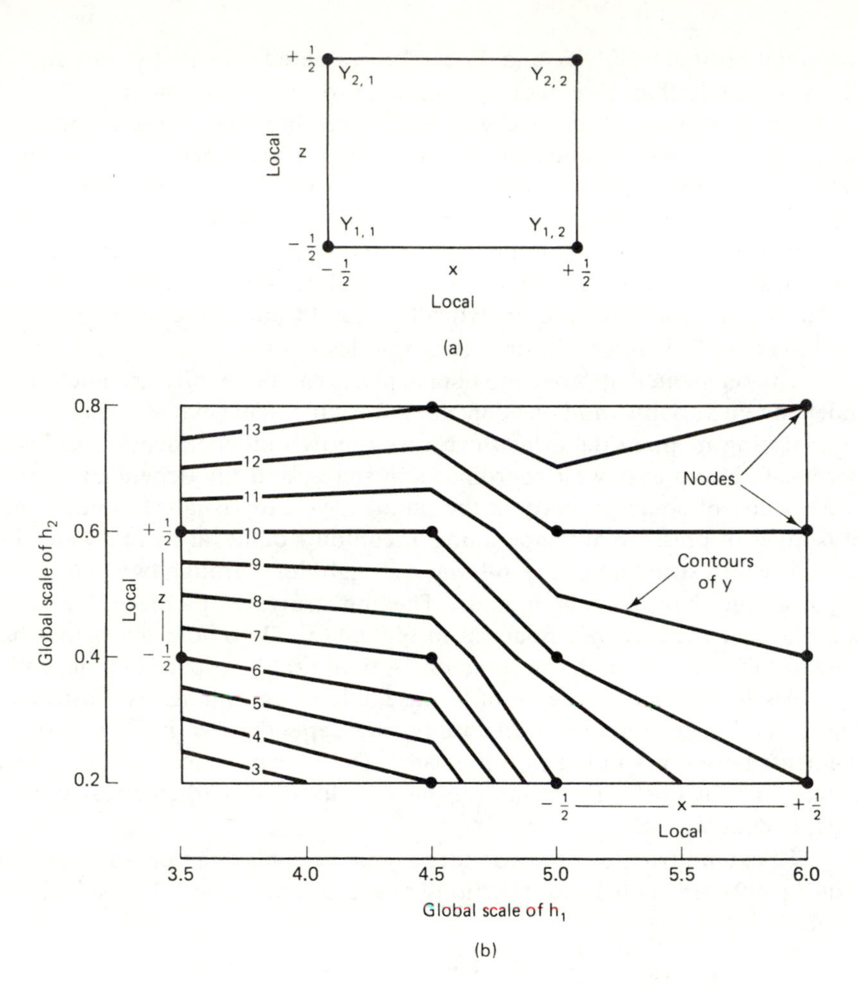

Figure 6-9 Schematic for segmented linear surfaces.

is still a linear structure that can be solved by least squares on a data set to provide "best" values of the nodal ordinates, Y_{ij}. Table 6-25 illustrates the arrangement of weights for data points at the center of two adjacent planes, and at the center of the common boundary between the planes.

Segmented functions and surfaces are convenient and versatile structures. We should note that the nodal values derived as regression coefficients are direct values of the function or surface at the nodes. Then Eq. 6-31, which gives the variance of a regression coefficient, can be used to compute the variance of the function or surface at the nodes directly. A further advantage of segmented functions is that data points only weight to the nearest nodes. This restricted weighting tends to localize the influence of any one data point upon the derived function.

TABLE 6-25 Example of Weights for Segmented Regression Surfaces

Data Point	Weights for Node						Data
	Y_{11}	Y_{21}	Y_{12}	Y_{22}	Y_{13}	Y_{23}	
1	0.25	0.25	0.25	0.25			y_1
2			0.25	0.25	0.25	0.25	y_2
3			0.50	0.50			y_3

A disadvantage of segmented structures is the requirement to solve relatively large sets of simultaneous equations. For example, Fig. 6-9(b) is based on 16 nodes, and their derivation by least squares would require solution of 16 simultaneous normal equations. Also, at least one data point must be located within each line segment or plane segment. If errors in the dependent variable are to be smoothed, a minimum of three or four points in each segment is desired. Nodal values that are not rational may result from deficiencies in data.

EXERCISES

For each of the following problems, use the data, which consist of a criterion variable and three predictor variables; **(a)** plot all combinations of variables (six graphs); **(b)** compute the correlation matrix; **(c)** discuss the degree of intercorrelation and predictor-criterion correlation and the potential for accurate predictions of Y.

6-1.

Y	X_1	X_2	X_3
1	1	3	5
1	2	6	3
4	4	1	4
4	5	2	2
6	6	1	3
2	3	5	1

6-2.

Y	X_1	X_2	X_3
0	1	3	5
5	3	1	2
3	3	4	7
7	2	6	3
10	5	4	4
4	4	2	6

6-3. Using the principle of least squares, derive the normal equations for the model $\hat{y} = b_1X_1 + b_2X_2 + b_3X_3$.

6-4. What is a partial regression coefficient? Standardized partial regression coefficient? What is the relationship between the two coefficients?

6-5. Using the standardized model $Z_y = \sum_{i=1}^{p} t_iZ_i$ and the equation for a standardized variable $Z = (X - \bar{X})/S$, derive the raw score model.

6-6. Find the determinant of each of the following matrices.

(a) $A = \begin{vmatrix} 1 & 0.2 \\ 0.2 & 1 \end{vmatrix}$ **(b)** $B = \begin{vmatrix} 1 & 0.8 \\ 0.8 & 1 \end{vmatrix}$ **(c)** $C = \begin{vmatrix} 1 & 0.2 & 0.2 \\ 0.2 & 1 & 0.2 \\ 0.2 & 0.2 & 1 \end{vmatrix}$

(d) $D = \begin{vmatrix} 1 & 0.2 & 0 \\ 0.2 & 1 & 0.2 \\ 0 & 0.2 & 1 \end{vmatrix}$ **(e)** $E = \begin{vmatrix} 1 & 0.2 & 0 \\ 0.2 & 1 & 0 \\ 0 & 0 & 1 \end{vmatrix}$ **(f)** $F = \begin{vmatrix} 1 & 0.8 & 0.8 \\ 0.8 & 1 & 0.8 \\ 0.8 & 0.8 & 1 \end{vmatrix}$

(g) $G = \begin{vmatrix} 1.00 & -0.22 & 0.58 & 0.82 \\ -0.22 & 1.00 & -0.26 & -0.30 \\ 0.58 & -0.26 & 1.00 & 0.75 \\ 0.82 & -0.30 & 0.75 & 1.00 \end{vmatrix}$

(h) $H = \begin{vmatrix} 1.00 & 0.34 & -0.17 & -0.44 \\ 0.34 & 1.00 & -0.05 & -0.19 \\ -0.17 & -0.05 & 1.00 & 0.07 \\ -0.44 & -0.19 & 0.07 & 1.00 \end{vmatrix}$

6-7. For the data base that follows, derive the least-squares estimates of the regression coefficients for the model $y = b_0 + b_1X_1 + b_2X_2$. (*Note:* The computations can be simplified by subtracting the mean value from each observation; if such a transformation is made, it is necessary to adjust the computed value of the intercept.)

Y	5	2	11	8	9
X_1	1	2	3	4	5
X_2	16	11	14	12	7

6-8. By creating a new predictor variable for the term involving the product X_1X_2, show the normal equations for the following model:

$$y = b_0 + b_1X_1 + b_2X_2 + b_3X_1X_2$$

Using the data base from Exercise 6-7, insert the values of the summations into the normal equations. (*Note:* It is not necessary to solve for the values of the unknowns.)

6-9. Compute the coefficient of multiple determination, the standard error of estimate, and the standardized partial regression coefficients for the regression model of Exercise 6-7. Use these values to provide an interpretation of the goodness of fit and the relative importance of the predictor variables.

For Exercises 6-10 to 6-15, do the following:

(a) Plot each pair of the variables and discuss (1) the degree of correlation between each pair (low, moderate, high); (2) the possible effect of intercorrelation between predictor variables; (3) the apparent relative importance of the predictor variables.
(b) Compute the correlation matrix.
(c) Compute the determinant of the matrix of intercorrelations between the predictor variables.
(d) Compute the partial and standardized partial regression coefficients.
(e) Compute the goodness-of-fit statistics (R, S_e) and discuss the implications.
(f) Discuss the relative importance of the predictor variables.
(g) Compute and analyze the residuals.

6-10. The peak discharge (Q) of a river is an important parameter for many engineering design problems, such as the design of dam spillways and levees. Accurate estimates can be obtained by relating Q to the watershed area (A) and the watershed slope (S). Perform a regression analysis using the following data.

A (sq mi)	S (ft/ft)	Q (cfs)
36	0.005	50
37	0.040	40
45	0.004	45
87	0.002	110
450	0.004	490
550	0.001	400
1200	0.002	650
4000	0.0005	1550

6-11. Sediment is a major water-quality problem in many rivers. Estimates are necessary for the design of sediment control programs. The sediment load (Y) in a stream is a function of the size of the contributing drainage area (A) and the average stream discharge (Z). Perform a regression analysis using the following data.

A ($\times 10^3$ sq mi)	Z (cfs)	Y (millions of tons/yr)
8	65	1.6
19	625	6.4
31	1,450	3.0
16	2,400	1.6
41	6,700	19.0
24	8,500	15.0
3	1,500	1.2
3	3,500	0.4
3	4,300	0.3
7	12,000	1.0

6-12. Snowmelt runoff rates (M), which serve as a source of water for irrigation, water supply, and power, are a function of the precipitation (P) incident to the snowpack and the mean daily temperature (T). Using the following data, derive a relationship for predicting M from P and T.

P (cm/day)	T (°C)	M (cm/day)	P (cm/day)	T (°C)	M (cm/day)
0.0	1	0.24	0.6	8	1.14
0.5	2	0.44	0.0	11	1.57
0.2	3	0.49	0.0	12	1.58
0.0	3	0.54	0.3	13	1.68
0.0	5	0.78	0.2	16	2.10
0.8	6	0.88	0.4	18	2.41
0.5	7	1.01			

6-13. Visitation to lakes for recreation purposes varies directly as the population of the nearest city or town and inversely as the distance between the lake and the city. Develop a regression equation to predict visitation (V) in person-days per year, using as predictors the population (P) of the nearby town and the distance (D) in miles between the lake and the city.

D	P	V
10	10,500	22,000
22	27,500	98,000
31	18,000	24,000
18	9,000	33,000
28	31,000	41,000
12	34,000	140,000
21	22,500	78,000
13	19,000	110,000
33	12,000	13,000

6-14. The stream reaeration coefficient (r) is a necessary parameter for computing the oxygen deficit sag curve for a stream reach. The coefficient is a function of the mean stream velocity in feet per second (X_1), water depth in feet (X_2), and the water temperature in degrees Celsius (X_3). Use the following data for the steps outlined above.

X_1	X_2	X_3	r	X_1	X_2	X_3	r
1.4	1.3	14	2.89	3.5	2.6	25	3.58
1.7	1.3	23	4.20	3.9	2.9	19	3.06
2.3	1.5	17	4.17	4.1	3.3	13	2.41
2.5	1.6	11	3.69	4.7	3.4	24	3.30
2.5	1.8	20	3.78	5.3	3.5	26	3.05
2.8	1.9	15	3.56	6.8	3.7	23	4.18
3.2	2.3	18	3.35	6.9	3.7	21	4.08
3.3	2.5	11	2.69				

6-15. Interception losses (I) vary with the rainfall rate (P) and the type of cover. For a small grain, the interception is also a function of the average crop height (h). The following data were measured during rainstorms from an experimental plot.

h (ft)	P (in.)	I (in.)	h (ft)	P (in.)	I (in.)
0.83	0.68	0.04	1.52	0.84	0.10
0.95	1.43	0.11	1.68	1.33	0.19
1.22	0.92	0.10	1.82	1.71	0.25
1.31	1.28	0.14	1.97	0.62	0.12
1.35	1.57	0.19	2.11	0.79	0.14

6-16. Find the least-squares estimates of the coefficients for the model $\hat{y} = b_1X_1 + b_2X_1X_2$. Using the data of Exercise 6-15, evaluate the model $I = b_1h + b_2hP$.

6-17. On the first step of a stepwise regression analysis, why are the partial correlation coefficients for each predictor variable equal to the predictor–criterion correlation?

6-18. Develop a table with column headings: (1) criteria, (2) indication of, (3) advantages, (4) disadvantages, (5) decision mode, and (6) rank. Under the column "criteria" list each of the criteria that are used in selecting a final model. In the column "rank" specify the order of importance of each of the criteria, with 1 given to the most important, 2 the next most important, and so on.

6-19. Using the data base of Exercise 6-14, perform a stepwise regression. At each step compute **(a)** partial F statistic to enter for each variable; **(b)** total F statistic; **(c)** partial F statistic for deletion; **(d)** multiple correlation coefficient; **(e)** standard error of estimate; **(f)** the regression equation; **(g)** standardized partial regression coefficients. Use a 5% level of significance to determine the final equation.

6-20. The results of a stepwise regression analysis are given in Tables 6-26 and 6-27, including the correlation coefficient matrix, the means and standard deviations of the six predictor variables, the partial F statistics for entering and deletion, the multiple correlation coefficient, the regression coefficients, and the standard error of the regression coefficients. Perform a complete regression analysis to select a model for prediction. State the reasons for selecting the model and make all additional computations that are necessary to support your decision. The analysis is based on a sample size of 71.

6-21. Transform the predictor variable for the model

$$y = a + b\sqrt{x}$$

so that the coefficients can be evaluated using multiple regression. Obtain estimates of the unknowns a and b using the data base

y	1.37	1.39	1.61	1.68	1.79	1.76
x	0.1	0.2	0.3	0.4	0.5	0.6

Compute the correlation coefficient and standard error of estimate, and compare the goodness-of-fit statistics with those for a linear model of y on x.

6-22. Transform the model

$$y = 10^{a+bx}$$

to linear form so that the coefficients can be evaluated using the principle of least squares. Obtain estimates of the unknowns a and b using the data base

y	2	3	5	8	13
x	1	2	3	4	5

Compute the correlation coefficient and the standard error of estimate.

6-23. Transform the model

$$y = e^{a+bx}$$

TABLE 6-26 Correlation Matrix and Statistics for Exercise 6-20 (n = 71)

		Correlation Matrix								Standard
		X_1	X_2	X_3	X_4	X_5	X_6	Y	Mean	Deviation
Variable	X_1	1.000	−0.139	0.181	−0.025	−0.142	−0.278	−0.300	38.72975	1.21257
	X_2		1.000	−0.601	0.048	0.159	0.239	0.124	169.54930	129.53981
	X_3			1.000	−0.040	−0.251	−0.103	0.133	24.93239	16.76210
	X_4				1.000	0.157	0.034	0.125	63.09859	19.46363
	X_5					1.000	0.433	0.134	25.64789	6.82872
	X_6						1.000	0.504	1.62113	0.28531
	Y							1.000	−0.09730	0.69231

TABLE 6-27 Results of Stepwise Regression for Exercise 6-20

	Step					
	1	2	3	4	5	6
Partial F to enter						
X_1		2.64	3.97	—	—	—
X_2		0.00	1.99	2.21	—	—
X_3		3.31	—	—	—	—
X_4		1.08	1.26	1.27	1.21	—
X_5		0.81	0.26	0.24	0.13	0.29
X_6		—	—	—	—	—
Correlation coefficient	0.5037	0.5369	0.5028	0.5915	0.6015	0.6039
Variable entered	X_6	X_3	X_1	X_2	X_4	X_5
Regression coefficient						
b_0	−2.078	−2.347	2.487	2.416	2.154	2.195
b_1	—	—	−0.1201	−0.1221	−0.1214	−0.1209
b_2	—	—	—	0.00101	0.00100	0.00096
b_3	—	0.00772	0.00907	0.01365	0.01373	0.01304
b_4	—	—	—	—	0.00390	0.00420
b_5	—	—	—	—	—	0.00630
b_6	1.222	1.269	1.135	1.0501	1.0442	1.1092
Standard error of b_i						
$S_e(b_1)$	—	—	0.0603	0.05978	0.0597	0.0600
$S_e(b_2)$	—	—	—	0.00068	0.00068	0.00069
$S_e(b_3)$	—	0.00425	0.00421	0.00519	0.00518	0.00537
$S_e(b_4)$	—	—	—	—	0.00353	0.00359
$S_e(b_5)$	—	—	—	—	—	0.01171
$S_e(b_6)$	0.252	0.2496	0.2534	0.2575	0.2571	0.2854
Partial F to delete						
X_1	—	—	3.97	4.17	4.14	4.06
X_2	—	—	—	2.21	2.14	1.95
X_3	—	3.30	4.63	6.92	7.02	5.90
X_4	—	—	—	—	1.22	1.36
X_5	—	—	—	—	—	0.29
X_6	23.45	25.84	20.07	16.63	16.49	15.11

to a linear form so that the coefficients can be evaluated using least squares. Obtain estimates for the regression coefficients a and b using the data base

y	20	24	30	36	40
x	1	1.5	2.0	2.5	3.0

Compute the correlation coefficient and standard error of estimate.

6-24. Using the data base for the drainage area (A) and discharge (Q) from Exercise 6-10, compute the regression coefficients for a bivariate power model with the discharge as the criterion variable. Determine the correlation coefficient and standard error of estimate for the linear and log-log spaces.

6-25. Using the following measurements of y and x, perform a bivariate polynomial regression analysis. Use a partial F test with a 5% level of significance to determine whether the model should be linear, quadratic, or cubic. Compute the correlation coefficient, standard error of estimate, and total F statistic for each model.

y	1.7	2.8	4.4	6.5	9.0
x	2	4	6	8	10

6-26. Using the following measurements of y and x, perform a regression analysis for a bivariate polynomial. Use a partial F test with a 1% level of significance to decide whether the model should be linear, quadratic, or cubic. Compute the correlation coefficient, standard error of estimate, and the total F statistic for each model.

y	−2.36	−2.37	−2.00	−1.43	−0.84
x	−2	−1	0	1	2

6-27. Given the values of y and x_2, determine the values of the coefficients for the model $y = b_0 + b_1 x + b_2 x$.

y	5	8	11	13	15	19
x	1	2	3	4	5	6

6-28. Place the equation $y = b_0(x_1 + 1)^{b_1} x_2^{b_2 - 1}$ in a form that can be calibrated using multiple regression. Show the equations for estimating the coefficients.

6-29. Use the principle of least squares to derive formulas for computing the partial regression coefficients for the nonlinear equation $y = b_1 x + b_2 x^2$.

6-30. Write a computer program to calculate the weights for data points for piecewise linear functions. The program should allow for varying numbers of points and varying numbers of nodes at varying locations. Test on the following data set.

	x	y
1	0.2	0.4
2	1.0	0.7
3	1.0	1.5
4	1.8	1.5
5	2.3	1.7
6	2.7	3.8
7	3.0	3.6
8	3.6	4.9
9	4.5	5.1
10	5.1	4.7
11	6.0	5.5

6-31. Use packaged software to compute the value of the nodes by least squares for the data weights calculated in Exercise 6-30. Plot the derived piecewise function and the data points.

6-32. Write a computer program to simulate the variability of prediction using the derived segments function of Exercise 6-31. Method: consider sweeping through the x range of the function n times. For each sweep calculate a random value for each node using its standard deviation. Next calculate y values on the randomly placed function at tenth points of the total range of x. Finally, add a random departure to each of the tenth point y values using the standard deviation of residuals. Plot the random tenth-point y values for all sweeps. Discuss the distribution of random nodal values and distribution of y values as compared to standard variance of prediction equations.

6-33. Write a computer program to calculate the weights for piecewise linear surfaces, similar to that for piecewise linear functions in Exercise 6-30.

6-34. Use packaged software to compute the values of the nodes by least squares for the data weights calculated in Exercise 6-33. Plot the derived nodal values at their grid points in $x - z$ space and then draw surface contours by interpolating between nodal values along the boundaries of the pieces.

REFERENCES AND SUGGESTED READINGS

ANDERSON, R. L., and T. A. BANCROFT, *Statistical Theory in Research*, McGraw-Hill, New York, 1952.

BENSON, M. A., *Factors Influencing the Occurrence of Floods in a Humid Region of Diverse Terrain*, U.S. Geological Survey Water-Supply Paper, 1580-B, U.S. Government Printing Office, Washington, D.C., 1964.

BENSON, M. A., *Factors Affecting the Occurrence of Floods in the Southwest*, U.S. Geological Survey Water-Supply Paper 1580-D, U.S. Government Printing Office, Washington, D.C., 1964.

BOX, G., Use and Abuse of Regression, *Technometrics*, Vol. 8, No. 4, pp. 625–29, November 1966.

BUCKETT, J., and F. OLIVER, Fitting the Pearson Type 3 Distribution in Practice, *Water Resources Research*, Vol. 13, No. 5, pp. 851–52, October 1977.

HAHN, G., The Hazards of Extrapolation in Regression Analysis, *Journal of Quality Technology*, Vol. 9, No. 4, pp. 159–65, October 1977.

RENARD, K. G., J. C. DRISSEL, and H. B. OSBORN, Flood Peaks from Small Southwest Range Watershed, *Journal of the Hydraulics Division, ASCE*, Vol. 96, No. HY3, Proc. Paper 7161, pp. 773–85, March 1970.

SNYDER, W. M., Hydrograph Analysis by the Method of Least Squares, *Proceedings of the ASCE*, Vol. 81, Paper No. 793, 1955.

SNYDER, W. M., Summary and Evaluation of Methods for Detecting Hydrologic Changes, *Journal of Hydrology*, Vol. 12, pp. 311–38, 1971.

STEDINGER, J. R., Confidence Intervals for Design Events, *Journal of Hydraulic Engineering, ASCE*, Vol. 109, No. 1, pp. 13–27, January 1983.

TASKER, G. D., Hydrologic Regression with Weighted Least Squares, *Water Resources Research*, Vol. 16, pp. 1107–13, December 1980.

7 Numerical Optimization

INTRODUCTION

Our discussions of multiple-variable models have been concerned with models in the linear form of Eq. 6-10 or with those that can be cast in that form through proper variate transforms. The technique that we have used to quantify our models by matching to data was the method of least squares. Before proceeding to nonlinear models we should develop a somewhat more comprehensive idea of what least squares actually is.

The basic definition of least squares is that the sum of the squares of the differences between observed values and values calculated from the model shall be a minimum. In other words, we want the values of the coefficients of our models to be such that this minimum is achieved. Achieving this minimum is the objective we have in mind in getting optimum correspondence between the model and the data. We should not think that this is the only objective we can have in achieving optimum correspondence. For example, we could state an objective of having the sum of the absolute values of the differences a minimum.

It should be clearly recognized that to this point we have only stated our objective. We have said nothing about how we will accomplish this objective numerically. It is not until we use the properties of the objective, as in Eqs. 6-8, that we state our approach to the objective. We are not limited to the approach expressed in Eqs. 6-8. We could follow some tedious trial-and-error procedure, changing values of our parameters until we assure ourselves that we have found the values giving the minimum residual sum of squares. Fortunately, in linear models we do not have to do this. The unique values of the parameters that accomplish our objective are provided by the simultaneous solution of equations such as Eqs. 6–11.

From this discussion we should retain two concepts in regard to model evaluation: (1) we must state an objective under which we will accept the values of our parameters as unique, and (2) we must state the numerical procedure by which we will accomplish our objective. These two concepts must carry over into our discussion of nonlinear models.

Many physical hydrologic processes cannot be adequately represented by linear models. We might, for example, wish to evaluate the parameters in a model taking a mathematical form as in the equation

$$Y = aX_1^b X_2^c \tag{7-1}$$

The shape parameters b and c cannot be evaluated by least squares. It is possible to transform Eq. 7-1 by logarithms to a linear form. However, our objective in evaluation must then be started in terms of log Y. This change of objective may produce strong distortion, deemphasizing large values of Y and emphasizing small values of Y.

Consider further that Eq. 7-1 has an extremely limited structure. When either X_1 or X_2, or both, are zero, Y must be zero. Stated in terms of process, the effect of either X on Y begins at the origin of measurement of the X. Unless we can categorically state this as a fact, we must allow the effect on Y to begin at some value other than the zero of X. We do this by writing

$$Y = a(X_1 - d)^b (X_2 - f)^c \tag{7-2}$$

Now d and f are thresholds of X_1 and X_2. They could also be called *shift parameters.*

It is not possible to linearize Eq. 7-2 through logarithmic transform. Even if we set the shape parameters (b and c) to unity, we cannot achieve a linear expression. The expansion of such a reduced form produces

$$Y = aX_1X_2 - adX_2 - afX_1 + adf \tag{7-3}$$

Equation 7-3 contains four terms but only three parameters. Although the equation does express a linear combination of terms, the products of coefficients are not linear. From the simple examples above, it is quite evident that we must develop some numerical procedures capable of handling models with nonlinear parameters. As stated earlier, trial-and-error procedures could be employed. However, this is impractical for any rational number of model parameters. If a model contains p parameters, and we try m levels of each parameter, the number of parameter combinations is m^p. Table 7-1 shows the fantastic number of trials needed to evaluate all combinations in order to find the best set. It is obvious that some other optimization process is needed, one

TABLE 7-1 Combinations of m^P

Parameters, P	Number of Levels, (m) 3	4	5
5	2.4×10^2	1.0×10^3	3.1×10^3
10	5.9×10^4	1.0×10^6	9.8×10^6
15	1.4×10^7	1.1×10^9	3.1×10^{10}
20	3.5×10^9	1.1×10^{12}	9.5×10^{13}

that will produce a continuity of parameter search. Such continuity must bridge across the vast number of discrete combinations in trial and error and produce near-optimal values with relatively few iterations. The field of numerical optimization provides methods for finding best fit values of the coefficients of models with nonlinear structures. Nonlinear least squares is probably the most widely used numerical optimization procedure and is sometimes referred to as the *method of differential correction.* Other numerical optimization methods use the basic concept of least squares but use strategies to increase the fitting efficiency.

THE PHASES OF NUMERICAL OPTIMIZATION

Before discussing methods of evaluating the coefficients of Eq. 7-2, it is important to put the problem in perspective. We have a nonlinear model, such as Eq. 7-2, that cannot be evaluated using the analytical derivation of the normal equations. But we seek a systematic method of evaluating the coefficients and still find some optimal value of an objective function. There is no reason to reject the principle of least squares solely because it cannot be evaluated analytically. One alternative is a numerical evaluation of the derivatives of the objective function. To reduce the computation effort, it is best to view the optimization process as a three-phased solution procedure. To perform the numerical optimization it is instructive to present the three phases as movement across a response surface.

A response surface is defined as the relationship between an objective function and the unknowns. For example, let's assume that the objective function, F, is a function of a single unknown, b:

$$F = 1.7b^2 - 3b + 2.4 \tag{7-4}$$

Also, let's assume that we cannot find the minimum of F using analytical techniques. If we graph the relationship between F and b (see Fig. 7-1) for values of b from 0 to 1.5 in increments of 0.25, we see that the resulting response surface has a minimum in the interval between 0.75 and 1.0. It is not unreasonable to believe that if we could start at some point on the response

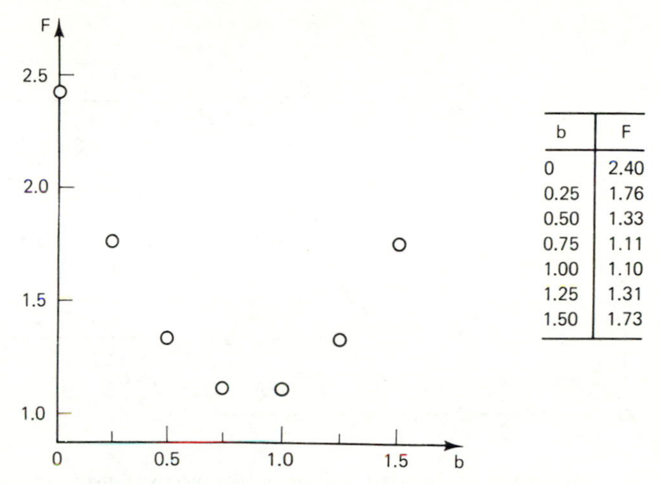

b	F
0	2.40
0.25	1.76
0.50	1.33
0.75	1.11
1.00	1.10
1.25	1.31
1.50	1.73

Figure 7-1 One-dimensional response surface for objective function of Eq. 7-4.

surface and move systematically along the surface, we could find the value of b that produces the minimum value of F. Thus we would need both a place to start and a systematic method for finding the minimum.

The objective function of Eq. 7-4 is neither a statistical function (i.e., it does not depend on data) nor a function of some model. Let's illustrate a response surface model that does both. Specifically, let's use the linear regression model and the principle of least squares. Thus the objective function is

$$F = \sum_{i=1}^{n} (\hat{Y}_i - Y_i)^2 \tag{7-5}$$

and the model is

$$\hat{Y}_i = b_0 + b_1 X_i \tag{7-6}$$

To show the response surface, we will use the following data, with $n = 3$:

Y	1	1	2
X	1	2	3

The response surface can be computed for values of b_0 and b_1 using the data above. For values of b_0 and b_1 in the range from 0 to 1, the response surface is plotted in Fig. 7-2. For Fig. 7-2 one could guess that the optimum is near $(b_0, b_1) = (0.5, 0.5)$, with the optimum value of b_0 probably slightly less than 0.5. In any case, we could move across the response surface if we had a place to start (i.e., values of b_0 and b_1) and a systematic procedure of moving from the initial point to the optimum. Numerical optimization is nothing more than a systematic procedure for moving across the response surface with the objective of approaching a stationary point (see Chapter 3).

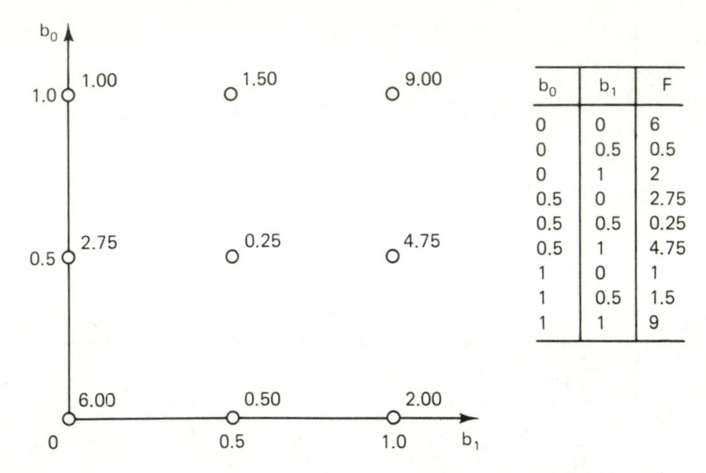

b_0	b_1	F
0	0	6
0	0.5	0.5
0	1	2
0.5	0	2.75
0.5	0.5	0.25
0.5	1	4.75
1	0	1
1	0.5	1.5
1	1	9

Figure 7-2 Two-dimensional response surface for the objective function of Eq. 7-5.

The three phases of the numerical optimization process seek to find a base point from which to begin the search procedure and explore the surface through either climbing to a maximum or descending to a minimum in search of a stationary point. The objectives of phase 1 are to gain an understanding of the surface, identify one or more points to begin the search procedure, and select a step size for movement. When examining the response surface in phase 1, the modeler should determine if the surface is unimodal (i.e., one optimum) or multimodal (i.e., one global optimum and one or more local optimum). For multimodal surfaces, more than one search (i.e., phase 2 and 3 analyses) may be necessary. Also, if the surface is not characterized by systematic variation, movement will have to occur using a small step size. The objective of phase 2 is to move rapidly across the response surface from the base point toward the stationary point. The third phase, which is not always necessary, seeks to move closer to the stationary point and to make sure that the optimum point identified in phase 2 is not a saddle point.

PHASE 1 SEARCH PROCEDURES

The most common phase 1 search procedure is a systematic search. Figs. 7-1 (one-dimensional) and 7-2 (two-dimensional) illustrate this type of search. The most likely range of each unknown was separated into an equally spaced grid, and the value of the objective function was evaluated at each grid point. This procedure is practical for a small number of unknowns, but as indicated by Table 7-1, for a large number of unknowns, the number of computations of the objective function can be large. If the minimum value in the phase 1 search lies at one of the grid nodes on the boundary of the search region, a second iteration of the phase 1 search may be conducted. It is often more economical to conduct a phase 1 exploration than it is to begin a phase 2 search at a base point that is far removed from the global optimum.

The output from a phase 1 analysis is a base point and a step size. For example, for the phase 1 analysis of Fig. 7-1, it would be prudent to begin the phase 2 search at $b = 0.75$. A step size, Δb, of 0.01 to 0.05 would be reasonable. For the problem of Fig. 7-2, the coordinates (0.5, 0.5) would be the base point for phase 2. It would be reasonable to use step sizes of 0.02 to 0.05 for both Δb_0 and Δb_1. The actual values selected for the step sizes would depend on the specific method used for the phase 2 search. For some numerical optimization methods, the step size is internal to the phase 2 search, and thus it is not necessary to evaluate this in phase 1.

Although the systematic grid is the easiest type of phase 1 search, it may be inadequate for models or objective functions that are characterized by a systematic cycle. In such cases, either a random or systematic random search might be used to conduct the phase 1 exploration of the response surface. With a random search, random numbers are used to generate values of the unknowns and the objective function is computed for each set. If we have an algorithm to generate random numbers having a uniform distribution with location and scale parameters of 0 and 1, respectively, and if we seek to explore the range for unknown b_i from b_{il} to b_{iu}, the following linear transformation can be used to randomly explore the range:

$$\hat{b}_i = b_{il} + (b_{iu} - b_{il})U \tag{7-7}$$

in which U is the random number. One disadvantage of the random search is that for small numbers of points, the distribution of points used to define the response surface may be very nonuniform, with large portions of the response surface being unexplored.

The systematic random search seeks to overcome the disadvantages of both the uniform grid and random searches while taking advantage of their positive characteristics. For the systematic random search the region of interest is separated into a uniform grid and one or more points are selected within each cell of the grid; the value of the objective function is computed at each of the points. Equation 7-7 is used to compute the value of each unknown for each grid cell. The values of b_{il} and b_{iu} are the lower and upper values of the grid cell.

Figure 7-3 shows the phase 1 response surfaces for random and systematic random searches. The same set of random numbers is used for both analyses. It is evident from Fig. 7-3(a) that the random search did not provide good coverage of the region above b of 0.5. The systematic random search of Fig. 7-3(b) shows better coverage than that for the random search but is not noticeably better than the coverage for the uniform grid search. The three phase 1 analysis methods would lead to approximately the same initial base point, with (0.37, 0.42) for the random search and (0.41, 0.46) for the systematic random search. It is reasonable to assume that the same minima would be

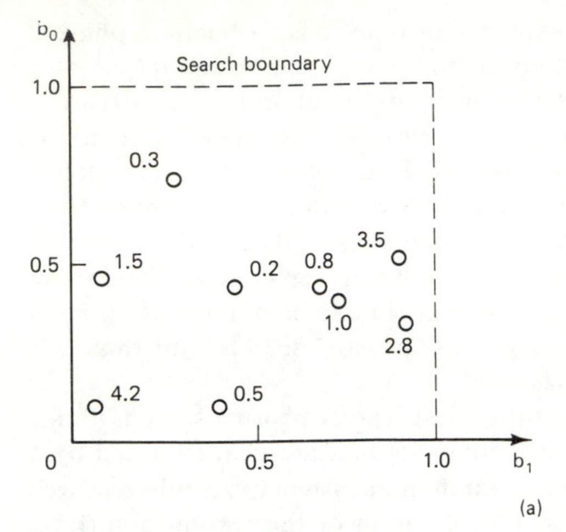

V_i	b_0	V_i	b_1	F
0.46	0.46	0.10	0.10	1.54
0.44	0.44	0.67	0.67	0.82
0.34	0.34	0.93	0.93	2.79
0.74	0.74	0.24	0.24	0.34
0.22	0.22	0.39	0.39	0.52
0.40	0.40	0.74	0.74	1.02
0.52	0.52	0.91	0.91	3.54
0.37	0.37	0.42	0.42	0.23
0.11	0.11	0.06	0.06	4.21

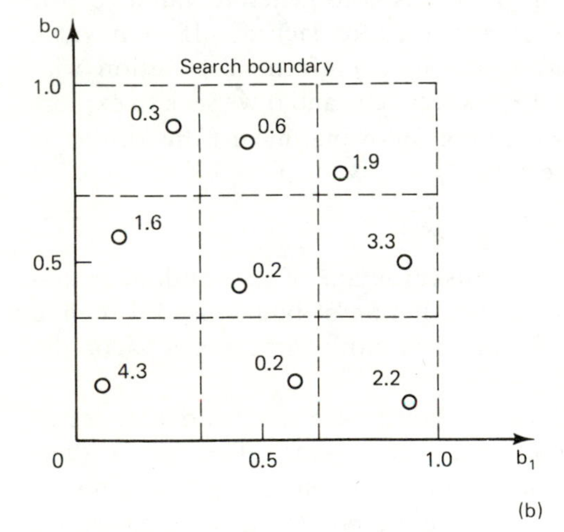

V_i	b_0	V_i	b_1	F
0.46	0.15	0.10	0.03	4.33
0.44	0.15	0.67	0.56	0.19
0.34	0.11	0.93	0.98	2.24
0.74	0.58	0.24	0.08	1.59
0.22	0.41	0.39	0.46	0.17
0.40	0.47	0.74	0.91	3.27
0.52	0.84	0.91	0.30	0.28
0.37	0.79	0.42	0.47	0.65
0.11	0.70	0.06	0.69	1.90

(b)

Figure 7-3 Response surface configuration for (a) random and (b) systematic random searches.

located in the phase 2 search, regardless of the phase I method used. However, this may not always be the case.

Before proceeding to a discussion of phase 2, it should be pointed out that quite frequently a phase 2 search is not preceded by a phase 1 analysis. This is acceptable if reasonable initial estimates of both the unknowns and the step sizes are available. In many cases, estimates of the unknowns can be obtained from either previous investigations or graphical analyses of the data. For example, if we plot the data ($n = 3$) used to derive the response surface of Figs. 7-2 and 7-3, one would find values of (0.2, 0.6) to be reasonable estimates of b_0 and b_1, respectively. The phase 1 search is especially useful when there are either constraints on the coefficients or model shift parameters that result in discontinuities in the response surface.

THE PHASE 2 SEARCH

Just as there is more than one phase 1 search procedure, there are numerous methods for conducting a phase 2 search. All the methods involve the same components: an objective function, a model, a data set, and an initial set of estimates for the unknowns. Additionally, all the methods are based on a numerical evaluation of the derivatives of the objective function with respect to the unknowns. They differ only in minor details of the search procedure. Most important, the interpretation of the output is always the same. Since least squares has been discussed extensively in previous chapters, the numerical method called nonlinear least squares will be selected for a detailed discussion. A brief review of characteristics of other methods will be given later.

Numerics for Nonlinear Least Squares

In developing procedures for nonlinear least squares, we will keep the same objective as we had in the linear situation. We wish to determine that unique set of values of the parameters such that the sum of the squares of the differences between the predicted and measured values of the criterion variable is a minimum. However, our approach to this objective will differ greatly from the linear situation.

The basic approach to nonlinear solutions is based on Taylor series expansion of the models to be fitted. Consider the extremely simple function

$$Y_1 = f(X, a) \tag{7-8}$$

where the value of Y_1 is specified by one variable X and one coefficient a. Since we wish to find an optimum value of the coefficient a for a specific data set, the X's are fixed for that set but a can change from one sample point to another within a data set. The value of the objective function for a slightly different value of a, say $a + h$, would be

$$Y_2 = f(X, a + h) \tag{7-9}$$

A Taylor series expansion allows us to write

$$Y_2 = Y_1 + f'(X, a)\frac{h}{1!} + f''(X, a)\frac{h^2}{2!} + f'''(X, a)\frac{h^3}{3!} + R_N \tag{7-10}$$

where

$$R_N = f^{(n)}(X, a_1)\frac{h^n}{n!} \qquad \text{and} \qquad a \le a_1 \le a + h$$

In practical numerical work we do not usually need the highly precise expansion given by Eq. 7-10. We are not interested in a single shift in our function from Y_1 to Y_2 caused by a single change h in the coefficient a. Rather having shifted from Y_1 to Y_2 with a change h, we can now consider shifting from Y_2 to a new value Y_3 by an additional parameter change h. Finally, if we keep h small, then h^2 and higher terms should be small enough to neglect in our successive shifts of the function. Hopefully, we can find some manner of shifting until we find a value of the function Y which will produce the smallest residual sum of squares.

A Taylor series expansion is not limited to the simple form of one coefficient and one independent variate given by Eq. 7-10. We might write the more general function as

$$Y_1 = f(X_1, X_2, \cdots, X_m, a_1, a_2, \cdots, a_k) \tag{7-11}$$

Since we are considering variations in the coefficients only, the X's are fixed. For simplicity, change Eq. 7-11 to

$$Y_1 = f(X_m, a_1, a_2, \ldots, a_k) \tag{7-12}$$

where X_m indicates the fixed value of the function due to m fixed independent variates. The Taylor series expansion of the multi-coefficient form is

$$\begin{aligned} Y_2 &= f(X_m, a_1 + h_1, a_2 + h_2, \ldots, a_k + h_k) \\ &= \sum_{h=0}^{N-1} \frac{1}{h!}\left(h_1 \frac{\partial}{\partial a_1} + h_2 \frac{\partial}{\partial a_2} + \cdots + h_k \frac{\partial}{\partial a_k}\right)^n f(X_m, a_1, a_2, \ldots, a_k) + R_N \end{aligned} \tag{7-13}$$

where

$$\left(h_1 \frac{\partial}{\partial a_1} + h_2 \frac{\partial}{\partial a_2} + \cdots + h_k \frac{\partial}{a_k}\right)^n$$

is a repeating differential operator on the function $f(X_m, a_1, a_2, \ldots, a_k)$ and

$$\begin{aligned} R_N = \frac{1}{N!}&\left(h_1 \frac{\partial}{\partial a_1} + h_2 \frac{\partial}{\partial a_2} + \cdots + h_k \frac{\partial}{\partial a_k}\right)^N \\ &\times f(X_m, a_1 + \Theta h_1, a_2 + \Theta h_2, \ldots, a_k + \Theta h_k) \end{aligned} \tag{7-14}$$

with $0 \leq \Theta \leq 1$.

If we again limit the expansion to just the first differential, we obtain

$$Y_2 = f(X_m, a, a_2, \ldots, a_k) + h_1 \frac{\partial f}{\partial a_1} + h_2 \frac{\partial f}{\partial a_2} + \cdots + h_k \frac{\partial f}{\partial a_k} \tag{7-15}$$

For example, if we had a function with four coefficients, we would write

$$Y_2 = Y_1 + h_1 \frac{\partial f}{\partial a_1} + h_2 \frac{\partial f}{\partial a_2} + h_3 \frac{\partial f}{\partial a_3} + h_4 \frac{\partial f}{\partial a_4} \tag{7-16}$$

A simple rearrangement gives

$$E = Y_2 - Y_1 = h_1 \frac{\partial f}{\partial a_1} + h_2 \frac{\partial f}{\partial a_2} + h_3 \frac{\partial f}{\partial a_3} + h_4 \frac{\partial f}{\partial a_4} \tag{7-17}$$

Now consider Y_2 to be an observed value of our function. Then Y_1 can be considered the value predicted by the function for some value of the four coefficients a_i. But Eq. 7-17 says that by changing each of our coefficients by the appropriate h_i we change the functional value from Y_1 to Y_2. This is the same as saying that we can adjust our coefficients to eliminate the error $Y_2 - Y_1$, or E_1.

The next step is to place the development and meaning of Eq. 7-17 within the context of a nonlinear least-squares procedure. Refer to Eq. 7-1. If we had some initial set of values for the coefficients, say a_0, b_0, and c_0 we could use them with given values of x_1 and x_2 to calculate $\hat{y}$. We cannot expect such an initial set of coefficients to be a "good" set for prediction of y, especially if we could only estimate the set. But Eq. 7-17 says we can improve the estimate by calculating the h_i. Of course we also need the values of $\partial f/\partial a_i$, but these can be readily calculated from Eq. 7-1.

If we had just three values of y the adjustment of the a_i by Eq. 7-17 would be absolute. But the usual problem in model evaluation is to produce a best estimate by some averaging process across many observed values y. The averaging process we use is the method of least squares.

We should now review quickly the evaluation of a linear equation by least squares. The basic form using four coefficients is repeated here for convenience:

$$Y = b_1x_1 + b_2x_2 + b_3x_3 + b_4x_4 \tag{7-18}$$

By least squares, using the simultaneous normal equations, we evaluated the regression coefficients b_i so that the sum of $(y - \hat{y})^2$ was a minimum.

Now compare Eqs. 7-18 and 7-17. They are identical in linear structure with $E = y - \hat{y}$ replacing $\hat{y}$, the $\partial f/\partial a_i$ replacing the x_i, and the h_i replacing the b_i. We now see that we can calculate the h_i as regression coefficients. The normal equations are formed on the $\partial f/\partial a_i$ and on E. Following evaluation of the h_i, our new values of the coefficients are

$$a_1 = a_0 + h_1 \tag{7-19a}$$

$$b_1 = b_0 + h_2 \tag{7-19b}$$

$$c_1 = c_0 + h_3 \tag{7-19c}$$

It should not be expected that a_1, b_1, and c_1 are final answers. Only if our model were linear would Eq. 7-17 be an exact solution. But we have ignored the higher order derivatives in Eq. 7-13, and, therefore, in a nonlinear model Eq. 7-17 is only approximate. Nevertheless, a_1, b_1, and c_1 are improvements. We simply repeat the whole process. We calculate the errors and partial derivatives using these new values of the coefficients. From these, a new set of h_i is derived. Finally, further improvement is produced by

$$a_2 = a_1 + h_{21} \tag{7-20a}$$

$$b_2 = b_1 + h_{22} \tag{7-20b}$$

$$c_2 = c_1 + h_{23} \tag{7-20c}$$

The entire process is continued until the changes h_i become very small, usually less than some preset test value. When the a_i can no longer be improved $\sum E^2$ is a minimum and we have accomplished a nonlinear least-squares fitting.

It has been demonstrated that Eq. 7-17 is a basic linear adjustment device by which, using least squares in repeated solutions, coefficients of nonlinear equations can be estimated to any desired precision. However, to use Eq. 7-17 it must be possible to form the partial derivatives $\partial f/\partial a_i$. This, in turn, usually means that our model is in the form of a comparatively simple equation that can be differentiated.

Many hydrologic models are not simple equations. A model using rainfall, for example, would require an explicit function representing rainfall intensity with time. But such explicit rainfall functions simply do not exist. The usual procedure is to work with quantities of rain falling in discrete increments of time. We need a procedure whereby we can operate on implicit functions, such as a time series of discrete rainfall amounts.

A simple procedure for operating on implicit functions can be called the method of partial finite divided differences. These divided differences replace the partial derivatives in Eq. 7-17.

The geometric definition of the partial divided finite difference and its relationship to the partial derivative are shown in Fig. 7-4. As the value of a coefficient of a model changes, the predicted value of Y will also change. This relationship of a and y is labeled $\hat{y} = f(a_i)$. Then $\partial f/\partial a$ is the tangent to this curve at point $(a_0, \hat{y})$. To get a slope by divided difference, change a_0 very slightly by adding Δa_0. At the new point $a_0 + \Delta a_0$ calculate a new predicted value $\hat{y}_a$, the subscript a indicating that it is a partial change, dependent only on Δa. Simple geometry gives the slope of the line between the points $(a_0 + \Delta a_0, \hat{y}_a)$ and $(a_0, \hat{y})$ as $(\hat{y}_a - \hat{y})/((a_0 + \Delta a_0) - a_0)$, or $(\hat{y}_a - \hat{y})/\Delta a$. If Δa is not large, the difference between $\partial f/\partial a$ and $(\hat{y}_a - \hat{y})/\Delta a$ should not be large. Also, it must be remembered that $\partial f/\partial a$ is not absolutely correct since the higher derivatives have been neglected. Furthermore, any error caused by the difference between the two slopes will tend to diminish in effect as the h_i of Eq. 7-17 get smaller with successive rounds. In practice, it has been found that the partial finite divided differences are a very satisfactory procedure for working with both explicit and implicit functions.

It is important to note the full implication of the use of divided differences. We have already mentioned that there is no longer a need to be able to compute and then numerically evaluate the $\partial f/\partial a_i$. But this means that as long as we

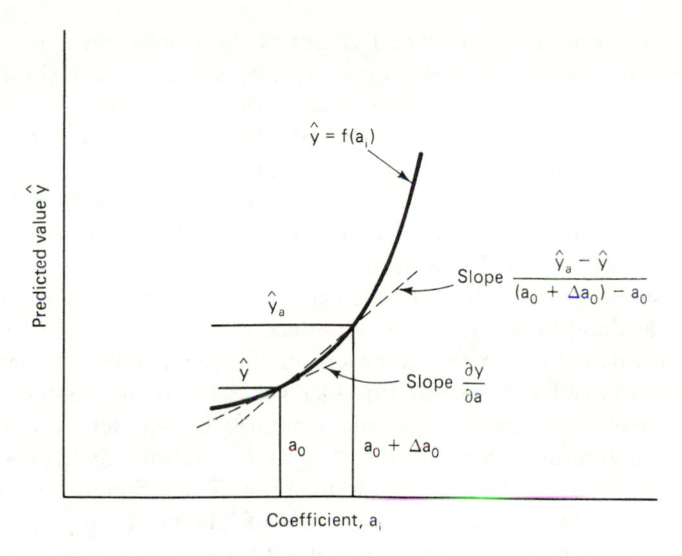

Figure 7-4 Definition of slope by divided finite difference.

have some system of calculation of $\hat{y}$ from given values of coefficients and variables, we can calculate the $(\hat{y}_a - \hat{y})/\Delta a$. Such systems can include table look-up, interpolation, parametric, and variate subfunctions—indeed, most of the tools of classical numerical analysis. The only requirement is that we be able to calculate the $\hat{y}$ in the transformed equation

$$E = y - \hat{y} = h_1 \frac{\hat{y}_{a1} - \hat{y}}{\Delta a_1} + h_2 \frac{\hat{y}_{a2} - \hat{y}}{\Delta a_2} + h_3 \frac{\hat{y}_{a3} - \hat{y}}{\Delta a_3} + \cdots \tag{7-21}$$

A word of caution is absolutely essential at this point. Equation 7-21 represents an extremely powerful tool for optimizing the correspondence between a model and an observational set of data. This does not mean that things can never go wrong. The builder of models must be extremely careful in the analysis phase. Numerical methods are prone to failure in some, usually rare, situations. Five of these are as follows.

The data set of the independent (x's) and dependent (y) variables must cover as wide a range as possible. Optimization can scarcely succeed in definition of a range of a function if all the data cluster at one point.

If the model consists of several sequential steps in calculation, each with variates and coefficients, it is possible that some of the parameters will interact with each other. In successive rounds each may overcompensate for an error in the other. An oscillatory situation may develop, instead of a simple, continuing decrease in values of the h_i.

Models of complex structure may have complex functions $\hat{y} = f(a)$ in Fig. 7-4. Particularly, there may be more than one point where $\hat{y}$ is such that $y - \hat{y} = E$ is a minimum. We must remember that we are really talking about

$\sum E^2$ being a minimum. But several values of the coefficient a may produce $\sum E^2$ all near the minimum, or even two separate minima. One should always seek assurance, usually through several solutions with different initial values, that a clear minimum of $\sum E^2$ was found. Minor, sometimes called local, minima of $\sum E^2$ should be rejected in favor of the global minimum. Often, a plot of $\sum E^2$ versus initial and all iterated values of the a_i will be helpful.

The $\partial f/\partial a_i$, or the $(\hat{y}_a - \hat{y})/\Delta a$, are called *sensitivity coefficients.* If these are large, we can see that h_i, given to some approximation by $E/(\partial f/\partial a)$, will tend to be small. If the sensitivity coefficient is small, h_i will tend to be large. Of course, the definition large or small for description of a sensitivity coefficient can only mean relative to the other sensitivity coefficients. Some difficulty may be experienced with use of Eq. 7-21 (or 7-17) if one of the sensitivity coefficients has an average value much greater or smaller (say a 100-fold difference of magnitude) than the others. In such situations again the successive values of the h_i may not be steadily diminishing. The difficulty can sometimes be corrected by an arbitrary scale change of the most directly associated independent variate. Sometimes a functional transform may be necessary.

The last difficulty is illustrated in Example 7-2 and will be discussed at that point. The sensitivity coefficients may be correlated in which case the values of the h_i given by Eq. 7-21 are questionable. Corrective measures are discussed with Example 7-2.

Such a number of cautionary statements might discourage a new user of nonlinear least squares. However, most problems of reasonable structure and complexity can be solved with no difficulty. The danger may lie more with the modeler than with the method. Given the tremendous potential of the method of nonlinear least squares, augmented by the freedom from functional form of the divided differences, one is tempted to put together highly exotic structures. Without sufficient and well-spread data, such temptation can lead to numerical disaster. Some simpler structure that did not overextend the data would be successful.

Example 7-1: Nonlinear Fit of a Simple Function

On the left side of Table 7-2 is a small set of observations on variables x, z, and y. Use nonlinear least squares to estimate values of the coefficients x_o, z_0, and k in a model of the form

$$y = k(x - x_0)(z - z_0) \tag{7-22}$$

This model is a simple form of Eq. 7-2. The six observational sets of x, z, and y are fictional data intended for use here only as a first computational illustration of the method of nonlinear least squares. Six observations on the variates is too small a number for practical use in hydrologic model building. However, the equation has practical form. The variable x might be storm rainfall on a watershed. The coefficient x_0 would represent the amount of rain before runoff begins. The variable z could be an index of soil moisture. The coefficient z_0 would represent the beginning effective soil moisture. The variable y would represent storm runoff.

TABLE 7-2 Iterative Divided Difference Solution for Model $y = k(x - x_0)(z - z_0)$

(a) First Round—Initial Values: $x_0 = 1.0$, $z_0 = 0.0$, $K = 0.12$; Parameter Changes: $\Delta x_0 = 0.2$, $\Delta z_0 = 0.2$, $\Delta k = 0.02$

x	z	y	$x - x_0$	$z - z_0$	$\hat{y}$	$x - (x_0 + \Delta x_0)$	$\hat{y}_x$	$z - (z_0 + \Delta z_0)$	$\hat{y}_z$	$\hat{y}_k$	$\frac{\hat{y}_x - \hat{y}}{\Delta x_0}$	$\frac{\hat{y}_2 - \hat{y}}{\Delta z_0}$	$\frac{\hat{y}_k - \hat{y}}{\Delta k}$	$y - \hat{y}$
2	1	0.054	1.0	1.0	0.12	0.8	0.096	0.8	0.096	0.14	−0.12	−0.12	1.0	−0.066
2	5	0.234	1.0	5.0	0.60	0.8	0.480	4.8	0.576	0.70	−0.60	−0.12	5.0	−0.366
5	2	0.693	4.0	2.0	0.96	3.8	0.912	1.8	0.864	1.12	−0.24	−0.48	8.0	−0.267
5	8	2.583	4.0	8.0	3.84	3.8	3.648	7.8	3.744	4.48	−0.96	−0.48	32.0	−1.257
10	3	2.448	9.0	3.0	3.24	8.8	3.168	2.8	3.024	3.78	−0.36	−1.08	27.0	−0.792
10	10	7.803	9.0	10.0	10.80	8.8	10.560	9.8	10.584	12.60	−1.20	−1.08	90.0	−2.997

Sums of products				Solution	Parameter corrections
h_x	h_z	h_k	E		
2.9232	2.3472	−153.48	5.37984	$h_x = 0.39$	$x_{10} = 1.0 + 3.9 = 1.39$
	2.8224	−146.28	4.87548	$h_z = -0.14$	$z_{10} = 0 - 0.14 = -0.14$
		9943.0	−335.370	$h_k = -0.03$	$k_1 = 0.12 - 0.03 = 0.09$

TABLE 7-2 *(cont.)*

(b) Second Round—Values: $x_0 = 1.39$, z_0: -0.14, $k = 0.09$; Parameter Changes: $\Delta x_0 = 0.2$, $\Delta z_0 = -0.2$, $\Delta k = 0.01$

x	z	y	$x - x_0$	$z - z_0$	$\hat{y}$	$x - (x + \Delta x_0)$	$\hat{y}_x$	$z - (z_0 + \Delta z_0)$	$\hat{y}_z$	$\hat{y}_k$	$\frac{\hat{y} - \hat{y}}{x_0}$	$\frac{\hat{y}_z - \hat{y}}{\Delta z_0}$	$\frac{\hat{y}_k - \hat{y}}{\Delta k}$	$y - \hat{y}$
2	1	0.054	0.61	1.14	0.0625	0.41	0.0421	1.34	0.0736	0.0695	−0.1020	−0.0555	0.70	−0.0085
2	5	0.234	0.61	5.14	0.2822	0.41	0.1897	5.34	0.2932	0.3135	−0.4625	−0.0555	3.13	−0.0482
5	2	0.693	3.61	2.14	0.6953	3.41	0.6568	2.34	0.7603	0.7725	−0.1925	−0.3250	7.72	−0.0023
5	8	2.583	3.61	8.14	2.6447	3.41	2.4982	8.34	2.7097	2.9383	−0.7325	−0.3250	29.38	−0.0617
10	3	2.448	8.61	3.14	2.4332	8.41	2.3767	3.34	2.5882	2.7035	−0.2825	−0.7750	27.03	−0.0148
10	10	7.803	8.61	10.14	7.8575	8.41	7.6750	10.34	8.0125	8.7305	−0.9125	−0.7750	87.30	−0.0545

Sums of products				Solution	Parameter corrections
h_x	h_z	h_k	E		
1.71038	1.25785	−111.8322	0.11439	$h_x = 0.111$	$x_{20} = 1.39 + 0.11 = 1.501$
	1.41861	−100.8742	0.05469	$h_z = -0.059$	$z_{20} = -0.14 - 0.059 = -0.199$
		9284.9806	−6.34512	$h_k = 0.000$	$k_2 = 0.090 + 0.000 = 0.090$

Table 7-2 shows two rounds of solution for the three coefficients. It is necessary to estimate initial values of the coefficients to get the iteration process going. These initial estimates are $x_0 = 1.0$, $z_0 = 0.0$, and $k = 0.12$. With these values, $\hat{y}$ can be computed for each observation. Then each coefficient must be incremented in turn, and for each, $\hat{y}_i$ must be computed for each observation. The divided differences and the residual error $y - \hat{y}$ can then be calculated for each of the six observations.

These last calculated values are shown to the right in Table 7-2. The next step is to consider the numbers in the last four columns as statistical variates and solve for the coefficients h_x, h_z, and h_k as though they were regression coefficients. The sums of products, the solution, and the parameter corrections are given immediately below.

A second round is now performed using the corrected coefficients in place of the initial values. Otherwise, all calculations proceed as in the first round. It should be noted that the residual errors, $\hat{y} - \hat{y}$, of the second round are much smaller than those of the first round, showing the effect of the first round corrections. The solution for the second round also shows substantially lower corrections.

A third round could now be performed. Corrections would be very small, however. Although it was not stated as part of the example, the fictitious observations were calculated using $x_0 = 1.5$, $z_o = -0.2$, and $k = 0.09$. The rapid convergence to these values from the initial estimates is partly explained by the lack of statistical error in y. These errors were not added in this first simple example. The method of computation would be identical for real data.

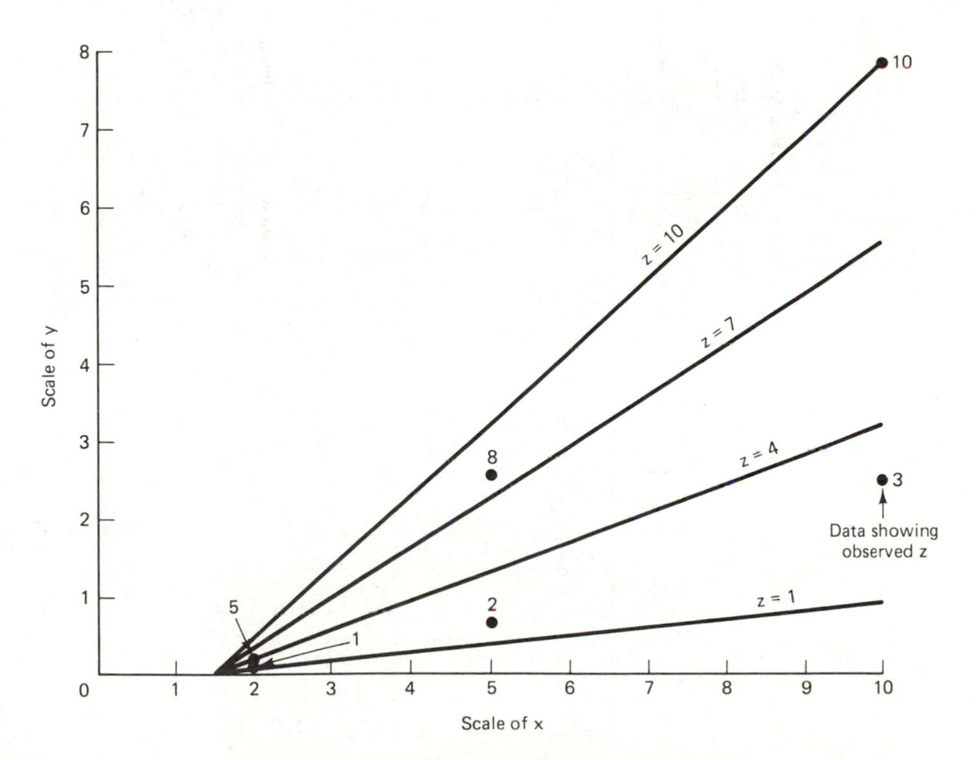

Figure 7-5 Verification plot of model for Example 7-1.

TABLE 7-3 Iterative Divided Difference for Recession Model

(a) First Round—Initial Values: Par. 1 = 20, Par. 2 = 8, Parameter Changes: $\Delta P1 = 1$, $\Delta P2 = -1$

Time	Q	$e^{-0.4T}$	Routing Function	Core Function	Core Function $\Delta P1$	Core Function $\Delta P2$	$\hat{Q}$	$\hat{Q}_1$	$\hat{Q}_2$	$\frac{\hat{Q}_1 - \hat{Q}}{\Delta P1}$	$\frac{\hat{Q}_2 - \hat{Q}}{\Delta P2}$	$Q - \hat{Q}$
0		1.0										
1	8.0	0.670	0.330	20.0	21.0	20.0	6.60	6.93	6.60	0.33	0.00	1.40
2	4.0	0.449	0.221	11.6	11.9	10.9	8.25	8.56	8.02	0.31	0.23	−4.25
3	2.0	0.301	0.148	10.4	10.6	9.6	8.96	9.24	8.54	0.28	0.42	−6.96
4	1.6	0.202	0.099	9.2	9.3	8.3	9.03	9.25	8.45	0.22	0.58	−8.53
5	1.4	0.135	0.067	8.0	8.0	7.0	8.70	8.85	7.98	0.15	0.72	−7.30
6	1.2	0.091	0.044	6.8	6.7	5.7	8.06	8.13	7.22	0.07	0.84	−6.86
7	1.2	0.061	0.030	5.6	5.4	4.4	7.25	7.23	6.29	−0.02	0.96	−6.05
8	1.0	0.041	0.020	4.4	4.1	3.1	6.31	6.20	5.24	−0.11	1.07	−5.31
9	1.0	0.027	0.014	3.2	2.8	1.8	5.30	5.09	4.11	−0.21	1.19	−4.30
10	1.0	0.018	0.009	2.0	1.5	0.5	4.21	3.90	2.92	−0.31	1.29	−3.21

Solution	Parameter corrections
$h_{p1} = -10.7$	Par. 1 = 20 − 10.7 = 9.3
$h_{p2} = -6.3$	Par. 2 = 8 − 6.3 = 1.7

TABLE 7-3 (*cont.*).

(b) Fourth Round—Parameter Values: Par. 1 = 10.3, Par. 2 = 2.5, Par. 3 = 0.4, Parameter Changes: $\Delta P1 = 0.5$, $\Delta P2 = 0.3$, $\Delta P3 = 0.3$

Time	Q	$e^{-0.4T}$	Routing Function	$e^{-0.7T}$	Routing Function $\Delta P3$	Core Function	Core Function $\Delta P1$	Core Function $\Delta P2$	$\frac{\hat{Q}_1 - \hat{Q}}{\Delta P1}$	$\frac{\hat{Q}_2 - \hat{Q}}{\Delta P2}$	$\frac{\hat{Q}_3 - \hat{Q}}{\Delta P3}$	$Q - \hat{Q}$
0	8.0	1.0		1.0								
1	4.0	0.670	0.330	0.497	0.503	10.3	10.8	10.3	0.330	0.000	5.940	4.601
2	2.0	0.449	0.221	0.247	0.250	4.06	4.16	4.30	0.288	0.263	3.337	0.384
3	1.6	0.301	0.148	0.122	0.125	3.28	3.33	3.55	0.226	0.473	1.493	−1.506
4	1.4	0.202	0.099	0.061	0.061	2.50	2.50	2.80	0.152	0.650	0.143	−1.570
5	1.2	0.135	0.067	0.030	0.031	1.72	1.67	2.05	0.068	0.793	−0.770	−1.298
6	1.2	0.091	0.044	0.015	0.015	0.94	0.84	1.30	−0.020	0.930	−1.380	−0.910
7	1.0	0.061	0.030	0.007	0.008	0.16	0.01	0.55	−0.112	1.050	−1.840	−0.270
8	1.0	0.041	0.020	0.004	0.003	−0.62	−0.82	−0.20	−0.208	1.167	−2.110	0.220
9	1.0	0.027	0.014	0.002	0.002	−1.40	−1.65	−0.95	−0.304	1.277	−2.367	0.933
10	1.0	0.018	0.009	0.001	0.001	−2.18	−2.48	−1.70	−0.402	1.383	−2.130	1.676

Three-parameter solution	Two-parameter solution	Two-parameter corrections
$h_{p1} = -11.9$	$h_{p2} = 0.53$	Par. 2 = 2.5 + 0.53 = 3.03
$h_{p2} = 0.3$	$h_{p3} = 0.41$	Par. 3 = 0.4 + 0.41 = 0.81
$h_{p3} = 1.3$		

The method of partial divided finite differences was used in Example 7-1. It would also be possible to use the partial derivatives because $\partial y/\partial x_0 = -k(z - z_0)$, $\partial y/\partial z_0 = -k(x - x_0)$, and $\partial y/\partial k = (x - x_0)(z - z_0)$. These would be evaluated for each of the six observations and, of course, would replace the divided differences. For this simple problem their use would shorten the calculations. For many practical problems in hydrology the method of divided differences must be used.

The given values of the variates and the derived family of lines are both plotted in Fig. 7-5. One could call this diverging family an "interaction" model, since the two variates x and z do not have a simple additive effect on y. The effect of z is much greater for a large x than for a small x.

Example 7-2: Streamflow Recession Analysis

A short segment of decreasing streamflow is given to the left in Table 7-3. These values are also plotted as the heavy, solid line in Fig. 7-6. This example illustrates the development of a recession model to represent the given streamflow and the estimation of the coefficients of this model by nonlinear least squares.

The recession of streamflow, which is the decrease from high to lower values of runoff as water drains from storage in a watershed, is frequently represented as having the mathematical form of the descending exponential. However, a different form will be used here. This form is more flexible than the simple descending exponential and can fit a variety of recession shapes. It is, therefore, no longer necessary to first separate recessions into arbitrary classifications of base flow, subsurface flow, and surface flow. This example has also been constructed to illustrate some of the pitfalls and some remedial measures in nonlinear least-squares methodology.

Great flexibility can be achieved in model building by specifying that the model is composed of elemental subfunctions. These subfunctions operate on each other to produce the model form. In this example two subfunctions were used, a core function and a routing function. Operation of the functions on each other is by convolution.

The *core function* is shown in Fig. 7-7. It is defined as two connecting linear segments. The unknown coefficients are defined as ordinates of the function at selected times. Nonlinear least squares will give the values of these ordinates. Construction of the function is shown in Table 7-4. The construction specifies each ordinate of the function by linear interpolation along the segments. This construction is somewhat arbitrary and was only used to hold the number of unknowns small for the sake of simplicity. Round 7 exhibits a slight increase in complexity with the number of parameters increasing from two to three. The core function will be used only at the discrete points of the integer time values for which the ordinates were calculated.

A simple descending exponential was used as the *routing function.* This function is here called a routing function only to identify it by analogy with the more common operation used in simple flood routing. While the descending exponential is a continuous mathematical function, it will be used only at the discrete points represented by the integer values of time from 1 to 10.

A descending exponential written in the form $r(t) = b\,e^{-bt}$ encloses an area mathematically equal to unity. If discrete routing coefficients are developed from fractional parts of this area, their total will also be unity. Define these routing coefficients

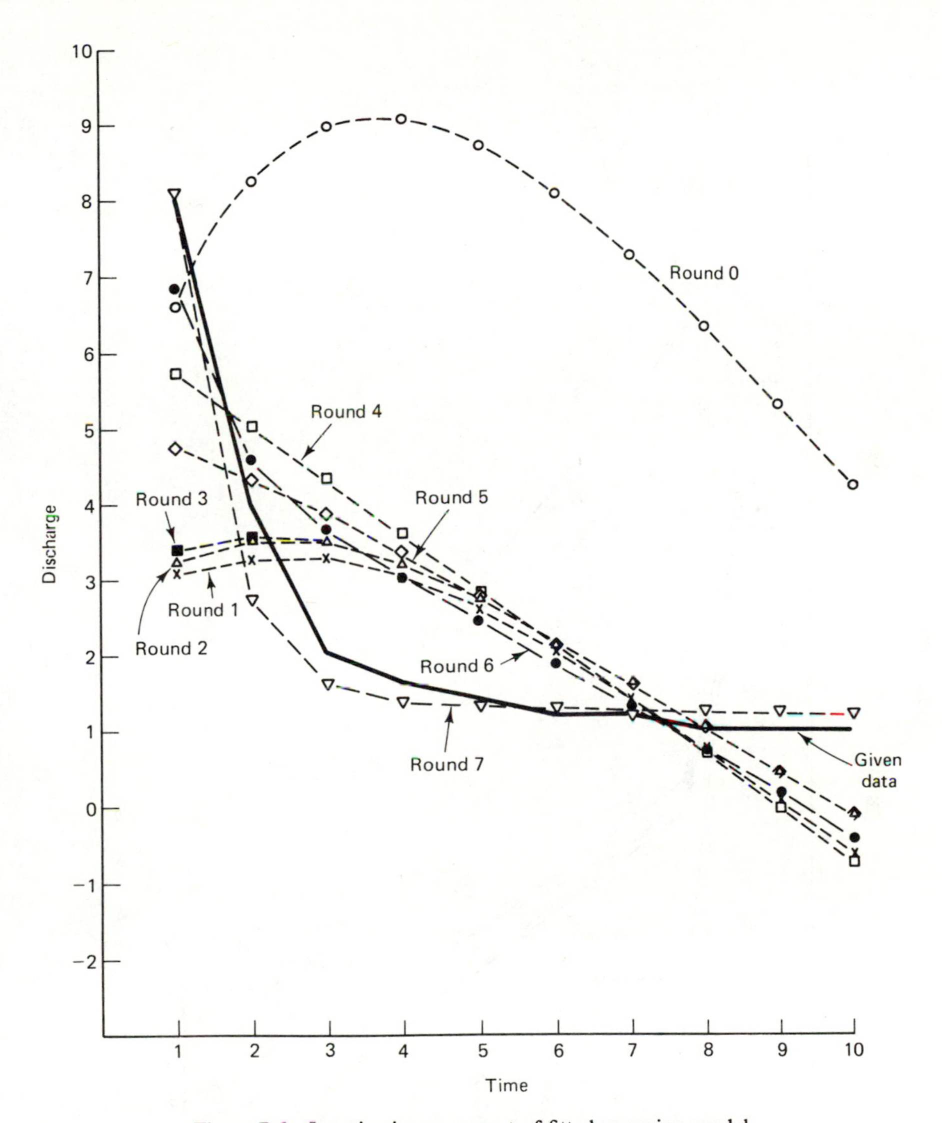

Figure 7-6 Iterative improvement of fitted recession model.

as follows:

$$r_1 = b \int_0^1 e^{-bt}\,dt = b\left(\frac{1}{-b}\,e^{-bt}\right)_0^1 = 1 - e^{-b} \tag{7-23a}$$

$$r_2 = b \int_1^2 e^{-bt}\,dt = b\left(\frac{1}{-b}\,e^{-bt}\right)_1^2 = e^{-b} - e^{-2b} \tag{7-23b}$$

$$\vdots$$

$$r_{10} = b \int_9^{10} e^{-bt}\,dt = b\left(\frac{1}{-b}\,e^{-bt}\right)_9^{10} = e^{-9b} - e^{-10b} \tag{7-23c}$$

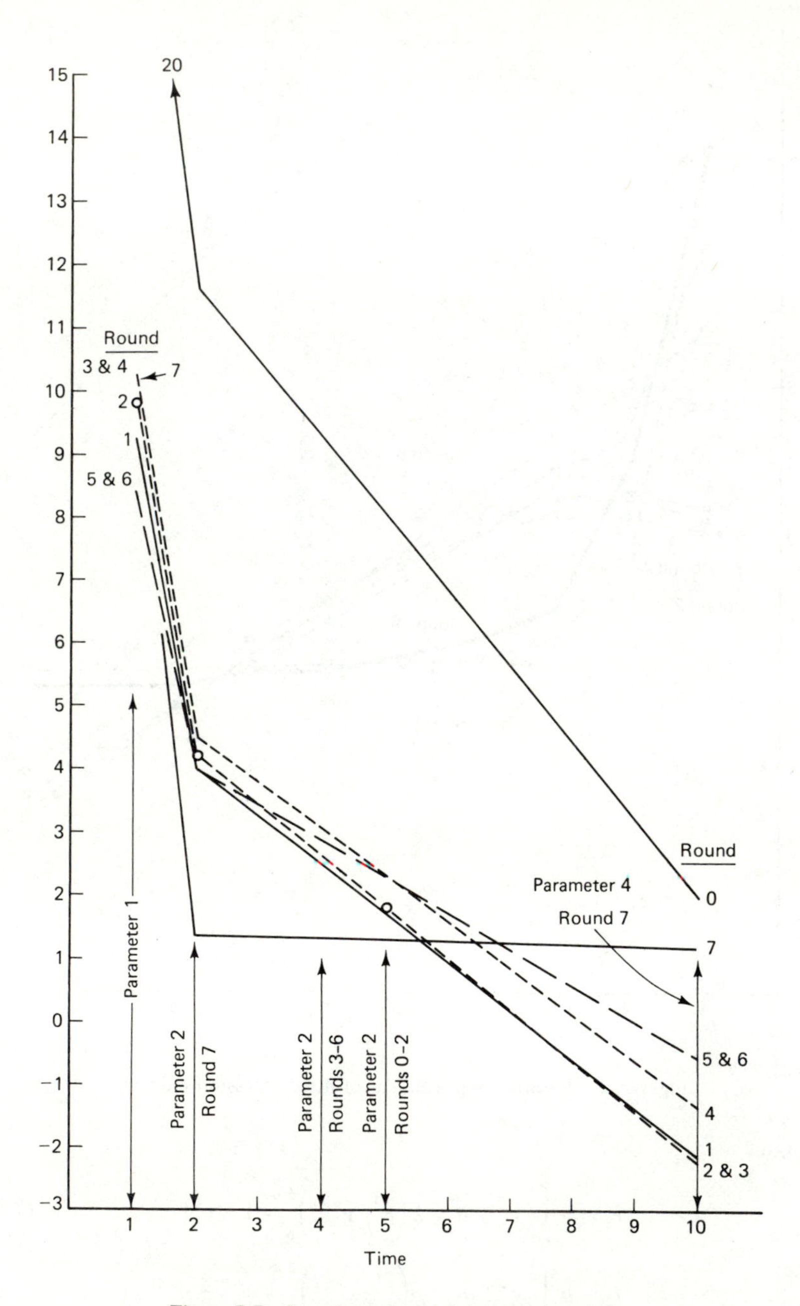

Figure 7-7 Iterative values of the core function.

TABLE 7-4 Construction of Core Function

Time	Rounds 0–2	Rounds 3–6	Round 7
1	$P1$	$P1$	$P1$
2	$0.3P1 + 0.7P2$	$0.2P1 + 0.8P2$	$P2$
3	$0.2P1 + 0.8P2$	$0.1P1 + 0.9P2$	$\frac{7P2 + 1P4}{8}$
4	$0.1P1 + 0.9P2$	$P2$	$\frac{6P2 + 2P4}{8}$
5	$P2$	$-0.1P1 + 1.1P2$	$\frac{5P2 + 3P4}{8}$
6	$-0.1P1 + 1.1P2$	$-0.2P1 + 1.2P2$	$\frac{4P2 + 4P4}{8}$
7	$-0.2P1 + 1.2P2$	$-0.3P1 + 1.3P2$	$\frac{3P2 + 5P4}{8}$
8	$-0.3P1 + 1.3P2$	$-0.4P1 + 1.4P2$	$\frac{2P2 + 6P4}{8}$
9	$-0.4P1 + 1.4P2$	$-0.5P1 + 1.5P2$	$\frac{1P2 + 7P4}{8}$
10	$-0.5P1 + 1.5P2$	$-0.6P1 + 1.6P2$	$P4$

Operating by convolution on the core function by the routing function can, therefore, only change the shape of the core function. The magnitude of the core function will not be changed.

A few points in Table 7-3, calculated by discrete convolution, are shown for illustration.

$$0.330 \times 20 = 6.60 = {}_1\hat{Q} \tag{7-24a}$$

$$(0.330 \times 11.6) + (0.221 \times 20) = 8.25 = {}_2\hat{Q} \tag{7-24b}$$

$$(0.330 \times 10.4) + (0.221 \times 11.6) + (0.148 \times 20) = 8.96 = {}_3\hat{Q} \tag{7-24c}$$

The entire set of 10 values of Q so calculated are plotted in Fig. 7-6 and labeled round 0. This is the prediction with the initial values of the unknowns before any adjustment by least squares is made.

Results of seven rounds of iteration are shown in Table 7-5. For rounds 1, 2, and 3 only unknowns 1 and 2 were adjusted. Figure 7-6 shows the predicted recession following each round. The adjustment by round 1 reduced the predictions drastically from those in round 0, computed by the initial values. However, further adjustment by rounds 2 and 3 is slight. Table 7-4 shows that the definition of unknown 2 was changed for round 3. Changing unknown 2 from time 5 to time 4 of the core function made little change in the prediction. We may say then that the prediction is not sensitive to such a change in the coefficient.

These adjustments for the first three rounds point out an important lesson. The method of nonlinear least squares is capable of changing the model coefficients so that the prediction dropped from the very high round 0 values to the nearly central

TABLE 7-5 Successive Values of Coefficients for Example 7-2

	Coefficient			
Round	1	2	3	4
0	20.0	8.0	0.4[a]	
1	9.3	1.7	0.4[a]	
2	9.8	1.8	0.4[a]	
3	10.3	2.5[b]	0.4[a]	
4	10.3[a]	3.03	0.81	
5	8.58	2.86	0.81[a]	
6	8.58[a]	2.865	1.61	
6A	8.58[a]	4.01[a,b]	1.61[a]	−0.564[a]
7	10.17	1.35	1.61[a]	1.20

[a]Coefficient held constant for this round.
[b]Redefinition of unknown coefficient.

values for rounds 1, 2, and 3. However, the model simply cannot take the shape of the given data. No amount of additional adjustment could accomplish this.

Part of the inflexibility mentioned above is due to the constant value of the b in the routing function. Table 7-5 shows the b as coefficient 3 with a constant value of 0.4 for the first three rounds. For round 4 coefficients 1, 2, and 3 were initially allowed to vary. The divided difference calculations are shown in Table 7-3(b); the solutions for corrections are shown at the bottom of this table. The correction value of −11.9 for coefficient 1 is unrealistic. Applying this amount of correction would produce a negative value of the coefficient.

This unrealistic value of the correction for coefficient 1 illustrates another of the pitfalls in nonlinear least squares. The difficulty is found in the correlation of two of the divided differences. The set of 10 values of $(\hat{Q}_1 - \hat{Q})/\Delta P1$ and $(\hat{Q}_3 - \hat{Q})/\Delta P3$ has a correlation coefficient in excess of 0.86. Now Eq. 7-21 shows that the divided differences serve as the statistical independent variates. Solution of Eq. 7-21 by least squares presupposes that these "independent" variates are uncorrelated. Although one can perform the computational mechanics when such correlations are present, the resulting values of the regression coefficients are not independent estimates of the relationship of the independent variates to the dependent variates.

This topic of correlations among the independent variates forms the central theme of the models in Chapter 11, where more will be said about nonlinear least squares. However, for purposes of solution in the present example, a simpler procedure is followed. Since the corrections for coefficients 1 and 3 are correlated we only adjust one of them, together with the remaining coefficient 2.

The adjustment restricted to coefficient 2 and 3 is shown in the bottom right of Table 7-3(b). Predictive results following this adjustment are shown as round 4 in Fig. 7-6. For round 5 coefficients 1 and 2 were allowed adjustment, and for round 6 coefficients 2 and 3 were allowed adjustment. Round 6 in

Fig. 7-6 shows very good improvement in the first part of the recession. However, the last part of the recession is completely unsatisfactory. Our model is still not flexible enough to fit the given data.

For round 7 the core function was redefined. Its construction is shown to the right in Table 7-4. Three coefficients are now used. The second coefficient has now been moved to time 2 and a new coefficient 4 has been added at time 10. The parameter values under round 6A in Table 7-5 show this redefinition based on round 6 values. The flexibility now added to the model is a capability of slope adjustment from time 2 to time 10. This should allow improvement of the tail of the recession in Fig. 7-6.

Coefficients 1, 2, and 4 were adjusted by round 7. The vast improvement is seen in the round 7 prediction in Fig. 7-6. The basic flexibility requirements have now been satisfied. Further adjustments could still be carried out. Round 8 could adjust coefficients 2, 3, and 4, and round 9 could further adjust coefficients 1, 2, and 4.

Variations on Nonlinear Least Squares

As indicated previously, there are a number of numerical optimization methods available; they are basically variations of the concepts underlying the nonlinear least-squares method. In each case, the derivatives of the objective function with respect to the unknowns are evaluated numerically.

The pattern search technique is a common variation. In an attempt to increase the efficiency of the search, the pattern search method attempts to minimize the number of times that the derivatives have to be evaluated by making the assumption that the derivatives are reevaluated only when an iteration fails to improve the value of the objective function. One might say that the philosophy of the pattern search method is: If it worked once, it is worth trying again. Thus movement across the response surface proceeds in a constant direction until the movement pattern does not lead to improvement in the objective function. When movement does not improve the value of the objective function, the derivatives are reevaluated and the pattern search continues. Since the numerical calculation of the derivatives is usually the most time-consuming element of the search, especially for large sample sizes and complex model structures, the pattern search method usually shows a significant reduction in computer time.

Methods can also use acceleration and deceleration to increase the computational efficiency of a numerical search. The philosophy underlying the acceleration concept is: If a movement of Δd improved the value of the objective function, why not try a movement of $2\,\Delta d$? Thus, if a pattern movement improved the value of the objective function, try the same pattern again, but increase the distance of movement by some factor, usually 2. Of course, we would expect to approach the stationary point with fewer iterations but as the step size continues to increase by a factor of 2, the possibility of overshooting the stationary point on the response surface increases. At the point when the value of the objective function does not improve or gets worse, the program reevaluates the pattern and returns to the original step size. For some programs, deceleration is also used to get closer to the stationary point. With deceleration, the step size is decreased by a preset amount, usually a fraction of 0.5, when the original step size is too large to approach the stationary point. The number of decreases in the step size per computer run is usually limited to some preset value, say 5. These variations on the numerical least-squares procedure increase the efficiency of the algorithm.

The Rosenbrock method is also used in calibrating hydrologic models. Although similar in many respects to the pattern search method, it differs from the other numerical optimization methods in that it finds a new orthogonal axis system in which to search the response surface. The principal axis is centered in the direction of the steepest ascent or descent of the response surface. A more detailed discussion of orthogonal rotation of axis systems is given in Chapter 11.

Before discussing phase 3 of the numerical optimization strategy, it may be worthwhile discussing an additional aspect of the phase 2 search. One must properly assess the output from the computer. The concepts in Chapter 3 on evaluating a regression equation apply also to numerically optimized models. Specifically, one should assess the rationality of the model coefficients, compute goodness-of-fit statistics (i.e., R, R^2, S_e), and evaluate the residuals to ensure that there are no local biases and that the residuals are homoscedastic and independent. Most computer programs list the values of the coefficients and the value of the objective function for each iteration. These should be reviewed to ensure that a stationary point was approached. For some numerical procedures that do not include deceleration, the step size is too large to approach the stationary point. In such cases, the program should be executed a second time using the final parameter estimates from the first execution as the initial parameter estimates for the second execution. This approach is also required when the magnitudes of the coefficients differ significantly. In such cases, some of the coefficients are not sufficiently sensitive and their values fail to get optimized. This can be overcome by optimizing the coefficients in sequence, with each computer run optimizing the coefficients having similar levels of sensitivity.

A PHASE 3 ANALYSIS

The first phase of the numerical optimization strategy has the objective of exploring the response surface and locating a base point to begin the phase 2 search. Phase 2 has the objective of climbing the response surface and approaching the stationary point. At a stationary point the first derivatives equal zero. Thus, in the phase 2 search, which is based on the first derivatives, the objective function becomes less sensitive to the very coefficients it is trying to optimize as the stationary point is approached. Not only does this make it difficult to find the optimum values, but the phase 2 search algorithm becomes less efficient as the optimum is approached.

In cases where the analysis has important consequences, it may be worthwhile conducting a phase 3 analysis. Phase 3 involves both exploration (like phase 1) and search (like phase 2). At the optimum identified in phase 2, the response surface should be explored. This can be accomplished graphically by constructing two-dimensional response surfaces for each pair of unknowns. Usually, a systematic grid with the phase 2 optimum point as the center node will be derived for each pair of points, with the cell width of the grid equal to some percentage (usually 2 to 10%) of the coefficient. The characteristics of the response surface can be assessed from these two-dimensional response functions. If isolines are constructed for the plots of the response surface, noninteracting coefficients will be indicated by a series of concentric ellipses in which the major and minor axes are parallel to the axis system for the coefficients. A set of concentric ellipses in which the major and minor axes are not normal with the axis system of the coefficients indicates that the coefficients interact with each other. A response surface that has a long, narrow ridge (maxima) or valley (minima) indicates that there are an array of values of the two coefficients that provide similar values of the objective function. If any attempt is to be made to relate the coefficients to potential causative factors, the situation where the response surface has a long, narrow valley should be given special consideration. It is highly likely that the resulting correlation will be an inaccurate indication of the true relationship. This situation may be best illustrated using an example. The response surface of Fig. 7-8 is for the watershed time of concentration and a unit hydrograph parameter for a single storm event. The long, narrow valley indicates that there are an array of values that provide almost identical values of the objective function. The direction of the valley indicates that the unit hydrograph parameter is highly correlated with the time of concentration. Thus, if one is going to attempt to relate the unit hydrograph parameter to watershed characteristics, it is important that the phase 2 search results be investigated to ensure that the true optimum is found. Otherwise, a small change in the time of concentration can lead to a change in the unit hydrograph parameter which may introduce significant errors into the relationship between the unit hydrograph parameters and the watershed characteristics. In summary, the phase 3 analysis can have important consequences; therefore, a response surface should, at the minimum, be investigated graphically.

In addition to the graphical exploration, the phase 3 analysis can also include a systematic search of the response surface. This systematic search attempts to reproduce the importance of the second derivatives in analytical optimization (see Chapter 3). Specifically, the phase 3 search uses a Taylor

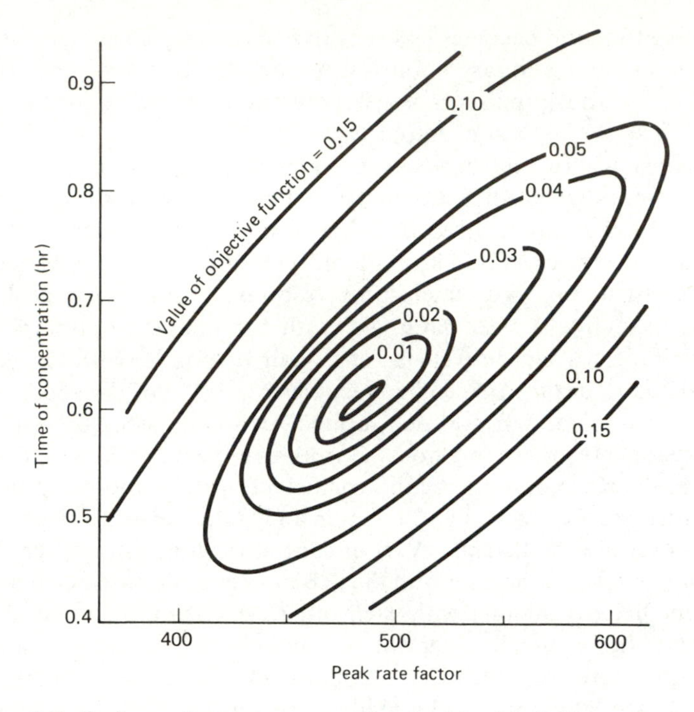

Figure 7-8 Peak rate factor response surface for May 31, 1962, storm event on Powells Creek watershed.

series expansion that includes nonlinear terms. This approach will account for some of the curvature in the response surface.

Let's consider the simple case where there is just one unknown. Assume that the phase 2 search indicated an optimum value for the unknown b of 0.95. The value of the objective function, F, is computed for values of b of 0.95, 0.96, and 0.97, with resulting values of F of 0.009875, 0.006336, and 0.003573, respectively. For the Taylor series expansion

$$F(b) = F(b_0) + (b - b_0)\frac{dF}{db} + \frac{(b - b_0)^2}{2!}\frac{d^2F}{db^2} \tag{7-25}$$

the values of the derivatives are unknown (if we knew these, the problem could be solved analytically). Letting Δ_1 and Δ_2 represent the first and second derivatives, respectively, we can form two simultaneous equations using the base point value of b of 0.95 and the two additional values of b at 0.96 and 0.97:

$$0.006336 = 0.009875 + (0.96 - 0.95)\Delta_1 + 0.5(0.96 - 0.95)^2\Delta_2 \tag{7-26a}$$

and

$$0.003573 = 0.009875 + (0.97 - 0.95)\Delta_1 + 0.5(0.97 - 0.95)^2\Delta_2 \tag{7-26b}$$

Solving for Δ_1 and Δ_2 yields $\Delta_1 = -0.3927$ and $\Delta_2 = 7.76$. This gives the following equation using Eq. 7-25:

$$F(b) = 0.009875 + (b - 0.95)(-0.3927) + 0.5(b - 0.95)^2(7.76) \tag{7-27}$$

If we differentiate Eq. 7-27 with respect to b and set the result equal to zero, we get the optimum value of b of 1.0006. In this case the true optimum was 1 since the objective function was known:

$$F(b) = b^3 + b^2 - 5b + 3 \tag{7-28}$$

Thus the phase 3 search, which is based on the second derivative, led to a better estimate of the true optimum.

This procedure can be extended to any number of unknowns. For two unknowns (b_1 and b_2) we have the following Taylor series expansion:

$$\begin{aligned} F(b_1, b_2) = {} & F(b_1^*, b_2^*) + (b_1 - b_1^*)\left.\frac{\partial F}{\partial b_1}\right|_* + (b_2 - b_2^*)\left.\frac{\partial F}{\partial b_2}\right|_* \\ & + \tfrac{1}{2}(b_1 - b_1^*)^2\left.\frac{\partial^2 F}{\partial b_1^2}\right|_* + \tfrac{1}{2}(b_2 - b_2^*)^2\left.\frac{\partial^2 F}{\partial b_2^2}\right|_* \\ & + (b_1 - b_1^*)(b_2 - b_2^*)\left.\frac{\partial^2 F}{\partial b_1\,\partial b_2}\right|_* \end{aligned} \tag{7-29}$$

Letting Δ_1, Δ_2, Δ_{11}, Δ_{22}, and Δ_{12} represent the respective derivatives, we have

$$\begin{aligned} F(b_1, b_2) = {} & F(b_1^*, b_2^*) + (b_1 - b_1^*)\Delta_1^* + (b_2 - b_2^*)\Delta_2^* \\ & + \tfrac{1}{2}(b_1 - b_1^*)^2\Delta_{11}^* + \tfrac{1}{2}(b_2 - b_2^*)^2\Delta_{22}^* + (b_1 - b_1^*)(b_2 - b_2^*)\Delta_{12}^* \end{aligned} \tag{7-30}$$

Equation 7-30 includes five unknowns (i.e., the five Δ's). We could form five simultaneous equations using six points on the response surface and solve for the five unknowns. However, the problem can be greatly simplified if we select five points and omit the unknown Δ_{12}. The five points should be selected using a center point and the end points for two orthogonal lines that intersect at the center point. This will reduce the problem to two sets of two simultaneous equations. The assumption is that the interaction term is critical only when the interaction is especially significant.

The procedure can be illustrated using the data of Fig. 7-9. The phase 2 search produced values of $b_1 = 3$ and $b_2 = 0.2$, with an objective function value of -0.98. Using $\Delta b_1 = 0.5$ and $\Delta b_2 = 0.05$, the values of the objective function for four points on the response surface are shown in Fig. 7-9. Substituting the values for the phase 3 base point into Eq. 7-30 (assuming the interaction term is not significant) yields

$$F = -0.98 + (b_1 - 3)\Delta_1 + (b_2 - 0.2)\Delta_2 + \tfrac{1}{2}(b_1 - 3)^2\Delta_{11} + \tfrac{1}{2}(b_2 - 0.2)^2\Delta_{22} \tag{7-31}$$

Since the lines in Fig. 7-9 are orthogonal, values for Δ_1 and Δ_{11} can be computed

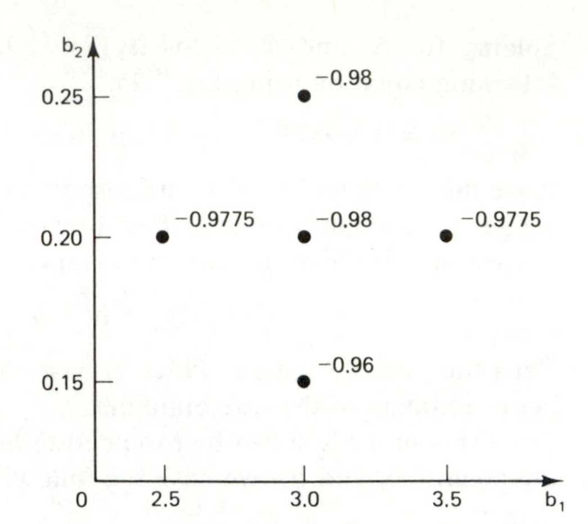

Figure 7-9 Delineation of response surface for phase 3 analysis.

with Eq. 7-31 using the three points with coordinates (2.5, 0.2), (3.0, 0.2), and (3.5, 0.2). Values for Δ_2 and Δ_{22} can be computed using the three points with coordinates (3.0, 0.15), (3.0, 0.2), and (3.0, 0.25) and Eq. 7-31. The resulting values are $\Delta_1 = 0$, $\Delta_{11} = 0.02$, $\Delta_2 = -0.2$, and $\Delta_{22} = 8$. This yields the following function:

$$F = -0.98 - 0.2(b_2 - 0.2) + \tfrac{1}{2}(0.02)(b_1 - 3)^2 + \tfrac{1}{2}(8)(b_2 - 0.2)^2 \tag{7-32}$$

Differentiating Eq. 7-32 twice, once with respect to b_1 and then with respect to b_2, and then setting the two derivatives equal to zero, yields values for b_1 and b_2 of 3.0 and 0.225, respectively. The corresponding value of the objective function is −0.9825, which is slightly better than the phase 2 optimum of −0.98.

In general, the true value is unknown. However, for this example, the objective function was a simple polynomial:

$$F = 0.01b_1^2 - 0.1b_1 + 4b_2^2 - 2.4b_2 - 0.57 + 0.2b_1b_2 \tag{7-33}$$

The minimum value for this function equals −0.983 when $b_1 = 2.667$ and $b_2 = 0.233$. An exact solution would have been found if the interaction term had not been omitted from Eq. 7-31. In general, the objective function is not a mathematical equation like Eq. 7-33, but a statistical function such as the least-squares function. In most cases, the phase 3 search will not produce an exact solution, but it should provide a better solution than the phase 2 search, which ignores the curvature around the stationary point.

APPLICATIONS IN HYDROLOGIC ANALYSIS

A number of problems that require numerical optimization arise frequently in hydrologic modeling. With the widespread availability of computers it is frequently necessary to fit functional forms to curves in order to develop software for hydrologic methods. For example, if one wished to write a program for the SCS peak discharge method that relates the unit peak discharge (cfm/in.) to the time of concentration, it would be necessary to calibrate a function to represent the nonlinear curve in the figure. When linear or log-linear models are not adequate, other forms are necessary. Unless the model structure can be transformed so that analytical least squares can be used, a numerical solution will be necessary. The problem of fitting a functional form to a curve will be considered in Example 7-3, where a logistic function will be fit to a curve representing the variation of a mean soil erosivity index with time over a 12-month period.

More frequently, the problem is to fit a model to a set of points. Actually, the problem of fitting a curve described in the preceding paragraph is solved by selecting a set of points and then fitting a structural function to the points. In either case, a significant deviation of the fitted model from either the known curve or the set of points would indicate a biased model, which implies that the model structure may be incorrect.

Numerical optimization is frequently used to fit functions in which the model structure has a discontinuity in either the model or one of its derivatives. The spherical semivariogram of Eqs. 5-11 is one such model. The existence of either a discontinuity or a constraint makes it especially difficult or impossible to fit the function using analytical methods. To illustrate this problem, a spherical semivariogram model will be fit to a set of regional rainfall data in Example 7-4.

Numerical optimization is especially suited to models that require intermediate calculations. This problem will be illustrated with Example 7-5, where the gamma probability function will be used as a unit hydrograph and the parameters will be fit using a set of rainfall-runoff data for a small watershed. Numerical optimization will be used since the solution requires convolution at each increment. Such numerical fitting would be an alternative to a method-of-moments solution to the problem.

Example 7-3: Distribution of the Erosivity Index

Soil erositivity is a function of a number of variables, including storm and soil characteristics as well as site characteristics, such as slope and land-cover type. Although erosivity will vary widely from storm to storm, it may be necessary to use some average value for either planning or hydrologic design. In such cases the average annual distribution may be adequate. Piest et al. (1965) provide a graph of the long-term average rainfall erosivity for a 74-acre, countour-planted, continuous-corn watershed near Treynor, Iowa. The distribution is shown as Fig. 7-10. The curve is noticeably nonlinear. Except for some local deviations that no simple nonlinear function could follow, the curve has the general shape of the *logistic function*:

$$y(t) = \frac{K_1}{1 + e^{-K_2(t-t_0)}} \tag{7-34}$$

in which K_1 and K_2 are scale parameters and t_0 is a location parameter. In addition to being referred to as a logistic function, Eq. 7-34 is referred to as the *biological growth curve* and the *autocatalytical function.* Values for K_1, K_2, and t_0 must be obtained through calibration.

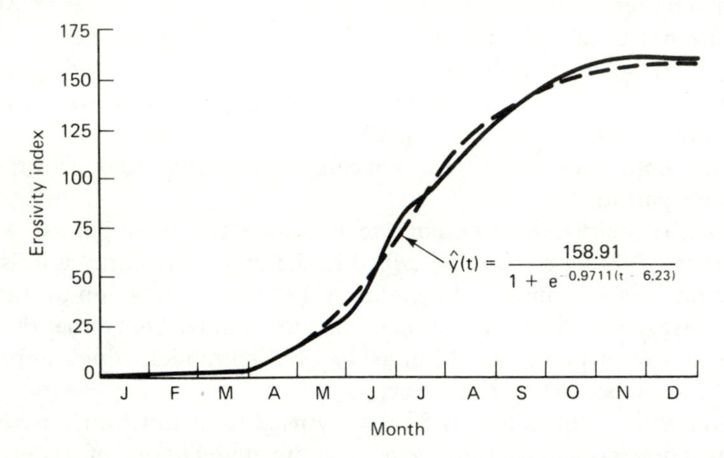

Figure 7-10 Monthly variation of erosivity index. (From Piest et al., 1965.)

The curve of Fig. 7-10 was reduced to a set of 13 data points. The points were selected using a nonconstant interval for the month, with values of 0 and 12 used to represent January 1 and December 31, respectively. The values for the 13 points are given in Table 7-6.

TABLE 7-6 Tabular Values for the Erosivity Index, $y(t)$

t	$y(t)$	$\hat{y}(t)$	$\hat{y}(t) - y(t)$
0	0	0.4	0.4
3.0	5	6.6	1.6
4.0	15	16.4	1.4
4.5	22	25.0	3.0
5.0	31	36.9	5.9
5.5	55	52.4	−2.6
6.0	80	70.6	−9.4
6.5	90	89.8	−0.2
7.0	105	107.8	2.8
8.0	130	134.7	4.7
9.0	147	148.8	1.8
10.0	158	154.9	−3.1
12.0	160	158.3	−1.7
			4.6

To use the numerical optimization method, initial estimates of the unknowns are required. These can be obtained using either a phase 1 exploration or from knowledge of the data and the model; the latter alternative was used for calibrating the model of Eq. 7-34 with the data of Table 7-6. It is known that K_1 represents the spread of the data; therefore, a value of 160 [i.e., $y(12) - y(0) = 160 - 0$] was used as the initial estimate of K_1. The value of t_0 is approximately the central estimate of the t, so a value of 6 was used as the initial estimate of t_0. The scale parameter K_2 reflects the curvature in the function. Using $K_1 = 160$ and $t_0 = 6$, an estimate of K_2 was obtained using the value of $y(t)$ at $t = 4$. Equation 7-34 becomes

$$y(t) = \frac{160}{1 + e^{-K_2(t-6)}} \tag{7-35}$$

Rearranging Eq. 7-34 and taking logarithms yields the following expression for K_2:

$$K_2 = \frac{\ln([K_1/y(t)] - 1)}{t_0 - t} \tag{7-36}$$

For $t = 4$ this becomes

$$K_2 = \frac{\ln(\frac{160}{15} - 1)}{6 - 4} = 1.13 \tag{7-37}$$

Therefore, a value of 1.0 was used as an initial estimate of K_2.

The data for t and $y(t)$ in Table 7-6 were used with the initial estimates of K_1, K_2, and t_0 to calibrate the logistic function of Eq. 7-34. A computerized numerical least-squares analysis was performed, with the following model as the best estimate:

$$\hat{y}(t) = \frac{158.91}{1 + e^{-0.9711(t-6.23)}} \tag{7-38}$$

The initial estimates of the unknowns were fairly accurate and convergence was achieved in a few rounds. The curve is shown in Fig. 7-10. Except for those parts of the curve that are highly nonlinear, such as near $t = 5$, $t = 6.25$, and $t = 7.5$, the logistic function provides a very reasonable approximation to the original curve.

In addition to the values of the coefficients, the numerical optimization computer program provided goodness-of-fit statistics. The predicted values of the erosivity index, $\hat{y}(t)$, are given in Table 7-6 together with values of the residuals, which are defined here as $\hat{y}(t) - y(t)$. The largest error, which was -9.4, occurred at $t = 7$; this reflects the inability of the logistic function to fit the relatively flat slope that is evident from Fig. 7-10 in the interval $6 < t < 8$. The sum of the residuals equals 4.6, which means that the average error equals 0.35; such a bias should not be considered significant. The logistic model of Eq. 7-38 provided a correlation coefficient of 0.92 (i.e., $R^2 = 0.848$) and a standard error of estimate of 4.35, which is small compared to the standard deviation of $y(t)$ of 59.7. Thus the goodness-of-fit statistics indicate that the logistic function provides an accurate, and unbiased, representation of the long-term average erosivity. A phase 3 analysis was performed but did not provide any increase in accuracy.

Example 7-4: Fitting a Spherical Semivariogram Model to Precipitation Data

Kriging was introduced in Chapter 5. The diversity of problems for which it has been used in hydrologic analysis suggests that it is an important tool for statistical estimation. To evaluate the weights for the kriging model, a semivariogram model must be selected. The spherical model appears to be the most generally applicable model.

Karlinger and Skrivan (1981) computed the sample semivariogram points using the mean annual precipitation data for stations in the Powder River basin, Montana and Wyoming. Mean annual precipitation is an important variable in regions where restoration of coal strip-mining lands is necessary. The sample points, which are shown in Fig. 7-11, are based on the mean annual precipitation at 60 stations in the region. The points show the trend that is typical of many semivariogram plots. The first three points have values of $\gamma(h)$ less than 4.1, while five of the last six values are greater than 4.5. Except for the value of $\gamma(h)$ at a separation distance of 187.5 miles, the sample values show little scatter about the trend. Additionally, it appears that the data would be best fit using a semivariogram model that includes an intercept (i.e., the nugget effect). While Karlinger and Skrivan used a linear semivariogram model, a spherical model will be used here:

$$\hat{\gamma}(h) = \begin{cases} \gamma_n + (\gamma_r - \gamma_n)\left[1.5\left(\frac{h}{r}\right) - 0.5\left(\frac{h}{r}\right)^3\right] & \text{for } h \le r \quad \text{(7-39a)} \\ \gamma_r & \text{for } h > r \quad \text{(7-39b)} \end{cases}$$

in which γ_n is the nugget effect, γ_r the sill, r the radius of influence, and h the separation distance. Values for γ_n, γ_r, and r must be estimated from data.

The numerical optimization strategy requires initial estimates of the unknowns. While a phase 1 exploration search could have been used, reasonable estimates of the three coefficients can be obtained from Fig. 7-11. Except for the point at a separation

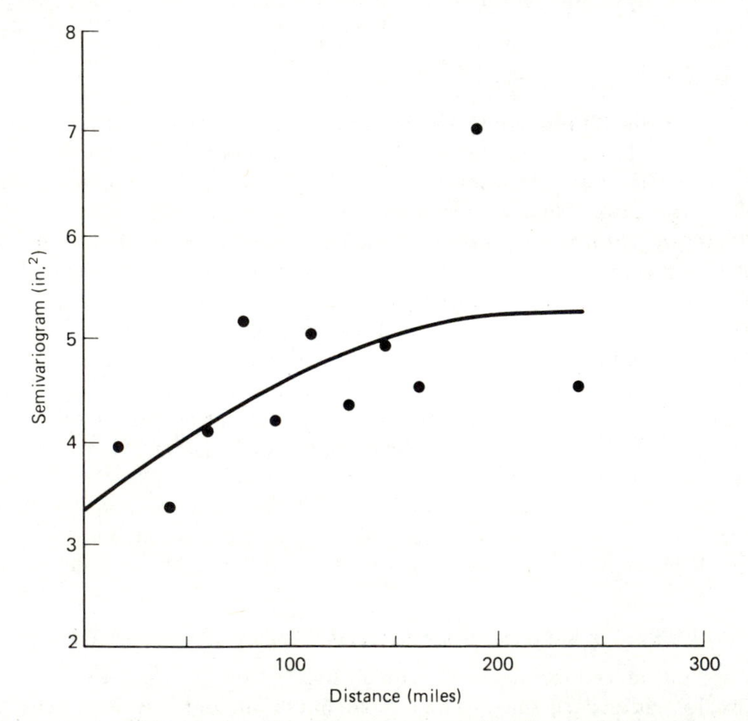

Figure 7-11 Semivariogram for mean annual precipitation in the Powder River basin.

distance of 187.5 miles, most of the points at large separation distances fall near a value of 5; thus a value of 5.0 in.^{2} will be used as the initial estimate of the sill, γ_r. If a line is drawn through the first few points and projected back to a separation distance of 0.0, a nugget effect of 2.5 is not unreasonable as an initial estimate. The same straight line intersects the sill at a separation distance of about 100 miles, which is used as an initial estimate of r.

TABLE 7-7 Data Base for Fitting a Semivariogram Model and the Distribution of Errors

Separation Distance (miles)	$\gamma(h)$ (in.^{2})	$\hat{\gamma}(h)$ (in.^{2})	$\hat{\gamma}(h) - \gamma(h)$ (in.^{2})
17.2	3.94	3.60	−0.34
42.5	3.38	3.95	0.57
59.4	4.06	4.17	0.11
76.6	5.16	4.38	−0.78
87.5	4.16	4.51	0.35
109.4	5.03	4.74	−0.29
128.1	4.34	4.92	0.58
145.3	4.94	5.06	0.12
160.9	4.53	5.17	0.64
187.5	7.05	5.29	−1.76
237.5	4.53	5.33	0.80
			0.00 Sum

The initial estimates of γ_r, γ_n, and r were used with the data of Table 7-7, which were estimated from a graph supplied by Karlinger and Skrivan, to calibrate Eqs. 7-39. A phase 2 analysis program based on nonlinear least squares converged after 50 rounds, with the values of γ_n, γ_r, and r shown in Table 7-8. The phase 2 analysis reduced the error sum of squares by about 9%. The nugget effect increased from 2.50 in.^{2} to 3.25 in.^{2}, while the estimate of the radius of influence increased from 100 miles to 210 miles. The sill was the least sensitive parameter, and its value did not change. The insensitivity of the sill occurs because there is just a single point beyond the radius of influenced of 210 miles.

TABLE 7-8 Summary Statistics for Numerical Optimization of a Spherical Semivariogram Model

Analysis	Nugget (in.^{2})	Radius of Influence (miles)	Sill (in.^{2})	Error Sum of Squares
Initial estimates	2.50	100	5.00	0.8372
Phase 2 results	3.25	210	5.00	0.7602
Phase 3 results	3.36	213	5.33	0.7589

Three iterations of a phase 3 analysis reduced the error further, with the error sum of squares decreasing by about 0.2%. The resulting parameters are given in Table 7-8. The final spherical semivariogram model is

$$\hat{\gamma}(h) = \begin{cases} 3.36 + (5.33 - 3.36)\left[1.5\left(\dfrac{h}{213}\right) - 0.5\left(\dfrac{h}{213}\right)^3\right] & \text{for } h \leq 213 \quad \text{(7-40a)} \\ 5.33 & \text{for } h > 213 \quad \text{(7-40b)} \end{cases}$$

The model is shown in Fig. 7-11, with the sample points. The predicted values and errors are given in Table 7-7. Although the model is unbiased, the error variation is still significant. The standard error of estimate equals 0.848 in.^{2}, while the standard deviation of the $\gamma(h)$ values equals 0.953 in.^{2}; thus the spherical model of Eqs. 7-40 reduced the error variation by only 11%. The large error variation may reflect either the small samples that were available to estimate the 11 sample points of Table 7-7 or the absence of a significant spatial structure of mean annual precipitation in the Powder River basin. Certainly, the observation at a separation distance of 187.5 miles is responsible for a large part of the error variation.

Example 7-5: Calibration of a Unit Hydrograph

Examples 7-3 and 7-4 differed little from statistical least-squares fitting, with the primary differences being the nonlinearity of both examples and the discontinuity in the spherical semivariogram of Example 7-4. The flexibility of numerical optimization is more apparent for problems where the model is more than just a nonlinear function. A unit hydrograph analysis is one such example because the criterion, or dependent, variable would be the values of the runoff hydrograph. Convolution is a necessary step in deriving the predicted runoff hydrograph. This makes it impossible to use the analytical approach.

One statistical approach to unit hydrograph derivation is method-of-moments estimation (MME). This method is sometimes used because it does not require the convolution integral to be used, except for showing the final predicted runoff hydrograph. However, MME, which is based on one or two of the statistical moments of the data, will not necessarily provide an accurate estimate of the entire runoff hydrograph because inherently it is less sensitive to the shape of the hydrograph than it is to the mean and variance.

Numerical optimization is an alternative to MME because it uses all the points that define the rainfall excess hyetograph and direct runoff hydrograph to derive the unit hydrograph. However, the solution procedure for numerical optimization is considerably more complex. Given the distributions of rainfall excess and direct runoff and assuming a form for the unit hydrograph, the following steps are used with the numerical optimization process:

1. Use estimates of the unit hydrograph parameters to compute the unit hydrograph.
2. Convolve the rainfall excess and the unit hydrograph to get an estimated runoff hydrograph.
3. Compare the predicted and measured runoff hydrographs and make any necessary adjustments in the model parameters using the numerical process.

The solution is iterative, with the above three steps executed until an optimal solution is found.

To illustrate the numerical optimization strategy in unit hydrograph analysis, a gamma distribution will be used to represent the unit hydrograph:

$$h(t) = \left(\frac{t}{b}\right)^{c-1} \frac{\exp(-t/b)}{b\Gamma(c)} \tag{7-41}$$

in which t is the time from beginning of rainfall excess, and b and c are the scale and shape parameter of the gamma probability function. While Eq. 7-41 is a distribution for a continuous random variable, it was discretized for the unit hydrograph analysis by computing $h(t)$ for each time value t and modifying the $h(t)$ values so that their sum equaled 1.0; this will ensure that the computed runoff volume equals the measured

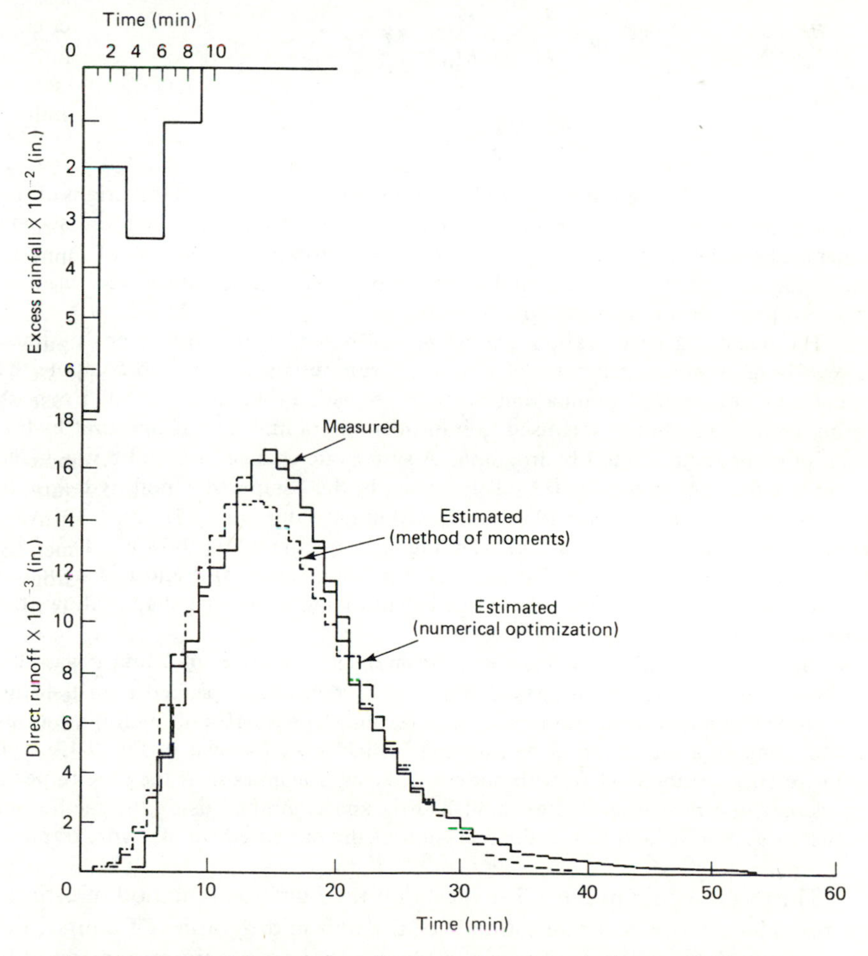

Figure 7-12 Rainfall excess and direct runoff for watershed W-6 at College Park, Maryland, for August 27, 1943.

runoff volume. To fit Eq. 7-41, values for b and c must be known, as well as the time base of the unit hydrograph.

Rainfall and runoff data for a 3.53 acre watershed in College Park, Maryland, were available on a 1-min. time increment. A storm for August 27, 1943, was used. Rainfall excess was separated using a loss function defined using an initial loss of 0.4 in. and a phi index abstraction of 0.036 in. The resulting distribution of rainfall excess is shown in Fig. 7-12. The direct runoff hydrograph is also shown in Fig. 7-12.

Initial estimates of the gamma unit hydrograph were obtained using the method of moments. When the gamma distribution is used as a unit hydrograph, it is often called the *Nash model*, because Nash was the first to make extensive use of the gamma distribution in hydrograph analysis. Nash showed that estimates of the shape and scale parameters can be obtained from the first and second moments of the rainfall excess and direct runoff distributions. Specifically, the scale parameter is defined as

$$b = \frac{M_{2Q} - M_{2p}}{M_{1Q} - M_{1p}} - M_{1Q} + M_{1P} \tag{7-42a}$$

and

$$c = \frac{M_{1Q} - M_{1P}}{b} \tag{7-42b}$$

in which M_{1P} and M_{2P} are the first and second moments about the time origin of the rainfall excess distribution, respectively, and M_{1Q} and M_{2Q} are the first and second moments about the time origin of the direct runoff distribution. The data of Fig. 7-12 yield values of b and c of 3.218 and 4.196, respectively; these values were used as initial estimates for a numerical optimization.

The numerical optimization procedure was based on the three steps outlined above. Using the initial estimates of b and c, a subroutine was used to compute the ordinates of a discretized gamma unit hydrograph, with a time increment of 1 min. A subroutine for convolution was used to transform the rainfall excess and unit hydrograph into a predicted runoff hydrograph. A systematic change in b and c was made to decrease the sum of the squares of the errors in the computed runoff hydrograph. The procedure converged after 195 rounds. Optimum values of 2.341 and 5.892 were found for b and c, respectively. The resulting unit hydrograph is shown in Fig. 7-13. The runoff hydrograph generated using these optimum values of b and c is shown in Fig. 7-12. The error in the peak discharge is less than 2% and the shape follows the shape of the measured runoff hydrograph.

The unit hydrograph for the method of moments estimates of b and c was also computed (see Fig. 7-13). The MME unit hydrograph is less peaked than the unit hydrograph computed using numerical optimization. The MME unit hydrograph was used to compute a direct runoff hydrograph, which is also shown in Fig. 7-12. The peak is in error by about 11%, with noticeable errors for times near the time to peak. Surprisingly, the recession of the runoff hydrograph computed using the method of moments was almost identical to the recession of the measured runoff hydrograph.

The results of Example 7-5 suggest that the numerical method of estimation provides a more accurate estimate of the unit hydrograph. Of course, the effort required to calibrate the unit hydrograph parameters is considerably greater.

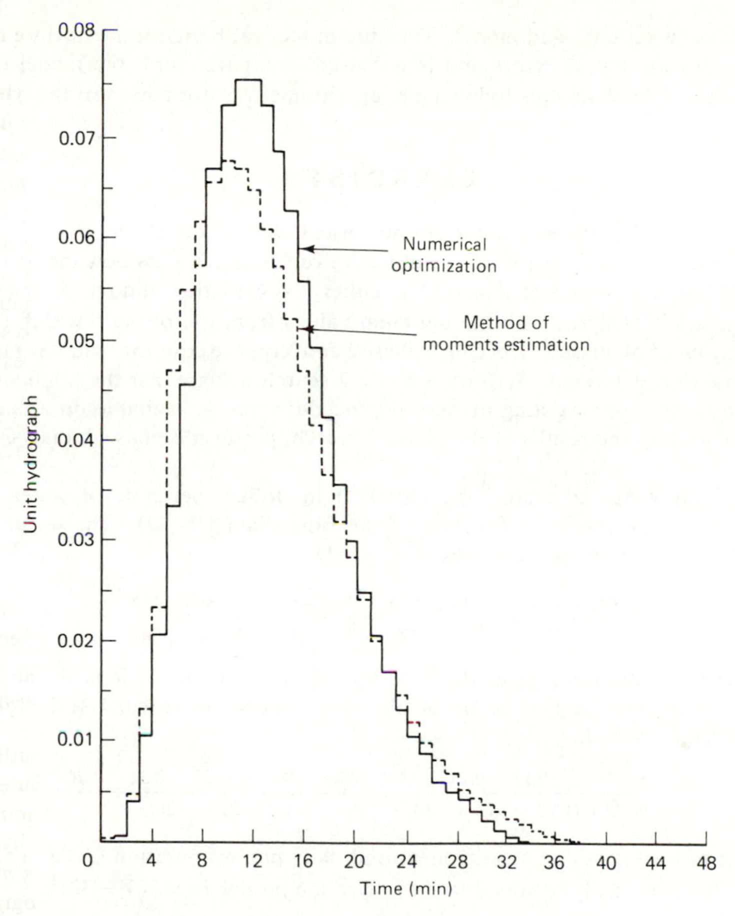

Figure 7-13 One-minute unit hydrograph for watershed W-6 at College Park, Maryland.

Example 7-6: Numerical Calibration of a Power Model

In Example 6-9 a power model that includes seven predictor variables was calibrated using a logarithmic transformation of the data. The resulting model showed a bias of −251 cfs, which is about 4.3% of the mean. The correlation coefficient of the logarithmic model, Eq. 6-76, could not be used as a measure of goodness of fit for the model (i.e., Eq. 6-77) that would be used to make predictions.

Numerical optimization provides an alternative method of calibrating a power model, with the added advantage of providing an unbiased model. Using the data of Appendix B-4, the coefficients of a multiple power model were computed:

$$\hat{Y} = 5.233X_1^{0.6965}X_2^{0.6565}X_3^{0.2230}X_4^{0.7009}X_5^{0.1497}X_6^{-0.1053}X_7^{1.1976} \tag{7-43}$$

The solution converged after 38 rounds, with the coefficients of Eq. 6-77 used as the initial estimates. The standard error of estimate equaled 1814 cfs, which is less than the standard error of Eq. 6-77 by about 15%. Also, the sum of the errors equaled zero, which indicates an unbiased model. Thus the model calibrated numerically provides a smaller distribution of errors and is unbiased, compared with the model derived using analytical least squares following a logarithmic transformation of the data.

EXERCISES

For each exercise, perform a complete numerical calibration of the model. Use the sum of the squares of the errors as the objective function. Discuss how the bounds on the phase 1 search were determined. Use either a systematic, random, or systematic-random phase 1 analysis, with the optimum values from the phase 1 search used as the starting value of phase 2. Develop a phase 2 convergence criterion, with the rationale for the criterion described. Perform a phase 2 search using either the nonliner least-squares method or an existing numerical procedure that is available on a computer system. Based on the results of the phase 2 search, perform a phase 3 analysis.

7-1. The following table gives the cost (C) in dollars per mile of water supply transmission lines as a function of the pipe diameter (D). The linear model without an intercept will be used: $\hat{C} = bD$.

C (\$10^3/mile)	48	107	173	245	320	398	482
D (in.)	6	12	18	24	30	36	42

7-2. The following table gives the time lag (t_L) for flow in various segments of a watershed as a function of the length (L) of the channel reach. Use the following model: $\hat{t}_L = b\sqrt{L}$.

t_L (hr)	0.13	0.21	0.18	0.22	0.23	0.27	0.31	0.26
L ($\times 10^2$ ft)	10	13	14	16	21	22	26	26

7-3. The following table gives the measured daily pan evaporation (E) as a function of the mean daily temperature (T). Use the model $\hat{E} = b(T - 32)^2$.

E (in./day)	0.05	0.12	0.24	0.10	0.18	0.19	0.09
T (°F)	56	74	91	67	79	83	71

7-4. The following table gives the measured interception storage (I) for nine storm events as a function of the storm rainfall (P) and the height of the crop (h). Use the model $\hat{I} = b_0 + b_1 h P^{0.9}$.

I (in.)	0.014	0.015	0.015	0.018	0.023	0.028	0.039	0.028	0.047
P (in.)	1.3	1.6	0.8	0.5	1.1	0.9	1.7	0.7	1.4
h (ft.)	0.5	0.8	1.4	1.8	2.9	3.3	3.9	4.6	5.2

7-5. To estimate consumptive use, the Blaney-Criddle method requires the monthly percentage of the annual daytime hours (p), which varies with the month. The table below gives the values of p for each month at a latitude of 40°. Use the model $\hat{p} = b_1 + b_2 \sin(30t - 75)$, in which t is the month ($t = 0$ for January, $t = 1$ for February etc.) and the sine is given in degrees.

t	0	1	2	3	4	5	6	7	8	9	10	11
p (%)	6.76	6.72	8.33	8.95	10.02	10.08	10.22	9.54	8.29	7.75	6.72	6.52

7-6. Where orographic effects can be significant, precipitation can vary with elevation. The table below gives the mean annual precipitation (P) for selected precipitation gages in a region as a function of the elevation (E) at the gaging station. Use the model

$$\hat{P} = \begin{cases} b_1 & \text{for } E \leq 5000 \text{ ft} \\ b_1 + b_2(E - 5000) & \text{for } E > 5000 \text{ ft} \end{cases}$$

P (in.)	21	18	24	18	20	22	28	28	37	35	41
E ($\times 10^2$ ft above MSL)	32	33	38	41	46	55	61	65	68	74	78

7-7. The average annual sediment production rate (S) was determined for eight watersheds varying in size. The table below gives the rate as a function of the drainage area (A). Use the model $\hat{S} = b_1 A^{b_2}$.

A (sq mi)	9	74	325	37	421	156	211	284
S (ac-ft/sq mi)	3.25	1.51	1.15	1.98	1.03	1.41	1.08	1.17

7-8. The storm runoff (Q) can be estimated from measurements of the storm rainfall (P) and a land-use factor (C). Using the data for seven watersheds given in the table below, estimate the values of the coefficients k and b for the model given by

$$\hat{Q} = \frac{(P - I)^b}{P + I} \qquad \text{where } I = k\left(\frac{1000}{C} - 10\right)$$

Q (in.)	1.9	2.0	2.7	1.7	2.7	2.4	2.5
P (in.)	4.1	3.7	5.2	2.9	5.5	3.4	4.6
C	75	83	78	86	72	89	80

7-9. Based on field measurements, the specific yield (Y) was computed as a function of depth (D). The data in the following table are for 11 tests for an area where bedrock occurs at a depth of about 60 ft. Derive the coefficients b_1 and b_2 for the model $\hat{Y} = 20[1 - b_1 e^{-b_2(60-D)}]$.

Y (1%)	19	16	18	17	18	16	14	13	9	8	4
D (ft)	1	3	4	9	11	17	26	34	38	47	55

7-10. Measurements of the moisture potential (P) and the moisture content (C) were obtained in a field under varying moisture contents. The data are given in the following table. Derive the coefficients b_1 and b_2 for the model $\hat{P} = b_1 e^{b_2 C}$.

P (cm water)	-10^6	-2×10^5	-10^4	-5×10^3	-4×10^2	-9×10^1	-2×10^1
C	0.05	0.06	0.1	0.22	0.22	0.35	0.41

7-11. The SCS TR-55 provides tabular data for a pond and swamp adjustment factor (F), which is a function of the storm frequency in years and the percentage of ponding and swampy area (P). The following table gives values of F and P for a 2-year storm event when the pond and swamp area is near the design point.

Evaluate the coefficients of the model $\hat{F} = 1 + b_1 P + b_2 P^2$.

F	1	0.92	0.86	0.8	0.74	0.69	0.64	0.59	0.57	0.53
P (%)	0	0.2	0.5	1.0	2.0	2.5	3.3	5.0	6.7	10.0

7-12. A nonstandard weir was constructed to measure small runoff rates (q) from an infiltration field plot. A narrow V notch was cut into a weir plate with bottom at head zero. To measure high flows, a second, wider V notch was cut with bottom at head of 0.75 ft. The compound rating curve (i.e., the relationship between depth, h, and q) will have the form

$$q = \begin{cases} Ah^B & h \le 0.75\text{ ft} \\ Ah^B + C(h - 0.75)^D & h > 0.75\text{ ft} \end{cases}$$

where A, B, C, and D are fitting coefficients. To provide accurate calibration of the compound weir, runoff during storm events is caught in a gaged tank so that volumes of flow are known. A recording gage also measures head on the weir continuously during the event. Write a computer program to perform phase 1 analysis for the following storm. Assume that the objective function is the sum of the squares of the differences between the computer and measured discharge rates (q).

Time (min)	0	1	2	3	4	5	6	7	8
Head (ft)	0	0.01	0.03	0.10	0.36	0.77	1.02	1.29	1.16
Volume (ft^3)	0	0.05	0.18	0.62	2.63	6.88	18.55	43.52	50.80

Time (min)	9	10	11	12	13	14	15	16
Head (ft)	0.98	0.72	0.47	0.31	0.11	0.08	0.02	0
Volume (ft^3)	33.52	16.58	7.15	4.88	2.51	1.04	0.57	0.08

7-13. Use an existing statistical software program to perform a phase 2 analysis for the compound rating curve and data in Exercise 7-12. Discuss this procedure in numerical optimization that calibrates a head-discharge function against volumes, the integral of discharge rate.

REFERENCES AND SUGGESTED READINGS

BEARD, L. R., Optimization Techniques for Hydrologic Engineering, *Water Resources Research*, Vol. 3, No. 3, pp. 809–15, 1967.

BETSON, R. P., and R. F. GREEN, *DIFCOR—A Program to Solve Nonlinear Equations*, Research Paper 6, Tennessee Valley Authority, Knoxville, Tenn., 1967.

BOX, G. E. P., The Exploration and Exploitation of Response Surfaces: Some General Considerations and Examples, *Biometrics*, Vol. 10, pp. 16–60, 1954.

BOX, G. E. P., and P. V. YOULE, The Exploration and Exploitation of Response Surfaces: An Example of the Link between the Fitted Surface and the Basic Mechanism of the System, *Biometrics*, Vol. 2, pp. 289–323, 1955.

BROMBERG, N. S., Maximization and Minimization of Complicated Multivariate Functions, *Communications and Electronics*, Vol. 58, pp. 725-30, 1962.

BROOKS, S. H., A Comparison of Maximum-Seeking Methods, *Operations Research*, Vol. 3, pp. 430-458, 1959.

CARRIGAN, P. H., Rosenbrock Technique for Determining Greatest or Least Value of a Function, *Geological Survey Computer Contribution*, 1972.

CROLEY, T. E., II, *Efficient Sequential Optimization in Water Resources*, Hydrology Paper 69, Colorado State University, Fort Collins, Colo., 1974.

DAWDY, D. R., and T. O'DONNELL, Mathematical Models of Catchment Behavior, *Journal of the Hydraulics Division*, Proc. Paper 4410, Vol. 91, No. HY4, pp. 123-37, 1965.

DECOURSEY, D. G., and W. M. SNYDER, Computer-Oriented Method of Optimizing Hydrologic Model Parameters, *Journal of Hydrology*, Vol. 9, pp. 34-56, 1969.

GREEN, R. F., *Optimization by the Pattern Search Method*, Research Paper 7, Tennessee Valley Authority, Knoxville, Tenn., January 1970.

IBBITT, R. P., and T. O'DONNELL, Fitting Methods for Conceptual Catchment Models, *Journal of the Hydraulics Division, ASCE*, Vol. 97, No. HY9, pp. 1331-42, 1971.

KARLINGER, M. R., and J. A. SKRIVAN, *Kriging Analysis of Mean Annual Precipitation, Powder River Basin, Montana and Wyoming*, USGS/WRD/WRI/81-050, Tacoma, Wash., May 1981.

MONRO, J. C., *Direct Search Optimization in Mathematical Modeling and a Watershed Model Application*, NOAA Tech. Memo, NWS HYDRO-12, Silver Spring, Md., April 1971.

OVERTON, D. E., and M. E. MEADOWS, *Stormwater Modeling*, Academic Press, New York, 1976.

PIEST, R. F., L. A. KRAMER, and H. HEINEMANN, Sediment Movement from Loessial Watersheds, in *Present and Prospective Technology for Predicting Sediment Yields and Sources*, ARS-S-40, pp. 131-41, June 1965.

ROSENBROCK, H. H., An Automatic Method of Finding the Greatest or Least Value of A Function, *The Computer Journal*, Vol. 3, p. 175, 1960.

SARMA, P., J. W. DELLEUR, and A. R. RAO, *A Program in Urban Hydrology*, Tech. Report 9, Water Resources Research Center, Purdue University, October 1969.

8 Subjective Optimization

INTRODUCTION

Two types of optimization have been discussed in Chapters 6 and 7. Analytical optimization was used to calibrate models having simple structures; for these types of models the performance criterion consisted of a single objective function, which could be differentiated analytically. The least-squares fit of a multiple regression model is one example where analytical optimization is used. When the model structure is of such complexity that it is difficult or impossible to evaluate analytically the derivatives of the single performance criterion, numerical optimization methods can be used. The derivatives are evaluated numerically. The nonlinear least-squares method is an example of numerical optimization.

There are optimization problems where neither analytical nor numerical optimization is appropriate. Specifically, the model may be so complex that even numerical analysis would be either too cumbersome or impossible to use. Additionally, there may be cases where there is more than one objective function. In either case, an optimal set of coefficients may be obtained using a third type of optimization, which is called *subjective optimization.* It differs from analytical and numerical optimization in that the derivatives of the objective function (or functions) are not computed for the direct purpose of modifying the coefficient values. Instead of finding derivatives either analytically or numerically, the user who is knowledgeable about the sensitivity of the model coefficients makes changes, somewhat subjectively, to the values on the basis of a comparison of the predicted and measured values of one or more criterion variables. In a sense, the derivatives are evaluated subjectively based on the modeler's knowledge of the sensitivity of the coefficients and are used to fit these coefficients.

The concept of subjective optimization is a significant departure from either analytical or numerical optimization because it does not provide reproducible solutions. That is, given the same model and input data, it is unlikely that two or more users would arrive at exactly the same solution. With analytical and numerical optimization, the solutions are usually reproducible.

AN INTRODUCTORY EXAMPLE

It may be instructive to introduce subjective optimization using a model and data set that could be optimized using either analytical or numerical optimization. Let's begin with the data of Table 8-1 and the linear bivariate model

$$\hat{y} = b_0 + b_1 x \tag{8-1}$$

in which b_0 and b_1 are unknown coefficients and y and x are the criterion and predictor variables, respectively. With subjective optimization we must provide initial estimates of the unknowns. Values for b_0 and b_1 of 0 and 1, respectively, are not unreasonable (see Fig. 8-1) because they provide a line that passes through the data. Using least squares as a criterion of fit, we can compute a

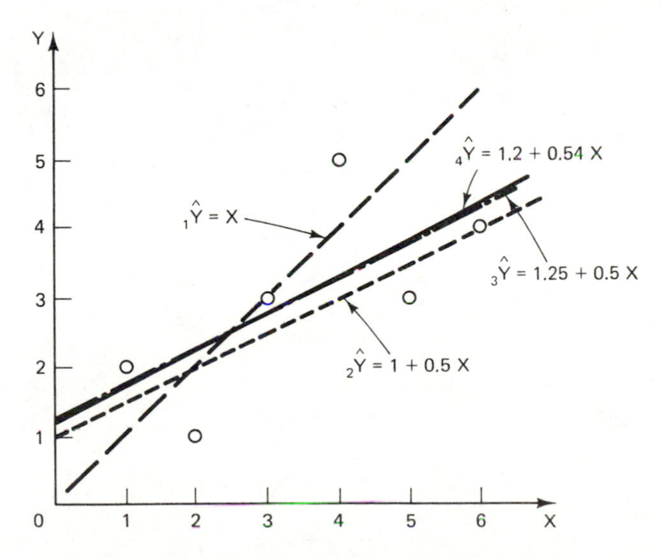

Figure 8-1 Graphical display of data for Table 8-1.

TABLE 8-1 Computations for the Introductory Example

i	X_i	y_i	${}_1\hat{Y}_i$	${}_1e_i$	${}_1e_i^2$	${}_2\hat{Y}_i$	${}_2e_i$	${}_2e_i^2$	${}_3\hat{Y}_i$	${}_3e_i$	${}_3e_i^2$	${}_4\hat{Y}_i$	${}_4e_i$	${}_4e_i^2$
1	1	2	1	−1	1	1.5	−0.5	0.25	1.75	−0.25	0.0625	1.714	−0.286	0.082
2	2	1	2	1	1	2.0	1.0	1.00	2.25	1.25	1.5625	2.228	1.228	1.508
3	3	3	3	0	0	2.5	−0.5	0.25	2.75	−0.25	0.0625	2.742	−0.258	0.067
4	4	5	4	−1	1	3.0	−2.0	4.00	3.25	−1.75	3.0625	3.256	−1.744	3.042
5	5	3	5	2	4	3.5	0.5	0.25	3.75	0.75	0.5625	3.770	0.770	0.593
6	6	4	6	2	4	4.0	0.0	0.00	4.25	0.25	0.0625	4.284	0.284	0.081
				3	11		−1.5	5.75		0.00	5.3750		−0.006	5.374

sum of squares for the model of Eq. 8-1 when $b_0 = 0$ and $b_1 = 1$:

$$_1\hat{Y} = 0 + 1X_i \tag{8-2}$$

With this model values of $_1\hat{Y}$ can be estimated, as shown in Table 8-1. Because the sum of the errors does not equal zero, the model is biased. The sum of the squares of the errors is also shown in Table 8-1.

Since the model is biased, we should seek values of b_0 and b_1 that provide a "better fit." We can adjust the values of b_0 and b_1 to achieve this. Since the bias is positive we may elect to decrease the value of b_1. The average error is 3/6 or 0.5, so let's try $b_1 = 0.5$, which equals the initial value of b_1 (1.0) minus the average error. If we have a slope coefficient of 0.5, we need to increase the value of b_0 so that the line passes through the data. In subjective optimization, the first few rounds involve a good deal of subjective assessment. In this case, from an examination of Fig. 8-1 a value of 1.0 may be a good estimate of b_0; at the very least it should be better than a value of zero. Thus we have a revised model:

$$_2\hat{Y} = 1 + 0.5X \tag{8-3}$$

The model of Eq. 8-3 is shown in Fig. 8-1. Using Eq. 8-3, we can calculate estimates of $_2\hat{Y}$ and compute the errors (see Table 8-1). The mean error has been reduced from 0.5 to −0.25 and the sum of the squares of the errors has been reduced by almost 50%. Thus we can assume that the model of Eq. 8-3 is better than the model of Eq. 8-2. But can we assume that Eq. 8-3 is the best possible model? Obviously, we may wish to have an unbiased model. We can get such a model by modifying b_0 alone. Specifically, since the mean error is −0.25, adding 0.25 to each predicted value would eliminate the bias. This is affected by increasing b_0 to 1.25, with the resulting model shown in Fig. 8-1:

$$_3\hat{Y} = 1.25 + 0.5X \tag{8-4}$$

In addition to providing an unbiased model, Eq. 8-4 also reduced the mean square error. We can again ask the question: Is this the best model we can get? We might also ask the question: Would continued adjustment of the coefficients result in a *significantly* better model? This illustrates one aspect of the subjectivity of the optimization process. One researcher may elect to use Eq. 8-4 as the final model while another researcher may elect to try to continue to reduce the mean square error. This is just one factor that leads to a lack of reproducibility in the final model.

Since the data and model have the form for which an analytical solution is possible, we can obtain the "true" values of b_0 and b_1 by bivariate linear regression:

$$_4\hat{Y} = 1.2 + 0.514X \tag{8-5}$$

The subjectively optimized model of Eq. 8-4 is surprisingly close to the true model of Eq. 8-5, as shown in Fig. 8-1. Except for round-off error, the model of Eq. 8-4 is unbiased (see Table 8-1), and the sum of the errors squared for Eq. 8-5 is only slightly smaller than the corresponding value for Eq. 8-4. It is evident that continued subjective adjustment of the coefficients of Eq. 8-4 would not have been worthwhile, at least if the best model is considered to be the model having the minimum sum of errors squared. But in practice the true set of coefficients is never known, so only continued iteration and a subjective assessment of the values of the objective function would indicate to the researcher when iteration should be stopped.

It should be evident from this simple example that subjective optimization is quite different from analytical optimization. First, subjective optimization is iterative, whereas analytical optimization requires one analysis. Second, initial estimates of the unknowns are required for a subjective analysis; initial estimates are not required for analytical optimization. Third, analytical optimization in the form of least squares uses the single objective function of the least-squares principle. Subjective optimization may involve more than one objective function. In the analysis above, an attempt was made to obtain a model that was unbiased and had a minimum mean square error. Fourth, subjective optimization involves a subjective decision on the best time to stop iteration. Other differences are identified below.

It should be apparent that the subjective analysis involves considerably greater effort than an analytical analysis. Thus, if the model structure and type of problem can be solved analytically, the analytical solution is preferable. But for many models used in hydrology the structure is sufficiently complex that derivatives for the least-squares solution either are not possible or are tedious to compute. Thus, subjective optimization is necessary.

CALIBRATION OF A SIMPLE WATERSHED MODEL

To illustrate the subjective optimization procedure, let's formulate a simple hydrologic model that cannot be optimized using analytical analysis. Probably the model type for which subjective optimization is most frequently used in hydrology is the conceptual rainfall–runoff model. Such a model attempts to simulate the translation of rainfall into runoff; the model is a conceptual simplification of the hydrologic cycle. For our purposes here, we can view the rainfall–runoff process as consisting of three parts: surface runoff, groundwater runoff, and evapotranspiration. The flowchart and schematic of Fig. 8-2 show a conceptualization of the rainfall–runoff process. It is evident from the schematic of Fig. 8-2 that a model structure having a form more complex than Eq. 8-1 will be required. Thus it may be worthwhile to get sidetracked from the subjective optimization process and confront the problem of model formulation.

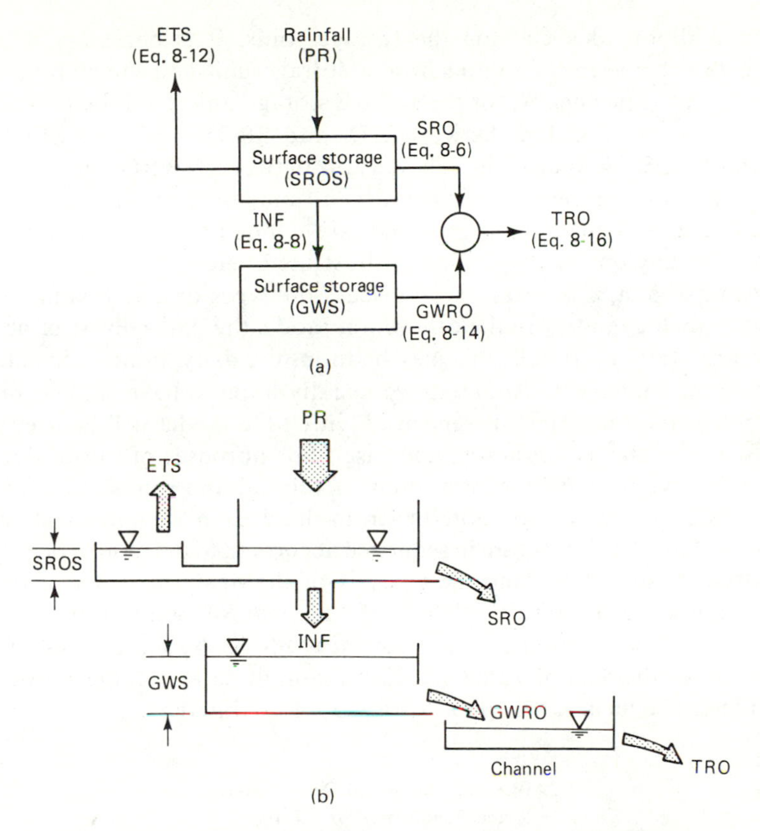

Figure 8-2 (a) Flowchart and (b) schematic of rainfall-runoff model.

Model Formulation

In formulating a complex model it is necessary to use as much theoretical and empirical evidence as possible. After collecting such evidence it will be necessary to make assumptions in order to reduce the problem to a manageable form. The first assumption was to simplify the rainfall-runoff process and include only those components shown in Fig. 8-2. A conceptual rainfall-runoff model consists of inputs, storage elements, and outputs. We can talk in terms of model input and outputs as well as input and output from a storage unit. The model of Fig. 8-1 includes two storage units: surface and subsurface (or groundwater) storage. The surface storage unit, which is denoted as SROS, has one input (rainfall, PR) and three outputs (evapotranspiration, ETS; infiltration, INF; and surface runoff, SRO). The subsurface storage unit, which is denoted as GWS, has one input (INF) and one output (groundwater runoff, GWRO). The model output is the sum of SRO and GWRO. It should be of interest to note that one output from SROS is the input to GWS.

In addition to specifying the storage units, it is necessary to define relationships between the output from a storage unit and the storage within a storage unit. Specifically, for the surface storage unit it will be necessary to formulate the relationships between ETS and SROS, INF and SROS, and SRO and SROS. Similarly, the relationship between GWRO and GWS must be formulated. Such relationships may be a function of coefficients that must be fit to data, as well as parameters that are defined by way of input but are constant for any one analysis. This is illustrated below.

At this point, it is necessary to discuss the types of data that are usually used with such a model. In the calibration mode, the data consist of observations on rainfall and runoff; this may be monthly, daily, hourly, or data on a variable time increment. Although we usually hope to have calibration data measured at the same time increment at which the model will be used in the forecast mode, this is not always the case. For purposes of illustration, let's assume that we have daily observations of rainfall and runoff, such as those of Table 8-2. These data are different from the data in the analytical analysis in that the data of Table 8-2 are in sequence according to the time of occurrence. With the data of Table 1 and the analysis of the data, the order of the data did not influence the results; that is, if the pairs of points were placed in reverse order, the solution of Eq. 8-5 would still be the same. This situation is not true for the data of Table 8-2; if the rainfall data of Table 8-2 occurred in a different sequence, the runoff data would be different.

TABLE 8-2 Measured Daily/Rainfall and Runoff (in.) for 6 Days

Day	Rainfall	Total Runoff
1	0.7	0.39
2	0.2	0.14
3	0.0	0.09
4	0.0	0.07
5	1.3	0.45
6	0.0	0.13

Before we can discuss further the calibration process, we need to define the relationships between the storage unit output and the coefficients of the storage. Let's assume that the surface runoff (SRO) is related to the surface-runoff storage (SROS) by the following:

$$\text{SRO} = \begin{cases} \dfrac{(\text{SROS} - \text{SROC} \times S)^2}{\text{SROS} + S - S \times \text{SROC}} & \text{if SROS} \geqslant 0.2 \times S \quad (8\text{-}6a) \\ 0 & \text{otherwise} \quad (8\text{-}6b) \end{cases}$$

in which SROC is the surface-runoff fitting coefficient and S is given by

$$S = \frac{1000}{\text{SROP}} - 10 \qquad \text{where } 0 < \text{SROP} \leq 100 \tag{8-7}$$

in which SROP is the surface-runoff parameter. In executing the model, we will assume that surface runoff is fulfilled before either infiltration or evaporation demands are met.

Water that does not appear as SRO remains in storage SROS and is used to fulfill infiltration and evapotranspiration. We can assume that the depth of water infiltrated is a function of the depths of water in both SROS and GWS, as well as a fitting coefficient and two input parameters. Specifically, we will assume that infiltration is given by

$$\text{INF} = \text{PINF} \times \text{SROS} \tag{8-8}$$

in which PINF is defined by

$$\text{PINF} = \begin{cases} \text{CINF}\left(1 - \dfrac{\text{GWS}}{\text{GWSM}}\right) & \text{if GWS} < \text{GWSM} \quad (8\text{-}9a) \\ 0 & \text{if GWS} \geq \text{GWSM} \quad (8\text{-}9b) \end{cases}$$

in which GWSM is an input parameter that depends on the subsurface storage available in the watershed and CINF is a coefficient that must be fitted for the particular watershed. Infiltration computed by Eq. 8-8 is used to increase the volume of subsurface storage:

$$\text{GWS} = \text{GWS} + \text{INF} \tag{8-10}$$

and deplete the volume of surface storage:

$$\text{SROS} = \text{SROS} - \text{INF} \tag{8-11}$$

Of course, if the SROS equals zero, there can be no infiltration.

The storage SROS is also used to satisfy evapotranspiration demand. Evapotranspiration loss (ETS) is a function of an input parameter (PETS) and the depth of surface-runoff storage (SROS) and is computed as

$$\text{ETS} = \begin{cases} \text{SROS} & \text{if SROS} < \text{PETS} \quad (8\text{-}12a) \\ \text{PETS} & \text{if SROS} \geq \text{PETS} \quad (8\text{-}12b) \end{cases}$$

The conceptual model for evapotranspiration loss (ETS) differs from the previous relationships in that it is not a function of a fitting coefficient; it depends only on a storage and an input parameter. Evapotranspiration losses must be subtracted from SROS:

$$\text{SROS} = \text{SROS} - \text{ETS} \tag{8-13}$$

The model assumes that surface-runoff and infiltration demands are fulfilled before the evapotranspiration demand.

After adding the depth of infiltration into groundwater storage (GWS), the depth of groundwater runoff (GWRO) can be computed. The following relationship is assumed to describe GWRO:

$$\text{GWRO} = \text{PGWRO}[1 - \exp(-\text{CGWS} \times \text{GWS})] \tag{8-14}$$

in which PGWRO is an input parameter that reflects the discharge capacity of the groundwater aquifer, and CGWS is a fitting coefficient, which reflects the rate at which GWRO occurs. After Eq. 8-14 is used to compute GWRO, the depth must be subtracted from GWS:

$$\text{GWS} = \text{GWS} - \text{GWRO} \tag{8-15}$$

The total runoff (TRO) is then the sum of surface and groundwater runoff:

$$\text{TRO} = \text{GWRO} + \text{SRO} \tag{8-16}$$

The computed value of TRO corresponds to $\hat{Y}$ in the bivariate regression model. The value of TRO should be compared with the measured runoff of Table 8-2.

An algorithm for the model of Fig. 8-2 is shown in Fig. 8-3. The algorithm must be used for each time increment for the calibration. Table 8-3 provides a summary of the input requirements for model calibration. In addition to rainfall and runoff data, values for the parameters must be input. The model also requires initial estimates of the storage units and the fitting coefficients. If we assume that values for the parameters are obtained from geomorphic analysis and are, therefore, constant, then fitting is achieved by varying the

```
SROS = SROS + PR
S = 1000.0/SROP - 10.0
SIA = SROC*S
SRO = (SROS - SIA)**2/(SROS + S - SIA)
IF(SROS.LT.SIA)SRO = 0.0
SROS = SROS - SRO
PINF = CINF*(1.0 - GWS/GWSM)
IF(GWS.GE.GWSM)PINF = 0.0
INF = PINF*SROS
SROS = SROS - INF
GWS = GWS + INF
ETS = SROS
IF(SROS.GE.PETS)ETS = PETS
SROS = SROS - ETS
GWRO = PGWRO*(1.0 - exp(-CGWS*GWS))
GWS = GWS - GWRO
TRO = SRO + GWRO
```

Figure 8-3 Algorithm for rainfall-runoff model.

values of the initial storages (SROS and GWS) and the coefficients (SROC, CINF, and CGWS).

TABLE 8-3 Data and Information Requirements for Rainfall–Runoff Model

Conceptual Component	Variable Name	Estimated Value for Trial		
		1	2	3
Input: rainfall	PR	—	—	—
Output: runoff	TRO	—	—	—
Parameters	SROP	70	75	75
	GWSM	12	12	12
	PETS	0.25	0.25	0.25
	PGWRO	0.20	0.20	0.20
Coefficients	SROC	0.20	0.05	0.02
	CINF	0.4	0.5	0.5
	CGWS	0.05	0.1	0.5
Storage	SROS	0.5	0.5	0.5
	GWS	6.0	6.0	1.5

Model Calibration

It should be evident that the model of Figs. 8-2 and 8-3 cannot be calibrated analytically. However, we must still provide one or more performance criteria that can be used to judge when the best fit has been achieved. One advantage of subjective optimization is that the calibration process can involve more than one performance criterion. For example, in fitting rainfall-runoff data such as that in Table 8-2 we might be interested in accurate fitting of the peak runoff depths as well as the volumes under the runoff depth hydrographs. Also, if measured evaporation data were available, we would want another objective function that assesses the fit between measured and predicted values of ETS. The different performance criteria reflect the greater complexity of the model and the greater understanding of the physical process being analyzed. For a simple model structure such as Eq. 8-1, we assume that either the true form of the relationship between Y and X is not known or the linear form should provide a sufficiently accurate estimate of Y. With the greater complexity of the model, we often need to use more than one performance criteria to ensure that the model is rational. Model rationality was previously shown to be the most important criterion in verifying a linear regression equation; the same rule applies with the more complex models, such as the model of Fig. 8-2.

We now have a model structure (Fig. 8-3), a data set (Table 8-2), and one or more performance criteria. This is the information that is necessary to calibrate the model. For the model of Fig. 8-3 we need values for the physical parameters (GWSM, PETS, SROP, and PGWRO), the fitting coefficients (SROC, CINF, and CGWS), and initial estimates of the storages (SROS and GWS). Initial estimates should be obtained from a rational analysis of the watershed; the ability to obtain rational estimates depends on the user's experience with the model and the user's familiarity with the watershed. In the absence of user experience with the model and knowledge of the watershed hydrology the user may have to resort to "educated guessing." Such a method usually leads to highly inaccurate predictions for the first few iterations and may ultimately lead to a nonoptimal final solution.

Assuming that values for the fitting coefficients and storages are obtained from a rational analysis, Table 8-3 provides a set of values that can be used for the first trial. Table 8-4(a) shows the calculations for these initial estimates of the coefficients and storages. The predicted daily runoff values, which are the sum of the computed daily values of surface and groundwater runoff, are shown next to the measured runoff values. It is evident that the predicted runoff is not sufficiently peaked and that the recession is too flat; for example, on days 1 and 5, the predicted runoff is about 20% of the measured runoff. Also, the computed volume of total runoff TRO for the 6-day period is much smaller than the measured volume (i.e., 0.39 in. versus 1.27 in.). The value of TRO could be increased by increasing either SRO or GWRO. One modeler might *subjectively* decide to change SROC to increase SRO while another modeler may believe CGWS should be changed to increase GWRO. Since the surface runoff is especially small, it would seem rational to change the values of the coefficients for the surface-runoff component of the model to increase the total volume of runoff. We could increase the surface runoff by decreasing SROC. But what change in SROC would lead to the desired end? At this point, if we do not have experience that would lead to a rational change in the coefficient, we must make a *subjective* estimate. A value of 0.05 will be used for the second trial. We could also increase the surface runoff by increasing SROP. While we usually set the values of the physical parameters from topographic and geomorphological data, quite frequently parameter values are changed because we know that physically based values are not exact.

In addition to changing the coefficients of the surface-runoff component, we must change at least one of the coefficients that control the groundwater runoff (GWRO). While the infiltration rate INF shows considerable daily variation, there is very little variation in GWRO. Therefore, it would seem more reasonable to change CGWS rather than CINF. However, there is merit in changing CINF to see how sensitive INF is to CINF. In any case, for the next trial, CINF was increased by 25% while CGWS was doubled (see Table 8-3). There was no indication that the initial storage estimates of 0.5 for SROS and 6.0 for GWS were unreasonable. Thus they were not changed. Table 8-3 gives the values for the parameters, coefficients, and storages for the second trial. The resulting calculations are shown in Table 8-4(b). In general, the performance criteria would suggest a much better fit than with trial 1. The computed volume increased from 0.39 in. to 1.16 in, but it is still less than the measured volume of 1.27 in. Thus there is a slight bias in the volume. The two computed peak values (i.e., 0.33 and 0.38 in.) on days 1 and 5 are still smaller than the measured peaks (0.39 and 0.45 in., respectively). Thus we need to make changes to increase the computed peaks. The computed recession also shows a bias in that it is much flatter than the measured recession. For example, the computed recession does not change from day 3 to day 4 while the measured runoff decreased by 0.02 in. In summary, for the third trial we need to increase the computed peaks and volumes as well as increasing the slope of the recession.

The biases that are evident in the computed runoff of Table 8-4(b) suggest that another trial is needed. Again, we are faced with the questions of which coefficients to change and by how much. Whereas the values used in trial 2 were largely guesses, we now have sufficient information to make more educated guesses. Specifically, we can use the same concepts of optimization as we used with analytical and numerical optimization to estimate the values of coefficients; specifically, we can compute derivatives to evaluate the change in a computed runoff value that results from a change in a fitting coefficient. Let's look at the problem of increasing the peaks.

When SROC was changed from 0.2 to 0.05, the peaks increased from 0.08 to 0.33 in. on day 1 and from 0.10 to 0.38 in. on day 5. This represents a rate of change of $(0.08 - 0.33)/(0.20 - 0.05) = -1.67$ for day 1 and $(0.10 - 0.38)/(0.20 - 0.05) = -1.87$ for day 5. If we use an average value of about -1.75 and we seek an additional increase in the peak of about 0.06 in., a Taylor series approximation would suggest a change of $0.06/-1.75 = -0.03$ in SROC. This would lead to a value of 0.02 for SROC in trial 3. It is important to note that the changes in the peak runoff from trial 1 to trial 2 resulted from changes in both SROP and SROC (also CINF and CGWS are not independent), while the above calculations assume that only SROC will be changed for trial 3. But the Taylor series approximation provides a good way to make estimates of the coefficients for the next trial. Increasing the peaks by way of changing SROC may also remove the bias from the volume. It will probably not have a significant effect on the bias in the groundwater-runoff recession.

It appears that the changes in CINF and CGWS from trial 1 to trial 2 did not change the shape of the recession, just the magnitude. A preliminary analysis of Eq. 8-14 suggests that we can get a greater rate of change in GWRO by increasing CGWS and decreasing GWS. Since a change in CGWS of 0.05 did not cause much of a change, an estimate of CGWS for the next trial will be more of a guess. Let's try 0.5 in trial 3. To get a magnitude of GWRO similar to the average value of trial 2, we will need a value of GWS of 1.5. Thus, let's use a value of 1.5 for the initial value of GWS in trial 3. Table 8-3 summarizes the input for the third trial.

The results for trial 3 are shown in Table 8-4(c). The changes caused a large change in the volume, with the total volume of computed runoff increasing

TABLE 8-4 Computations for the Iteration of the Rainfall–Runoff Model

(a) First Round

Day	PR	SROS	SRO	SROS	PINF	INF	SROS	GWS	ETS	SROS	GWRO	GWS	TRO Computed	TRO Measured
0	—	—	—	—	—	—	—	—	—	0.50	—	6.00	—	—
1	0.70	1.20	0.03	1.17	0.20	0.23	0.94	6.23	0.25	0.69	0.05	6.18	0.08	0.39
2	0.20	0.89	0.00	0.89	0.19	0.17	0.72	6.35	0.25	0.47	0.05	6.30	0.05	0.14
3	0.00	0.47	0.00	0.47	0.19	0.09	0.38	6.39	0.25	0.13	0.05	6.34	0.05	0.09
4	0.00	0.13	0.00	0.13	0.19	0.02	0.11	6.36	0.11	0.00	0.05	6.29	0.05	0.07
5	1.30	1.30	0.04	1.26	0.19	0.24	1.02	6.53	0.25	0.77	0.06	6.47	0.10	0.45
6	0.00	0.77	0.00	0.77	0.18	0.14	0.63	6.61	0.25	0.38	0.06	6.55	0.06	0.13
													0.39	1.27

(b) Second Round

Day	PR	SROS	SRO	SROS	PINF	INF	SROS	GWS	ETS	SROS	GWRO	GWS	TRO Computed	TRO Measured
0	—	—	—	—	—	—	—	—	—	0.50	—	6.00	—	—
1	0.70	1.20	0.24	0.96	0.25	0.24	0.72	6.24	0.25	0.47	0.09	6.15	0.33	0.39
2	0.20	0.67	0.06	0.61	0.24	0.15	0.46	6.30	0.25	0.21	0.09	6.21	0.15	0.14
3	0.00	0.21	0.00	0.21	0.24	0.05	0.16	6.26	0.16	0.00	0.09	6.17	0.09	0.09
4	0.00	0.00	0.00	0.00	—	0.00	0.00	6.17	0.00	0.00	0.09	6.08	0.09	0.07
5	1.30	1.30	0.29	1.01	0.25	0.25	0.76	6.33	0.25	0.51	0.09	6.24	0.38	0.45
6	0.00	0.51	0.03	0.48	0.24	0.12	0.36	6.36	0.25	0.11	0.09	6.27	0.12	0.13
													1.16	1.27

TABLE 8-4 (*cont.*).

(c) Third Round

Day	PR	SROS	SRO	SROS	PINF	INF	SROS	GWS	ETS	SROS	GWRO	GWS	TRO Computed	TRO Measured
0	—	—	—	—	—	—	—	—	—	0.50	—	1.50	—	—
1	0.70	1.20	0.29	0.91	0.44	0.40	0.51	1.90	0.25	0.26	0.12	1.78	0.41	0.39
2	0.20	0.46	0.04	0.42	0.43	0.18	0.24	1.96	0.24	0.00	0.12	1.84	0.16	0.14
3	0.00	0.00	0.00	0.00	—	0.00	0.00	1.84	0.00	0.00	0.12	1.72	0.12	0.09
4	0.00	0.00	0.00	0.00	—	0.00	0.00	1.72	0.00	0.00	0.12	1.60	0.12	0.07
5	1.30	1.30	0.33	0.97	0.43	0.42	0.55	2.02	0.25	0.30	0.11	1.91	0.44	0.45
6	0.00	0.30	0.02	0.28	0.43	0.12	0.16	2.03	0.16	0.00	0.11	1.92	0.13	0.13
													1.38	1.27

by 0.22 in., to 1.38 in. Thus, instead of underestimating by 0.11 in., the computed runoff is overestimated by 0.11 in. There is little bias in the computed peak values, with errors on days 1 and 5 of 0.02 and −0.01, respectively. Also, the groundwater runoff appears to be too high. Again, the values of GWRO show very little variation, which may suggest that they do not adequately reflect a normal recession curve. This may result from either poor values for the coefficients or an inadequate model structure.

We are now confronted by the question: Is the accuracy sufficient that we can conclude that the coefficients can be considered optimal? If one accepts this statement, the surface-runoff component must be slightly underestimating the surface-runoff effect because if we reduced the magnitude of GWRO, the computed values of TRO would also be smaller on days 1 and 5. But just as we had to decide whether Eq. 8-4 was our best solution for that problem, we face a similar problem with the results of trial 3 for the watershed model. Unfortunately, there is no true answer. One person may decide to stop the iteration process, whereas another may go on to trial 4. It depends on their *subjective* assessment of whether the lack of total fit will cause errors in design.

Measures of Performance

Performance criteria have been discussed in a qualitative sense. In practice, one must deal with values of quantitative indices. Thus it is worthwhile to examine some of the more common performance criteria. They have previously been called goodness-of-fit indices. Either name is acceptable. It is important to recall the discussion of goodness-of-fit statistics when dealing with regression analysis (see Chapters 3 and 6) because with only a few exceptions, the goodness-of-fit criteria that were used to assess the quality of a regression model can also be used to assess the quality of a complex model that has been calibrated subjectively. Indices such as a correlation coefficient and standard error of estimate can still serve as measures of goodness of fit. However, with some models, such as the time-dependent rainfall–runoff model discussed previously, the indices should be viewed only as a goodness-of-fit index and not a statistical measure. For example, the correlation coefficient for the bivariate model of Eq. 8-1 equals the square root of the ratio of the explained variation to the total variation. It has a statistical base that assumes that (1) the model is unbiased, (2) there are $(n - 2)$ degrees of freedom, and (3) the observations in the data set are independent. With a time-dependent model such as the rainfall–runoff model discussed previously, the runoff values are not independent. The runoff (either the measured or predicted) for day 2 is not independent of the runoff for day 1. In general, the runoff for day (or time period) t depends at least on the runoff for day $t - 1$. It may also depend on the runoff for days $t - 2$, $t - 3$, and so on. Actually, the dependence is not necessarily on the runoff but on factors such as the storages that control runoff rates. For example, there was very little change in the storage GWS (see Table

8-4) over the 6-day period, which resulted in values of GWRO that were essentially the same. This causes a correlation (often called *serial correlation*) between both the measured and observed runoff with time. Since the values are not independent, the statistical assumptions that underlie the correlation coefficient are not valid and thus it cannot be used as a statistical quantity in measuring goodness of fit. However, it provides a very good index of the fit.

The violation of a statistical assumption does not mean that the correlation coefficient cannot be used as a goodness-of-fit statistic. It is still a measure of the explained variance. The violation of the assumption only means that, in practice, statistical tests of hypotheses cannot be applied in their statistical framework. Because the data points are not independent, one is really dealing with a data set containing far fewer "independent" data points; that is, the effective record length is really much less than the actual record length. Although some research has tried to address the question of effective record length, it is not widely used. In summary, statistical indices such as the correlation coefficient can be used as an indicator of fit, but they should not be used with standard tests of significance.

The principle of least squares with a linear model provides an unbiased model. Subjective optimization does not ensure unbiasedness, although the user may adjust the coefficients so that biases are not apparent. However, for large data bases, it may be very difficult to assess biasedness subjectively. For this reason there is a need to adopt one or more single-valued indices that can detect biases. There are a number of such indices that can be used. In developing appropriate indices, the user should ensure that the index is capable of measuring what it is intended to measure. In any case, a comparison of mean values may detect a bias. For example, any difference between the computed and measured mean daily runoffs would reflect a bias in the coefficients that control the outflows from the surface and groundwater storages of Fig. 8-2. But biases measured by mean values are subject to fluctuations due to extreme events. For example, with small samples one large error in a computed peak discharge may affect a computed bias. In addition to mean values of the total runoff, one may be interested in biases of either the peak discharges or the recession. For such cases, means can be computed for parts of the data base. Although this is probably done subjectively in evaluating the coefficients, it is still worthwhile to develop algorithms that can be used to detect biases that can be associated with specific components and/or coefficients of the model. For example, if peak values of SRO are controlled largely by the value of SROC, any bias in the peak values of SRO can be reduced by adjusting the value of SROC. Thus it is important that a set of performance criteria be developed and evaluated for each iteration of the model. The information supplied for each execution can be used to systematically adjust the coefficients for the next execution. As the values of the performance criteria approach stable values, one can assume that the optimum set of coefficients is being approached. The discussion of performance criteria will be continued when the case study is discussed in the next section.

APPLICATION WITH HYDROLOGIC DATA

The purpose of the example provided in the preceding section was solely to introduce the topic of subjective optimization in hydrology. Both the data and the model were greatly simplified in order to highlight the fundamentals of subjective optimization. In practice, watershed models are much more complex, with 20 to 40 fitting coefficients not uncommon. The input usually consists of several years of rainfall, runoff, and evaporation data. The nature of the data depends on the intended use. Specifically, the data base for a model to be used for irrigation scheduling would be different from the data requirements for a real-time flood forecast model. Daily totals may be acceptable for the former case, while 5-min to hourly data may be required for calibrating a real-time flood forecast model. Although the required data base is a function of the intended use of the model, the data available may limit the accuracy of the model.

Because of the simplicity of the model and data base in the preceding section, several characteristics of subjective optimization could not be adequately demonstrated. The objective here is to formulate and calibrate a slightly more complex model, with one year of daily rainfall and runoff data to be used in calibrating the model.

Model Formulation

A model having three storage components will be used. The algorithm, which is provided in Appendix D-1, is purposely kept simple so that the model can be used in some of the exercises. A schematic of the model is provided in Fig. 8-4. The rainfall is divided into two parts, depending on the value of the variable PINF. Part of the rainfall enters surface storage while the remaining part enters the groundwater system, which consists of two storages (GWS and SUBFLO).

Water that enters surface storage is discharged directly to the channel using a unit hydrograph having a time base of four time periods. The fraction of the daily discharge depends on the parameter SROP. The unit hydrograph for surface runoff is defined by

$$h_i = \frac{(\text{SROP})^i}{\sum_{j=1}^{4} (\text{SROP})^j} \qquad \text{for } i = 1, 2, 3, 4 \tag{8-17}$$

where SROP is the fitting coefficient and can vary from 0 to 1.0. For example, if SROP equals 0.3, the unit hydrograph would have the ordinates (0.706,

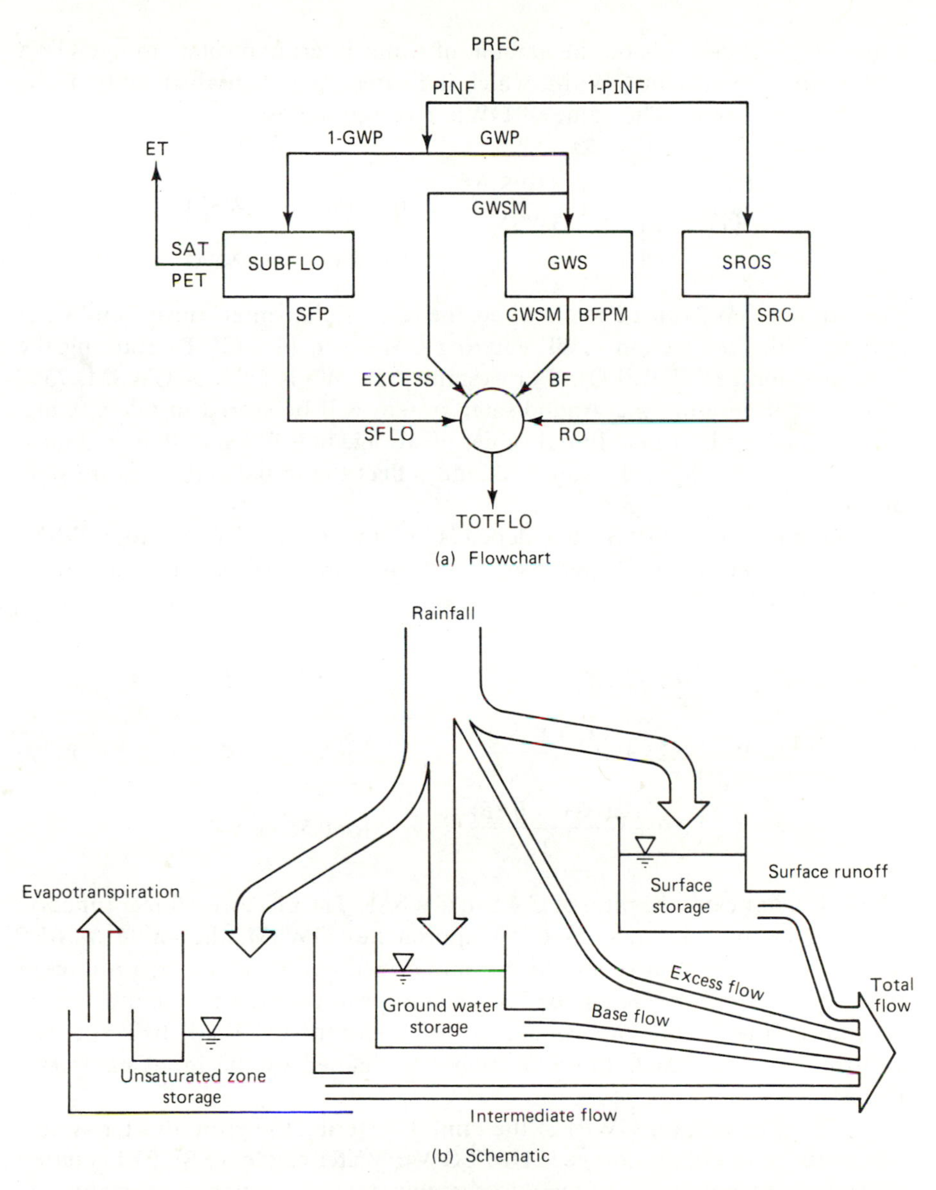

(a) Flowchart

(b) Schematic

Figure 8-4

0.212, 0.064, 0.018); this would reflect a rapid response of a small urban watershed. For SROP equal to 0.6, the unit hydrograph has a more uniform distribution with time (0.460, 0.276, 0.165, 0.099); this may be more appropriate for a large rural watershed.

Rainfall that goes directly to the groundwater system may enter one of two storages, depending on the value of an internal parameter GWP. The value of GWP depends on the amount of water in groundwater storage GWS and a fitting coefficient GWSM, which indicates the potential capacity of the storage component. The value of GWP is computed as

$$\text{GWP} = \begin{cases} 1 - \left(\dfrac{\text{GWS}}{\text{GWSM}}\right)^2 & \text{for GWS} \le \text{GWSM} \\ 0 & \text{for GWS} > \text{GWSM} \end{cases} \tag{8-18}$$

The variable GWP is used to distribute the rainfall that enters the groundwater system, with the fraction GWP entering GWS and (1 − GWP) entering the interflow storage (SUBFLO). For example, if GWS is 50% of GWSM, 75% of the water entering the groundwater system will be stored in GWS, while 25% will enter SUBFLO. Initial values of storage in GWS and SUBFLO must be input to the model; the values should reflect the initial soil moisture state of the watershed.

Outflow from GWS also depends on the ratio of GWS to GWSM. Specifically, an internal parameter BFP is computed as a function of GWS/GWSM by

$$\text{BFP} = \begin{cases} 0.5 \times XX & \text{for } XX \le 0.5 \\ 0.25 + \dfrac{13(XX - 0.5)}{7} & \text{for } 0.5 < XX \le 0.85 \\ 0.9 + \dfrac{2(XX - 0.85)}{3} & \text{for } 0.85 < XX \end{cases} \tag{8-19}$$

where XX indicates the ratio of GWS to GWSM. The piecewise linear function of Eq. 8-19 indicates that as GWS approaches GWSM, the value of BFP increases. The amount of baseflow from GWS is computed as the product of BFP and a fitting coefficient BFPM, which represents the maximum rate of baseflow. Thus when GWS is equal to or greater than GWSM, BFP equals 1 and the baseflow equals the maximum rate defined by the input parameter BFPM.

The fraction (1 − GWP) of the rainfall entering the groundwater system is directed to interflow storage (SUBFLOW). Water can leave SUBFLO either by way of evaporation or flow into the channel. The latter is controlled by the fitting coefficient SFP, with the interflow given by

$$\text{SFLO} = \text{SFP} \times \text{SUBFLO} \tag{8-20}$$

Water leaving SUBFLO storage by way of evaporation is controlled by two fitting coefficients, PET and SAT. The evaporation losses (ET) equal

$$\text{ET} = \text{ETCO} \times \text{PET} \tag{8-21}$$

where PET is the potential rate and ETCO is an internal parameter defined by

$$\text{ETCO} = \begin{cases} 0.75 \times YY & \text{for } YY \le \frac{1}{3} \\ 0.25 + 1.5(YY - \frac{1}{3}) & \text{for } \frac{1}{3} < YY \le \frac{2}{3} \\ 0.75 + 0.75(YY - \frac{2}{3}) & \text{for } \frac{2}{3} < YY \end{cases} \tag{8-22}$$

in which YY is the ratio of the volume of interflow storage (SUBFLO) to a fitting parameter SAT, which reflects saturated conditions. As the volume in storage approaches saturation, the evapotranspiration losses increase. Flows from the two groundwater components, GWS and SUBFLO, are added to give the total groundwater runoff.

In summary, the model was formulated with three storage components and seven fitting coefficients (PINF, SROP, GWSM, BFPM, SFP, PET, and SAT). Values for two of the storage elements (GWS and SUBFLO) must be input. As specified by the program in Appendix D-1, the program is set up for daily values of rainfall and runoff. The rainfall serves as the driving function, while the runoff is used solely to assess the goodness of fit.

Assessing Model Performance

Before calibrating the model using hydrologic data, it may be worthwhile discussing the computer output, since this will be used in calibrating the model. The output includes a summary of the input, which serves mainly as documentation and a means of checking to ensure that the values used were the values intended to be used. The values of the surface runoff unit hydrograph are listed. Annual totals are given for the following: precipitation, observed runoff, and the predicted values of the total runoff, surface runoff, base flow, interflow (or subsurface flow), and evaporation. Two goodness-of-fit statistics are provided; the correlation and modified correlation coefficients reflect the proportion of the variance in the measured runoff that is associated with the predicted runoff. The modified correlation coefficient accounts for differences in the scale of the measured and predicted runoff, while the correlation coefficient is insensitive to differences in scale. The end-of-year storages are given for the three storage components. An annual water balance is computed as

$$P - Q - \text{ET} \tag{8-23}$$

where P is the annual rainfall, Q the predicted total runoff, and ET the predicted evapotranspiration. Finally, a plot of the rainfall and both the measured and predicted runoff is provided; the rainfall is plotted on the upper axis while the measured and predicted runoff hydrographs are plotted on the lower axis. All of this information can be used to calibrate the fitting coefficients.

One of the advantages of subjective optimization is that more than one objective function can be considered. This is especially important in watershed modeling. In calibrating a model to a set of data, the following are some of the questions that must be answered:

1. Does the total predicted runoff adequately approximate the total measured runoff?
2. Do the predicted peak discharges for the storm events closely approximate the measured peak discharges? Or is there a bias in the predicted peak discharges?
3. Is the timing of the predicted runoff an adequate reflection of timing in the measured runoff?
4. Do the assumed initial storages provide an accurate assessment of the soil moisture state at the beginning of the record?
5. Do the recessions of the predicted storm event hydrographs closely match the recessions of the measured hydrographs?
6. Is there a bias in the flow during periods when rainfall has not occurred?

Each of these questions represents an objective function that can be used in calibrating the model. The importance of the objective function will depend on the intended use of the model. For example, question 6 may be given more consideration when the model is being used for analyzing water-quality conditions during periods of low flow, whereas question 2 may be more important for maximum flood flow analysis. In practice, each of these questions must be given consideration during the calibration process.

Model Calibration

The calibration process will be demonstrated using a set of data for the Anacostia River watershed. The rainfall consists of daily total depths (area-inches) measured at the Washington National Airport weather station for the period October 1, 1964, to September 30, 1965. A total depth of 31.8 in. was recorded during the water year. The runoff data were the mean daily runoff at the Riverdale gaging station. The drainage area above the gage was 72.6 square miles.

The rainfall data are used with the values of the fitting coefficients to derive predicted values of the total runoff. The predicted total runoff hydrograph is computed and compared with the measured runoff hydrograph. The following steps summarize the fitting procedure:

1. Using the initial estimates of the two storage elements and the seven fitting coefficients, derive a first estimate of the total runoff for each day of the year.
2. By comparing the predicted and measured annual runoff hydrographs and examining the annual totals, the major biases should be assessed for each of the objective functions discussed previously.

3. Based on one's knowledge of the model, attempt to eliminate the biases by modifying the values of the coefficients that are expected to have the greatest potential for eliminating the biases.
4. Using the revised estimates of the storage and the coefficients, recompute the total predicted runoff hydrograph and repeat these four steps until a best fit is achieved.

In general, the procedure is designed to eliminate biases in the predicted hydrograph and reduce the imprecision to a minimum.

Trial 1. Values for the seven coefficients were selected based on knowledge of the model and experience with the model on similar watersheds. The values selected are listed as the initial parameter estimates for trial 1 in Table 8-5. The rainfall data and values for the input resulted in the predicted runoff

TABLE 8-5 Initial Parameter Estimates, Annual Totals, Goodness-of-Fit Statistics, and Final Storages for Each Trial of the Rainfall–Runoff Model

	Trial			
	1	2	3	4
Initial parameter estimate				
GWS	12.0	12.0	14.0	14.5
SUBFLO	5.0	4.0	1.50	1.00
PINF	0.70	0.80	0.87	0.80
SROP	0.80	0.50	0.30	0.30
GWSM	15.0	15.0	15.0	15.0
BFPM	0.01	0.01	0.007	0.001
SFP	0.01	0.01	0.01	0.008
PET	0.25	0.25	0.29	0.30
SAT	5.00	5.00	5.00	5.00
Annual totals				
PREC	31.80			
OBS RO	10.48			
PRED RO	17.03	14.23	11.26	10.17
SURF RO	9.54	6.36	4.13	6.36
BASE FLOW	3.33	3.36	2.76	0.36
SUB FLOW	4.16	4.51	4.38	3.45
EVAP	17.23	18.72	20.89	21.34
Goodness of fit				
R	0.68	0.74	0.74	0.73
R_M	0.55	0.49	0.39	0.57
Final storages				
GWS	13.72	13.90	14.24	14.88
SUBFLO	0.83	0.95	0.90	0.91
Water balance	−2.45	−1.16	−0.36	0.29

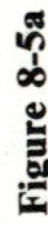

Figure 8-5a

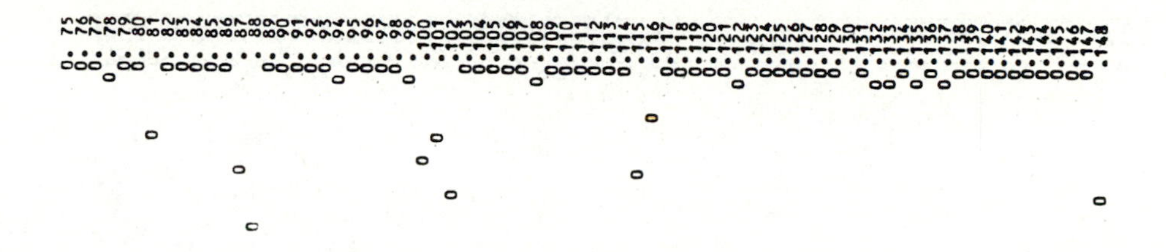

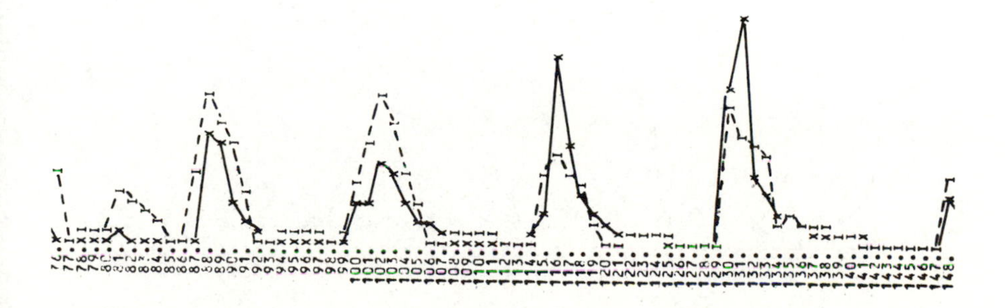

Figure 8-5a (*cont.*).

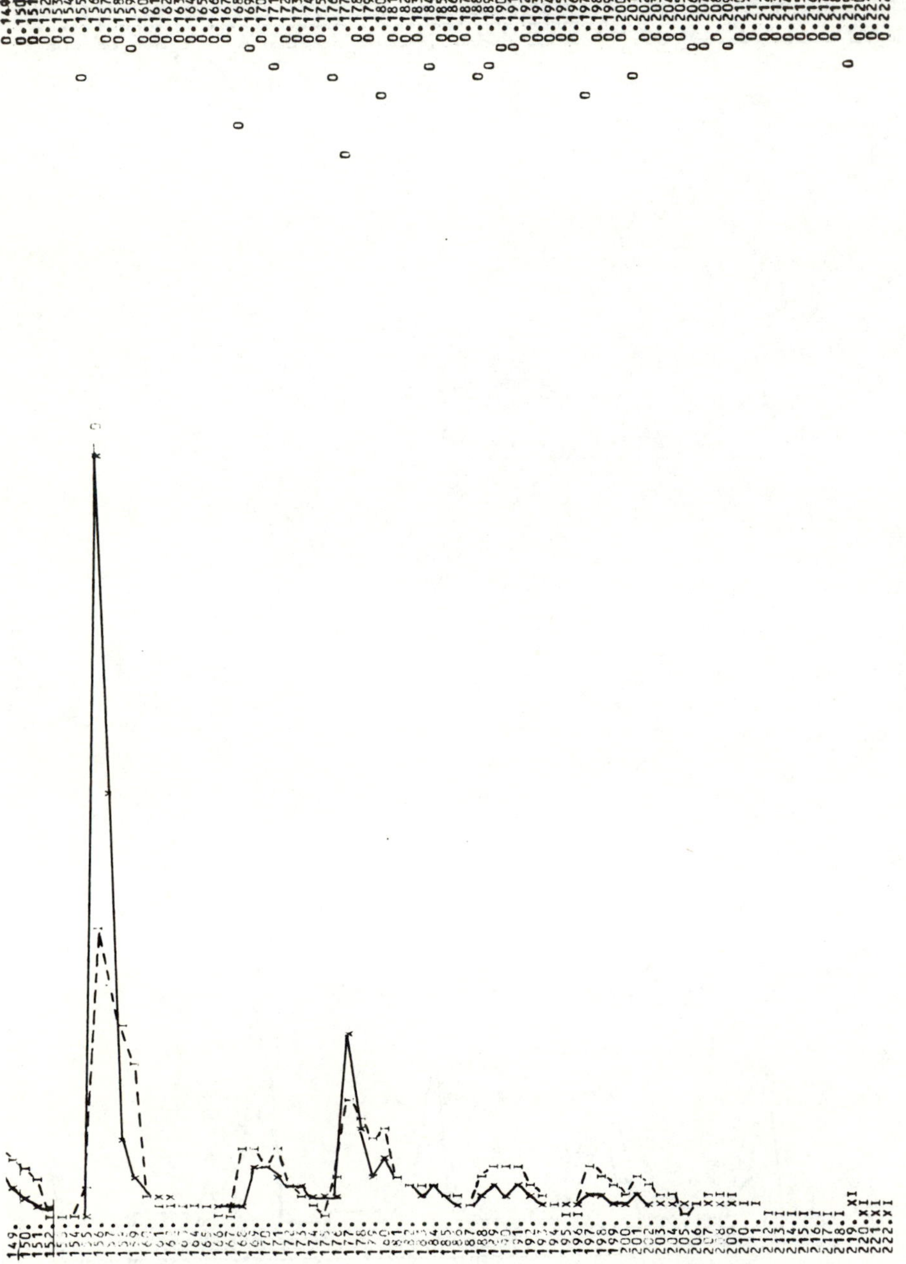

Figure 8-5a (cont.).

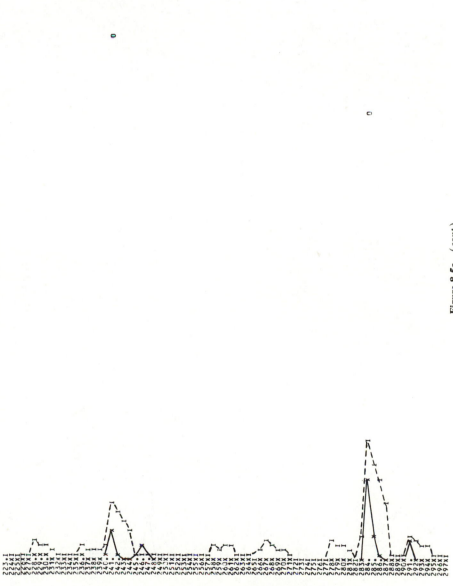

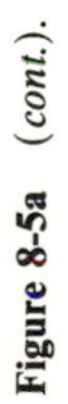

Figure 8-5a (*cont.*).

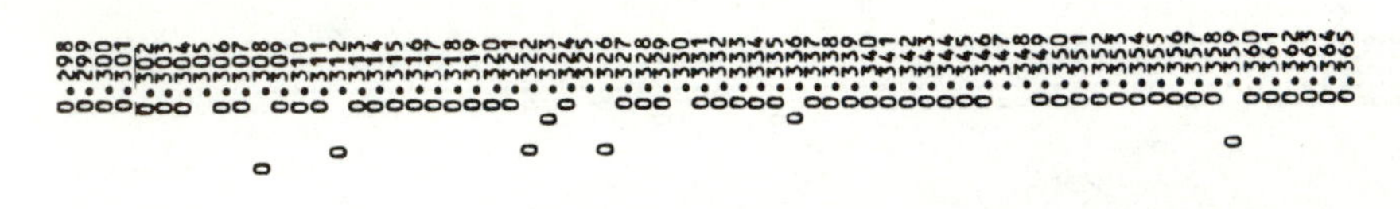

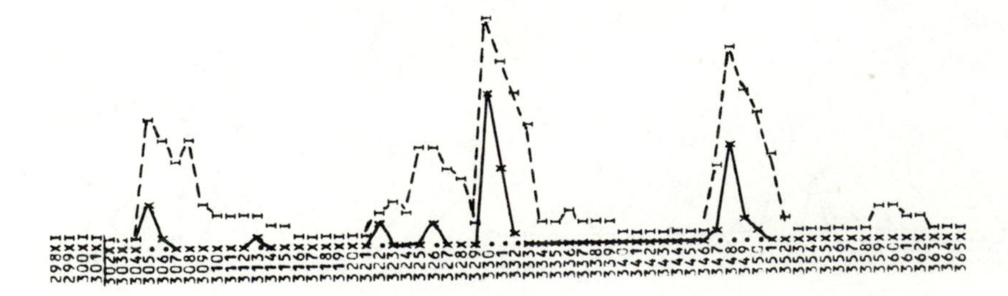

Figure 8-5a (*cont.*).

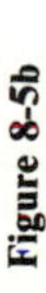

Figure 8-5b

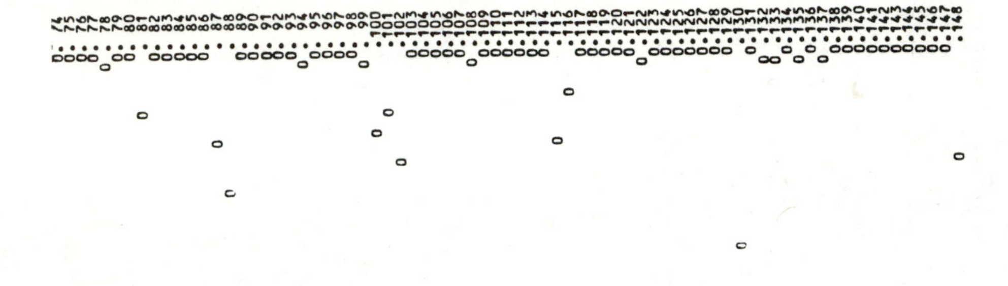

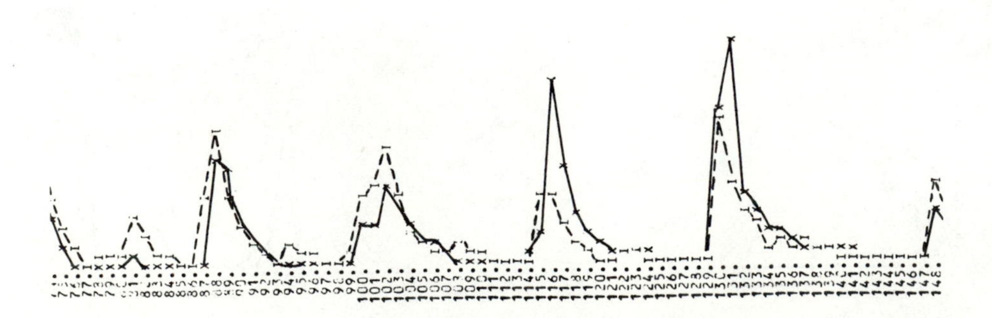

Figure 8-5b (*cont.*).

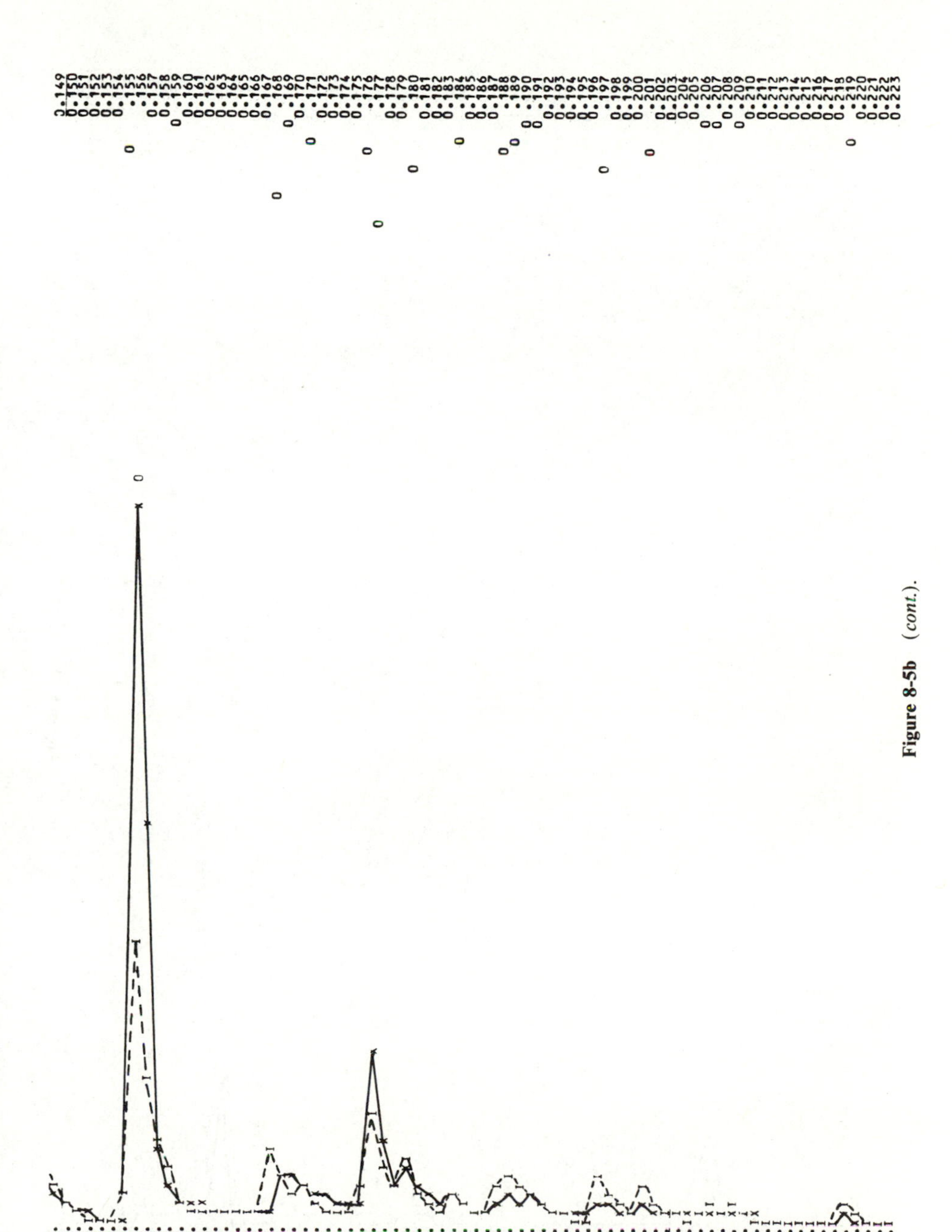

Figure 8-5b (*cont.*).

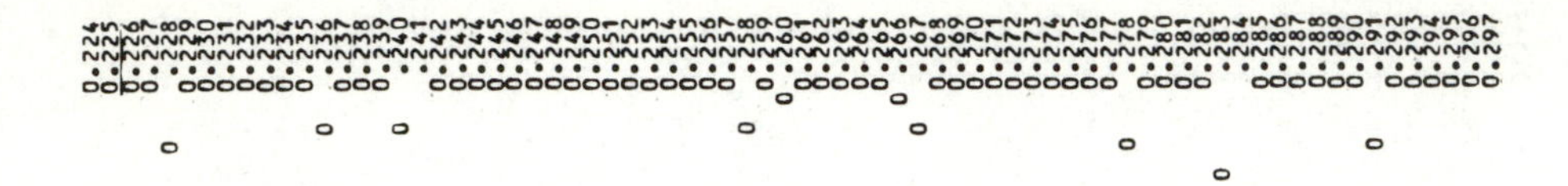

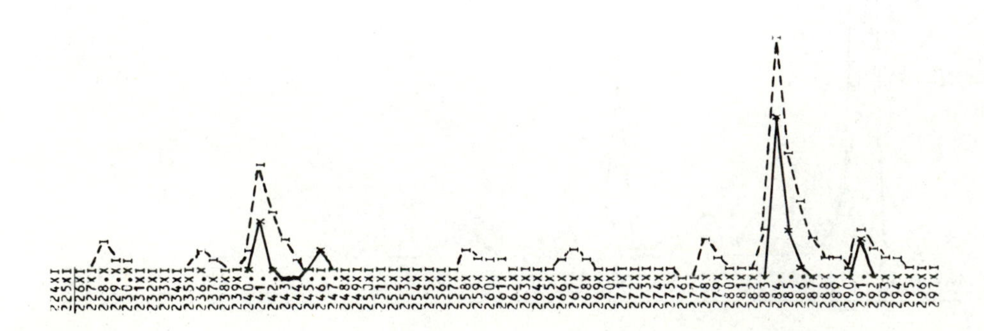

Figure 8-5b (*cont.*).

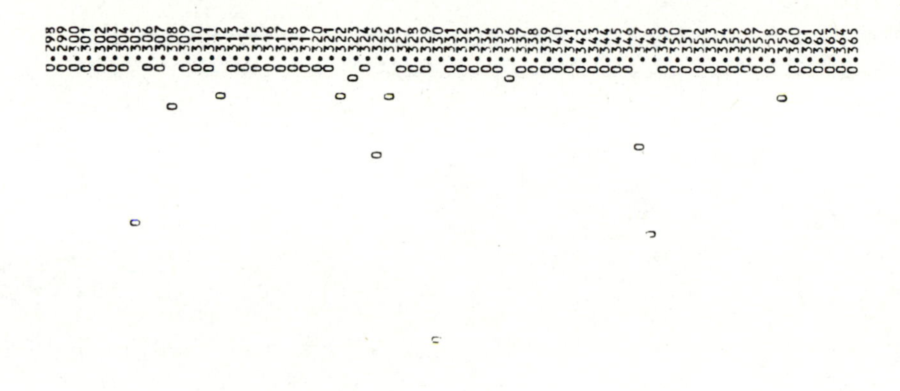

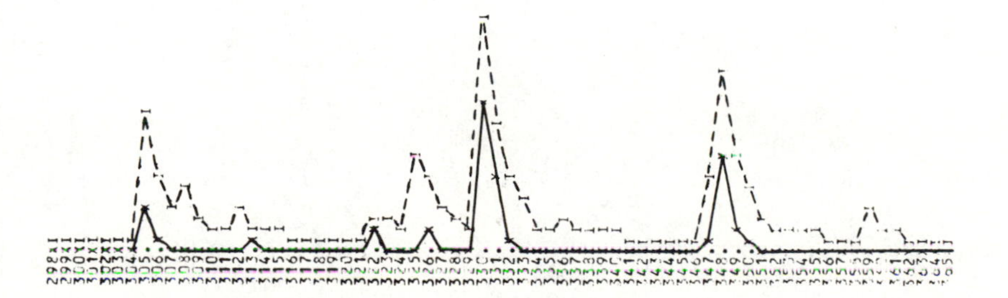

Figure 8-5b (*cont.*).

Figure 8-5c

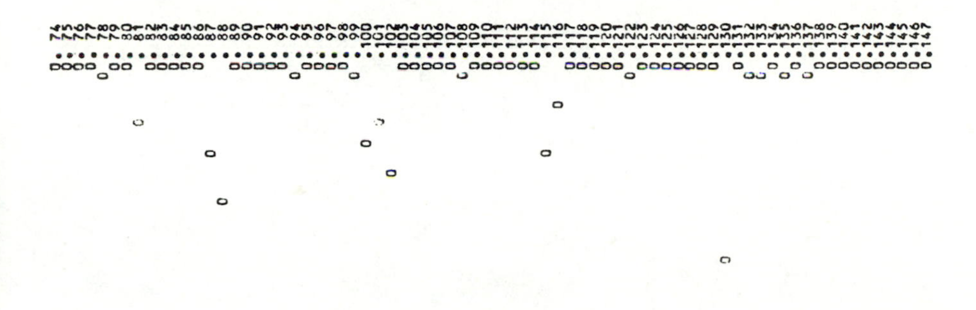

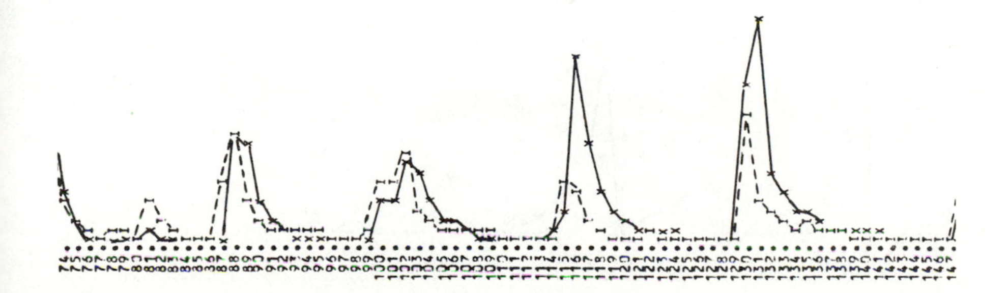

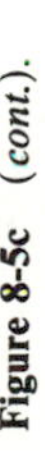

Figure 8-5c *(cont.)*.

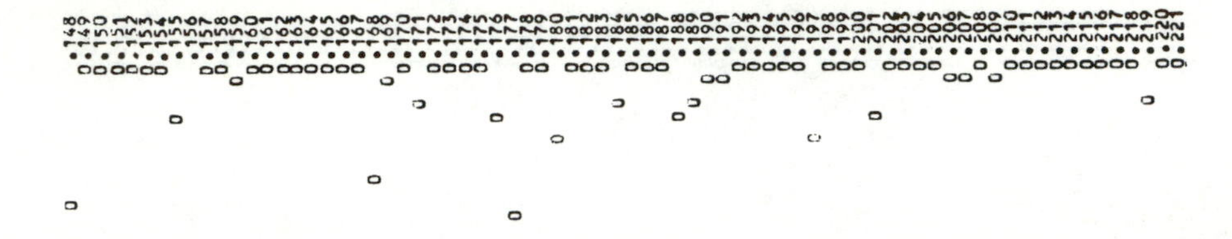

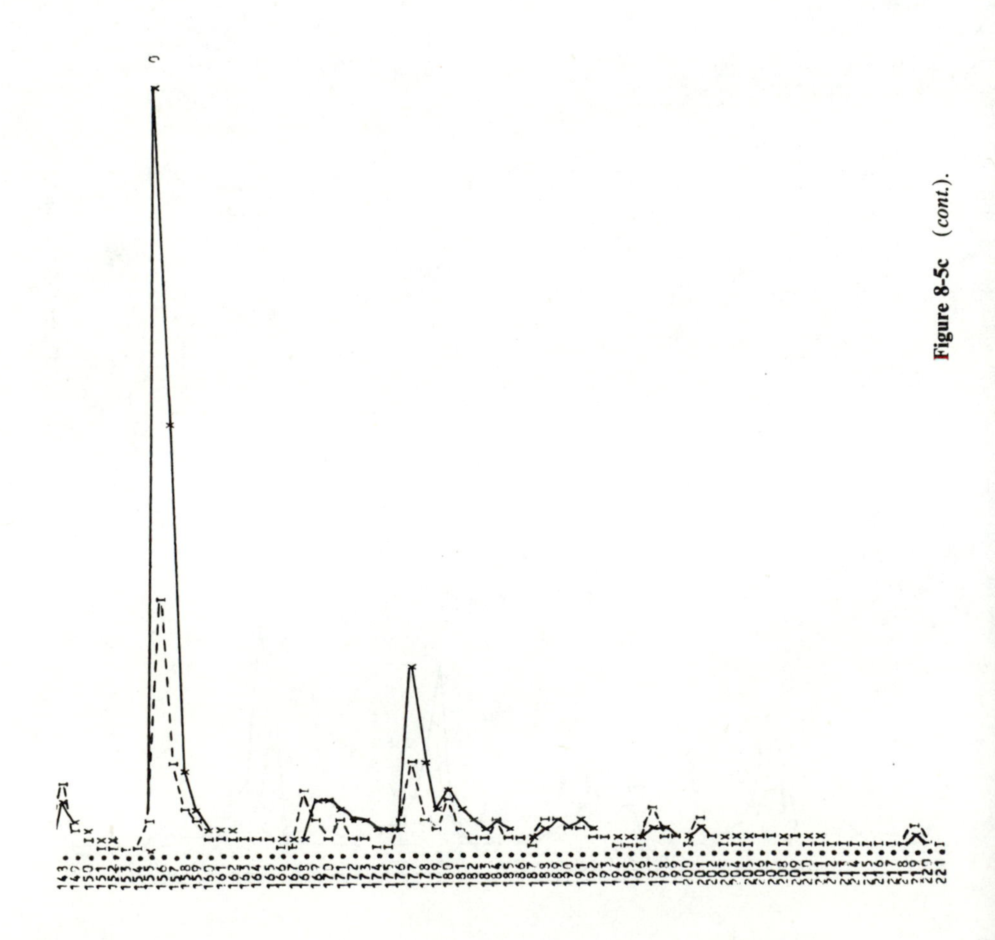

Figure 8-5c *(cont.)*.

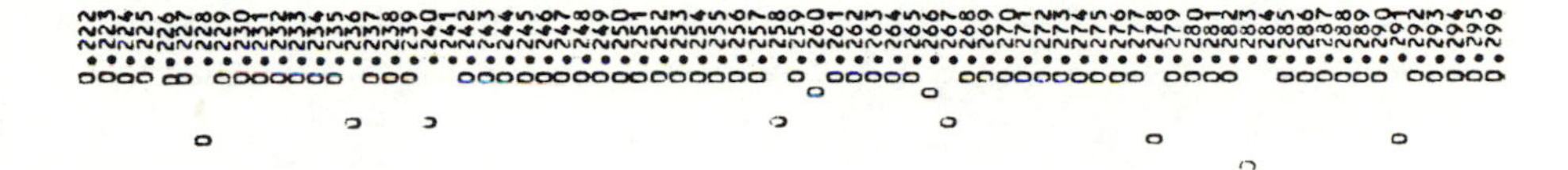

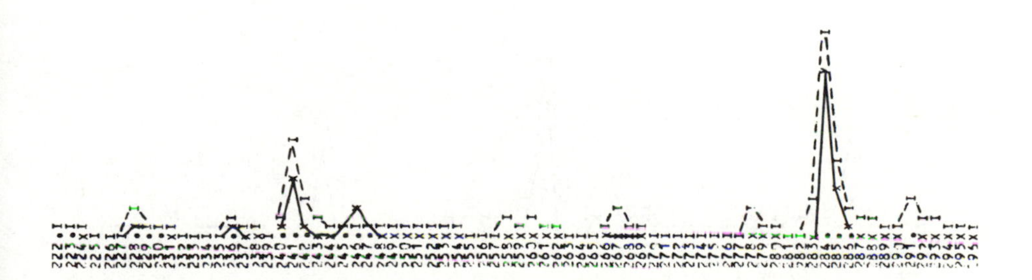

Figure 8-5c *(cont.)*.

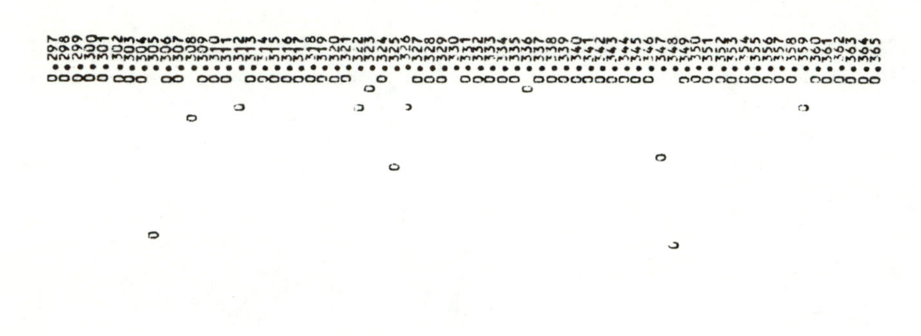

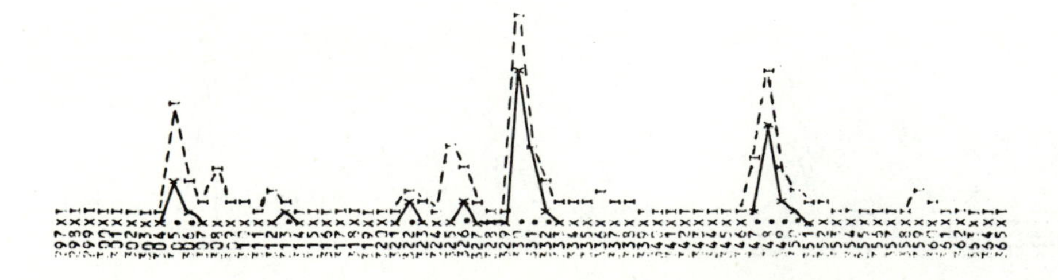

Figure 8-5c (*cont.*).

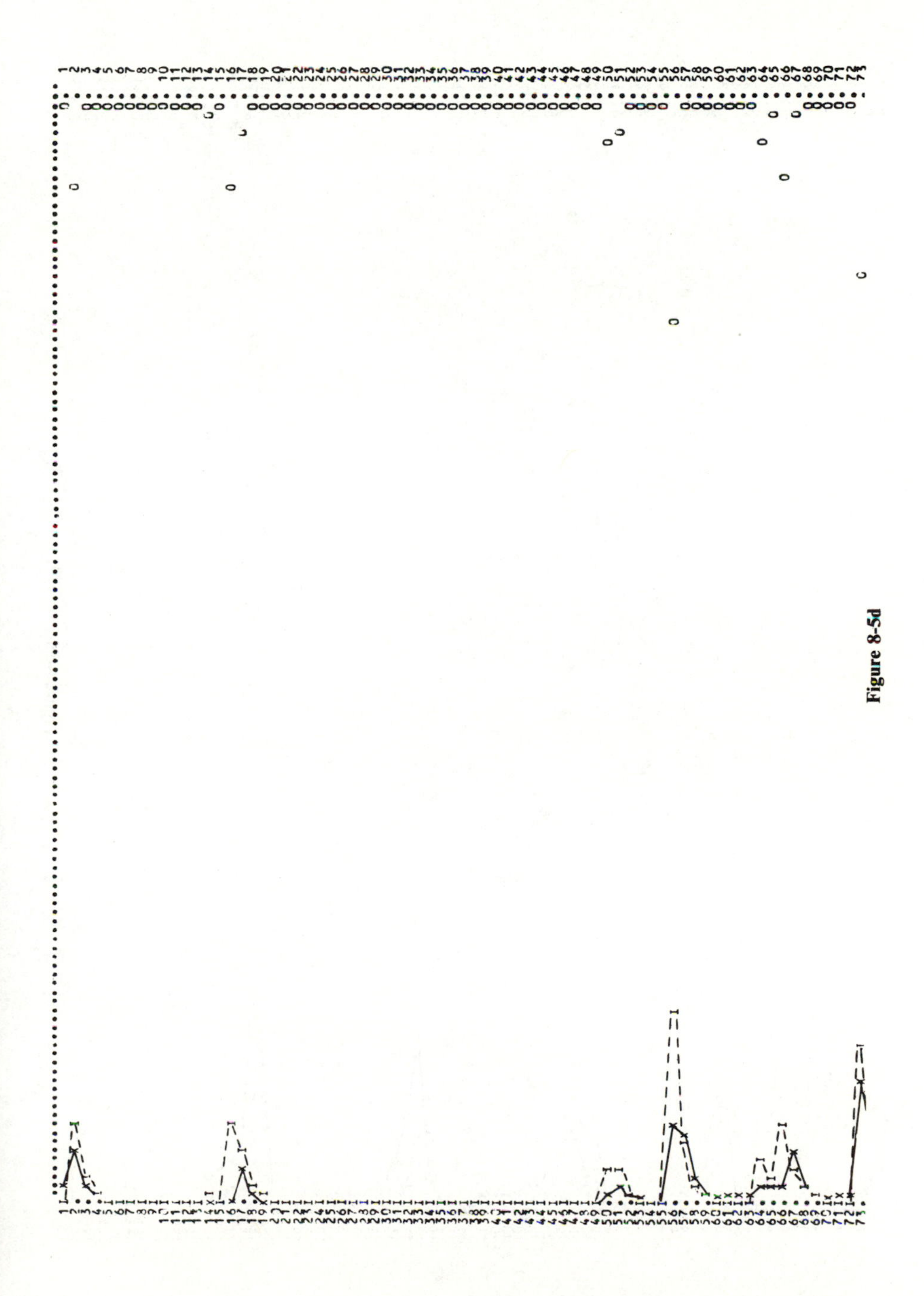

Figure 8-5d

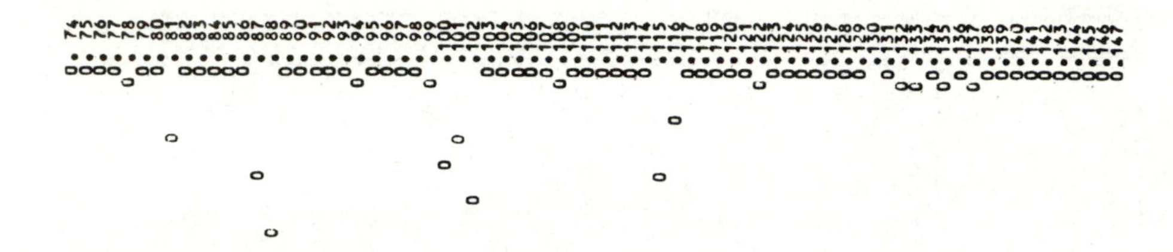

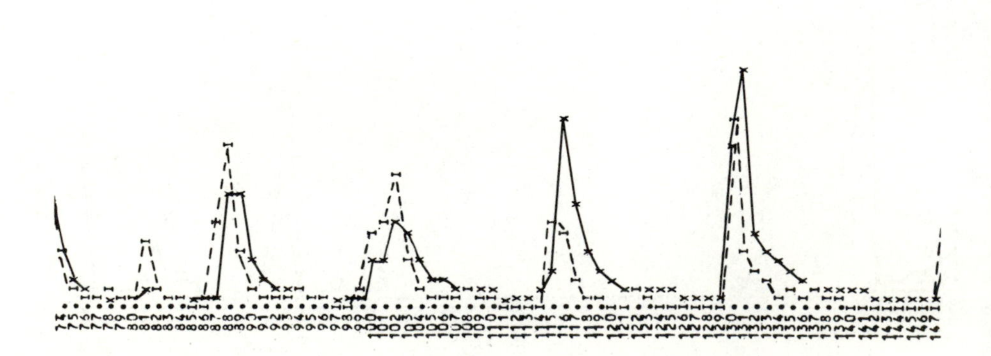

Figure 8-5d *(cont.)*.

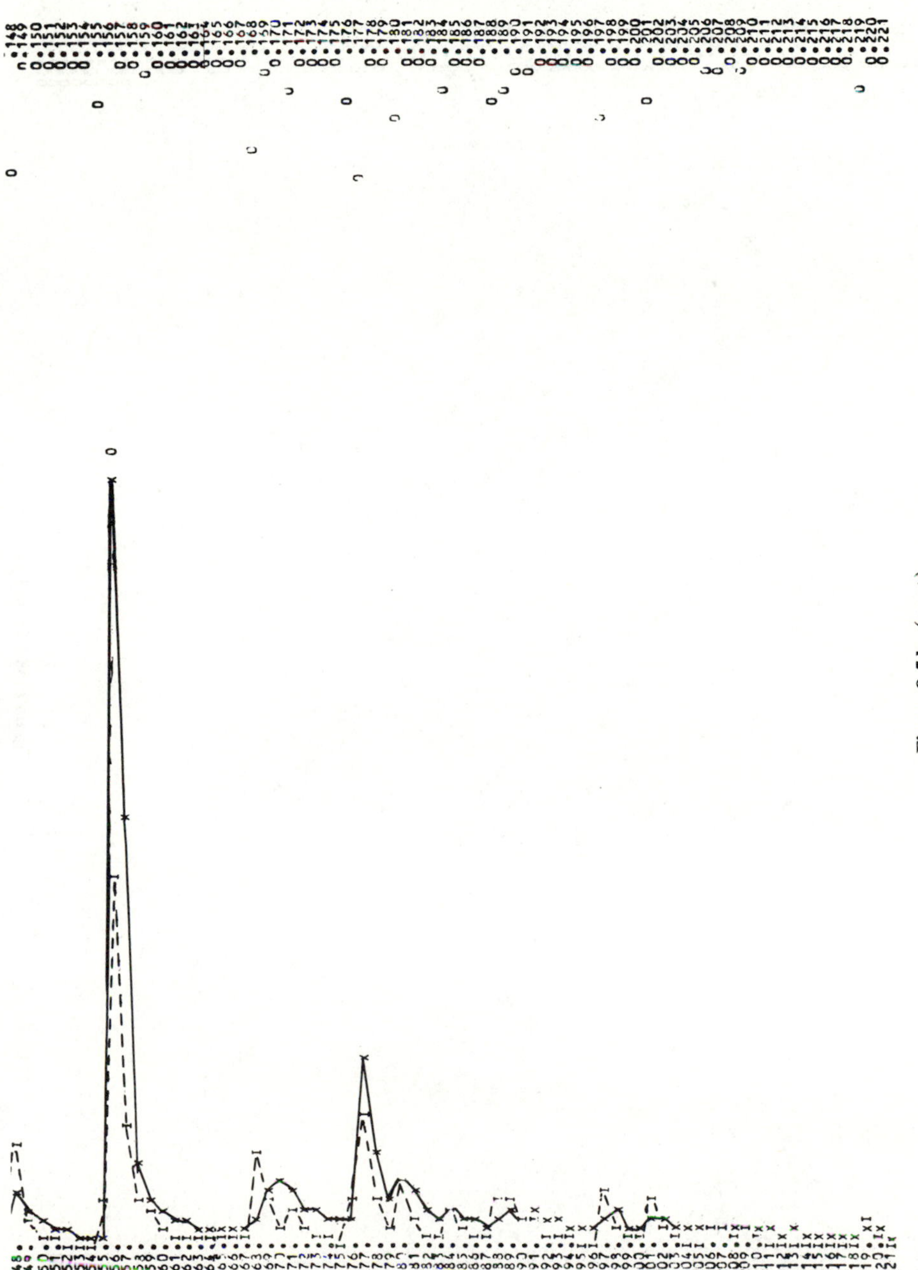

Figure 8-5d (*cont.*).

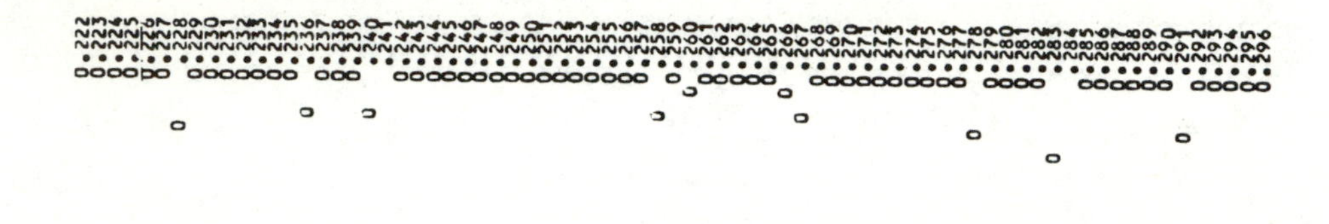

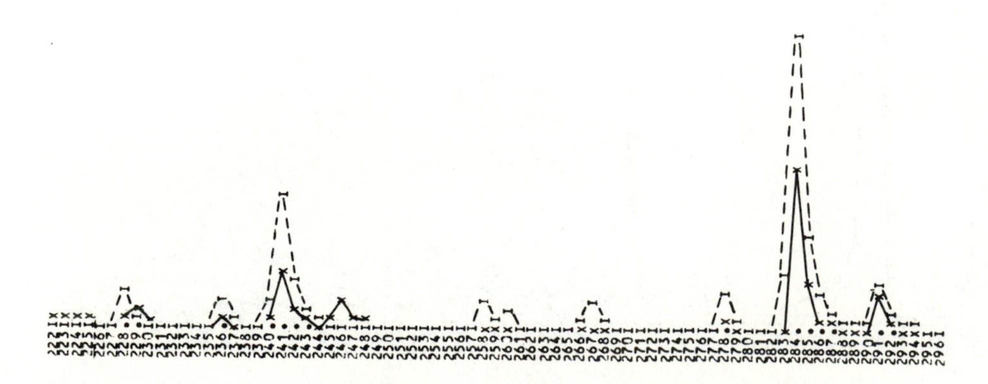

Figure 8-5d (*cont.*).

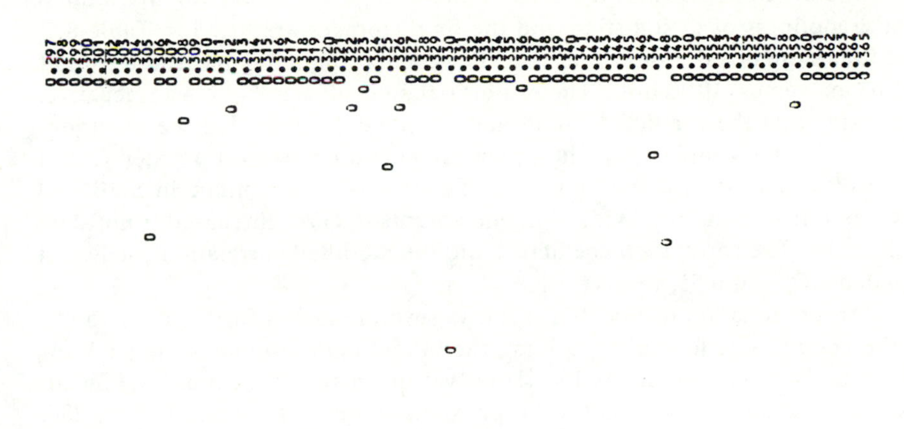

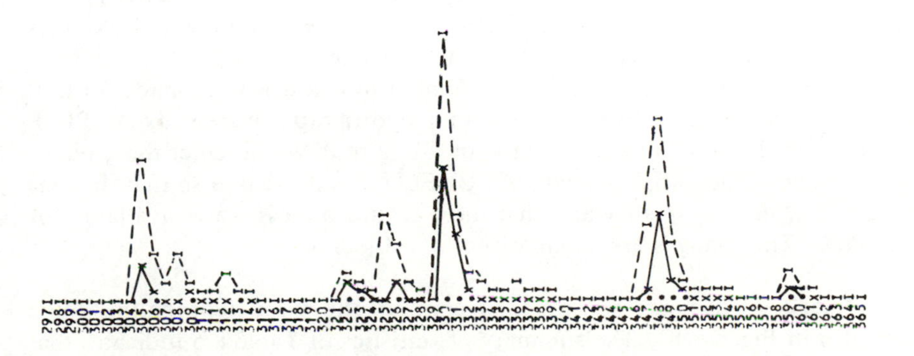

Figure 8-5d (*cont.*).

hydrograph shown in Fig. 8-5(a) and the computed values for the annual totals, goodness-of-fit statistics, and the final storages are listed in Table 8-5. The total predicted runoff was 6.55 in. greater than the measured runoff (17.03 in. versus 10.48 in.). The water balance of Eq. 8-23 was negative, indicating that the predicted runoff and evaporation exceeded the precipitation; while one would expect the water balance to be zero for a water year, a value of 2.45 in. is probably not a significant bias. The volume in SUBFLO decreased from 5 in. to 0.83 in, while the volume in GWS increased from 12 in. to 13.72 in. The correlation coefficient and the modified correlation coefficient equaled 0.68 and 0.55, respectively.

An examination of Fig. 8-5(a) shows several reasons for the poor values of the goodness-of-fit statistics. First, the model underestimates the peak for the largest storm event on day 156. However, for many of the other significant events, the model overestimates the peak discharge (e.g., days 56, 102, 284, 330, and 348). Second, the computed recessions are usually less steep than the recessions of the measured storm hydrographs, especially for storms on days 56, 73, 88, 116, 131, 156, 177, 284, and 330. This would suggest that the unit hydrograph parameter SROP should be decreased so that the unit hydrograph would be more peaked. However, increasing the peakedness of the unit hydrograph would increase the computed peak discharge, which would compound the problem previously noted of overestimating peak discharges. Thus a parameter would have to be changed simultaneously to counterbalance the problem of overestimating peak discharges. For example, PINF could be increased, which would result in a smaller fraction of the rainfall entering surface storage. A third observation is that the computed flow on day 1 is higher than the measured flow. Although the difference is not great, it does suggest that either the initial groundwater storages are too high or that the coefficients that control the groundwater flow rate are causing too much water to be released on a daily basis. The difference between the computed and measured groundwater rates is also evident from day 21 to day 44 in the early part of the water year and from day 295 to day 303 in the latter part of the water year. Thus changes to both the coefficients and initial storages of the two groundwater components may be needed.

Based on these considerations, a number of changes were made for trial 2. SROP was decreased to make the unit hydrograph more peaked. PINF was increased so that a larger fraction of the rainfall would enter the groundwater system. The initial estimate of SUBFLO was decreased so that the bias on day 1 would be smaller and that there would be less water available for interflow. The changes are summarized in Table 8-5.

Trial 2. The model was used to recompute the runoff hydrograph, which is shown in Fig. 8-5(b). The summary of statistics of Table 8-5 indicates that the changes decreased the predicted total runoff by 2.80 in., which was not sufficient to make the predicted (14.23 in.) equal the measured (10.48 in.). The water balance reduced to -1.16 in., while the evaporation increased by 1.49 in. The correlation coefficient indicated an improved fit, while the modified coefficient indicated a poorer fit. The implications of these changes will be discussed below. The end-of-year storages were similar to those values from the first trial, so the initial estimates for the next trial should be changed appropriately (i.e., GWS should be increased and SUBFLO should be decreased). The water balance was closer to zero, although a value of zero would not necessarily be optimum for any one year.

The change in SROP improved the shapes of many hydrograph recessions, although it still appears that they can be improved further. In comparing Fig. 8-5(a) and (b), it is evident that the recessions starting on days 56, 88, and 330 are much improved; however, the recessions starting on days 16, 156, 241, and 284 are examples where the computed recessions are still too flat and further improvement may be possible by decreasing the value of SROP.

It appears that the simultaneous changes to SROP and PINF tended to cancel their individual effects on the peak discharges since there is little change in the peak discharges between the two trials. If SROP is to be decreased further on trial 3, it will be necessary to increase PINF further to maintain the fit of the peak discharges.

The total computed groundwater runoff was not improved from trial 1 to trial 2. The periods during which rainfall did not occur still show differences between the computed and measured values. For example, for the time periods of days 248 to 257, 269 to 275, 295 to 303, 314 to 321, and 341 to 346, the computed flow is greater than the measured flow. Thus changes in parameters that control groundwater flow rates are needed. Since the computed is greater than the measured, changes to the coefficients should be made to decrease the computed.

The total predicted evaporation increased by 1.49 in. from trial 1 to trial 2 even though neither of the parameters (PET and SAT) that control evaporation was changed. This results from the change in PINF that caused increased water entering SUBFLO storage. The increased evapotranspiration is necessary to decrease the bias in the total runoff and thus improve the water balance.

In summary, values of the coefficients should be selected for trial 3 such that (1) the recessions of the predicted storm hydrographs need to be steeper, which is possible by decreasing SROP; (2) the predicted groundwater flow rates for the first month are much higher than the measured flow; therefore, the initial value of SUBFLO can be reduced; (3) the predicted groundwater flow rates near the end of the water year are still too high, so a parameter such as BFPM needs to be reduced to lower the groundwater flow rates; (4) to maintain the water balance, evaporation rates should be increased to offset the higher storage of groundwater; thus PET can be increased; and (5) if SROP is to be decreased significantly, PINF should be increased to maintain reasonable estimates of the peak discharges for most of the smaller storm events.

Trial 3. Based on the previous discussion, values for four coefficients and both initial storages (GWS and SUBFLO) were modified, as evident from Table 8-5. The resulting statistics are given in Table 8-5 and the computed and measured annual hydrograph is shown in Fig. 8-5(c).

The results of the changes for trial 3 are mixed. While the total predicted runoff was improved, the predicted annual hydrograph does not provide as good a fit, as evidenced by the lower value of the modified correlation coefficient. The total water balance was improved and the final groundwater storages were not much different from the initial storages.

In comparing the hydrographs of Fig. 8-5(b) and (c), it is evident that the initial state of the watershed is much improved in trial 3. The predicted hydrograph for the first 54 days is a much better representation of the measured runoff; however, the predicted values are still slightly larger during this period.

Again, there are large differences between many of the peaks for individual storm events, especially for the largest storm event on day 156. The computed peak is lower for trial 3 than for trial 2; this is the primary reason for the decrease in the modified correlation coefficient. The value of R_M is sensitive to differences in scale, while the value of the correlation coefficient is insensitive to differences in scale. Because the computed and measured peak discharges do not show close agreement, the value of R_M is far below the value of R. The value of R_M is relatively insensitive to the values of base flow. Thus there is still a need to improve the prediction of the peak discharges of the 10 or 12 major runoff events.

The value of SROP was reduced to enhance the fit of the recessions. For the most part, the computed storm event recessions closely followed the shape of the measured recessions. Thus the unit hydrograph defined by a value of 0.30 for SROP appears to be optimum.

For non-storm-event periods, the computed runoff hydrographs show only slightly better agreement with the measured values. This is especially evident in the last 150 days of the water year, which is the drier period when the effect of the initial storages would be nil. Although there are some small changes, it appears that it will be necessary to make much larger changes for the base flow coefficients than were made for trial 3.

Trial 4. The analysis of trial 3 indicated a need for (1) better estimates of peak discharges, especially the larger events; and (2) better prediction of the low flows during the last 150 days of the year, which is the relatively dry period during the late spring and summer. The total predicted runoff also needs to be reduced toward the measured value of 10.48 in. Changes were made to four coefficients and both ground water storages, although only the changes to BFPM and PINF were relatively large changes. PINF was decreased to increase the peak discharges, and BFPM was reduced to decrease the baseflow. The value of PET was increased in an attempt to try to increase evaporation and thus improve the water balance. The value of SFP was also reduced to decrease groundwater flow following storm events; for example, on days 288–289, 334–338, and 352 the predicted runoff is much higher than the measured. The value of SFP will have a major effect on runoff after surface runoff has been depleted (i.e., after the 4 days defined by the distribution graph). The resulting initial estimates for trial 4 are shown in Table 8-5.

The annual hydrograph for trial 4 is shown in Fig. 8-5(d), and the annual totals are given in Table 8-5. The modified correlation coefficient increased significantly, with virtually no change in *R*. The predicted runoff of 10.17 in. is only 0.31 in. less than the measured value and the water balance was less than the 1% of the annual rainfall. These annual statistics would suggest a very good fit.

In comparing Fig. 8-5(c) and (d), it is evident that trial 4 shows less agreement of peak discharges than trial 3, except for the largest event, which occurs on day 156. The significant improvement for this one event is the primary reason for the improvement in the value of the modified correlation coefficient. Although the change on most storms was not an improvement, the differences on most storm events is not necessarily cause for concern. After all, there is considerable variation in flow over the 365-day period and very few coefficients to adjust, in addition to the constraints imposed by the model structure. With respect to peak discharge estimation for the minor events, trial 3 may be considered as a better solution than trial 4. It depends on which of the objective functions is given the most weight in selecting an optimum.

Trial 4 provides a better representation of the base flow characteristics for the entire year than do the results of trial 3. The errors on days 5 to 13 and 20 to 48 are smaller for trial 4 than for trial 3; this indicates better estimation of both initial storages and the base flow coefficients. The errors during the summer months are also smaller, such as for days 249–254, 295–303, 314–320, 340–345, and 362–365. Again, this indicates that the groundwater coefficients are more accurate for trial 4, especially for the GWS component.

Discussion

Does trial 4 represent the optimum solution? Probably not! Additional trials could be made and a better solution found. For example, the computed peak discharge for the largest event (day 156) could be increased by continuing to reduce PINF. Although this would cause the correlation coefficients to increase, it would lead to larger errors in the peak discharges for other storm events. Whether one considers it optimum depends on which of the objective functions is given the greatest weight.

At this point it may be worthwhile reviewing the title of this chapter, "Subjective Optimization." In reviewing the changes made from one trial to the next, good arguments could be made for other changes rather than those listed in Table 8-5. This is the nature of subjective optimization. But for the purpose of this chapter, a better solution than Fig. 8-5(d) is not important. It is much more important to keep in mind the objectives of this example. First, the trial-and-error nature of subjective optimization was demonstrated for a complex model. A solution could not have been found using either analytical or numerical optimization methods, although a numerical procedure could be used for a partial optimization. Second, the importance of recognizing the multiobjective nature of subjective optimization was demonstrated. The weight given to each of the objective functions will depend on the purpose of the model and data analysis. Finally, the use of goodness-of-fit criteria in assessing the adequacy of a model was demonstrated. Statistical criteria such as a correlation coefficient are useful summary statistics; that is, they are a means of reducing 365 errors into a single number. They have both advantages and limitations.

The concept of using a first-order Taylor series expansion for calibrating a complex model was demonstrated in our example, "Calibration of a Simple Watershed Model"; however, it was not used with this example. In practice, it should be used, especially as one fine-tunes the coefficients. It was not used for this example because there were only four trials and three or more coefficients were changed on each trial. Because of the interaction between coefficients, such as between PINF and SROP or between SFP and PET, it is often difficult to get meaningful estimates of individual sensitivities when multiple coefficient changes are made. If trials beyond trial 4 were to be made, the numerical procedure may prove useful in adjusting the coefficients when only one or two coefficients are adjusted per trial.

In comparison to many textbook examples where individual storm events are fitted, one might be alarmed at the relatively large scatter in Fig. 8-5(d). Such scatter is not uncommon when fitting rainfall–runoff models to actual data. Even when there is only 1 year of record and daily totals are being used as the data base, there are 365 values to fit and only 5 or 10 coefficients that can be adjusted. When the record consists of 5 years of data, the degrees of freedom are even greater. The model cannot hope to fit the variety of watershed conditions that might exist over such a time span. This should not be discouraging. The fitting process seeks only to represent the average of the watershed processes, especially as they would exist under design conditions.

In discussing the means of evaluating a regression problem (Chapters 3 and 6) the analysis of the residuals was emphasized. Specifically, it was stated that the residuals should be checked for randomness and any biases in the residuals should be explained. Furthermore, it was shown that biases often suggest an incorrect model structure. These same concepts should be considered when evaluating a more complex model, such as a rainfall–runoff model. The plot for an optimum fit should be analyzed to detect trends in the residuals. Trends in the residuals may suggest either an incorrect model structure or a nonoptimum solution.

The structure of the model presented here was purposely kept simple; thus biases are to be expected. Two sources of bias will be identified to illustrate this point. First, there are a number of days when rainfall occurred but there was no measured direct runoff (e.g., days 206, 258, 266, and 336). There are even more days where rainfall occurred with little resulting surface runoff (e.g., days 50, 81, 228, 236, 278, and 359). In each of these cases, the computed runoff was several times larger than the measured runoff. This may reflect the need for an initial abstraction component for the model, such that no surface runoff occurs until a specified volume of rainfall occurs. Use of an initial abstraction in this model would probably eliminate this source of bias. Second, there appears to be a seasonal bias, with the summer storm events showing a large overprediction and the winter and spring storm events showing better agreement. This could possibly reflect the lack of a realistic evaporation component. The existing evaporation component is simply constant over the year, with variation in evaporation rates depending only on the availability of water. If evaporation rates computed by the model varied over the year, a better fit would probably result. Although these two biases are most evident, others may exist in such a simple model. The important point is that the residuals may contain important information about either the model or the data base, and the modeler should not stop the analysis after an apparent optimum has been found. The residuals should be analyzed for biases.

CALIBRATION OF A RETENTION FUNCTION

Thomas et al. (1981) used subjective optimization to partition rainfall to surface runoff and an increment to soil water storage. With dual model outputs of runoff and soil water, no single criterion could be used for numerical optimization. The immediate objective was to partition the daily rainfall and assess the results against 3 years of field data. The wider objective was to initiate development of a methodology to increase the effectiveness of management of water for agriculture.

Partitioning of rainfall on small homogeneous research plots has been studied extensively. Accurate modeling results can be obtained with a high level of characterization of rainfall and detailed soils information. Integration of such results over heterogeneous segments of the landscape has not been as successful. Therefore, at present a semiempirical approach is required. The rainfall partitioning device used in the study is the retention function described previously by Snyder (1971).

Briefly, the retention function starts from the general differential equation

$$\frac{dr}{dt} = -b[r(t) - r_c] \qquad (8\text{-}24)$$

Integrating Eq. 8-24 gives the following expression, which has the form of the familiar Horton infiltration equation:

$$r(t) = (r_0 - r_c)\, e^{-bt} + r_c \tag{8-25}$$

The variable, r, is the rate of intake of water into storage. Such storages are all waters that do not become "runoff." The variable, r_c, is the minimum retention or saturated intake rate, and r_0 is the initial intake at time, t, equal to zero. The mathematical parameter, b, controls the depletion rate of unfilled storage. The retention function represents processes occurring over a large heterogeneous area as randomly interconnected storages for water at and below the soil surface are depleted. As such storages are depleted during rainfall, more and more surface water is deflected downslope to become runoff.

Equations 8-24 and 8-25 describe a depletion of storage that is not explicitly dependent on the rainfall, and yet changes of storages in time must be dependent on the rate of input. Also, Eq. 8-25 is monotonic downward, with $r(t)$ approaching r_c asymptotically. A retention function is needed to describe more specifically the dynamics of storage when a variable rainfall occurs. To accomplish this, the parameter b in Eqs. 8-24 and 8-25 was made a state-dependent variable and written as (Snyder, 1971)

$$b = a_1\left(\frac{P_{\Delta t} + r_u - r_t}{P_{\Delta t} + r_u - r_c}\right)\frac{P_{\Delta t} - r_c}{P_{\Delta t} + r_c} \tag{8-26}$$

where $P_{\Delta t}$ is the rainfall intensity during time step Δt, r_u the maximum or upper limit retention, r_c the minimum retention, r_t the retention at time t, a_1 the parameter.

To compute $r(t)$, the quantity of rainfall during sequential time periods, Δt, must be supplied. Equation 8-26 was substituted into the explicit finite difference form of Eq. 8-24, which may be written as follows:

$$r_{t+\Delta t} = r_t - b(r_t - r_c)\,\Delta t \tag{8-27}$$

The time step, Δt, in Eq. 8-27 is identical to the time period, Δt, for rainfall increments. Forward solution of Eq. 8-27 would amount to projecting straight-line segments. The solution is improved by using Eq. 8-25 to calculate r_t sequentially over short intervals $\delta t = \Delta t/5$. Since $P_{\Delta t}$ and r_t in Eq. 8-26 are held constant during δt, b is constant. Equation 8-25 may be written as follows:

$$r_{t+\delta t} = (r_t - r_c)\, e^{-b\delta t} + r_c \tag{8-28}$$

By this procedure r_t over interval Δt is represented by five points obtained by projecting short exponential arcs in the form of Eq. 8-28.

Equation 8-26 yields values for state variable b when the retention function is rain driven. Basically, r_t approaches r_c, but can reverse and increase in value when the rainfall is less than r_c. During prolonged periods of no rainfall the value of r_t approaches r_u because of the water used by evapotranspiration. Rescaled pan evaporation was used to drive the retention function for periods of no rain. The retention function was changed simply by computing the state variable b as in

$$b = -a_2\left(\frac{r_u - r_t}{r_u - r_c}\right)\frac{\text{PAN}'_{\Delta t}}{\text{PAN}'_{\Delta t} + r_c} \tag{8-29}$$

where $\text{PAN}'_{\Delta t}$ is the rescaled pan evaporation in Eq. 8-30 below and a_2 is a parameter.

Initially, pan evaporation was used in Eq. 8-29. When results were checked against measured data, however, recovery of the value of retention during the winter dormant season was too great. When the grass was dormant it acted as mulch, a thermal shield against incoming energy, suppressing soil evaporation. After the grasses began growth in the spring, their hydrologic role changed. They changed from shields to wicks, readily removing water from soil storage through transpiration. Consequently, a plant-activity index was developed ranging from unity, when recovery of retention is described by pan evaporation, down to a small fraction during the dormant season. The plant-activity index values were used to scale pan evaporation as in the equation

$$\text{PAN}'_{\Delta t} = I(\text{PAN}_{\Delta t}) \tag{8-30}$$

where $\text{PAN}_{\Delta t}$ is daily pan evaporation and I is the index value.

Figure 8-6 illustrates the retention function separating that portion of rainfall that becomes runoff. During periods of rainfall the retention function is controlled between an upper and lower limit by rainfall. During periods of no rainfall the retention function is controlled by pan evaporation rescaled by the plant-activity index. The pan-driven retention function approaches the upper retention limit asymptotically. Such a structure is thus an alternative to a changing empirical ratio between potential evapotranspiration and actual evapotranspiration.

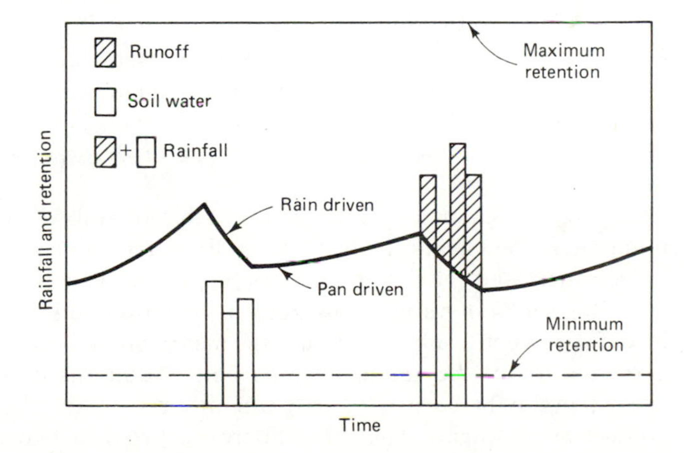

Figure 8-6 Operation of retention function during rain and nonrain periods.

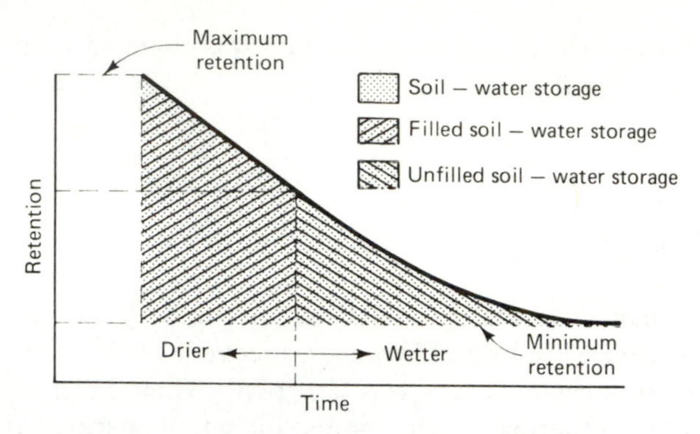

Figure 8-7 Relationship between retention function and soil-water storage of experimental watershed.

The retention function also evaluates soil-water storage. A unique mathematical relationship between retention and unfilled soil-water storage was developed. This relationship is defined in Fig. 8-7.

If the retention function is driven downward by rainfall, from some initial value to near the final value, it encloses an area on the graph. The value of this area, which is the integral of intake rate over time, must be the volume of water entering the soil or being stored on it. The incipient ponding case was assumed where the value of retention is equivalent to the value of the rainfall intensity. Therefore, the surface storage is zero and the integral of the retention function is the volume of water entering soil storage. The shape of the curve will be controlled by the parameter, a_1, in Eq. 8-26. The integration thus produces a unique relationship between initial retention intake, r_0, final retention intake, r_c, maximum storage, and the parameter, a_1. Given any three, the fourth can be determined.

If the initial value of retention, r_0, is set at the maximum retention, r_u, the volume is maximum and is thus the maximum storage of the soil. For a given value of parameter a_1, the value of retention relates uniquely to the unfilled soil-water storage. The consequence of this relationship is that the state of the watershed system with regard to its capacity to take in water should be expressed in terms of unfilled soil-water storage or dryness rather than filled storage or wetness.

The model was programmed to compute a table of soil-water storages versus retention rate. The integration is based on an initial value of r_0 slightly in excess of an anticipated maximum value of r_u, a final value, r_c, and the parameter a_1. This series of values constitutes a retention versus soil-water storage table, or figure, such as Fig. 8-8. Given the maximum soil-water storage, one can interpolate to get the maximum retention. Additionally, given the retention at, say, midnight each day, one can interpolate for the unfilled soil-water storage at midnight. Then the difference between two midnight values is the soil-water storage change for the day.

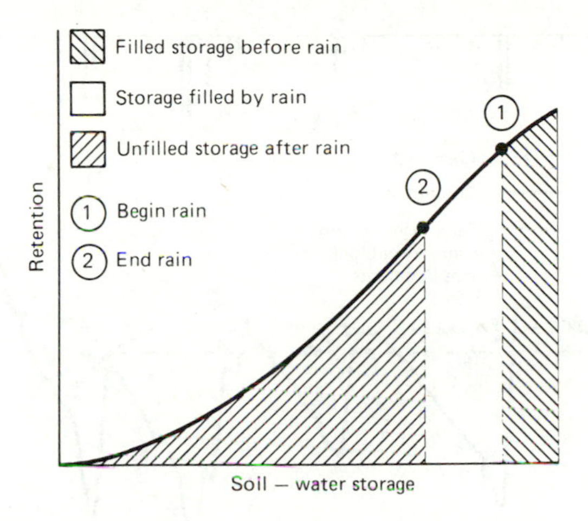

Figure 8-8 Relationship between retention and soil-water storage as affected by rainfall.

The retention model was checked against runoff and soil-moisture data from a small 1.3-ha watershed on the southern Piedmont Conservation Research Center at Watkinsville, Georgia. Rainfall and runoff were continuously recorded by gages at the site. Soil moisture was measured by a vertical set of electrical resistance blocks at each of 15 sites in the watershed. The block resistance was read daily and converted to soil-water pressure by calibration curve. Laboratory determined curves of pressure versus water content yielded available storage for water. Soil depth used for storage was 18 to 71 cm of sandy clay loam above a sandy clay or clay layer.

Hourly rainfall was used with the rain-drive retention function to compute daily runoff from hourly increments. Computed values of available storage were obtained from the retention values by interpolation in the constructed retention storage table. These calculated daily values were compared with measured values for a 3-year period.

Parameters a_1 and a_2 were varied and comparisons checked. The action of the parameters is not independent. However, since a_2, which controls the recovery of storage, sets the initial value of retention for a rain event, it relates strongly to the calculation of runoff. Parameter a_1 relates more strongly to the calculation of soil water. It can be made site specific if the soil characteristics of the intake rate and storage volume are known.

Figure 8-9 shows a comparison between computed and measured unfilled or available storage during a winter wet season. The electrical resistance blocks can sense soil-water pressure between −0.25 and −15 bar. They do not respond to soil-water pressures greater than about −0.25 bar and are insensitive in very wet soils. The retention model computes unfilled storage down to near zero, while the moisture blocks do not register much below 1.5 cm of unfilled or available storage.

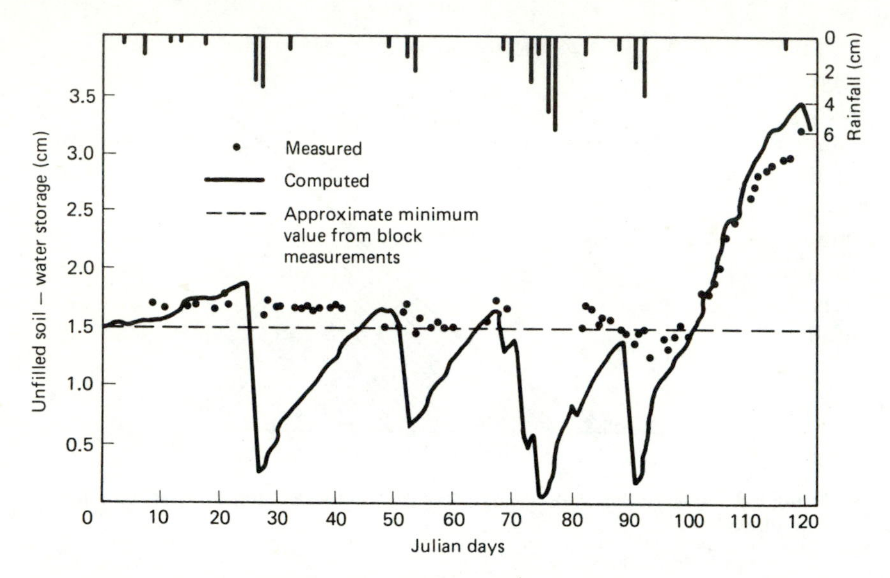

Figure 8-9 Comparison of measured and computed unfilled soil-water storage for 120 days in 1976.

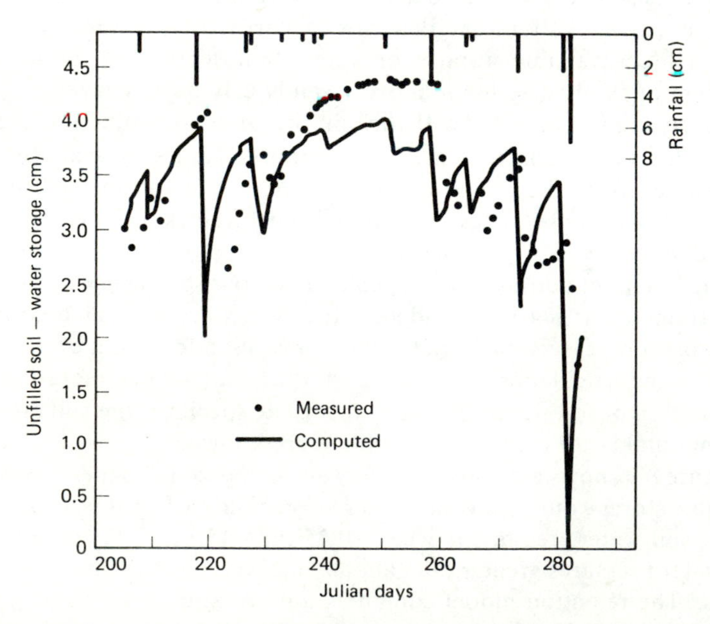

Figure 8-10 Comparison of measured and computed unfilled soil-water storage for days 205–285 in 1976.

Figure 8-10 shows a comparison of computed and measured unfilled storage in the summer dry season. During this season the computed values tended to respond quickly to wetting and drying. This is rational since electrical resistance blocks require time to respond to changes in matric soil water.

Runoff comparisons are summarized in Table 8-6. Annual totals of measured and computed runoff for a range of wet to dry years are similar. Numbers of days with runoff were in agreement. However, the computed value of the annual maximum daily runoff was always less than the measured. This is most likely due to use of hourly rainfall. Other work has shown that rainfall intensity by 5-min periods is necessary on small watersheds.

TABLE 8-6 Annual Summary for Observed and Computed Values of Surface Runoff and Soil-Water Storage[a]

	Runoff			
Year	Total (cm/yr)	Maximum (cm/day)	Number of Days with Runoff	Maximum Unfilled Soil-Water Storage (cm)
1976 (wet)	10.4(13.0)	3.8(2.0)	29(25)	4.3(4.0)
1977 (avge)	8.4(9.4)	4.6(2.5)	18(15)	4.3(4.1)
1978 (dry)	5.1(5.1)	3.8(2.3)	12(10)	4.3(4.2)

[a]Values in parentheses were computed.

EXERCISES

For Exercises 8-1 to 8-3, fit a linear bivariate model $\hat{Y} = b_0 + b_1 X$ using subjective optimization and an objective function of minimizing the sum of the squares of the errors. Use the data sets from Exercise 3-7 and compare the results with the analytical solutions. In each case, plot the resulting line and explain why each change was made.

8-1. Use data set 1, with $b_0 = 3$ and $b_1 = 0$ as initial estimates.

8-2. Use data set 2, with $b_0 = 0$ and $b_1 = 0.6$ as initial estimates.

8-3. Use data set 3, with $b_0 = 12$ and $b_1 = -1$ as initial estimates.

For Exercises 8-4 to 8-14, fit the models indicated in the exercises of Chapter 7 using subjective optimization and an objective function of minimizing the sum of the squares of the errors. In each case, plot the resulting model for each iteration and explain the reasons for changes made to the fitting coefficients.

8-4. Exercise 7-1 with $b = 10$ as the initial estimate.

8-5. Exercise 7-2 with $b = 0.05$ as the initial estimate.

8-6. Exercise 7-3 with $b = 0.000075$ as the initial estimate.

8-7. Exercise 7-4 with $b_0 = 0.01$ and $b_1 = 0.1$ as initial estimates.

8-8. Exercise 7-5 with $b_1 = 8.5$ and $b_2 = 2$ as initial estimates.

8-9. Exercise 7-6 with $b_1 = 20$ and $b_2 = 0.007$ as initial estimates.

8-10. Exercise 7-7 with $b_1 = 5$ and $b_2 = -0.2$ as initial estimates.

8-11. Exercise 7-8 with $k = 0.25$ and $b = 2.5$ as initial estimates.

8-12. Exercise 7-9 with $b_1 = 0.8$ and $b_2 = 0.075$ as initial estimates.

8-13. Exercise 7-10 with $b_1 = -3 \times 10^6$ and $b_2 = -40$ as initial estimates.

8-14. Using the model of Figs. 8-2 and 8-3 and the data given below, find the values of SROC, CINF, CGWS, SROS, and GWS that result in the minimum sum of squares of the differences between the computed and measured runoff. Assume that SROP = 65, GWSM = 15, PETS = 0.3, and PGWRO = 0.15.

Day	1	2	3	4	5	6	7	8
P (in.)	0.8	0.4	0	0	1.2	0.9	0.7	0
TRO (in.)	0.23	0.15	0.10	0.08	0.42	0.37	0.31	0.20

8-15. Two years of rainfall and runoff data for the Anacostia River at Colesville, Maryland (area = 21.1 square miles), are provided in Appendix D-2. Using the computer model of Appendix D-1, calibrate the model using the data for the 1963 water year. Then test the model using the data for the 1964 water year. Discuss the results.

8-16. Using the rainfall–runoff model (see Appendix D-1), develop an initial abstraction component and calibrate the model using the data for the Anacostia River at Colesville for the 1963 water year (Appendix D-2). One possible initial abstraction model would be

$$\text{FTGW} = \begin{cases} P & \text{for } P \le I_a \\ I_a + \text{PINF}(P - I_a) & \text{for } P > I_a \end{cases}$$

where FTGW is the flow to the groundwater system, I_a the initial abstraction, P the daily rainfall depth, and PINF the infiltration coefficient of the existing model. The value of I_a could then be varied in fitting the model to the data. Test the model using the data for the 1964 water year. Discuss the results.

8-17. Using the rainfall–runoff model (see Appendix D-1), develop an evaporation component that shows daily variation of the evaporation rate and use it in place of the existing component. Calibrate the model using the data for the Anacostia River at Colesville for the 1963 water year (Appendix D-2). Test the model with the data for the 1964 water year, and discuss the results.

8-18. **(a)** Using a water balance approach, formulate a daily water-yield model for forecasting snowmelt runoff. Make the model spatially distributed so that it can use data for subareas. Formulate the model so that it can use the following input for each subarea: drainage area, the daily precipitation (inches), the percentage of snow-covered area, and a temperature index (degree-days).

(b) Using the data of Appendix C-4, calibrate the model for the single snowmelt season by comparing the predicted and measured streamflow (cfs). The data are for the Conejos River watershed at Magote, Colorado, which is a 282-

square-mile watershed. The watershed is divided into three subareas, with areas of 57, 129, and 96 square miles.

8-19. Read the following three articles:

1. J. Nash and J. Sutcliffe, *J. Hydrology*, Vol. 10, pp. 282-90, 1970.
2. P. O'Connell et al., *J. Hydrology*, Vol. 10, pp. 317-27, 1970.
3. A. Mandeville et al., *J. Hydrology*, Vol. 11, p. 109, 1971.

Obtain a set of data from local climatological records and reproduce the experiment discussed in these three articles. Evaluate and discuss the change in accuracy as the model complexity is changed.

REFERENCES AND SUGGESTED READINGS

BEARD, L. R., Use of Interrelated Records to Simulate Stream Flow, *Journal of the Hydraulics Division, ASCE*, Vol. 91, No. HY5, pp. 13-19, 1965.

BEARD, L. R., Hydrologic Simulation in Water-Yield Analysis, *Journal of the Irrigation and Drainage Division, ASCE*, Vol. 93, No. IR1, pp. 33-42, March 1967.

BEVEN, K. J., and G. M. HORNBERGER, Assessing the Effect of Spatial Pattern of Precipitation in Modeling Stream Flow Hydrographs, *Water Resources Bulletin*, Vol. 18, No. 5, pp. 823-29, 1982.

BEVEN, K. J., M. J. KIRKBY, N. SCHOLFIELD, and A. F. TAGG, Testing a Physically-Based Flood Forecasting Model for Three UK Catchments, *Journal of Hydrology*, Vol. 69, pp. 119-44, 1984.

CAMPBELL, K. L., and H. P. JOHNSON, Hydrologic Simulation of Watersheds with Artificial Drainage, *Water Resources Research*, Vol. 11, No. 1, pp. 120-26, 1975.

CHEN, C., and R. SHUBINSKI, Computer Simulation of Urban Storm Water Runoff, *Journal of the Hydraulics Division, Proceedings of the ASCE*, Vol. 97, pp. 289-301, February 1971.

CRAWFORD, N. H., and R. K. LINSLEY, *Digital Simulation in Hydrology: Stanford Watershed Model 4*, Tech. Report 39, Department of Civil Engineering, Stanford University, 1966.

CROLEY, T. E., II, Great Lakes Basins Runoff Modeling, *Journal of Hydrology*, Vol. 64, pp. 135-58, 1983.

CUNDY, T. W., and K. N. BROOKS, Calibrating and Verifying the SSARR Model: Missouri River Watershed Study, *Water Resources Bulletin*, Vol. 17, No. 5, pp. 775-82, 1981.

DAWDY, D. R., and J. M. BERGMANN, Effect of Rainfall Variability on Streamflow Simulation, *Water Resources Research*, Vol. 5, No. 5, pp. 958-66, 1969.

DAWDY, D. R., R. W. LICHTY, and J. M. BERGMANN, *A Rainfall-Runoff Simulation Model for Estimation of Flood Peaks for Small Drainage Basins*, U.S. Geological Survey Professional Paper 506-B, 1972.

DEBO, T. N., and A. M. LUMB, Role of Computer Simulation in Administering an Urban Drainage Ordinance, *Simulation Network Newsletter, HYDROCOMP*, Vol. 7, No. 2, February 1975.

DISKIN, M. H., and E. SIMON, A Procedure for the Selection of Objective Functions for Hydrologic Simulation Models, *Journal of Hydrology*, Vol. 34, pp. 129-49, 1977.

DISKIN, M. H., and E. SIMON, The Relationship between the Time Bases of Simulation Models and Their Structure, *Water Resources Bulletin*, Vol. 15, No. 6, pp. 1716-32, 1979.

FERRARA, R. A., and A. HILDICK-SMITH, A Modeling Approach for Storm Water Quantity and Quality Control via Detention Basins, *Water Resources Bulletin*, Vol. 18, No. 6, 975-81, 1982.

GLENNE, B., Simulation of Water Pollution Generation and Abatement on Suburban Watersheds, *Water Resources Bulletin*, Vol. 20, No. 2, pp. 211-17, 1984.

HUGGINS, L. F., and E. J. MONKE, A Mathematical Model for Simulating the Hydrologic Response of a Watershed, *Water Resources Research*, Vol. 4, pp. 529-40, 1968.

JAMES, L. D., Using a Digital Computer to Estimate the Effects of Urban Development on Flood Peaks, *Water Resources Research*, Vol. 1, No. 2, 1965.

KHANBILVARDI, R. M., A. S. ROGOWSKI, and A. C. MILLER, Modeling Upland Erosion, *Water Resources Bulletin*, Vol. 19, No. 1, pp. 29-35, 1983.

LANGFORD, K. J., and J. L. MCGUINNESS, A Comparison of Modeling and Statistical Evaluation of Hydrologic Change, *Water Resources Research*, Vol. 12, No. 6, pp. 1322-24, 1976.

LEAF, C. F., and G. E. BRINK, *Computer Simulation of Snowmelt within a Colorado Subalpine Watershed*, USDA Forest Service Research Paper RM-99, February 1973.

LINSLEY, R., Continuous Simulation Models in Urban Hydrology, *Geophysical Research Letters*, Vol. 1, No. 1, pp. 59-62, May 1974.

LITCHTY, R. W., D. R. DAWDY, and J. M. BERGMANN, Rainfall Runoff Model for Small Basin Flood Hydrograph Simulation, *Proceedings of the Symposium on the Use of Analog and Digital Computers in Hydrology*, Tucson, Ariz., pp. 356-67, 1968.

MAGATTE, W. L., V. O. SHANHOLTZ, and J. C. CARR, Estimating Selected Parameters for the Kentucky Watershed Model from Watershed Characteristics, *Water Resources Research*, Vol. 12, No. 3, pp. 472-76, 1976.

MEDINA, M. A., Jr., and J. BUZUN, Continuous Simulation of Receiving Water Quality Transients, *Water Resources Bulletin*, Vol. 17, No. 4, pp. 549-57, 1981.

MURPHY, C. E., and K. R. KNOERR, The Evaporation of Intercepted Rainfall from a Forest Stand: An Analysis by Simulation, *Water Resources Research*, Vol. 11, No. 2, pp. 273-80, 1975.

NOZDRYN-PLOTNICKIN, W. W., Assessment of Fitting Techniques for the Log Pearson Type 3 Distribution Using Monte Carlo Simulation, *Water Resources Research*, Vol. 15, No. 3, pp. 714-18, June 1979.

PATRY, G. G., and M. A. MARINO, Nonlinear Runoff Modeling: Parameter Identification, *Journal of the Hydraulics Division, ASCE*, Vol. 109, No. HY6, pp. 865-60, 1983.

PORTER, J. W., and T. A. MCMAHON, A Model for the Simulation of Streamflow Data from Climatic Records, *Journal of Hydrology*, Vol. 13, No. 4, pp. 297-324, 1971.

SHANHOLTZ, V. O., and J. H. LILLARD, *Simulations of Watershed Hydrology on Agricultural Watersheds in Virginia with the Stanford Model*, Dept. of Agricultural

Engineering, Virginia Polytechnic Institute, Blacksburg, Va., NTIS PB-237-508, September 1971.

SNYDER, W. M., A Proposed Watershed Retention Function, *Proceedings of the ASCE*, Vol. 97, No. (IR1), pp. 193–201, 1971.

STRZEPEK, K. M., and D. H. MARKS, River Basin Simulation Models: Guidelines for Their Uses in Water Resources Planning, *Water Resources Bulletin*, Vol. 17, No. 1, pp. 10–15, 1981.

THOMAS, A. W., R. R. BRUCE, and W. M. SNYDER, Daily Partitioning of Rainfall to Surface Runoff and Soil Water on Complex Landscapes, *Transactions of the ASAE*, Vol. 24, No. 5, pp. 1191–98, 1981.

WARD, R. C., Hypothesis Testing by Modeling Catchment Response, *Journal of Hydrology*, Vol. 67, pp. 281–305, 1984.

WOOD, E. F., An Analysis of the Effects of Parameter Uncertainty in Deterministic Hydrologic Models, *Water Resources Research*, Vol. 12, No. 5, pp. 925-40, 1976.

WOOD, K. B., L. BOERSMA, and L. N. STONE, Dynamic Simulation Model of the Transpiration Process, *Water Resources Research*, Vol. 2, No. 1, pp. 85-97, 1966.

9 Time-Series Analysis and Synthesis

INTRODUCTION

Simply stated, *time-series analysis* is defined as the analysis of data in which time is an independent variable. A time series is analyzed for the purpose of formulating and calibrating a model that can be used both to describe the time-dependent characteristics of a hydrologic variable and to predict (i.e., forecast) future values of the time-dependent variable. Methods used to analyze time series are also being used to analyze spatial data of hydrologic systems, such as soil moisture within a watershed or the transport of pollutants in groundwater systems.

In the following portions of this chapter we will refer frequently to time as an independent variable. In spatial variability we may similarly refer to space as an independent variable. As long as we clearly recognize that we are using time and space in a sense of relationship, we may regard them as mathematical or statistical independent variables. However, physical, chemical, and biological processes take place *in* space or time. Such processes do not take place because *of* space or time. For example, acceleration of a particle of mass is conveniently expressed as length divided by time squared. But acceleration is caused by a force acting on the particle; it is not caused by time.

Time and space are not causal properties. They would be regarded as convenient parameters by which we bring true cause and effect into proper relationships. As another example, evapotranspiration is normally highest in June. This maximum is not caused by June, but because insolution is highest in June. The seasonal time of June is a parameter relating evapotranspiration and insolation.

Time-series analyses, particularly autocorrelation methods to be studied in this chapter, are sometimes considered a necessary step prior to power spectral analysis. Spectral analysis is useful for detection of unknown periods or cycles in data. It is not too useful when the predominant cycles are known, as they are in hydrology. Annual cycles predominate at intermediate time scale because of seasonal changes in insolation. Diurnal cycles control short time scales, but their analysis has not proven very useful in applied hydrology. Climatic megacycles exist; glacial and interglacial periods in earth climate are documented. Irregular 10- to 20-year cycles can sometimes be detected in climatic records. Portions of such cycles may exist as time trends in data, but their periods are too long for conventional time-series analysis.

The annual cycle predominates in hydrologic time-series analysis. In much of hydrology we deal with a sequence of observations in time. The year may be broken into 12 months, 73 five-day periods, or 365 days. All of these observational periods have been used. Monthly data have been used more frequently than daily data, because of reduced data handling and because of accessibility in published records.

In its most basic form, time-series analysis is a bivariate analysis in which time is the independent or predictor variable. However, it differs from the bivariate form of regression in that regression assumes independence among the residuals. Time-series analysis recognizes a time dependence and attempts to explain the dependence. More specifically, time series are analyzed to separate the systematic and nonsystematic variation and characterize the time dependence within the systematic component. In addition to the difference based on the time-dependence characteristic, regression analysis is usually applied to unordered data, while the order in a time series is an important characteristic that must be considered in handling the data. Actually, it may not be fair to compare regression with time-series analysis because regression is a method of calibrating the coefficients of an explicit function, while time-series analysis is much broader and refers to data analysis techniques that are used to handle data in which the independent variable is time (or space). The principle of least squares is used in time-series analysis to calibrate the coefficients of explicit time-dependent models.

The statistical methods that are used to analyze time series can also be used to analyze other types of correlated variables. For example, spatial data often exhibit a dependence between adjacent points, lines, or fields. For some types of spatial data, estimation is based on kriging and semivariogram analysis; for other types of spatial data, the methods presented here for the analysis of time series can be used to characterize the spatial dependence.

A time series consists of two general types of variation, systematic and nonsystematic. Both types of variation must be analyzed and characterized in order to formulate a model that can be used to predict or synthesize expected values and future events. The objective of the analysis phase of time-series modeling is to decompose the data so that the types of variation can be characterized. The objective of the synthesis phase is to formulate a model that reflects the characteristics of the systematic and nonsystematic variation.

COMPONENTS OF A TIME SERIES

In the decomposition of a time series, there are five general components, all of which may or may not be present in any one time series. The three components that can be characterized as systematic are secular, periodic, and cyclical trends. The names of these components are probably based on the importance of time trends within business forecasting. Nonsystematic variation may be based on episodic events and random variation. Different methods are used to identify each of these components. The process of time-series analysis must be viewed as a process of separating the total variation in measured data into five components.

By definition, a *secular trend* is a tendency to increase or decrease continuously for an extended period of time in a systematic manner. If urbanization of a watershed occurs over an extended period of time, the progressive increase in peak discharge characteristics may be viewed as a secular trend. Secular trends are often separated from the other components by fitting a functional form such as a line or polynomial. The coefficients of the equation can be evaluated using regression analysis; however, it must be recognized that the assumption of independence of the errors will most likely be violated. At the minimum, the violation of the assumption of independence means that the significance test on either the regression coefficients or the correlation coefficient may not be meaningful.

Periodic trends are very common in hydrologic time series. Rainfall, runoff, and evaporation rates show periodic trends with an annual period. Seasonal trends may also be apparent in hydrologic data. Where a periodic trend is expected, the period of the trend can be identified using a moving-average analysis, a correlation analysis of adjacent values of the criterion (i.e., an autocorrelation analysis), or a spectral analysis. Once a periodic trend has been shown to exist in the data, a functional form can be used to represent the trend. The calibration of trigonometric transforms was discussed in Chapter 6. Quite frequently, one or more sine functions are used to represent the trend:

$$f(t) = A \sin (2\pi f_0 t + \theta) \tag{9-1}$$

in which A is the amplitude of the trend, f_0 the frequency, θ the phase angle, and t the time measured from some zero point. The frequency is the reciprocal of the period of the trend, with the units depending on the units of t. The phase angle is necessary to adjust the trend so that the sine function crosses the mean of the trend at the appropriate time. In addition to the method outlined in Chapter 6, the values of A, f_0, and θ can be optimized using a numerical optimization method such as nonlinear least squares. As the number of sine waves increases, the complexity of the optimization also increases.

Unlike periodic trends, *cyclical trends* occur with an irregular period. Business cycles are the classic example. In general, these types of trends are less common in hydrology; however, some say that meteorological characteristics change with sunspot activity, which has an irregular period. The cyclical trend is considered part of the systematic variation because it provides for some local correlation; that is, given the start of a time series with a dominant cyclical trend, the nonsystematic variation may be minor compared with the variation due to the cyclical trend. However, because the period of the cycle is not deterministic, there is a certain element of indeterminism with cyclical trends. Cyclical trends can often be detected using the same methods as periodic trends. Sometimes, they can be identified only by eliminating other components of the time series.

Episodic variation results from events that occur on a "one-shot" basis. In a long record, there may only be one or two such events. Extreme meteorological events, such as monsoons or hurricanes, may cause episodic variation in hydrological data. The change in the location of a recording gage may also act as an episodic event. The identification of an episodic event usually requires supplementary information. Although extreme changes may appear in a time series, one should be cautious about labeling the variation as an episodic event without the supporting data. It must be remembered that extreme events can be observed in any set of measurements on a random variable. If the supporting data do not provide the basis for evaluating the characteristics of the episodic event, one must characterize the remaining components of the time series and use the residual to define the characteristics of the episodic event.

Random fluctuations are often a dominant source of variation in time series. This source of variation results from physical occurrences that are not measurable; these are sometimes called *environmental factors* since they are considered to be external to the physical processes that compose the system. The objective of the analysis phase is to characterize the random variation by a known probability function and the values of its parameters (see Chapter 4).

MOVING-AVERAGE FILTERING

Moving-average filtering is one method for separating the systematic and nonsystematic variations. The term *filter* is the result of use by electrical engineers of statistical methods to "filter" the systematic variation from the total time series. It is based on the premise that the systematic components exhibit some *autocorrelation* (i.e., correlation between adjacent and nearby measurements) while the random fluctuations are not autocorrelated. Therefore, the averaging of adjacent measurements will eliminate the random fluctuations, with the result converging to a description of the systematic trend.

In general, the moving-average computation uses a weighted average of adjacent observations to produce a new time series that consists of the systematic trend. Given a time series Y_t, the filtered series $\hat{Y}_t$ is derived by

$$\hat{Y}_i = \sum_{j=1}^{m} w_j Y_{i-k+j-1} \qquad \text{for } i = (k+1), (k+2), \ldots, (n-k) \tag{9-2}$$

in which n is the length of the measured time series, m the length of the smoothing interval, w_j the weight given to the jth value in the smoothing interval, and k is given by

$$k = \frac{m-1}{2} \tag{9-3}$$

The smoothing interval should be an odd integer, and a total of $2k$ observations are lost; that is, while the length of the measured time series equals n, the smoothed series, $\hat{Y}_i$, has $n - 2k$ values. The simplest weighting scheme would be the arithmetic mean (i.e., $w_j = 1/m$). Other weighting schemes give the greatest weight to the central point in the interval, with successively smaller weights given to points farther removed from the central point. (The similarities with kriging estimation should be recognized.) The filter is essentially a plot of the weights versus the separation distance from the central point. Most filters assume equally spaced measurements in the time series. Periodic variations will be eliminated when the smoothing period equals the period of the periodic component; thus the choice of a smoothing interval is not an arbitrary choice. If the smoothing interval is less than the period, the periodic component will be dampened.

There are several disadvantages to moving-average filtering. First, $2k$, or $m - 1$, observations are lost, which may be a disadvantage for short record lengths. Second, a moving-average filter is not itself a mathematical representation, and thus forecasting with the filter is not possible; a structural form must still be calibrated to forecast any systematic trend identified by the filtering. Third, the choice of the smoothing interval is not always obvious, and it is often necessary to try several values in order to provide the best separation of systematic and nonsystematic variation. Fourth, if the smoothing interval is not properly selected, it is possible to eliminate both the systematic and the nonsystematic variation.

It is important to emphasize the role of moving-average filtering in time-series analysis. The filtering process is used to separate the systematic component(s) from the total measured values. Moving-average calculations can be used when we know the filter, that is, when we know the weights, w_j, in Eq. 9-2. When we do not know the filter we usually resort to least-squares methods. By such methods we smooth the data and estimate the weights simultaneously. After identifying the systematic component(s), all components must be calibrated using a known functional form. The values predicted for the systematic variation can then be subtracted from the total measured value to obtain the residuals, which represent the nonsystematic variation. Then a known probability function (e.g., normal or uniform) must be calibrated with the residuals. The sum of the functional forms used to represent the systematic components and the probability function used to represent the nonsystematic variation then serves as the forecasting or simulation model. Thus the modeling process consists of two analysis steps, one for the systematic and one for the nonsystematic.

Example 9-1: Effect of Urbanization on Peak Discharges

A common problem in hydrologic modeling is the evaluation of the effect of urban development on runoff characteristics, especially the peak discharge. It is difficult to determine the hydrologic effects of urbanization with time because in a large watershed

TABLE 9-1 Annual Flood Series and Smoothed Series for the Pond Creek Watershed, 1945–1968 (cfs)

Year	Annual Maximum	Smoothed Series
1945	2000	
1946	1740	
1947	1460	
1948	2060	1720
1949	1530	1640
1950	1590	1580
1951	1690	1460
1952	1420	1360
1953	1330	1380
1954	607	1480
1955	1380	1610
1956	1660	1870
1957	2290	2040
1958	2590	2390
1959	3260	2560
1960	2490	2800
1961	3080	3620
1962	2520	3860
1963	3360	4020
1964	8020	4130
1965	4310	4300
1966	4380	
1967	3220	
1968	4320	

development occurs gradually and there are many other factors that can cause variation in runoff characteristics, such as rainfall.

The objective of this example is to demonstrate the use of moving-average smoothing for detecting a secular trend in data. The data consist of the annual flood series for the 24-year period from 1945 through 1968 for the Pond Creek watershed, which is a 64-square-mile watershed in north-central Kentucky. Between 1946 and 1966 the percentage of urbanization increased from 2.3 to 13.3 while the degree of channelization increased from 18.6% to 56.7% with most of the changes occurring after 1954. The annual flood series for the 24-year period is given in Table 9-1.

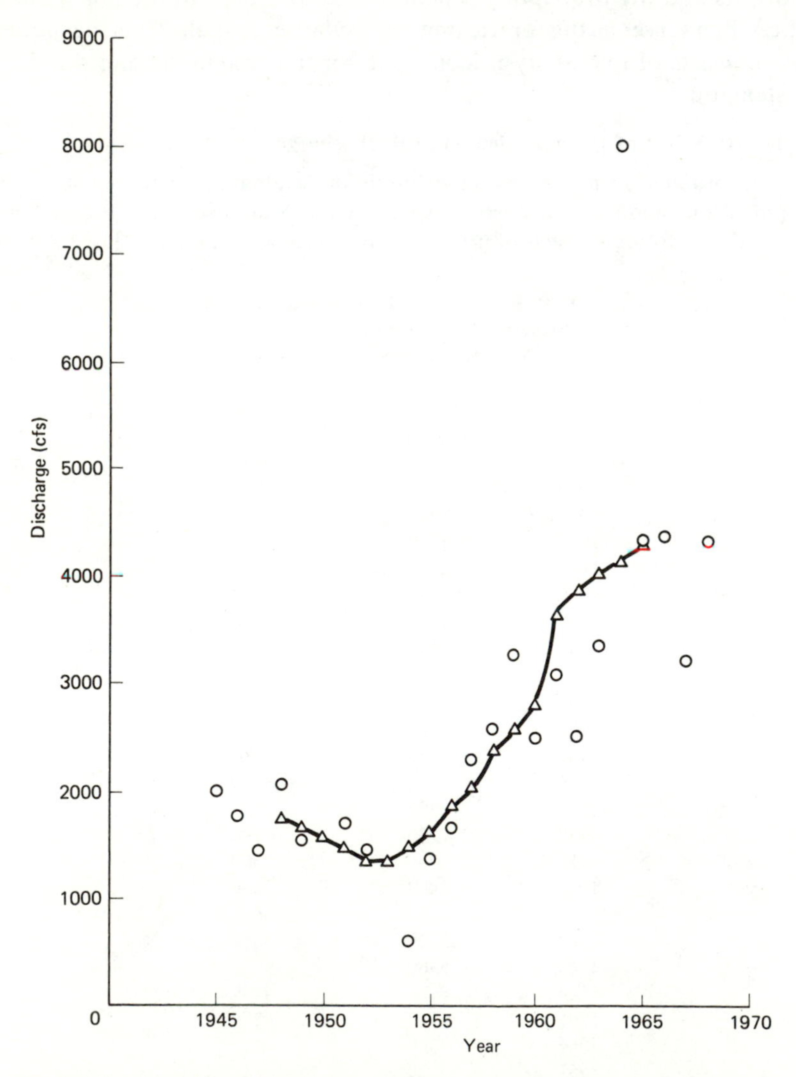

Figure 9-1 Annual flood series and smoothed series for the Pond Creek watershed, 1945–1968. ⊙, annual flood series; △, smoothed series.

The data were subjected to a moving-average smoothing with a smoothing interval of 7 years. Shorter smoothing intervals were tried but did not show the secular trend as well as did the 7-year interval. The smoothed series is shown in Fig. 9-1. The smoothed series has a length of 18 years because three values are lost at each end of the series for a smoothing interval of 7 years. From Fig. 9-1 it is evident there is a trend in the smoothed series. There is relatively little variation in the smoothed series prior to the mid-1950s; this variation can be considered random variation. After development became significant in the mid-1950s, the flood peaks appear to be larger, as evident from the nearly linear trend in the smoothed series.

It is important to emphasize two points. First, the moving-average smoothing does not provide a forecast equation. After the systematic trend has been identified as an almost linear secular trend, it would be necessary to fit a representative equation to the data. Second, there is no assurance that the trend evident in the smoothed series is the result of urban development. There is some chance that it is due either to randomness or to another causal factor, such as increased rainfall. Therefore, once a trend has been found to exist and a structure hypothesized, a reason for the trend should be identified. The reasons may suggest a type of model to be used to represent the data.

Example 9-2: Periodic Trends in Rainfall and Runoff Data

To formulate an accurate water-yield time-series model, it is necessary to identify the structure of the variation of water yield with time. The moving-average filtering method can be used in identifying the components of a water-yield time series. For a watershed that has not undergone significant changes in land use, one would not expect a secular trend. Therefore, one would expect the dominant element of the systematic variation to be a periodic trend that is correlated with the seasonal variation in meteorological conditions. For such a component, it would be necessary to identify the period (probably one year) and both the amplitude and phase of the periodic trend. A moving-average filtering of the time series will enable the modeler to get reasonably accurate initial estimates of these elements so that a formal model structure can be fit using numerical optimization (see Chapter 7). For locations where two flooding seasons occur per year, two periodic trends may be indicated by the moving-average filtering. Once the periodic trends have been identified and subtracted from the time series, the methods of Chapter 4 can be used to identify the population that underlies the nonsystematic variation of the residuals.

A record of 18 years of monthly runoff data for the Chestuee Creek near Dial, Georgia, was subjected to a moving-average filtering. The data are given in Appendix C-2, and both the time series and the smoothed series for a smoothing interval of 3 months are shown in Fig. 9-2. The data were also filtered using smoothing intervals of 5, 7, 9, 11, and 13 months. As the smoothing interval was increased, the variation decreased significantly, with little variation evident in the smoothed series for smoothing intervals of 11 and 13 months. Table 9-2 provides the sums of squares of both the smoothed and truncated series, which is defined as the time series with the truncated points at each end of the time series excluded. Table 9-2 also gives the variation of the smoothed series as a percentage of the variation in the truncated series; this percentage reflects the amount of variation that remains in the smoothed series. It is evident from Table 9-2 that a significant portion of the nonsystematic variation is eliminated even for a smoothing interval of 3 months. Figure 9-2 shows that the monthly

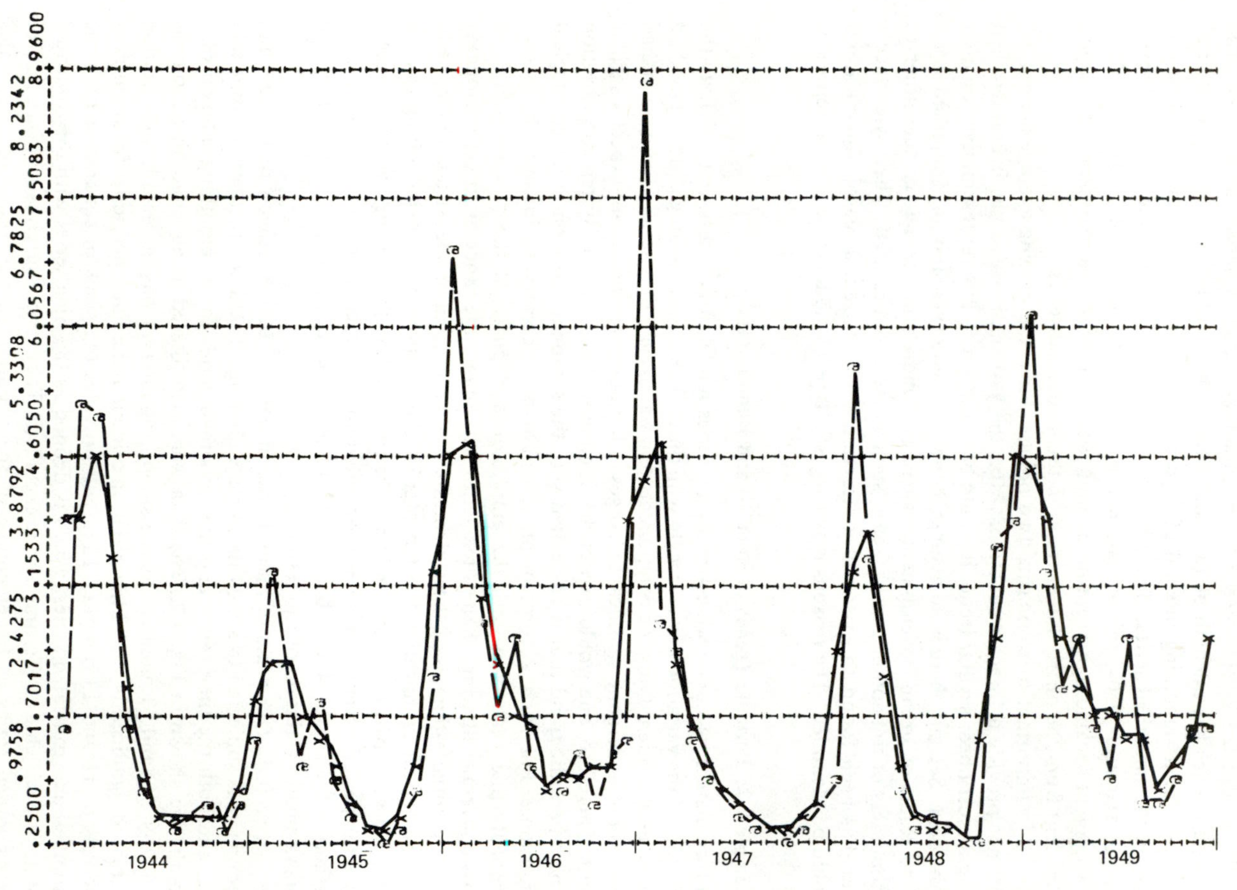

Figure 9-2 Monthly water yield and moving-average filtered series for the Chestuee Creek watershed, 1944–1961. Solid line, moving-average filtered series; dashed line, monthly water yield.

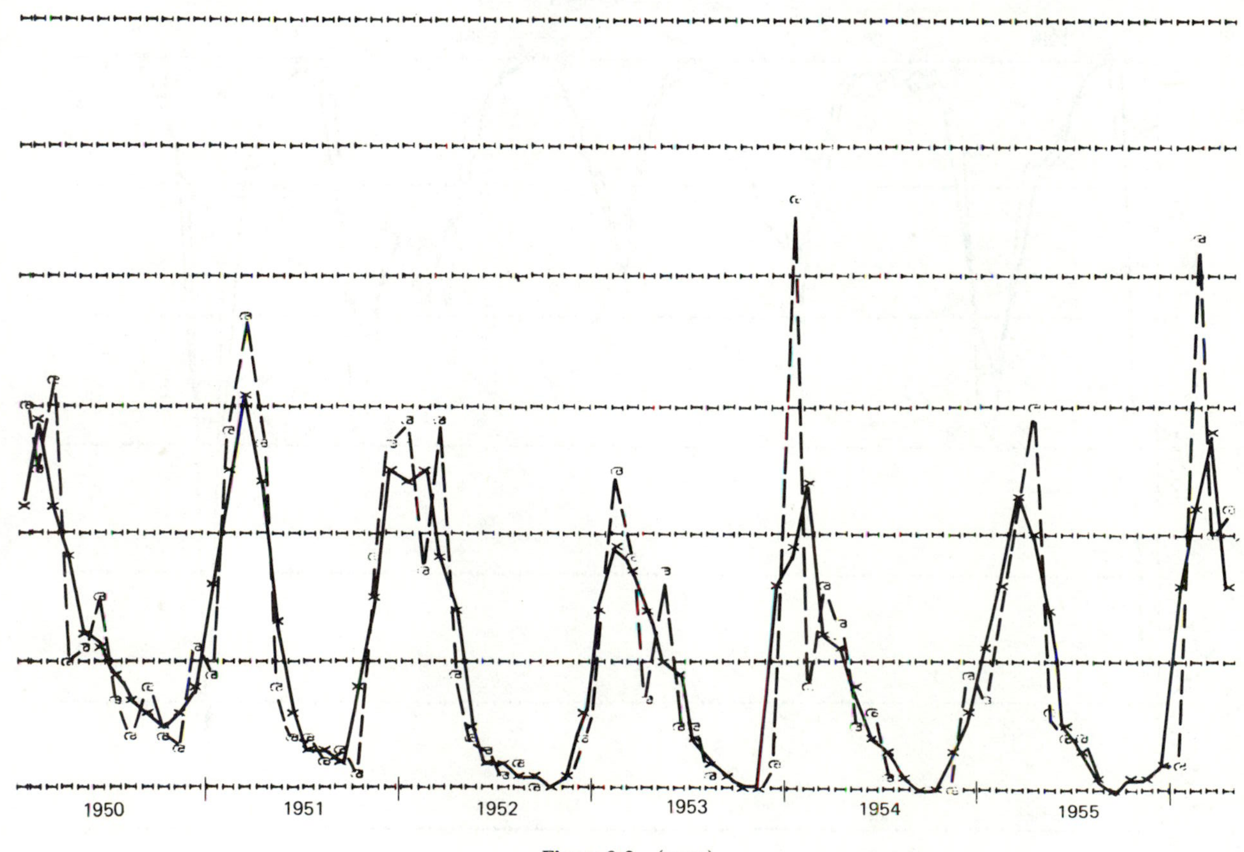

Figure 9-2 (*cont.*).

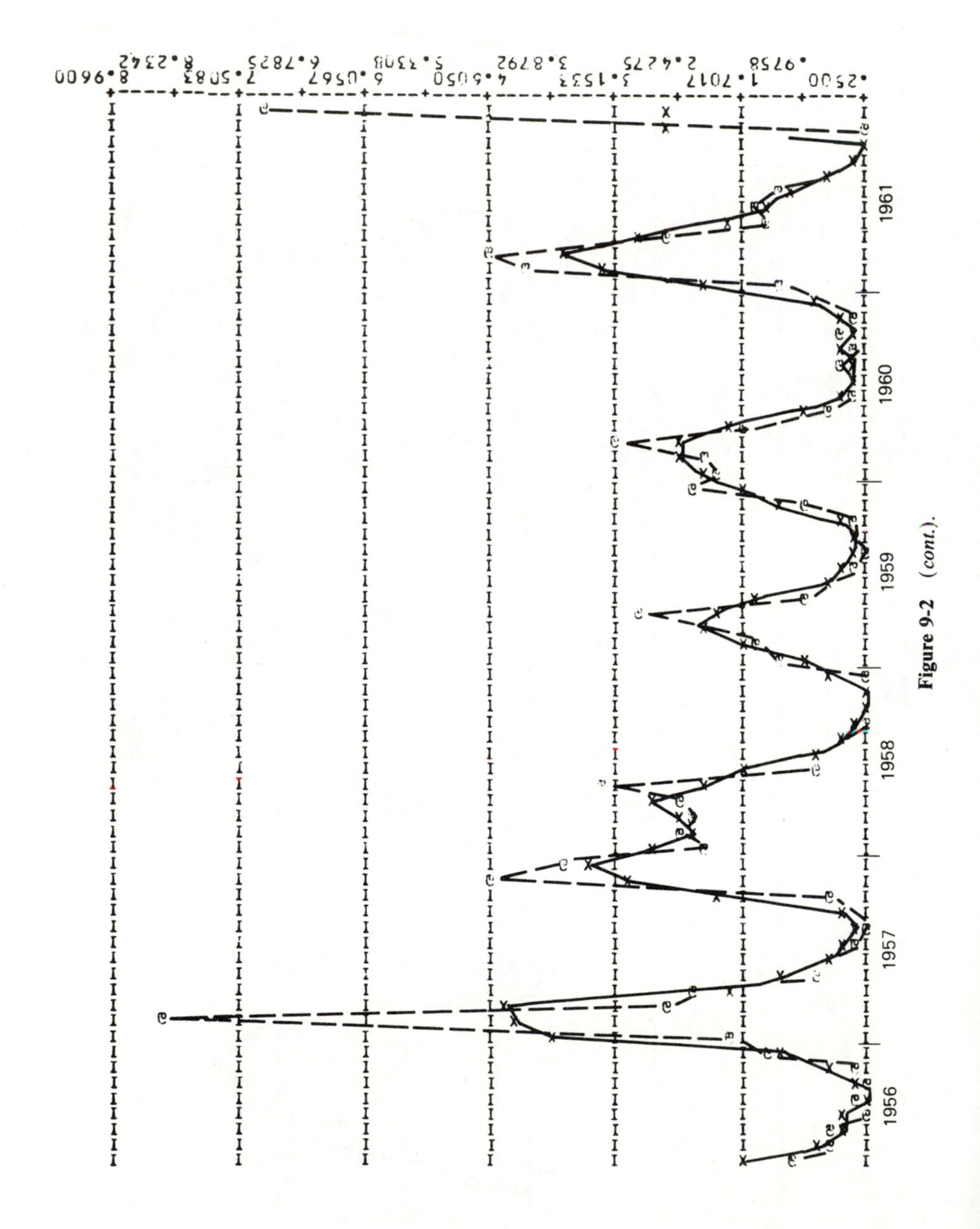

Figure 9-2 *(cont.)*.

TABLE 9-2 Statistics for Moving-Average Filtering of Chestuee Creek Data

Smoothing Interval (months)	Sums of Squares		Goodness of Fit (%)
	Truncated Data[a]	Smoothed Data	
3	563.8	331.4	58.8
5	550.0	201.7	36.7
7	536.2	114.3	21.3
9	531.4	59.4	11.2
11	530.3	34.2	6.5
13	529.3	30.3	5.7

[a]Sum of squares of time series = 594.7.

water yield data are characterized by a dominant annual cycle, with a fairly constant base flow at the trough between each pair of peaks. The smoothed series closely approximates the actual water yield during the dry period of the year. However, there is considerable nonsystematic variation around the peaks. For example, the smoothed peaks for 1944, 1946, 1947, and 1949 are nearly equal, even though the actual monthly peaks for those years show considerable variation.

The smoothed series suggests that the systematic variation can be represented by a sine series with a mean of about 2.5, a period of 12 months, and a phase angle of about 0 months. The residuals would suggest that the nonsystematic variation has a nonconstant variance and thus the population of the nonsystematic variation would be time dependent.

AUTOCORRELATION ANALYSIS

Many time series are characterized by high serial correlation. That is, adjacent values in a time series may be correlated. For example, when the daily streamflow from a large river system is high above the mean flow, we should expect high flows for one or more days following; there is very little chance of having a flow of 100 cfs if the flow the preceding day was 10,000 cfs. This serial correlation is sometimes called *persistence*. As will be indicated later, monthly water yields on larger streams show a high correlation between flows measured in adjacent months. If this serial correlation is sufficiently high, one can take advantage of it in forecasting water yield. The depletion of snow cover is another hydrologic variable that exhibits serial correlation for time intervals of a month or more, depending on the size of the watershed, the amount of snow cover, and the amounts and time distribution of precipitation. In any case, many snowmelt forecast models use this serial correlation to improve the accuracy of forecasts.

The computational objective of autocorrelation analysis is to analyze a time series and determine the degree of correlation in adjacent values. Actually, the analysis usually examines the change in correlation as the separation distance increases. The separation distance is often called the *lag*, and a plot of the correlation coefficient versus time lag is called a *correlogram.*

Computationally, the correlogram is computed by finding the value of the Pearson product-moment correlation coefficient (Eq. 3-23) for lags from 1 time unit to a maximum lag of approximately 10% of the record length. For large samples, the autocovariance function is computed first:

$$V(\tau) = \frac{1}{N-\tau} \sum_{i=1}^{N-\tau} X_i X_{i+\tau} \tag{9-4}$$

in which X_i is the value of the time series for time i, τ the time lag, N the record length, and $V(\tau)$ the autocovariance function. The autocorrelation function, $R(\tau)$, is determined by dividing the autocovariance function by the variance of the time series, S_X^2:

$$R(\tau) = \frac{V(\tau)}{S_X^2} \tag{9-5}$$

For small sample sizes (i.e., small N or large τ) the variance will not be constant with τ. In such cases the variance must be recomputed for each τ. The autocorrelation function is then computed by

$$R(\tau) = \frac{\sum_{i=1}^{N-\tau} X_i X_{i+\tau} - \frac{1}{N-\tau}\left(\sum_{i=1}^{N-\tau} X_i\right)\left(\sum_{i=\tau+1}^{N} X_i\right)}{\left[\sum_{i=1}^{N-\tau} X_i^2 - \frac{1}{N-\tau}\left(\sum_{i=1}^{N-\tau} X_i\right)^2\right]^{0.5}\left[\sum_{i=\tau+1}^{N} X_i^2 - \frac{1}{N-\tau}\left(\sum_{i=\tau+1}^{N} X_i\right)^2\right]^{0.5}} \tag{9-6}$$

Obviously, for τ equal to zero, $R(\tau)$ equals 1. The graph of $R(\tau)$ versus τ is the correlogram. As τ increases, the number of values used to compute $R(\tau)$ decreases, and the correlogram may begin to oscillate. This is the reason for the empirical rule of thumb that suggests limiting the maximum value of τ to approximately 10% of N.

It is instructive to view the computation of $R(\tau)$ schematically. If a time series is offset from itself by one time unit, the correlation coefficient can be used to compute the *one-lag correlation.* This is shown in Fig. 9-3(a). For the one-lag correlation coefficient, one observation is lost (i.e., $N_1 = N - 1$). For computing the two-lag correlation the time series is offset from itself by two time lags (i.e., $\tau = 2$ [Fig. 9-3(b)]). In this case two values are lost, so the record length for $R(2)$ is $N_2 = N - 2$. The values used to compute $R(\tau)$ are shown within the dashed lines in Fig. 9-3. The lagging process could be presented for any τ.

The correlogram shows the degree of persistence in a time series. Thus the correlogram indicates the degree to which past measurements will help forecast future values. If $R(\tau)$ is near zero for all τ, a time-series model

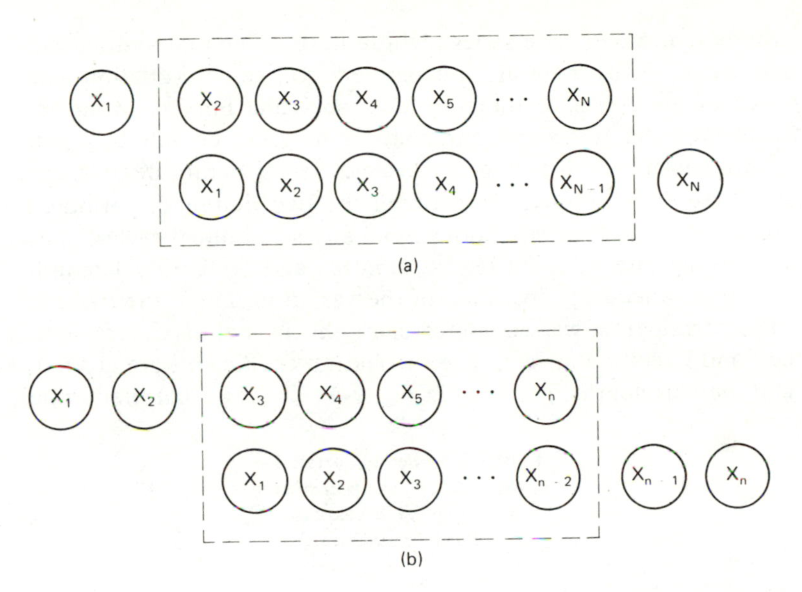

Figure 9-3 Schematics for the computation of the autocorrelation function: (a) one-lag correlation; (b) two-lag correlation.

probably will not provide accurate forecasts. This implies that physical processes that are time dependent are not the ones that have a significant affect on the variation in the criterion variable. If only $R(1)$ is sufficiently high, a one-lag model will provide the best forecast model. When more than one lag is significant, a multilag may be appropriate; however, one must recognize that there is correlation between the one-lag and two-lag values, and thus the intercorrelation may prevent the model based on two lags from being significantly better than a model based on a single lag. It is also possible to have $R(1)$ near zero and yet have $R(\tau)$ highly significant for some other τ. For example, if the time series includes a dominant annual cycle, one would expect $R(12)$ to be high for monthly data. Also, $R(6)$ may be near -1 for the half-year cycle.

As indicated previously, the feasible number of lags depends on the length of the time series, with a 10% maximum being a rule of thumb. Figure 9-3 indicates that as the number of lags increases, the number of observations decreases. From our knowledge of confidence intervals, we know that the width of the confidence interval increases as the number of observations decreases. Thus our confidence in the value of $R(\tau)$ decreases as τ increases. Also, as the length of the time series increases, the accuracy of the correlogram should improve because of the larger number of values used to compute each point of the correlogram. Obviously, for extremely long time series the width of the confidence interval will not change much as τ increases, even beyond 10% of the record length.

Many hydrologic time series include both an annual and a semiannual periodic trend. Thus there are independent sources of variation associated with each of the cycles. If there are an insufficient number of lags between the two cycles, the autocorrelation analysis may not be able to separate the variation associated with each of the cycles. As a general rule, it is preferable to have three or more intervals between the two dominant periodic trends. Thus for hydrologic data with both annual and semiannual cycles, one should use a sampling interval of at least 2 months, and preferably 1 month. This will ensure an adequate separation of the variation due to the two cycles.

The correlogram was computed using the annual maximum flood series for the Pond Creek watershed. The correlation coefficients are given in Table 9-3, and the correlogram is shown in Fig. 9-4. The lag-1 correlation indicates

TABLE 9-3 Serial Correlation Coefficients, (R) as a Function of Time Lag (τ) in Years for Pond Creek Data

τ	R	τ	R
0	1.00	4	0.43
1	0.61	5	0.33
2	0.55	6	0.24
3	0.43		

that about 36% of the variation in an annual maximum discharge can be explained by the value from the previous year. Of course, this is an irrational interpretation because the events are actually independent of each other. Instead, the serial correlation of 0.6 is indicative of the secular trend that occurs in the annual maximum series because of the increase in urbanization. Although correlograms are usually used to detect periodic trends in time series, this example indicates that secular trends in data can also be the cause of significant serial correlations. In this case the significant lag-1 correlation should not be viewed as an indication that a forecast model with time as the predictor variable can be calibrated to predict the annual maximum floods. Instead, the significant correlation only indicates that a third variable (i.e.,

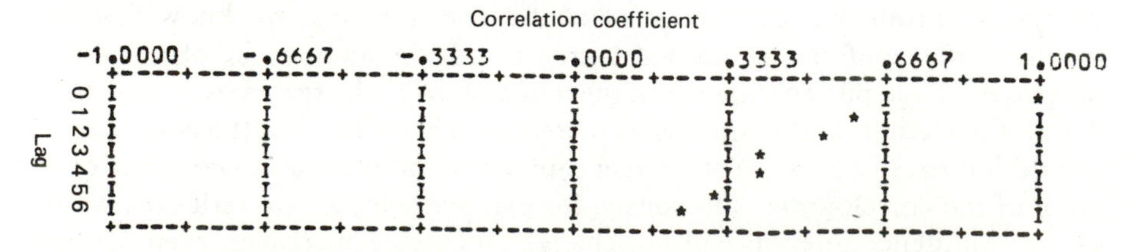

Figure 9-4 Correlogram for the Pond Creek annual maximum flood series.

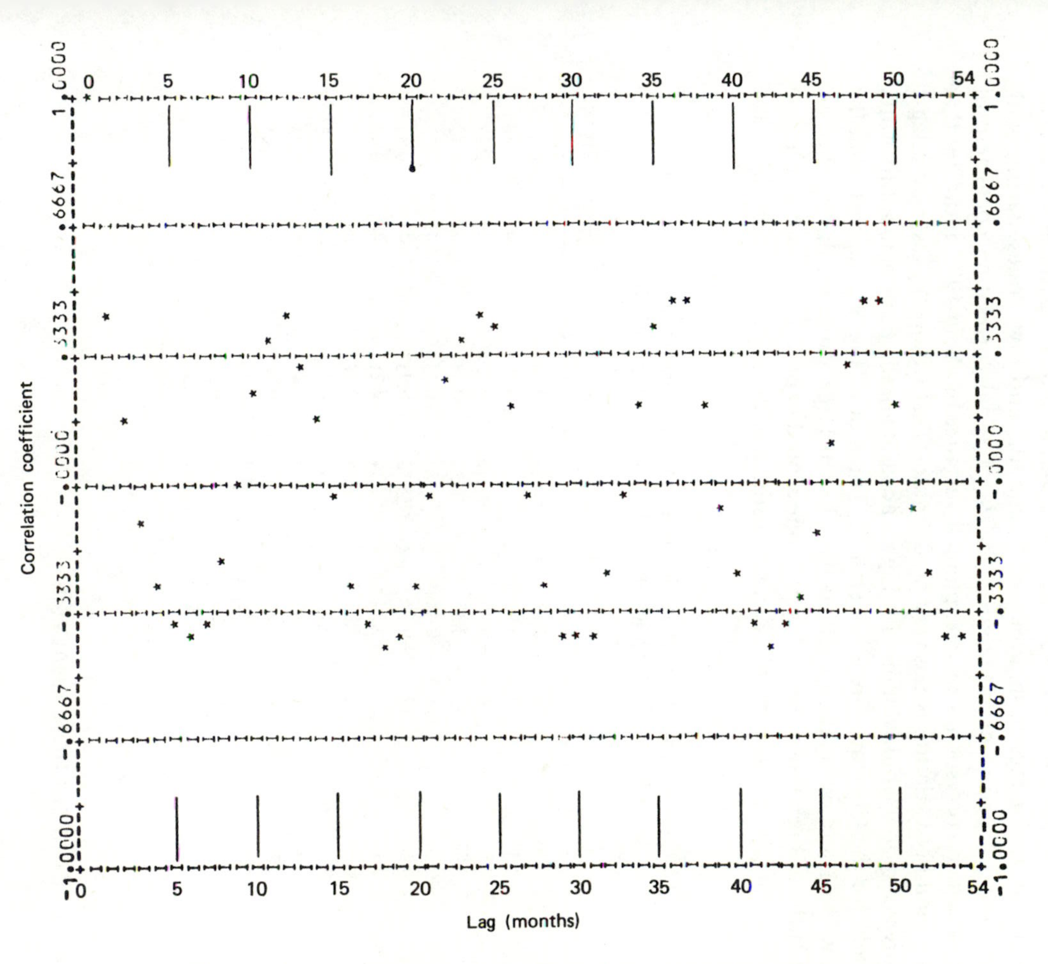

Figure 9-5 Correlogram for the Chestuee Creek monthly streamflow.

urbanization) that is changing systematically with time causes the variation in the annual maximum flood data. If a forecast model is developed, urbanization should be used as the predictor variable rather than time.

The Chestuee Creek water-yield data were also subjected to a serial correlation analysis. The correlogram is shown in Fig. 9-5, with the correlation coefficients given in Table 9-4. The correlogram shows a periodic trend, with high values for lags in multiples of 12 months and low values starting with lag 6 and a period of 12 months. The lag-1 serial correlation coefficient of 0.44, which corresponds to an explained variance of about 19%, indicates that the monthly variation is not very predictable. This would suggest that a lag-1 forecast model would not provide highly accurate predictions of monthly water yield for the Chestuee Creek watershed. The plot of Fig. 9-2 suggests that the low lag-1 autocorrelation coefficient is due to large variation during the few months of high flows. From Fig. 9-2, one would expect a lag-1 autoregressive model to provide reasonable accurate estimates during the months with low flows.

TABLE 9-4 Serial Correlation Coefficients (R) as a Function of Time Lag (τ) in Months for the Chestuee Creek Monthly Water-Yield Data

τ	$R(\tau)$	τ	$R(\tau)$	τ	$R(\tau)$	τ	$R(\tau)$
0	1.00	14	0.18	28	−0.24	42	−0.41
1	0.44	15	−0.03	29	−0.38	43	−0.36
2	0.19	16	−0.26	30	−0.39	44	−0.27
3	−0.08	17	−0.35	31	−0.38	45	−0.12
4	−0.24	18	−0.41	32	−0.23	46	0.13
5	−0.34	19	−0.37	33	−0.03	47	0.33
6	−0.37	20	−0.24	34	0.21	48	0.49
7	−0.34	21	−0.03	35	0.42	49	0.49
8	−0.17	22	0.27	36	0.47	50	0.22
9	0.01	23	0.39	37	0.47	51	−0.04
10	0.23	24	0.43	38	0.20	52	−0.23
11	0.39	25	0.41	39	−0.06	53	−0.37
12	0.47	26	0.20	40	−0.22	54	−0.39
13	0.32	27	−0.01	41	−0.36		

CROSS-CORRELATION ANALYSIS

In some situations, one may wish to compute or forecast values for one time series using values from a second time series. For example, we may wish to use rainfall for time t (i.e., P_t) to predict runoff at time t (i.e., RO_t). Or we may wish to predict the runoff at a downstream site (i.e., Y_t) from the runoff at an upstream site for the preceding time period (i.e., X_{t-1}). The objective of a cross-correlation analysis is to identify the correlation, and thus the predictability, between two time series. Cross-correlation analysis is computationally similar to autocorrelation analysis except that two time series are involved rather than one time series offset from itself. This produces two distinct differences between auto- and cross-correlation. First, whereas the autocorrelation coefficient for lag 0 must be 1, the cross-correlation coefficient for lag 0 can take on any value between −1 and 1. In fact, the peak of the correlogram for two time series may peak at a value other than lag zero, especially when there is a definite lag between the two time series. For example, the correlogram between stream gages on the same river would probably have a peak on the correlogram corresponding to the time lag equal to the travel time of flows between the two gages. A second distinguishing characteristic of the cross-correlogram is that one must compute the correlations for both positive and negative lags. While the autocorrelation function is a mirror image about lag $\tau = 0$, the cross-correlation function is not a mirror image. This is evident from Fig. 9-6. For a lag of 1, the values of Y_1 and X_n are not

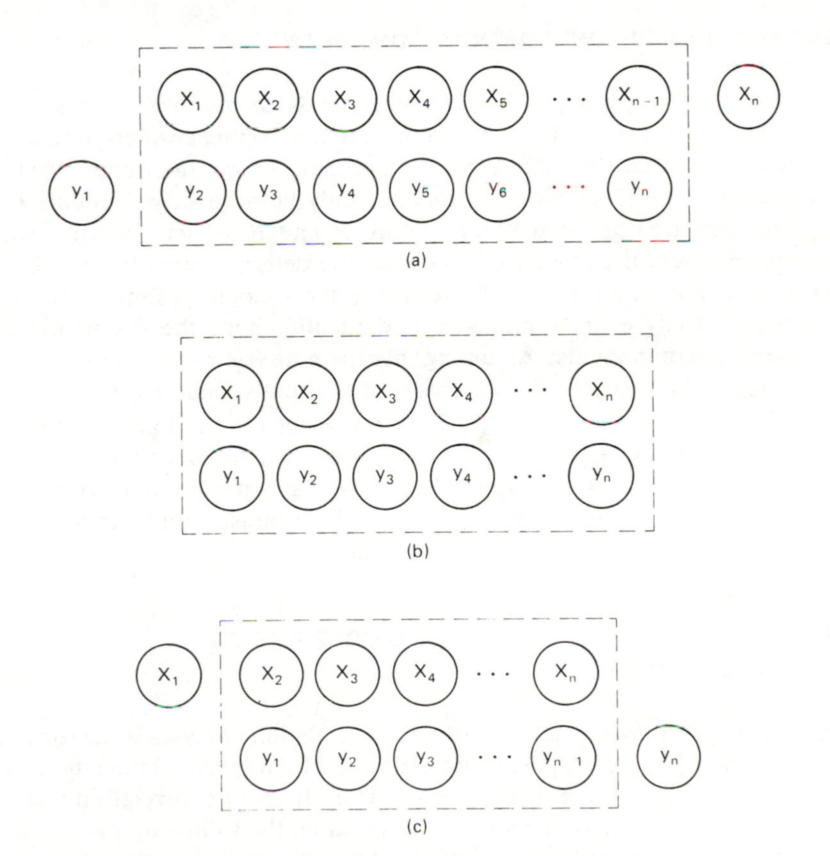

Figure 9-6 Schematics for calculating the cross-correlation coefficient: (a) $\tau = +1$; (b) $\tau = 0$; (c) $\tau = -1$.

used and X_i is paired with Y_{i+1}. For a lag of -1, the values of X_1 and Y_n are not used and X_i is paired with Y_{i-1}.

The cross-correlation coefficient $r(\tau)$ can be computed with Eq. 9-6 by substituting $Y_{i+\tau}$ for $X_{i+\tau}$:

$$r(\tau) = \frac{\sum_{i=1}^{N-|\tau|} X_i Y_{i+\tau} - \frac{1}{N-|\tau|}\left(\sum_{i=1}^{N-|\tau|} X_i\right)\left(\sum_{i=\tau+1}^{N} Y_i\right)}{\left[\sum_{i=1}^{N-|\tau|} X_i^2 - \frac{1}{N-|\tau|}\left(\sum_{i=1}^{N-|\tau|} X_i\right)^2\right]^{0.5}\left[\sum_{i=1+|\tau|}^{N} Y_i^2 - \frac{1}{N-|\tau|}\left(\sum_{i=1+|\tau|}^{N} Y_i\right)^2\right]^{0.5}} \tag{9-7}$$

It is important to note that the absolute value of τ is used in Eq. 9-7 and that the computations must be performed so that they correspond to those shown in Fig. 9-6.

IDENTIFICATION OF THE RANDOM COMPONENT

For some types of problems it is necessary only to identify the systematic components of the time series. For example, in developing a water-yield model that relates the volume of runoff for any month to the rainfall depths for both the current and previous months, it would only be necessary to identify the systematic variation between RO_t and both P_t and P_{t-1}. The resulting water yield model would then be based solely on the deterministic elements of the time series. For other types of problems, most notably simulation, it is necessary to analyze the time series and identify both the systematic and nonsystematic components. Assuming that the nonsystematic variation is due entirely to random variation (i.e., noise), it is usually sufficient to identify a probability function that can be used to represent the random variation. A probability function can be identified using either a frequency analysis or a moment-ratio analysis (see Chapter 4). In some cases it is possible to calibrate simultaneously both the coefficients of the deterministic and random components using a numerical optimization strategy.

FORMULATION OF AUTO- AND CROSS-CORRELATION WATER-YIELD MODELS

A primary use of autocorrelation and cross-correlation analyses is the formulation of water-yield models. The volume of runoff for a specified time increment (e.g., hour or month) is the dependent variable. If an autocorrelation analysis indicates that there is a significant lag correlation, the following model might be a reasonable representation of the deterministic component:

$$RO_t = b_0 + b_1 RO_{t-1} \tag{9-8}$$

in which b_0 and b_1 are coefficients that must be fit to measured data, and RO_t and RO_{t-1} are the volumes of runoff for time periods t and $t-1$, respectively. Equation 9-8 assumes that the characteristics of RO for time periods t and $t-1$ are the same. If t represents monthly values, it is possible that the mean and standard deviation of runoff volumes may vary with the month. If so, it is necessary to generalize the model of Eq. 9-8. If we assume that the deviation of RO_t from the mean runoff for that month, $\overline{RO}_t$, is proportional to the deviation of RO_{t-1} from the mean runoff for that month, Eq. 9-8 can be generalized as

$$RO_t = \overline{RO}_t + b_j(RO_{t-1} - \overline{RO}_{t-1}) \tag{9-9}$$

Whereas Eq. 9-8 has two fitting coefficients, Eq. 9-9 uses the additional information of the means but only a single fitting coefficient, b_j. With Eq. 9-9 it is necessary to evaluate b_j for each time period; for example, if t is a period of one month, there would be 12 values of b_j, in addition to the 12 monthly means, $\overline{RO}_t$. It is important to ensure that the accuracy of the distributed model of Eq. 9-9 is significantly greater than that of the lumped model of Eq. 9-8 before electing to use the more complex model. If the lag correlations for lags greater than one are significant, a multilag model may be more accurate. For example, the two-lag model alternative to Eq. 9-8 would be

$$RO_t = b_0 + b_1 RO_{t-1} + b_2 RO_{t-2} \tag{9-10}$$

The inclusion of the additional term increases the number of coefficients that must be fit, as well as introducing the potential for intercorrelation between RO_{t-1} and RO_{t-2}. The inclusion of the RO_{t-2} term can be tested for statistical significance using a partial F test.

The same concepts can lead to cross-correlation water-yield models. For example, for smaller watersheds, the monthly runoff volumes may not exhibit high serial correlation. Therefore, a model based on the correlation between rainfall and runoff may provide more accurate estimates of monthly yield:

$$RO_t = b_0 + b_1 P_t \tag{9-11}$$

A multilag model may also prove to be more accurate:

$$RO_t = b_0 + b_1 P_t + b_2 P_{t-1} \tag{9-12}$$

Also, it may be necessary to use a time-distributed model to make accurate projections:

$$RO_t = \overline{RO}_t + b_j(P_t - \bar{P}_{tj}) \tag{9-13}$$

in which the subscript j indicates the month, or season, of the values. Again, the time-distributed model of Eq. 9-13 requires sufficient data for calibrating the coefficient b_j for each time period.

AUTOREGRESSION AND CROSS-REGRESSION

Having formulated a model based on either autocorrelation or cross-correlation analysis, it is necessary to estimate, or fit, the unknowns. Given that regression theory is widely accepted and that unbiased models are desirable, calibration usually involves a regression analysis of the data. For lumped models, such as Eqs. 9-8, 9-10, 9-11, and 9-12, the entire data set is used to calibrate the unknowns. For the time-distributed models, such as Eqs. 9-9 and 9-13, it is necessary to do a separate regression analysis for each time period, or else specify a functional relation among the time periods, thus pooling the data for evaluation of a seasonally continuous model. In addition to the linear models of Eqs. 9-8 to 9-13, logarithmic models (i.e., $\log RO_i$ replacing RO_i and $\log P_i$ replacing P_i) may prove to be better representations of the causal processes. When a model contains more than one predictor variable (e.g., Eqs. 9-10 and 9-12), partial F tests can be used to test the statistical significance of the terms.

Computational Example

Assume that the values in Table 9-5 represent a hydrologic variable such as water yield. The hypothetical data consist of 5 years of record, with values given for three seasons. The one-lag and two-lag correlation coefficients are 0.193 and 0.023, respectively. The one-lag autocorrelation and autoregression are based on 14 pairs, while the two-lag autocorrelation and autoregression are based on only 13 pairs. The one-lag autoregression equation, which is based on the general form of Eq. 9-9, is

$$\hat{X}_{t+1} = 7.929 + 0.190(X_t + 8) \tag{9-14}$$

The intercept of 7.929 does not equal the mean of the 15 values of Table 9-5 because the first value (i.e., the season 1 value in 1970) was not used as a value of X_{t+1} in deriving the regression equation. The one-lag autocorrelation coefficient of 0.193 indicates that the deterministic model of Eq. 9-14 will not provide accurate predictions. The same impression would be evident if X_{t+1} were plotted versus X_t.

TABLE 9-5 Data Base for Seasonally Distributed Hydrologic Variable

	Year						
Season	1970	1971	1972	1973	1974	Mean	Variance
1	9	7	9	10	5	8	4
2	6	6	8	9	6	7	2
3	10	8	11	8	8	9	2
						8	2.286

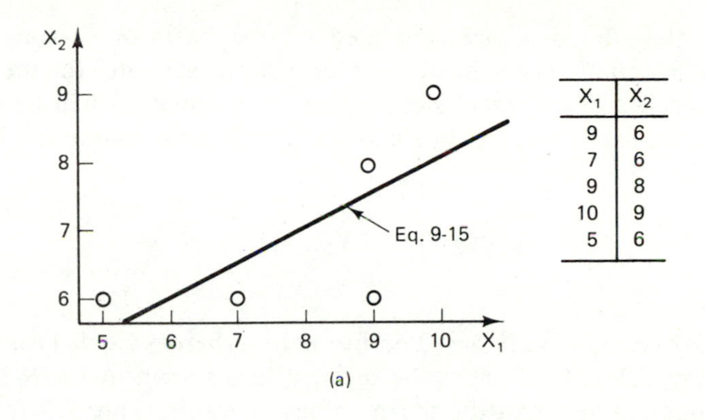

X_1	X_2
9	6
7	6
9	8
10	9
5	6

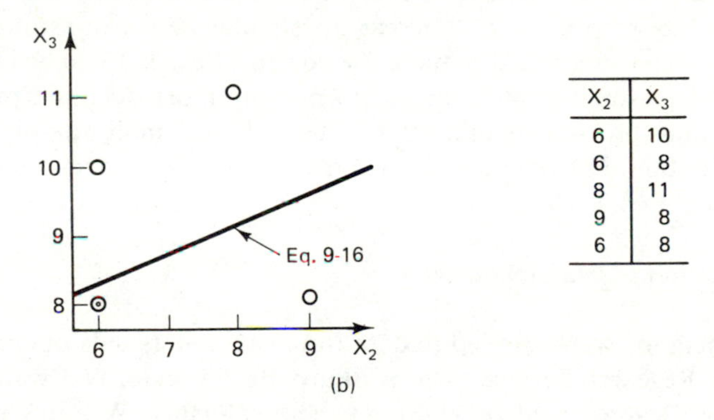

X_2	X_3
6	10
6	8
8	11
9	8
6	8

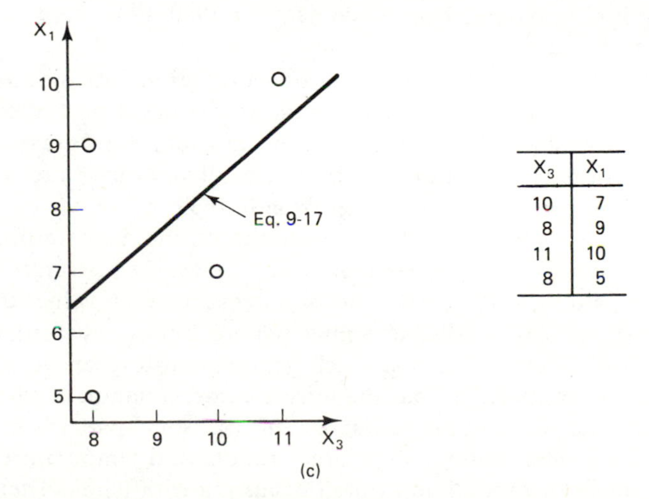

X_3	X_1
10	7
8	9
11	10
8	5

Figure 9-7 Autoregression relationships for seasonal model: (a) X_2 versus X_1; (b) X_3 versus X_2; (c) X_1 versus X_3.

Given that the data are separated on the basis of season, separate autoregression equations can be derived for making estimates for each season. However, because of the small sample size, each equation will be based on an even smaller sample size than Eq. 9-14. The autoregression models are

$$\hat{X}_2 = 7 + 0.5(X_1 - 8) \tag{9-15}$$

$$\hat{X}_3 = 9 + 0.125(X_2 - 7) \tag{9-16}$$

$$\hat{X}_1 = 7.75 + 0.7778(X_3 - 9.25) \tag{9-17}$$

Equations 9-15 and 9-16 are based on five pairs, whereas Eq. 9-17 is based on only four pairs (Fig. 9-7). Thus the mean values shown in Eq. 9-17 are for the four values rather than the mean values given in Table 9-5 for the five points. Equations 9-15, 9-16, and 9-17 had correlation coefficients of 0.707, 0.125, and 0.526, respectively. Whereas the standard error of estimate for Eq. 9-14 is 1.741, the combined standard error for Eqs. 9-15 to 9-17 is 1.317. Although the seasonal models appear to provide more accurate predictions, the correlations are not significant because of the small sample sizes. In practice, one should be concerned when the sample sizes are as small as those used to derive Eqs. 9-15 to 9-17.

Application to Watershed Data

The methods were applied to data from two watersheds operated by the Agricultural Research Service. The Watkinsville, Georgia, W-1 watershed has an area of 19.2 acres, and the Taylor Creek watershed W-2 in Vero Beach, Florida, has an area of 98.6 square miles. Data for 1956–1966 were available for the Taylor Creek watershed, while data for 1940–1964 were available for the Watkinsville watershed.

Figure 9-8 shows the auto- and cross-correlations for monthly rainfall and runoff for the Taylor Creek watershed. In the lower part of Fig. 9-8 the autocorrelation coefficient for monthly mean temperature and cross-correlation coefficients between temperature and both rainfall and runoff are shown. The coefficients were computed for all lags from 0 to 39.

The strong seasonal pattern of rainfall, runoff, and temperature is evident from the sinusoidal waves with a period of 12 lags. Temperature shows the most distinct pattern, with correlations oscillating between values of −0.9 and +0.9. Rainfall also shows a distinct annual pattern, but values oscillate between −0.4 and +0.4, which indicates a much greater year-to-year variability than temperature. Monthly runoff has the lowest autocorrelation coefficients, and hence the greatest year-to-year variability. The seasonal patterns of the cross-correlation coefficients show that rainfall, runoff, and temperature are all in phase, with highs of rainfall and runoff occurring with highs of temperature. This is an indication of the semitropical climate of Florida, where the summer season is the wet season of the year.

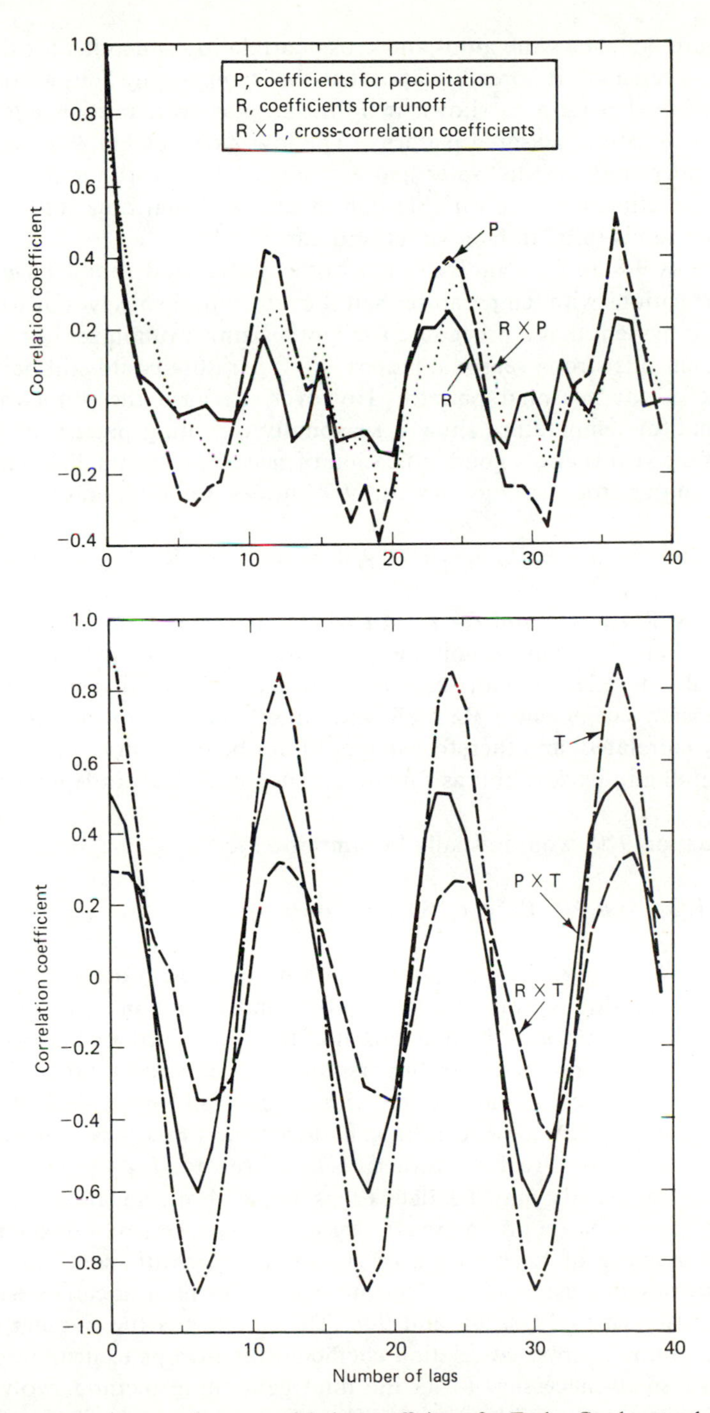

Figure 9-8 Auto- and cross-correlation coefficients for Taylor Creek watershed.

Figure 9-9 shows the auto- and cross-correlation functions for the Watkinsville watershed. It is readily apparent that temperature follows the same strong seasonal pattern as shown in Florida. However, rainfall and runoff show only indistinct seasonal patterns. The lower part of Fig. 9-9 shows that rainfall and runoff for this watershed are not in phase with temperature. In a temperate climate under variable continental and maritime influence, the wet season tends to be in late winter and early spring.

Figures 9-8 and 9-9 indicate that both rainfall and runoff have higher serial correlations with temperature than they do with themselves or with each other. It is evident that a procedure for synthesizing a runoff series must take into account this strong seasonal pattern. Temperature could still be used as an index of the seasonal pattern. However, because the autocorrelation coefficients for temperature show a seasonally repeating pattern of 0.9, the month of the year is also a good indication of season and is much easier to use.

The information in Figs. 9-8 and 9-9 implies a regression of the type

$$\mathrm{RO}_M = a + b_1 P_M + c_1\mathrm{RO}_{M-1} + b_2 P_{M-1} + c_2\mathrm{RO}_{M-2} + \cdots \tag{9-18}$$

In Eq. 9-18, RO_M is runoff for month M, P_M is precipitation for month M, and RO_{M-1} and P_{M-1} are runoff and precipitation for the antecedent month. Additional antecedent months can be included. The coefficients a, b, and c are regression coefficients. Rainfall and runoff during any month are significantly correlated, and therefore it should not be necessary to include both back rainfall and back runoff as functional (not statistical) independent variables.

Equation 9-18 would usually be simplified to

$$\mathrm{RO}_M = a + b_1 P_M + c_1\mathrm{RO}_{M-1} + c_2\mathrm{RO}_{M-2} + c_3\mathrm{RO}_{M-3} + \cdots \tag{9-19}$$

Equation 9-19 implies an operating position for prediction as follows. A rainfall record, P_M, is available by months, and values of runoff for some number of months prior to the first month of rainfall record are also available. Runoff can then be predicted for the first month of the rainfall record. Using this synthesized runoff as back runoff with the rainfall for the second month, runoff for the second month can be generated; the process is then repeated for the full length of a real or synthetic rainfall record, P_M.

Small watersheds may be defined as those drainage areas for which soil-water processes are predominant over channel processes. For such watersheds the number of back values of runoff that can influence runoff of a succeeding month must be small. The number of back months can be estimated from graphs such as Figs. 9-8 and 9-9. The number is the number of lags before the auto- or cross-correlation coefficient first drops to near zero.

It is usually necessary to let the final generating method evolve from systematic examination of the data set. Therefore, all the regression coefficients

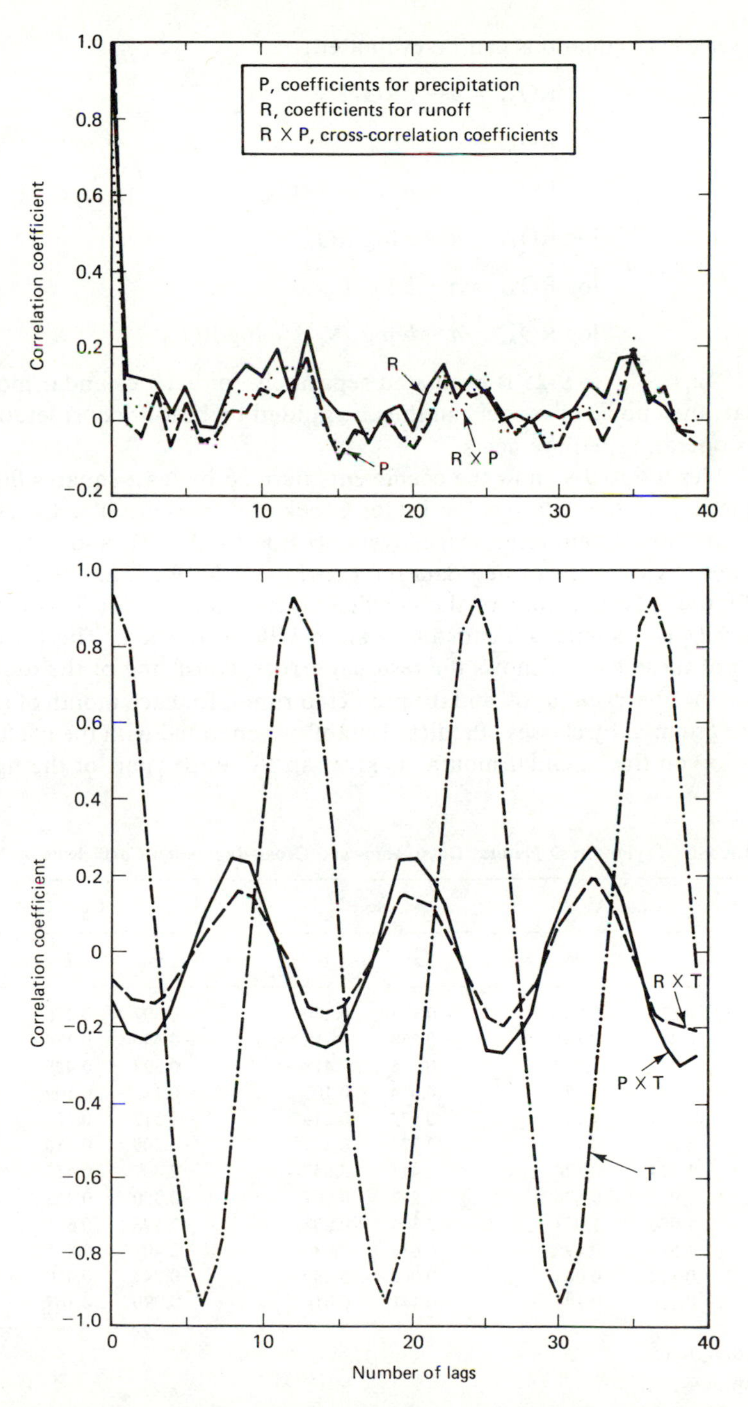

Figure 9-9 Auto- and cross-correlation coefficients for Watkinsville watershed.

in the set of six equations can be evaluated:

$$\mathrm{RO}_M = a + c\mathrm{RO}_{M-1} \tag{9-20}$$

$$\mathrm{RO}_M = a + bP_M \tag{9-21}$$

$$\mathrm{RO}_M = a + bP_M + c\mathrm{RO}_{M-1} \tag{9-22}$$

$$\log \mathrm{RO}_M = a + c \log \mathrm{RO}_{M-1} \tag{9-23}$$

$$\log \mathrm{RO}_M = a + b \log P_M \tag{9-24}$$

$$\log \mathrm{RO}_M = a + b \log P_M + c \log \mathrm{RO}_{M-1} \tag{9-25}$$

Each of Eqs. 9-20 to 9-25 is evaluated separately for each calendar month of the year; the choice of model can thus be guided by best-fit considerations as well as operating requirements.

Tables 9-6 to 9-9 show the coefficients derived by least-squares fitting of the equations to the data sets for Taylor Creek and Watkinsville. Cases A, B, and C for natural data refer, respectively, to Eqs. 9-20, 9-21, and 9-22. Cases A, B, and C with transformed data refer to Eqs. 9-23, 9-24, and 9-25.

Figure 9-10 is a plot of the coefficients of Eq. 9-22 for Taylor Creek. Figure 9-11 is a similar plot for the Watkinsville watershed. The lower part of each of these figures shows the residual errors, consisting of the difference between the observed runoff and the predicted runoff for each month of record; they are grouped by classes. Predicted runoff is generated with the coefficients depending on the calendar month, as given in the upper part of the figure.

TABLE 9-6 Taylor Creek Natural Data: Auto- and Cross-Regression Coefficients by Month

	Case A[a]		Case B[b]		Case C[c]		
Month	a	c	a	b	a	b	c
Jan.	0.339	0.935	−0.105	0.270	−0.307	0.276	1.101
Feb.	0.481	0.100	−0.268	0.313	−0.446	0.334	0.279
Mar.	0.387	1.022	−0.625	0.478	−0.894	0.428	0.819
Apr.	0.148	0.078	−0.034	0.102	−0.133	0.109	0.088
May	−0.061	1.914	−0.572	0.219	−0.512	0.124	1.561
June	1.696	−0.114	−2.771	0.579	−3.308	0.616	0.710
July	1.474	0.124	−2.411	0.657	−2.407	0.655	0.006
Aug.	0.937	0.676	−1.257	0.507	−0.950	0.352	0.422
Sept.	1.000	1.067	−2.365	0.802	−2.598	0.676	0.535
Oct.	1.531	0.195	−0.636	0.699	−2.401	0.805	0.418
Nov.	0.177	0.061	−0.060	0.289	−0.245	0.310	0.073
Dec.	0.116	0.149	0.141	0.014	0.090	0.167	0.151

[a]Equation 9-20.

[b]Equation 9-21.

[c]Equation 9-22.

TABLE 9-7 Taylor Creek Log-Transformed Data: Auto- and Cross-Regression Coefficients by Month

	Case A[a]		Case B[b]		Case C[c]		
Month	a	c	a	b	a	b	c
Jan.	−0.359	0.547	−1.935	0.999	−0.883	0.974	0.495
Feb.	−0.248	0.815	−2.364	1.168	−1.067	1.015	0.728
Mar.	−0.089	0.811	−2.831	1.536	−1.547	1.561	0.824
Apr.	−1.337	0.600	−2.728	0.702	−1.932	0.747	0.605
May	−1.005	0.410	−3.465	1.156	−2.575	1.145	0.405
June	0.263	0.469	−7.190	3.335	−6.281	3.302	0.446
July	0.161	0.643	−7.231	3.888	−5.646	3.101	0.271
Aug.	−0.009	1.029	−5.875	3.075	−3.221	1.732	0.837
Sept.	0.531	0.437	−3.559	2.178	−3.005	1.917	0.302
Oct.	−0.440	0.678	−1.069	0.832	−1.528	0.969	0.734
Nov.	−1.678	0.420	−1.752	0.361	−1.682	0.552	0.478
Dec.	−1.207	0.502	−2.127	0.396	−1.316	0.334	0.461

[a]Equation 9-23.
[b]Equation 9-24.
[c]Equation 9-25.

TABLE 9-8 Watkinsville Natural Data: Auto- and Cross-Regression Coefficients by Month

	Case A[a]		Case B[b]		Case C[c]		
Month	a	c	a	b	a	b	c
Jan.	0.164	1.410	−1.021	0.326	−1.014	0.272	1.065
Feb.	0.444	−0.061	−0.482	0.195	−0.552	0.203	0.060
Mar.	0.417	0.522	−1.440	0.349	−1.451	0.337	0.193
Apr.	0.243	0.457	−0.632	0.265	−0.695	0.233	0.327
May	0.053	0.302	−0.202	0.118	−0.432	0.134	0.331
June	0.272	−0.105	−0.774	0.281	−0.771	0.281	−0.011
July	0.439	0.143	−1.210	0.334	−1.214	0.333	0.030
Aug.	0.422	−0.077	−0.734	0.302	−0.735	0.303	0.002
Sept.	0.012	0.033	0.005	0.007	−0.036	0.015	0.040
Oct.	0.061	0.553	−0.102	0.069	−0.107	0.068	0.337
Nov.	0.352	−0.513	−0.822	0.346	−0.921	0.358	0.792
Dec.	0.254	−0.037	−0.406	0.147	−0.394	0.146	−0.031

[a]Equation 9-20.
[b]Equation 9-21.
[c]Equation 9-22.

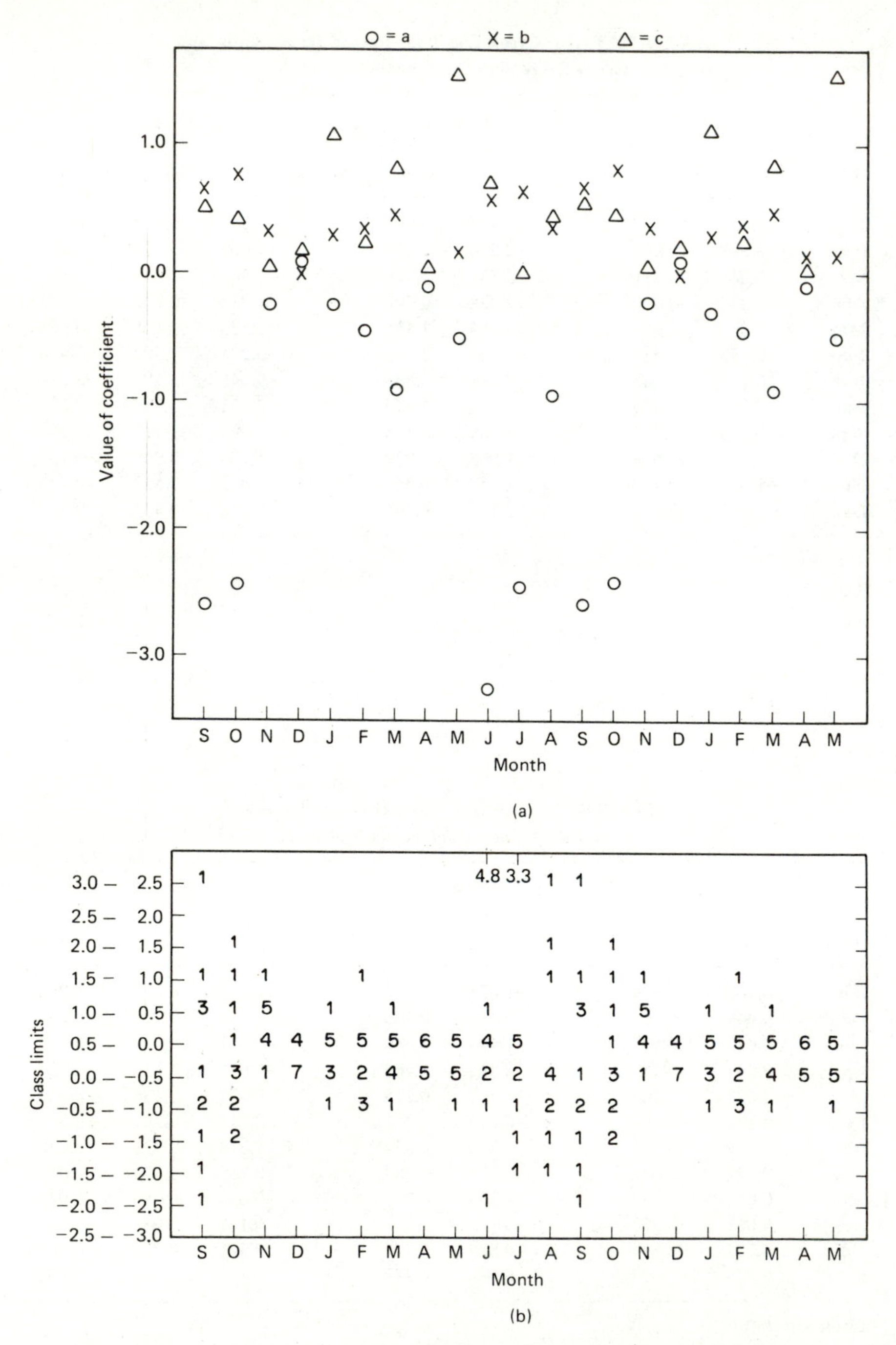

Figure 9-10 Monthly (a) regression coefficients and (b) residual errors by classes for Taylor Creek watershed.

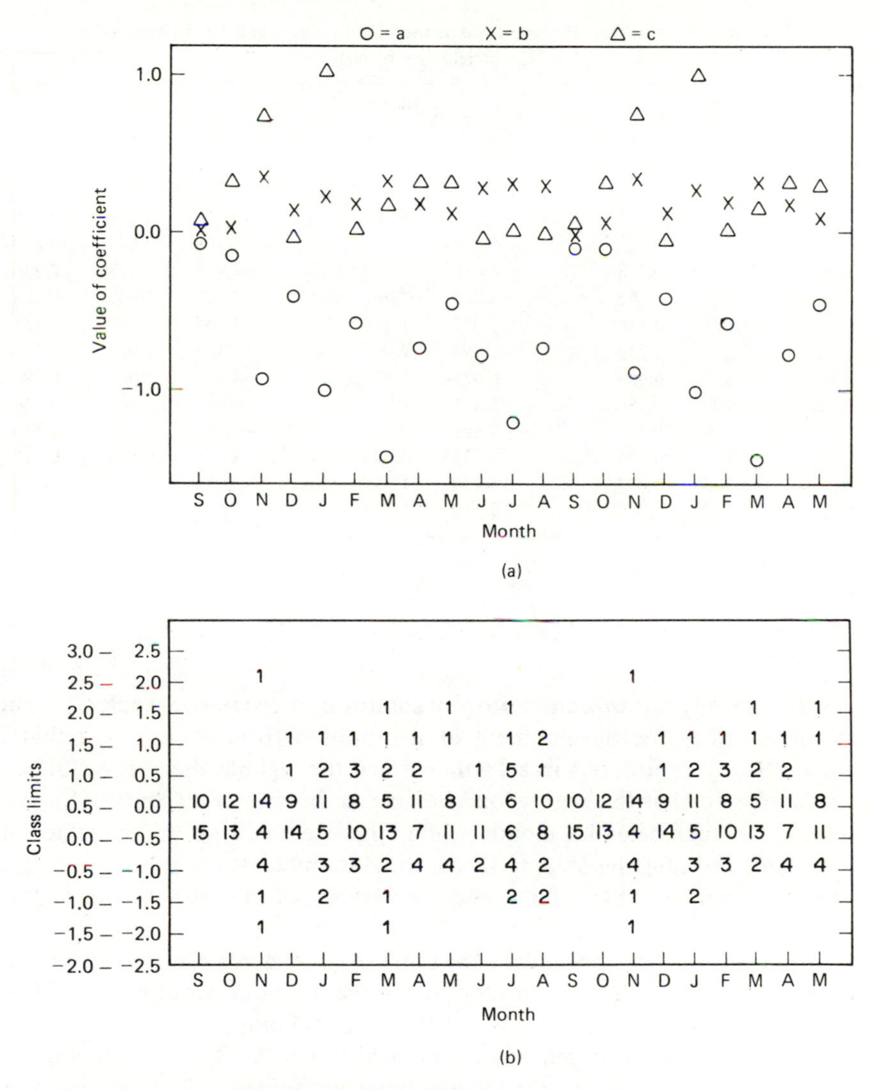

Figure 9-11 Monthly (a) regression coefficients and (b) residual errors by classes for Watkinsville watershed.

The coefficients in Figs. 9-10 and 9-11 exhibit in a crude way the seasonal patterns of Figs. 9-4 and 9-5. The pattern is irregular because the statistical averaging in Figs. 9-8 and 9-9 is over only approximately one-twelfth of the data points in Figs. 9-8 and 9-9. For example, a lag-1 correlation coefficient is based on all pairs January–February, February–March, March–April, and so on, through the years of record. The monthly regression coefficients for a calendar month, however, are based only on observations for that calendar month in the record.

TABLE 9-9 Watkinsville Log-Transformed Data: Auto- and Cross-Regression Coefficients by Month

	Case A[a]		Case B[b]		Case C[c]		
Month	*a*	*c*	*a*	*b*	*a*	*b*	*c*
Jan.	−1.416	0.374	−7.532	3.402	−6.556	3.258	0.238
Feb.	−1.825	0.183	−5.819	2.478	−5.306	2.699	0.320
Mar.	−1.209	0.318	−8.656	3.922	−7.914	3.863	0.279
Apr.	−1.763	0.242	−4.768	1.965	−4.337	1.935	0.202
May	−3.228	0.130	−5.057	1.364	−4.765	1.365	0.131
June	−2.785	0.219	−5.894	2.027	−5.163	2.014	0.203
July	−0.654	0.669	−6.673	2.437	−4.248	2.004	0.500
Aug.	−2.902	0.180	−5.411	1.822	−4.952	1.802	0.144
Sept.	−3.439	0.187	−4.595	0.532	−3.884	1.006	0.338
Oct.	−3.115	0.225	−4.333	0.593	−3.475	0.589	0.210
Nov.	−4.143	−0.097	−5.150	1.515	−3.957	1.815	0.364
Dec.	−3.212	−0.020	−7.180	2.911	−7.604	2.938	−0.103

[a]Equation 9-23.
[b]Equation 9-24.
[c]Equation 9-25.

Seasonally continuous autoregression and cross-regression. This section describes the development of a modified form of Eq. 9-22 that is mathematically continuous in a cyclic pattern throughout the year. It should be remembered that the regression coefficients in Figs. 9-10 and 9-11 are in discrete sets, each calendar month producing a set. A total of 36 coefficients are needed, and must be obtained from the data by least squares. There is no continuity from month to month, and the patterns of the coefficients are quite irregular.

Equation 9-22 can be modified by making coefficients *a*, *b*, and *c* functions of calendar month. In such a modified equation, one would normally be required to estimate the parameters of the three calendar-month functions. A simpler and more direct method of establishing continuity of the coefficients is by fitting an interpolation function by least squares. The interpolating function used here was continuous parabolic interpolation modified to a cyclic form.

Consider the arrangement of monthly values of the coefficient *a* of Eq. 9-22, shown in Table 9-10. Base values are located as shown, one every 4 months. Continuous parabolic interpolation requires four base values for an interpolated value. The values for a_2, a_3, and a_4 are based on a_9, a_1, a_5, and a_9. It can be seen that a seasonally continuous cyclic system of interpolation is produced.

Tables 9-11 and 9-12 are listings of the base values, defined in Table 9-10, for the three seasonal functions for the coefficients of Eq. 9-22. Several

TABLE 9-10 Schematic for Seasonally Continuous Interpolation

Month	Base Value	Interpolated Value
Sept.	a_9	
Oct.		
Nov.		
Dec.		
Jan.	a_1	
Feb.		a_2
Mar.		a_3
Apr.		a_4
May	a_5	
June		a_6
July		a_7
Aug.		a_8
Sept.	a_9	
Oct.		a_{10}
Nov.		a_{11}
Dec.		a_{12}
Jan.	a_1	
Feb.		
Mar.		
Apr.		
May	a_5	

different determinations of these seasonal functions were made, using different fitting criteria. The criteria for and deviations of coefficients are discussed below.

The coefficients for natural data, unrestrained, in Tables 9-11 and 9-12 are the seasonally continuous versions of the discrete sets of coefficients in Figs. 9-10 and 9-11. In further explanation, in Table 9-12 the coefficients for natural data, unrestrained, are expanded to include all calendar months. These expanded coefficients are plotted as continuous lines in Fig. 9-12 for both Taylor Creek and Watkinsville. For convenient comparison, the discrete coefficients from Figs. 9-10 and 9-11 are also plotted. It can be seen that the cyclically continuous coefficients form a smooth average of the discretely derived coefficients. By evaluating only nine parameters from the historical record, instead of the 36 in monthly discrete analysis, a greater degree of averaging is imposed. This averaging tends to smooth the irregularities in the discrete coefficients.

The comparisons evident in Fig. 9-12 are based on Eq. 9-22 as the predictor model. Similar comparisons could be made for all other models implied in Eqs. 9-20 to 9-25. The researcher may need to make such additional comparisons before choosing the predictor model for a specific watershed.

TABLE 9-11 Taylor Creek Seasonally Cyclic Coefficients in $RO_M = a_M + b_M P_M + c_M RO_{M-1}$

	Fitting Criteria				
Coefficient[a]	[b]	[c]	Natural, RO ≥ 0	Natural, $b \leq 1$, RO ≥ 0	Estimated Natural for Simulation
a_1	−0.6763	−0.1540	−12.1178	−24.7986	−3.50
a_5	−3.0677	−1.0969	−3.6173	−4.9686	−1.71
a_9	−2.5040	−2.1945	−2.6206	−3.2856	−2.34
b_1	0.7140	0.3653	2.2952	1.0000	0.85
b_5	1.7859	0.3411	0.7240	0.8315	0.46
b_9	1.5078	0.6773	0.6865	0.7178	0.68
c_1	0.6396	0.2367	1.2124	2.0082	0.49
c_5	0.5463	0.0302	0.1887	0.1050	0.07
c_9	0.5400	0.3246	0.4268	0.5433	0.35
S_e	0.8819	1.0965	0.9487	1.5490	1.05
$\hat{Y}$	−1.1115	1.1455	1.0458	1.1709	1.1439
$\bar{Y}$	−1.1115	1.1439	1.1439	1.1439	—

[a] S_e standard deviation of residual errors; $\hat{Y}$, mean of calculated values; $\bar{Y}$, mean of observed values.

[b] Log-transformation of data—unrestrained.

[c] Natural data—unrestrained.

TABLE 9-12 Watkinsville Seasonally Cyclic Coefficients in $RO_M = a_M + b_M P_M + c_M RO_{M-1}$

	Fitting Criteria		
Coefficient[a]	[b]	[c]	Natural, RO>0
a_1	−5.5750	−0.8104	−3.1206
a_5	−5.0523	−0.8910	−2.0072
a_9	−3.6056	−0.5464	−5.0934
b_1	2.5766	0.2556	0.5482
b_5	2.1564	0.2684	0.4032
b_9	0.7799	0.2355	0.7652
c_1	0.2144	0.1165	0.2612
c_5	0.2632	0.2769	0.4624
c_9	0.2432	−0.0124	0.1258
S_e	1.2789	0.5868	0.4022
$\hat{Y}$	−3.1350	0.3380	0.3104
$\bar{Y}$	−3.1350	0.3380	0.3380

[a] S_e, standard deviation of residual errors; $\hat{Y}$, mean of calculated values; $\bar{Y}$, mean of observed values.

[b] Log-transformation of data—unrestrained.

[c] Natural data—unrestrained.

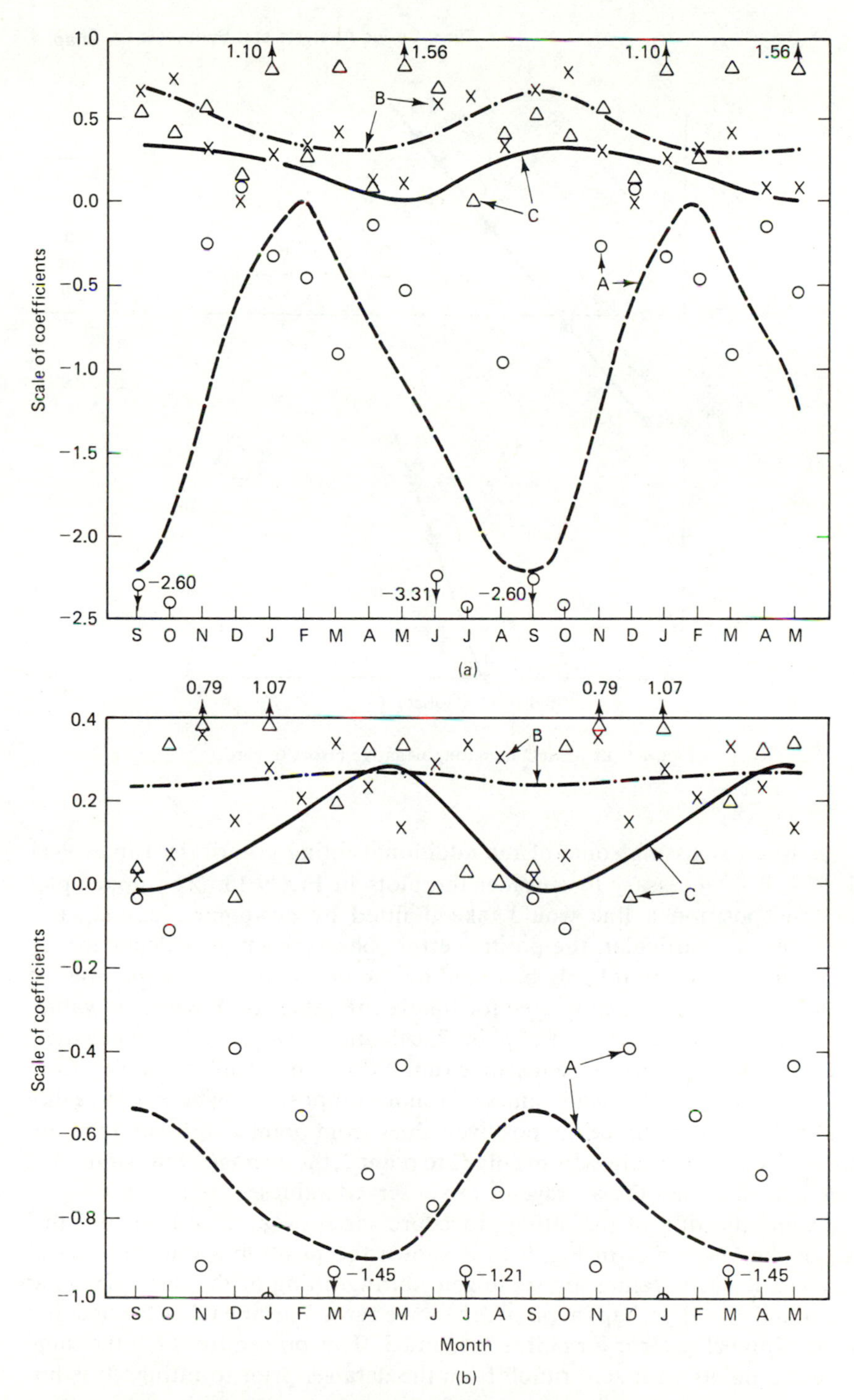

Figure 9-12 Seasonally continuous regression coefficients: (a) Taylor Creek; (b) Watkinsville.

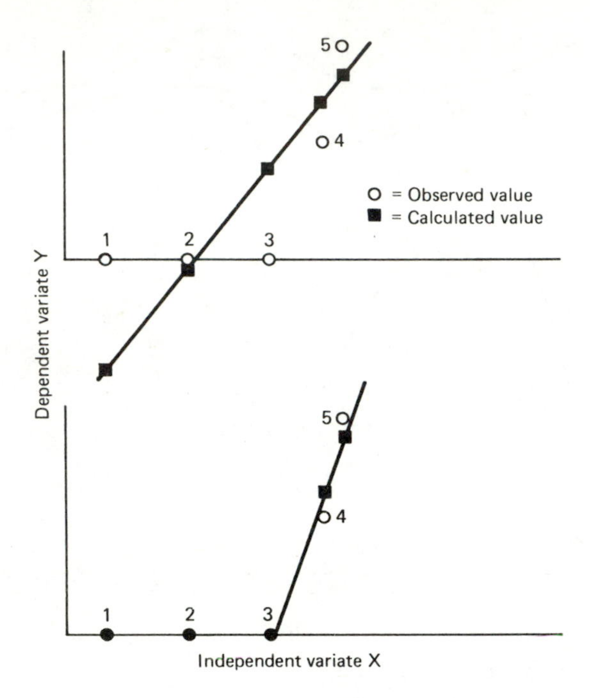

Figure 9-13 Effect of setting negative errors to zero.

Before discussing some of the additional fitting criteria in Tables 9-11 and 9-12, it is necessary to consider the plots in Fig. 9-13. The upper plot shows the position a line would take if fitted by customary least-squares procedures. In particular, the positive error (observed minus calculated) for data point 1 would be roughly balanced by the negative error for point 3. If this fitted line is subsequently used for simulating values of Y based on values of X, a bias results. To the left of point 2, calculated values of Y are negative. These can be set to zero, since negative runoff does not occur. However, from point 2 to point 3, calculated values of runoff are positive, whereas in reality they should be zero. Since the positive values from point 2 to point 3 are not balanced by negative values from point 1 to point 2, the average of the simulated values is higher than the average of the observed values.

A modification of the fitting procedure should significantly reduce this bias. In the lower plot in Fig. 9-13 is shown the position a line will take if negative calculated values are set to zero during fitting of the line. Errors at points 1 and 2, and perhaps at point 3, become zero. The line takes the position shown, minimizing error for points 3, 4, and 5. This procedure is not the same as deleting points with zero runoff from the data set prior to fitting. It is not known ahead of time whether point 3 would have a negative or positive calculated value; consequently, it is not known ahead of time whether point

3 should be deleted or retained. In a real data set many points could be expected to lie near the critical value of point 3.

The line shown in the lower plot of Fig. 9-13 cannot be fitted by customary least squares. It can, however, be routinely fitted by the iterative procedures of nonlinear least squares (see Chapter 7). The coefficients for natural data, under the restraint that predicted runoff cannot be negative, are shown in the third column of Tables 9-11 and 9-12.

Table 9-12 shows that for Watkinsville the incorporation of the runoff restraint tended to increase the slopes of the two lines (increase in coefficient b and in coefficient c) and to decrease significantly the value of the intercept coefficient a. This follows the results expected from the sketches in Fig. 9-13. Table 9-11 shows that the single restraint did not produce satisfactory results for Taylor Creek. Coefficient b_1 is 2.2952. Physically, this value may not be more than 1.0; otherwise, runoff at some value of rainfall would become greater

TABLE 9-13 Examples of Complete Sets of Seasonally Cyclic Coefficients

Watershed and Month	Coefficient		
	a	b	c
Taylor Creek			
Jan.	−0.1540	0.3653	0.2367
Feb.	−0.0222	0.3306	0.1817
Mar.	−0.4293	0.3127	0.1095
Apr.	−0.7804	0.3151	0.0493
May	−1.0969	0.3411	0.0302
June	−1.4340	0.4150	0.0775
July	−1.8322	0.5272	0.1700
Aug.	−2.1372	0.6304	0.2661
Sept.	−2.1945	0.6773	0.3246
Oct.	−1.8351	0.6381	0.3323
Nov.	−1.1839	0.5438	0.3120
Dec.	−0.5279	0.4383	0.2760
Watkinsville			
Jan.	−0.8104	0.2556	0.1165
Feb.	−0.8534	0.2604	0.1675
Mar.	−0.8887	0.2653	0.2228
Apr.	−0.9050	0.2686	0.2676
May	−0.8910	0.2684	0.2769
June	−0.8204	0.2621	0.2018
July	−0.7072	0.2515	0.1342
Aug.	−0.5997	0.2410	0.0410
Sept.	−0.5464	0.2355	−0.0124
Oct.	−0.5739	0.2370	−0.0103
Nov.	−0.6518	0.2427	0.0239
Dec.	−0.7430	0.2499	0.0723

than the rainfall. The Taylor Creek data were rerun under the additional restraint that the coefficients b could not be greater than 1.0. The results in Table 9-13 under this fitting criterion show that this procedure is numerically feasible. However, the change in the coefficients is quite large. The distortion is due to some extreme values of rainfall occurring in the short record of 11 years. It is expected that a longer record would produce better averaging.

EXERCISES

9-1. Compute and graph the smoothed series of the Pond Creek data (Table 9-1) using smoothing intervals of 3 and 5 years. Discuss the rationality of using a moving-average filter for such data. Fit a regression equation to the time series of Table 9-1 and analyze the residuals. Explain why the regression equation and the series defined by the moving-average filter differ.

9-2. Using the snow-covered-area data for zone 1 (see Appendix C-4), use a moving-average filter to separate the systematic and nonsystematic variation. Define the underlying population for the data, including both the systematic and nonsystematic components.

9-3. Using the degree-day data for zone 1 (see Appendix C-4), use a moving-average filter to separate the systematic and nonsystematic variation. Formulate a model that uses time as the predictor variable to represent the systematic variation and fit the coefficients using either a numerical or subjective optimization (see Chapters 7 and 8). Use frequency analysis (Chapter 4) to analyze the nonsystematic variation.

9-4. Discuss the applicability of moving-average filtering for analyzing the daily precipitation data of Appendix C-4.

9-5. The data in the table below include Manning's n, the tree density d (i.e., the average number of trees per foot for both sides within the reach), and the average channel width (w) for various sections of the Kankakee River in Indiana. Using moving-average filtering, separate the systematic and nonsystematic variation. Fit a regression equation (linear or nonlinear) for each of the three variables using the river mile as the predictor variable. Analyze the nonsystematic variation and discuss its importance. Analyze the correlations between n, d, and w.

River Mile	n	d	w
70	0.032	0.10	208
72	0.034	0.11	198
74	0.035	0.10	196
76	0.034	0.10	191
78	0.036	0.09	194
80	0.035	0.11	190
82	0.035	0.12	192
84	0.037	0.14	200
86	0.039	0.17	172

River Mile	n	d	w
88	0.040	0.16	177
90	0.039	0.14	165
92	0.040	0.15	162
94	0.041	0.14	153
96	0.043	0.17	131
98	0.046	0.19	142
100	0.050	0.21	139
102	0.051	0.20	132
104	0.050	0.16	117
106	0.045	0.20	124
108	0.050	0.26	120
110	0.050	0.25	127
112	0.055	0.30	102
114	0.053	0.26	109
116	0.053	0.17	111
118	0.060	0.44	110
120	0.055	0.30	112
122	0.050	0.25	102
124	0.048	0.18	98
126	0.055	0.20	101
128	0.050	0.21	100

9-6. If a long time series consisted of numbers generated using a random number generator, what smoothing interval would be required to support statistically that there is no systematic trend in the data? Would this vary with the probability distribution of the random numbers?

9-7. Using the daily streamflow data for the Conejos River (see Appendix C-4), compute the correlogram and develop an autoregressive model for predicting the daily streamflow. Analyze the residuals and qualitatively compare them with the precipitation and degree-day data of Appendix C-4.

9-8. Compute the correlogram for the zone 1 degree-day data of Appendix C-4 and develop an autoregression prediction model. Compare the accuracy of the model with the model developed in Exercise 9-3.

9-9. Using the precipitation (R) data of Appendix C-4, estimate the following conditional probabilities:

		Day $t-1$: $R = 0$	Day $t-1$: $R > 0$
Day t	$R = 0$	$P(R_t = 0 \mid R_{t-1} = 0)$	$P(R_t = 0 \mid R_{t-1} > 0)$
	$R > 0$	$P(R_t > 0 \mid R_{t-1} = 0)$	$P(R_t > 0 \mid R_{t-1} > 0)$

Then perform a frequency analysis of the rainfall amounts on those days where $R > 0$. Using the probability table above and the frequency curve, formulate a model that could be used to simulate a rainfall sequence. Using a table of random numbers or a random number generator on a computer, generate a sequence of 3 months of daily rainfall data. Compare the statistical characteristics of the generated data with the sample data.

9-10. Compute the cross-correlograms for the daily streamflow, precipitation, degree-day, and snow-covered-area data of Appendix C-4. Formulate a set of regression models for forecasting streamflow using the best predictor variables. Analyze the data using a stepwise regression analysis and select the best forecast model.

9-11. Using the rainfall generator of Exercise 9-9, simulate one year of rainfall and compute the streamflow using the optimized model of Fig. 8-3 and Table 8-3.

9-12. Write a simple computer program to do smoothing by moving averages. Use the following set of linear weights to smooth the data series, and plot.

Linear weights:

Time	0	1	2	3	4
Weight	0	0.25	0.5	0.25	0

Data series:

Time	Data	Time	Data	Time	Data
1	0.93	7	−9.74	13	−0.46
2	4.11	8	−5.97	14	−4.30
3	9.69	9	−0.52	15	−9.93
4	5.40	10	5.85	16	−5.92
5	0.83	11	9.34	17	0.73
6	−4.87	12	4.94		

9-13. Smooth the data series of Exercise 9-12 with the following set of linear weights.

Time	0	1	2	3	4	5	6
Weight	0	0.1111	0.2222	0.3334	0.2222	0.1111	0

Plot the results and compare the results with the results of Exercise 9-12. Which set of weights appears to provide the better model? Is either set fully adequate? Why?

9-14. Use packaged software to compute auto- and cross-correlation coefficients for lags 0 through 7 of the following series:

P	2	1	2	3	4	5	4	3	2	1	1	2	3	4	5	4	3	2
R	3	2	1	1	2	3	4	5	4	3	2	1	1	2	3	4	5	4

9-15. Use packaged software to compute the auto- and cross-correlation coefficients for monthly rainfall and runoff for the Chestuee River near Dahlonega, Georgia, for the years 1944–1961. The data are given in Appendix C-2. Calculate for lags from 0 through 18 and plot the correlograms. Why does the autocorrelation of runoff show a stronger 12-month pattern than that shown by precipitation? Are the positive coefficients of cross-correlation for lags 0 through 3 consistent with the distribution of runoff?

9-16. Use packaged software to construct correlograms of monthly rainfall and runoff for the Chattooga River near Clayton, Georgia. Use at least 30 lags. The data are given in Appendix C-1.

9-17. Repeat Exercise 9-16 for rainfall and monthly average temperature from Appendix C-1.

9-18. Repeat Exercise 9-16 for runoff and monthly average temperature from Appendix C-1.

9-19. Discuss the auto- and cross-correlogram of Exercises 9-16 to 9-18. Note any phase shifts in the cross-correlations. Are such phase shifts rational?

9-20. Use packaged software to evaluate the coefficients of the monthly water-yield model (Equation 1-4) for the data of Exercise 9-16. Are the coefficients consistent with your results from Exercises 9-16 to 9-19? Study the residual errors. Do they show seasonal patterns?

9-21. Repeat Exercise 9-16 using the data for the Wildcat Creek watershed (Appendix C-3).

REFERENCES AND SUGGESTED READINGS

BENSON, M. A., and N. C. MATALAS, Synthetic Hydrology Based on Regional Statistical Parameters, *Water Resources Research*, Vol. 3, No. 4, 1967.

BONNÉ, J., Stochastic Simulation of Monthly Streamflow by a Multiple Regression Model Utilizing Precipitation Data, *Journal of Hydrology*, Vol. 12, No. 4, pp. 285–310, 1971.

CLARKE, R. T., Extension of Annual Streamflow Record by Correlation with Precipitation Subject to Heterogeneous Errors, *Water Resources Research*, Vol. 15, No. 5, pp. 1081–88, 1979.

DELLEUR, J. W., P. C. TAO, and M. C. KAVVAS, An Evaluation of the Practicality and Complexity of Some Rainfall and Runoff Time Series Models, *Water Resources Research*, Vol. 12, No. 5, pp. 953–70, 1976.

DRACUP, J. A., K. S. LEE, and E. G. PAULSON, On the Statistical Characteristics of Drought Events, *Water Resources Research*, Vol. 16, No. 2, pp. 289–96, 1980.

DUFFY, C. J., L. W. GELHAR, and P. J. WIERENGA, Stochastic Models in Agricultural Watersheds, *Journal of Hydrology*, Vol. 69, pp. 145–62, 1984.

HANES, W. T., M. FOGEL, and L. DUCKSTEIN, Forecasting Snowmelt Runoff: Probabilistic Model, *Journal of the Irrigation and Drainage Division, ASCE*, Vol. IR3, pp. 343–55, September 1977.

HIEMSTRA, L., and R. CREESE, Synthetic Generalization of Seasonal Precipitation, *Journal of Hydrology*, Vol. 11, pp. 30–46, 1970.

JULIAN, P. R., Variance Spectrum Analysis, *Water Resources Research*, Vol. 3, No. 3, pp. 831–45, 1967.

KISIEL, C. C., Time Series Analysis of Hydrologic Data, in *Advances in Hydroscience*, Vol. 5, V. T. Chow (ed.), Academic Press, New York, 1969.

KUCZERA, G., On the Relationship between the Reliability of Parameter Estimates and Hydrologic Time Series Data Used in Calibration, *Water Resources Research,* Vol. 18, No. 1, pp. 146–54, 1982.

LETTENMAIER, D. P., and S. J. BURGES, Operational Assessment of Hydrologic Models of Long-Term Persistence, *Water Resources Research,* Vol. 13, No. 1, pp. 113–24, 1977.

MATALAS, N. C., Time Series Analysis, *Water Resources Research,* Vol. 3, No. 3, pp. 817–29, 1967.

MCLEOD, A. I., K. W. HIPEL, and F. COMANCHO, Trend Assessment of Water Quality Time Series, *Water Resources Bulletin,* Vol. 19, No. 4, pp. 537–48, 1983.

MEHTA, B., and R. AHLERT, Stochastic Variation of Water Quality of the Passaic River, *Water Resources Research,* Vol. 11, No. 2, pp. 300–308, April 1975.

PHIEN, H. N., and M. A. KHAN, Comparison of Two Autoregressive Models for Monthly Stream Flow Generation, *Water Resources Bulletin,* Vol. 17, No. 6, pp. 1035–41, 1981.

QUIMPO, R. G., Stochastic Model of Daily River Flow Sequences, Hydrologic Paper 18, Colorado State University, Fort Collins, Colo., February 1967.

RODRIQUEZ, I., The Application of Cross Spectral Analysis to Hydrologic Time Series, Hydrologic Paper 24, Colorado State University, Fort Collins, Colo., September 1967.

SEN, Z., Effect of Periodic Parameters on the Autocorrelation Structure of Hydrologic Series, *Water Resources Research,* Vol. 15, No. 6, pp. 1639–42, 1979.

SHAW, S. R., An Investigation of the Cellular Structure of Storms Using Correlation Techniques, *Journal of Hydrology,* Vol. 62, No. 1, pp. 63–79, 1983.

SRIKANTHAN, R., and T. A. MCMAHON, Stochastic Generation of Monthly Streamflows, *Journal of the Hydraulics Division, ASCE,* Vol. 108, No. HY2, pp. 419–42, 1982.

STEDINGER, J. R., Estimating Correlations in Multivariate Streamflow Models, *Water Resources Research,* Vol. 17, No. 1, pp. 200–208, 1981.

STEDINGER, J. R., and M. R. TAYLOR, Synthetic Streamflow Generation, *Water Resources Research,* Vol. 18, No. 4, pp. 909–24, 1982.

TASKER, G. D., Approximate Sampling Distribution of the Serial Correlation Coefficient for Small Samples, *Water Resources Research,* Vol. 19, No. 2, pp. 579–82, 1983.

WASTLER, T. A., *Application of Spectral Analysis to Stream and Estuary Field Surveys,* Public Health Service Publ. 999-WP-7, Cincinnati, Ohio, November 1963.

YEVJEVICH, V., *Structural Analysis of Hydrologic Time Series,* Hydrology Paper 56, Colorado State University, Fort Collins, Colo., 1972.

10 Sensitivity Analysis and Probabilistic Modeling

INTRODUCTION

Statistical hydrologic modeling is a process by which simplified functional algorithms that represent elements of the hydrologic cycle are formulated and fit to measured data. To this point of the book, the emphasis has been toward the use of data in formulating, calibrating, and verifying hydrologic models. This emphasis should not suggest that all aspects of modeling require data. In Chapter 6 we indicated that a rational analysis of a model was the first and most important element of verifying a model. Simulation was shown in Chapter 2 to be a useful statistical tool for decision making. Two additional modeling tools will be introduced in this chapter, sensitivity analysis and probabilistic modeling. These two tools do not use data in the way that data have been used for model formulation and calibration. Instead, the calibrated models are used to generate data that indicate the response of the system to various specified states of the system. That is, the model is used to generate data that reflect the sensitivity of the hydrologic process being modeled to the inputs to the system. Sensitivity analysis and probabilistic modeling provide a systematic means of examining the response of a hydrologic model in a way that is free of the error variation that exists when dealing with measured data. This freedom from error variation makes it possible to assess more easily the rationality of the model, as well as examining the effect of error in the input.

MATHEMATICAL FOUNDATIONS OF SENSITIVITY ANALYSIS

Definition

Sensitivity is the rate of change in one factor with respect to change in another factor. Although such a definition is vague in terms of the factors involved, nevertheless it implies a quotient of two differentials. Stressing the nebulosity of the definition is important because, in practice, the sensitivity of model parameters is rarely recognized as a special case of the concept of sensitivity. The failure to recognize the generality of sensitivity has been partially responsible for the limited use of sensitivity as a tool for the design and analysis of hydrologic models.

The Sensitivity Equation

The general definition of sensitivity can be expressed in mathematical form by considering a Taylor series expansion of the explicit function

$$O = f(F_1, F_2, \ldots, F_n) \tag{10-1}$$

The factor O is often a model output or the output of one component of a model. The change in factor O resulting from change in a factor F_i is given by

$$f(F_i + \Delta F_i, F_{j|j \neq i}) = O_0 + \frac{\partial O_0}{\partial F_i} \Delta F_i + \frac{1}{2!} \frac{\partial^2 O_0}{\partial F_i^2} \Delta F_i^2 + \cdots \tag{10-2}$$

in which O_0 is the value of O at some specified level of each F_i. If the nonlinear terms are small in comparison with the linear terms, Eq. 10-2 reduces to

$$f(F_i + \Delta F_i, F_{j|j \neq i}) = O_0 + \frac{\partial O_0}{\partial F_i} \Delta F_i \tag{10-3}$$

Thus

$$\Delta O_0 = f(F_i + \Delta F_i, F_{j|j \neq i}) - O_0 = \left(\frac{\partial O_0}{\partial Fi} \right) \Delta F_i \tag{10-4}$$

Equation 10-4 is referred to herein as the *linearized sensitivity equation.* It measures the change in factor O that results from change in factor F_i. The linearized sensitivity equation can be extended to the case where more than one parameter is changed simultaneously. The general definition of sensitivity is derived from Eqs. 10-1 and 10-4:

$$S = \frac{\partial O_0}{\partial F_i} = \frac{f(F_i + \Delta F_i, F_{j|j \neq i}) - f(F_1, F_2, \ldots, F_n)}{\Delta F_i} \tag{10-5}$$

Computational Methods

The general definition of sensitivity, which is expressed in mathematical form by Eq. 10-5, suggests two methods of computation. The left-hand side of Eq. 10-5 suggests that the sensitivity of O to changes in factor F_i can be estimated by differentiating the explicit relationship of Eq. 10-1 with respect to factor F_i:

$$S = \frac{\partial O_0}{\partial F_i} \tag{10-6}$$

Analytical differentiation has not been used extensively for analyzing hydrologic models because the mathematical framework of sensitivity has not been sufficiently developed. It will be used even less frequently as hydrologic models become more complex.

The method of *factor perturbation*, which is the second computational method suggested by Eq. 10-5, is the more commonly used method in hydrologic analysis. The right-hand side of Eq. 10-5 indicates that the sensitivity of O to change in F_i can be derived by incrementing F_i and computing the resulting change in the solution O. The sensitivity is the ratio of the two differentials and can be expressed in finite difference form:

$$S = \frac{\Delta O_0}{\Delta F_i} = \frac{f(F_i + \Delta F_i, F_{j|j \neq i}) - f(F_1, F_2, \ldots, F_n)}{\Delta F_i} \tag{10-7}$$

However, use of the method of parameter perturbation is often impractical for a complete sensitivity analysis of multiparameter systems because of the extensive computational effort required for complex models and because the sensitivity state depends on the data base used to calibrate the model.

Parametric and Component Sensitivity

A simplified system or a component of a more complex system is described by three functions: the input function, the output function, and the system response function. The response function is the component(s) of the system that transforms the input function into the output function and is often defined by a distribution function that depends on one or more parameters. In the past, sensitivity analyses of models have been limited to measuring the effect of parametric variations on the output. Such analyses focus on the output and response functions. Using the form of Eq. 10-5 parametric sensitivity can be mathematically expressed as

$$S_{pi} = \frac{\partial O}{\partial P_i} = \frac{f(P_i + \Delta P_i; P_{j|j \neq i}) - f(P_1, P_2, \ldots, P_n)}{\Delta P_i} \tag{10-8}$$

where O represents the output function and P_i is the parameter of the system response function under consideration.

Unfortunately, the general concept of sensitivity has been overshadowed by parametric sensitivity. As hydrologic models have become more complex, the derivation of parametric sensitivity estimates has become increasingly more difficult and, most often, impossible, to compute. However, by considering the input and output functions, the general definition of sensitivity (Eq. 10-5) can be used to define another form of sensitivity. Specifically, component sensitivity measures the effect of variation in the input function I on the output function:

$$S_c \equiv \frac{\partial O}{\partial I} = \frac{\Delta O}{\Delta I} \tag{10-9}$$

Combining component and parameter sensitivity functions makes it feasible to estimate the sensitivity of parameters of complex hydrologic models.

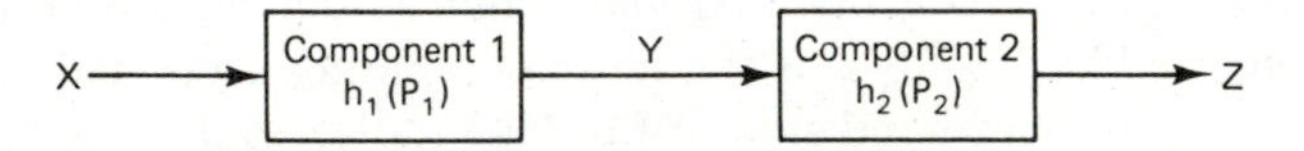

Figure 10-1 Two-component model.

For example, in the simplified two component model of Fig. 10-1, the sensitivity of Y to variation in P_1 and the sensitivity of Z to variation in P_2 are readily computed using sensitivity as defined by Eq. 10-6:

$$S_1 = \frac{\partial Y}{\partial P_1} \quad \text{and} \quad S_2 = \frac{\partial Z}{\partial P_2} \tag{10-10}$$

However, the sensitivity of the output from component 2 to change in the parameter of component 1 cannot always be estimated directly from the differential $\partial Z/\partial P_1$. In such cases, the component sensitivity function of component 2 can be used with the parametric sensitivity function S_1 to estimate the sensitivity of Z to change in P_1. Specifically, the sensitivity of $\partial Z/\partial P_1$ equals the product of the component sensitivity function $\partial Z/\partial Y$ and the parametric sensitivity function $\partial Y/\partial P_1$:

$$\frac{\partial Z}{\partial P_1} = \frac{\partial Z}{\partial Y}\frac{\partial Y}{\partial P_1} \tag{10-11}$$

Whereas the differentials $\partial Z/\partial Y$ and $\partial Y/\partial P_1$ are often easily derived, an explicit sensitivity function $\partial Z/\partial P_1$ can be computed only for very simple models. When a solution cannot be obtained analytically, the numerical method of Eq. 10-7 must be used.

Absolute and Relative Sensitivity

Sensitivity can be expressed in two forms: absolute and relative. The form in which sensitivity values are presented depends on the intended use.

Sensitivity values computed with the definition of Eq. 10-5 are in absolute form. Such a definition is inappropriate for the comparison of sensitivity values because values computed using Eq. 10-5 are not invariant to the magnitude of either factor O or F_i. Dividing the numerator of Eq. 10-5 by O_0 and the denominator by F_i provides an estimate of the relative change in O with respect to a relative change in F_i:

$$R_s = \frac{\partial O/O_0}{\partial F_i/F_i} = \frac{\partial O}{\partial F_i}\frac{F_i}{O_0} \tag{10-12}$$

Relative sensitivity values are invariant to the magnitude of O and F_i and thus provide a valid means for comparing factor sensitivity.

A Correspondence between Sensitivity and Correlation

The Pearson product-moment correlation coefficient R is used as a measure of the linear association between two random variables:

$$R = \frac{\sum xy}{\sqrt{\sum x^2 \sum y^2}} \tag{10-13}$$

where x is the difference between the random variable X and the mean value $\bar{X}$, and y is the difference between Y and $\bar{Y}$. The square of the correlation coefficient represents the proportion of the variance in Y that can be attributed to its linear regression on X.

Least-squares regression analysis is commonly used to derive a linear relationship between two random variables. There is a strong structural similarity between correlation analysis and regression analysis (Snyder, 1971). The linear regression coefficient b can be determined from the correlation coefficient and the standard deviations of X and Y:

$$b = \frac{\sum xy}{\sum x^2} = \frac{RS_y}{S_x} \tag{10-14}$$

Equation 10-14 suggests that for values of R near 1, changes in X will produce comparatively large changes in Y.

For a linear regression equation relating two random variables (see Eq. 3-17), the sensitivity of the dependent variate Y to variation in the independent variate X can be determined by differentiating the regression equation with respect to X:

$$\frac{\partial Y}{\partial X} = b \tag{10-15}$$

Equation 10-15 indicates that the regression coefficient represents the rate of change in Y with respect to change in X. Furthermore, Eqs. 10-14 and 10-15 suggest that there is a direct correspondence between correlation and sensitivity. Such a correspondence is supported by the results of empirical investigations with hydrologic simulation models. For example, Dawdy and O'Donnell (1965) found that the more sensitive parameters of the model were in better agreement with the true values than were the less sensitive parameters.

Example 10-1: Linear Rainfall–Runoff Model

The data of Table 6-1 provide values of the runoff in the month of March (RO_M), with values of the rainfall for both March (P_M) and February (P_F) as predictor variables. A linear regression analysis of the data resulted in Eq. 6-22:

$$\widehat{RO}_M = -0.0346 + 0.5880P_M + 0.1238P_F$$

The model explained 78.4% (R^2) of the variation in RO_M, which suggests that estimates of RO_M should be reasonably accurate. Taking the derivatives of the prediction equation results in the following values of absolute sensitivity:

$$\frac{\partial RO_M}{\partial P_M} = 0.5880$$

and

$$\frac{\partial RO_M}{\partial P_F} = 0.1238$$

The mean values of P_M and P_F are 4.85 in. and 4.98 in., respectively, and the standard deviations are 2.076 in. and 2.615 in., respectively.

The absolute sensitivities can be used to evaluate the effect of errors in the rainfall measurements. There are several sources of errors that we may wish to consider. For example, errors in estimates of RO_M can occur because of either instrumental errors in measurement of P_M and P_F or the inability of the gage to reflect the rainfall over the drainage area. If either of these sources of error amounts to 5% of the estimate of P_M or P_F, the error in RO_M that results from the error in the rainfall estimate can be computed from the absolute sensitivity and the linear sensitivity equation (Eq. 10-4); for example, if the rainfall is 5 in. in any one month, the error in the resulting runoff would be

$$\Delta RO_M = \frac{\partial RO_M}{\partial P_M}\Delta P_M = 0.5880[(0.05)(5)] = 0.147 \text{ inch}$$

If 5 in. of rainfall fell in both February and March, the resulting runoff (from Eq. 6-22) would be 3.52 in. Thus the error in runoff resulting from the 5% error in rainfall would be about 4% of the runoff.

The relative sensitivities of the inputs can also be computed. At the mean values the relative sensitivity of RO_M to P_M and P_F would be, respectively,

$$R_{SM} = \frac{\partial RO_M}{\partial P_M}\frac{\bar{P}_M}{\overline{RO}_M} = 0.5880\left(\frac{4.85}{3.43}\right) = 0.831$$

$$R_{SF} = \frac{\partial RO_M}{\partial P_F}\frac{\bar{P}_F}{\overline{RO}_M} = 0.1238\left(\frac{4.98}{3.43}\right) = 0.180$$

As was suggested in the discussion of Example 6-1, the relative sensitivity values indicate that the rainfall during March has a greater effect on the runoff in March than on the rainfall in February.

Time Variation of Sensitivity

Example 10-1 provided constant values of both the absolute and relative sensitivities. Sensitivity will actually vary depending on the values selected for the variables. This general statement implies that the sensitivity of a hydrologic model to errors in the input will vary with the state of the system (i.e., the values of the coefficients and variables). This is easily illustrated using a model of Example 6-4. The three-coefficient model for estimating the mean monthly temperature had the form

$$T = b_0 + b_1 \sin \omega M + b_2 \cos \omega M$$

The three-coefficient model was selected because the five-coefficient model was not significantly better than the simpler three-coefficient model. The least-squares fitting provided the following values for the fitting coefficients:

$$T = 55.967 - 10.050 \sin \omega M - 17.637 \cos \omega M$$

It was illustrated in Chapter 6 that the values of the coefficients (i.e., b_0, b_1, and b_2) are values of random variables and thus are not exact. Therefore, we would probably be interested in the sensitivity of T to error in the values of the coefficients. Taking the derivatives of the prediction equation with respect to b_i yields

$$\frac{\partial T}{\partial b_0} = 1$$

$$\frac{\partial T}{\partial b_1} = \sin \omega T$$

$$\frac{\partial T}{\partial b_2} = \cos \omega T$$

These derivatives indicate that the sensitivity is not necessarily a constant. Error in b_0 is linearly related to b_0 since a unit change in b_0 causes a unit change in T, which is evident from the sensitivity of 1 for $\partial T/\partial b_0$. However, the sensitivities of T to b_1 and b_2 are not constant and will vary with the month. When ωM equals either 0° or 180°, error in b_1 will not cause any error in T, while the effect of error in b_2 will be at its greatest. Similarly, when ωM equals either 90° or 270°, the effect of error in b_1 will be greatest while error in b_2 will have no effect on T.

This example illustrates that model sensitivity is not always constant. For the time-dependent model of this example, the sensitivity varied with time. This would be true of time-dependent rainfall-runoff models such as those discussed in Chapters 8 and 9. While sensitivity analyses of hydrologic models are usually performed using the mean values of the coefficients and input variables, it is quite likely that the sensitivity estimates should be computed for the watershed conditions at which a design will be made. For example, if a hydrologic model is being used for design at extreme flood conditions, such as at the 100-year rainfall event, the sensitivity analysis should be performed at the conditions that would exist at that time rather than at average watershed conditions. For example, a high soil-moisture state would be more appropriate than the average soil-moisture state. One can reasonably expect the relative sensitivities, and therefore the relative importance of the model elements, to vary with the state of the watershed. In summary, while the mathematical development of sensitivity was presented in its basic form, it is important to recognize that sensitivity analyses of hydrologic models must consider the time and spatial characteristics of the watershed conditions.

SENSITIVITY IN MODEL FORMULATION

The general definition of sensitivity, Eq. 10-5, indicates that the sensitivity of one factor depends, in the general case, on the magnitude of all factors of the system. For a dynamic system that is not in steady state, such as a watershed, the output and sensitivity functions will also be time dependent. The unit response function proposed by Nash (1957) will be used to demonstrate the dynamic and parametric nature of sensitivity. The gamma distribution proposed by Nash (1957) has been used by others as a conceptual representation of the response of a watershed (Nash, 1959; Sarma et al., 1969):

$$h = h(n, K; t) = \frac{t^{n-1} e^{-t/K}}{K^n \Gamma(n)} \tag{10-16}$$

where n represents the number of equivalent reservoirs whose hydrologic output is the same as that of the drainage basin, K is the constant of proportionality relating outflow rate and storage, t represents the time from the impulse input, and $\Gamma(n)$ is the gamma function with argument n. Equation 10-6 can be used to derive the sensitivity functions for the parameters n and K. The sensitivity function for n is

$$S_n = \frac{\partial h}{\partial n} = \frac{t^{n-1} e^{-t/K}[\ln_e (t/K) - \Gamma'(n)/\Gamma(n)]}{K^n \Gamma(n)} \tag{10-17}$$

where $\Gamma'(n)$ is the derivative of $\Gamma(n)$ with respect to n. The sensitivity function for K is

$$S_K = \frac{\partial h}{\partial K} = \frac{t^{n-1} e^{-t/K}(t - nK)}{K^{n+2} \Gamma(n)} \tag{10-18}$$

It is evident from Eqs. 10-17 and 10-18 that both of the sensitivity functions depend on the value of n and K and also that the sensitivity of a parameter

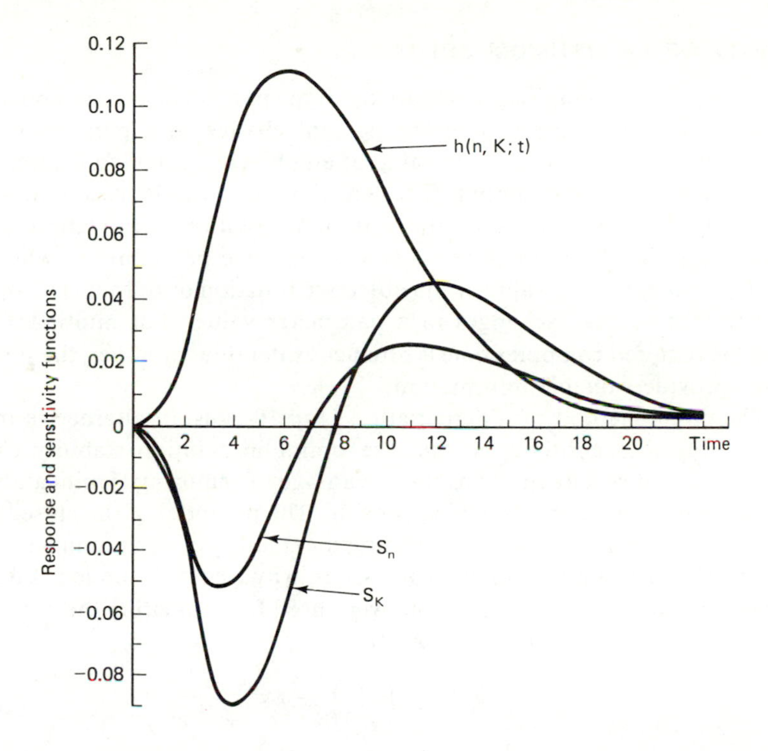

Figure 10-2 Nash model sensitivity: $n = 4.0$; $K = 2.0$.

varies with time *t*. In Fig. 10-2 both the unit response function and the two parametric sensitivity functions are plotted as a function of time. The sensitivity plot for the parameter *n* indicates that the rising limb of the unit response function is most sensitive to change in the parameter *n*. Furthermore, the change of the peak of the response function is approximately one-third of the maximum change. Similarly, change in the value of *K* has the greatest effect on the rising limb of the response function. Although such information concerning the effect of parameter variation is only qualitative it can be of considerable value in the process of deriving relationships between the optimal parameter values and watershed or storm properties. Also, parameters such as *n* and *K* that are highly intercorrelated may confound the optimization technique and thus prevent the identification of the true values. Failure to identify the true parameter values would adversely affect relationships derived between model parameters and physical properties of the watershed. The effect of parameter interaction on relationships between optimal model parameters and watershed properties can be reduced if the similarities of the parameter sensitivity functions are considered when formulating the model structure.

STABILITY OF AN OPTIMUM SOLUTION

A sensitivity plot is a graphical comparison of the percent change in one factor, which is usually the output, and the percent change in a parameter value (Dawdy, 1969). The change in the value of an objective function is often used to represent the change in output. The sensitivity plot can be used to examine the stability of a parameter of the optimum solution. Derivation of the sensitivity plot has traditionally involved an iterative procedure in which the percentage change in the value of the objective function or output is computed for different percentage changes in a parameter value. For multiparameter models the required computer time is often considerable; however, the sensitivity plots provide valuable information.

The direct method of differentiation (Eq. 10-6) is an alternative means of computing the sensitivity so that the optimum solution stability can be examined. The direct method has the advantage of requiring significantly less computer time than the numerical approach. The potential of the closed-form sensitivity function as a means of indicating the stability of an optimum solution is demonstrated using the double-routing unit hydrograph model. Effective precipitation and observed runoff data were used for estimating the parameter K of the unit response function $h(K; t)$:

$$h(K; t) = \left(\frac{t}{K^2}\right) e^{-t/K} \tag{10-19}$$

The optimum value of K was determined by minimizing the sum of the squares of the differences between observed and computed runoff. The closed-form sensitivity function $S(K; t)$ can be derived by differentiation of the unit response function with respect to K:

$$S(K; t) = \frac{dh(K; t)}{dK} = \frac{t(t - 2K)\, e^{-t/K}}{K^4} \tag{10-20}$$

To simplify graphical presentation the sensitivity function of Eq. 10-20 will be evaluated at the time of the peak of the response function. Such a simplification is not unreasonable because the ordinates of the sensitivity function are proportional to the sensitivity at the peak for a given value of the parameter K. For the double-routing response function the time of the peak response t_p equals the value of the parameter K. Thus the sensitivity at the peak is

$$S(K; t = t_p) = \frac{-0.36788}{K^2} \tag{10-21}$$

A plot of $S(K; t = t_p)$ versus a percent change in the parameter value is given in Fig. 10-3. Both the sensitive plot of Fig. 10-4 and the graph of the sensitivity function in Fig. 10-3 indicate that an increase in the parameter value K would cause significantly less change in the output than an equal percentage decrease in the parameter value. The sensitivity function approach is strictly valid only

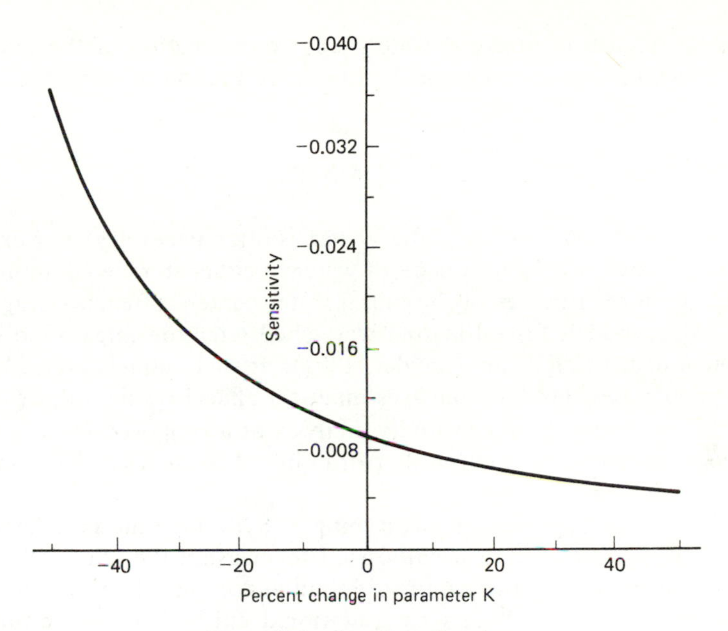

Figure 10-3 Sensitivity function.

when the nonlinear terms of Eq. 10-2 are insignificant. However, even for large changes in a parameter value the sensitivity function approach provides a reliable qualitative indication of the stability of the optimum solution. Furthermore, the sensitivity function approach (Fig. 10-3) requires considerably less computational effort than the derivation of a sensitivity plot (Fig. 10-4). The sensitivity function thus indicates that if it is necessary to estimate

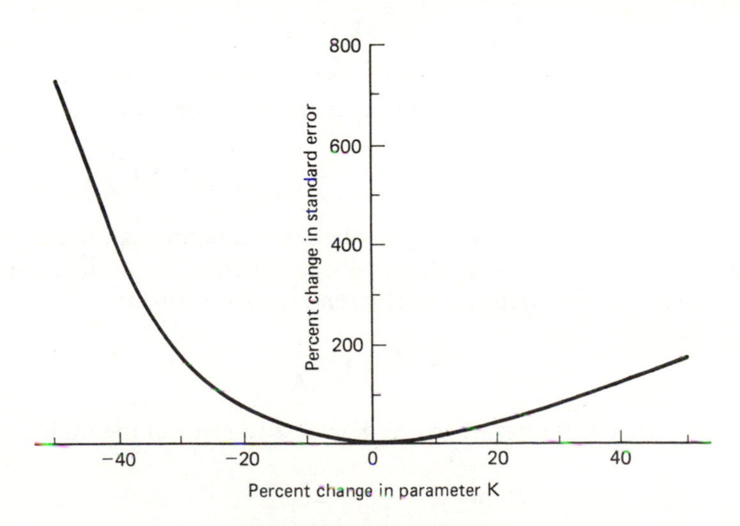

Figure 10-4 Sensitivity plot.

the value of K for an ungaged watershed, overestimation of the parameter value K would result in less error than underestimation of its value.

SENSITIVITY OF INITIAL VALUE ESTIMATES

An estimate of the initial state of the system is often necessary. For example, simulation models involving storage of water in either storage components or soil zones require estimates of the initial water content of each storage component. Also, models for estimating watershed retention rates often require an estimate of the initial rate (Snyder, 1971). If such models are to be used for design purposes, it is desirable to estimate the effect in initial value estimates on the design variables. A sensitivity analysis of a proposed initial estimate can be used to estimate the effect of error in initial estimates on the computed output.

When designing or planning with complex hydrologic models, the method of parameter perturbation is usually used to measure the effect of error in initial value estimates. However, the differential approach to estimating sensitivity requires considerably less computational effort than the perturbation technique, especially for complex models. The linear storage model (Zoch, 1934–1937) will be used to demonstrate the differential approach to estimating initial value sensitivity. The equation of continuity relates input X, output Y, and the time rate of change in storage S:

$$X - Y = \frac{dS}{dt} \tag{10-22}$$

Storage is assumed to be a linear function of outflow:

$$S = KY \tag{10-23}$$

where K is a constant. Substituting the time differential of Eq. 10-23 into Eq. 10-22 and expressing the result in finite difference form yields

$$Y_{t+\Delta t} = Y_t - \frac{(Y_t - X_t)\,\Delta t}{K} = Y_t\left(1 - \frac{\Delta t}{K}\right) + X_t\left(\frac{\Delta t}{K}\right) \tag{10-24}$$

The model given by Eq. 10-24 can be solved by iteration. An initial estimate Y_0 is required. The sensitivity of the output at time $t + \Delta t$ to error in the output at time t can be derived by differentiating Eq. 10-24:

$$\frac{dY_{t+\Delta t}}{dY_t} = 1 - \frac{\Delta t}{K} \tag{10-25}$$

After n time periods the sensitivity of $Y_{n\Delta t}$ due to error in the initial estimate Y_0 is

$$\frac{dY_{n\Delta t}}{dY_0} = \left(1 - \frac{\Delta t}{K}\right)^{n\Delta t} \tag{10-26}$$

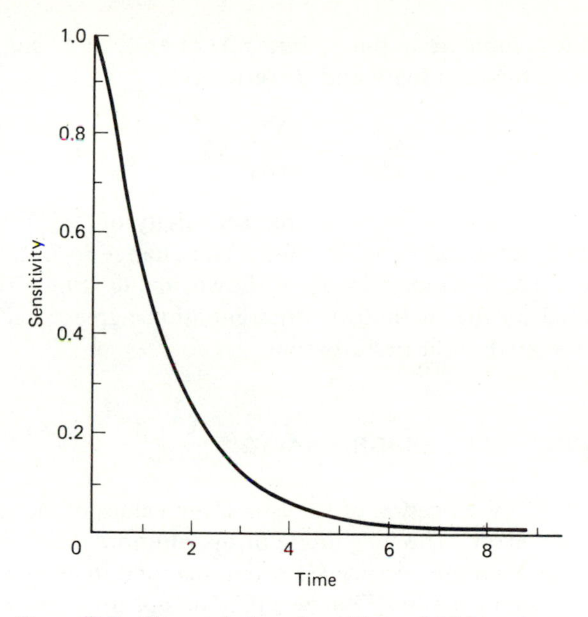

Figure 10-5 Time variation of sensitivity of initial estimate.

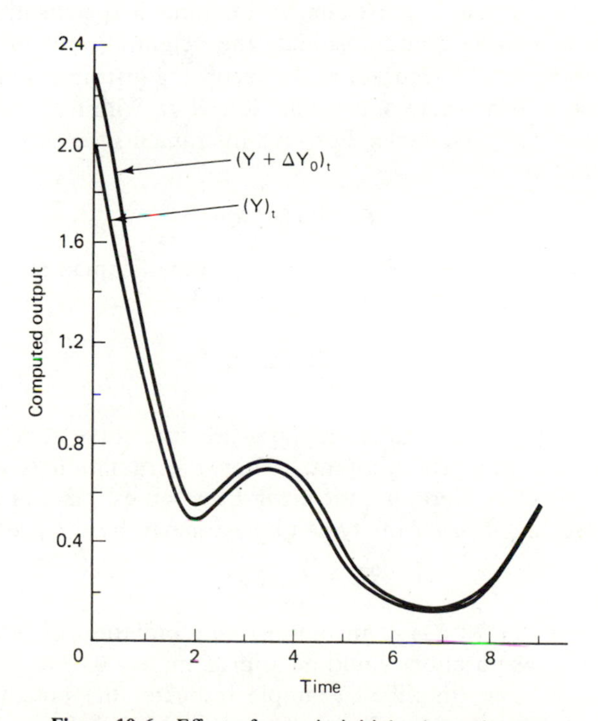

Figure 10-6 Effect of error in initial value estimates.

The effect on the computed output at time $n\,\Delta t$ of error in Y_0 can be estimated by the product of the sensitivity and the error ΔY_0:

$$\Delta Y_{n\Delta t} = \frac{dY_{n\Delta t}}{dY_0}\Delta Y_0 \tag{10-27}$$

Equation 10-26 was used to compute the sensitivity of $Y_{n\Delta t}$ with respect to Y_0 and is shown graphically in Fig. 10-5. The change in output, Eq. 10-27 resulting from a 12.5% change in Y_0 is shown in Fig. 10-6. The change in output computed by the method of differentiation agrees exactly with that computed by the method of perturbation.

SENSITIVITY AND DATA ERROR ANALYSIS

Data used for model verification or for estimating values of model parameters invariably contain error. The magnitude of distribution of error often cannot be evaluated. For example, airspeed is often included in evaporation models as a measure of air instability. But whether or not airspeed measurements quantitatively measure the effect of air instability on evaporation rates is difficult to assess. However, the effects of data error from other sources can be quantitatively evaluated. For example, instrument specifications supplied by manufacturers can be used to estimate the potential error in an observed measurement due to the inaccuracy of the recording instrument. Radiometers that are expected to measure solar radiation R to within 0.5% of the true value are commercially available. For such instruments the expected error E_R can be estimated by

$$E_R = 0.005R \tag{10-28}$$

The sensitivity of evaporation E computed with the Penman model (Penman, 1948, 1956) to change in radiation is estimated by

$$S_R = \frac{\partial E}{\partial R} = \frac{\Delta}{\Delta + \gamma} \tag{10-29}$$

where Δ is the slope of the saturation vapor pressure curve at air temperature and γ is the psychrometric constant. When the radiation is expressed in equivalent inches of evaporation, the error E_E in an estimate of daily evaporation from instrument insensitivity can be estimated by the product

$$E_E = E_R(S_R) \tag{10-30}$$

For a temperature of 70°F and incoming solar radiation of 500 ly/day the expected error in evaporation would be approximately 0.01 in., or 4% of the total evaporation. The simplified example indicates the potential value of using a closed-form sensitivity function to estimate the effect of data error.

The method of parameter perturbation would require considerably more computer time to estimate the effect of data error on complex models.

Reproducibility is a measure of how well different designers can estimate the value of a property at a site with the same input information and procedure. The lack of reproducibility is a source of random error. Sensitivity and error analyses are a means of examining the effect of the lack of reproducibility on a hydrologic design variable, such as peak discharge.

Example 10-2: Sensitivity of a Peak-Discharge Model

The SCS TR-55 graphical method can be used to demonstrate the use of sensitivity and error analysis. The graphical method gives the peak discharge (q_p) as a function of the unit peak discharge (q_u) in cfs/mile^2 per inch of runoff, the drainage area (A) in square miles, and the depth of runoff (Q) in inches:

$$q_p = q_u A Q \tag{10-31}$$

The unit peak discharge is a function of the time of concentration (t_c) in hours; for t_c greater than 0.7 hr the relationship is

$$q_u = 321.48 t_c^{-0.74946} \tag{10-32}$$

The runoff depth can be estimated as

$$Q = \frac{(P - 0.2S)^2}{P + 0.8S} \tag{10-33}$$

in which P is the 24-hour rainfall depth in inches for the selected exceedence probability, and S is the maximum retention, which is given by

$$S = \frac{1000}{\text{CN}} - 10 \tag{10-34}$$

in which CN is the runoff curve number. While the velocity method of computing t_c is recommended, for purposes of illustration the SCS lag method will be used:

$$t_c = \frac{1.67\text{HL}^{0.8}(S + 1)^{0.7}}{1900\, Y^{0.5}} \tag{10-35}$$

in which HL is the hydraulic length in feet and Y is the watershed slope in percent. The hydraulic length can be computed from the drainage area (A_a) in acres:

$$\text{HL} = 209 A_a^{0.6} \tag{10-36}$$

Using the equations above, the following sensitivity functions can be derived for the input variables of P, A, t_c, and CN:

$$\frac{\partial q_p}{\partial P} = q_p \left(\frac{2}{P - 0.2S} - \frac{1}{P + 0.8S} \right) \tag{10-37}$$

$$\frac{\partial q_p}{\partial A_a} = \frac{q_p}{A_a}(1 + 0.48b) \tag{10-38}$$

$$\frac{\partial q_p}{\partial t_c} = \frac{b q_p}{t_c} \tag{10-39}$$

$$\frac{\partial q_p}{\partial \text{CN}} = \frac{100 q_p}{\text{CN}^2}\left(\frac{-7\text{b}}{S+1} + \frac{4}{P-0.2S} + \frac{8}{P+0.8S}\right) \tag{10-40}$$

in which b is the exponent of Eq. 10-32. The sensitivity functions of Eqs. 10-37 to 10-40 are a function of the peak discharge (q_p) and the input variables (A_a, CN, and P). These sensitivity functions can be used to compute relative sensitivity functions, as follows:

$$R_p = \frac{\partial q_p}{\partial P}\frac{P}{q_p} = P\left(\frac{2}{P-0.2S} - \frac{1}{P+0.8S}\right) \tag{10-41}$$

$$R_{A_a} = \frac{\partial q_p}{\partial A_a}\frac{A_a}{q_p} = 1 + 0.48b \tag{10-42}$$

$$R_{t_c} = \frac{\partial q_p}{\partial t_c}\frac{t_c}{q_p} = b \tag{10-43}$$

$$R_{\text{CN}} = \frac{\partial q_p}{\partial \text{CN}}\frac{\text{CN}}{q_p} = \frac{100}{\text{CN}}\left(\frac{-7b}{S+1} + \frac{4}{P-0.2S} + \frac{8}{P+0.8S}\right) \tag{10-44}$$

in which R_p, R_A, R_{t_c}, and R_{CN} are the relative sensitivity functions for rainfall, drainage area, the time of concentration, and the curve number, respectively. These are a function only of the rainfall and curve number, with R_A and R_{t_c} being constant for a given value of b. They will vary with P and CN. The relative sensitivity functions were computed for rainfall depths of 3 to 5 in. and CN values of 65 and 80, with the results plotted in Fig. 10-7. The curves suggest that the CN is the most important input, with the CN having little effect on the relative sensitivity of the CN. Rainfall is the second most important variable, with the effect of CN on the relative sensitivity much greater than for the relative sensitivity of the CN. Drainage area and the time of concentration have similar relative sensitivities. It should be emphasized that the plots of Fig. 10-7 do not indicate that error in the CN will have the greatest effect, only the relative error will have the greatest effect when the relative errors are equal. For unequal relative errors, the relative error in the computed discharge will vary according to the relative sensitivity functions in Fig. 10-7.

The sensitivity functions of Eqs. 10-37 to 10-40 can be used with Eq. 10-4 to perform an error analysis:

$$\Delta R = \frac{\partial q_p}{\partial X}\Delta X \tag{10-45}$$

in which R_p, R_A, R_{t_c}, and R_{CN} are the relative sensitivity functions for rainfall, drainage input variable X, $\partial q_p/\partial X$ is one of the derivatives of Eqs. 10-37 to 10-40, and ΔX is a measure of the reproducibility. The WRC report (1981) reported coefficients of variation of 0.03, 0.02, 0.70, and 0.06 for P, A, t_c, and CN, respectively. The coefficient of variation can be converted to an average error by multiplying the coefficient of variation (C_{vx}) by the mean ($\bar{X}$) of the variable X:

$$\Delta X = C_{\text{vx}}\bar{X} \tag{10-46}$$

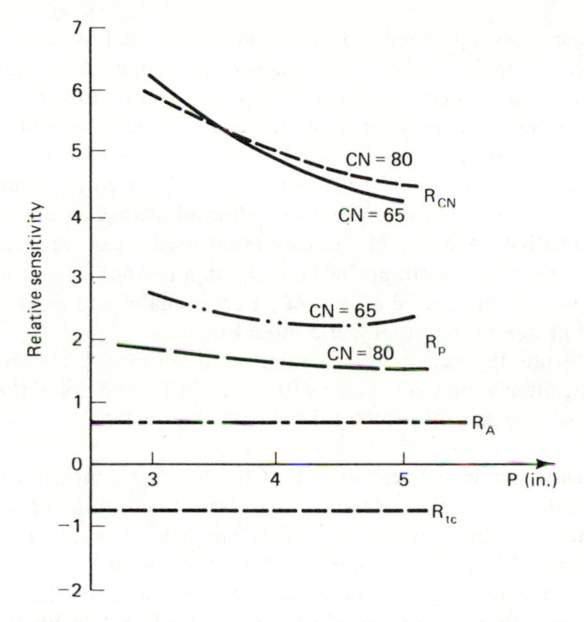

Figure 10-7 Relative sensitivity plots for the curve number (S_{CN}), rainfall (S_p), area (S_A), and time of concentration (S_{tc}) as a function of rainfall (P).

The effect of reproducibility errors on peak discharge can be computed with the equations above. For example, if we assume that $A_a = 500$ acres, $P = 5$ in., and $\text{CN} = 80$, we can compute $t_c = 1.498$ hr from Eq. 10-35 and the derivatives of Eqs. 10-37 to 10-40. The values of the reproducibility error using Eq. 10-41 are

$$\Delta R_P = \frac{\partial q_p}{\partial P}\Delta P = 161.9[0.03(5)] = 24 \text{ cfs} \tag{10-47}$$

$$\Delta R_A = \frac{\partial q_p}{\partial A_a}\Delta A_a = 0.687[0.02(500)] = 7 \text{ cfs} \tag{10-48}$$

$$\Delta R_{t_c} = \frac{\partial q_p}{\partial t_c}\Delta t_c = -268.6[0.7(1.498)] = -282 \text{ cfs} \tag{10-49}$$

$$\Delta R_{\text{CN}} = \frac{\partial q_p}{\partial \text{CN}}\Delta \text{CN} = 29.6[0.06(80)] = 142 \text{ cfs} \tag{10-50}$$

The values of Eqs. 10-47 to 10-50 indicate that the lack of reproducibility in the drainage area and rainfall cause little error in peak discharge but a very large error due to the lack of reproducibility of t_c. Drainage and stormwater management policies should be designed to minimize the problem associated with t_c.

Example 10-3: Sensitivity Analysis of a Rainfall–Runoff Model

Sensitivity analyses become more important as the complexity of the model becomes greater. Whereas there is little need for sensitivity analyses of peak discharge formulas such as the rational equation, every rainfall-runoff model should be subjected to a sensitivity analysis. As suggested before, a sensitivity analysis can be used in the formulation, calibration, and verification stages of modeling. A sensitivity analysis of a rainfall-runoff model might suggest changes to the structure of the model. An understanding of the sensitivity of a model can increase the efficiency of model calibration when the model must be calibrated subjectively (see Chapter 8); the sensitivity functions for the model coefficients would be used to estimate the change in a coefficient that is necessary to produce a desired change in one of the objective functions. A sensitivity analysis of the calibrated model can be used to assess the rationality of the relative importance of both the input variables and the components of the model; also, it can be used as part of an error analysis to assess the inaccuracy of a model output due to inaccuracy of a model input.

To demonstrate the sensitivity of a rainfall-runoff model, the model of Fig. 8-2 will be used. The fitted values associated with trial 3 in Table 8-3 and the computations of Table 8-4(c) will be assumed to be the calibrated model.

Use of sensitivity in calibration. In Chapter 8 the model coefficients were adjusted, with little explanation provided as to how changes in the coefficients were determined. The sensitivity equation (Eq. 10-4) can be used in subjective optimization to make an estimate of the needed change. For this use, ΔO_o of Eq. 10-4 would be the desired change in the objective function, $\partial O_o/\partial F_i$ would be the sensitivity of the objective function with respect to coefficient F_i, and ΔF_i would be necessary change in F_i. Thus ΔF_i is computed by

$$\Delta F_i = \frac{\Delta O_o}{\partial O_o/\partial F_i} \tag{10-51}$$

The sensitivity can be determined using values of O and F_i from two iterations.

The use of Eq. 10-51 can be illustrated using the data of Table 8-4. The value of the coefficient SROC was changed from 0.20 to 0.05 between trials 1 and 2. If we use the total computed runoff as the objective function, the sensitivity would be

$$\frac{\partial O_o}{\partial F_i} = \frac{\text{TRO}_2 - \text{TRO}_1}{\text{SROC}_2 - \text{SROC}_1} = \frac{1.16 - 0.39}{0.05 - 0.20} = -5.133 \tag{10-52}$$

in which the subscripts on TRO and SROC indicate the trial number. The value of -5.133 indicates an increase of 1.0 in SROC will cause TRO to decrease by 5.133. If we set as our objective to increase the computed total runoff so that it equals the total measured runoff, we would want to change SROC by an amount that would produce an increase in TRO of 0.11, which equals 1.27 minus 1.16. Using Eq. 10-51 with $\partial O_o/\partial F_i$ equal to -5.133 and ΔO_o equal to 0.11, we get a change in SROC of

$$\Delta\text{SROC} = \frac{0.11}{-5.133} = -0.021 \tag{10-53}$$

It is important to emphasize two points. First, the linear sensitivity function of Eq. 10-4 assumes that just a single variable F_i is changed, while all other F_j remain constant. In going from trial 1 to trial 2 in Table 8-3 several coefficients were changed. Thus Eq. 10-53 is only an approximation. Second, Eq. 10-4 assumed that the nonlinear terms were not important. Although it is not easy to determine whether or not this assumption is valid, it is very likely that it is not necessarily a good assumption because of the very large change in the objective function from trial 1 to trial 2 and the nonlinearity of the model. In spite of these problems, Eq. 10-53 would still provide a reasonably good "guesstimate" of the change that should be made in SROC to achieve the objective of making the total computed runoff equal to the total measured runoff.

One of the reasons for using subjective optimization is that more than one objective function can be used in reaching a "best fit." If instead of matching the total measured and computed runoffs, we were interested in matching the computed and measured peak discharges, we could use Eq. 10-31 with 0 being the value of the peak discharge. Thus we might be interested in knowing the change in SROC that would be necessary to make the computed peak discharge on day 5 match the measured peak discharge on day 5. Thus our objective function is different than that used in Eq. 10-53. Using the values from trials 1 and 2 [Table 8-4(a) and (b)], we can compute the sensitivity of the day 5 peak discharge to SROC:

$$\frac{\partial O_o}{\partial F_i} = \frac{Q_2 - Q_1}{\mathrm{SROC}_2 - \mathrm{SROC}_1} = \frac{0.38 - 0.10}{0.05 - 0.20} = -1.867 \tag{10-54}$$

We would like the two discharges to be equal, so we would want to increase the peak discharge by 0.07 (i.e., 0.45 − 0.38). Thus, using Eqs. 10-51 and 10-54, the desired change in SROC is

$$\Delta\mathrm{SROC} = \frac{0.07}{-1.867} = -0.0375 \tag{10-55}$$

The change indicated by Eq. 10-55 is about 75% greater than the change indicated by Eq. 10-53. Thus a value of −0.03 was used as a compromise.

In summary, the foregoing computations were made to show how sensitivity analysis can be used in calibrating a complex rainfall-runoff model. The sensitivity approach can increase the efficiency of the calibration process.

Sensitivity of input error. It is common in using measured rainfall and runoff data with hydrologic models to find either large runoffs occurring even when the rainfall was light or small runoffs when the rainfall was heavy. This occurs when the volume of rainfall measured at a raingage is not representative of the volume (or depth) of rain that fell on the watershed. For this reason it is good to have an idea of the sensitivity of runoff to rainfall so that adjustments are not made to the coefficients to compensate for the unrepresentativeness of the catch at the raingage.

The sensitivity of the runoff to error in the rainfall can be computed by incrementing a rainfall depth by some small amount and computing the resulting change in runoff. For example, if we assume the calculations of Table 8-4c to be the optimum conditions, we can adjust the rainfall on day 5 by 0.05 in. and compute the runoffs on days 5 and 6. These calculations are given in Table 10-1. The increase in rainfall on day 5 caused an increase in runoff on day 5 of 0.05 and on day 6, an increase of 0.02 in. Thus the sensitivity of runoff to rainfall on day 5 was 1 while on day 6 the sensitivity was 0.4. These sensitivities indicate that any error in rainfall measurements will translate directly into error in runoff on the same day and a lesser effect on following days.

Sensitivity to error in input parameters. As indicated previously, error in the values of input parameters may result in errors in the predicted output of the model.

TABLE 10-1 Computations to Estimate Sensitivity of Runoff to Rainfall

Day	PR	SROS	SRO	SROS	PINF	INF	SROS	GWS	ETS	SROS	GWRO	GWS	TRO	
													Computed	Measured
5	1.35	1.35	0.36	0.99	0.43	0.43	0.56	2.03	0.25	0.31	0.13	1.90	0.49	0.45
6	0.00	0.31	0.02	0.29	0.42	0.12	0.17	2.02	0.17	0.00	0.13	1.89	0.15	0.13

The inaccuracy of the predicted values will depend on the sensitivity of the output to the parameter, which according to Eq. 10-3 is a function of the values of all other parameter values, as well as the value of the parameter of interest. A linear sensitivity function can be used to assess the sensitivity of either the output or the value of an objective function to a parameter, and the sensitivity equation (Eq. 10-4) can be used to assess the effect of error in the parameter.

The sensitivity and error analysis of a rainfall–runoff watershed model can be demonstrated using the parameter SROP. Using the parameters for the third trial of Table 8-3 as the optimum solution, with the results shown in Table 8-4(c), the sensitivity can be determined by incrementing SROP by a small amount and measuring the change in the output or an objective function. The value of SROP was increased by 3.0 to 78 and the output recomputed for the data of Table 8-2; the resulting values of the computed runoff are shown in Table 10-2. If fitting the total runoff is the objective, the sensitivity of the objective function to SROP can be computed using Eq. 10-5:

$$\frac{\partial \text{TRO}}{\partial \text{SROP}} = \frac{\text{TRO}_4 - \text{TRO}_3}{\text{SROP}_4 - \text{SROP}_3} = \frac{1.52 - 1.38}{78 - 75} = 0.0467 \tag{10-56}$$

in which the subscripts 3 and 4 refer to the optimum trial [i.e., trial 3 of Table 8-4(c)] and the incremented trial of Table 10-2, respectively. If we assume that the estimated value of SROP is accurate to within ± 1, Eq. 10-4 can be used to estimate the effect of the error in SROP on the total runoff:

$$\Delta\text{TRO} = \frac{\partial \text{TRO}}{\partial \text{SROP}} \Delta\text{SROP} = 0.0467(1) = 0.0467 \text{ in.} \tag{10-57}$$

In some cases, the estimated sensitivity can be computed from the results of the computer runs for the last few trials of the calibration process. However, the sensitivities may only be approximations if the last few trials involved simultaneous changes to two or more fitting coefficients.

It is important to remember that parametric sensitivities can be computed for more than one objective function. If our objective was to provide accurate fits of the peak discharges, we might be interested in the sensitivity of the peak discharge to the accuracy of input parameters. For the rainfall–runoff model we can use the results of Tables 8-6 and 10-2 to compute the sensitivity of the discharge on day 5 to SROP:

$$\frac{\partial Q_p}{\partial \text{SROP}} = \frac{0.50 - 0.44}{78 - 75} = 0.02 \tag{10-58}$$

The effect of the probable error in SROP on the peak discharge can be computed using Eqs. 10-4 and 10-58:

$$\Delta Q_p = \frac{\partial Q_p}{\partial \text{SROP}} \Delta\text{SROP} = 0.02(1) = 0.02 \tag{10-59}$$

Thus an error of 1.0 in SROP would produce an error of 0.02 in the computed peak discharge.

Component sensitivity of the rainfall–runoff model. The sensitivity of a model component can be computed to examine both the importance of the component and the rationality of the structure of the component. Although the data of Table 8-2 are

TABLE 10-2 Sensitivity of Model Output to Input Parameter SROP

Day	PR	SROS	SRO	SROS	PINF	INF	SROS	GWS	ETS	SROS	GWRO	GWS	TRO Computed	TRO Measured
0	—	—	—	—	—	—	—	—	—	0.50	—	1.50	—	—
1	0.70	1.20	0.33	0.87	0.44	0.38	0.49	1.88	0.25	0.24	0.12	1.76	0.45	0.39
2	0.20	0.44	0.08	0.36	0.43	0.15	0.21	1.91	0.21	0.00	0.12	1.79	0.20	0.14
3	0.00	0.00	0.00	0.00	0.43	0.00	0.00	1.79	0.00	0.00	0.12	1.67	0.12	0.09
4	0.00	0.00	0.00	0.00	0.43	0.00	0.00	1.67	0.00	0.00	0.11	1.56	0.11	0.07
5	1.30	1.30	0.38	0.92	0.44	0.40	0.52	1.96	0.25	0.27	0.12	1.84	0.50	0.45
6	0.00	0.27	0.02	0.25	0.44	0.11	0.14	1.95	0.14	0.00	0.12	1.83	0.14	0.13
													1.52	1.27

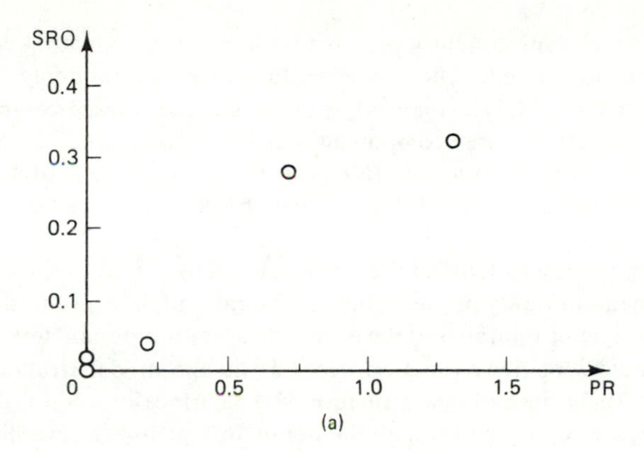

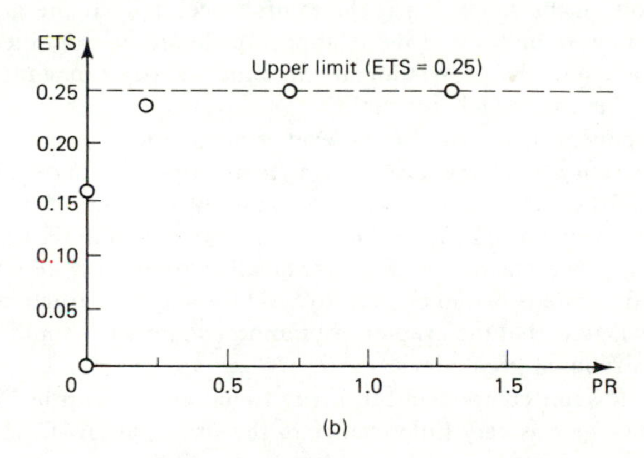

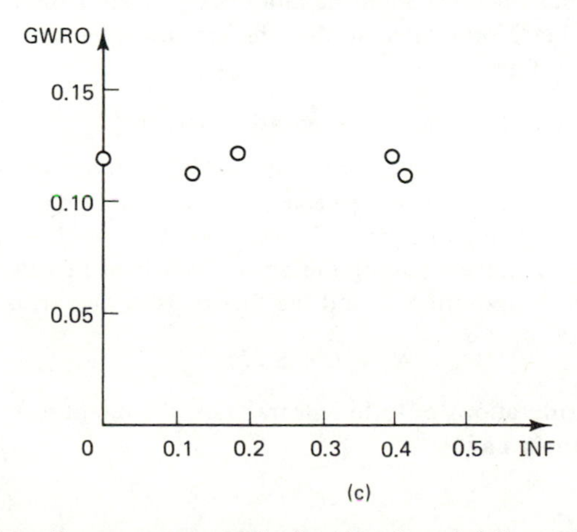

Figure 10-8 Component sensitivity functions for rainfall-runoff model: (a) surface runoff (SRO) versus precipitation (PR); (b) evapotranspiration (ETS) versus precipitation (PR); (c) ground-water runoff (GRWO) versus infiltration (INF).

too sparse for developing conclusions, a methodology for evaluating component sensitivity can be demonstrated. The model includes two components (i.e., the surface SROS and groundwater GWS storages), with the surface storage component including two outputs. Therefore, three-component sensitivity analyses (i.e., SRO versus PR, ETS versus PR, and GWRO versus INF) can be made. Figure 10-8 shows the input versus output for the data from Table 8-6, which was considered to be the calibrated solution.

The component sensitivity of the surface runoff to rainfall is shown in Fig. 10-8(a). The values depend not only on the values of the rainfall, but on the storage that exists in SROS at the time of rainfall and the evapotranspiration parameters. Also, the shape of the component sensitivity function depends on the model structure; in this case, Eqs. 10-33 and 10-34 describe the structure of the surface runoff-rainfall relationship. If we fit by eye a straight line through the points in Fig. 10-8(a), we find that the slope of the function is approximately 0.3. This indicates that 30% of the rainfall appears as surface runoff on the same day as the rainfall occurred. If the graph were based on a larger data base, the form of the relationship should be examined for rationality in both magnitude and shape. Irrationality in either shape or magnitude provides the basis for modifying the model structure.

The component sensitivity of the evapotranspiration output is shown in Fig. 10-8(b). The function is characterized by a steep slope for small rainfalls, with the upper limit of ETS controlling the shape for rainfalls of about 0.5 in. and above. For large rainfalls the value of ETS is completely insensitive to rainfall, with evapotranspiration occurring at the maximum. For very small rainfalls (i.e., less than 0.2 in.) the evapotranspiration rate is extremely sensitive to rainfall. The structure appears highly irrational and suggests that the evapotranspiration component should be reformulated so that it yields rational predictions.

The groundwater component sensitivity function is shown in Fig. 10-8(c). The points show that there is very little change in the daily outflow from the component as the input changes. This may be considered rational for very large watersheds, but it is probably not rational for most small or moderately-sized watersheds. Thus the model structure should be reformulated so that the groundwater component shows a rational response to the infiltration.

Example 10-4: Sensitivity of the Penman Evaporation Model

Both empirically derived prediction equations and model structures that were synthesized from theoretical considerations have been used for estimating evaporation rates from open water surfaces. Penman proposed an evaporation model that was derived from a simplified energy balance equating the net radiant heat flux density R_n to the sum of the latent heat of evaporation E and the flux density of sensible heat S:

$$R_n = E + S \tag{10-60}$$

From aerodynamic considerations of turbulent transfer, the latent heat flux density is given by a form of Dalton's Law

$$E = f_o(v)(e_o - e_a) \tag{10-61}$$

and the sensible heat transfer is given by

$$S = \gamma f_o(V)(T_o - T_a) \tag{10-62}$$

where γ is the psychrometric constant, $f_o(V)$ is a wind function, e is the vapor pressure, T is the temperature, and the subscripts o and a refer to water surface and air quantities, respectively. Equations 10-60, 10-61, and 10-62 provide the basis for the Penman Model which is given by

$$E_o = \frac{\Delta R_n + \gamma E_{ao}}{\Delta + \gamma} \tag{10-63}$$

where E_o is the evaporation estimate, Δ is the slope of the saturation vapor pressure curve at air temperature T_a, and E_{ao} is an empirical representation of the latent heat flux density given by

$$E_{ao} = 0.35\left(0.5 + \frac{V}{100}\right)(e_o - e_a) \tag{10-64}$$

where V is the windspeed in miles per day.

Equation 10-6 can be used to derive sensitivity functions for the Penman evaporation procedure. The relative sensitivity defined by Eq. 10-12 can then be computed using the sensitivity functions and the mean values of the evaporation and the meteorological factors.

Both the slope of the saturation vapor pressure curve and the vapor pressure deficit $(e_o - e_a)$ are a function of the air temperature T_a. Therefore, the sensitivity with respect to the air temperature T_a is derived from the following total derivative of Eq. 10-63:

$$S_T = \frac{dE_o}{dT_a} = \frac{dE_o}{d\Delta} \cdot \frac{d\Delta}{dT_a} + \frac{dE_o}{d\bar{e}} \cdot \frac{d\bar{e}}{dT_a} \tag{10-65}$$

where $\bar{e}$ is used to represent the vapor pressure deficit $(e_o - e_a)$. The derivative $d\bar{e}/dT_a$ is the quantity Δ and the derivative $d\Delta/dT_a$ is then $d^2\bar{e}/dT_a^2$. For the range of air temperatures used, the second derivative $d^2\bar{e}/dT_a^2$ and the product $(dE_o/d\Delta)(d\Delta/dT_a)$ are very small. Thus, the sensitivity function S_T can be approximated by

$$S_T = \Delta \cdot \frac{dE_o}{d\bar{e}} \tag{10-66}$$

where $dE_o/d\bar{e}$ represents the rate of change of evaporation with respect to change in the vapor pressure deficit, thus,

$$S_T = \frac{0.35\Delta\gamma(1 - 0.01H)(0.5 + 0.01V)}{\Delta + \gamma} \tag{10-67}$$

Similarly, the derivative approach can be used to derive the sensitivity functions for net radiation, air speed and humidity, respectively:

$$S_{R_n} = \frac{\Delta}{2758(\Delta + \gamma)} \tag{10-68}$$

$$S_v = \frac{0.35\,\gamma e(1 - 0.01H)}{100(\Delta + \gamma)} \tag{10-69}$$

$$S_H = \frac{-0.35\,\gamma e(0.5 + 0.01V)}{100(\Delta + \gamma)} \tag{10-70}$$

where e is the saturation vapor pressure.

Since Δ is a function of air temperature, the sensitivity function S_{R_n} is a function of only air temperature. The air speed sensitivity function depends on the levels of humidity and air temperature while the sensitivity function of humidity is a function of air temperature and air speed. The temperature sensitivity function depends on the humidity H, air speed V, and air temperature.

Sources of error in the meteorological data used for estimating evaporation from open water surfaces include: (1) inaccuracy of the recording instrument; (2) recording instrument calibration error; (3) observational error; (4) data processing error; and (5) the inability of the recording instrument to provide a quantitative indication of the intended meteorological characteristics, e.g., the horizontal speed of air masses is used to represent air turbulence. By maintaining the recording instruments in proper calibration and by using consistent practices in the collection and processing of the data, the primary source of error can be limited to the inaccuracies of the recording instruments. Estimates of error contained in the recorded measurements of meteorological factors resulting from the inaccuracy of the instruments can be obtained from the instrument specifications supplied by the manufacturer. From such sources of information, instrument error functions were developed for the meteorological factors required by the Penman method.

Radiometers, which are commonly used for the measurement of radiation, provide measurements that are expected to be within 0.5% of the true value. Thus, the following is assumed to be a reasonable error function:

$$E_R = 0.005R \tag{10-71}$$

where E_R is the expected error in the measurement of radiation R.

Air speed is measured by instruments called anemometers, of which there are several types. Such instruments are often less accurate at low air speeds due to the frictional resistance threshold of the instruments. Thus, manufacturers' specifications often specify a two-phase error function. The following is assumed to be a reasonable error function for air speed measurements:

$$E_V = \begin{cases} 0.15 \text{ mph} & \text{for } V \le 5 \text{ mph} \\ 0.03\ V & \text{for } V > 5 \text{ mph} \end{cases} \tag{10-72}$$

where V is the air speed in miles per hour.

Temperature measurements are often reported to the nearest degree of Fahrenheit even though values can be read to the nearest tenth of a degree. Thus, there is a potential error of one-half of a degree in any single observation:

$$E_T = 0.5°\text{F} \tag{10-73}$$

will be used as the error function for observations of temperature.

Psychrometers, instruments used for measuring humidity, can be calibrated such that a selected range of the relative humidity scale will provide more accurate measurements than measurements outside the specified range. If the majority of the measurements will be between 20 and 80 percent relative humidity, the psychrometer can be calibrated such that the following is a reasonable error function:

$$E_H = \begin{cases} 1\% & 20\% \le H \le 80\% \\ 3\% & \text{otherwise} \end{cases} \tag{10-74}$$

Equation 10-4 provides a means for estimating the effect on evaporation estimates of data error that results from imprecise measurements of the meteorological factors. The sensitivity and error functions previously defined can be used in Eq. 10-4. Values for the sensitivity and error functions were determined using information from a *Climatic Atlas of the United States.* Average values of the meteorological factors for the May–October period were used because the evaporation is greatest during that period. The potential errors, as indicated using Eq. 10-4, for five geographic locations are given in Table 10-3. The evaporation error values indicate that the meteorological factor measurement error is very small compared with the mean daily evaporation for the May–October period. For the Penman model, the largest error is less than 5% of the daily evaporation.

TABLE 10-3 Error in Evaporation Estimates Resulting from Data Measurement Error

	Error in evaporation estimate				
Location	$S_H E_H$ (in./day)	$S_V E_V$ (in./day)	$S_R E_R$ (in./day)	$S_T E_T$ (in./day)	Mean daily evaporation (in./day)
Miami, Fla.	−0.00521	0.000623	0.000676	0.000694	0.178
El Paso, Texas	−0.00101	0.00174	0.000722	0.00234	0.285
Rapid City, S.D.	−0.00129	0.00104	0.000594	0.00157	0.208
Elkins, W.V.	−0.00100	0.000598	0.000501	0.000583	0.133
Boston, Mass.	−0.00162	0.000769	0.000485	0.00116	0.164

Rates of evaporation from open water surfaces vary, depending in part on meteorological conditions. The relative sensitivity of the meteorological factors used in the estimation of evaporation rates provides a means of quantitatively examining the relative influence of changes in the level of the meteorological factors on computed evaporation rates. In order to estimate the relative sensitivity of the various meteorological factors, it is necessary to specify a model from which values of relative sensitivity can be computed. The sensitivity function, and thus the relative sensitivity, are computed directly from the functional representation of the model. Thus, the model should be structured from consideration of the underlying physical principles. Such a model will provide more realistic estimates of factor sensitivity. An arbitrarily selected mathematical equation or an oversimplified model can provide an unrealistic indication of factor importance even though the equation provides good estimates of evaporation rates.

Using mean values of four meteorological factors (T_a, R, H and V) for the period from May-to-October, values of the sensitivity functions were computed and used to derive values of relative sensitivity. Meteorological data from thirteen locations in the United States were used. The resulting relative sensitivity values are given in Table 10-4. The relative sensitivity values of Table 10-4 indicate that humidity, or vapor pressure deficit, is the most sensitive meteorological factor for Miami, Elkins, Liberty, Raleigh, Charleston, and Boston. These six cities are located near the Atlantic Ocean. Due to the storage and release of heat, one would expect such a body of water to significantly influence evaporation rates. The availability of water vapor from the

TABLE 10-4 Relative Sensitivities of Meteorological Factors for Selected Cities

Location	Meteorological factors			
	Humidity	Wind	Radiation	Temperature
Miami, Fla.	−0.98	0.19	0.76	0.63
Ft. Worth, Tex.	−0.71	0.35	0.58	1.05
El Paso, Tex.	−0.35	0.41	0.51	1.24
Wichita, Kans.	−0.79	0.36	0.58	1.04
Rapid City, S.D.	−0.62	0.36	0.57	0.95
Peoria, Ill.	−0.78	0.28	0.65	0.81
Nashville, Tenn.	−0.70	0.22	0.70	0.73
Cincinnati, Ohio	−0.62	0.20	0.73	0.66
Elkins, W.V.	−0.75	0.18	0.75	0.53
Liberty, N.Y.	−0.79	0.24	0.69	0.66
Boston, Mass.	−0.99	0.34	0.59	0.93
Raleigh, N.C.	−0.79	0.22	0.72	0.68
Charleston, S.C.	−0.91	0.21	0.73	0.67

Atlantic Ocean serves as a barrier to increased evaporation rates and thus, variation in the level of humidity greatly influences variation in computer evaporation rates.

For Boston, air temperature is the second most sensitive factor, while for the other five previously mentioned locations, radiation is more sensitive than temperature. From the values of relative sensitivity, it appears that radiation becomes less influential with increasing latitude. The simultaneous increase in sensitivity of wind and temperature in Boston could be an indication of an increase in importance of the sensible heat flux density.

The sensitivity of evaporation rates to changes in humidity is less in the midwestern states than in the eastern states. Temperature is a more sensitive factor than humidity for the cities of Fort Worth, Wichita, Rapid City, Peoria, Nashville, and Cincinnati. Although high evaporation rates result from a large vapor pressure deficit, the increased sensitivity to temperature indicates that the variation in evaporation rates stems from the variation in temperature levels.

For comparatively arid climates, such as that of El Paso, variation in temperature produces the greatest variation in evaporation rates. The temperature relative sensitivity value of 1.24, which is the largest value in Table 10-4, is an indication of the great influence of temperature variation on computed evaporation rates. Although the relative sensitivity value of radiation indicates that radiation is more sensitive than wind or humidity, it should be stressed that the value of 0.51 for radiation and 0.41 for wind are the lowest and highest respectively, for the thirteen cities included in the analysis. Also, a relative sensitivity value of −0.345 for humidity is the lowest humidity relative sensitivity for the thirteen locations. Thus, variations in evaporation rates in an arid climate appear to be controlled by temperature with wind speed being a more influential factor than in the more humid regions.

Wind is the least sensitive factor in twelve of the thirteen locations. In general, on a long term basis, a change of ten percent in wind speed results in a change in evaporation of one to three percent. Such changes would correspond to relative sensitivity values of 0.1 and 0.3. The values in Table 10-4 vary from 0.18 to 0.41 with a mean value of 0.27. Thus, the relative sensitivity values derived from the Penman Model appear to be rational and provide a good indication of the relative effect of changes in the meteorological factors.

LIMITATIONS OF SENSITIVITY ANALYSIS

Sensitivity analysis was shown to be a useful tool for all phases of modeling (formulation, calibration, and verification), as well as part of an error analysis for decision making. However, it has some limitations. First, sensitivity analysis is usually applied using the linear sensitivity equation (Eq. 10-4); however, the linear form is valid only over a limited range of the variable in the denominator. Since most hydrologic models are nonlinear, the sensitivity changes over the range of each parameter. In most applications, sensitivity coefficients are usually computed using the means of the variables as a base point. However, the sensitivity at the extreme values of the physical conditions is often of primary interest in hydrologic design.

The univariate nature of sensitivity is a second limitation. For the case of the hydrograph model of Fig. 10-2, the sensitivity functions for n and K were derived while holding the other variable constant at some base-point value. Similarly, for the error analysis of Eqs. 10-47 to 10-50, the reproducibility errors ΔR_i represent the "independent" effects of the four input variables. That is, the reproducibility errors assume error only in the single variable of interest. An underlying assumption is that there is no interaction between the variables. This is obviously not true because the time of concentration depends on both the drainage area and the curve number (Eqs. 10-35 and 10-39). Also, we cannot assume that any user on one design will make an error on only one variable. However, because of the interaction between variables one cannot just simply add the reproducibility errors. For example, even if a user made an error of one standard error in each of the variables, the net effect would not be the sum of the reproducibility errors of Eqs. 10-47 to 10-50. For the data given, the peak discharge would be 537 cfs. If we use a change of one standard deviation, as with Eqs. 10-47 to 10.50, we would have changes in P, A, and CN of 0.15 in., 10 acres, and 4.8, respectively. The change in CN tends to decrease the t_c, while the change in drainage area tends to increase the t_c. The net effect is a decrease in t_c from 1.50 hr to 1.29 hr. The changes in P, A, and CN cause the peak discharge to increase from 537 cfs to 738 cfs, an increase of 201 cfs. Adding the reproducibility errors for P, A, and CN indicates a change of 173 cfs, which is an error of −14%. While the difference is relatively small, it is only because the coefficients of variation of P, A, and CN are small. If the time of concentration was computed by another method such as the velocity method, the large coefficient of variation of 0.7 would result in a very large error in the computed peak discharge. Thus the univariate nature of sensitivity is a limiting factor for large errors or deviations in the variables.

The third, and probably the most important, limitation of sensitivity analysis is that it provides only a single-valued indication of the effect on the criterion or dependent variable. Ideally, one would like to have some idea of the distribution of the design variable. One could approximate this by using the linearized sensitivity equation with the distribution of the error or variation of the independent or input variable, but this approach is limited in usefulness because of the first two limitations (i.e., linearity and univariate). A method that circumvents the limitations of sensitivity analysis would be an improvement for many analyses of design methods. This does not imply that sensitivity is not of value, only that advanced forms of sensitivity analysis are needed. The linear, univariate form of sensitivity analysis will still be of value for analyses described previously.

PROBABILISTIC MODELING

Probabilistic modeling is an extension of sensitivity analysis with the important distinctions that (1) it is not limited to the univariate form, (2) it allows for the interaction of the input variables, and (3) it provides the distribution of the design variable (i.e., dependent or criterion variable), not just a single-valued measure of the dispersion of the design variable. Of course, there is a price to pay for the additional information that is obtained from a probabilistic analysis. First, we must know the distributions of the input variables, and second, the computational requirements for a probabilistic analysis are much greater than those required for a linear, univariate sensitivity analysis.

A probabilistic analysis is based on an iterative analysis of the design model for a sufficient number of conditions so that the distribution of the design variable is identified. While a numerical analysis of sensitivity requires two iterations of the design model in order to define the sensitivity by Eq. 10-7, the probabilistic approach involves numerous solutions of the design model. While the sensitivity analysis of Eq. 10-7 is univariate, with each variable being perturbed independently of the others, probabilistic analysis uses a simultaneous perturbation of all the design variables. While the sensitivity analysis requires only a measure of the dispersion or error variation of the input variables, a probabilistic analysis requires knowledge of the entire distribution function of the input variables.

Actually, probabilistic analysis is quite simple. Given the distribution functions of the input variables, values of the input variables are generated using values of random variables having the specified distribution functions and used to compute a single value of the design variable. This process is repeated until a sufficient number of values of the design variables have been generated to define the distribution function of the design variable.

A simple computational example can be used to illustrate the methodology. Let's assume that a design process consists of two input variables, X_1 and X_2, which are both normally distributed. The variable X_1 has a mean and a standard deviation of 5 and 1, respectively. Given that both are normal, values of X_1 and X_2 can be generated using a random normal number generator with the appropriate statistics. A sample of 25 was generated for each of the input variables (see Table 10-5). The value of a design variable Y is related to X_1 and X_2 by

$$Y = 7.0 + 0.6X_1 + 1.6X_2 \tag{10-75}$$

The 25 values of Y generated with Eq. 10-75 are also given in Table 10-5. It is known from theory that the sum of m independent random variables, each normally distributed, is also normally distributed for a linear combination. Thus, since Eq. 10-75 is a linear equation and X_1 and X_2 are normally distributed, Y must also be normally distributed. It can also be shown from theory that the mean and variance of Y are sums of the linear combinations of the input variables X_1 and X_2. However, the important point here is that we used randomly generated values of the input variables to generate the

TABLE 10-5 Generated Values of a Design Variable (Y) as a Function of Two Input Variables (X_1 and X_2)

i	X_1	X_2	Y
1	5.080	4.889	17.870
2	1.874	4.884	15.939
3	2.574	5.948	18.061
4	4.796	4.309	16.772
5	2.726	5.952	18.159
6	0.830	3.422	12.973
7	−0.736	5.677	15.642
8	3.122	3.740	14.857
9	1.932	3.270	13.391
10	3.508	5.421	17.778
11	2.640	3.702	14.507
12	2.618	5.705	17.699
13	3.322	4.956	16.923
14	2.754	4.740	16.236
15	2.570	4.842	16.289
16	5.506	5.321	18.817
17	2.708	5.285	17.081
18	0.414	4.444	14.359
19	2.468	5.920	17.953
20	−0.638	5.245	15.009
21	4.338	4.239	16.385
22	4.620	5.461	18.510
23	0.608	3.746	13.358
24	4.422	6.241	19.639
25	2.264	2.892	12.986

values of the design variable, from which an estimate of the underlying distribution of Y was determined. In the example used, the true distribution could have been derived from theory only because the input variables had a normal distribution and Eq. 10-75 defines Y to be a linear combination of X_1 and X_2. In practice, the input variables are frequently nonnormal and the relationships between Y and the input variables are not linear. In that case it would be impossible to derive the distribution of the design variable Y from theory. For example, the equations relating the design variable q_p of Eq. 10-31 and the input variables P, CN, and A are not linear. Therefore, theory could not be used to derive the distribution function of q_p; however, a probabilistic analysis could be used.

In most cases, sample sizes of 1000 or more are required to generate an accurate estimate of the probability distribution of the design variable. Therefore, probabilistic analyses are conducted on a computer. To illustrate the design of a probabilistic analysis and an interpretation, an application for the design of a two-stage riser of a stormwater detention basin will be used.

Example 10-5: Design Accuracy of Detention Basins

Detention basins are used in developing areas to control flood runoff. Figure 10-9 shows a schematic cross section of a detention basin. Numerous design methods are available for sizing the outlet structure.

Probably the greatest source of inaccuracy in detention outlet design is the hydrologic model that is used to estimate the design discharge rates. Most hydrologic models that are used to compute peak discharges for storm-detention-basin design are based on four general types of input: (1) rainfall; (2) land use; (3) watershed area; and (4) timing of runoff, which is usually expressed as the time of concentration. Rainfall is often characterized by a design intensity (i.e., the T-year, X-hour rainfall depth). Land use can be represented by an index such as the runoff coefficient for the rational method or the SCS runoff curve number. Each of these inputs is subject to error, and error in an input parameter causes inaccuracy in the design. Error in an input parameter can be due to inconsistent evaluation of the underlying data. If the

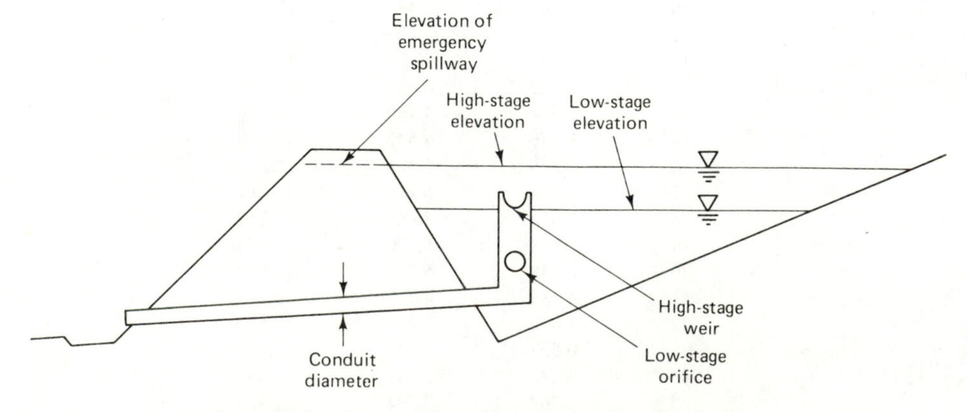

Figure 10-9 Cross section of stormwater detention basin.

rainfall depth is obtained from a rainfall intensity–duration–frequency graph, reading errors cause variation; this error is usually small. The WRC report indicates a mean coefficient of variation of approximately 3%. Errors in the runoff curve number can result during the integration of the soils and land-use data. The mean coefficient of variation for the runoff curve number is approximately 5%. Delineation of the watershed area introduces a random error that can be characterized by a coefficient of variation of 5%. The time of concentration is the input that has the greatest potential for error since it requires delineation of a flow path, measurement of slopes and lengths, estimation of the flow velocities, and is dependent on the topographic and site maps. The WRC reported a coefficient variation of 20 to 70% for the time of concentration estimates with the variation dependent on the method used to estimate the time of concentration.

Errors in input affect designs. The error in a design will depend on the magnitude of the errors in the input parameters. Although errors in different input variables may compensate each other, they can also compound each other. To examine the effect of errors in the input to hydrologic models on stormwater detention basin design, a probabilistic approach was used to simulate the probability distributions of five design characteristics for multistage outlet facilities: the conduit diameter, the area of the low-stage orifice, the length of the high-stage weir, and both the low-stage and high-stage storage volumes. A drainage area of 50 acres was used, with 24-hour rainfall depths of 3.2 and 4.8 in.; these rainfall depths correspond to the 2-year and 10-year events. A corregated metal riser was assumed with a friction factor that varied with the pipe diameter. The SCS two-stage riser design method was used in computing values of the five design characteristics. The error variations discussed previously were used to reflect variation expected in actual design. The true values of the design characteristics are given in Table 10-6; the minimum, maximum, mean, and standard deviation of the simulated values are also given. The difference between the mean and the true value represents the best estimate of the bias. The distribution of the design characteristics is given in Table 10-7 and Figs. 10-10, 10-11, and 10-12. It is evident from Table 10-6 that the storage volumes are unbiased but have a large error variation. The large error variation is due to the high coefficient of variation for the peak discharge estimates. The riser diameter and the high-stage weir length are highly dependent on the magnitude of the high-stage before-development peak discharge. The low-stage orifice area is highly dependent on the low-stage before-development peak discharge. The storage

TABLE 10-6 Summary of Statistical Characteristics of Design Characteristics

Design Characteristic	True Value	Mean Value	Bias	Std. Dev. of Errors	Minimum Value	Maximum Value
Riser diameter (ft)	2.25	2.65	0.40	1.16	1.25	6.5
Low-stage orifice (ft^2)	0.959	2.09	1.13	3.10	0.415	14.3
Low-stage storage volume (ac-ft)	5.555	5.545	−0.01	0.30	4.784	6.73
High-stage storage volume (ac-ft)	7.258	7.200	−0.06	0.42	5.971	8.682
High-stage weir length (ft)	3.436	5.937	2.50	6.62	1.039	33.70

TABLE 10-7 Computed Probability Distribution of Pipe Diameters with a True Diameter of 21 Inches, with Different Coefficients of Variation for Time of Concentration Estimates

Pipe Diameter (in.)	Probability for $C_V = 0.7$	Probability for $C_V = 0.2$
15	0.001	0.000
18	0.027	0.000
21	0.164	0.018
24	0.204	0.275
27	0.189	0.445
30	0.118	0.209
33	0.071	0.044
36	0.050	0.007
42	0.048	0.002
48	0.023	
54	0.015	
60	0.012	
66	0.022	
72	0.044	
78	0.012	
	1.000	1.000

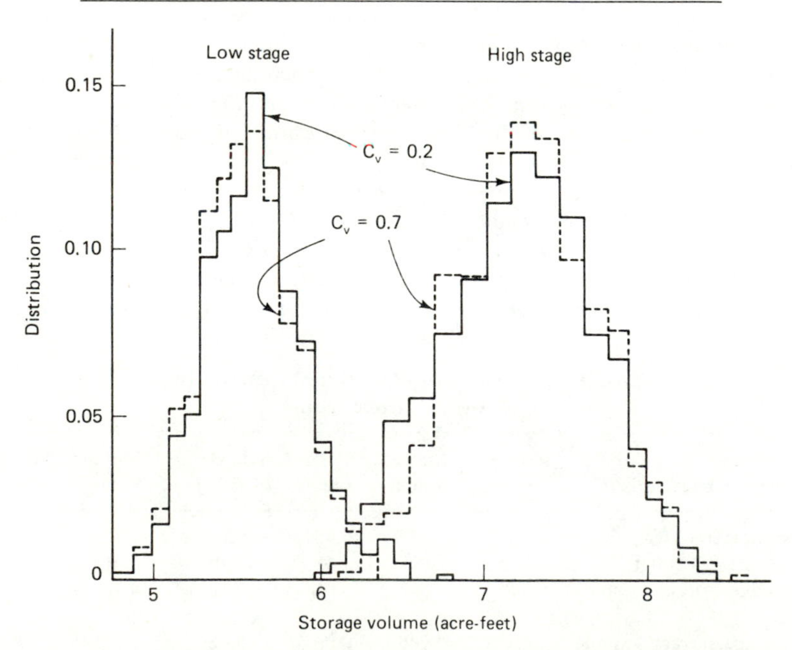

Figure 10-10 Distributions of high-stage and low-stage storage volumes due to variations in inputs for peak discharge estimation with selected coefficients of variation (C_v) for the time of concentration.

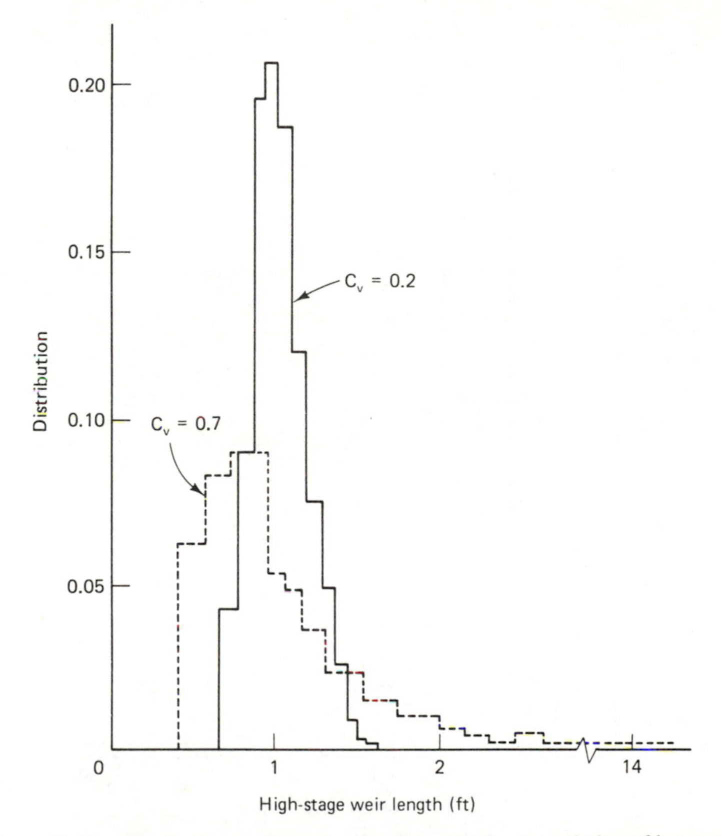

Figure 10-11 Distributions of high-stage weir length due to variation of inputs for peak-discharge estimation with selected coefficients of variation (C_v) for the time of concentration.

volumes depend more on the change in land use, which was represented here by the change in the curve number.

The coefficient of variation for time of concentration estimates was at least 14 times as great as the error variations of the other inputs; therefore, one might conclude that the wide distributions of Table 10-7 and Figs. 10-10, 10-11, and 10-12 are due to the variation in estimates of the time of concentration. The accuracy of t_c values depends highly on the method used to estimate t_c, and numerous methods are available for estimating t_c. The WRC report (1981) indicates that the mean coefficient of variation for t_c can be as small as 0.2. Table 10-7 and Figs. 10-10, 10-11, and 10-12 show the distributions of the design parameters for coefficients of variation of 0.2 and 0.7. It is evident that increased accuracy in t_c estimates could reduce substantially the errors in three of the design parameters: the riser diameter, the low-stage orifice area, and the high-stage weir length. Figure 10-10 indicates that it has no effect on the volumes because the volumes are more highly dependent on the change in runoff characteristics rather than the before or after peak discharge. But the other distributions indicate that SWM policies can affect the accuracy of designs by requiring a consistent method of

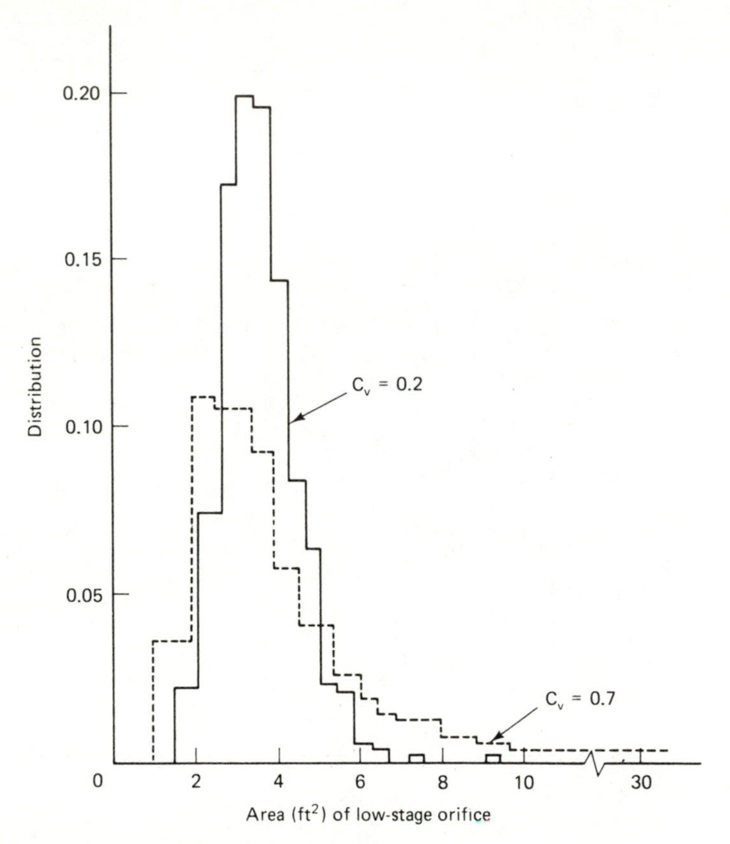

Figure 10-12 Distributions of low-stage orifice area due to variation in inputs for peak-discharge estimation with selected coefficients of variation (C_v) for the time of concentration.

estimating t_c and providing for adequate evaluation of t_c estimates in the project plan review stage.

EXERCISES

10-1. Based on Eqs. 10-2 and 10-3, graphically show the error introduced into the linear sensitivity equation when nonlinear terms are ignored. Use the function

$$O = 2F^3 - 3F^2 + 5F - 7$$

and evaluate the sensitivity at $F = 2$ and $\Delta F = \pm 0.2, \pm 0.5$.

10-2. For the sensitivity of SCS runoff model (Eq. 10-31) to the curve number, evaluate the magnitude of the second-order term of Eq. 10-2 for the case when CN = 75, $P = 5$ in., $A = 0.01$ square mile, and a change in CN of 0.5. Compare the magnitude of the second-order term with that of the linear term.

10-3. Using a Taylor series expansion with two independent variables (F_1 and F_2), derive a linear function for the change in the output function as a function of change in F_1 and F_2 of ΔF_1 and ΔF_2, respectively. Using the total runoff (TRO) as the objective function, compute the expected change in TRO for the changes in SROP and SROC between trial 1 [Table 8-4(a)] and trial 2 [Table 8-4(b)]. Compare this with the actual change and explain the difference.

10.4. Derive the absolute and component sensitivity function for the depth of runoff (Q) in Eq. 10-33 as a function of the rainfall (P) and the curve number (CN). Derive relative sensitivity plots for rainfall depths from 3 to 5 in. and CN values of 65 and 80. What do the plots suggest about the accuracy of the two inputs? Compare the relative sensitivity plots for Q with those for q_p (Fig. 10-7).

10-5. Compute the relative sensitivities of the four predictor variables in the sediment yield model of Table 6-6. Compare the relative sensitivities with the values of the standardized partial regression coefficients (Table 6-6) as measures of the relative importance of the predictor variables.

10-6. Compute the relative sensitivities of the four predictor variables in the evaporation model of Table 6-8. Compare the relative sensitivities with the values of the standardized partial regression coefficients as measures of the relative importance of the predictor variables.

10-7. Using the evaporation data from Table 6-8, compute the relative sensitivity for each predictor variable for each step of the stepwise regression analysis. Explain why the values change with each step.

10-8. Using the model calibrated for Exercise 7-4, compute the sensitivity functions for P and L. Also, plot the relative sensitivity functions for storm rainfalls from 0 to 3 in. and plant heights of 1 and 6 ft.

10-9. Using the model calibrated in Exercise 7-6, show the variation of the absolute sensitivity of P with respect to E for evaluations from 4000 to 8000 ft.

10-10. Using the linear sensitivity function, compare the expected errors in the runoff depth (Q) of Eq. 10-33 for errors of $\Delta P = \pm 0.2$ in. and ΔCN of ± 0.5.

10-11. Using the linear sensitivity equation and the data of Table 8-4(c), evaluate the error in the peak discharge on day 5 due to an error in PGWRO of 0.05.

REFERENCES AND SUGGESTED READINGS

AGUADO, E., N. SITAR, and I. REMSON, Sensitivity Analysis in Aquifer Studies, *Water Resources Research*, Vol. 13, No. 4, pp. 733-37, 1977.

BETSON, R. P., and R. F. GREEN, Analytically Derived Unit Graph and Runoff, *Journal of the Hydraulics Division, ASCE*, Vol. 94, No. HY6, pp. 1489-1505, 1968.

CARLSON, R. F., and P. FOX, A Northern Snowmelt-Flood Frequency Model, *Water Resources Research*, Vol. 12, No. 4, pp. 786-94, 1976.

COLEMAN, G., and D. G. DECOURSEY, Sensitivity and Model Variance Analysis Applied to Some Evaporation and Evapotranspiration Models, *Water Resources Research*, Vol. 12, No. 5, pp. 873-79, 1976.

CORDOVA, J. R., and R. L. BRAS, Physically Based Probabilistic Models of Infiltration, Soil Moisture, and Actual Evapotranspiration, *Water Resources Research*, Vol. 17, No. 1, pp. 93–106, 1981.

DAVIS, D. R., and W. DVORANCHIK, Evaluation of the Worth of Additional Data, *Water Resources Bulletin*, Vol. 7, No. 4, pp. 700–707, 1971.

DAVIS, D. R., C. C. KISIEL, and L. DUCKSTEIN, Bayesian Decision Theory Applied to Design in Hydrology, *Water Resources Research*, Vol. 8, No. 1, pp. 33–41, 1972.

DAWDY, D. R., Considerations Involved in Evaluating Mathematical Modeling of Urban Hydrologic Systems, *Urban Water Resources Research*, Amer. Society of Civil Engineers, New York (Appendix A, Chapter 6), 1969.

DAWDY, D. R., and T. O'DONNELL, Mathematical Models of Catchment Behavior, *Journal of the Hydraulics Division, ASCE*, Vol. 91, No. HY4, 1965.

DECOURSEY, D. G., and W. M. SNYDER, Computer-Oriented Method of Optimizing Hydrologic Model Parameters, *Journal of Hydrology*, Vol. 9, pp. 34–56, 1969.

HANES, W. T., M. M. FOGEL, and L. DUCKSTEIN, Forecasting Snowmelt Runoff: Probabilistic Model, *Journal of the Irrigation and Drainage Division, ASCE*, Vol. 103, No. IR3, pp. 343–56, 1977.

LIND, N., Formulation of Probabilistic Design, *Journal of the Engineering Mechanics Division, Proceedings of the ASCE*, pp. 273–84, April 1977.

MCBEATH, B. C., and R. ELIASSEN, Sensitivity Analysis of Activated Sludge Economics, *Journal of the Sanitary Engineering Division, ASCE*, Vol. 92, No. SA2, pp. 147–67, April 1966.

MEIER, W., A. WEISS, C. PUENTES, and J. MOSELEY, Sensitivity Analysis: A Necessity in Water Planning, *Water Resources Bulletin*, Vol. 7, No. 3, pp. 529–41, 1971.

MOSS, M. E., and D. R. DAWDY, Application of Sensitivity Analysis to Reservoir Design and Worth of Streamflow Data, Symposium on Statistical Hydrology, Tucson, Ariz., 1971.

NASH, J. E., The Form of the Instantaneous Unit Hydrograph, *International Association of Scientific Hydrology*, Vol. 45, pp. 114–21, 1957.

NASH, J. E., Systematic Determination of Unit Hydrograph Parameters, *Journal of Geophysical Research*, Vol. 64(1), pp. 111–15, 1959.

PENMAN, H. L., Estimating Evaporation, *American Geophysical Union Transactions*, Vol. 37, pp. 43–46, 1956.

PENMAN, H. L., Natural Evaporation from Open Water, Bare Soil, and Grass, Proc. Royal Society (London), A, Vol. 193, pp. 120–45, 1948.

PINGOUD, K., Sensitivity Analysis of a Lumped-Parameter Model for Infiltration, *Journal of Hydrology*, Vol. 67, pp. 97–113, 1984.

PLINSTON, D. T., Parameter Sensitivity and Interdependence in Hydrological Models, in *Mathematical Models in Ecology*, J. N. R. Jerffers (ed.), Blackwell Scientific, Oxford, 1972, pp. 237–47.

SARMA, P. B., J. W. DELLEUR, and A. R. RAO, *A Program in Urban Hydrology: Part II*, Purdue Univ., Water Resources Research Center, Technical Report No. 9, 1969.

SAXTON, K. E., Sensitivity Analysis of the Combination Evapotranspiration Equation, *Agricultural Meteorology*, Vol. 15, pp. 343–53, 1975.

SINGH, K. P., and A. SNORRASON, Sensitivity of Outflow Peaks and Flood States to the Selection of Dam Breach Parameters and Simulation Models, *Journal of Hydrology*, Vol. 68, pp. 295–310, 1984.

SINGH, V. P., Sensitivity of Some Runoff Models to Errors in Rainfall Excess, *Journal of Hydrology*, Vol. 33, pp. 301–18, 1977.

SNYDER, W. M., Summary and Evaluation of Methods for Detecting Hydrologic Changes, *Journal of Hydrology*, Vol. 12, pp. 311–38, 1971.

SOROOSHIAN, S., and F. ARFI, Response Surface Parameter Sensitivity Analysis Methods for Postcalibration Studies, *Water Resources Research*, Vol. 18, No. 5, pp. 1531–38, 1982.

TERJUNG, W. H., J. T. HAYES, P. A. O'ROURKE, J. E. BURT, and P. E. TODHUNTER, Consumptive Water Use Response of Maize to Changes in Environmental Management Practices: Sensitivity Analysis of a Model, *Water Resources Research*, Vol. 18, No. 5, pp. 1539–50, 1982.

THORNTON, K. W., and A. S. LESSEM, *Sensitivity Analysis of the Water Quality for River-Reservoir Systems Model*, U.S. Army Engineer Waterways Exp. Station, Misc. Paper Y-76-4, 1976.

VEMURI, V., J. A. DRACUP, and R. C. ERDMANN, Sensitivity Analysis Method of System Identification and Its Potential in Hydrologic Research, *Water Resources Research*, Vol. 5, No. 2, pp. 341–49, April 1969.

WALKER, W. W., A Sensitivity and Error Analysis Framework for Lake Entrophication Modeling, *Water Resources Research*, Vol. 18, No. 1, pp. 53–60, 1982.

Water Resources Council, *Estimates Peak Flow Frequencies for Natural Ungaged Watersheds*, Hydrology Committee Report, Washington, D.C., 346 pp., 1981.

YOUNG, G. K., R. W. SCHRECONGOST, and W. FITCH, Design Sensitivity of Pollution Control Reservoirs, *Journal of the Sanitary and Engineering Division, ASCE*, Vol. 94, No. SA5, pp. 829–40, October 1968.

YOUNG, G. K., M. T. TSENG, and R. S. TAYLOR, Estuary Water Temperature Sensitivity to Meteorological Conditions, *Water Resources Research*, Vol. 7, No. 5, pp. 1173–81, October 1971.

ZOCH, R. T., On the Relation Between Rainfall and Stream Flow, *Monthly Weather Review*, Vol. 62 (315–22), Vol. 64 (105–21), Vol. 65 (135–47), 1934–1937.

11 Multivariate Models

INTRODUCTION

In some previous chapters we have used the terms *multiple variable* and *multivariate* almost interchangeably. These terms convey the meaning that more than one independent variable stands in relation to the dependent or criterion variable. In Chapter 6 multiple and stepwise regression were discussed as tools for optimizing the model structure relating the variables. In Chapter 7 techniques were presented for optimization when the model structure contains nonlinear coefficients. All such previous analyses are appropriate when the functional independent variables are also statistically independent of each other. A host of problems can be solved where the correlation among the independents is not large. For some problems, however, such intercorrelations produce an irrational evaluation of the model coefficients when least squares is the optimizing principle.

In this chapter we introduce just a few concepts from multivariate statistics that have been found useful for problems with intercorrelated variables. When we say "multivariate statistics" we place a unique definition upon "multivariate." This particular meaning will become clearer as the chapter develops.

Figure 11-1 shows how the contents of Chapter 11 are related to Chapters 6 and 7. Methodology for both linear and nonlinear model coefficients will be developed. Eigenvalue analysis of a matrix will replace least squares.

	Model Coefficient Structure		
Intercorrelation	Linear	Nonlinear	Required mathematics
Independent variables not correlated	Chapter 6	Chapter 7	Matrix inversion
	Stepwise regression	Numerical optimization	
	Multiple regression	Differential multiple regression	
Independent variables correlated	Chapter 11	Chapter 11	Eigenvalue-eigenvector matrix analysis
	Components regression	Numerical optimization	
	Factor analysis	Differential components regression	

Figure 11-1 Relationship of methodologies.

Although the treatment of multivariate statistics could be considered extensive, it is but a sampling of this topic. The chapter was designed as a model-oriented introduction to the topic for advanced students.

The linear or linearizable models that we have optimally matched to data by least squares had only one random variate. This random variate is the functional dependent variable or criterion variable. The functional independent variables are not considered statistical variates in least-squares fitting. The meaning of this will be clearer if one remembers that correlation, the measure of association in regression models, is based on the reduction in deviation of only the dependent variable. The variability of the independent variables does not enter. All adjustment of deviation, and the standard deviates residual to fitting, are in the scale of the dependent variable. We call the independent variables *fixed variates*, not statistical variates. Thus we say that regression models have multiple variables, one dependent and one or more independent. But they have only one variate, the dependent variable. Hence, in a strict sense, they are multiple variable but not multivariate.

We can now take up the need for multivariate models in certain circumstances. These circumstances are simply the presence of correlations among the functional independent variables of a regression-type model. When such correlations are present one can still perform mechanically the numerical process of a least-squares fitting. The difficulty arises in the interpretation of the results of fitting. This is one distinction between blind least-squares fitting and rational linear model building. After optimizing the correspondence between our models and observational data, the results of the optimization should make sense. Considering a simple linear model, we want the regression coefficients to measure structurally the relationship between the dependent variable and the appropriate independent variables. We want to recognize the regression coefficient as the slope of the line of relationship between the functional independent and dependent variables. If simple least squares produces results that we cannot interpret rationally, we must reject the resultant model structure and search for other means of optimization.

THE EFFECT OF CROSS-CORRELATION

A partial explanation of the failure of regression models in the presence of correlations among the functional independent variables can be found by consideration of the degrees of freedom in the model. Consider the multiple regression equation

$$y = b_0 + b_1x_1 + b_2x_2 + b_3x_3 + {}_1\epsilon \tag{11-1}$$

Here the term ${}_1\epsilon$ means the random deviation between the predictive component $\hat{y} = b_0 + b_1x_1 + b_2x_2 + b_3x_3$ and the observed value y. Now assume a correlation between x_2 and x_3 so that we can write a regression relationship as in

$$x_3 = \alpha + \gamma x_2 + {}_2\epsilon \tag{11-2}$$

Substituting Eq. 11-2 for x_3 in Eq. 11-1 produces

$$y = (b_0 + b_3\alpha) + b_1x_1 + (b_2 + b_3\gamma)x_2 + b_3\,{}_2\epsilon + {}_1\epsilon \tag{11-3}$$

Comparing Eq. 11-1 and 11-3 we can see immediately that certain of the regression coefficients are suspect if the condition in Eq. 11-2 exists. We fit Eq. 11-1 and hope to interpret the b_i on this basis. But actually we are fitting Eq. 11-3. For example, what we might regard as the b_2 of Eq. 11-1 is actually $b_2 + b_3\gamma$ of Eq. 11-3. Note also that in Eq. 11-3, b_3 associates with the error term ${}_2\epsilon$. Nonrigorously, we can say that b_3 in Eq. 11-1 is an excess degree of freedom if condition 11-2 exists. Since it is excess, it can associate with residual errors rather than with the variable x_3.

One cannot predict deterministically in any one sample how b_3 will evaluate. Its resultant value from application of least squares is random, depending on the relative strength of correlation expressed by Eqs. 11-1 and 11-2, and also on the correlation between residual errors ${}_2\epsilon$ and ${}_1\epsilon$. Since b_3 is random, we cannot regard it as true structural expression of our model.

THE BASIS OF MULTIVARIATE ANALYSIS

Our problem now can be simply stated. We want a rational value for b_3 in Eq. 11-1 even if the relationship in Eq. 11-2 exists. We will proceed according to the following logic. Since x_2 and x_3 are related as in Eq. 11-2, we may say that they are both functions of some other, and at the moment, abstract, variable. Then this abstract variable is both a function of x_2 and x_3. Express this as a statistical relationship:

$$z_2 = k_{21} + k_{22}x_2 + k_{23}x_3 \tag{11-4}$$

Now to be complete, we should also consider that x_1 may be a function of some other abstract variable z_1. Express this as

$$z_1 = k_{11} + k_{12}x_1 \tag{11-5}$$

We next consider a redefinition of our problem. Instead of the linear regression model (Eq. 11-1), write the model in terms of the abstract variables z_1 and z_2:

$$\hat{y} = \alpha_0 + \alpha_1 z_1 + \alpha_2 z_2 \tag{11-6}$$

Although it is true that we do not know what z_1 and z_2 are, we know that we can express them in terms of x_1, x_2, and x_3. Placing Eqs. 11-5 and 11-4 in Eq. 11-6, we have

$$y = (\alpha_0 + \alpha_1 k_{11} + \alpha_2 k_{21}) + \alpha_1 k_{12}x_1 + \alpha_2 k_{22}x_2 + \alpha_2 k_{23}x_3 + {}_3\epsilon \tag{11-7}$$

Equation 11-7 is similar in structure to Eq. 11-1. However, the definition of the coefficients has changed.

Procedurally, we already know how to compute the α's of Eq. 11-6. We can define them as regression coefficients and apply the principle of least squares to evaluate them. What remains to be specified is some method for obtaining the k's of Eqs. 11-4 and 11-5. Finding values such as these k's is known as *components analysis* in multivariate statistics. It is the first major topic in multivariate analysis that we will explore.

In optimizing multiple-variable models by least squares we depended on the solution of simultaneous equations by matrix inversion. In gaining familiarity with the computational aspects of components analysis we will have to learn eigenvalue–eigenvector analysis of a matrix. These procedures are developed in the following sections of this chapter.

Multivariate analysis is not limited to components analysis. However, the derivation of statistical component variates and their use in regression modeling forms the simplest and most utilitarian bridge between ordinary statistics and multivariate statistics as tools for modeling.

The theoretical bases of multivariate statistics are beyond the scope of this book. This chapter on multivariate models is intended only as a computational introduction to the topic as it occurs in statistical model building. The references at the end of this chapter may be consulted for the more theoretical aspects.

THE SIMPLE TWO-VARIABLE RELATIONSHIP

Before proceeding to the general case of components analysis of sets of statistical variates, we will treat the simple two-variable case. This relationship is simple linear but differs markedly from simple linear regression.

In regression the simple linear relationship is expressed as

$$y = a + bx \tag{11-8}$$

with the a and b to be derived by least squares. All the error of fitting is associated with y.

We may write Eq. 11-8 in the form

$$x = -\frac{a}{b} + \frac{1}{b}y \tag{11-9}$$

If we solve Eq. 11-9 by least squares, the resultant values of a and b will not be the same as those found for Eq. 11-8. The reason is that the errors associated with x are now minimized, producing a different solution than when the y errors are minimized. We can summarize by saying that the solutions are not reversible when the positions of the dependent and independent variable are exchanged in simple linear regression.

Many hydrologic models should be reversible. Suppose, for example, that we are dealing with the familiar problem of estimating missing data. The stream gage for station y may malfunction. A gage at station x nearby may have continued to operate. One way to estimate a missing day at y is to predict it statistically from station x. This can be done by evaluating Eq. 11-8 from the record of flows at both stations. But suppose that station x gage malfunctions. Should we then evaluate Eq. 11-9? Intuitively, one would say that two equations, 11-8 and 11-9, are incorrect. If a linear relationship exists between the two gages its evaluation should not depend on which gage happens to malfunction. Errors do exist in streamflow data. They probably exist about equally for station x and station y. Therefore, our relationship should recognize errors at both stations. For this reason we need a simple linear relationship with statistical error along both the x and y axes.

Kendall (1957) discusses the linear relationship of two variables and gives a solution for equal error variance along the x and y axes. Snyder (1962) presented an example based on Kendall's solution. In a bivariate relation, the value for slope in Eq. 11-8 is given by

$$b = W + \sqrt{1 + W^2} \tag{11-10}$$

where

$$W = \frac{\sum y^2 - \sum x^2}{2 \sum xy}$$

In the definition of W; x, and y are variates corrected for their means. Thus, with $y = bx$, we get

$$y - \bar{y} = b(x - \bar{x}) \tag{11-11}$$

Rearranging Eq. 11-11 yields

$$y = (\bar{y} - b\bar{x}) + bx \tag{11-12}$$

If one reverses the position of x and y in Eq. 11-12, it can be shown simply with Eq. 11-10 that the interchanged variates have a multivariate slope equal to $1/b$. That is, the relationship does not change and is reversible with interchanging variates.

Using 12 years of annual runoff data for two small streams, Snyder (1962) developed the relationships shown in Fig. 11-2. Note that the multivariate relationship lies between the two possible regression relationships. The square of the correlation coefficient was 0.70 for this example. Correlation based on percentage reduction in variation of y in Fig. 11-2 for the multivariate relationship was reduced to 0.63. We must recognize that least squares, by definition, will produce the best fit if we limit our objective to reduction of variance in

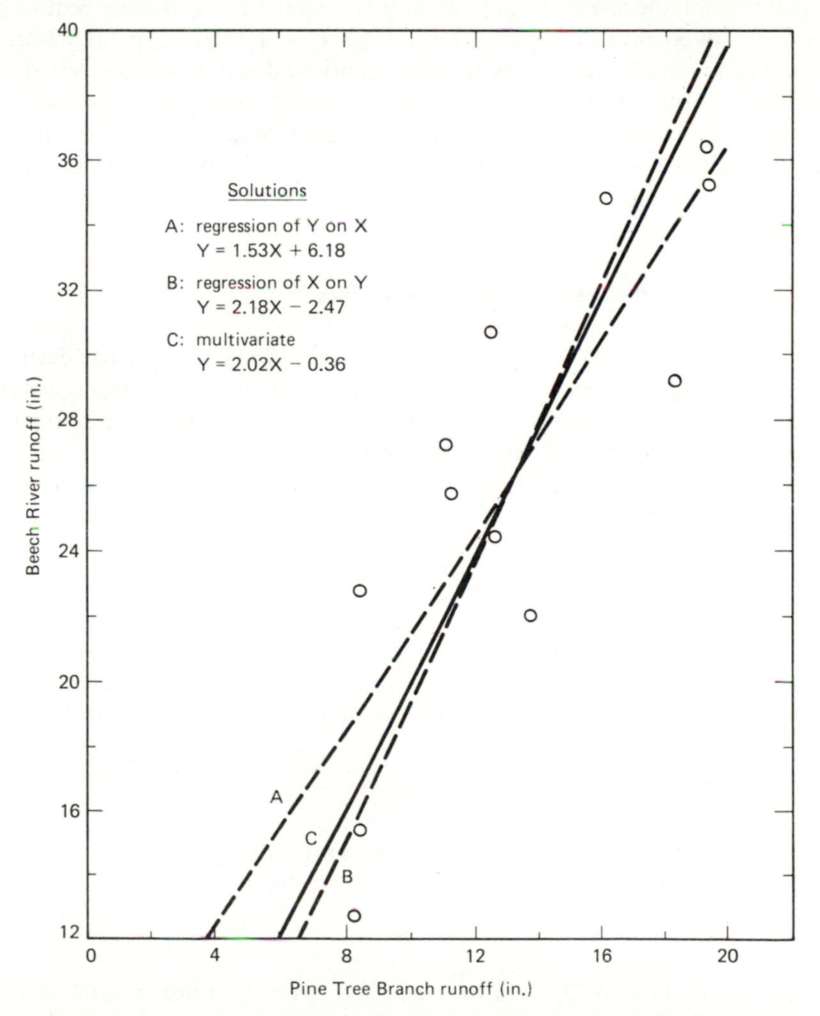

Figure 11-2 Three models of simple linear relationships. (From Snyder, 1962.)

y. This should not be taken to mean that least squares automatically produces the best set of optimized parameters. Distinctions are illustrated by the following two questions.

If we ask the question "What is the relationship of least error?", the answer must be "That derived by least squares." But if we ask the question "What is the reversible relationship of least error?", the answer is "That derived by the multivariate method."

MULTIPLE-VARIABLE RELATIONSHIPS

The solution of the general case of multiple statistical variables requires a knowledge of eigenvalue–eigenvector analysis of a matrix. In geometrical terms the problem of components analysis mentioned earlier may be considered a problem of rotation of axes. Our original variates, the things we observe and measure, are located in our familiar system of axes. But our abstract variables, such as expressed in Eq. 11-4, are measured along some other axes. The k values of Eq. 11-4 could be considered the mathematical devices for changing from one axial set to another.

Eigenvalue–Eigenvector Analysis of a Matrix

Before proceeding to the rotation of axes we will need to learn the methods for eigenvalue–eigenvector analysis of matrices. Although some rotations can be performed without such analysis, the rotation to a special set of axes, called *principal axes*, does require it.

The eigenvalue problem arises when it is required to solve a problem expressed in the form

$$\boldsymbol{A}\mathbf{X} = \lambda \mathbf{X} \tag{11-13}$$

In Eq. 11-13 we state that a matrix $\boldsymbol{A}$ multiplied by a vector $\mathbf{X}$ equals that vector multiplied by a scalar quantity λ. Our problem is to find both the value or values of λ for which the equation is true and the unknown vector $\mathbf{X}$.

We can modify Eq. 11-13 by transposing all terms to the left of the equal sign, getting

$$\boldsymbol{A}\mathbf{X} - \lambda \mathbf{X} = 0 \tag{11-14}$$

Now $\lambda \mathbf{X}$ is the same as $\lambda \boldsymbol{I}\mathbf{X}$, since $\boldsymbol{I}$ is the unity, or identity, matrix. If we substitute $\lambda \boldsymbol{I}\mathbf{X}$ for $\lambda \mathbf{X}$ and then factor out $\mathbf{X}$, we have

$$(\boldsymbol{A} - \lambda \boldsymbol{I})\mathbf{X} = 0 \tag{11-15}$$

In Eq. 11-15 either $\mathbf{X} = 0$ or $(\boldsymbol{A} - \lambda \boldsymbol{I}) = 0$. The condition $\mathbf{X} = 0$ is trivial; therefore, we are interested in the alternative solution. If we write Eq. 11-15 in terms of its elements, we get, for an order 3 matrix,

$$(A_{11} - \lambda)X_1 + A_{12}X_2 + A_{13}X_3 = 0 \qquad (11\text{-}16a)$$

$$A_{21}X_1 + (A_{22} - \lambda)X_2 + A_{23}X_3 = 0 \qquad (11\text{-}16b)$$

$$A_{31}X_1 + A_{32}X_2 + (A_{33} - \lambda)X_3 = 0 \qquad (11\text{-}16c)$$

For Eq. 11-16 to be true without the X_i being equal to zero, the determinant of $(\boldsymbol{A} - \lambda\boldsymbol{I})$ must equal zero. Thus

$$|\boldsymbol{A} - \lambda\boldsymbol{I}| = \begin{vmatrix} A_{11} - \lambda & A_{12} & A_{13} \\ A_{21} & A_{22} - \lambda & A_{23} \\ A_{31} & A_{32} & A_{33} - \lambda \end{vmatrix} = 0 \qquad (11\text{-}17)$$

Equation 11-17 is known as the *characteristic equation.* When the elements A_{ij} are known, it provides a means for solution of λ. Expansion of the determinant shows that the characteristic equation will contain a λ^3. Thus, there are three roots to the equation, or three values of λ, for which Eq. 11-13 will be true. The number of roots corresponds to the order of the matrix. For a symmetric matrix these roots are real numbers. It is these roots that are known as *eigenvalues.*

If, after determining the λ's, or eigenvalues, we substitute each in turn into Eq. 11-16, we will get for each λ a set of values for X_1, X_2, and X_3. These sets can be considered vectors, and when so derived they are called *eigenvectors.* There is an eigenvector corresponding to each eigenvalue. The eigenvalues and eigenvectors form the complete solution to Eq. 11-16, or to the more general equation 11-13.

Direct Solution of the Characteristic Equation

Two methods are available for obtaining the numerical values of eigenvalues and eigenvectors. These two methods are the direct solution of the characteristic equation and an iterative numerical method. We will take up the direct solution first.

For an explanatory description of the direct solution we will find the eigenvalues and eigenvectors of the matrix

$$\begin{pmatrix} 1 - \lambda & 0.13 & 0.18 \\ 0.13 & 1 - \lambda & 0.95 \\ 0.18 & 0.95 & 1 - \lambda \end{pmatrix}$$

We make use of the fact, expressed in Eq. 11-17, that the determinant of this matrix must be zero for nontrivial solutions to exist. Then

$$\begin{vmatrix} 1 - \lambda & 0.13 & 0.18 \\ 0.13 & 1 - \lambda & 0.95 \\ 0.18 & 0.95 & 1 - \lambda \end{vmatrix} = 0$$

Expanding the determinant, we get

$$(1-\lambda)\begin{vmatrix} 1-\lambda & 0.95 \\ 0.95 & 1-\lambda \end{vmatrix} - 0.13\begin{vmatrix} 0.13 & 0.95 \\ 0.18 & 1-\lambda \end{vmatrix} + 0.18\begin{vmatrix} 0.13 & 1-\lambda \\ 0.18 & 0.95 \end{vmatrix} = 0$$

Further expansion gives

$$(1-\lambda)(1-\lambda)(1-\lambda) - (1-\lambda)(0.95)(0.95) - (0.13)(0.13)(1-\lambda) + (0.13)(0.95)(0.18) + (0.18)(0.13)(0.95) - (0.18)(1-\lambda)(0.18) = 0$$

which results in the cubic equation

$$\lambda^3 - 3\lambda^2 + 2.0482\lambda - 0.09266 = 0 \tag{11-18}$$

There are thus three roots, or three values of λ. For a symmetric matrix these roots will always be real numbers, although they may not always be separate. The roots of Eq. 11-18 may be found by any of the usual methods for finding the roots of a polynomial. They are $\lambda_1 = 1.9982$, $\lambda_2 = 0.9532$, and $\lambda_3 = 0.0486$.

For each of these eigenvalues there is an eigenvector. To obtain the eigenvectors we write the three equations

$$(1-\lambda)X_1 + 0.13X_2 + 0.18X_3 = 0 \tag{11-19a}$$

$$0.13X_1 + (1-\lambda)X_2 + 0.95X_3 = 0 \tag{11-19b}$$

$$0.18X_1 + 0.95X_2 + (1-\lambda)X_3 = 0 \tag{11-19c}$$

These are specific expressions of the more general equations 11-16. Equations 11-19 form a set of *homogeneous equations.* For such a set, no complete numerical solution exists. However, there does exist a consistent solution between two of the variables. To find this solution, one of the variables to the right is algebraically moved across the equal sign. Then one of the two variables on the left can be eliminated by subtraction following standardization of its coefficients. One may eliminate this variable by subtracting the second equation from the first, and also by subtracting the third equation from the second. If the two resulting reduced equations are a consistent relationship between the remaining two variables, the solution is found. If the relationship is inconsistent, one moves one of the other variables to the right of the equal sign. Again, two reduced equations are scrutinized for a consistent relationship between the remaining two variables. The computational process is illustrated below.

Substituting λ_1 into Eqs. 11-19, we get

$$-0.9982\text{X}_1 + 0.13\text{X}_2 + 0.18\text{X}_3 = 0 \tag{11-20a}$$

$$0.13X_1 - 0.9982X_2 + 0.95X_3 = 0 \tag{11-20b}$$

$$0.18X_1 + 0.95X_2 - 0.9982X_3 = 0 \tag{11-20c}$$

We can move X_2 to the right and form the equations with the coefficient of X_1 standardized at unity:

$$X_1 - 0.180325X_3 = 0.130234X_2 \tag{11-21a}$$

$$X_1 + 7.307692X_3 = 7.678462X_2 \tag{11-21b}$$

$$X_1 - 5.545556X_3 = -5.277778X_2 \tag{11-21c}$$

Subtracting the first equation from the second and the third from the second gives the set

$$7.488017X_3 = 7.548228X_2 \tag{11-22a}$$

$$12.853248X_3 = 12.956240X_2 \tag{11-22b}$$

which further standardizes to

$$X_3 = 1.00804X_2 \tag{11-23a}$$

$$X_3 = 1.00801X_2 \tag{11-23b}$$

Allowing for round-off error, this is a consistent relationship between X_3 and X_2. Using an average value of the coefficient of 1.00803, we can eliminate X_3 from Eq. 11-21a and find $X_1 = 0.312002X_2$.

We now can write our solution to the homogenous set of Eqs. 11-20. This solution is a vector in terms of X_2 written as

$$\begin{pmatrix} X_1 \\ X_2 \\ X_3 \end{pmatrix} = X_2 \begin{pmatrix} 0.31200 \\ 1.00000 \\ 1.00803 \end{pmatrix}$$

Now we can eliminate the scalar X_2 by standardizing the numerical elements of this vector. A convenient standardization is to normalize on the root-mean-square value of the elements. This normalization is performed by dividing the elements by the square root of the sum of their squares.

The sum of squares of the three elements is 2.11347 and the square root of this sum is 1.45378. Dividing each of the vector elements by this value, we get the vector

$$\begin{pmatrix} 0.21461 \\ 0.68786 \\ 0.69339 \end{pmatrix}$$

This is the eigenvector corresponding to the first eigenvalue.

By substituting the second eigenvalue into Eq. 11-19, we get the homogeneous set

$$0.0468X_1 + 0.13X_2 + 0.18X_3 = 0 \tag{11-24a}$$

$$0.13X_1 + 0.0468X_2 + 0.95X_3 = 0 \tag{11-24b}$$

$$0.18X_1 + 0.95X_2 + 0.0468X_3 = 0 \tag{11-24c}$$

The solution in terms of X_1 is

$$X_1 \begin{pmatrix} 1.00000 \\ -0.18302 \\ -0.12782 \end{pmatrix}$$

which normalizes to

$$\begin{pmatrix} 0.97598 \\ -0.17862 \\ -0\ 12475 \end{pmatrix}$$

Similarly, the third set of equations

$$0.9514X_1 + 0.013X_2 + 0.18X_3 = 0 \qquad (11\text{-}25a)$$

$$0.13X_1 + 0.9514X_2 + 0.95X_3 = 0 \qquad (11\text{-}25b)$$

$$0.18X_1 + 0.95X_2 + 0.9514X_3 = 0 \qquad (11\text{-}25c)$$

has a solution

$$X_3 \begin{pmatrix} -0.05375 \\ -0.99124 \\ 1.00000 \end{pmatrix}$$

which normalizes to

$$\begin{pmatrix} -0.03815 \\ -0.70348 \\ 0.70969 \end{pmatrix}$$

We may now collect the three normalized eigenvectors into a matrix we will define as $\boldsymbol{Q}$:

$$\boldsymbol{Q} = \begin{pmatrix} 0.21461 & 0.97598 & -0.03815 \\ 0.68786 & -0.17862 & -0.70348 \\ 0.69339 & -0.12475 & 0.70969 \end{pmatrix} \qquad (11\text{-}26)$$

If we take the sum of the squares of the elements of each eigenvector we obtain unity, which indicates that each eigenvector is normalized. If we take the sum of the products of elements of any two of the eigenvectors, we obtain zero. The vectors are thus orthogonal to each other. We call such sets of vectors *orthonormal*, meaning both orthogonal and normalized.

Iterative Numerical Method

An iterative numerical method can also be used to calculate the eigenvalues and eigenvectors of a matrix. A number of numerical methods are described in standard texts on numerical methods. The two methods yield the same eigenstructure.

An important difference should be noted between the two methods of computing eigenvalues. When using the direct method the roots of the characteristic equation may be found in any order. However, with the iterative method the largest eigenvalue always appears first. Operation on the reduced matrix then produces the second-largest eigenvalue. The eigenvalues will always appear in this order of largest first, and successively smaller values following. This order of appearance of the roots will have an important statistical implication to be brought out later. Because of this statistical implication, and because of the relative ease of programming the iterative method, this is the one usually used in eigenvalue–eigenvector calculation. Computer library programs are readily available.

Rotation of Axes

Gere and Weaver (1965) present a direct and understandable discussion of eigenvalue–eigenvector solutions in terms of rotation of axes from one system of coordinates to another. We recall from geometry that a vector in one set of coordinate axes can be expressed as a vector in another set of coordinate axes through the direction cosines of the angles between the axes. This may be expressed as

$$Y_1 = l_{11}X_1 + l_{12}X_2 + l_{13}X_3 \tag{11-27a}$$

$$Y_2 = l_{21}X_1 + l_{22}X_2 + l_{23}X_3 \tag{11-27b}$$

$$Y_3 = l_{31}X_1 + l_{32}X_2 + l_{33}X_3 \tag{11-27c}$$

In Eqs. 11-27 the X_i represent the vector in the original coordinate system, and the Y_i represent the vector in the rotated system. The l_{ij} are direction cosines between the original and rotated axes, with the subscript i referring to the new axes and the subscript j to the old axes.

In matrix form, Eqs. 11-27 become

$$\mathbf{Y} = \boldsymbol{R}\mathbf{X} \tag{11-28}$$

where $\boldsymbol{R}$ is the rotation matrix made up of the l_{ij}. $\boldsymbol{R}$ is an orthogonal matrix if the axes are at right angles to each other, which is the case for customary systems of coordinates.

Suppose now that we have established a relationship between two vectors in our original set of coordinates. How do we express this relationship in the rotated axes? Express the original relationship as the matrix vector relation

$$\boldsymbol{A}\mathbf{X}_1 = \mathbf{X}_2 \tag{11-29}$$

We may compute the two vectors in the rotated system by multiplying by the rotation matrix:

$$\mathbf{Y}_1 = \boldsymbol{R}\mathbf{X}_1 \tag{11-30a}$$

$$\mathbf{Y}_2 = \boldsymbol{R}\mathbf{X}_2 \tag{11-30b}$$

Also, multiplying Eq. 11-29 by $\mathbf{R}$ gives

$$\boldsymbol{RA}\mathbf{X}_1 = \boldsymbol{R}\mathbf{X}_2 \tag{11-31}$$

Now insert $\boldsymbol{R}^{-1}\boldsymbol{R}$ into Eq. 11-31. This does not change the equation since this term ιo be inserted is the unity matrix and produces no change:

$$\boldsymbol{RAR}^{-1}\boldsymbol{R}\mathbf{X}_1 = \boldsymbol{R}\mathbf{X}_2 \tag{11-32}$$

If we define $\boldsymbol{B} = \boldsymbol{RAR}^{-1}$, Eq. 11-32 becomes

$$\boldsymbol{B}\mathbf{Y}_1 = \mathbf{Y}_2 \tag{11-33}$$

We have now converted our matrix of relationship, $\boldsymbol{A}$ in the original coordinate system, to the matrix $\boldsymbol{B}$ in the rotated system. $\mathbf{Y}_1$ and $\mathbf{X}_1$ are one and the same vector, but $\mathbf{X}_1$ is its expression in the old system, $\mathbf{Y}_1$ its expression in the new. $\mathbf{Y}_2$ and $\mathbf{X}_2$ are also identical vectors in space. The operation $\boldsymbol{RAR}^{-1}$ simply changes the matrix of relationship to the new coordinate system. A special set of rotated axes results when the new relationship matrix, $\mathbf{B}$, is a diagonal matrix. These special new axes are called *principal axes.*

It can be shown that any matrix, $\boldsymbol{A}$, can be converted to a diagonal matrix, $\boldsymbol{S}$, by the operation

$$\boldsymbol{S} = \boldsymbol{Q}^{-1}\boldsymbol{AQ} = \boldsymbol{Q}'\boldsymbol{AQ} \tag{11-34}$$

In Eq. 11-34, $\boldsymbol{S}$ is called the *spectral matrix.* It consists of the eigenvalues of $\boldsymbol{A}$ on the main diagonal, and zeros for all off-diagonal elements. $\boldsymbol{Q}$ is the matrix of normalized eigenvectors of $\boldsymbol{A}$. We can write the double form of Eq. 11-34 because the inverse and the transpose of an orthogonal matrix are equal. We can write

$$\boldsymbol{B} = \boldsymbol{RAR}^{-1} = \boldsymbol{RAR}' \tag{11-35a}$$

and

$$\boldsymbol{S} = \boldsymbol{Q}^{-1}\boldsymbol{AQ} = \boldsymbol{Q}'\boldsymbol{AQ} \tag{11-35b}$$

From Eq. 11-35 we can see that if we use $\boldsymbol{Q}'$ for our rotation matrix $\boldsymbol{R}$, then $\boldsymbol{B}$ becomes the diagonal matrix $\boldsymbol{S}$.

In summary, we can rotate our coordinate system in which we have specified a vector relationship given by Eq. 11-29. In any general new coordinate system this relationship is given by Eq. 11-33. However, we can specify rotation to a special new set of axes, called principal axes, by making the new matrix of relationship, $\boldsymbol{B}$, equal to the spectral matrix, $\boldsymbol{S}$. Therefore, we need only compute the eigenvalues and eigenvectors of A, then use them as specified in Eqs. 11-35, and we have accomplished rotation to the special set of principal axes.

Call the vectors in the space of the principal axes $\mathbf{P}_1$ and $\mathbf{P}_2$ instead of the more general $\mathbf{Y}_1$ and $\mathbf{Y}_2$ of an unspecified rotation. Then

$$\mathbf{P}_1 = \begin{pmatrix} P_{11} \\ P_{12} \\ P_{13} \end{pmatrix} \tag{11-36a}$$

and

$$\mathbf{P}_2 = \begin{pmatrix} P_{21} \\ P_{22} \\ P_{23} \end{pmatrix} \tag{11-36b}$$

Now it is the special property of principal axes that

$$\frac{P_{21}}{P_{11}} = \lambda_1 \tag{11-37a}$$

$$\frac{P_{22}}{P_{12}} = \lambda_2 \tag{11-37b}$$

$$\frac{P_{23}}{P_{13}} = \lambda_3 \tag{11-37c}$$

Equations 11-37 simply state that in the space of the principal axes the relationship between the vector elements becomes a simple ratio equal to the eigenvalues of matrix $\boldsymbol{A}$ in the original space. The relationship of the first element of the two vectors is a ratio equal to the first eigenvalue, and so on.

In our original coordinate system, designated as axes X_1, X_2, X_3, and so on, our relationship was expressed by Eq. 11-29. If we expand this equation for a matrix of order 3 we get

$$A_{11}X_{11} + A_{12}X_{12} + A_{13}X_{13} = X_{21} \tag{11-38a}$$

$$A_{21}X_{11} + A_{22}X_{12} + A_{23}X_{13} = X_{22} \tag{11-38b}$$

$$A_{31}X_{11} + A_{23}X_{12} + A_{33}X_{13} = X_{23} \tag{11-38c}$$

However, if we rotate our axes to the principal axes, Eqs. 11-37 hold, and our relationship between vectors in this new coordinate space becomes

$$\lambda_1 P_{11} \qquad\qquad\qquad = P_{21} \tag{11-39a}$$

$$\qquad \lambda_2 P_{12} \qquad\quad = P_{22} \tag{11-39b}$$

$$\qquad\qquad \lambda_3 P_{13} = P_{23} \tag{11-39c}$$

Because our matrix of relationship is a diagonal matrix in the space of the principal axes, we see that we no longer have any interactions between the elements of the vectors.

Now consider Eqs. 11-38 as representing a linear hydrologic model. We wish to predict vector $\mathbf{X}_2$. There is no simple way to do this since each element of vector $\mathbf{X}_1$ contributes to each element of vector $\mathbf{X}_2$. Now vector $\mathbf{P}_1$ in the rotated system is the same as vector $\mathbf{X}_1$ in the old system. Equations 11-39 tell us that we have found a system of axes in which the elements of the first vector contribute only to the respective elements of the second vector. This rotation to the principal axes forms the very heart of multivariate model building. We will see in a later section that the elimination of the interaction between elements of the vectors of relationship is the very device we use to eliminate correlations among the functional independent variables in our linear hydrologic models.

Components Analysis

It will be recalled that one of the very important reasons for studying multivariate statistics is the occasional failure of ordinary least squares. This failure occurs when the functional independent variables in a regression model are not statistically independent. We now introduce the concept that this statistical independence has the same meaning as orthogonality of vectors. If two vectors are orthogonal, their elements are uncorrelated, which we will take to mean independent. We used this calculation to demonstrate orthogonality of the eigenvectors in Eq. 11-26.

Consider the vector relationship in Eqs. 11-38. We can compute the three elements of vector $\mathbf{X}_2$ from the three elements of vector $\mathbf{X}_1$. But suppose that X_{12} and X_{13} are measures of hydrologic data and are highly correlated. Then we could say that X_{13} is proportional to X_{12}, and our order 3 matrix would reduce to order 2. We could then calculate only two elements of $\mathbf{X}_2$, since the third element would be a repetition of one of the other two. In a sense we have reduced the three elements of our original hydrologic data to two elements by eliminating the relationship of X_{12} and X_{13}.

We may state the reduction in the number of elements in another way. We can specify that we wish to find a special matrix of relationship, the A_{ij} of Eqs. 11-36. This special matrix must be of such form that the elements of $\mathbf{X}_2$ are orthogonal. Now if some of the elements of $\mathbf{X}_1$ are correlated, some of the elements of X_2, all being orthogonal, must be essentially zero. It is this transform from our original data vector $\mathbf{X}_1$ to some other vector of orthogonal elements, $\mathbf{X}_2$, that we call *components analysis.*

Kendall (1957) shows that the statistical problem of eliminating correlations is equivalent to the eigenvalue–eigenvector problem of rotation to principal axes. If we specify rotation to the principal axes, we attempt to establish the principal components of the matrix of hydrologic variables.

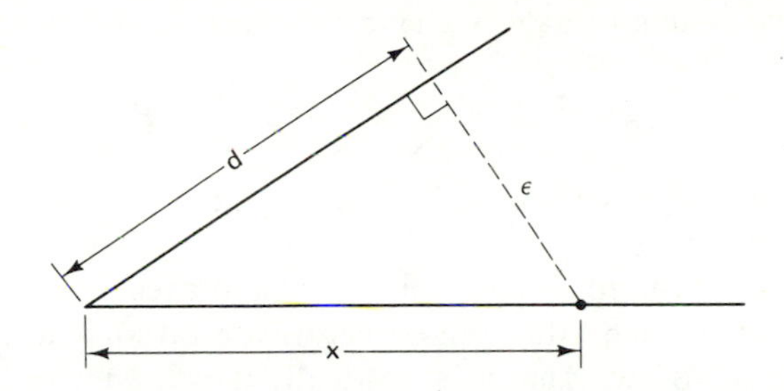

Figure 11-3 Definition of rotated axes.

Kendall bases his solution on the principle of establishing new, rotated axes so that the distances from the hydrologic data points perpendicular to the new axes have a minimum squared value. In Fig. 11-3 x represents some hydrologic measurement. The distance d is the hydrologic variable measured along a rotated axis to the point intercepted by the perpendicular from the point on the x axis. This perpendicular distance to the new axis is ϵ. The rotated axis has a direction cosine $d/x = l$ from the x axis. It is evident that

$$\epsilon^2 = x^2 - d^2 = x^2 - (lx)^2 \tag{11-40}$$

In most instances we have several simultaneous hydrologic measurements. This number is the same as the number of elements of $\mathbf{X}_1$ in Eqs. 11-38. Call the number of elements P. Then our total of the deviation squared from point to axis must represent the sum across all vector elements:

$$\sum \epsilon^2 = \sum_{i=1}^{P} x_i^2 - \left(\sum_{i=1}^{P} l_i x_i \right)^2 \tag{11-41}$$

Finally, we must sum these deviations across n, the number of observations we have on the elements x_i. The total deviations squared is then

$$S = \sum_{j=1}^{n} \left[\sum_{i=1}^{P} x_{ij}^2 - \left(\sum_{i=1}^{P} l_i x_{ij} \right)^2 \right] \tag{11-42}$$

Kendall points out that if Eq. 11-42 is differentiated with respect to each l_i and the resulting expressions set to zero to establish the minimum squared distances from the points to the new axes, the matrix equation

$$(\boldsymbol{r} - \lambda \boldsymbol{I})\mathbf{l} = 0 \tag{11-43}$$

results where $\boldsymbol{r}$ is the matrix of correlation coefficients of our hydrologic variates, x_i, and $\mathbf{l}$ is the vector of direction cosines.

Briefly, we can establish the matrix of correlation coefficients of a set of hydrologic variates. Eigenvalue–eigenvector analysis rotates our system of data, in x-axes measurement, to some new axes given by the eigenvectors, which are also the direction cosines.

In our new system of data we have new variates defined by

$$\zeta_k = \sum_{j=1}^{P} l_{kj} x_j \qquad k = 1, 2, \ldots = P \tag{11-44}$$

There are as many of these new variates ζ_k as P, the number of items in our set of hydrologic variates.

The ζx_k are measured along our new, rotated axes. Since these axes are all at right angles to each other (eigenvectors are orthogonal), the ζ_k are at right angles to each other. That is, statistically they are uncorrelated.

An important property of these ζ_k is that their respective variances are λ_k. Stated simply, the variances of the orthogonal variates are the eigenvalues. The first, largest, eigenvalue is the variance of ζ_1. The second eigenvalue is the variance of ζ_2, and so on. A further property is that the sum of the eigenvalues is equal to P, the order of the r matrix, which is, of course, the same as the number of original hydrologic variates. This second property is of immense practical importance. By the iterative method of solution we always derive the eigenvalues in order of magnitude. We can sum these eigenvalues and check against the known total, P. If, for example, for a 4×4 matrix we find that the sum of the first three eigenvalues is 3.99, these three eigenvalues contain a $3.99/4 = 0.9975$ proportion of the total variance. The remaining eigenvalue is nearly zero and we would lose very little information in ignoring it. By ignoring this very small eigenvalue we have reduced our hydrologic set of four variates to three new orthogonal variates. These new orthogonal variates must initially be regarded as mathematical abstractions. However, sometimes they can be given physical meaning, as will be illustrated below.

Illustration of Principal Components Analysis

In a preceding section illustrating the method of computing eigenvalues and eigenvectors, we analyzed the matrix

$$\begin{pmatrix} 1.00 & 0.13 & 0.18 \\ 0.13 & 1.00 & 0.95 \\ 0.18 & 0.95 & 1.00 \end{pmatrix}$$

If we consider this as a matrix of correlation coefficients of three hydrologic variates, we see that X_2 and X_3 have a high correlation of 0.95. X_1 and X_2 are correlated only slightly, having a coefficient of 0.13. X_1 and X_3 with a coefficient of 0.18 are also only slightly correlated. We also derived the $\boldsymbol{Q}$ matrix of eigenvectors:

$$\boldsymbol{Q} = \begin{pmatrix} 0.21461 & 0.97598 & -0.03815 \\ 0.68786 & -0.17862 & -0.70348 \\ 0.69339 & -0.12475 & 0.70969 \end{pmatrix}$$

Now by Eq. 11-44,

$$\zeta_1 = 0.21416x_1 + 0.68786x_2 + 0.69339x_3 \quad (11\text{-}45a)$$

$$\zeta_2 = 0.97598x_1 - 0.17862x_2 - 0.12475x_3 \quad (11\text{-}45b)$$

$$\zeta_3 = -0.03815x_1 - 0.70348x_2 + 0.70969x_3 \quad (11\text{-}45c)$$

In Eqs. 11-45 the x_i are the standardized values of x_i. The ζ_k are the new orthogonal variates.

Studying the set of Eqs. 11-45, we see that ζ_1 is strongly dependent on x_2 and x_3 and only moderately dependent on x_1. ζ_2, on the other hand, is very strongly dependent on x_1 and only slightly dependent on x_2 and x_3. ζ_3 must be very small, since it has near zero dependence on x_1 and canceling effects on x_2 and x_3. The canceling effect is present because x_2 and x_3 in standardized form must be nearly equal since they are highly correlated. The sum of the first two eigenvalues is $1.9982 + 0.9532 = 2.9514$. These two eigenvalues thus account for about $(2.9514/3) \times 100 = 98.4\%$ of the total variance. Thus very little information is contained in the third eigenvalue. Also, we have already established that the value of ζ_3 is small. We can thus ignore ζ_3 and we have reduced our original three hydrologic variates, X_1, X_2, and X_3, to two new variates ζ_1 and ζ_2.

It is convenient to call the ζ_k components. Using this terminology, we have reduced three hydrologic variates to two orthogonal components. The first component identifies strongly with X_2 and X_3, which are highly correlated. The second component identifies with X_1, which is nearly independent of X_2 and X_3.

Components Regression

To this point in our study of multivariate statistics we have seen how to convert a set of correlated hydrologic variables, X_i, into a set of orthogonal variates, ζ_k, where $k \leq i$. We are now ready to take a fresh look at the problem of multiple regression. It will be recalled that it was stated earlier that one of the very strong reasons for studying multivariate statistics was to develop a substitute for multiple regression. The reason given was that multiple regression, or least squares, does not always give the answers we need in our model evaluation. This happens when our X_i are correlated. We can now state a simple solution to the problem. Instead of stating our linear model as a relationship of Y to the X_i, we state it as a relationship of Y to the ζ_k. Specifically, we can write

$$y = \alpha_1\zeta_1 + \alpha_2\zeta_2 + d_3\zeta_3 + \cdots + \alpha_k\zeta_k \quad (11\text{-}46)$$

which simply states that we "regress" y on the ζ_k instead of on the x_1. Use of the lowercase y and x_i only means we have first standardized our variates to zero mean and variance of unity. The α_k of Eq. 11-46 have the same meaning as the b_i of the multiple regression equation. They are the partial regression coefficients.

Equation 11-46 can be expressed in terms of the standardized hydrologic variables x_i by making use of Eq. 11-44. Expanding this equation, we can write

$$\zeta_1 = l_{11}x_1 + l_{21}x_2 + l_{31}x_3 + \cdots + l_{i1}x_i$$
$$\zeta_2 = l_{12}x_1 + l_{22}x_2 + l_{32}x_3 + \cdots + l_{i2}x_i$$
$$\vdots$$
$$\zeta_k = l_{1k}x_1 + l_{2k}x_2 + l_{3k}x_3 + \cdots + l_{ik}x_i$$

The contribution of ζ_1 to y is $\alpha_1 \zeta_1$. We can write this as a separate contribution since there is no interaction of the variates ζ_k. This is a consequence of principal axes as we saw in Equation 11-39. Then from the expansion of Eq. 11-44 and from Eq. 11-46,

$${}_1y = \alpha_1 l_{11}x_1 + \alpha_1 l_{21}x_2 + \alpha_1 l_{31}x_3 + \cdots + \alpha_1 l_{i1}x_i \tag{11-47a}$$
$${}_2y = \alpha_2 l_{12}x_1 + \alpha_2 l_{22}x_2 + \alpha_2 l_{32}x_3 + \cdots + \alpha_2 l_{i2}x_i \tag{11-47b}$$
$$\vdots$$
$${}_ky = \alpha_k l_{1k}x_1 + \alpha_k l_{2k}x_2 + \alpha_k l_{3k}x_3 + \cdots + \alpha_k l_{ik}x_i \tag{11-47k}$$

In Eqs. 11-47, ${}_1y$ is the independent contribution of ζ_1, ${}_2y$ is the independent contribution of ζ_2, and so on. Since the contributions to the total relationship of y to the x_i are independent, we may add. A relationship on two components is given by

$$\begin{aligned} {}_1y + {}_2y = y_{12} = {} & (\alpha_1 l_{11} + \alpha_2 l_{12})x_1 \\ & + (\alpha_1 l_{21} + \alpha_2 l_{22})x_2 \\ & + (\alpha_1 l_{31} + \alpha_2 l_{32})x_3 + \cdots \\ & + (\alpha_1 l_{i1} + \alpha_2 l_{i2})x_i \end{aligned} \tag{11-48}$$

A relationship on three components is given by

$$\begin{aligned} {}_1y + {}_2y + {}_3y = y_{123} = {} & (\alpha_1 l_{11} + \alpha_2 l_{12} + \alpha_3 l_{13})x_1 \\ & + (\alpha_1 l_{21} + \alpha_2 l_{22} + \alpha_3 l_{23})x_2 \\ & + (\alpha_1 l_{31} + \alpha_2 l_{32} + \alpha_3 l_{33})x_3 + \cdots \\ & + (\alpha_1 l_{i1} + \alpha_2 l_{i2} + \alpha_3 l_{i3})x_i \end{aligned} \tag{11-49}$$

Looking at Eqs. 11-48 and 11-49, we can see that we can express our relationship of y to the x_i if we can calculate the numerical values of the quantities in parentheses. We have already seen how to compute l_{ij}, since these compose the eigenvector matrix of the correlation matrix of the x_i. Then all that remains is to compute the α_k. This calculation is very simple. Since Eq. 11-46 is a regression equation, we can write the simultaneous normal equations for solution of the α_k in a manner identical to Eqs. 6-11.

$$\alpha_1 \sum \zeta_1^2 + \alpha_2 \sum \zeta_1\zeta_2 + \alpha_3 \sum \zeta_1\zeta_3 + \cdots + \alpha_k \sum \zeta_1\zeta_k = \sum \zeta_1 y \qquad (11\text{-}50a)$$

$$\alpha_1 \sum \zeta_1\zeta_2 + \alpha_2 \sum \zeta_2^2 + \alpha_3 \sum \zeta_2\zeta_3 + \cdots + \alpha_k \sum \zeta_2\zeta_k = \sum \zeta_2 y \qquad (11\text{-}50b)$$

$$\alpha_1 \sum \zeta_1\zeta_3 + \alpha_2 \sum \zeta_2\zeta_3 + \alpha_3 \sum \zeta_3^2 + \cdots + \alpha_k \sum \zeta_3\zeta_k = \sum \zeta_3 y \qquad (11\text{-}50c)$$

$$\vdots$$

$$\alpha_1 \sum \zeta_1\zeta_k + \alpha_2 \sum \zeta_2\zeta_k + \alpha_3 \sum \zeta_3\zeta_k + \cdots + \alpha_k \sum \zeta_k^2 = \sum \zeta_k y \qquad (11\text{-}50k)$$

However, in Eqs. 11-50 all the $\sum \zeta_i\zeta_k$ terms for which $i \neq k$ are zero since the ζ_i are orthogonal. Dropping all these zero terms, there results

$$\alpha_1 = \frac{\sum \zeta_1 y}{\sum \zeta_1^2} = \frac{\sum \zeta_1 y}{\lambda_1} \qquad (11\text{-}51a)$$

$$\alpha_2 = \frac{\sum \zeta_2 y}{\sum \zeta_2^2} = \frac{\sum \zeta_2 y}{\lambda_2} \qquad (11\text{-}51b)$$

$$\alpha_3 = \frac{\sum \zeta_3 y}{\sum \zeta_3^2} = \frac{\sum \zeta_3 y}{\lambda_3} \qquad (11\text{-}51c)$$

$$\vdots$$

$$\alpha_k = \frac{\sum \zeta_k y}{\sum \zeta_k^2} = \frac{\sum \zeta_k y}{\lambda_k} \qquad (11\text{-}51k)$$

The denominator substitution in Eqs. 11-51 can be made because $\sum \zeta_i^2$ is by definition the variance of ζ_i and it was stated earlier that this is the eigenvalue, λ_i. Making use of the definition of ζ_i in terms of the χ_i yields

$$\sum \zeta_k y = l_{1k} \sum x_1 y + l_{2k} \sum x_2 y - l_{3k} \sum x_3 y + \cdots + l_{ik} \sum x_i y \qquad (11\text{-}52)$$

Making use of Eqs. 11-52 and 11-51 and noting that $\sum x_i y = r_{x_i y}$ for a standardized variate, we get

$$\alpha_1 = (1/\lambda_1)(l_{11}r_{x_1y} + l_{21}r_{x_2y} + l_{31}r_{x_3y} + \cdots + l_{ix}r_{x_iy}) \qquad (11\text{-}53a)$$

$$\alpha_2 = (1/\lambda_2)(l_{12}r_{x_1y} + l_{22}r_{x_2y} + l_{32}r_{x_3y} + \cdots + l_{i2}r_{x_1y}) \qquad (11\text{-}53b)$$

$$\vdots$$

$$\alpha_k = (1/\lambda_k)(l_{1k}r_{x_1y} + l_{2k}r_{x_2y} + l_{3k}r_{x_3y} + \cdots + l_{ik}r_{x_1y}) \qquad (11\text{-}53k)$$

Kendall (1957) shows that the correlation coefficient for Eq. 11-46 is the sum of the correlations of the ζ_k. These separate correlation coefficients are

$$R_1^2 = \lambda_1 \alpha_1^2 \tag{11-54a}$$

$$R_2^2 = \lambda_2 \alpha_2^2 \tag{11-54b}$$

$$\vdots$$

$$R_k^2 = \lambda_k \alpha_k^2 \tag{11-54k}$$

Then the total correlation given by the first j orthogonal variates is

$$R_j^2 = \sum_{k=1}^{j} \lambda_k \alpha_k^2 \tag{11-55}$$

In summary, we need only Eqs. 11-49, 11-53, and 11-55 to derive a components regression equation. The steps are as follows:

1. Compute the correlation matrix of the X_i and the correlation vector for each X_i with Y.
2. Calculate the eigenvalues and eigenvectors of the correlation matrix of the X_i.
3. Use Eqs. 11-53 to calculate the α_k.
4. Use Eq. 11-55 to calculate the correlation coefficient.
5. Use Eq. 11-49 to calculate the coefficients of the X_i in the linear model.

Usually, we will want our model expressed in the original variates of measurement, X_i and Y, rather than x_1 and y. Remembering that

$$x_i = \frac{X_i - \bar{X}_i}{S_x} \qquad \text{and} \qquad y = \frac{Y - \bar{Y}}{S_y}$$

we can substitute these standardizing expressions in Eq. 11-49 and solve for Y in terms of the X_i. Equation 11-49 obviously must be extended to the proper number of X_i and ζ_k for the model under study.

Example 11-1: Analysis of Synthetic Data

In Table 11-1 is shown a data set of 15 observations on four functional independent variates and a functional dependent variate. This is a synthetic data set. The dependent variate was computed from the equation $\hat{Y} = 0.2X_1 + 0.1X_2 + 0.05X_3 + 0.35X_4$. A random error was then added to $\hat{Y}$ to produce the variate Y in Table 11-1. We will evaluate a linear relationship of these variates by components regression and compare with the linear relationship evaluated by least squares. At the bottom of Table 11-1 are shown the means and standard deviations of the variates. These will be needed in conversion of components regression coefficients in standardized variate form to original variate form as shown following Eq. 11-55.

TABLE 11-1 Data Set for Example 11-1

Observation Vector	Variate X_1	X_2	X_3	X_4	Y
1	1.0	1.0	8.3	6.9	3.23
2	2.0	1.0	7.0	9.9	4.22
3	2.3	2.0	7.0	8.5	4.01
4	3.0	1.8	5.5	3.3	2.40
5	3.5	2.5	6.5	9.4	4.56
6	4.0	2.3	4.5	2.8	2.37
7	4.5	1.8	3.0	7.6	4.07
8	4.5	3.0	5.0	5.9	3.71
9	5.8	2.5	2.5	1.2	1.94
10	5.8	3.3	3.3	4.8	3.71
11	7.0	2.8	1.5	8.6	4.59
12	7.0	3.3	1.8	5.7	3.98
13	1.0	2.0	8.5	1.0	1.15
14	3.0	2.8	6.5	0.1	1.54
15	1.8	2.5	7.3	3.4	3.66
$\sum X$	56.2	34.6	78.2	79.1	49.14
$\sum X^2$	266.56	86.98	484.06	561.03	178.3468
$\bar{X}$	3.7467	2.3067	5.2133	5.2733	3.2760
S_X	3.7332	0.4780	5.0918	9.5939	1.1576

The matrix of correlation coefficients is shown in Table 11-2. The correlations of each X_1 with Y are included. These are needed as given by Eqs. 11-53. However, it should be clear that eigenvalue-eigenvector analysis is performed using only the 4×4 matrix of correlation coefficients of the X_i. We should note in Table 11-2 that X_1 and X_3 are very highly correlated with correlation coefficient of -0.970. Also X_1 and X_2 are correlated with a coefficient of 0.706. X_4 is not correlated with the other X_i but is highly correlated with Y.

TABLE 11-2 Matrix of Correlation Coefficients for Example 11-1

	X_1	X_2	X_3	X_4	Y
X_1	1.00000	0.70648	−0.97035	0.07752	0.30626
X_2	0.70648	1.00000	−0.57484	−0.27669	0.05274
X_3	−0.97035	−0.57484	1.00000	−0.05899	−0.27059
X_4	0.07752	−0.27669	−0.05899	1.00000	0.90627

Table 11-13 gives the results of eigenvalue–eigenvector analysis of the correlation matrix of the X_i. Only three components are meaningful. The sum of the three eigenvalues is about 3.99 out of a possible total of 4 for a 4×4 matrix. The fourth component therefore does not contain any meaningful amount of information concerning the structure of the X_i correlation matrix.

TABLE 11-3 Eigenvalues and Eigenvectors of the *X* Correlation Matrix for Example 11-1

		Component 1	Component 2	Component 3
			Eigenvalues	
		2.51623	1.12177	0.35279
			Eigenvectors	
Variate	X_1	0.61731	0.15702	−0.15268
	X_2	0.51869	−0.32094	0.76620
	X_3	−0.59015	−0.18016	0.48718
	X_4	−0.04013	0.91645	0.39025

The elements of the eigenvectors show that component 2 is identified almost exclusively with variate X_4. We would anticipate such an eigenvector since the correlation matrix of Table 11-2 showed that X_4 is not correlated with the other X_i. The eigenvector number one indicates that component number 1 is identified fairly strongly with variates X_1 and X_3 and to a lesser degree with X_2. Component number 3 identifies strongly with variate X_2.

Table 11-4 shows the completed components regression. The regression coefficients were computed using Eqs. 11-53 and 11-47. As the first step the vector products indicated in Eqs. 11-53 are computed. The first product is $(0.306255 \times 0.61731) + (0.052736 \times 0.51869) + (-0.270590 \times -0.59015) + (0.006274 \times -0.04013) = 0.33973$. Then α_1 of Eqs. 11-53 is $0.33973/2.51623 = 0.13502$. Further, α_1 times the elements of eigenvector 1 are the regression coefficients for component 1 as in Eqs. 11-47. For example, $0.13502\,(0.61731) = 0.08335$. This is b_1 for component 1. Similarly, b_2 for component 1 is $0.13502 \times 0.51869 = 0.07003$. The remaining regression coefficients are computed in the same way. The second vector product of the correlations

TABLE 11-4 Solution on Successive Components for Example 11-1

Component	Regression Coefficient b_1	b_2	b_3	b_4	Correlation, R^2	X Variance, λ
1	0.08335	0.07003	−0.07968	−0.00542	0.0459	2.51623
2	0.12744	−0.26049	−0.14623	0.74383	0.7390	1.12177
3	−0.09326	0.46802	0.29758	0.23838	0.1316	0.35279
1 + 2	0.21079	−0.19046	−0.22591	0.73841	0.7849	3.63800
1 + 2 + 3	0.11753	0.27756	0.07167	0.97679	0.9165	3.99079

Three component equation in original variates	Corr.
$Y = 0.06705 + 0.06544X_1 + 0.43195X_2 + 0.03418X_3 + 0.33930X_4$	0.9573
Equation by method of least squares	
$Y = 4.42650 - 0.75260X_1 + 1.04386X_2 - 0.53870X_3 + 0.39251X_4$	0.9755

with Y and elements of the second eigenvector is 0.91047. Then α_2 is $0.91047/1.12122 = 0.81164$.

Calculation of the correlation coefficient for each component is by Eqs. 11-54. For example, R^2 due to component 1 is $2.51623 \times 0.13502^2 = 0.0459$. Table 11-4 shows that component 2 has the largest correlation with Y. Components 1 and 3 have considerably lower correlation.

Equations 11-49 and 11-55 indicate that the regression coefficients and the square of the correlation coefficient can be summed by components. This feature of components regression is one of the extremely valuable procedures in model evaluation. It allows the structure of the model to be studied as components are brought into the correlation with Y. The regression coefficients of each of the individual components in Table 11-4 contain some negative values. Now, the full relationship of the X_i to Y is not necessarily given by any one component. Each component simply represents a relation of one independent, or orthogonal, element in the X_i to Y. The full relationship of all the independent elements of the X_i to Y is given by the sum of all the nontrivial components. The λ's are the measure, on a variance scale, of the information content of the components. As stated earlier, three λ's total to about 3.99. The remaining λ of 0.01 is a trivial level of information. Our linear model is then given by the sum of the first three components. Note that the regression coefficients for components $1+2+3$ are all positive. They are a reasonable reproduction of the data generating equation given at the beginning of this example, when due allowance is made for the random error added to the generated variate $\hat{Y}$. The strong consequence of our use of components regression is that b_1 represents an effect of X_1 on Y which is independent of any relationship of X_1 to X_2, to X_3 or to X_4. We are then able to judge b_1 as to its rationality as a structural element of our linear model.

The independent nature of the components regression coefficients is further illustrated by comparison with coefficients derived by least squares. For ease in making the comparison the two equations are shown at the bottom of Table 11-4. Before making the comparison we need to note that if

$$\hat{y} = b_1 x_1 + b_2 x_2$$

is a linear relationship of standardized variates, then

$$\hat{Y} = \left(\bar{Y} - b_1 \frac{S_y}{S_{x_1}} \bar{X}_1 - b_2 \frac{S_y}{S_{x_2}} \bar{X}_2\right) + \left(b_1 \frac{S_y}{S_{x_1}}\right) X_1 + \left(b_2 \frac{S_y}{S_{x_2}}\right) X_2$$

is the equivalent relationship of the original hydrologic variates.

We can note that the regression coefficients for X_1, X_2, and X_3 are vastly different in the two equations shown at the bottom of Table 11-4. These are of course, the variates having strong correlations, as shown in Table 11-2. The coefficient for X_4, on the other hand, is not so greatly different in the two equations. This again is a consequence of the independence of X_4 as evidenced by the correlation coefficients of Table 11-2.

The correlation coefficient of 0.9573 for the components regression is slightly lower than the value of 0.9755 for the least-squares equation. This is to be expected. We have used only three components to express four variates and there is some loss in the level to which we can reproduce Y with our linear model. It is necessary to note, however, that the X information in this omitted component is small. We should regard the omitted component as representing the random portions of the X_i. By omitting this random portion of the X_i, and thereby eliminating random relationships to the Y, we have gained structural rationality of the regression coefficients at the cost of a slight reduction in correlation. But we can say this another way. We have derived rational coefficients in the presence of error and in the presence of nonindependence of our original variates. We have prevented our coefficients from being influenced by the random errors. The reduced correlation coefficient then becomes a more meaningful measure of fit of a rational set of coefficients.

Example 11-2: Analysis of Synthetic Data

In this example a linear model will be evaluated by components regression using the synthetic data set in Table 11-5. The matrix of correlation coefficients and the results of eigenvalue-eigenvector analysis are shown in Tables 11-6 and 11-7, respectively.

TABLE 11-5 Synthetic Data Set for Example 11-2

Observation	Variate						
Vector	X_1	X_2	X_3	X_4	X_5	X_6	Y
1	1.15	1.25	11.72	0.41	1.56	7.74	4.19
2	1.64	1.47	9.86	0.27	1.91	6.76	4.52
3	1.60	0.50	9.31	0.44	2.01	2.56	3.03
4	2.10	0.90	10.48	0.28	1.18	5.59	2.43
5	1.58	2.20	13.31	0.46	2.66	6.22	6.64
6	1.20	2.30	8.75	0.27	2.57	4.42	3.13
7	2.03	2.54	6.12	0.86	3.40	3.01	2.16
8	1.91	2.76	10.85	0.42	1.33	3.21	1.29
9	1.71	2.89	11.12	0.91	2.62	7.60	2.37
10	1.42	3.04	10.77	0.20	1.62	8.18	3.73
11	2.35	3.03	4.38	0.42	2.77	6.62	2.19
12	1.11	3.26	8.88	0.64	1.75	3.42	4.16
13	2.16	3.30	3.02	0.28	2.44	2.36	1.29
14	2.61	3.40	2.17	0.89	3.06	7.85	3.88
15	1.42	3.51	7.34	0.51	7.85	0.29	3.37
16	2.45	3.63	6.08	0.93	7.11	6.62	3.53
17	1.09	3.72	10.09	0.20	3.92	9.49	2.74
18	2.50	4.10	8.00	0.43	4.53	6.08	4.84
19	1.23	4.11	9.20	0.85	2.08	1.16	0.93
20	3.26	7.02	5.18	0.72	7.74	8.34	4.89
21	2.59	6.69	3.43	0.29	3.72	4.57	4.26
22	3.11	5.61	1.89	0.92	6.53	7.42	5.42
23	2.08	5.28	1.50	0.61	5.89	0.39	1.06
24	2.80	5.06	1.78	0.95	6.01	9.94	4.65
25	3.17	6.44	0.48	0.20	3.37	6.90	2.79
26	1.59	4.80	8.67	0.44	13.47	1.27	4.25
27	3.20	4.49	1.26	0.81	5.75	5.24	2.81
28	1.86	5.79	9.50	0.30	10.29	8.23	5.25
$\bar{X}$	2.0329	3.6818	6.9693	0.5325	4.2550	5.4100	3.4214
S_X	0.6700	1.6851	3.7207	0.2587	2.9183	2.7387	1.4011

TABLE 11-6 Matrix of Correlation Coefficients for Example 11-2

	X_1	X_2	X_3	X_4	X_5	X_6	Y
X_1	1.0000	0.5968	−0.7796	0.3448	0.2605	0.2943	0.1383
X_2	0.5968	1.0000	−0.6135	0.1748	0.5918	0.1233	0.1785
X_3	−0.7796	−0.6135	1.0000	−0.3343	−0.2527	0.0093	0.1427
X_4	0.3448	0.1748	−0.3343	1.0000	0.1963	0.0242	0.0043
X_5	0.2605	0.5918	−0.2527	0.1963	1.0000	−0.0645	0.3332
X_6	0.2943	0.1233	0.0093	0.0242	−0.0645	1.0000	0.4943

TABLE 11-7 Eigenvalues and Eigenvectors of the *X* Correlation Matrix for Example 11-2

	Component					
	1	2	3	4	5	6
			Eigenvalues			
	2.76656	1.12036	0.93196	0.76733	0.25830	0.15550
			Eigenvectors			
1	0.52075	0.28280	−0.08847	−0.23428	−0.43680	0.62877
2	0.50487	−0.13996	0.36195	−0.04727	0.74441	0.19527
3	−0.50862	−0.01098	0.22206	0.46078	0.08387	0.68739
4	0.28784	−0.00357	−0.71889	0.60407	0.18745	0.01738
5	0.34661	−0.49299	0.41774	0.47654	−0.46138	−0.14951
6	0.11063	0.81072	0.34720	0.37225	0.00227	−0.26715

In interpreting the evaluation of this model it should first be noted that exceedingly high correlations are present among the X variates. The highest value is −0.7796 between X_1 and X_3. The least correlated variates are X_4 and X_6, but even with these, moderate values of 0.3448 and 0.2943 are found.

The eigenvalues and eigenvectors reflect the diffuse pattern of correlations among the X_i. Six eigenvalues are shown since even the smallest contains some small but important portion of the total. This smallest eigenvector contains $0.1555/6 = 2.6\%$ of the total variance of the X_i, compared to $0.01/4 = 0.25\%$ for the smallest eigenvalue in Example 11-1. The elements of the eigenvectors show that component 1 is related to variates X_1, X_2, and X_3. Component 2 identifies clearly with variate X_6, and component 3 identifies with X_4. Variates X_3, X_4, X_5, and X_6 are all loosely associated in component 4. Variates X_1, X_2, and X_5 have the highest elements in component 5, while X_1 and X_2 are identified in component 6. In summary, a diffuse pattern of linkages between the variates and the components is found. No simple structure, possible of clear and concise interpretation, is present.

Table 11-8 shows the regression coefficients given by adding sequentially the components from highest to lowest. The changes in the correlation coefficient and the X variance caused by the addition of sequential components are also shown.

In this example of diffuse correlation, reasons for omitting some of the components are less clear than in Example 11-1. The regression coefficients and the correlation coefficients for four components most likely form the best solution for Example 11-2. With four components the sum of the eigenvalues

TABLE 11-8 Solution on Successive Components for Example 11-2

Number of Components	Regression Coefficients						Correlation			X Variance		
	b_1	b_2	b_3	b_4	b_5	b_6	$R_c^{2\,a}$	$R_T^{2\,b}$	R_T	λ	λ_T	$\frac{\lambda_T}{b}$
1	0.0491	0.0476	−0.0480	0.0272	0.0327	0.0104	0.0246	0.0246	0.157	2.7666	2.7666	0.461
2	0.1120	0.0165	−0.0504	0.0264	−0.0769	0.1907	0.0554	0.0800	0.283	1.1204	3.8870	0.648
3	0.0748	0.1687	0.0429	−0.2758	0.0987	0.3366	0.1647	0.2447	0.495	0.9320	4.8190	0.803
4	−0.0382	0.1459	0.2653	0.0157	0.3287	0.5162	0.1787	0.4233	0.651	0.7673	5.5863	0.931
5	0.0757	−0.0483	0.2434	−0.0332	0.4490	0.5156	0.0176	0.4409	0.664	0.2583	5.8446	0.974
6	0.2297	−0.0004	0.4118	−0.0290	0.4124	0.4502	0.0093	0.4503	0.671	0.1555	6.0000	1.000

[a]Correlation due to last component added.

[b]Total correlation due to all components added.

is 93% of the total variance. The correlation coefficient increases only a small amount with addition of components 5 and 6. Also, all regression coefficients but one are positive, and this is in agreement with the positive y correlations in Table 11-6. The single negative regression coefficient is small. The regression coefficients shown in Table 11-8 are for standardized variates. Values for the original variates are not shown but could be computed using the variate means and standard deviations at the bottom of Table 11-5.

Example 11-3: Effect of Antecedent Precipitation on Streamflow

We wish to investigate the influence of antecedent precipitation on streamflow during the dormant season of the year. We wish further that this influence be independent of any correlation patterns in the rainfall. A linear model will be evaluated by components regression to establish this influence.

Monthly runoff data for the calendar months January, February, and March for the years 1954–1962 were taken from Tennessee Valley Authority (1963). Monthly rainfall values for the same months, plus 12 antecedent months, were also copied. We thus have data to evaluate the linear model.

$$a + b_{13}P_{13} + b_{12}P_{12} + b_{11}P_{11} + \cdots + b_2P_2 + b_1P_1 = \mathrm{R}O \tag{11-56}$$

In summary, in Eq. 11-56, RO stands for runoff in each of the months January, February, and March. P_1 is the precipitation in the same month, P_2 is the precipitation in the previous month, and so on, back to and including precipitation in the same calendar month of the previous year.

Table 11-9 gives the full matrix of correlation coefficients computed from the rainfall and runoff data. We should first note that no extremely strong correlations are found among the monthly rainfall values. On the other hand, several nontrivial correlations, in the range of magnitude from 0.3 to 0.6, can be found. We thus have a diffuse picture, somewhat like Example 11-2. A strong correlation of about 0.84 is found between rainfall and runoff in the same month. Correlations with runoff drop off irregularly as we go back through antecedent months, but remain positive for five antecedent months.

Table 11-10 gives the eigenvalues and eigenvectors for the 13×13 matrix of monthly rainfall correlation coefficients. The eigenvalues can be seen to drop off slowly and steadily from 2.6339 to 0.1408. No abrupt break occurs beyond which small values could be omitted from our solution. We can note, however, that eigenvalues 9 and 10 are not too dissimilar in magnitude, with eigenvalue 11 dropping to about two-thirds of their value.

Some relationship can be found between the variates and the components in the elements of the eigenvectors. The rainfall of the month concurrent with our runoff shows a clear relationship to component 6 with a loading of 0.7769. Rainfall in the month previous to the month of runoff associates with component 7 with a loading of 0.6469. However, rainfall for 7 months previous to runoff also associates with component 7 with a loading of −0.6091. Other such diffuse relationships can be seen. We must conclude that our eigenvalue-eigenvector analysis gives us no clear indication of how many components to use in our regression. We see again that to this point Example 11-3 is similar to Example 11-2.

TABLE 11-9 Matrix of Correlation Coefficients for Example 11-3

	For precipitation during month													
	JFM	FMA	MAM	AMJ	MJJ	JJA	JAS	ASO	SON	OND	NDJ	DJF	JFM	Jan-Feb-Mar RO
JFM	1.0000	−0.5035	0.2695	−0.1872	0.2648	0.1603	−0.3376	0.0346	−0.3270	0.2345	0.0306	0.0290	0.0393	−0.0203
FMA	−0.5035	1.0000	−0.2749	0.3378	−0.3717	−0.0577	0.3501	−0.0355	0.4553	−0.1350	−0.1162	0.0019	−0.1164	−0.0833
MAM	0.2695	−0.2749	1.0000	−0.1468	0.0431	−0.2700	−0.1102	0.3804	−0.0699	0.3102	−0.0864	−0.3313	−0.0697	0.0040
AMJ	−0.1872	0.3378	−0.1468	1.0000	−0.1794	0.2700	−0.3067	−0.0898	0.6840	0.2976	0.1463	0.0610	−0.2150	−0.1236
MJJ	0.2648	−0.3717	0.0431	−0.1794	1.0000	−0.1843	0.1108	−0.3291	−0.2558	0.4337	0.2685	0.0214	−0.0681	−0.0810
JJA	0.1603	−0.0577	−0.2700	0.2722	−0.1843	1.0000	−0.4286	0.2247	0.0040	−0.0029	0.2730	0.2329	0.0309	0.0735
JAS	−0.3376	0.3501	−0.1102	−0.3067	0.1108	−0.4286	1.0000	−0.2623	0.0042	0.1111	−0.0075	−0.1270	−0.0560	−0.0013
ASO	0.0346	−0.0355	0.3804	−0.0898	−0.3291	0.2247	−0.2623	1.0000	−0.2615	−0.0708	0.0460	0.1818	−0.1047	0.0506
SON	−0.3270	0.4553	−0.0699	0.6840	−0.2558	0.0040	0.0042	−0.2615	1.0000	0.2277	−0.0667	−0.1789	−0.0142	0.1492
OND	0.2345	−0.1350	0.3102	0.2976	0.4337	−0.0029	0.1111	−0.0708	0.2277	1.0000	0.4270	−0.0434	−0.2161	0.0011
NDJ	0.0306	−0.1162	−0.0864	0.1463	0.2685	0.2730	−0.0075	0.0460	−0.0667	0.4270	1.0000	0.3266	0.0484	0.3455
DJF	0.0290	0.0019	−0.3313	0.0610	0.0214	0.2329	−0.1270	0.1818	−0.1789	−0.0434	0.3266	1.0000	0.0207	0.2392
JFM	0.0393	−0.1164	−0.0697	−0.2150	−0.0681	0.0309	−0.0560	−0.1047	−0.0142	−0.02161	0.0484	0.0207	1.0000	0.8392

TABLE 11-10 Eigenvalues and Eigenvectors of Correlation Matrix for Example 11-3

	Component												
	1	2	3	4	5	6	7	8	9	10	11	12	13
						Eigenvalues							
	2.6339	2.1172	1.9649	1.6423	1.2480	0.9426	0.6510	0.5659	0.3641	0.3368	0.2139	0.1787	0.1408
						Eigenvectors							
JFM	0.4329	0.1462	0.0403	−0.1158	−0.2589	−0.1614	−0.0053	0.7001	−0.0987	−0.3084	−0.1160	0.1844	0.1977
FMA	−0.5022	−0.0580	−0.0120	0.0209	0.2380	0.0302	−0.0492	0.3821	0.4550	−0.4951	0.1418	−0.0850	−0.2416
MAM	0.2596	−0.0921	0.1154	−0.5844	0.1730	0.2850	0.1400	−0.0095	0.0181	−0.0435	0.6001	−0.2337	0.1463
AMJ	−0.3526	0.4323	0.2033	−0.1761	−0.1261	−0.0853	0.1807	−0.0945	0.0307	−0.0962	−0.2646	−0.4316	0.5339
MJJ	0.3092	−0.0500	0.4447	0.2556	−0.0608	−0.1620	0.0993	−0.2256	0.6797	−0.0115	−0.0925	0.1895	0.2037
JJA	0.0060	0.5107	−0.2180	0.0886	−0.0934	−0.0452	−0.6091	0.1045	0.2098	0.3201	0.3666	−0.0829	0.0582
JAS	−0.1497	−0.4269	0.2440	0.2435	0.3384	0.1323	−0.2652	0.3356	−0.1363	0.2921	−0.0165	−0.0379	0.5060
ASO	0.1337	0.1808	−0.3470	−0.3170	0.5328	0.1776	0.0025	−0.0302	0.3012	0.1017	−0.4288	0.3176	0.1627
SON	−0.4495	0.1483	0.2452	−0.2195	−0.2368	0.1690	0.1532	0.0745	−0.0635	0.2218	0.1845	0.6822	0.0364
OND	0.1128	0.2218	0.5808	−0.1261	0.1490	0.1486	−0.0536	0.2166	0.0188	0.3165	−0.2840	−0.2169	−0.5125
NDJ	0.1128	0.3604	0.2435	0.3156	0.2675	0.3873	−0.1894	−0.2334	−0.3232	−0.4909	0.0772	0.1806	0.0561
DJF	0.0422	0.3098	−0.1591	0.4190	0.2952	−0.0545	0.6469	0.2465	−0.0621	0.2385	0.2608	−0.0571	−0.0089
JFM	0.0582	−0.0681	−0.1903	0.2036	−0.4275	0.7769	0.1159	0.1085	0.2334	0.0565	−0.1434	−0.1516	0.0394

A difference between Example 11-3 and Example 11-2 emerges if we study Table 11-11. This table shows the regression coefficients for the cumulative number of components we use in the regression, starting with the first component. The main pattern we should notice is that regression coefficients 1 through 7 gradually build to positive values which decrease backward in time. Our pattern of influence of antecedent rainfall on runoff is thus beginning to emerge.

The regression coefficients based on the first 10 components seem to present the smoothest picture of diminishing influence backward through time. These coefficients were plotted in Fig. 11-4(a). We can see here that influence of antecedent rain probably extends through 7 months. For comparison, the regression coefficients based on all 13 components are also shown in the figure. We can see that this pattern is much more irregular. The coefficients based on 10 components seem to fit better our a priori ideas of what diminishing influence should look like.

We cannot leave Example 11-3 without making special note that our choice of solution is, to a certain degree, based on our preconceived ideas. We could find no clear statistical basis for selecting the number of components on which to base the regression coefficients. We used as a guide our knowledge that rainfall influence on runoff must diminish as we go backward in time. We also used the idea that influence should diminish smoothly. Admittedly, we make a subjective choice of solution; but our choice is guided by experience, judgment, and knowledge of the physical process. We do not need to choose a solution that is obviously nonsensical. Additionally, we can always design additional tests on other hydrologic data to test our choice of solution.

We could not reason logically in Example 11-2, because this was a synthetic data set with no controlling physical process present. We had no alternative but to choose as the solution the coefficients based on all six components. In Example 11-2 we do have a concept of physical process that we can use as a guide in choice of solution.

Example 11-4: Effect of Rainfall on Water Yield

We will perform a rainfall influence analysis, similar to Example 11-3, but using different data. In this example we will study diminishing influence of rainfall on runoff for the months of August, September, and October. Data will be from Parker Branch watershed in western North Carolina, the same area that provided the data for Example 11-3. The correlation matrix for this period of the year again yielded a pattern of numerous but not strong relationships between runoff and rainfall. The general impression is a strong similarity to the data for Example 11-3.

As in Example 11-3, the regression coefficients gradually build to a positive pattern for the rainfall of the more recent months. The pattern only extends through three or four antecedent months, considerably less than for the runoff of the winter months. The solution based on the first 11 components offers a reasonably smooth picture of diminishing influence. These coefficients are plotted in Fig. 11-4(b). The solution based on all 13 components is also plotted.

The 11-component solution is a more regular set of coefficients than the 13 component solution. We make the same subjective judgment as in Example 11-3. Our

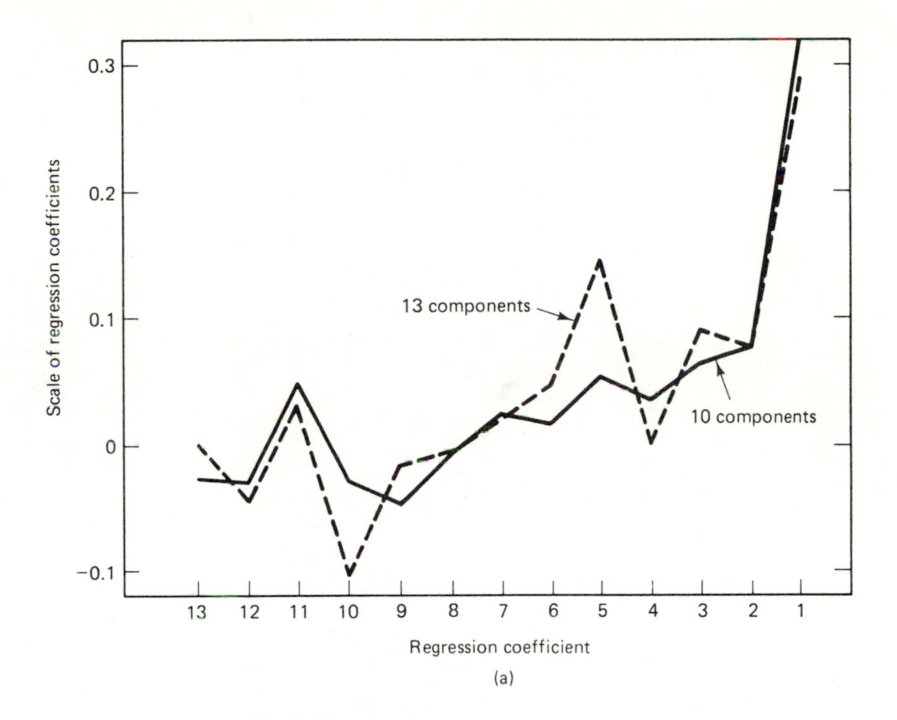

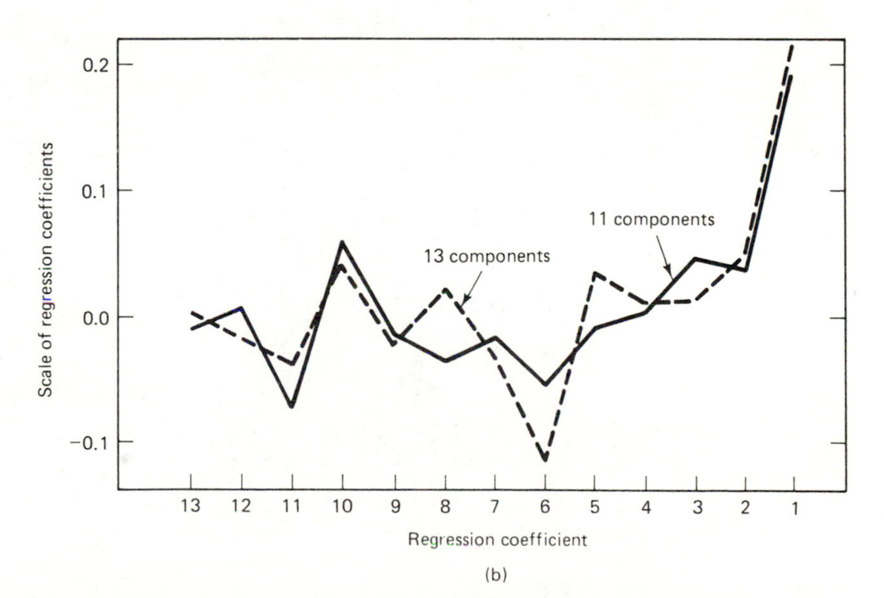

Figure 11-4 Influence of antecedent precipitation on streamflow: (a) January-February-March runoff; (b) August-September-October runoff.

TABLE 11-11 Solution on Successive Components for Example 11-3

Cumulative Number of Components	Regression Coefficient														Cumulative R^2	Precipitation Variance	
	a	13	12	11	10	9	8	7	6	5	4	3	2	1		Cumulative	Percent of Total
1	1.1171	0.0057	−0.0069	0.0037	−0.0045	0.0039	0.0001	−0.0014	0.0013	−0.0045	0.0012	0.0011	0.0005	0.0008	0.0031	2.6339	20.3
2	0.9979	0.0100	−0.0087	0.0008	0.0079	0.0025	0.0115	−0.0103	0.0052	−0.0012	0.0065	0.0093	0.0083	−0.0012	0.0157	4.7511	36.5
3	1.1007	0.0087	−0.0083	−0.0034	0.0013	−0.0116	0.0171	−0.0161	0.0136	−0.0074	−0.0091	0.0031	0.0128	0.0050	0.0306	6.7160	51.7
4	1.1142	−0.0004	−0.0066	−0.0530	−0.0121	0.0075	0.0224	−0.0025	−0.0045	−0.0204	−0.0170	0.0220	0.0409	0.0206	0.0996	8.3583	64.3
5	1.2128	0.0158	−0.0221	−0.0648	−0.0044	0.0112	0.0268	−0.0176	−0.0289	−0.0092	−0.0245	0.0092	0.0251	0.0468	0.1332	9.6063	73.9
6	−0.2422	−0.0383	−0.0115	0.0391	−0.0323	−0.0410	0.0152	0.0140	0.0147	0.0340	0.0157	0.1088	0.0094	0.3024	0.8627	10.5489	81.1
7	−0.3107	−0.0387	−0.0157	0.0515	−0.0179	−0.0332	−0.0228	−0.0014	0.0148	0.0436	0.0121	0.0969	0.0547	0.3117	0.8926	11.1999	86.2
8	−0.5424	−0.0056	0.0032	0.0510	−0.0223	−0.0434	−0.0190	0.0099	0.0138	0.0462	0.0204	0.0884	0.0647	0.3167	0.9013	11.7658	90.5
9	−0.5249	−0.0052	0.0013	0.0510	−0.0224	−0.0460	−0.0197	0.0103	0.0129	0.0464	0.0204	0.0894	0.0649	0.3159	0.9013	12.1299	93.3
10	−0.4871	−0.0250	−0.0318	0.0479	−0.0284	−0.0467	−0.0039	0.0236	0.0177	0.0573	0.0368	0.0652	0.0780	0.3194	0.9109	12.4667	95.9
11	−0.5539	−0.0290	−0.0267	0.0704	−0.0373	−0.0436	0.0057	0.0232	0.0069	0.0621	0.0289	0.0673	0.0857	0.3146	0.9127	12.6806	97.5
12	−0.5274	0.0005	−0.0406	0.0306	−0.1031	−0.0152	−0.0042	0.0190	0.0433	0.1435	0.0015	0.0890	0.0780	0.2913	0.9428	12.8593	98.9
13	−0.5393	0.0011	−0.0415	0.0311	−0.1013	−0.0145	−0.0040	0.0202	0.0437	0.1436	0.0001	0.0891	0.0780	0.2914	0.9428	13.0001	100.0

position is strengthened, however, because we note that the shorter period of antecedent influence of rainfall is hydrologically correct. During the summer months the soil profile of a watershed normally dries. The base flow of a stream diminishes during this drying period. Rain falling during the summer months usually provides mainly surface runoff, but little recharge of water deep in the soil profile. Antecedent rainfall thus has a shorter period of influence on runoff in late summer and early autumn.

We note finally that the rainfall influence lines in Fig. 11-4 are not runoff distribution curves. The regression coefficients plotted here are conditioned by two hydrologic effects. First, they do contain some part of a time-of-travel function, showing the delayed delivery of water to streamflow caused by travel time through the soil profile. But they also represent initial losses that are abstracted from rainfall before contribution to streamflow begins.

Example 11-15: Calibration of a Peak Discharge Model

Benson (1962, 1964) has studied the relationship between watershed characteristics and peak discharges of various recurrence intervals. In these studies he described the relationship by an equation of the form

$$Q_T = b_0 X_1^{b_1} X_2^{b_2} X_3^{b_3} \cdots X_p^{b_p}$$

where Q_T is the flood for recurrence interval T, the X_i are watershed characteristics, and the b_i are the parameters to be evaluated. Benson evaluated the parameters using least squares after linearizing the equation by a logarithmic transform. We will evaluate a linear relationship of the characteristics to the flood discharge using components regression.

Watersheds were selected from Tables 1 and 2 of Benson (1962). All watersheds with drainage areas between 1,000 and 10,000 square miles were used. A total of 63 watersheds fell in this range. The definition of the variables, and the matrix of correlation coefficients are given in Table 11-12. We can note immediately in this table several strong correlations among the watershed characteristics. Area and length of the main channel are both measures of watershed size and correlate at 0.86. High correlations are also found among the hydrologic characteristics of mean annual precipitation, mean annual runoff, and the ratio of runoff to precipitation. Another set of relations is found among slope, basin rise, and altitude. We should suspect, then, that certain advantages would be derived by reducing the problem from 11 interrelated characteristics to some lesser number of orthogonal components.

The eigenvalue-eigenvector analysis of the correlation matrix of the watershed characteristics is shown in Table 11-13. Note first that seven components explain 98.1% of the variance in 11 characteristics. Component 1 appears to be a general component moderately associated with all characteristics except area, length, and area in ponds and lakes. Component 2 is a watershed size component and relates to drainage area and channel length. Component 3 is identified with the number of thunderstorm days. Component 4 shows a strong relationship to hydraulic storage, as expressed by area in ponds and lakes. Component 5 is most strongly related to slope. Component 6 appears a weak conglomerate expression of the area, the number of thunderstorm days, and the ratio of runoff to rainfall. Component 7 relates to rainfall intensity.

Regression coefficients of our linear model relating floods to watershed characteristics are given in Table 11-14. The coefficients based on seven components appear rational considering the weak correlation shown by an R^2 of only 0.41. Coefficients

TABLE 11-12 Matrix of Correlation Coefficients for Example 11-5[a]

Variate

	A	S	S_t	E	L	H	P	I	N	R	R_a	$Q_{2.33}$
A	1.0000	−0.3341	0.1798	−0.2334	0.8626	−0.0451	0.1302	0.2347	0.0320	0.1762	0.1053	0.4787
S	−0.3341	1.0000	−0.1832	0.8212	−0.3744	0.8331	−0.5705	−0.7006	−0.1451	−0.5630	−0.4276	−0.3502
S_t	0.1798	−0.1832	1.0000	−0.1433	0.1197	−0.2569	0.1550	0.1261	0.3053	0.2832	0.1193	0.0341
E	−0.2334	0.8212	−0.1433	1.0000	−0.3079	0.7948	−0.7224	−0.8700	−0.0589	−0.6400	−0.5186	−0.3511
L	0.8626	−0.3744	0.1197	−0.3079	1.0000	−0.0371	0.2604	0.3185	0.1070	0.2910	0.2028	0.4179
H	−0.0451	0.8331	−0.2569	0.7948	−0.0371	1.0000	−0.6573	−0.6805	−0.2139	−0.6255	−0.5358	−0.2743
P	0.1302	−0.5705	0.1550	−0.7224	0.2604	−0.6573	1.0000	0.7935	0.6243	0.9532	0.9251	0.3965
I	0.2347	−0.7006	0.1261	−0.8700	0.3185	−0.6805	0.7935	1.0000	0.1931	0.7027	0.6373	0.4115
N	0.0320	−0.1451	0.3053	−0.0589	0.1070	−0.2139	0.6243	0.1931	1.0000	0.6759	0.7029	0.2301
R	0.1762	−0.5630	0.2832	−0.6400	0.2910	−0.6255	0.9532	0.7027	0.6759	1.0000	0.9285	0.4339
R_a	0.1053	−0.4276	0.1193	−0.5186	0.2028	−0.5358	0.9251	0.6373	0.7029	0.9285	1.0000	0.4643

[a]Definition of variates: A, drainage area in square miles; S, main channel slope in feet per mile; S_t, percentage of area in lakes and ponds plus 1%; E, altitude index in feet above mean sea level; L, length of main channel in miles; H, basin rise in feet; P, mean annual precipitation in inches; I, 10-year, 24-hour rainfall intensity in inches; N, mean annual number of thunderstorm days; R, ratio of runoff to precipitation during months when annual peak discharges occur; R_a, mean annual runoff in inches; $Q_{2.33}$, peak discharge at 2.33-year recurrence interval.

TABLE 11-13 Eigenvalues and Eigenvectors of Correlation Matrix for Example 11-5

		Component						
		1	2	3	4	5	6	7
					Eigenvalues			
		5.7400	1.8437	1.5117	0.9889	0.3990	0.1706	0.1354
					Eigenvectors			
	A	0.1308	0.6151	0.2945	0.0301	−0.0966	0.4229	0.4756
	S	−0.3302	−0.1990	0.2936	0.1324	0.5292	0.3052	0.0623
	S_t	0.1185	0.0243	0.2401	−0.8948	0.3069	0.0099	−0.0712
	E	−0.3521	−0.1258	0.3545	0.0030	−0.2105	0.1208	0.1871
	L	0.1673	0.5747	0.3136	0.1566	0.0060	−0.2628	−0.4639
Variable	H	−0.3318	0.0690	0.3528	0.2408	0.3933	−0.3740	−0.0794
	P	0.3901	−0.1762	0.0873	0.1731	0.1623	0.0184	−0.0858
	I	0.3600	0.0641	−0.2163	0.1115	0.4902	−0.3204	0.5430
	N	0.2144	−0.3257	0.5314	−0.0374	−0.3770	−0.4514	0.3070
	R	0.3838	−0.1674	0.1859	0.0555	0.0784	0.1812	−0.3310
	R_a	0.3504	−0.2477	0.2224	0.2271	0.0496	0.4059	−0.0481

b_1 and b_2 show that floods are positively related to the size of drainage area and the slope of the channel. These are rationally correct because large areas and steep slopes both cause relatively high flood peaks. Coefficient b_3 is negative, showing the effect of storage in ponds and lakes in reducing peak discharge. Coefficient b_4 shows a positive relationship between floods and altitude. No clear direct reason is apparent. However, altitude may be an expression of steep slopes, or short channel lengths, or even indirectly of shallow upland soils with little water storage capacity. Coefficient b_5 shows a slight negative relation to channel legnth. This is a hydrologically correct relationship because flood waves are attenuated as they travel down-channel. Coefficient b_6 shows a negative relationship to basin rise. This seems at first thought illogical. However, basin rise is simply a difference in elevation of two channel points. A small steeply sloped, high-lying watershed could have the same rise as a large watershed, low lying, with small channel slope. Basin rise may be an indistinct characteristic. Coefficients 7 and 8 show positive relationship to annual rainfall and to rainfall intensity. These are logical. Coefficient b_9 shows a negative relationship between floods and the number of thunderstorm days. This is probably correct for the size of drainage area in our sample of 63 watersheds. Drainage areas of this size most likely do not have maximum floods caused by high-intensity thunderstorm rains, but rather by prolonged cyclonic rains during the dormant season of the year. Coefficients b_{10} and b_{11} show a correct positive relationship between floods and both the ratio of runoff to rainfall and annual precipitation.

Our seven components thus establish a logical linear relationship to the 11 watershed characteristics.

Factor Analysis

In components analysis we had a single, clear definition of meaning. We determined orthonormal components of a correlation matrix through eigenvalue-eigenvector analysis of the matrix. In factor analysis we do not have such a single clear meaning. Factor analysis has come to mean any one of

TABLE 11-14 Solution on Successive Components for Example 11-5

Cumulative Number of Components	Regression Coefficient												Cumulative R^2	Component Variance	
	a	b_1	b_2	b_3	b_4	b_5	b_6	b_7	b_8	b_9	b_{10}	b_{11}		Cumulative	Percent of Total
1	8.9821	0.1229	−0.0617	1.1425	−0.3560	0.0039	−0.0924	0.0637	0.4759	0.0429	3.3879	0.1500	0.2297	5.7400	52.2
2	9.1180	0.6313	−0.0943	1.3484	−0.4679	0.0155	−0.0755	0.0384	0.5504	−0.0144	2.0888	0.0567	0.2867	7.5837	68.9
3	1.5895	0.8218	−0.0566	2.9410	−0.2213	0.0205	−0.0079	0.0482	0.3536	0.0587	3.2176	0.1223	0.3153	9.0954	82.7
4	7.3310	0.8463	−0.0352	−4.5317	−0.2187	0.0237	0.0502	0.0727	0.4813	0.0522	3.6416	0.2064	0.3450	10.0843	91.7
5	8.4972	0.8648	−0.0553	−5.1344	−0.1753	0.0236	0.0279	0.0673	0.3493	0.0676	3.5008	0.2021	0.3457	10.4833	95.3
6	19.9092	1.8541	0.0864	−4.8965	0.1287	0.0085	−0.2313	0.0748	−0.7044	−0.1570	7.4795	0.6344	0.3879	10.6539	96.9
7	10.7583	2.8409	0.1121	−6.4125	0.5462	−0.0151	−0.2801	0.0439	0.8800	−0.0215	1.0299	0.5890	0.4143	10.7893	98.1

several methods of analyzing the matrix of correlation coefficients of a set of variates. We shall follow Cooley and Lohnes (1971), who advocate principal components as a primary procedure for factoring a matrix. A broader discussion of factoring procedures can be found in Harman (1960).

Components analysis was used to establish the independent elements present in a matrix. Factor analysis has different purposes. We will consider two such purposes. First, we will try to obtain better understanding and interpretability of the structure of the correlation matrix. Second, we will attempt to screen our data sets of hydrologic variables in order to eliminate redundant ones and thus produce simpler linear models.

Our method of factoring a matrix will be based on components analysis, as stated above. Therefore, we will first discuss the transition from components to factors. In Eq. 11-44 we can see that the l_{kj} are really regression coefficients that relate the hydrologic variates to the components. Remember also that these l_{kj} are the elements of our eigenvectors. Now, in factoring based on principal components we simply change the eigenvector elements to correlation coefficients. We will then consider that these modified eigenvectors define factors rather than components.

The numerical basis for converting the eigenvector elements from regression coefficients to correlation coefficients can be derived very easily. Consider a simple linear regression equation $\hat{Y} = a + bX$. Now we know from basic statistics that this linear regression equation can also be cast in the form

$$\hat{Y} = \left[\bar{Y} - r\left(\frac{S_Y}{S_X}\right)\bar{X}\right] + r\left(\frac{S_Y}{S_X}\right)X \tag{11-57}$$

In Eq. 11-57, r is the simple linear correlation coefficient and S_y and S_x are the standard deviations of Y and X, respectively. We note specifically that the regression coefficient and the correlation coefficient are related by

$$b = r\frac{S_Y}{S_X} \tag{11-58}$$

which means that the intercept coefficient, a, equals $(\bar{Y} - b\bar{X})$.

Now write Eq. 11-44 as the matrix-vector product

$$\boldsymbol{\zeta} = \boldsymbol{L}\mathbf{X} \tag{11-59}$$

Premultiply Eq. 11-59 by the inverse of $\boldsymbol{L}$:

$$\boldsymbol{L}^{-1}\boldsymbol{\zeta} = \boldsymbol{L}^{-1}\boldsymbol{L}\mathbf{X} = \mathbf{X} \tag{11-60}$$

Now, the inverse and transpose of an orthogonal matrix are equal; therefore, we can write

$$\boldsymbol{L}^{\mathrm{T}}\boldsymbol{\zeta} = \mathbf{X} \tag{11-61}$$

Expanding Eq. 11-61, we get

$$X_1 = l_{11}\zeta_1 + l_{12}\zeta_2 + l_{13}\zeta_3 + \cdots \quad (11\text{-}62a)$$

$$X_2 = l_{21}\zeta_1 + l_{22}\zeta_2 + l_{23}\zeta_3 + \cdots \quad (11\text{-}62b)$$

$$X_3 = l_{31}\zeta_1 + l_{32}\zeta_2 + l_{33}\zeta_3 + \cdots \quad (11\text{-}62c)$$

We write our relationship in the form of Eq. 11-62 because we are interested in interpreting our variates in terms of the components. For any simple linear relationship between a variate and a component we can write the form equivalent to Eq. 11-58:

$$l = r\frac{S_x}{S_\zeta} \quad (11\text{-}63)$$

Solving for r_x yields

$$r = l\frac{S_\zeta}{S_x} = l\sqrt{\lambda} \quad (11\text{-}64)$$

Equation 11-64 follows because S_ζ, the standard deviation of a component, is the square root of its variance. We learned earlier that the variance of a component is the eigenvalue defining that component. Also, since x is a standardized variate, its variance and standard deviation are 1. Equation 11-64 tells us that we can convert our eigenvector elements from regression coefficients to correlation coefficients simply by multiplying by the square root of the respective eigenvalue.

Earlier in this chapter we performed components analysis of a 3×3 matrix to illustrate the computations. We can now convert to a factor solution. The matrix of eigenvectors and the new factor matrix are given side by side for ease of comparison. We make use of the square roots of the eigenvalues 1.9982, 0.9532, and 0.0486.

Eigenvectors			*Factor matrix*		
0.2146	0.9760	−0.0382	0.3034	0.9529	−0.0084
0.6879	−0.1786	−0.7035	0.9723	−0.1744	−0.1551
0.6934	−0.1248	0.7097	0.9802	−0.1218	0.1564

We should note several properties of the factor matrix. If we take the sum of products of any two columns we get zero, showing that the factors are orthogonal. However, the sum of squares of a column is not unity, but equal to the eigenvalue. Therefore, the column elements, called *factor loadings*, give the proportioning of the factor variance to the variates. The sum of squares of any row totals unity. Therefore, the factor loadings by rows give the distribution of the variance of the variates to the factors.

It can readily be seen that the factor matrix is easier to interpret than the eigenvector matrix. Variates X_2 and X_3 have high loadings in the first factor. Variate X_1 has high loading in the second factor. Loadings in the third factor are all trivial. The variance of X_1 in the first two factors is $0.3034^2 + 0.9529^2 \approx 1.000$. Most of the information in X_1 is therefore contained in the first two factors. The variances of X_2 and X_3 in the first two factors are 0.9759 and 0.9756, also illustrating that no significant information is contained in the third factor.

We may recall that in components analysis we rotated the axes in which we measured our hydrologic variates to new positions, called principal axes, in order to achieve orthogonality. When we first compute factor loadings from the principal components we are still in this orthogonal frame of reference. In factor analysis there is no particular reason for maintaining this component frame of reference. Since we wish to interpret our variates in terms of the factors, we can consider any new system of axes that will aid in interpretability.

A system of rotation in common usage is called *varimax rotation* (Cooley and Lohnes, 1971). The principle of rotation used here is that the variance of each factor is maximized. This maximization is accomplished when the large elements in the factor loadings are made as large as possible and the small elements are made as small as possible. Since the loadings are correlation coefficients, this rescaling by rotation to new axes tends to magnify the larger elements toward plus or minus unity and the smaller elements are squeezed toward zero.

The varimax computation will not be further illustrated here. Actual computation is feasible only on a computer, and library programs are available. The technique will be used in the examples of factor analysis to follow.

In the examples we will see one other difference between factor analysis and component analysis leading to components regression. In regression we analyze only the functional independent variates and then relate the components to the dependent variate. In factor analysis we usually include the dependent variate in the eigenvalue–eigenvector analysis. When we convert the components to factors, we can then study the role of the dependent variate as well as the independent variates in the factor patterns. In this way we can, hopefully, find factors composed of the dependent and one or more of the independent variates. Any independents that do not have high loadings in a factor in which the dependent also has high loadings must be considered a candidate for deletion from the set of hydrologic variates.

Example 11-6: Factor Analysis of Synthetic Data

We will perform a factor analysis on the same data used in Example 11-1. The 5×5 matrix of the independent variates plus the dependent variate will be analyzed. The derived factors are shown in Table 11-15. It should be pointed out that the eigenvalue–eigenvector solution of the 5×5 matrix is obviously not the same solution as for the 4×4 matrix of Example 11-1.

Four significant components were found in the 5×5 matrix. The factor loadings derived from these four components are shown at the top of Table 11-15. We see that X_1, X_2, and X_3 have high loadings in the first factor. Y has only moderate loading in

TABLE 11-15 Factor Analysis of Data in Table 11-1 for Example 11-6

Variates	Factor 1	Factor 2	Factor 3	Factor 4	Variance
	Square root of eigenvalues				
	1.6204	1.3795	0.6502	0.2088	
	Loadings				
	Factor loadings from principal components				
X_1	0.9725	−0.1578	−0.1503	0.0667	0.998
X_2	0.7274	−0.4657	0.5028	0.0303	1.000
X_3	−0.9273	0.1506	0.3357	0.0566	0.998
X_4	0.2286	0.9643	0.0032	0.1323	1.000
Y	0.4886	0.8419	0.1862	−0.1324	1.000
	Factor loadings from two rotated vectors				
X_1	0.9708	0.1678			0.971
X_2	0.8394	−0.2031			0.746
X_3	−0.9257	−0.1598			0.883
X_4	−0.0981	0.9861			0.982
Y	0.1875	0.9552			0.948
	Factor loadings from three rotated vectors				
X_1	0.9150	0.1318	0.3723		0.993
X_2	0.4192	−0.1045	0.9012		0.999
X_3	−0.9742	−0.0881	−0.1961		0.995
X_4	0.0221	0.9683	−0.2099		0.982
Y	0.1625	0.9725	0.1004		0.982
	Factor loadings from four rotated vectors				
X_1	0.9132	0.1314	0.3788	0.0553	0.998
X_2	0.4158	−0.1040	0.9032	−0.0153	1.000
X_3	−0.9742	−0.0873	−0.1963	0.0574	0.998
X_4	0.0223	0.9682	−0.2032	0.1428	1.000
Y	0.1645	0.9723	0.0925	−0.1372	1.000

this factor. Variates X_4 and Y have high loadings in factor 2. Loadings are small in factors 3 and 4, except for the moderate loading of X_2 in factor 3.

When we rotate our factors by the varimax procedure, it is not necessary to rotate all four. We can select any grouping of the vectors for rotation. Since factor 1 unrotated is derived from component 1, factor 2 from component 2, and so on, and since the components are derived in the order of highest variances, it is logical to rotate successive groups of the factors starting with the first. For example, Table 11-15 shows results of rotation of factors 1 and 2; of factors 1, 2, and 3; and of factors 1, 2, 3, and 4.

Rotation of the first two factors shows clearly that factor 1 is composed of variates X_1, X_2, and X_3. Factor 2 is clearly made up of variates X_4 and Y. If we rotate three factors, the only change is that variate X_2 separates from factor 1 and identifies with factor 3. The structure does not change when four factors are rotated.

In simple models it is sometimes helpful to plot the variates as vectors in the factor domains. Wong (1963) reanalyzed Benson's (1962) New England flood data and used this technique to show the patterns of relationships among the watershed characteristics. We can plot the variates in only two factors on ordinary graph paper. This plotting is shown in Fig. 11-5 for this example. Two rays of vectors appear in this figure. The first is made up of X_4 and Y, the second of X_1, X_2, and X_3. A second order detail may be noticed. X_2 is not quite in line with X_1 and X_3. Also, Y is not quite parallel to X_4. These details can have a bearing on our selection of variates to use as predictors of Y. For example, Fig. 11-5 should be kept in mind in the following discussion of a two-variate versus a three-variate linear model.

We can now summarize our findings. First, we reject the four-factor solution. The fourth factor contains no usable information relative to the problem of prediction of the variate Y. Second, we have a choice between the three-factor and two-factor solution. Consider the three-factor solution. The highest loading for Y is in the second factor. We select X_4, also with high loading, as the independent variate to represent this factor, but also as a predictor for Y in a linear model. Our next highest loading for Y is the first factor. The loading is not strong for Y, but is strong for X_1 and X_3. We should select either X_1 or X_3 as an independent variate to predict Y. We would probably not use both X_1 and X_3 since they tend to duplicate each other.

If one of these variates is easy and economical to measure and the other is difficult and expensive, we would naturally select the cheaper one and delete the

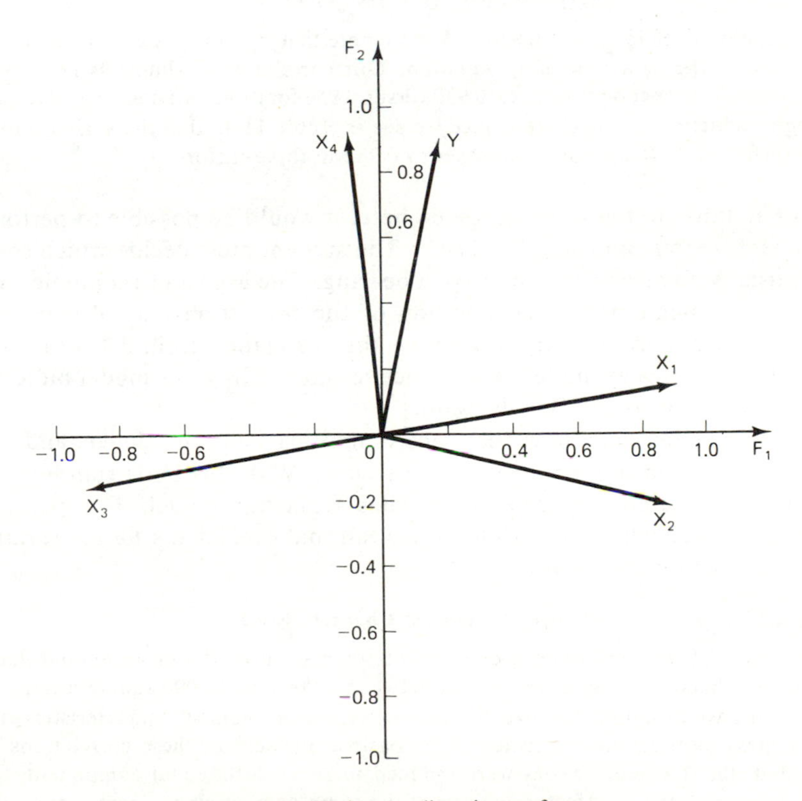

Figure 11-5 Variate loadings in two factors.

expensive one from our predictor variates. If outside reasons do not control, then we can select the variate with the highest loading, X_3 in this case. From factor 3 we select X_2 as a predictor for Y.

Our linear model is thus

$$\hat{Y} = a + b_2X_2 + b_3X_3 + b_4X_4 \tag{11-65}$$

Equation 11-65 was evaluated by least squares. The resultant equation was

$$\hat{Y} = 0.462 + 0.475X_2 - 0.018X_3 + 0.343X_4$$

The correlation coefficient was 0.956. We now note two things. The coefficient of X_3 is negative and the coefficient of X_2 is relatively large. These coefficients are at odds with the generating equation given for the data in Example 11-1.

We next try a solution based on the two-factor rotation. We select variates X_1 and X_4 as predictor variates, these having the highest factor loadings. Then our linear model is

$$\hat{Y} = a + b_1X_1 + b_4X_4 \tag{11-66}$$

A least-squares fitting gave

$$\hat{Y} = 1.154 + 0.132X_1 + 0.308X_4$$

The correlation coefficient was 0.930. We now note that b_1 and b_4 are very nearly the values used in the data-generating equation. Our correlation is almost as good with use of only two independent variates, 0.930 versus 0.956 for three variates. We anticipate this slight reduction in correlation since we see in Table 11-15 that the variance of Y in two factors is 0.948 and only increases to 0.982 in three factors.

In addition to the solutions given here, it would be possible to perform components regression using Eq. 11-65. The student must decide which solution to use. Many elements may have a bearing. The aspect of economics has already been mentioned. The precision of the prediction may also have a bearing. Is it important to try to increase the correlation from 0.930 to 0.956 as we can do in this example? The objectives controlling our model building play an important role in this decision.

It should be clear now that factor analysis helps us understand the patterns of relationship among all our variates. With this understanding we can consider retaining or deleting variates from our model. Components regression is used when we wish to derive rational coefficients *for the variates we have chosen for our model.*

Example 11-7: Factor Analysis of Watershed Characteristics

In Example 11-5 we performed a components regression of the mean annual flood, $Q_{2.33}$, on 11 characteristics of watersheds between 1,000 and 10,000 square miles. We found relationships among the size characteristics, meteorological characteristics, and the steepness–altitude characteristics of the basins. Because of these correlations we found that the 11 characteristics were reduced to seven orthogonal components. In this example we perform a factor analysis of the same data, attempt clearer definitions of the underlying physical factors, and then select a reduced number of characteristics as predictors of $Q_{2.33}$.

Table 11-16 gives the loadings for the derived and rotated factors. Factors were derived from the first five components. More than 80% of the variance of all characteristics is contained in these first five factors, and for 8 of the 12 the variance is above 90%. Note particularly that nearly all the variance of the mean annual flood is contained in these five factors. It is immediately evident that the factors give a clearer picture of relationships than the components in Table 11-13. The loadings show that factor 1 is a composite of the meteorological and slope–altitude characteristics. Factor 2 remains the size factor. Factor 3 relates only to the number of thunderstorm days. Factor 4 is a unique expression of hydraulic storage as measured by the percentage of area in lakes and ponds. Factor 5 has only one strong loading, which is our desired dependent variate of mean annual floods. This last factor indicates that we will probably not have a good model to predict $Q_{2.33}$. Since much of $Q_{2.33}$ variance is in factor 5, with no strong loadings from any of our potential independent variates, we must expect that there is a substantial portion of $Q_{2.33}$ variation which we cannot explain with our proposed linear model.

When we rotate the first two factors to vector positions having maximum variance, we find little change. Essentially, the makeup of these two factors is enhanced by higher loadings. The variance content of the characteristics in these two factors is, for the most part, considerably reduced. Consequently, we examine the pattern produced by rotation of the first three factors.

In the three-factor rotation factor 2 remains the size factor. However, the meteorological and slope–altitude factors have now separated. We may now interpret factor 1 as an expression of altitude as composed of altitude itself, slope, basin rise, and rainfall intensity. We can only speculate that intensity relates to altitude through some orographic effect on rainfall. Factor 3 is now made up very clearly of the characteristics of rainfall, number of thunderstorm days, ratio of runoff to rainfall, and annual runoff. The mean annual flood is seen to be related more to the size factors than to the meteorological or altitude factors. We note finally that the variance content of $Q_{2.33}$ in three factors is still low, although for some of the other characteristics it is substantially improved.

In the four-factor rotation we produce little change in the first three factors. Factor 4 simply identified with our unique characteristic of hydraulic storage. However, we see that mean annual flood loads nearly as heavily in this factor as in the altitude factor, factor 1. The variance of $Q_{2.33}$ is still small compared to the value of 0.9927 in the unrotated five factors.

The rotation of all five factors to vector positions of maximum variance does little more than emphasize the uniqueness of the mean annual flood. Although we have regained our variance of this variate, we have lost the strength of loadings in the other factors where our intended predictors have good loadings. Consequently, we backtrack to the four-factor rotation.

We are now in a position to select those characteristics that could serve as predictors for the mean annual flood. Generally, the number of predictors should be kept small. This keeps our model simple. It can also reduce future costs, since the determination of long lists of watershed characteristics is both tedious and expensive. Certainly, we wish to eliminate as many redundancies as possible. The strongest loading for the mean annual flood is the 0.5947 value in factor 2. We have a choice of either

TABLE 11-16 Factor Analysis of Data in Table 11-12 for Example 11-7

	Factor					
	1	2	3	4	5	
	Square root of eigenvalues					
	2.4507	1.3913	1.2446	1.0217	0.7591	
Variates	Loadings					Variance
	Factor loadings from principal components					
A	0.3526	0.8668	0.2116	−0.0756	−0.0692	0.9309
S	−0.7859	−0.1606	0.4288	0.1382	0.0024	0.8464
S_t	0.2725	0.0006	0.2151	−0.9006	0.1621	0.9579
E	−0.8348	−0.0565	0.4793	0.0043	0.0596	0.9334
L	0.4302	0.7971	0.2324	0.0048	−0.2825	0.9543
H	−0.7782	0.2031	0.4373	0.1860	−0.1417	0.8928
P	0.9252	−0.2575	0.1287	0.1326	−0.1414	0.9764
I	0.8588	0.0119	−0.2885	0.1003	−0.0774	0.8370
N	0.5094	−0.3622	0.6972	−0.0884	−0.0564	0.8878
R	0.9146	−0.2267	0.2434	0.0193	−0.0711	0.9526
R_a	0.8410	−0.2986	0.3283	0.2097	−0.0215	0.9487
$Q_{2.33}$	0.5523	0.3749	0.2153	0.3061	0.6380	0.9927
	Factor loadings from two rotated vectors					
A	−0.0194	0.9356				0.8757
S	−0.6581	−0.4586				0.6434
S_t	0.2500	0.1084				0.0742
E	−0.7442	−0.3825				0.7001
L	0.0794	0.9022				0.8203
H	−0.7951	−0.1217				0.6470
P	0.9516	0.1298				0.9224
I	0.7839	0.3510				0.7377
N	0.6112	−0.1309				0.3907
R	0.9296	0.1540				0.8879
R_a	0.8905	0.0588				0.7964
$Q_{2.33}$	0.3587	0.5629				0.4455
	Factor loadings from three rotated vectors					
A	0.0714	0.9567	0.0103			0.9205
S	−0.8614	−0.2749	−0.0982			0.8272
S_t	0.0612	0.1406	0.3114			0.1205
E	−0.9376	−0.1815	−0.1331			0.9298
L	0.1198	0.9222	0.0985			0.8745
H	−0.8832	0.0623	−0.2330			0.8382
P	0.6217	0.0818	0.7387			0.9389
I	0.8320	0.1982	0.2991			0.8210
N	−0.0589	0.0081	0.9345			0.8768
R	0.5360	0.1369	0.8007			0.9472
R_a	0.4292	0.0708	0.8456			0.9043
$Q_{2.33}$	0.2477	0.5688	0.3271			0.4919

Table 11-16 (*cont.*)

	Factor loadings from four rotated vectors					
A	0.0892	0.9479	−0.0328	−0.1362		0.9261
S	−0.8693	−0.2593	−0.0986	0.1172		0.8464
S_t	0.1163	0.0800	0.1415	−0.9442		0.9315
E	−0.9337	−0.1761	−0.1634	−0.0186		0.9299
L	0.1288	0.9209	0.0714	−0.0684		0.8744
H	−0.8857	0.0778	−0.2347	0.1646		0.8727
P	0.5899	0.1021	0.7729	0.0275		0.9565
I	0.8164	0.2048	0.3402	0.0825		0.8310
N	−0.0802	0.0218	0.8991	−0.2631		0.8845
R	0.5112	0.1507	0.8079	−0.1040		0.9476
R_a	0.3893	0.1006	0.8840	0.0717		0.9483
$Q_{2.33}$	0.2238	0.5947	0.3676	0.2164		0.5857
	Factor loadings from five rotated vectors					
A	0.1007	0.9249	−0.0198	−0.1164	0.2270	0.9311
S	−0.8709	−0.2318	−0.0999	0.1223	−0.0964	0.8464
S_t	0.1105	0.0924	0.1397	−0.9575	−0.0281	0.9579
E	−0.9358	−0.1556	−0.1701	−0.0222	−0.0628	0.9333
L	0.1457	0.9563	0.1201	0.0020	0.0638	0.9542
H	−0.8789	0.1362	−0.2101	0.2067	−0.1222	0.8928
P	0.5905	0.0823	0.7839	0.0451	0.0671	0.9765
I	0.8185	0.1746	0.3447	0.0866	0.1009	0.8369
N	−0.0835	0.0094	0.9021	−0.2536	0.0519	0.8879
R	0.5100	0.1157	0.8093	−0.1000	0.1191	0.9526
R_a	0.3859	0.0352	0.8693	0.0569	0.1990	0.9487
$Q_{2.33}$	0.2091	0.3055	0.2325	0.0354	0.8946	0.9927

area or channel length as the predictor. The logical choice would be area, since this characteristic is usually determined for many purposes, and therefore should be readily available. Our next best factor is number 3. Here we can choose one of the meteorological factors. Annual runoff could be used if a streamflow record is available. But if we wish to predict mean annual flood for an ungaged drainage area, mean annual runoff would not be available. Consequently, we choose N, the number of thunderstorm days, since we can estimate this characteristic from meteorological records. From factor 1 we choose altitude as a predictor. This characteristic has the highest loading, and it is also fairly easy to calculate. From factor 4 we obviously have to select hydraulic storage.

We now have a minimum model made up of drainage area, number of thunderstorm days, altitude, and hydraulic storage in ponds and lakes. The beauty and simplicity of this model lie in our ability to predict mean annual flood for ungaged drainage areas. Certainly, such prediction should be done with caution and should be limited to the area for which Benson prepared the data.

Example 11-8: Principal Components Regression of Watershed Data

We will perform a components regression of $Q_{2.33}$ on the watershed characteristics of area, percent ponds and lakes, altitude, and number of thunderstorm days. These were the promising variates identified in Example 11-7. We will use the same watershed data as used in previous examples.

The eigenvalue–eigenvector structure of the 4 × 4 matrix of watershed characteristics is shown in Table 11-17. Four components are needed to explain the variance of these characteristics. We can see that none of the eigenvalues is small. There is no clear pattern of the relationship of characteristics to components to be noted in the elements of the eigenvectors.

TABLE 11-17 Eigenvalues and Eigenvectors for Example 11-8

	Component			
	1	2	3	4
		Eigenvalues		
	1.4867	1.0800	0.7746	0.6586
		Eigenvectors		
1	0.4713	−0.5164	0.6234	0.3501
2	0.5970	0.3130	0.2063	−0.7093
3	−0.4575	0.5009	0.7331	0.0492
4	0.4606	0.6201	−0.1771	0.6099

Table 11-18 gives the regression coefficients as successive principal components are added into the linear relationship. The most noteworthy feature is that component 3 adds little to the relationship. Addition of this component produced only slight changes in the regression coefficients and virtually zero change in correlation. Consequently, we should delete component 3 from the summations. The regression coefficients rescaled to the original hydrologic variates are listed at the bottom of Table 11-18. We can see that deleting component 3 makes very little change.

Recall that the four characteristics used in this example were also used as part of a larger set of characteristics in Example 11-5. The regression coefficients corresponding to the four characteristics of this example are also noted at the bottom of Table 11-18 for ready comparison. The values of the coefficients for area and for ponds and lakes have changed little. However, the coefficients for altitude and number of thunderstorm days are very different. The coefficient for altitude has taken on a significantly negative value, whereas it had been positive in Table 11-14. The coefficient for number of thunderstorm days has taken on a small positive value. There is no way of knowing absolutely which regression coefficients are "best." Such an answer would have to depend in part on additional study of the physiography and hydrometeorology of these drainage areas. Without such extra information we should probably prefer to use our four-variate model rather than the 11-variate model of Example 11-5. The simplicity of the four-variate model has value, and we note that the square of the correlation coefficient has dropped only from 0.41 to 0.35. In a situation of such weak correlation the use of a large number of variates is not justified.

TABLE 11-18 Solution on Successive Components for Example 11-8

Number of Components	a	Regression Coefficient b_1	b_2	b_3	b_4	Cumulative R^2	Cumulative Variance
			Standardized variates				
1		0.1624	0.2058	−0.1577	0.1588	0.1767	1.4867
2		0.2914	0.1277	−0.2828	0.0040	0.2440	2.5667
3		0.2973	0.1296	−0.2759	0.0023	0.2441	3.3413
4		0.4390	−0.1573	−0.2560	0.2491	0.3520	4.0000
			Original variates				
1 to 4	10.7823	2.0633	−7.5860	1.2932	0.2490		
1 + 2 + 4	10.8858	2.0404	−7.6758	1.2934	0.2507	0.3519	3.2253
			Corresponding coefficients from Table 11-14				
		2.8409	−6.4125	0.5462	−0.0215	0.4143	
Variate		Area	Percent Ponds and Lakes	Altitude	Number of Thunderstorm Days		

Before leaving this example we should review what we have accomplished. Through factor analysis in Example 11-7 we screened a large number of correlated watershed characteristics. We were able to select four characteristics from a larger set of 11. Each of these four was chosen from among others as a representative predictor of mean annual flood. A strong advantage of our subset of four is that all can be determined for an ungaged drainage area. Finally, it should be noted that components regression was still necessary for analysis of the relationship of the four characteristics to the mean annual flood. This analysis showed us that while there are four orthogonal elements in our four variates, only three of them have meaningful correlation with the mean annual flood.

Example 11-9: Factor Analysis of Geomorphic Data

Overton (1969) used factor analysis to study geomorphic interrelations on small agricultural watersheds. He determined nine geomorphic characteristics on 37 widely dispersed experimental watersheds gaged by the Agricultural Research Service. His characteristics were defined as follows:

DA *drainage area*: the area of the basin drained in acres.

HY *hypsometric integral*: the area beneath the hypsometric curve of the basin.

SD *stream density*: the number of stream segments per unit drainage area in acres.

WSL *average watershed slope*: determined by calculating slope from contours on a topographic map at each point of a grid overlay. Watershed slope is the average of the grid-point slopes and is in percent.

LWR *length/width ratio*: calculated by dividing the maximum length of the watershed by its maximum width. Units are feet per foot.

FPR *floodplain ratio*: the total area of the floodplain in alluvial depositional material divided by the drainage area. Units are acre per acre.

CSL *average channel slope*: determined by using the same grid system as for watershed slope.

RR *relief ratio*: the total elevation drop in the watershed divided by the total length of the watershed. Units are feet per foot.

ELD *total elevation drop*: units are feet.

The matrix of correlations among the nine characteristics is given in Table 11-19. Only two strong relationships are noted: 0.73 between drainage area and elevation drop, and 0.83 between relief ratio and channel slope.

Overton identified four components following eigenvalue–eigenvector analysis of the correlation matrix. The first component was a slope variate and was made up of the hypsometric integral, watershed slope, channel slope, relief ratio, and elevation drop. Component 2 was a size variate made up of drainage area and basin drop. Component 3 was a drainage variate with significant loadings from the length/width ratio and the floodplain ratio. Component 4 identified singly with the length/width ratio. Additional components had weak loadings in all characteristics and could not be identified.

TABLE 11-19 Matrix of Correlations among Characteristics of Example 11-9

	DA	HY	SD	WSL	LWR	FPR	CSL	RR	ELD
DA	1.00								
HY	−0.25	1.00							
SD	−0.28	−0.48	1.00						
WSL	0.21	−0.64	0.41	1.00					
LWR	−0.14	−0.24	0.22	−0.08	1.00				
FPR	−0.11	−0.20	0.27	0.12	0.54	1.00			
CSL	−0.02	−0.23	0.24	0.35	−0.16	−0.33	1.00		
RR	−0.09	−0.28	0.12	0.32	−0.13	−0.35	0.83	1.00	
ELD	0.73	−0.57	−0.13	0.43	0.12	−0.02	0.20	0.16	1.00

Source: From Overton (1969).

Overton was interested primarily in finding redundancies in his data and in the possibility of predicting certain characteristics from others selected by examination of the factor loadings. He kindly provided the data in Table 11-20, which were not reported in his 1969 research. This table shows factor loadings for seven factors rotated to vector position of maximum variance. Factor 2 in Table 11-20 is composed primarily of CSL (average channel slope), and RR (the relief ratio). Since the relief ratio may also be considered a slope, factor 2 may be identified with hydraulic slope of the watershed. Factor 3 is made up primarily of DA (drainage area) and ELD (total elevation drop). A large watershed can be expected to have a greater drop in elevation, on the average, than a small watershed. Consequently, total drop is also a measure of size. Therefore, factor 3 is a size factor.

All other factors in Table 11-20 have strong loadings in only one watershed characteristic. All of these are therefore one-to-one correspondences between unique watershed characteristics and factors. Several points should be noted in summarizing Table 11-20. Two redundancies are present; either relief ratio or channel slope could be deleted. Also, either total elevation drop or drainage area could be deleted. Probably, we should keep drainage area in our set of watershed characteristics since this is so

TABLE 11-20 Factor Loadings from Seven Rotated Vectors for Example 11-9

	Factor						
Variable	1	2	3	4	5	6	7
DA	0.081	−0.085	0.961	0.146	0.026	−0.013	0.031
HY	0.322	−0.163	−0.307	0.155	0.105	0.786	−0.329
SD	−0.914	0.110	−0.200	−0.114	−0.118	−0.187	0.202
WSL	−0.231	0.220	0.203	0.092	−0.103	−0.232	0.882
LWR	−0.104	−0.080	−0.013	−0.945	−0.257	−0.097	−0.069
FPR	−0.122	−0.254	−0.057	−0.301	−0.900	−0.069	0.101
CSL	−0.198	0.936	0.084	0.053	0.122	0.066	0.134
RR	0.054	0.928	−0.066	0.046	0.140	−0.229	0.104
ELD	0.141	0.149	0.831	−0.200	0.044	−0.313	0.259

Source: Data from D. E. Overton, personal communication, 1975.

fundamental a characteristic of a drainage area. We can thus reduce our original tentative list of nine watershed characteristics to seven fairly uncorrelated characteristics. Overton's set of characteristics, for the watershed sample used, is thus a good descriptive set with few redundancies. It should be a good set to use as independent predictors of hydrologic characteristics such as peak rate of discharge, volume of storm flow, or even recession curve parameters.

NONLINEAR ANALYSIS

In Chapter 6 we first developed the concepts of evaluation of linear multiple-variable models through the methods of least squares. In Chapter 7 we explored multiple-variable models which were nonlinear in the coefficients to be evaluated from data. It will be remembered that an iterative technique of nonlinear least squares was presented. This technique had as its essential basis the use of the first derivatives of the model as a linear correction system for improving initial estimates of our unknown coefficients.

In Chapter 11 we have developed multiple-variate, or multivariate models. One of our primary reasons for taking up multivariate techniques was a failure of least squares, or more popularly, multiple regression, to produce rational coefficients in a particular circumstance. This particular circumstance was shown to be statistical nonindependence of the functional independent variables. To summarize more concisely, in the linear equation

$$y = a + b_1x_1 + b_2x_2 + \cdots + b_nX_n$$

our evaluation of the b_i by ordinary least squares does not give good estimates of the effect of x_i on y when the x_i are significantly correlated.

We now apply these considerations to nonlinear models. For convenience we repeat Eq. 7-17:

$$y_2 - y_1 = h_1\frac{\partial f}{\partial a_1} + h_2\frac{\partial f}{\partial a_2} + h_3\frac{\partial f}{\partial a_3} + h_4\frac{\partial f}{\partial a_4} \tag{7-17}$$

A quick review of iterative nonlinear least squares is in order. We replaced $y_2 - y$, by $y - \hat{y} = E$, in which E represents the error between the observed value of the dependent variable, y, and its predicted value, $\hat{y}$. Then Eq. 7-17 is a linear correction scheme. The partial derivatives of Eq. 7-17 take the place of the x_i in a simple linear model. The h_i, which are corrections to the initial estimates of the unknown equation parameters, take the place of regression coefficients. Given any set of errors, E, based on some values of our parameters, we can compute through simple least squares, corrections to this set of parameters so that E is reduced, and with successive corrections, should approach a minimum.

We must now recall our reason for studying multivariate techniques. If the x_i are correlated, the regression coefficients are questionable expressions of relationship. Applying this to Eq. 7-17, if the partial derivatives are correlated, the h_i are questionable expressions of relationship between the derivatives and the errors. It follows that they are questionable corrections to the set of parameters on which the errors are based. We cannot expect efficient convergence to stable and optimum values of the parameters if our analysis scheme produces faulty corrections.

The solution postulated for evaluation of linear models was the use of components regression instead of ordinary least squares. We now make the simple proposal of substituting components regression for ordinary least squares in our iteration scheme for nonlinear models. This means that in use of Eq. 7-17 we first perform components analysis of our correlation matrix of first derivatives. We then regress E on these components just as we regressed y on the components of the x_i of our linear models. Such components regression then produces more realistic values of the parameter corrections, h_i.

Although we have discussed the wedding of nonlinear least squares and components regression in terms of Eq. 7-17, the same argument for such a wedding of numerical methods also applies to Eq. 7-21. This equation is also repeated for convenience:

$$E = h_1 \frac{\hat{y}_{a1} - \hat{y}}{\Delta a_1} + h_2 \frac{\hat{y}_{a2} - \hat{y}}{\Delta a_2} + h_3 \frac{\hat{y}_{a3} - \hat{y}}{\Delta a_3} + h_4 \frac{\hat{y}_{a4} - \hat{y}}{\Delta a_4} \tag{7-21}$$

The only difference, it will be recalled, was the substitution of partial divided finite differences for partial derivatives. We can perform components analysis on the correlation matrix of partial divided finite differences, and then regress E on the resultant components to get our parameter corrections h_i.

The combination of nonlinear least squares and components regression was suggested by Snyder (1962). Since that time the method has been additionally treated (Betson and Green, 1967; DeCoursey and Snyder, 1969; Snyder et al., 1970; Snyder et al., 1971). It is obvious that the amount of computation required for nonlinear least squares–components regression is too large to be performed on microcomputers. The composite technique is very efficient for high-speed mainframe computers. The doubly iterative techniques of nonlinear least squares and eigenvalue–eigenvector analysis of a matrix lend themselves to recycling through the same computer steps time after time until a stable set of parameter values is reached.

Example 11-10: Nonlinear Water Yield Model

In this example we construct a fairly simple water-yield model based on monthly values of precipitation and runoff and an average seasonal value expressing seepage and evapotranspiration. The basic principle will be a partitioning of monthly precipitation into two components, seepage and evapotranspiration, and streamflow. Water yield analysis deals with the long-term delivery of water to the channel system. We are interested in the pattern of volumes of flow per month over several months, rather than the detailed pattern of changing rates of discharge during a storm event. Streamflow must therefore be separated into two components, a portion that is rapid response to rainfall, and a delayed response, some of which will become streamflow in the months following that in which the rain fell. The basic partitioning of the precipitation into components is illustrated in Fig. 11-6.

Runoff passing a gaging point in any month at time t in a sequence of months is the sum of rapid response to rainfall that month, plus a portion of the delayed response to rainfall of that month, plus the delayed response to rainfall in previous months. We may write the general equation

$$\begin{aligned} \mathrm{RO}(t) = {} & \mathrm{RR}(t) + \mathrm{RC}(1) \times \mathrm{DR}(t) + \mathrm{RC}(2) \times \mathrm{DR}(t-1) \\ & + \mathrm{RC}(3) \times \mathrm{DR}(t-2) + \mathrm{RC}(4) \times \mathrm{DR}(t-3) + \mathrm{RC}(5) \times \mathrm{DR}(t-4) \end{aligned} \tag{11-67}$$

In Eq. 11-67 RO(t) is runoff for the month, RR(t) is rapid response to rainfall that month, DR(t) is delayed response to rainfall that month, and DR($t-1$) is delayed response to rainfall of the preceding month. All units are inches depth on the watershed. Four antecedent months are specified in Eq. 11-67. The RC(T), $T = 1, 5$, in the equation are recession coefficients. These coefficients will schedule the delivery of delayed response from rainfall of the current month plus the four antecedent months.

For this model the RC(T) were computed by

$$\mathrm{RC}(T) = \frac{\mathrm{EXP}(-\mathrm{PAR}(1) \times (T-1))}{\sum_T \mathrm{EXP}(-\mathrm{PAR}(1) \times (T-1))} \qquad T = 1, 5 \tag{11-68}$$

In Eq. 11-68, T is a time variable defined as equal to 1 for the current month and running backward in the time sequence. PAR(1) is a parameter which will be evaluated from recorded data. Equation 11-68 produces five recession coefficients whose sum is equal to unity.

Streamflow as defined in Fig. 11-6 is precipitation less seepage and evapotranspiration. This can be written as

$$\mathrm{ST}(t) = \mathrm{PP}(t) - \mathrm{SE}(t) \tag{11-69}$$

In Eq. 11-69, ST(t) is streamflow for month t, PP(t) is the precipitation in that month, and SE(t) is seepage and evaporation during the month. It is necessary to define some

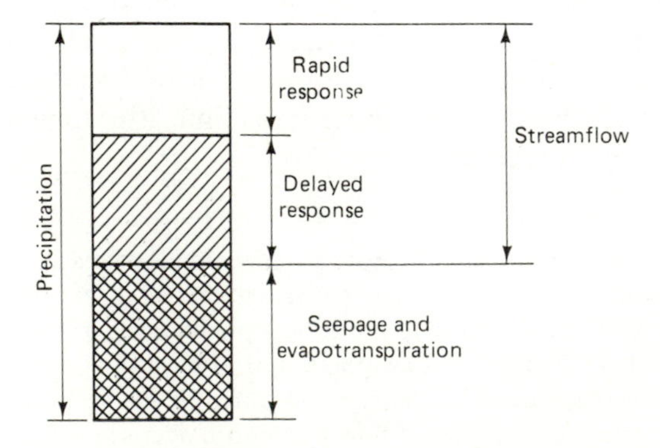

Figure 11-6 Partitioning for simple water-yield model.

function to specify $SE(t)$. As a first approximation let seepage and evapotranspiration vary as a trigonometric sine wave expressing season of the year, as in

$$SE(M) = PAR(2) + PAR(3) \times SIN((30 \times M) + PAR(4)) \quad (M = 1, 12; \text{July} = 1) \tag{11-70}$$

$SE(M)$ will take the form of a sine wave imposed on some average value. $SE(m)$ will be the same for each January in a record, for each February in a record, and so on. PAR(2) establishes an average value of the sine wave. A wave of amplitude equal to PAR(3) is set to run above and below this average value. The seasonal index M is set to 1 for July, but PAR(4), expressed in circular units, allows a seasonal phase-shift of the wave.

Streamflow is partitioned into rapid response and delayed response which may be written as Equation 11-71.

$$ST(t) = RR(t) + DR(t) \tag{11-71}$$

Now define a proportioning coefficient, k, such that

$$RR(t) = kST(t) \tag{11-72a}$$

and

$$DR(t) = (1 - k)ST(t) \tag{11-72b}$$

The proportioning coefficient is probably not constant. For example, during a month with excessive rainfall more than a normal amount of the streamflow would be rapid response. As a first approximation we can make the proportioning coefficient a linear function of streamflow as in Equation 11-73.

$$k(t) = PAR(5) + PAR(6) \times ST(t) \tag{11-73}$$

PAR(5) and PAR(6) are to be specified by evaluation from data. Equation 11-67 can now be written more explicitly by substituting Eqs. 11-72 for $RR(t)$ and $DR(t)$.

$$\begin{aligned} RO(t) = [k(1) \times ST(t)] &+ [RC(1) \times (1 - k(1)) \times ST(t)] \\ &+ [RC(2) \times (1 - k(2)) \times ST(t-1)] \\ &+ [RC(3) \times (1 - k(3)) \times ST(t-2)] \\ &+ [RC(4) \times (1 - k(4)) \times ST(t-3)] \\ &+ [RC(5) \times (1 - k(5)) \times ST(t-4)] \end{aligned} \tag{11-74}$$

In Eq. 11-74 the term $[k(1) \times ST(t)]$ is the rapid response to the rainfall of the month and the remaining terms, shown in vertical arrangement, are the delayed responses, routed through the soil profile by the coefficients $RC(T)$. $ST(t)$ in Eq. 11-74 is computed from Eq. 11-69, with $SE(t)$, in turn, computed from Eq. 11-70. The $RC(T)$, T running backward in time, are computed from Eq. 11-68, and the proportioning coefficients, $k(T)$, are computed from Eq. 11-73. Note particularly that the $RC(T)$ form a constant set but that the $k(T)$ vary with the streamflow of the antecedent months. Equation 11-74 is structurally dependent on the six parameters indicated in Eqs. 11-68, 11-70, and 11-73. These six parameters can be evaluated from historical

TABLE 11-21 Illustration of Runoff Calculations for Water-Yield Model of Muddy Creek, North Carolina

	Month					
	Sept.	Oct.	Nov.	Dec.	Jan.	
January 1935						
Precipitation	8.71	1.86	4.48	2.88	1.30	
Seepage and evapotranspiration	3.62	2.76	1.82	1.07	0.70	
Streamflow	5.09	−0.90	2.66	1.81	0.60	
Prop. coefficient, k	0.302	0.068	0.307	0.174	0.126	
Rapid response						0.08
$1-k$	0.698	0.932	0.793	0.826	0.874	
Delayed response	3.55	−0.84	2.11	1.50	0.52	
Recession coefficient, RC	0.050	0.087	0.150	0.261	0.452	
Delayed cont. to Jan. runoff	0.18	−0.07	0.32	0.39	0.24	
Jan. total delayed runoff						1.06
Jan. total calculated runoff						1.14
	Oct.	Nov.	Dec.	Jan.	Feb.	
February 1935						
Precipitation					2.92	
Seepage and evapotranspiration					0.81	
Streamflow					2.11	
Prop. coefficient, k					0.185	
Rapid response						0.39
$1-k$					0.815	
Delayed response	−0.84	2.11	1.50	0.52	1.72	
Recession coefficient, RC	0.050	0.087	0.150	0.261	0.452	
Delayed cont. to Feb. runoff	−0.04	0.18	0.22	0.14	0.78	
Feb. total delayed runoff						1.28
Feb. total calculated runoff						1.67

data. The method of evaluation can be based on Eq. 7-21, but with components regression substituted for ordinary least squares.

Table 11-21 shows the calculation of runoff for the months of January and February 1935 for Muddy Creek watershed at High Point, North Carolina. The numerical values of the parameters used in these illustrations are the initial values used in the optimization for the Muddy Creek watershed. These initial values are given as round 0 in Table 11-22. For example, seepage and evapotranspiration for the month of September calculated by Eq. 11-70 is

$$SE(\text{Sept}) = 2.49 + 1.81 \sin[(3 \times 90 \times 0.01745) + 0.9] = 3.62$$

The seasonal angle is calculated in radians.

The proportioning coefficient for September calculated by Eq. 11-73 is $k = 0.103 + 0.0390(8.71 - 3.62) = 0.302$ and $1 - k = 0.698$.

TABLE 11-22 Muddy Creek Watershed, High Point, North Carolina: Iterated Values of Parameters

	Parameter					
Round	1	2	3	4	5	6
0	0.5500	2.4900	1.8100	0.9000	0.1030	0.0390
1	0.1957	2.4878	1.9219	1.0914	0.0492	0.0618
2	0.3294	2.4862	1.9042	1.0700	0.0	0.0598
3	0.3186	2.4867	1.9015	1.0830	0.0	0.0600
4	0.3210	2.4866	1.9012	1.0818	0.0	0.0598
5	0.3207	2.4866	1.9011	1.0821	0.0	0.0598

The routing coefficients are calculated from Eq. 11-68 as follows:

T	$0.55T$	EXP $(-0.55T)$	EXP $(-0.55T)/1.2766$
1	0.55	0.5770	0.452
2	1.10	0.3329	0.261
3	1.65	0.1920	0.150
4	2.20	0.1108	0.087
5	2.75	0.0639	0.050
		1.2766	1.000

Other values in Table 11-22 are computed similarly.

On the basis of these numerical illustrations it should now be possible to visualize the procedure of calculating runoff for each month of record. These calculated runoff values are the $\bar{y}$ of Eq. 7-21. One can also now visualize the process of incrementing one parameter and again calculating runoff for each month of record. These runoff values would form the $\hat{y}_{a1}$ of Eq. 7-21 if the first parameter were incremented. The parameter increment Δa_1 is arbitrary but small. A value of 0.5% of the parameter value appears to work well in most cases. Care must be taken when initial or intermediate values of the parameters are zero. In this case a value of 0.0005 seems workable.

All parameters must be incremented and the partial divided finite differences calculated for each month of the record. Table 11-23 shows the results of all the basic calculations for the first 18 months. In this table the observed minus predicted runoff is the E of Eq. 7-21 and assumes the role of the dependent variable in the components regression. The partial divided finite differences assume the role of the independent variables in the regression.

Table 11-24 gives the matrix of correlation coefficients for the partial divided finite differences. The correlation between the parameter 1 difference and the parameter 5 difference is high. The correlation between the parameter 5 difference and the parameter 6 difference is moderately high. We should suspect that simple regression analysis to get the parameter corrections would be unsatisfactory.

TABLE 11-23 Simple Water-Yield Model, Muddy Creek Watershed, High Point, North Carolina: Data for Initial Estimates of Parameters

Observation	Precipitation (in.)	Observed Runoff (in.)	Predicted Runoff (in.)	Observed Minus Predicted (in.)	Partial Divided Differences for Parameter					
					1	2	3	4	5	6
1	1.30	1.78	1.125	0.655	−0.446	−0.944	0.693	0.721	−0.719	−3.085
2	2.92	1.45	1.672	−0.222	0.341	−1.042	0.880	−0.028	0.544	1.210
3	6.67	3.65	4.062	−0.412	1.054	−1.154	0.825	−1.007	1.976	13.569
4	3.90	3.05	2.246	0.804	0.098	−0.928	0.368	−1.230	−0.926	−6.667
5	4.13	1.19	1.606	−0.416	−0.518	−0.929	−0.027	−1.418	−0.914	−4.851
6	1.50	0.22	−0.268	0.488	−2.186	−0.813	−0.319	−1.106	−2.392	−0.159
7	5.08	0.42	0.228	0.192	−0.599	−1.040	−0.779	−0.821	0.508	−2.957
8	1.69	0.03	−1.167	1.197	−1.135	−0.893	−0.738	−0.159	−1.376	2.118
9	4.50	0.12	−0.238	0.358	0.522	−1.091	−0.817	0.759	1.172	−1.826
10	1.52	0.04	−0.827	0.867	0.064	−0.962	−0.468	1.153	−0.483	−0.641
11	3.14	0.28	0.336	−0.056	0.715	−1.085	−0.037	1.643	1.085	−0.135
12	2.62	0.26	0.895	−0.635	1.012	−1.058	0.409	1.399	0.739	0.256
13	7.18	5.88	4.634	1.246	1.857	−1.235	0.941	0.847	3.013	21.960
14	3.76	4.00	3.009	0.991	0.881	−0.972	0.801	−0.062	−0.356	−6.757
15	5.13	3.10	3.615	−0.515	0.315	−1.009	0.703	−0.909	1.183	−1.117
16	5.84	4.92	3.628	1.292	0.030	−0.993	0.352	−1.425	−0.085	−1.610
17	0.13	0.13	0.205	−0.075	−2.835	−0.708	0.014	−1.087	−3.748	−3.385
18	4.81	0.30	0.447	−0.147	−0.887	−1.021	−0.460	−1.373	0.265	−5.574

TABLE 11-24 Muddy Creek Watershed, High Point, North Carolina: Matrix of Correlation Coefficients for First Set of Partial Divided Differences

	Parameter					
	1	2	3	4	5	6
1	1.0000	−0.0674	0.2613	0.3092	0.8692	0.5335
2	−0.0674	1,0000	−0.0127	−0.0261	−0.1007	−0.0760
3	0.2613	−0.0127	1.0000	0.0404	0.1133	0.0808
4	0.3092	−0.0261	0.0403	1.0000	0.2508	0.1149
5	0.8692	−0.1007	0.1133	0.2508	1.0000	0.6968
6	0.5335	−0.0760	0.0808	0.1149	0.6968	1.0000

Table 11-25 shows the eigenvalues and eigenvectors of the matrix of correlation coefficients of the partial divided finite differences. We should note that six eigenvalues were found. There is thus no reduction of dimension, and our computed corrections to the regression coefficients will be the same as if we had used simple multiple regression in Eq. 7-21. However, the eigenvalue-eigenvector analysis that has been performed should not be regarded as wasted effort. Until this analysis was performed we had no way of knowing that the high correlations in Table 11-24 were not sufficient to cause reduction in dimension of the matrix. Also, we have no way of knowing how the matrix of correlation coefficients will change with changing values of the parameters in subsequent rounds.

We should note additionally that none of the eigenvectors is trivial. Even vectors 5 and 6 with relatively weak eigenvalues still each show one strong element. Vector 5 associates with the correction to parameter 6 and vector 6 associates with the correction to vector 5. Referrring to Eq. 11-73, we can see that these two vectors will exercise strong control on the solution for the proportioning coefficient.

We have now established all basic processes of calculation necessary to optimize the correspondence between our simple water yield model and our historical data. Although not specifically mentioned, the step from the eigenvalue-eigenvector solution

TABLE 11-25 Muddy Creek Watershed, High Point, North Carolina: Eigenvalues and Eigenvectors of Correlation Matrix

Component					
1	2	3	4	5	6
		Eigenvalues			
2.5721	1.0003	0.9704	0.9248	0.4405	0.0920
		Eigenvectors			
0.5691	0.0925	−0.0101	−0.0001	−0.5275	−0.6239
−0.0909	0.7982	0.5518	−0.2228	0.0072	0.0205
0.1688	0.5709	−0.7740	0.0747	0.1731	0.1049
0.2467	0.0913	0.2656	0.8951	0.2410	0.0304
0.5866	−0.0650	0.1089	−0.1427	−0.2595	0.7431
0.4842	−0.1258	0.1181	−0.3510	0.7525	−0.2151

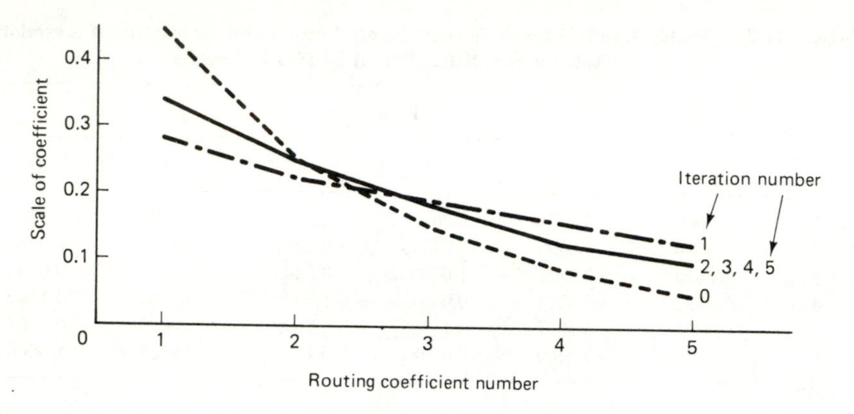

Figure 11-7 Derived routing coefficients.

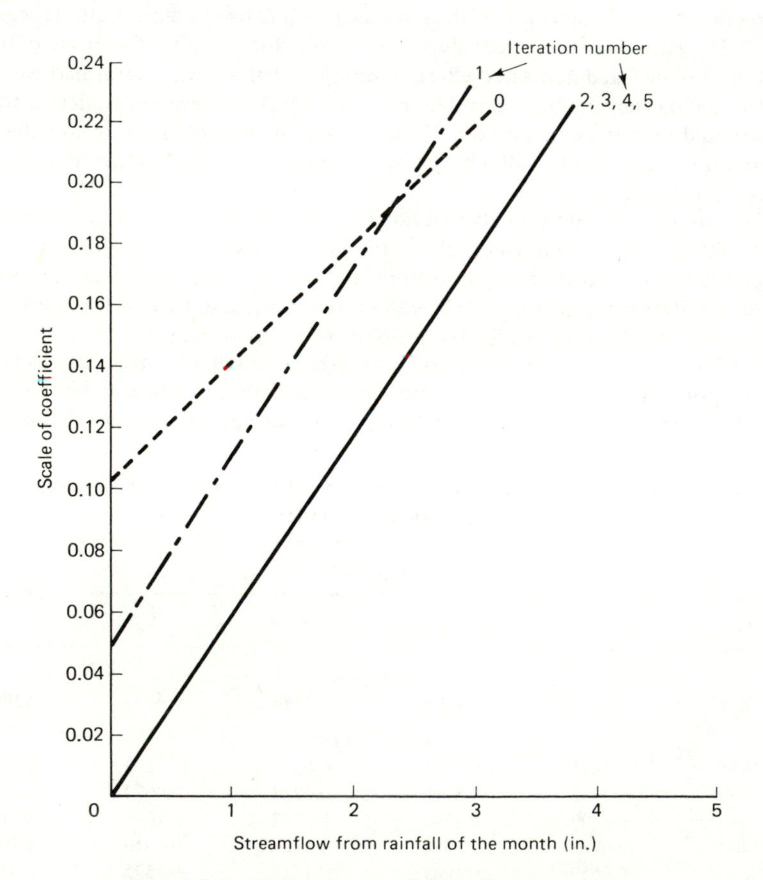

Figure 11-8 Partitioning coefficient of rapid response.

to calculation of the parameter corrections, the h_i of Eq. 7-17, is identical to the calculations already developed in Eq. 11-49.

Table 11-22 shows the values of the parameters for five rounds of solution. The corrections calculated from the eigenvalue–eigenvector analysis in Table 11-25 produce the changes in the parameters from round 0 to round 1. The steps of calculation between each round are, of course, identical. The values of the parameters change, but not the iterative procedure. The change between round 4 and 5 is small for every parameter. The optimization is thus considered to have converged to a stable value.

Parameter 1 controls the recession coefficient which schedules the delivery of delayed response of streamflow to watershed monthly runoff. The coefficients computed by Eq. 11-68 are plotted in Fig. 11-7 for the different values of parameter 1 during successive rounds. It can be noted that after only two adjustments the changes were too small to show in the graph.

Parameters 5 and 6 relate the value of the proportioning coefficient to streamflow. This relationship is plotted in Fig. 11-8. Again, only two rounds of adjustment were needed before further changes are lost in the scale of the graph.

The seepage and evapotranspiration component of our water-yield model is given in Fig. 11-9. This component is dependent on parameters 2, 3, and 4 in Eq. 11-70. This component of the model also shifted very slightly during optimization. The general level, controlled by parameter 2, the wave amplitude, controlled by parameter 3, and the seasonal shift, controlled by parameter 4, all changed very little.

Figure 11-10 gives a comparison of monthly runoff versus the monthly runoff computed by the water yield model. The correlation coefficient for the model is 0.89. The standard deviation of the residuals to fitting is 0.65 in. The overall fit of the model to data is satisfactory but not exceptional. Probably the most severe failure of the model is in representation of months of very low flow. Prediction for months with low flow tends to be negative. Figure 11-10 shows this negative prediction, and also suggests that some compensation occurs by overprediction when monthly flows are between 0.5 and 1.5 in.

The total pattern of residual errors approaches a normal distribution, as shown in Fig. 11-11. Our optimizing principle of least-squares residual error obviously requires a central tendency of the errors, but the approximate symmetry exhibited in Fig. 11-11 gives us some assurance that our nonlinear iterative process has converged to a position giving balanced errors.

Figure 11-12 shows the time series of observed and predicted streamflow. In this figure the seasonal pattern of flow is very evident, with the time of seasonal low flows varying generally within the late summer and autumn period of the year. We can see now that the model underpredicts, the predictions going almost always negative, following a major recession of streamflow. Stated another way, when streamflow falls off rapidly it tends to stabilize after a few months of decline, at some sustained low flow level. The water-yield model, on the other hand, predicts a continued recession of flow well below the sustained low flow, and into the negative.

In conclusion of this example of a simple water-yield model, these points should be noted:

1. The model is a stable form for optimization. Few rounds of nonlinear components regression were required.

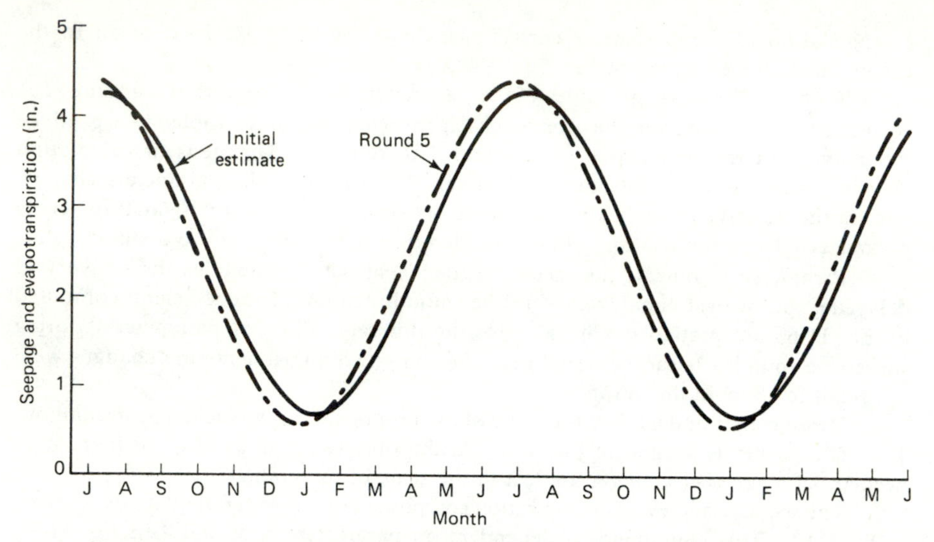

Figure 11-9 Derived seasonal seepage and evapotranspiration, Muddy Creek watershed, High Point, North Carolina.

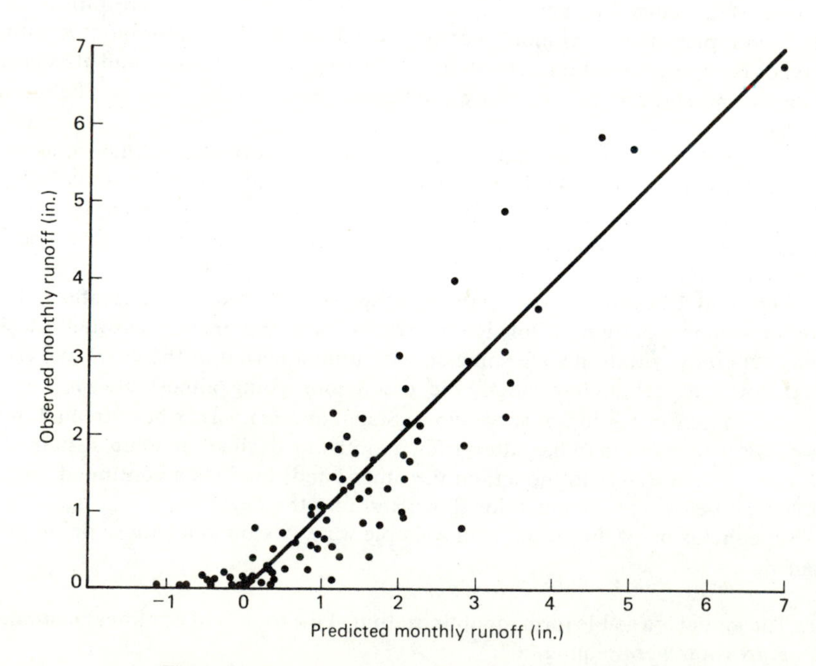

Figure 11-10 Comparison of model and observed data.

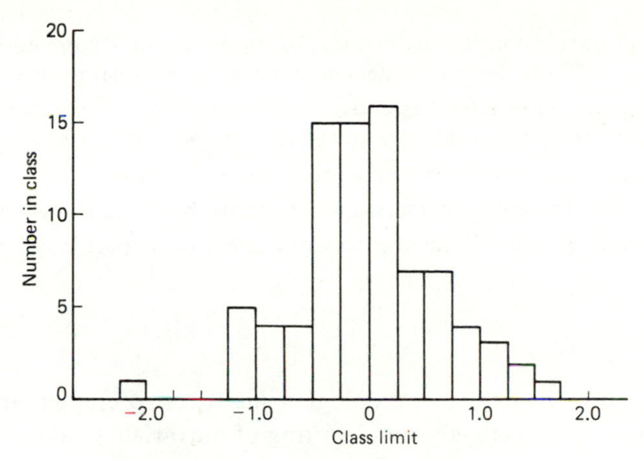

Figure 11-11 Frequency of residual errors.

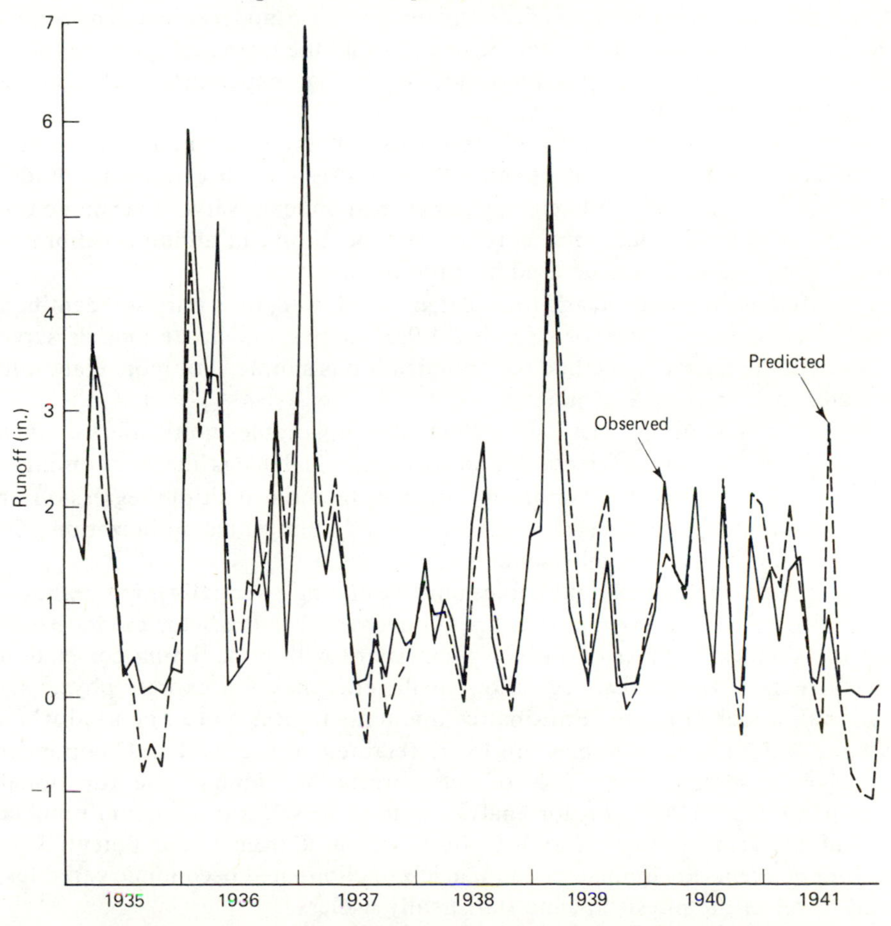

Figure 11-12 Time series of observed and predicted runoff for Muddy Creek.

2. The model fits historical data in a statistical sense, with a correlation coefficient of nearly 0.9. The derived model components are rational in structure.
3. Components regression gives added assurance of the rationality of the model structure, showing that iterative corrections are not correlated.
4. The model is unsatisfactory in the prediction of low flows.
5. To improve the model the recession curve must be extended beyond 5 months.
6. More water must be allocated to delayed runoff to support low flows.

SUMMARY

A spectrum of numerical methodologies has been presented in this text. Methods range from relatively simple fitting of univariate models to elaborate numerics for differential components regression. An attempt was made throughout to retain a common denominator for these model-oriented methods. This common denominator is the perception that the methodologies are tools. It takes practice to develop skills with tools. It takes experience to choose the proper tool for the job at hand.

The numerical tools for optimization are intended for a particular function. This function is to produce some "best" correspondence between a model and experimental data. Achieving good correspondence serves to critique the model, validate the basic concepts of the model, and in addition calibrates the model so that it may be used as a predictor.

Models are designed and formulated, or selected, for a purpose identified by the student, engineer, or scientist. When simple univariate models serve the need, the choice of method for optimization is simple. For more elaborate models the choices are complex.

The first choice of method for the multiple-variable/multivariate models depends on whether the scientist has designed a model that is linear or nonlinear in its coefficients. If it is nonlinear, either differential multiple regression or differential components regression can be used. Subjective optimization will be necessary for very cumbersome models.

If the model is linear, the choice must be among essentially three methods. Stepwise regression can be used to produce a linear predictor model based on the smallest number of variables consistent with a maximum correlation coefficient. The resultant regression coefficients may not express physically rational model structure. Principal components regression can be used when all the independent variables are to be retained in the model. Component analysis suppresses the effect of intercorrelations among the functional independent variables. Factor analysis is used to select a minimum number of independent variables. In this analysis we purge from a large potential set of independents to eliminate redundancies, to eliminate uneconomic variables, and to reject the physically and statistically useless.

EXERCISES

11-1. Obtain a data set of hydrologic, meteorologic, geomorphic, and/or cultural information and use packaged software to perform a factor analysis of the data. Identify the highest factor loadings for the criterion variable. Do any factors take on describable identity through their variable associations? (Data suggestions: daily, monthly, or weekly records of rainfall, runoff, temperature, or radiation at several stations in a geographic region, watershed land-use and geomorphic data, information from *Agricultural Index of the U.S.*, etc.)

11-2. Perform a components regression on the data set of Exercise 11-1 and quantify the regression coefficients in the equation to predict your criterion variable. Contrast the solutions and the component-factor identities.

11-3. Screen your data set of Exercise 11-1 of redundant or useless predictor variables. Use packaged software to perform multiple regression analysis on the reduced set of variables. Discuss the structure of your predictor model. Do the regression coefficients provide a physically rational model?

11-4. Upgrade the water-yield model of Example 11-10. Write a computer subprogram of the model to optimize with prepared software for nonlinear differential components regression. The following model elements could be used;

Seepage and evapotranspiration for calendar month m.

$$\mathrm{SE}(m) = a + b \sin(\omega m + c)$$

Potential runoff is rainfall minus seepage and evapotranspiration.

$$\mathrm{PRO}(m) = P(m) - \mathrm{SE}(m) \qquad \mathrm{PRO}(m) \geq 0$$

Rapid runoff is excess of potential runoff over some threshold d.

$$\mathrm{RRO}(m) = \mathrm{PRO}(m) - d \qquad \mathrm{RRO}(m) \geq 0$$

Delayed runoff is difference between potential runoff and rapid runoff.

$$\mathrm{DRO}(m) = \mathrm{PRO}(m) - \mathrm{RRO}(m) \qquad \mathrm{DRO}(m) \geq 0$$

Construct a recession function to distribute the delayed runoff in current and future months. Predicted runoff will be the sum of rapid runoff of the month plus the scheduled amounts of delayed runoff. Parameters to be optimized will be a, b, c, d, and those in the recession function.

11-5. Run your upgraded water-yield model on at least two data sets from watersheds with some known differences in characteristics. Discuss the rationality of the two resultant optimized models. Do any differences in the model properly reflect the known physical differences in the watersheds?

REFERENCES AND SUGGESTED READINGS

ABRAHAMS, A. D., Factor Analysis of Drainage Basin Properties: Evidence for Stream Abstraction Accompanying the Degradation of Relief, *Water Resources Research*, Vol. 8, No. 3, pp. 624–33, 1972.

ANDERSON, T. W., *Introduction to Multivariate Statistical Analysis*, Wiley, New York, 1958.

ANDERSON, T. W., The Use of Factor Analysis in the Statistical Analysis of Multiple Time Series, *Psychometrika*, Vol. 28, pp. 1-25, 1963.

BEAUDOIN, P., and J. ROUSSELLE, A Study of Space Variations of Precipitation by Factor Analysis, *Journal of Hydrology*, Vol. 59, No. 1/2, pp. 123-38, 1982.

BENSON, M. A., Factors Influencing the Occurrence of Floods in a Humid Region of Diverse Terrain, *Water Supply Paper 1580-B*, United States Geological Survey, Washington, 1962.

BENSON, M. A., Factors Affecting the Occurrences of Floods in the Southwest, *Water Supply Paper 1580-D*, United States Geological Survey, Washington, 1964.

BETSON, R. P., and R. F. GREEN, *DIFCOR—A Program to Solve Nonlinear Equations*, Research Paper 6, Tennessee Valley Authority, Knoxville, Tenn., 1967.

COOLEY, W. W., and P. R. LOHNES, *Multivariate Data Analysis*, Wiley, New York, 1971, p. 137.

COOPER, D. M., and E. F. WOOD, Identification of Multivariate Time Series and Multivariate Input-Output Models, *Water Resources Research*, Vol. 18, No. 4, pp. 937-46, 1982.

DECOURSEY, D. G., and W. M. SNYDER, Computer-Oriented Method of Optimizing Hydrologic Model Parameters, *Journal of Hydrology*, Vol. 9, 1969.

GERE, J. M., and W. WEAVER, JR., *Matrix Algebra for Engineers*, D. Van Nostrand, Princeton, N.J., 1965.

HARMAN, H. H., *Modern Factor Analysis*, University of Chicago Press, Chicago, 1960.

JEFFERS, J. N. R., Two Case Studies in the Application of Principal Components Analysis, *Applied Statistics*, Vol. 16, No. 3, pp. 225-36, 1967.

KENDALL, M. G., *A Course in Multivariate Analysis*, Hafner, New York, 1957.

KISIEL, C. C., Applications of Principal Components, Canonical Correlation, and Factor Analysis in Hydrology, *Institute of Applications of Stochastic Methods in Civil Engineering*, Fort Collins, Colo., 1972.

KNISEL, W. G., A Factor Analysis of Reservoir Losses, *Water Resources Research*, Vol. 6, No. 2, pp. 491-98, April 1970.

OVERTON, D. E., Eigenvector Analysis of Geomorphic Interrelations of Small Agricultural Watersheds, *Transactions of the ASAE*, Vol. 12, No. 4, 1969.

RAO, C. R., *Advanced Statistical Methods in Biometric Research*, Wiley, New York, 1952.

SNYDER, W. M., Some Possibilities for Multivariate Analysis in Hydrologic Studies, *Journal of Geophysical Research*, Vol. 67, No. 2, February 1962.

SNYDER, W. M., W. C. MILLS, and J. C. STEPHENS, A Method of Derivation of Non-Constant Watershed Response Functions, *Water Resources Research*, Vol. 6, No. 1, February 1970.

SNYDER, W. M., W. C. MILLS, and J. C. STEPHENS, A Three-Component, Nonlinear Water Yield Model, *Systems Approach to Hydrology*, Proceedings of First Bilateral U.S.-Japan Seminar in Hydrology, Honolulu, January 1971.

Tennessee Valley Authority, *Parker Branch Research Watershed Project Report 1953–1962.* In cooperation with North Carolina State College of Agriculture and Engineering. Knoxville, Tennessee. 1963.

WONG, S. T., A Multivariate Statistical Model for Predicting Mean Annual Flood in New England, *Annals of the Association of American Geographers*, Vol. 53, No. 3, 1963.

Appendix

TABLE A-1	Normal Distribution Probabilities
TABLE A-2	t-Distribution Probabilities
TABLE A-3	χ^2 Distribution
TABLE A-4	F Distribution
TABLE A-5	Binomial Distribution Probabilities
TABLE A-6	Tolerance Limit Factors
TABLE A-7	Critical Values of the Kolmogorov-Smirnov One-Sample Test
TABLE A-8	Pearson Type III Deviates
TABLE B-1	Sediment Yield Data
TABLE B-2	Evaporation Data
TABLE B-3	French Broad River Water Yield Data
TABLE B-4	Indiana Flood Data
TABLE C-1	Chattooga River Watershed Data
TABLE C-2	Chestuee Creek Watershed Data
TABLE C-3	Wildcat Creek Watershed Data
TABLE C-4	Conejos River Watershed Data
TABLE D-1	Algorithm for Rainfall/Runoff Model of Chapter 8
TABLE D-2	Rainfall and Runoff Data for the Anacostia River at Colesville
TABLE D-3	Rainfall and Runoff Data for the Anacostia River at Riverdale

TABLE A-1 Normal Distribution Probabilities

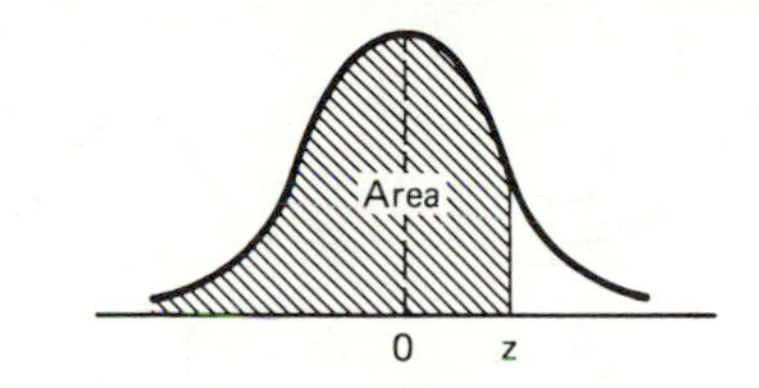

z	0.00	0.01	0.02	0.03	0.04	0.05	0.06	0.07	0.08	0.09
-3.4	.0003	.0003	.0003	.0003	.0003	.0003	.0003	.0003	.0003	.0002
-3.3	.0005	.0005	.0005	.0004	.0004	.0004	.0004	.0004	.0004	.0003
-3.2	.0007	.0007	.0006	.0006	.0006	.0006	.0006	.0005	.0005	.0005
-3.1	.0010	.0009	.0009	.0009	.0008	.0008	.0008	.0008	.0007	.0007
-3.0	.0013	.0013	.0013	.0012	.0012	.0011	.0011	.0011	.0010	.0010
-2.9	.0019	.0018	.0018	.0017	.0016	.0016	.0015	.0015	.0014	.0014
-2.8	.0026	.0025	.0024	.0023	.0023	.0022	.0021	.0021	.0020	.0019
-2.7	.0035	.0034	.0033	.0032	.0031	.0030	.0029	.0028	.0027	.0026
-2.6	.0047	.0045	.0044	.0043	.0041	.0040	.0039	.0038	.0037	.0036
-2.5	.0062	.0060	.0059	.0057	.0055	.0054	.0052	.0051	.0049	.0048
-2.4	.0082	.0080	.0078	.0075	.0073	.0071	.0069	.0068	.0066	.0064
-2.3	.0107	.0104	.0102	.0099	.0096	.0094	.0091	.0089	.0087	.0084
-2.2	.0139	.0136	.0132	.0129	.0125	.0122	.0119	.0116	.0113	.0110
-2.1	.0179	.0174	.0170	.0166	.0162	.0158	.0154	.0150	.0146	.0143
-2.0	.0228	.0222	.0217	.0212	.0207	.0202	.0197	.0192	.0188	.0183
-1.9	.0287	.0281	.0274	.0268	.0262	.0256	.0250	.0244	.0239	.0233
-1.8	.0359	.0351	.0344	.0336	.0329	.0322	.0314	.0307	.0301	.0294
-1.7	.0446	.0436	.0427	.0418	.0409	.0401	.0392	.0384	.0375	.0367
-1.6	.0548	.0537	.0526	.0516	.0505	.0495	.0485	.0475	.0465	.0455
-1.5	.0668	.0655	.0643	.0630	.0618	.0606	.0594	.0582	.0571	.0559
-1.4	.0808	.0793	.0778	.0764	.0749	.0735	.0721	.0708	.0694	.0681
-1.3	.0968	.0951	.0934	.0918	.0901	.0885	.0869	.0853	.0838	.0823
-1.2	.1151	.1131	.1112	.1093	.1075	.1056	.1038	.1020	.1003	.0985
-1.1	.1357	.1335	.1314	.1292	.1271	.1251	.1230	.1210	.1190	.1170
-1.0	.1587	.1562	.1539	.1515	.1492	.1469	.1446	.1423	.1401	.1379
-.9	.1841	.1814	.1788	.1762	.1736	.1711	.1685	.1660	.1635	.1611
-.8	.2119	.2090	.2061	.2033	.2005	.1977	.1949	.1922	.1894	.1867
-.7	.2420	.2389	.2358	.2327	.2296	.2266	.2236	.2206	.2177	.2148
-.6	.2743	.2709	.2676	.2643	.2611	.2578	.2546	.2514	.2483	.2451
-.5	.3085	.3050	.3015	.2981	.2946	.2912	.2877	.2843	.2810	.2776
-.4	.3446	.3409	.3372	.3336	.3300	.3264	.3228	.3192	.3156	.3121
-.3	.3821	.3783	.3745	.3707	.3669	.3632	.3594	.3557	.3520	.3483
-.2	.4207	.4168	.4129	.4090	.4052	.4013	.3974	.3936	.3897	.3859
-.1	.4602	.4562	.4522	.4483	.4443	.4404	.4364	.4325	.4286	.4247
-.0	.5000	.4960	.4920	.4880	.4840	.4801	.4761	.4721	.4681	.4641
.0	.5000	.5040	.5080	.5120	.5160	.5199	.5239	.5279	.5319	.5359
.1	.5398	.5438	.5478	.5517	.5557	.5596	.5636	.5675	.5714	.5753
.2	.5793	.5832	.5871	.5910	.5948	.5987	.6026	.6064	.6103	.6141
.3	.6179	.6217	.6255	.6293	.6331	.6368	.6406	.6443	.6480	.6517
.4	.6554	.6591	.6628	.6664	.6700	.6736	.6772	.6808	.6844	.6879
.5	.6915	.6950	.6985	.7019	.7054	.7088	.7123	.7157	.7190	.7224
.6	.7257	.7291	.7324	.7357	.7389	.7422	.7454	.7486	.7517	.7549
.7	.7580	.7611	.7642	.7673	.7704	.7734	.7764	.7794	.7823	.7852
.8	.7881	.7910	.7939	.7967	.7995	.8023	.8051	.8078	.8106	.8133
.9	.8159	.8186	.8212	.8238	.8264	.8289	.8315	.8340	.8365	.8389
1.0	.8413	.8438	.8461	.8485	.8508	.8531	.8554	.8577	.8599	.8621
1.1	.8643	.8665	.8686	.8708	.8729	.8749	.8770	.8790	.8810	.8830
1.2	.8849	.8869	.8888	.8907	.8925	.8944	.8962	.8980	.8997	.9015
1.3	.9032	.9049	.9066	.9082	.9099	.9115	.9131	.9147	.9162	.9177
1.4	.9192	.9207	.9222	.9236	.9251	.9265	.9279	.9292	.9306	.9319
1.5	.9332	.9345	.9357	.9370	.9382	.9394	.9406	.9418	.9429	.9441
1.6	.9452	.9463	.9474	.9484	.9495	.9505	.9515	.9525	.9535	.9545
1.7	.9554	.9564	.9573	.9582	.9591	.9599	.9608	.9616	.9625	.9633
1.8	.9641	.9649	.9656	.9664	.9671	.9678	.9686	.9693	.9699	.9706
1.9	.9713	.9719	.9726	.9732	.9738	.9744	.9750	.9756	.9761	.9767
2.0	.9772	.9778	.9783	.9788	.9793	.9798	.9803	.9808	.9812	.9817
2.1	.9821	.9826	.9830	.9834	.9838	.9842	.9846	.9850	.9854	.9857
2.2	.9861	.9864	.9868	.9871	.9875	.9878	.9881	.9884	.9887	.9890
2.3	.9893	.9896	.9898	.9901	.9904	.9906	.9909	.9911	.9913	.9916
2.4	.9918	.9920	.9922	.9925	.9927	.9929	.9931	.9932	.9934	.9936
2.5	.9938	.9940	.9941	.9943	.9945	.9946	.9948	.9949	.9951	.9952
2.6	.9953	.9955	.9956	.9957	.9959	.9960	.9961	.9962	.9963	.9964
2.7	.9965	.9966	.9967	.9968	.9969	.9970	.9971	.9972	.9973	.9974
2.8	.9974	.9975	.9976	.9977	.9977	.9978	.9979	.9979	.9980	.9981
2.9	.9981	.9982	.9982	.9983	.9984	.9984	.9985	.9985	.9986	.9986
3.0	.9987	.9987	.9987	.9988	.9988	.9989	.9989	.9989	.9990	.9990
3.1	.9990	.9991	.9991	.9991	.9992	.9992	.9992	.9992	.9993	.9993
3.2	.9993	.9993	.9994	.9994	.9994	.9994	.9994	.9995	.9995	.9995
3.3	.9995	.9995	.9995	.9996	.9996	.9996	.9996	.9996	.9996	.9997
3.4	.9997	.9997	.9997	.9997	.9997	.9997	.9997	.9997	.9997	.9998

TABLE A-2 *t*-Distribution Probabilities

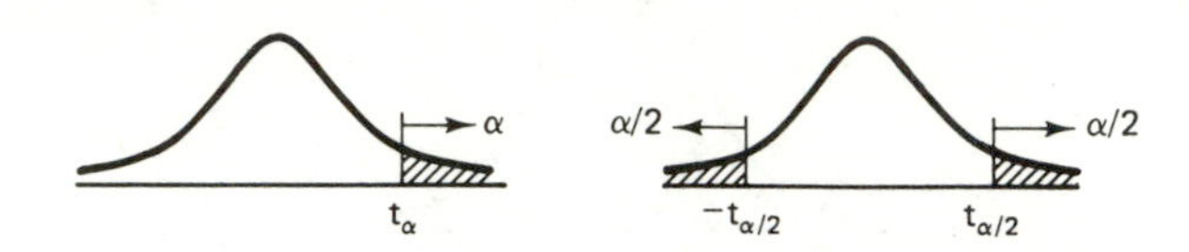

	Level of Significance for One-Tailed Test							
	.250	.100	.050	.025	.010	.005	.0025	.0005
	Level of Significance for a Two-Tailed Test							
Degrees of Freedom	.500	.200	.100	.050	.020	.010	.005	.001
1.	1.000	3.078	6.314	12.706	31.821	63.657	27.321	536.627
2.	.816	1.886	2.920	4.303	6.965	9.925	14.089	31.599
3.	.765	1.638	2.353	3.182	4.541	5.841	7.453	12.924
4.	.741	1.533	2.132	2.776	3.747	4.604	5.598	8.610
5.	.727	1.476	2.015	2.571	3.365	4.032	4.773	6.869
6.	.718	1.440	1.943	2.447	3.143	3.707	4.317	5.959
7.	.711	1.415	1.895	2.365	2.998	3.499	4.029	5.408
8.	.706	1.397	1.860	2.306	2.896	3.355	3.833	5.041
9.	.703	1.383	1.833	2.262	2.821	3.250	3.690	4.781
10.	.700	1.372	1.812	2.228	2.764	3.169	3.581	4.587
11.	.697	1.363	1.796	2.201	2.718	3.106	3.497	4.437
12.	.695	1.356	1.782	2.179	2.681	3.055	3.428	4.318
13.	.694	1.350	1.771	2.160	2.650	3.012	3.372	4.221
14.	.692	1.345	1.761	2.145	2.624	2.977	3.326	4.140
15.	.691	1.341	1.753	2.131	2.602	2.947	3.286	4.073
16.	.690	1.337	1.746	2.120	2.583	2.921	3.252	4.015
17.	.689	1.333	1.740	2.110	2.567	2.898	3.222	3.965
18.	.688	1.330	1.734	2.101	2.552	2.878	3.197	3.922
19.	.688	1.328	1.729	2.093	2.539	2.861	3.174	3.883
20.	.687	1.325	1.725	2.086	2.528	2.845	3.153	3.850
21.	.686	1.323	1.721	2.080	2.518	2.831	3.135	3.819
22.	.686	1.321	1.717	2.074	2.508	2.819	3.119	3.792
23.	.685	1.319	1.714	2.069	2.500	2.807	3.104	3.768
24.	.685	1.318	1.711	2.064	2.492	2.797	3.091	3.745
25.	.684	1.316	1.708	2.060	2.485	2.787	3.078	3.725
26.	.684	1.315	1.706	2.056	2.479	2.779	3.067	3.707
27.	.684	1.314	1.703	2.052	2.473	2.771	3.057	3.690
28.	.683	1.313	1.701	2.048	2.467	2.763	3.047	3.674
29.	.683	1.311	1.699	2.045	2.462	2.756	3.038	3.659
30.	.683	1.310	1.697	2.042	2.457	2.750	3.030	3.646
35.	.682	1.306	1.690	2.030	2.438	2.724	2.996	3.591
40.	.681	1.303	1.684	2.021	2.423	2.704	2.971	3.551
45.	.680	1.301	1.679	2.014	2.412	2.690	2.952	3.520
50.	.679	1.299	1.676	2.009	2.403	2.678	2.937	3.496
55.	.679	1.297	1.673	2.004	2.396	2.668	2.925	3.476
60.	.679	1.296	1.671	2.000	2.390	2.660	2.915	3.460
65.	.678	1.295	1.669	1.997	2.385	2.654	2.906	3.447
70.	.678	1.294	1.667	1.994	2.381	2.648	2.899	3.435
80.	.678	1.292	1.664	1.990	2.374	2.639	2.887	3.416
90.	.677	1.291	1.662	1.987	2.368	2.632	2.878	3.402
100.	.677	1.290	1.660	1.984	2.364	2.626	2.871	3.390
125.	.676	1.288	1.657	1.979	2.357	2.616	2.858	3.370
150.	.676	1.287	1.655	1.976	2.351	2.609	2.849	3.357
200.	.676	1.286	1.653	1.972	2.345	2.601	2.839	3.340
∞	.6745	1. 2816	1. 6448	1. 9600	2. 3267	2. 5758	2. 8070	3. 2905

TABLE A-3 χ^2 Distribution

Degrees of Freedom \ α	.999	.995	.990	.975	.950	.900	.700	.500	.200	.100	.050	.020	.010	.005	.001
1.	.000	.000	.000	.001	.004	.016	.148	.455	1.642	2.706	3.842	5.405	6.637	7.905	10.809
2.	.002	.010	.020	.051	.102	.211	.713	1.386	3.219	4.604	5.995	7.822	9.221	10.589	13.691
3.	.024	.071	.115	.216	.352	.584	1.424	2.366	4.642	6.252	7.817	9.841	11.325	12.819	16.292
4.	.090	.205	.297	.484	.711	1.064	2.195	3.357	5.989	7.782	9.492	11.660	13.280	14.824	18.432
5.	.209	.411	.553	.831	1.145	1.610	3.000	4.352	7.291	9.237	11.073	13.385	15.088	16.762	20.751
6.	.377	.673	.871	1.236	1.635	2.204	3.828	5.349	8.559	10.646	12.596	15.033	16.810	18.550	22.677
7.	.597	.988	1.237	1.688	2.167	2.833	4.671	6.346	9.804	12.020	14.070	16.624	18.471	20.270	24.527
8.	.850	1.341	1.642	2.179	2.732	3.489	5.527	7.344	11.031	13.363	15.512	18.171	20.082	21.938	26.318
9.	1.135	1.728	2.086	2.699	3.324	4.168	6.393	8.343	12.243	14.686	16.925	19.683	21.654	23.563	28.061
10.	1.446	2.152	2.555	3.244	3.938	4.864	7.266	9.342	13.443	15.990	18.311	21.165	23.194	25.154	29.763
11.	1.819	2.597	3.047	3.815	4.574	5.576	8.148	10.341	14.633	17.278	19.681	22.623	24.755	26.714	31.431
12.	2.188	3.064	3.568	4.402	5.225	6.303	9.034	11.340	15.813	18.551	21.030	24.059	26.246	28.249	33.070
13.	2.577	3.560	4.102	5.006	5.890	7.041	9.926	12.340	16.986	19.814	22.367	25.477	27.717	29.878	34.683
14.	3.018	4.066	4.653	5.624	6.568	7.789	10.821	13.339	18.152	21.067	23.691	26.879	29.169	31.376	36.272
15.	3.449	4.588	5.226	6.260	7.260	8.546	11.721	14.339	19.312	22.310	25.000	28.266	30.605	32.857	37.842
16.	3.894	5.135	5.807	6.905	7.960	9.311	12.624	15.339	20.466	23.546	26.301	29.640	32.027	34.321	39.392
17.	4.350	5.687	6.400	7.560	8.670	10.083	13.530	16.338	21.616	24.771	27.593	31.002	33.435	35.771	40.926
18.	4.864	6.251	7.004	8.225	9.388	10.864	14.440	17.337	22.761	25.992	28.877	32.353	34.831	37.208	42.444
19.	5.351	6.825	7.627	8.904	10.114	11.650	15.351	18.338	23.902	27.206	30.148	33.694	36.216	38.633	43.949
20.	5.848	7.422	8.252	9.587	10.849	12.442	16.265	19.337	25.039	28.415	31.416	35.026	37.591	40.046	45.440
21.	6.398	8.018	8.886	10.278	11.590	13.238	17.182	20.337	26.173	29.619	32.678	36.350	38.957	41.449	46.919
22.	6.919	8.622	9.528	10.976	12.336	14.040	18.100	21.337	27.304	30.817	33.933	37.666	40.314	42.843	48.387
23.	7.447	9.247	10.187	11.685	13.088	14.846	19.020	22.337	28.431	32.012	35.178	38.975	41.662	44.228	49.845
24.	8.027	9.869	10.846	12.397	13.845	15.657	19.943	23.337	29.556	33.199	36.421	40.277	43.004	45.604	51.293
25.	8.576	10.498	11.510	13.115	14.607	16.471	20.866	24.337	30.678	34.384	37.660	41.573	44.338	46.973	52.732
26.	9.130	11.132	12.190	13.837	15.377	17.291	21.792	25.337	31.796	35.566	38.894	42.863	45.665	48.334	54.162
27.	9.735	11.789	12.868	14.565	16.149	18.113	22.718	26.336	32.913	36.745	40.119	44.147	46.986	49.688	55.584
28.	10.306	12.438	13.551	15.304	16.925	18.938	23.646	27.336	34.028	37.920	41.344	45.426	48.301	51.036	56.998
29.	10.882	13.092	14.240	16.042	17.705	19.766	24.576	28.336	35.140	39.092	42.565	46.699	49.610	52.378	58.405
30.	11.509	13.767	14.943	16.784	18.488	20.598	25.507	29.336	36.251	40.261	43.782	47.968	50.914	53.713	59.805
35.	14.554	17.160	18.494	20.562	22.461	24.793	30.176	34.336	41.780	46.063	49.811	54.250	57.363	60.313	66.714
40.	17.846	20.669	22.139	24.423	26.508	29.055	34.879	39.337	47.261	51.796	55.753	60.443	63.710	66.802	73.490
45.	21.183	24.275	25.878	28.357	30.611	33.355	39.591	44.337	52.721	57.497	61.652	66.562	69.976	73.201	80.160
50.	24.609	27.957	29.685	32.349	34.763	37.693	44.319	49.336	58.157	63.159	67.501	72.619	76.172	79.523	86.740
55.	28.111	31.702	33.549	36.390	38.957	42.064	49.061	54.336	63.571	68.789	73.308	78.625	82.309	85.780	93.243
60.	31.678	35.503	37.465	40.474	43.187	46.463	53.814	59.336	68.966	74.390	79.078	84.586	88.396	91.982	99.679
65.	35.303	39.353	41.424	44.596	47.449	50.887	58.578	64.336	74.344	79.966	84.817	90.507	94.438	98.134	106.057
70.	38.980	43.246	45.423	48.750	51.739	55.333	63.351	69.335	79.709	85.521	90.528	96.393	100.441	104.243	112.383
80.	46.466	51.145	53.523	57.147	60.391	64.282	72.920	79.335	90.400	96.572	101.876	108.075	112.344	116.348	124.901
90.	54.104	59.171	61.738	65.641	69.126	73.295	82.515	89.335	101.048	107.559	113.143	119.654	124.130	128.324	137.267
100.	61.869	67.303	70.049	74.216	77.929	82.362	92.133	99.335	111.662	118.493	124.340	131.147	135.820	140.193	149.505
125.	81.726	88.007	91.166	95.941	100.178	105.217	116.254	124.335	138.071	145.638	152.092	159.580	164.706	169.493	179.653
150.	102.073	109.122	112.655	117.980	122.692	128.278	140.460	149.334	164.345	172.577	179.579	187.683	193.219	198.380	209.310
200.	143.807	152.224	156.421	162.724	168.279	174.838	189.051	199.334	216.605	226.017	233.993	243.191	249.455	255.281	267.579

TABLE A-4 ***F* Distribution**

ν_2 \ ν_1 (Degrees of Freedom, ν_1)	1	2	3	4	5	6	7	8	9	10	11	12	13	14	15
1.	161.	200.	216.	225.	230.	234.	237.	239.	241.	242.	243.	244.	245.	245.	246.
1.	4052.	5000.	5403.	5625.	5764.	5859.	5928.	5981.	6022.	6056.	6083.	6106.	6126.	6143.	6157.
2.	18.51	19.00	19.16	19.25	19.30	19.33	19.35	19.37	19.38	19.40	19.40	19.41	19.42	19.42	19.43
2.	98.50	99.00	99.17	99.25	99.30	99.33	99.36	99.37	99.39	99.40	99.41	99.42	99.42	99.43	99.43
3.	10.13	9.55	9.28	9.12	9.01	8.94	8.89	8.85	8.81	8.79	8.76	8.74	8.73	8.71	8.70
3.	34.12	30.82	29.46	28.71	28.24	27.91	27.67	27.49	27.34	27.23	27.13	27.05	26.99	26.93	26.87
4.	7.71	6.94	6.59	6.39	6.26	6.16	6.09	6.04	6.00	5.96	5.94	5.91	5.89	5.87	5.86
4.	21.20	18.00	16.69	15.98	15.52	15.21	14.98	14.80	14.66	14.55	14.45	14.37	14.31	14.25	14.20
5.	6.61	5.79	5.41	5.19	5.05	4.95	4.88	4.82	4.77	4.74	4.70	4.68	4.66	4.64	4.62
5.	16.26	13.27	12.06	11.39	10.97	10.67	10.46	10.29	10.16	10.05	9.96	9.89	9.82	9.77	9.72
6.	5.99	5.14	4.76	4.53	4.39	4.28	4.21	4.15	4.10	4.06	4.03	4.00	3.98	3.96	3.94
6.	13.75	10.92	9.78	9.15	8.75	8.47	8.26	8.10	7.98	7.87	7.79	7.72	7.66	7.60	7.56
7.	5.59	4.74	4.35	4.12	3.97	3.87	3.79	3.73	3.68	3.64	3.60	3.57	3.55	3.53	3.51
7.	12.25	9.55	8.45	7.85	7.46	7.19	6.99	6.84	6.72	6.62	6.54	6.47	6.41	6.36	6.31
8.	5.32	4.46	4.07	3.84	3.69	3.58	3.50	3.44	3.39	3.35	3.31	3.28	3.26	3.24	3.22
8.	11.26	8.65	7.59	7.01	6.63	6.37	6.18	6.03	5.91	5.81	5.73	5.67	5.61	5.56	5.52
9.	5.12	4.26	3.86	3.63	3.48	3.37	3.29	3.23	3.18	3.14	3.10	3.07	3.05	3.03	3.01
9.	10.56	8.02	6.99	6.42	6.06	5.80	5.61	5.47	5.35	5.26	5.18	5.11	5.05	5.01	4.96
10.	4.96	4.10	3.71	3.48	3.33	3.22	3.14	3.07	3.02	2.98	2.94	2.91	2.89	2.86	2.85
10.	10.04	7.56	6.55	5.99	5.64	5.39	5.20	5.06	4.94	4.85	4.77	4.71	4.65	4.60	4.56
11.	4.84	3.98	3.59	3.36	3.20	3.09	3.01	2.95	2.90	2.85	2.82	2.79	2.76	2.74	2.72
11.	9.65	7.21	6.22	5.67	5.32	5.07	4.89	4.74	4.63	4.54	4.46	4.40	4.34	4.29	4.25

* For degrees of freedom ν_1 and ν_2 the upper and lower values in the table are $F_{.05}$ and $F_{.01}$.

TABLE A-4 (*cont.*)

Degrees of Freedom, ν_1

ν_2 \ ν_1	1	2	3	4	5	6	7	8	9	10	11	12	13	14	15
12.	4.75	3.89	3.49	3.26	3.11	3.00	2.91	2.85	2.80	2.75	2.72	2.69	2.66	2.64	2.62
12.	9.33	6.93	5.95	5.41	5.06	4.82	4.64	4.50	4.39	4.30	4.22	4.16	4.10	4.05	4.01
13.	4.67	3.81	3.41	3.18	3.03	2.92	2.83	2.77	2.71	2.67	2.63	2.60	2.58	2.55	2.53
13.	9.07	6.70	5.74	5.21	4.86	4.62	4.44	4.30	4.19	4.10	4.02	3.96	3.91	3.86	3.82
14.	4.60	3.74	3.34	3.11	2.96	2.85	2.76	2.70	2.65	2.60	2.57	2.53	2.51	2.48	2.46
14.	8.86	6.51	5.56	5.04	4.69	4.46	4.28	4.14	4.03	3.94	3.86	3.80	3.75	3.70	3.66
15.	4.54	3.68	3.29	3.06	2.90	2.79	2.71	2.64	2.59	2.54	2.51	2.48	2.45	2.42	2.40
15.	8.68	6.36	5.42	4.89	4.56	4.32	4.14	4.00	3.89	3.80	3.73	3.67	3.61	3.56	3.52
17.	4.45	3.59	3.20	2.96	2.81	2.70	2.61	2.55	2.49	2.45	2.41	2.38	2.35	2.33	2.31
17.	8.40	6.11	5.19	4.67	4.34	4.10	3.93	3.79	3.68	3.59	3.52	3.46	3.40	3.35	3.31
20.	4.35	3.49	3.10	2.87	2.71	2.60	2.51	2.45	2.39	2.35	2.31	2.28	2.25	2.22	2.20
20.	8.10	5.85	4.94	4.43	4.10	3.87	3.70	3.56	3.46	3.37	3.29	3.23	3.18	3.13	3.09
25.	4.24	3.39	2.99	2.76	2.60	2.49	2.40	2.34	2.28	2.24	2.20	2.16	2.14	2.11	2.09
25.	7.77	5.57	4.68	4.18	3.85	3.63	3.46	3.32	3.22	3.13	3.06	2.99	2.94	2.89	2.85
30.	4.17	3.32	2.92	2.69	2.53	2.42	2.33	2.27	2.21	2.16	2.13	2.09	2.06	2.04	2.01
30.	7.56	5.39	4.51	4.02	3.70	3.47	3.30	3.17	3.07	2.98	2.91	2.84	2.79	2.74	2.70
40.	4.08	3.23	2.84	2.61	2.45	2.34	2.25	2.18	2.12	2.08	2.04	2.00	1.97	1.95	1.92
40.	7.31	5.18	4.31	3.83	3.51	3.29	3.12	2.99	2.89	2.80	2.73	2.66	2.61	2.56	2.52
50.	4.03	3.18	2.79	2.56	2.40	2.29	2.20	2.13	2.07	2.03	1.99	1.95	1.92	1.89	1.87
50.	7.17	5.06	4.20	3.72	3.41	3.19	3.02	2.89	2.78	2.70	2.63	2.56	2.51	2.46	2.42
75.	3.97	3.12	2.73	2.49	2.34	2.22	2.13	2.06	2.01	1.96	1.92	1.88	1.85	1.83	1.80
75.	6.99	4.90	4.05	3.58	3.27	3.05	2.89	2.76	2.65	2.57	2.49	2.43	2.38	2.33	2.29
100.	3.94	3.09	2.70	2.46	2.31	2.19	2.10	2.03	1.97	1.93	1.89	1.85	1.82	1.79	1.77
100.	6.90	4.82	3.98	3.51	3.21	2.99	2.82	2.69	2.59	2.50	2.43	2.37	2.31	2.27	2.22
200.	3.89	3.04	2.65	2.42	2.26	2.14	2.06	1.98	1.93	1.88	1.84	1.80	1.77	1.74	1.72
200.	6.76	4.71	3.88	3.41	3.11	2.89	2.73	2.60	2.50	2.41	2.34	2.27	2.22	2.17	2.13
500.	3.86	3.01	2.62	2.39	2.23	2.12	2.03	1.96	1.90	1.85	1.81	1.77	1.74	1.71	1.69
500.	6.69	4.65	3.82	3.36	3.05	2.84	2.68	2.55	2.44	2.36	2.28	2.22	2.17	2.12	2.07
1000.	3.85	3.00	2.61	2.38	2.22	2.11	2.02	1.95	1.89	1.84	1.80	1.76	1.73	1.70	1.68
1000.	6.66	4.63	3.80	3.34	3.04	2.82	2.66	2.53	2.43	2.34	2.27	2.20	2.15	2.10	2.06
10000.	3.84	3.00	2.61	2.37	2.21	2.10	2.01	1.94	1.88	1.83	1.79	1.75	1.72	1.69	1.67
10000.	6.64	4.61	3.78	3.32	3.02	2.80	2.64	2.51	2.41	2.32	2.25	2.19	2.13	2.08	2.04

TABLE A-4 *(cont.)*

Degrees of Freedom, ν_1

ν_2 \ ν_1	17	20	25	30	40	50	75	100	200	500	1000	10000
1. 1.	247. 6181.	248. 6209.	249. 6240.	250. 6261.	251. 6287.	252. 6303.	253. 6324.	253. 6334.	254. 6350.	254. 6360.	254. 6363.	254. 6364.
2. 2.	19.44 99.44	19.45 99.45	19.46 99.46	19.46 99.47	19.47 99.47	19.48 99.48	19.48 99.49	19.49 99.49	19.49 99.49	19.49 99.50	19.49 99.50	19.49 99.49
3. 3.	8.68 26.79	8.66 26.69	8.63 26.58	8.62 26.50	8.59 26.41	8.58 26.35	8.56 26.28	8.55 26.24	8.54 26.18	8.53 26.15	8.53 26.14	8.52 26.12
4. 4.	5.83 14.11	5.80 14.02	5.77 13.91	5.75 13.84	5.72 13.75	5.70 13.69	5.68 13.61	5.66 13.58	5.65 13.52	5.64 13.49	5.63 13.47	5.63 13.46
5. 5.	4.59 9.64	4.56 9.55	4.52 9.45	4.50 9.38	4.46 9.29	4.44 9.24	4.42 9.17	4.41 9.13	4.39 9.08	4.37 9.04	4.37 9.03	4.37 9.02
6. 6.	3.91 7.48	3.87 7.40	3.83 7.30	3.81 7.23	3.77 7.14	3.75 7.09	3.73 7.02	3.71 6.99	3.69 6.93	3.68 6.90	3.67 6.89	3.67 6.88
7. 7.	3.48 6.24	3.44 6.16	3.40 6.06	3.38 5.99	3.34 5.91	3.32 5.86	3.29 5.79	3.27 5.75	3.25 5.70	3.24 5.67	3.23 5.66	3.23 5.65
8. 8.	3.19 5.44	3.15 5.36	3.11 5.26	3.08 5.20	3.04 5.12	3.02 5.07	2.99 5.00	2.97 4.96	2.95 4.91	2.94 4.88	2.93 4.87	2.93 4.86
9. 9.	2.97 4.89	2.94 4.81	2.89 4.71	2.86 4.65	2.83 4.57	2.80 4.52	2.77 4.45	2.76 4.41	2.73 4.36	2.72 4.33	2.71 4.32	2.71 4.31
10. 10.	2.81 4.49	2.77 4.41	2.73 4.31	2.70 4.25	2.66 4.17	2.64 4.12	2.60 4.05	2.59 4.01	2.56 3.96	2.55 3.93	2.54 3.92	2.54 3.91
11. 11.	2.69 4.18	2.65 4.10	2.60 4.01	2.57 3.94	2.53 3.86	2.51 3.81	2.47 3.74	2.46 3.71	2.43 3.66	2.42 3.62	2.41 3.61	2.41 3.60
12. 12.	2.58 3.94	2.54 3.86	2.50 3.76	2.47 3.70	2.43 3.62	2.40 3.57	2.37 3.50	2.35 3.47	2.32 3.41	2.31 3.38	2.30 3.37	2.30 3.36
13. 13.	2.50 3.75	2.46 3.66	2.41 3.57	2.38 3.51	2.34 3.43	2.31 3.38	2.28 3.31	2.26 3.27	2.23 3.22	2.22 3.19	2.21 3.18	2.21 3.17
14. 14.	2.43 3.59	2.39 3.51	2.34 3.41	2.31 3.35	2.27 3.27	2.24 3.22	2.21 3.15	2.19 3.11	2.16 3.06	2.14 3.03	2.14 3.02	2.13 3.00
15. 15.	2.37 3.45	2.33 3.37	2.28 3.28	2.25 3.21	2.20 3.13	2.18 3.08	2.14 3.01	2.12 2.98	2.10 2.92	2.08 2.89	2.07 2.88	2.07 2.87
17. 17.	2.27 3.24	2.23 3.16	2.18 3.07	2.15 3.00	2.10 2.92	2.08 2.87	2.04 2.80	2.02 2.76	1.99 2.71	1.97 2.68	1.97 2.66	1.96 2.65
20. 20.	2.17 3.02	2.12 2.94	2.07 2.84	2.04 2.78	1.99 2.69	1.97 2.64	1.93 2.57	1.91 2.54	1.88 2.48	1.86 2.44	1.85 2.43	1.85 2.42
25. 25.	2.05 2.78	2.01 2.70	1.96 2.60	1.92 2.54	1.87 2.45	1.84 2.40	1.80 2.33	1.78 2.29	1.75 2.23	1.73 2.19	1.72 2.18	1.72 2.17
30. 30.	1.98 2.63	1.93 2.55	1.88 2.45	1.84 2.39	1.79 2.30	1.76 2.25	1.72 2.17	1.70 2.13	1.66 2.07	1.64 2.03	1.63 2.02	1.62 2.01
40. 40.	1.89 2.45	1.84 2.37	1.78 2.27	1.74 2.20	1.69 2.11	1.66 2.06	1.61 1.98	1.59 1.94	1.55 1.87	1.53 1.83	1.52 1.82	1.51 1.82
50. 50.	1.83 2.35	1.78 2.27	1.73 2.17	1.69 2.10	1.63 2.01	1.60 1.95	1.55 1.87	1.52 1.82	1.48 1.76	1.46 1.71	1.45 1.70	1.44 1.68
75. 75.	1.76 2.22	1.71 2.13	1.65 2.03	1.61 1.96	1.55 1.87	1.52 1.81	1.47 1.72	1.44 1.67	1.39 1.60	1.36 1.55	1.35 1.53	1.34 1.52
100. 100.	1.73 2.15	1.68 2.07	1.62 1.97	1.57 1.89	1.52 1.80	1.48 1.74	1.42 1.65	1.39 1.60	1.34 1.52	1.31 1.47	1.30 1.45	1.28 1.43
200. 200.	1.67 2.06	1.62 1.97	1.56 1.87	1.52 1.79	1.46 1.69	1.41 1.63	1.35 1.53	1.32 1.48	1.26 1.39	1.22 1.33	1.21 1.30	1.19 1.28
500. 500.	1.64 2.00	1.59 1.92	1.53 1.81	1.48 1.74	1.42 1.63	1.38 1.57	1.31 1.47	1.28 1.41	1.21 1.31	1.16 1.23	1.14 1.20	1.12 1.17
1000. 1000.	1.63 1.98	1.58 1.90	1.52 1.79	1.47 1.72	1.41 1.61	1.36 1.54	1.30 1.44	1.26 1.38	1.19 1.28	1.13 1.19	1.11 1.16	1.08 1.12
10000. 10000.	1.62 1.97	1.57 1.88	1.51 1.77	1.46 1.70	1.40 1.59	1.35 1.53	1.28 1.42	1.25 1.36	1.17 1.25	1.11 1.16	1.08 1.11	1.03 1.05

TABLE A-5 Binomial Distribution Probabilities

n	x	0.05	0.10	0.15	0.20	0.25	0.30	0.35	0.40	0.45	0.50
2	0	.9025	.8100	.7225	.6400	.5625	.4900	.4225	.3600	.3025	.2500
	1	.9975	.9900	.9775	.9600	.9375	.9100	.8775	.8400	.7975	.7500
3	0	.8574	.7290	.6141	.5120	.4219	.3430	.2746	.2160	.1664	.1250
	1	.9928	.9720	.9393	.8960	.8438	.7840	.7182	.6480	.5748	.5000
	2	.9999	.9990	.9966	.9920	.9844	.9730	.9571	.9360	.9089	.8750
4	0	.8145	.6561	.5220	.4096	.3164	.2401	.1785	.1296	.0915	.0625
	1	.9860	.9477	.8905	.8192	.7383	.6517	.5630	.4752	.3910	.3125
	2	.9995	.9963	.9880	.9728	.9492	.9163	.8735	.8208	.7585	.6875
	3	1.0000	.9999	.9995	.9984	.9961	.9919	.9850	.9744	.9590	.9375
5	0	.7738	.5905	.4437	.3277	.2373	.1681	.1160	.0778	.0503	.0313
	1	.9774	.9185	.8352	.7373	.6328	.5282	.4284	.3370	.2562	.1875
	2	.9988	.9914	.9734	.9421	.8965	.8369	.7648	.6826	.5931	.5000
	3	1.0000	.9995	.9978	.9933	.9844	.9692	.9460	.9130	.8688	.8125
	4	1.0000	1.0000	.9999	.9997	.9990	.9976	.9947	.9898	.9815	.9688
6	0	.7351	.5314	.3771	.2621	.1780	.1176	.0754	.0467	.0277	.0156
	1	.9672	.8857	.7765	.6554	.5339	.4202	.3191	.2333	.1636	.1094
	2	.9978	.9842	.9527	.9011	.8306	.7443	.6471	.5443	.4415	.3438
	3	.9999	.9987	.9941	.9830	.9624	.9295	.8826	.8208	.7447	.6563
	4	1.0000	.9999	.9996	.9984	.9954	.9891	.9777	.9590	.9308	.8906
	5	1.0000	1.0000	1.0000	.9999	.9998	.9993	.9982	.9959	.9917	.9844
7	0	.6983	.4783	.3206	.2097	.1335	.0824	.0490	.0280	.0152	.0078
	1	.9556	.8503	.7166	.5767	.4449	.3294	.2338	.1586	.1024	.0625
	2	.9962	.9743	.9262	.8520	.7564	.6471	.5323	.4199	.3164	.2266
	3	.9998	.9973	.9879	.9667	.9294	.8740	.8002	.7102	.6083	.5000
	4	1.0000	.9998	.9988	.9953	.9871	.9712	.9444	.9037	.8471	.7734
	5	1.0000	1.0000	.9999	.9996	.9987	.9962	.9910	.9812	.9643	.9375
	6	1.0000	1.0000	1.0000	1.0000	.9999	.9998	.9994	.9984	.9963	.9922
8	0	.6634	.4305	.2725	.1678	.1001	.0576	.0319	.0168	.0084	.0039
	1	.9428	.8131	.6572	.5033	.3671	.2553	.1691	.1064	.0632	.0352
	2	.9942	.9619	.8948	.7969	.6785	.5518	.4278	.3154	.2201	.1445
	3	.9996	.9950	.9786	.9437	.8862	.8059	.7064	.5941	.4770	.3633
	4	1.0000	.9996	.9971	.9896	.9727	.9420	.8939	.8263	.7396	.6367
	5	1.0000	1.0000	.9998	.9988	.9958	.9887	.9747	.9502	.9115	.8555
	6	1.0000	1.0000	1.0000	.9999	.9996	.9987	.9964	.9915	.9819	.9648
	7	1.0000	1.0000	1.0000	1.0000	1.0000	.9999	.9998	.9993	.9983	.9961
9	0	.6302	.3874	.2316	.1342	.0751	.0404	.0207	.0101	.0046	.0020
	1	.9288	.7748	.5995	.4362	.3003	.1960	.1211	.0705	.0385	.0195
	2	.9916	.9470	.8591	.7382	.6007	.4628	.3373	.2318	.1495	.0898
	3	.9994	.9917	.9661	.9144	.8343	.7297	.6089	.4826	.3614	.2539
	4	1.0000	.9991	.9944	.9804	.9511	.9012	.8283	.7334	.6214	.5000
	5	1.0000	.9999	.9994	.9969	.9900	.9747	.9464	.9006	.8342	.7461
	6	1.0000	1.0000	1.0000	.9997	.9987	.9957	.9888	.9750	.9502	.9102
	7	1.0000	1.0000	1.0000	1.0000	.9999	.9996	.9986	.9962	.9909	.9805
	8	1.0000	1.0000	1.0000	1.0000	1.0000	1.0000	.9999	.9997	.9992	.9980
10	0	.5987	.3487	.1969	.1074	.0563	.0282	.0135	.0060	.0025	.0010
	1	.9139	.7361	.5443	.3758	.2440	.1493	.0860	.0464	.0233	.0107
	2	.9885	.9298	.8202	.6778	.5256	.3828	.2616	.1673	.0996	.0547
	3	.9990	.9872	.9500	.8791	.7759	.6496	.5138	.3823	.2660	.1719
	4	.9999	.9984	.9901	.9672	.9219	.8497	.7515	.6331	.5044	.3770
	5	1.0000	.9999	.9986	.9936	.9803	.9527	.9051	.8338	.7384	.6230
	6	1.0000	1.0000	.9999	.9991	.9965	.9894	.9740	.9452	.8980	.8281
	7	1.0000	1.0000	1.0000	.9999	.9996	.9984	.9952	.9877	.9726	.9453
	8	1.0000	1.0000	1.0000	1.0000	1.0000	.9999	.9995	.9983	.9955	.9893
	9	1.0000	1.0000	1.0000	1.0000	1.0000	1.0000	1.0000	.9999	.9997	.9990
11	0	.5688	.3138	.1673	.0859	.0422	.0198	.0088	.0036	.0014	.0005
	1	.8981	.6974	.4922	.3221	.1971	.1130	.0606	.0302	.0139	.0059
	2	.9848	.9104	.7788	.6174	.4552	.3127	.2001	.1189	.0652	.0327
	3	.9984	.9815	.9306	.8389	.7133	.5696	.4256	.2963	.1911	.1133
	4	.9999	.9972	.9841	.9496	.8854	.7897	.6683	.5328	.3971	.2744
	5	1.0000	.9997	.9973	.9883	.9657	.9218	.8513	.7535	.6331	.5000
	6	1.0000	1.0000	.9997	.9980	.9924	.9784	.9499	.9006	.8262	.7256
	7	1.0000	1.0000	1.0000	.9998	.9988	.9957	.9878	.9707	.9390	.8867
	8	1.0000	1.0000	1.0000	1.0000	.9999	.9994	.9980	.9941	.9852	.9673
	9	1.0000	1.0000	1.0000	1.0000	1.0000	1.0000	.9998	.9993	.9978	.9941
	10	1.0000	1.0000	1.0000	1.0000	1.0000	1.0000	1.0000	1.0000	.9998	.9995
12	0	.5404	.2824	.1422	.0687	.0317	.0138	.0057	.0022	.0008	.0002
	1	.8816	.6590	.4435	.2749	.1584	.0850	.0424	.0196	.0083	.0032
	2	.9804	.8891	.7358	.5583	.3907	.2528	.1513	.0834	.0421	.0193
	3	.9978	.9744	.9078	.7946	.6488	.4925	.3467	.2253	.1345	.0730
	4	.9998	.9957	.9761	.9274	.8424	.7237	.5833	.4382	.3044	.1938
	5	1.0000	.9995	.9954	.9806	.9456	.8822	.7873	.6652	.5269	.3872
	6	1.0000	.9999	.9993	.9961	.9857	.9614	.9154	.8418	.7393	.6128
	7	1.0000	1.0000	.9999	.9994	.9972	.9905	.9745	.9427	.8883	.8062
	8	1.0000	1.0000	1.0000	.9999	.9996	.9983	.9944	.9847	.9644	.9270
	9	1.0000	1.0000	1.0000	1.0000	1.0000	.9998	.9992	.9972	.9921	.9807
	10	1.0000	1.0000	1.0000	1.0000	1.0000	1.0000	.9999	.9997	.9989	.9968
	11	1.0000	1.0000	1.0000	1.0000	1.0000	1.0000	1.0000	1.0000	.9999	.9998

TABLE A-5 (*cont.*)

n	x	0.05	0.10	0.15	0.20	0.25	0.30	0.35	0.40	0.45	0.50
13	0	.5133	.2542	.1209	.0550	.0238	.0097	.0037	.0013	.0004	.0001
	1	.8646	.6213	.3983	.2336	.1267	.0637	.0296	.0126	.0049	.0017
	2	.9755	.8661	.6920	.5017	.3326	.2025	.1132	.0579	.0269	.0112
	3	.9969	.9658	.8820	.7473	.5843	.4206	.2783	.1686	.0929	.0461
	4	.9997	.9935	.9658	.9009	.7940	.6543	.5005	.3530	.2279	.1334
	5	1.0000	.9991	.9925	.9700	.9198	.8346	.7159	.5744	.4268	.2905
	6	1.0000	.9999	.9987	.9930	.9757	.9376	.8705	.7712	.6437	.5000
	7	1.0000	1.0000	.9998	.9988	.9944	.9818	.9538	.9023	.8212	.7095
	8	1.0000	1.0000	1.0000	.9998	.9990	.9960	.9874	.9679	.9302	.8666
	9	1.0000	1.0000	1.0000	1.0000	.9999	.9993	.9975	.9922	.9797	.9539
	10	1.0000	1.0000	1.0000	1.0000	1.0000	.9999	.9997	.9987	.9959	.9888
	11	1.0000	1.0000	1.0000	1.0000	1.0000	1.0000	1.0000	.9999	.9995	.9983
	12	1.0000	1.0000	1.0000	1.0000	1.0000	1.0000	1.0000	1.0000	1.0000	.9999
14	0	.4877	.2288	.1028	.0440	.0178	.0068	.0024	.0008	.0002	.0001
	1	.8470	.5846	.3567	.1979	.1010	.0475	.0205	.0081	.0029	.0009
	2	.9699	.8416	.6479	.4481	.2811	.1608	.0839	.0398	.0170	.0065
	3	.9958	.9559	.8535	.6982	.5213	.3552	.2205	.1243	.0632	.0287
	4	.9996	.9908	.9533	.8702	.7415	.5842	.4227	.2793	.1672	.0898
	5	1.0000	.9985	.9885	.9561	.8883	.7805	.6405	.4859	.3373	.2120
	6	1.0000	.9998	.9978	.9884	.9617	.9067	.8164	.6925	.5461	.3953
	7	1.0000	1.0000	.9997	.9976	.9897	.9685	.9247	.8499	.7414	.6047
	8	1.0000	1.0000	1.0000	.9996	.9978	.9917	.9757	.9417	.8811	.7880
	9	1.0000	1.0000	1.0000	1.0000	.9997	.9983	.9940	.9825	.9574	.9102
	10	1.0000	1.0000	1.0000	1.0000	1.0000	.9998	.9989	.9961	.9886	.9713
	11	1.0000	1.0000	1.0000	1.0000	1.0000	1.0000	.9999	.9994	.9978	.9935
	12	1.0000	1.0000	1.0000	1.0000	1.0000	1.0000	1.0000	.9999	.9997	.9991
	13	1.0000	1.0000	1.0000	1.0000	1.0000	1.0000	1.0000	1.0000	1.0000	.9999
15	0	.4633	.2059	.0874	.0352	.0134	.0047	.0016	.0005	.0001	.0000
	1	.8290	.5490	.3186	.1671	.0802	.0353	.0142	.0052	.0017	.0005
	2	.9638	.8159	.6042	.3980	.2361	.1268	.0617	.0271	.0107	.0037
	3	.9945	.9444	.8227	.6482	.4613	.2969	.1727	.0905	.0424	.0176
	4	.9994	.9873	.9383	.8358	.6865	.5155	.3519	.2173	.1204	.0592
	5	.9999	.9978	.9832	.9389	.8516	.7216	.5643	.4032	.2608	.1509
	6	1.0000	.9997	.9964	.9819	.9434	.8689	.7548	.6098	.4522	.3036
	7	1.0000	1.0000	.9994	.9958	.9827	.9500	.8868	.7869	.6535	.5000
	8	1.0000	1.0000	.9999	.9992	.9958	.9848	.9578	.9050	.8182	.6964
	9	1.0000	1.0000	1.0000	.9999	.9992	.9963	.9876	.9662	.9231	.8491
	10	1.0000	1.0000	1.0000	1.0000	.9999	.9993	.9972	.9907	.9745	.9408
	11	1.0000	1.0000	1.0000	1.0000	1.0000	.9999	.9995	.9981	.9937	.9824
	12	1.0000	1.0000	1.0000	1.0000	1.0000	1.0000	.9999	.9997	.9989	.9963
	13	1.0000	1.0000	1.0000	1.0000	1.0000	1.0000	1.0000	1.0000	.9999	.9995
	14	1.0000	1.0000	1.0000	1.0000	1.0000	1.0000	1.0000	1.0000	1.0000	1.0000
16	0	.4401	.1853	.0743	.0281	.0100	.0033	.0010	.0003	.0001	.0000
	1	.8108	.5147	.2839	.1407	.0635	.0261	.0098	.0033	.0010	.0003
	2	.9571	.7892	.5614	.3518	.1971	.0994	.0451	.0183	.0066	.0021
	3	.9930	.9316	.7899	.5981	.4050	.2459	.1339	.0651	.0281	.0106
	4	.9991	.9830	.9209	.7982	.6302	.4499	.2892	.1666	.0853	.0384
	5	.9999	.9967	.9765	.9183	.8103	.6598	.4900	.3288	.1976	.1051
	6	1.0000	.9995	.9944	.9733	.9204	.8247	.6881	.5272	.3660	.2272
	7	1.0000	.9999	.9989	.9930	.9729	.9256	.8406	.7161	.5629	.4018
	8	1.0000	1.0000	.9998	.9985	.9925	.9743	.9329	.8577	.7441	.5982
	9	1.0000	1.0000	1.0000	.9998	.9984	.9929	.9771	.9417	.8759	.7728
	10	1.0000	1.0000	1.0000	1.0000	.9997	.9984	.9938	.9809	.9514	.8949
	11	1.0000	1.0000	1.0000	1.0000	1.0000	.9997	.9987	.9951	.9851	.9616
	12	1.0000	1.0000	1.0000	1.0000	1.0000	1.0000	.9998	.9991	.9965	.9894
	13	1.0000	1.0000	1.0000	1.0000	1.0000	1.0000	1.0000	.9999	.9994	.9979
	14	1.0000	1.0000	1.0000	1.0000	1.0000	1.0000	1.0000	1.0000	.9999	.9997
	15	1.0000	1.0000	1.0000	1.0000	1.0000	1.0000	1.0000	1.0000	1.0000	1.0000
17	0	.4181	.1668	.0631	.0225	.0075	.0023	.0007	.0002	.0000	.0000
	1	.7922	.4818	.2525	.1182	.0501	.0193	.0067	.0021	.0006	.0001
	2	.9497	.7618	.5198	.3096	.1637	.0774	.0327	.0123	.0041	.0012
	3	.9912	.9174	.7556	.5489	.3530	.2019	.1028	.0464	.0184	.0064
	4	.9988	.9779	.9013	.7582	.5739	.3887	.2348	.1260	.0596	.0245
	5	.9999	.9953	.9681	.8943	.7653	.5968	.4197	.2639	.1471	.0717
	6	1.0000	.9992	.9917	.9623	.8929	.7752	.6188	.4478	.2902	.1662
	7	1.0000	.9999	.9983	.9891	.9598	.8954	.7872	.6405	.4743	.3145
	8	1.0000	1.0000	.9997	.9974	.9876	.9597	.9006	.8011	.6626	.5000
	9	1.0000	1.0000	1.0000	.9995	.9969	.9873	.9617	.9081	.8166	.6855
	10	1.0000	1.0000	1.0000	.9999	.9994	.9968	.9880	.9652	.9174	.8338
	11	1.0000	1.0000	1.0000	1.0000	.9999	.9993	.9970	.9894	.9699	.9283
	12	1.0000	1.0000	1.0000	1.0000	1.0000	.9999	.9994	.9975	.9914	.9755
	13	1.0000	1.0000	1.0000	1.0000	1.0000	1.0000	.9999	.9995	.9981	.9936
	14	1.0000	1.0000	1.0000	1.0000	1.0000	1.0000	1.0000	.9999	.9997	.9988
	15	1.0000	1.0000	1.0000	1.0000	1.0000	1.0000	1.0000	1.0000	1.0000	.9999
	16	1.0000	1.0000	1.0000	1.0000	1.0000	1.0000	1.0000	1.0000	1.0000	1.0000
18	0	.3972	.1501	.0536	.0180	.0056	.0016	.0004	.0001	.0000	.0000
	1	.7735	.4503	.2241	.0991	.0395	.0142	.0046	.0013	.0003	.0001
	2	.9419	.7338	.4797	.2713	.1353	.0600	.0236	.0082	.0025	.0007
	3	.9891	.9018	.7202	.5010	.3057	.1646	.0783	.0328	.0120	.0038
	4	.9985	.9718	.8794	.7164	.5187	.3327	.1886	.0942	.0411	.0154
	5	.9998	.9936	.9581	.8671	.7175	.5344	.3550	.2088	.1077	.0481
	6	1.0000	.9988	.9882	.9487	.8610	.7217	.5491	.3743	.2258	.1189
	7	1.0000	.9998	.9973	.9837	.9431	.8593	.7283	.5634	.3915	.2403
	8	1.0000	1.0000	.9995	.9957	.9807	.9404	.8609	.7368	.5778	.4073
	9	1.0000	1.0000	.9999	.9991	.9946	.9790	.9403	.8653	.7473	.5927
	10	1.0000	1.0000	1.0000	.9998	.9988	.9939	.9788	.9424	.8720	.7597
	11	1.0000	1.0000	1.0000	1.0000	.9998	.9986	.9938	.9797	.9463	.8811
	12	1.0000	1.0000	1.0000	1.0000	1.0000	.9997	.9986	.9942	.9817	.9519
	13	1.0000	1.0000	1.0000	1.0000	1.0000	1.0000	.9997	.9987	.9951	.9846
	14	1.0000	1.0000	1.0000	1.0000	1.0000	1.0000	1.0000	.9998	.9990	.9962
	15	1.0000	1.0000	1.0000	1.0000	1.0000	1.0000	1.0000	1.0000	.9999	.9993
	16	1.0000	1.0000	1.0000	1.0000	1.0000	1.0000	1.0000	1.0000	1.0000	.9999
	17	1.0000	1.0000	1.0000	1.0000	1.0000	1.0000	1.0000	1.0000	1.0000	1.0000

TABLE A-5 (*cont.*)

n	x	0.05	0.10	0.15	0.20	0.25	0.30	0.35	0.40	0.45	0.50
19	0	.3774	.1351	.0456	.0144	.0042	.0011	.0003	.0001	.0000	.0000
	1	.7547	.4203	.1985	.0829	.0310	.0104	.0031	.0008	.0002	.0000
	2	.9335	.7054	.4413	.2369	.1113	.0462	.0170	.0055	.0015	.0004
	3	.9868	.8850	.6841	.4551	.2631	.1332	.0591	.0230	.0077	.0022
	4	.9980	.9648	.8556	.6733	.4654	.2822	.1500	.0696	.0280	.0096
	5	.9998	.9914	.9463	.8369	.6678	.4739	.2968	.1629	.0777	.0318
	6	1.0000	.9983	.9837	.9324	.8251	.6655	.4812	.3081	.1727	.0835
	7	1.0000	.9997	.9959	.9767	.9225	.8180	.6656	.4878	.3169	.1796
	8	1.0000	1.0000	.9992	.9933	.9713	.9161	.8145	.6675	.4940	.3238
	9	1.0000	1.0000	.9999	.9984	.9911	.9674	.9125	.8139	.6710	.5000
	10	1.0000	1.0000	1.0000	.9997	.9977	.9895	.9653	.9115	.8159	.6762
	11	1.0000	1.0000	1.0000	1.0000	.9995	.9972	.9886	.9648	.9129	.8204
	12	1.0000	1.0000	1.0000	1.0000	.9999	.9994	.9969	.9884	.9658	.9165
	13	1.0000	1.0000	1.0000	1.0000	1.0000	.9999	.9993	.9969	.9891	.9682
	14	1.0000	1.0000	1.0000	1.0000	1.0000	1.0000	.9999	.9994	.9972	.9904
	15	1.0000	1.0000	1.0000	1.0000	1.0000	1.0000	1.0000	.9999	.9995	.9978
	16	1.0000	1.0000	1.0000	1.0000	1.0000	1.0000	1.0000	1.0000	.9999	.9996
	17	1.0000	1.0000	1.0000	1.0000	1.0000	1.0000	1.0000	1.0000	1.0000	1.0000
	18	1.0000	1.0000	1.0000	1.0000	1.0000	1.0000	1.0000	1.0000	1.0000	1.0000
20	0	.3585	.1216	.0388	.0115	.0032	.0008	.0002	.0000	.0000	.0000
	1	.7358	.3917	.1756	.0692	.0243	.0076	.0021	.0005	.0001	.0000
	2	.9245	.6769	.4049	.2061	.0913	.0355	.0121	.0036	.0009	.0002
	3	.9841	.8670	.6477	.4114	.2252	.1071	.0444	.0160	.0049	.0013
	4	.9974	.9568	.8298	.6296	.4148	.2375	.1182	.0510	.0189	.0059
	5	.9997	.9887	.9327	.8042	.6172	.4164	.2454	.1256	.0553	.0207
	6	1.0000	.9976	.9781	.9133	.7858	.6080	.4166	.2500	.1299	.0577
	7	1.0000	.9996	.9941	.9679	.8982	.7723	.6010	.4159	.2520	.1316
	8	1.0000	.9999	.9987	.9900	.9591	.8867	.7624	.5956	.4143	.2517
	9	1.0000	1.0000	.9998	.9974	.9861	.9520	.8782	.7553	.5914	.4119
	10	1.0000	1.0000	1.0000	.9994	.9961	.9829	.9468	.8725	.7507	.5881
	11	1.0000	1.0000	1.0000	.9999	.9991	.9949	.9804	.9435	.8692	.7483
	12	1.0000	1.0000	1.0000	1.0000	.9998	.9987	.9940	.9790	.9420	.8684
	13	1.0000	1.0000	1.0000	1.0000	1.0000	.9997	.9985	.9935	.9786	.9423
	14	1.0000	1.0000	1.0000	1.0000	1.0000	1.0000	.9997	.9984	.9936	.9793
	15	1.0000	1.0000	1.0000	1.0000	1.0000	1.0000	.9999	.9997	.9985	.9941
	16	1.0000	1.0000	1.0000	1.0000	1.0000	1.0000	1.0000	1.0000	.9997	.9987
	17	1.0000	1.0000	1.0000	1.0000	1.0000	1.0000	1.0000	1.0000	1.0000	.9998
	18	1.0000	1.0000	1.0000	1.0000	1.0000	1.0000	1.0000	1.0000	1.0000	1.0000
	19	1.0000	1.0000	1.0000	1.0000	1.0000	1.0000	1.0000	1.0000	1.0000	1.0000
21	0	.3406	.1094	.0329	.0092	.0024	.0006	.0001	.0000	.0000	.0000
	1	.7170	.3647	.1550	.0576	.0190	.0056	.0014	.0003	.0001	.0000
	2	.9151	.6484	.3705	.1787	.0745	.0271	.0086	.0024	.0006	.0001
	3	.9811	.8480	.6113	.3704	.1917	.0856	.0331	.0110	.0031	.0007
	4	.9968	.9478	.8025	.5860	.3674	.1984	.0924	.0370	.0126	.0036
	5	.9996	.9856	.9173	.7693	.5666	.3627	.2009	.0957	.0389	.0133
	6	1.0000	.9967	.9713	.8915	.7436	.5505	.3567	.2002	.0964	.0392
	7	1.0000	.9994	.9917	.9569	.8701	.7230	.5365	.3495	.1971	.0946
	8	1.0000	.9999	.9980	.9856	.9439	.8523	.7059	.5237	.3413	.1917
	9	1.0000	1.0000	.9996	.9959	.9794	.9324	.8377	.6914	.5117	.3318
	10	1.0000	1.0000	.9999	.9990	.9936	.9736	.9228	.8256	.6790	.5000
	11	1.0000	1.0000	1.0000	.9998	.9983	.9913	.9687	.9151	.8159	.6682
	12	1.0000	1.0000	1.0000	1.0000	.9996	.9976	.9892	.9648	.9092	.8083
	13	1.0000	1.0000	1.0000	1.0000	.9999	.9994	.9969	.9877	.9621	.9054
	14	1.0000	1.0000	1.0000	1.0000	1.0000	.9999	.9993	.9964	.9868	.9608
	15	1.0000	1.0000	1.0000	1.0000	1.0000	1.0000	.9999	.9992	.9963	.9867
	16	1.0000	1.0000	1.0000	1.0000	1.0000	1.0000	1.0000	.9998	.9992	.9964
	17	1.0000	1.0000	1.0000	1.0000	1.0000	1.0000	1.0000	1.0000	.9999	.9993
	18	1.0000	1.0000	1.0000	1.0000	1.0000	1.0000	1.0000	1.0000	1.0000	.9999
	19	1.0000	1.0000	1.0000	1.0000	1.0000	1.0000	1.0000	1.0000	1.0000	1.0000
	20	1.0000	1.0000	1.0000	1.0000	1.0000	1.0000	1.0000	1.0000	1.0000	1.0000
22	0	.3235	.0985	.0280	.0074	.0018	.0004	.0001	.0000	.0000	.0000
	1	.6982	.3392	.1367	.0480	.0149	.0041	.0010	.0002	.0000	.0000
	2	.9052	.6200	.3382	.1545	.0606	.0207	.0061	.0016	.0003	.0001
	3	.9778	.8281	.5752	.3320	.1624	.0681	.0245	.0076	.0020	.0004
	4	.9960	.9379	.7738	.5429	.3235	.1645	.0716	.0266	.0083	.0022
	5	.9994	.9818	.9001	.7326	.5168	.3134	.1629	.0722	.0271	.0085
	6	.9999	.9956	.9632	.8670	.6994	.4942	.3022	.1584	.0705	.0262
	7	1.0000	.9991	.9886	.9439	.8385	.6713	.4736	.2898	.1518	.0669
	8	1.0000	.9999	.9970	.9799	.9254	.8135	.6466	.4540	.2764	.1431
	9	1.0000	1.0000	.9993	.9939	.9705	.9084	.7916	.6244	.4350	.2617
	10	1.0000	1.0000	.9999	.9984	.9900	.9613	.8930	.7720	.6037	.4159
	11	1.0000	1.0000	1.0000	.9997	.9971	.9860	.9526	.8793	.7543	.5841
	12	1.0000	1.0000	1.0000	.9999	.9993	.9957	.9820	.9449	.8672	.7383
	13	1.0000	1.0000	1.0000	1.0000	.9999	.9989	.9942	.9785	.9383	.8569
	14	1.0000	1.0000	1.0000	1.0000	1.0000	.9998	.9984	.9930	.9757	.9331
	15	1.0000	1.0000	1.0000	1.0000	1.0000	1.0000	.9997	.9981	.9920	.9738
	16	1.0000	1.0000	1.0000	1.0000	1.0000	1.0000	.9999	.9996	.9979	.9915
	17	1.0000	1.0000	1.0000	1.0000	1.0000	1.0000	1.0000	.9999	.9995	.9978
	18	1.0000	1.0000	1.0000	1.0000	1.0000	1.0000	1.0000	1.0000	.9999	.9996
	19	1.0000	1.0000	1.0000	1.0000	1.0000	1.0000	1.0000	1.0000	1.0000	.9999
	20	1.0000	1.0000	1.0000	1.0000	1.0000	1.0000	1.0000	1.0000	1.0000	1.0000
	21	1.0000	1.0000	1.0000	1.0000	1.0000	1.0000	1.0000	1.0000	1.0000	1.0000

TABLE A-5 (*cont.*)

n	x	0.05	0.10	0.15	0.20	0.25	0.30	0.35	0.40	0.45	0.50
23	0	.3074	.0886	.0238	.0059	.0013	.0003	.0000	.0000	.0000	.0000
	1	.6794	.3151	.1204	.0398	.0116	.0030	.0007	.0001	.0000	.0000
	2	.8948	.5920	.3080	.1332	.0492	.0157	.0043	.0010	.0002	.0000
	3	.9742	.8073	.5396	.2965	.1370	.0538	.0181	.0052	.0012	.0002
	4	.9951	.9269	.7440	.5007	.2832	.1356	.0551	.0190	.0055	.0013
	5	.9992	.9774	.8811	.6947	.4685	.2688	.1309	.0540	.0186	.0053
	6	.9999	.9942	.9537	.8402	.6537	.4399	.2534	.1240	.0510	.0173
	7	1.0000	.9988	.9848	.9285	.8037	.6181	.4136	.2373	.1152	.0466
	8	1.0000	.9998	.9958	.9727	.9037	.7709	.5860	.3884	.2203	.1050
	9	1.0000	1.0000	.9990	.9911	.9592	.8799	.7408	.5562	.3636	.2024
	10	1.0000	1.0000	.9998	.9975	.9851	.9454	.8575	.7129	.5278	.3388
	11	1.0000	1.0000	1.0000	.9994	.9954	.9786	.9318	.8364	.6865	.5000
	12	1.0000	1.0000	1.0000	.9999	.9988	.9928	.9717	.9187	.8164	.6612
	13	1.0000	1.0000	1.0000	1.0000	.9997	.9979	.9900	.9651	.9063	.7976
	14	1.0000	1.0000	1.0000	1.0000	.9999	.9995	.9970	.9872	.9589	.8950
	15	1.0000	1.0000	1.0000	1.0000	1.0000	.9999	.9992	.9960	.9847	.9534
	16	1.0000	1.0000	1.0000	1.0000	1.0000	1.0000	.9998	.9990	.9952	.9827
	17	1.0000	1.0000	1.0000	1.0000	1.0000	1.0000	1.0000	.9998	.9988	.9947
	18	1.0000	1.0000	1.0000	1.0000	1.0000	1.0000	1.0000	1.0000	.9998	.9987
	19	1.0000	1.0000	1.0000	1.0000	1.0000	1.0000	1.0000	1.0000	1.0000	.9998
	20	1.0000	1.0000	1.0000	1.0000	1.0000	1.0000	1.0000	1.0000	1.0000	1.0000
	21	1.0000	1.0000	1.0000	1.0000	1.0000	1.0000	1.0000	1.0000	1.0000	1.0000
	22	1.0000	1.0000	1.0000	1.0000	1.0000	1.0000	1.0000	1.0000	1.0000	1.0000
24	0	.2920	.0798	.0202	.0047	.0010	.0002	.0000	.0000	.0000	.0000
	1	.6608	.2925	.1059	.0331	.0090	.0022	.0005	.0001	.0000	.0000
	2	.8841	.5643	.2798	.1145	.0398	.0119	.0030	.0007	.0001	.0000
	3	.9702	.7857	.5049	.2639	.1150	.0424	.0133	.0035	.0008	.0001
	4	.9940	.9149	.7134	.4599	.2466	.1111	.0422	.0134	.0036	.0008
	5	.9990	.9723	.8606	.6559	.4222	.2288	.1044	.0400	.0127	.0033
	6	.9999	.9925	.9428	.8111	.6074	.3886	.2106	.0960	.0364	.0113
	7	1.0000	.9983	.9801	.9108	.7662	.5647	.3575	.1919	.0863	.0320
	8	1.0000	.9997	.9941	.9638	.8787	.7250	.5257	.3279	.1730	.0758
	9	1.0000	.9999	.9985	.9874	.9453	.8472	.6866	.4891	.2991	.1537
	10	1.0000	1.0000	.9997	.9962	.9787	.9258	.8167	.6502	.4539	.2706
	11	1.0000	1.0000	.9999	.9990	.9928	.9686	.9058	.7870	.6151	.4194
	12	1.0000	1.0000	1.0000	.9998	.9979	.9885	.9577	.8857	.7580	.5806
	13	1.0000	1.0000	1.0000	1.0000	.9995	.9964	.9836	.9465	.8659	.7294
	14	1.0000	1.0000	1.0000	1.0000	.9999	.9990	.9945	.9783	.9352	.8463
	15	1.0000	1.0000	1.0000	1.0000	1.0000	.9998	.9984	.9925	.9731	.9242
	16	1.0000	1.0000	1.0000	1.0000	1.0000	1.0000	.9996	.9978	.9905	.9680
	17	1.0000	1.0000	1.0000	1.0000	1.0000	1.0000	.9999	.9995	.9972	.9887
	18	1.0000	1.0000	1.0000	1.0000	1.0000	1.0000	1.0000	.9999	.9993	.9967
	19	1.0000	1.0000	1.0000	1.0000	1.0000	1.0000	1.0000	1.0000	.9999	.9992
	20	1.0000	1.0000	1.0000	1.0000	1.0000	1.0000	1.0000	1.0000	1.0000	.9999
	21	1.0000	1.0000	1.0000	1.0000	1.0000	1.0000	1.0000	1.0000	1.0000	1.0000
	22	1.0000	1.0000	1.0000	1.0000	1.0000	1.0000	1.0000	1.0000	1.0000	1.0000
	23	1.0000	1.0000	1.0000	1.0000	1.0000	1.0000	1.0000	1.0000	1.0000	1.0000
25	0	.2774	.0718	.0172	.0038	.0008	.0001	.0000	.0000	.0000	.0000
	1	.6424	.2712	.0931	.0274	.0070	.0016	.0003	.0001	.0000	.0000
	2	.8729	.5371	.2537	.0982	.0321	.0090	.0021	.0004	.0001	.0000
	3	.9659	.7636	.4711	.2340	.0962	.0332	.0097	.0024	.0005	.0001
	4	.9928	.9020	.6821	.4207	.2137	.0905	.0320	.0095	.0023	.0005
	5	.9988	.9666	.8385	.6167	.3783	.1935	.0826	.0294	.0086	.0020
	6	.9998	.9905	.9305	.7800	.5611	.3407	.1734	.0736	.0258	.0073
	7	1.0000	.9977	.9745	.8909	.7265	.5118	.3061	.1536	.0639	.0216
	8	1.0000	.9995	.9920	.9532	.8506	.6769	.4668	.2735	.1340	.0539
	9	1.0000	.9999	.9979	.9827	.9287	.8106	.6303	.4246	.2424	.1148
	10	1.0000	1.0000	.9995	.9944	.9703	.9022	.7712	.5858	.3843	.2122
	11	1.0000	1.0000	.9999	.9985	.9893	.9558	.8746	.7323	.5426	.3450
	12	1.0000	1.0000	1.0000	.9996	.9966	.9825	.9396	.8462	.6937	.5000
	13	1.0000	1.0000	1.0000	.9999	.9991	.9940	.9745	.9222	.8173	.6550
	14	1.0000	1.0000	1.0000	1.0000	.9998	.9982	.9907	.9656	.9040	.7878
	15	1.0000	1.0000	1.0000	1.0000	1.0000	.9995	.9971	.9868	.9560	.8852
	16	1.0000	1.0000	1.0000	1.0000	1.0000	.9999	.9992	.9957	.9826	.9461
	17	1.0000	1.0000	1.0000	1.0000	1.0000	1.0000	.9998	.9988	.9942	.9784
	18	1.0000	1.0000	1.0000	1.0000	1.0000	1.0000	1.0000	.9997	.9984	.9927
	19	1.0000	1.0000	1.0000	1.0000	1.0000	1.0000	1.0000	.9999	.9996	.9980
	20	1.0000	1.0000	1.0000	1.0000	1.0000	1.0000	1.0000	1.0000	.9999	.9995
	21	1.0000	1.0000	1.0000	1.0000	1.0000	1.0000	1.0000	1.0000	1.0000	.9999
	22	1.0000	1.0000	1.0000	1.0000	1.0000	1.0000	1.0000	1.0000	1.0000	1.0000
	23	1.0000	1.0000	1.0000	1.0000	1.0000	1.0000	1.0000	1.0000	1.0000	1.0000
	24	1.0000	1.0000	1.0000	1.0000	1.0000	1.0000	1.0000	1.0000	1.0000	1.0000

TABLE A-6a Factors for Two-sided Tolerance Limits for Normal Distributions[a]

Factors K such that the probability is γ that at least a proportion P of the distribution will be included between $\bar{X} \pm Ks$, where $\bar{X}$ and s are estimates of the mean and the standard deviation computed from a sample size of n.

	γ = 0.75					γ = 0.90				
n \ P	0.75	0.90	0.95	0.99	0.999	0.75	0.90	0.95	0.99	0.999
2	4.498	6.301	7.414	9.531	11.920	11.407	15.978	18.800	24.167	30.227
3	2.501	3.538	4.187	5.431	6.844	4.132	5.847	6.919	8.974	11.309
4	2.035	2.892	3.431	4.471	5.657	2.932	4.166	4.943	6.440	8.149
5	1.825	2.599	3.088	4.033	5.117	2.454	3.494	4.152	5.423	6.879
6	1.704	2.429	2.889	3.779	4.802	2.196	3.131	3.723	4.870	6.188
7	1.624	2.318	2.757	3.611	4.593	2.034	2.902	3.452	4.521	5.750
8	1.568	2.238	2.663	3.491	4.444	1.921	2.743	3.264	4.278	5.446
9	1.525	2.178	2.593	3.400	4.330	1.839	2.626	3.125	4.098	5.220
10	1.492	2.131	2.537	3.328	4.241	1.775	2.535	3.018	3.959	5.046
11	1.465	2.093	2.493	3.271	4.169	1.724	2.463	2.933	3.849	4.906
12	1.443	2.062	2.456	3.223	4.110	1.683	2.404	2.863	3.758	4.792
13	1.425	2.036	2.424	3.183	4.059	1.648	2.355	2.805	3.682	4.697
14	1.409	2.013	2.398	3.148	4.016	1.619	2.314	2.756	3.618	4.615
15	1.395	1.994	2.375	3.118	3.979	1.594	2.278	2.713	3.562	4.545
16	1.383	1.977	2.355	3.092	3.946	1.572	2.246	2.676	3.514	4.484
17	1.372	1.962	2.337	3.069	3.917	1.552	2.219	2.643	3.471	4.430
18	1.363	1.948	2.321	3.048	3.891	1.535	2.194	2.614	3.433	4.382
19	1.355	1.936	2.307	3.030	3.867	1.520	2.172	2.588	3.399	4.339
20	1.347	1.925	2.294	3.013	3.846	1.506	2.152	2.564	3.368	4.300
21	1.340	1.915	2.282	2.998	3.827	1.493	2.135	2.543	3.340	4.264
22	1.334	1.906	2.271	2.984	3.809	1.482	2.118	2.524	3.315	4.232
23	1.328	1.898	2.261	2.971	3.793	1.471	2.103	2.506	3.292	4.203
24	1.322	1.891	2.252	2.959	3.778	1.462	2.089	2.489	3.270	4.176
25	1.317	1.883	2.244	2.948	3.764	1.453	2.077	2.474	3.251	4.151
26	1.313	1.877	2.236	2.938	3.751	1.444	2.065	2.460	3.232	4.127
27	1.309	1.871	2.229	2.929	3.740	1.437	2.054	2.447	3.215	4.106

[a] Adapted by permission from *Techniques of Statistical Analysis* by C. Eisenhart, M. W. Hastay, and W. A. Wallis; copyright 1947, McGraw-Hill Book Company.

TABLE A-6a (*cont.*)

	$\gamma = 0.95$					$\gamma = 0.99$				
n \ P	0.75	0.90	0.95	0.99	0.999	0.75	0.90	0.95	0.99	0.999
2	22.858	32.019	37.674	48.430	60.573	114.363	160.193	188.491	242.300	303.054
3	5.922	8.380	9.916	12.861	16.208	13.378	18.930	22.401	29.055	36.616
4	3.779	5.369	6.370	8.299	10.502	6.614	9.398	11.150	14.527	18.383
5	3.002	4.275	5.079	6.634	8.415	4.643	6.612	7.855	10.260	13.015
6	2.604	3.712	4.414	5.775	7.337	3.743	5.337	6.345	8.301	10.548
7	2.361	3.369	4.007	5.248	6.676	3.233	4.613	5.488	7.187	9.142
8	2.197	3.136	3.732	4.891	6.226	2.905	4.147	4.936	6.468	8.234
9	2.078	2.967	3.532	4.631	5.899	2.677	3.822	4.550	5.966	7.600
10	1.987	2.839	3.379	4.433	5.649	2.508	3.582	4.265	5.594	7.129
11	1.916	2.737	3.259	4.277	5.452	2.378	3.397	4.045	5.308	6.766
12	1.858	2.655	3.162	4.150	5.291	2.274	3.250	3.870	5.079	6.477
13	1.810	2.587	3.081	4.044	5.158	2.190	3.130	3.727	4.893	6.240
14	1.770	2.529	3.012	3.955	5.045	2.120	3.029	3.608	4.737	6.043
15	1.735	2.480	2.954	3.878	4.949	2.060	2.945	3.507	4.605	5.876
16	1.705	2.437	2.903	3.812	4.865	2.009	2.872	3.421	4.492	5.732
17	1.679	2.400	2.858	3.754	4.791	1.965	2.808	3.345	4.393	5.607
18	1.655	2.366	2.819	3.702	4.725	1.926	2.753	3.279	4.307	5.497
19	1.635	2.337	2.784	3.656	4.667	1.891	2.703	3.221	4.230	5.399
20	1.616	2.310	2.752	3.615	4.614	1.860	2.659	3.168	4.161	5.312
21	1.599	2.286	2.723	3.577	4.567	1.833	2.620	3.121	4.100	5.234
22	1.584	2.264	2.697	3.543	4.523	1.808	2.584	3.078	4.044	5.163
23	1.570	2.244	2.673	3.512	4.484	1.785	2.551	3.040	3.993	5.098
24	1.557	2.225	2.651	3.483	4.447	1.764	2.522	3.004	3.947	5.039
25	1.545	2.208	2.631	3.457	4.413	1.745	2.494	2.972	3.904	4.985
26	1.534	2.193	2.612	3.432	4.382	1.727	2.469	2.941	3.865	4.935
27	1.523	2.178	2.595	3.409	4.353	1.711	2.446	2.914	3.828	4.888

TABLE A-6a (*cont.*)

	γ = 0.75					γ = 0.90				
n \ P	0.75	0.90	0.95	0.99	0.999	0.75	0.90	0.95	0.99	0.999
30	1.297	1.855	2.210	2.904	3.708	1.417	2.025	2.413	3.170	4.049
35	1.283	1.834	2.185	2.871	3.667	1.390	1.988	2.368	3.112	3.974
40	1.271	1.818	2.166	2.846	3.635	1.370	1.959	2.334	3.066	3.917
45	1.262	1.805	2.150	2.826	3.609	1.354	1.935	2.306	3.030	3.871
50	1.255	1.794	2.138	2.809	3.588	1.340	1.916	2.284	3.001	3.833
55	1.249	1.785	2.127	2.795	3.571	1.329	1.901	2.265	2.976	3.801
60	1.243	1.778	2.118	2.784	3.556	1.320	1.887	2.248	2.955	3.774
65	1.239	1.771	2.110	2.773	3.543	1.312	1.875	2.235	2.937	3.751
70	1.235	1.765	2.104	2.764	3.531	1.304	1.865	2.222	2.920	3.730
75	1.231	1.760	2.098	2.757	3.521	1.298	1.856	2.211	2.906	3.712
80	1.228	1.756	2.092	2.749	3.512	1.292	1.848	2.202	2.894	3.696
85	1.225	1.752	2.087	2.743	3.504	1.287	1.841	2.193	2.882	3.682
90	1.223	1.748	2.083	2.737	3.497	1.283	1.834	2.185	2.872	3.669
95	1.220	1.745	2.079	2.732	3.490	1.278	1.828	2.178	2.863	3.657
100	1.218	1.742	2.075	2.727	3.484	1.275	1.822	2.172	2.854	3.646
110	1.214	1.736	2.069	2.719	3.473	1.268	1.813	2.160	2.839	3.626
120	1.211	1.732	2.063	2.712	3.464	1.262	1.804	2.150	2.826	3.610
130	1.208	1.728	2.059	2.705	3.456	1.257	1.797	2.141	2.814	3.595
140	1.206	1.724	2.054	2.700	3.449	1.252	1.791	2.134	2.804	3.582
150	1.204	1.721	2.051	2.695	3.443	1.248	1.785	2.127	2.795	3.571
160	1.202	1.718	2.047	2.691	3.437	1.245	1.780	2.121	2.787	3.561
170	1.200	1.716	2.044	2.687	3.432	1.242	1.775	2.116	2.780	3.552
180	1.198	1.713	2.042	2.683	3.427	1.239	1.771	2.111	2.774	3.543
190	1.197	1.711	2.039	2.680	3.423	1.236	1.767	2.106	2.768	3.536
200	1.195	1.709	2.037	2.677	3.419	1.234	1.764	2.102	2.762	3.529
250	1.190	1.702	2.028	2.665	3.404	1.224	1.750	2.085	2.740	3.501
300	1.186	1.696	2.021	2.656	3.393	1.217	1.740	2.073	2.725	3.481
400	1.181	1.688	2.012	2.644	3.378	1.207	1.726	2.057	2.703	3.453
500	1.177	1.683	2.006	2.636	3.368	1.201	1.717	2.046	2.689	3.434
600	1.175	1.680	2.002	2.631	3.360	1.196	1.710	2.038	2.678	3.421
700	1.173	1.677	1.998	2.626	3.355	1.192	1.705	2.032	2.670	3.411
800	1.171	1.675	1.996	2.623	3.350	1.189	1.701	2.027	2.663	3.402
900	1.170	1.673	1.993	2.620	3.347	1.187	1.697	2.023	2.658	3.396
1000	1.169	1.671	1.992	2.617	3.344	1.185	1.695	2.019	2.654	3.390
∞	1.150	1.645	1.960	2.576	3.291	1.150	1.645	1.960	2.576	3.291

TABLE A-6a (*cont.*)

n \ P	γ = 0.95					γ = 0.99				
	0.75	0.90	0.95	0.99	0.999	0.75	0.90	0.95	0.99	0.999
30	1.497	2.140	2.549	3.350	4.278	1.668	2.385	2.841	3.733	4.768
35	1.462	2.090	2.490	3.272	4.179	1.613	2.306	2.748	3.611	4.611
40	1.435	2.052	2.445	3.213	4.104	1.571	2.247	2.677	3.518	4.493
45	1.414	2.021	2.408	3.165	4.042	1.539	2.200	2.621	3.444	4.399
50	1.396	1.996	2.379	3.126	3.993	1.512	2.162	2.576	3.385	4.323
55	1.382	1.976	2.354	3.094	3.951	1.490	2.130	2.538	3.335	4.260
60	1.369	1.958	2.333	3.066	3.916	1.471	2.103	2.506	3.293	4.206
65	1.359	1.943	2.315	3.042	3.886	1.455	2.080	2.478	3.257	4.160
70	1.349	1.929	2.299	3.021	3.859	1.440	2.060	2.454	3.225	4.120
75	1.341	1.917	2.285	3.002	3.835	1.428	2.042	2.433	3.197	4.084
80	1.334	1.907	2.272	2.986	3.814	1.417	2.026	2.414	3.173	4.053
85	1.327	1.897	2.261	2.971	3.795	1.407	2.012	2.397	3.150	4.024
90	1.321	1.889	2.251	2.958	3.778	1.398	1.999	2.382	3.130	3.999
95	1.315	1.881	2.241	2.945	3.763	1.390	1.987	2.368	3.112	3.976
100	1.311	1.874	2.233	2.934	3.748	1.383	1.977	2.355	3.096	3.954
110	1.302	1.861	2.218	2.915	3.723	1.369	1.958	2.333	3.066	3.917
120	1.294	1.850	2.205	2.898	3.702	1.358	1.942	2.314	3.041	3.885
130	1.288	1.841	2.194	2.883	3.683	1.349	1.928	2.298	3.019	3.857
140	1.282	1.833	2.184	2.870	3.666	1.340	1.916	2.283	3.000	3.833
150	1.277	1.825	2.175	2.859	3.652	1.332	1.905	2.270	2.983	3.811
160	1.272	1.819	2.167	2.848	3.638	1.326	1.896	2.259	2.968	3.792
170	1.268	1.813	2.160	2.839	3.527	1.320	1.887	2.248	2.955	3.774
180	1.264	1.808	2.154	2.831	3.616	1.314	1.879	2.239	2.942	3.759
190	1.261	1.803	2.148	2.823	3.606	1.309	1.872	2.230	2.931	3.744
200	1.258	1.798	2.143	2.816	3.597	1.304	1.865	2.222	2.921	3.731
250	1.245	1.780	2.121	2.788	3.561	1.286	1.839	2.191	2.880	3.678
300	1.236	1.767	2.106	2.767	3.535	1.273	1.820	2.169	2.850	3.641
400	1.223	1.749	2.084	2.739	3.499	1.255	1.794	2.138	2.809	3.589
500	1.215	1.737	2.070	2.721	3.475	1.243	1.777	2.117	2.783	3.555
600	1.209	1.729	2.060	2.707	3.458	1.234	1.764	2.102	2.763	3.530
700	1.204	1.722	2.052	2.697	3.445	1.227	1.755	2.091	2.748	3.511
800	1.201	1.717	2.046	2.688	3.434	1.222	1.747	2.082	2.736	3.495
900	1.198	1.712	2.040	2.682	3.426	1.218	1.741	2.075	2.726	3.483
1000	1.195	1.709	2.036	2.676	3.418	1.214	1.736	2.068	2.718	3.472
∞	1.150	1.645	1.960	2.576	3.291	1.150	1.645	1.960	2.576	3.291

TABLE A-6b Factors for One-sided Tolerance Limits for Normal Distributions[a]

Factors K such that the probability is γ that at least a proportion P of the distribution will be less than $\bar{X} + Ks$ (or greater than $\bar{X} - Ks$), where $\bar{X}$ and s are estimates of the mean and the standard deviation computed from a sample size of n.

	$\gamma = 0.75$					$\gamma = 0.90$				
n \ P	0.75	0.90	0.95	0.99	0.999	0.75	0.90	0.95	0.99	0.999
3	1.464	2.501	3.152	4.396	5.805	2.602	4.258	5.310	7.340	9.651
4	1.256	2.134	2.680	3.726	4.910	1.972	3.187	3.957	5.437	7.128
5	1.152	1.961	2.463	3.421	4.507	1.698	2.742	3.400	4.666	6.112
6	1.087	1.860	2.336	3.243	4.273	1.540	2.494	3.091	4.242	5.556
7	1.043	1.791	2.250	3.126	4.118	1.435	2.333	2.894	3.972	5.201
8	1.010	1.740	2.190	3.042	4.008	1.360	2.219	2.755	3.783	4.955
9	0.984	1.702	2.141	2.977	3.924	1.302	2.133	2.649	3.641	4.772
10	0.964	1.671	2.103	2.927	3.858	1.257	2.065	2.568	3.532	4.629
11	0.947	1.646	2.073	2.885	3.804	1.219	2.012	2.503	3.444	4.515
12	0.933	1.624	2.048	2.851	3.760	1.188	1.966	2.448	3.371	4.420
13	0.919	1.606	2.026	2.822	3.722	1.162	1.928	2.403	3.310	4.341
14	0.909	1.591	2.007	2.796	3.690	1.139	1.895	2.363	3.257	4.274
15	0.899	1.577	1.991	2.776	3.661	1.119	1.866	2.329	3.212	4.215
16	0.891	1.566	1.977	2.756	3.637	1.101	1.842	2.299	3.172	4.164
17	0.883	1.554	1.964	2.739	3.615	1.085	1.820	2.272	3.136	4.118
18	0.876	1.544	1.951	2.723	3.595	1.071	1.800	2.249	3.106	4.078
19	0.870	1.536	1.942	2.710	3.577	1.058	1.781	2.228	3.078	4.041
20	0.865	1.528	1.933	2.697	3.561	1.046	1.765	2.208	3.052	4.009
21	0.859	1.520	1.923	2.686	3.545	1.035	1.750	2.190	3.028	3.979
22	0.854	1.514	1.916	2.675	3.532	1.025	1.736	2.174	3.007	3.952
23	0.849	1.508	1.907	2.665	3.520	1.016	1.724	2.159	2.987	3.927
24	0.845	1.502	1.901	2.656	3.509	1.007	1.712	2.145	2.969	3.904
25	0.842	1.496	1.895	2.647	3.497	0.999	1.702	2.132	2.952	3.882
30	0.825	1.475	1.869	2.613	3.454	0.966	1.657	2.080	2.884	3.794
35	0.812	1.458	1.849	2.588	3.421	0.942	1.623	2.041	2.833	3.730
40	0.803	1.445	1.834	2.568	3.395	0.923	1.598	2.010	2.793	3.679
45	0.795	1.435	1.821	2.552	3.375	0.908	1.577	1.986	2.762	3.638
50	0.788	1.426	1.811	2.538	3.358	0.894	1.560	1.965	2.735	3.604

[a] Adapted by permission from *Industrial Quality Control*, Vol. XIV, No. 10 (April 1958), from the article "Tables for One-Sided Statistical Tolerance Limits," by G. J. Lieberman. Permission granted by the American Society for Quality Control.

TABLE A-6b *(cont.)*

* The two starred values have been corrected to the values given by D. B. Owen in "Factors for One-Sided Tolerance Limits and for Variables Sampling Plans," *Sandia Corporation Monograph SCR-607*, available from the Clearing House for Federal Scientific and Technical Information, U.S. Department of Commerce, Springfield, VA 22151. The Owen Tables indicate other errors in the table below, not exceeding 4 in the last digit.

	γ = 0.95					γ = 0.99				
n \ P	0.75	0.90	0.95	0.99	0.999	0.75	0.90	0.95	0.99	0.999
3	3.804	6.158	7.655	10.552	13.857	—	—	—	—	—
4	2.619	4.163	5.145	7.042	9.215	—	—	—	—	—
5	2.149	3.407	4.202	5.741	7.501	—	—	—	—	—
6	1.895	3.006	3.707	5.062	6.612	2.849	4.408	5.409	7.334	9.550*
7	1.732	2.755	3.399	4.641	6.061	2.490	3.856	4.730	6.411	8.348
8	1.617	2.582	3.188	4.353	5.686	2.252	3.496	4.287	5.811	7.566
9	1.532	2.454	3.031	4.143	5.414	2.085	3.242	3.971	5.389	7.014
10	1.465	2.355	2.911	3.981	5.203	1.954	3.048	3.739	5.075	6.603
11	1.411	2.275	2.815	3.852	5.036	1.854	2.897	3.557	4.828	6.284
12	1.366	2.210	2.736	3.747	4.900	1.771	2.773	3.410	4.633	6.032
13	1.329	2.155	2.670	3.659	4.787	1.702	2.677	3.290	4.472	5.826
14	1.296	2.108	2.614	3.585	4.690	1.645	2.592	3.189	4.336	5.651
15	1.268	2.068	2.566	3.520	4.607	1.596	2.521	3.102	4.224	5.507
16	1.242	2.032	2.523	3.463	4.534	1.553	2.458	3.028	4.124	5.374
17	1.220	2.001	2.486	3.415	4.471	1.514	2.405	2.962	4.038	5.268
18	1.200	1.974	2.453	3.370	4.415	1.481	2.357	2.906	3.961	5.167
19	1.183	1.949	2.423	3.331	4.364	1.450	2.315	2.855	3.893	5.078
20	1.167	1.926	2.396	3.295	4.319	1.424	2.275	2.807	3.832	5.003
21	1.152	1.905	2.371	3.262	4.276	1.397	2.241	2.768	3.776	4.932
22	1.138	1.887	2.350	3.233	4.238	1.376	2.208	2.729	3.727	4.866
23	1.126	1.869	2.329	3.206	4.204	1.355	2.179	2.693	3.680	4.806
24	1.114	1.853	2.309	3.181	4.171	1.336	2.154	2.663	3.638	4.755
25	1.103	1.838	2.292	3.158	4.143	1.319	2.129	2.632	3.601	4.706
30	1.059	1.778	2.220	3.064	4.022	1.249	2.029	2.516	3.446	4.508
35	1.025	1.732	2.166	2.994	3.934	1.195	1.957	2.431	3.334	4.364
40	0.999	1.697	2.126	2.941	3.866	1.154	1.902	2.365	3.250	4.255
45	0.978	1.669	2.092	2.897	3.811	1.122	1.857	2.313	3.181	4.168
50	0.961	1.646	2.065	2.863	3.766	1.096	1.821	2.269*	3.124	4.096

TABLE A-7 Critical Values for the Kolmogorov–Smirnov One-Sample Test[a]

This table provides the critical value, D, for sample size N and level of significance α.

Sample size (N)	Level of Significance α				
	.20	.15	.10	.05	.01
1	.900	.925	.950	.975	.995
2	.684	.726	.776	.842	.929
3	.565	.597	.642	.708	.828
4	.494	.525	.564	.624	.733
5	.446	.474	.510	.565	.669
6	.410	.436	.470	.521	.618
7	.381	.405	.438	.486	.577
8	.358	.381	.411	.457	.543
9	.339	.360	.388	.432	.514
10	.322	.342	.368	.410	.490
11	.307	.326	.352	.391	.468
12	.295	.313	.338	.375	.450
13	.284	.302	.325	.361	.433
14	.274	.292	.314	.349	.418
15	.266	.283	.304	.338	.404
16	.258	.274	.295	.328	.392
17	.250	.266	.286	.318	.381
18	.244	.259	.278	.309	.371
19	.237	.252	.272	.301	.363
20	.231	.246	.264	.294	.356
25	.21	.22	.24	.27	.32
30	.19	.20	.22	.24	.29
35	.18	.19	.21	.23	.27
Over 35	$\frac{1.07}{\sqrt{N}}$	$\frac{1.14}{\sqrt{N}}$	$\frac{1.22}{\sqrt{N}}$	$\frac{1.36}{\sqrt{N}}$	$\frac{1.63}{\sqrt{N}}$

[a] Adapted from F. J. Massey, Jr., "The Kolmogorov–Smirnov Test for Goodness of Fit," *Journal of the American Statistical Association*, Vol. 46 (1951), p. 70, with the permission of the publisher.

TABLE A-8 Pearson Type III Deviates (K) for Selected Skew Coefficients (G) and Excedence Probabilities (P)

P	G =0.0	G =0.1	G =0.2	G =0.3	G =0.4	G =0.5	G =0.6
0.9999	-3.71902	-3.50703	-3.29921	-3.09631	-2.89907	-2.70836	-2.52507
0.9995	-3.29053	-3.12767	-2.96698	-2.80889	-2.65390	-2.50257	-2.35549
0.9990	-3.09023	-2.94834	-2.80786	-2.66915	-2.53261	-2.39867	-2.26780
0.9980	-2.87816	-2.75706	-2.63672	-2.51741	-2.39942	-2.28311	-2.16884
0.9950	-2.57583	-2.48187	-2.38795	-2.29423	-2.20092	-2.10825	-2.01644
0.9900	-2.32635	-2.25258	-2.17840	-2.10394	-2.02933	-1.95472	-1.88029
0.9800	-2.05375	-1.99973	-1.94499	-1.88959	-1.83361	-1.77716	-1.72033
0.9750	-1.95996	-1.91219	-1.86360	-1.81427	-1.76427	-1.71366	-1.66253
0.9600	-1.75069	-1.71580	-1.67999	-1.64329	-1.60574	-1.56740	-1.52830
0.9500	-1.64485	-1.61594	-1.58607	-1.55527	-1.52357	-1.49101	-1.45762
0.9000	-1.28155	-1.27037	-1.25824	-1.24516	-1.23114	-1.21618	-1.20028
0.8000	-0.84162	-0.84611	-0.84986	-0.85285	-0.85508	-0.85653	-0.85718
0.7000	-0.52440	-0.53624	-0.54757	-0.55839	-0.56867	-0.57840	-0.58757
0.6000	-0.25335	-0.26882	-0.28403	-0.29897	-0.31362	-0.32796	-0.34198
0.5704	-0.17733	-0.19339	-0.20925	-0.22492	-0.24037	-0.25558	-0.27047
0.5000	0.0	-0.01662	-0.03325	-0.04993	-0.06651	-0.08302	-0.09945
0.4296	0.17733	0.16111	0.14472	0.12820	0.11154	0.09478	0.07791
0.4000	0.25335	0.23763	0.22168	0.20552	0.18916	0.17261	0.15589
0.3000	0.52440	0.51207	0.49927	0.48600	0.47228	0.45812	0.44352
0.2000	0.84162	0.83639	0.83044	0.82377	0.81638	0.80829	0.79950
0.1000	1.28155	1.29178	1.30105	1.30936	1.31671	1.32309	1.32850
0.0500	1.64485	1.67279	1.69971	1.72562	1.75048	1.77428	1.79701
0.0400	1.75069	1.78462	1.81756	1.84949	1.88039	1.91022	1.93896
0.0250	1.95996	2.00688	2.05290	2.09795	2.14202	2.18505	2.22702
0.0200	2.05375	2.10697	2.15935	2.21081	2.26133	2.31084	2.35931
0.0100	2.32635	2.39961	2.47226	2.54421	2.61539	2.68572	2.75514
0.0050	2.57583	2.66965	2.76321	2.85636	2.94900	3.04102	3.13232
0.0020	2.87816	2.99978	3.12169	3.24371	3.36566	3.48737	3.60872
0.0010	3.09023	3.23322	3.37703	3.52139	3.66608	3.81090	3.95567
0.0005	3.29053	3.45513	3.62113	3.78820	3.95605	4.12443	4.29311
0.0001	3.71902	3.93453	4.15301	4.37394	4.59687	4.82141	5.04718

P	G =0.7	G =0.8	G =0.9	G =1.0	G =1.1	G =1.2	G =1.3
0.9999	-2.35015	-2.18448	-2.02891	-1.88410	-1.75053	-1.62838	-1.51752
0.9995	-2.21328	-2.07661	-1.94611	-1.82241	-1.70603	-1.59738	-1.49673
0.9990	-2.14053	-2.01739	-1.89894	-1.78572	-1.67825	-1.57695	-1.48216
0.9980	-2.05701	-1.94806	-1.84244	-1.74062	-1.64305	-1.55016	-1.46232
0.9950	-1.92580	-1.83660	-1.74919	-1.66390	-1.58110	-1.50114	-1.42439
0.9900	-1.80621	-1.73271	-1.66001	-1.58838	-1.51808	-1.44942	-1.38267
0.9800	-1.66325	-1.60604	-1.54886	-1.49188	-1.43529	-1.37929	-1.32412
0.9750	-1.61099	-1.55914	-1.50712	-1.45507	-1.40314	-1.35153	-1.30042
0.9600	-1.48852	-1.44813	-1.40720	-1.36584	-1.32414	-1.28225	-1.24028
0.9500	-1.42345	-1.38855	-1.35299	-1.31684	-1.28019	-1.24313	-1.20578
0.9000	-1.18347	-1.16574	-1.14712	-1.12762	-1.10726	-1.08608	-1.06413
0.8000	-0.85703	-0.85607	-0.85426	-0.85161	-0.84809	-0.84369	-0.83841
0.7000	-0.59615	-0.60412	-0.61146	-0.61815	-0.62415	-0.62944	-0.63400
0.6000	-0.35565	-0.36889	-0.38186	-0.39434	-0.40638	-0.41794	-0.42899
0.5704	-0.28516	-0.29961	-0.31368	-0.32740	-0.34075	-0.35370	-0.36620
0.5000	-0.11578	-0.13199	-0.14807	-0.16397	-0.17968	-0.19517	-0.21040
0.4296	0.06097	0.04397	0.02693	0.00987	-0.00719	-0.02421	-0.04116
0.4000	0.13901	0.12199	0.10486	0.08763	0.07032	0.05297	0.03560
0.3000	0.42851	0.41309	0.39729	0.38111	0.36458	0.34772	0.33054
0.2000	0.79002	0.77986	0.76902	0.75752	0.74537	0.73257	0.71915
0.1000	1.33294	1.33640	1.33889	1.34039	1.34092	1.34047	1.33904
0.0500	1.81864	1.83916	1.85856	1.87683	1.89395	1.90992	1.92472
0.0400	1.96660	1.99311	2.01848	2.04269	2.06573	2.08758	2.10823
0.0250	2.26790	2.30764	2.34623	2.38364	2.41984	2.45482	2.48855
0.0200	2.40670	2.45298	2.49811	2.54206	2.58480	2.62631	2.66657
0.0100	2.82359	2.89101	2.95735	3.02256	3.08660	3.14944	3.21103
0.0050	3.22281	3.31243	3.40109	3.48874	3.57530	3.66073	3.74497
0.0020	3.72957	3.84981	3.96932	4.08802	4.20582	4.32263	4.43839
0.0010	4.10022	4.24439	4.38807	4.53112	4.67344	4.81492	4.95549
0.0005	4.46189	4.63057	4.79899	4.96701	5.13449	5.30130	5.46735
0.0001	5.27389	5.50124	5.72899	5.95691	6.18480	6.41249	6.63980

TABLE A-8 *(cont.)*

P	G =1.4	G =1.5	G =1.6	G =1.7	G =1.8	G =1.9	G =2.0
0.9999	-1.41753	-1.32774	-1.24728	-1.17520	-1.11054	-1.05239	-0.99990
0.9995	-1.40413	-1.31944	-1.24235	-1.17240	-1.10901	-1.05159	-0.99950
0.9990	-1.39408	-1.31275	-1.23805	-1.16974	-1.10743	-1.05068	-0.99900
0.9980	-1.37981	-1.30279	-1.23132	-1.16534	-1.10465	-1.04898	-0.99800
0.9950	-1.35114	-1.28167	-1.21618	-1.15477	-1.09749	-1.04427	-0.99499
0.9900	-1.31815	-1.25611	-1.19680	-1.14042	-1.08711	-1.03695	-0.98995
0.9800	-1.26999	-1.21716	-1.16584	-1.11628	-1.06864	-1.02311	-0.97980
0.9750	-1.25004	-1.20059	-1.15229	-1.10537	-1.06001	-1.01640	-0.97468
0.9600	-1.19842	-1.15682	-1.11566	-1.07513	-1.03543	-0.99672	-0.95918
0.9500	-1.16827	-1.13075	-1.09338	-1.05631	-1.01973	-0.98381	-0.94871
0.9000	-1.04144	-1.01810	-0.99418	-0.96977	-0.94496	-0.91988	-0.89464
0.8000	-0.83223	-0.82516	-0.81720	-0.80837	-0.79868	-0.78816	-0.77686
0.7000	-0.63779	-0.64080	-0.64300	-0.64436	-0.64488	-0.64453	-0.64333
0.6000	-0.43949	-0.44942	-0.45873	-0.46739	-0.47538	-0.48265	-0.48917
0.5704	-0.37824	-0.38977	-0.40075	-0.41116	-0.42095	-0.43008	-0.43854
0.5000	-0.22535	-0.23996	-0.25422	-0.26808	-0.28150	-0.29443	-0.30685
0.4296	-0.05803	-0.07476	-0.09132	-0.10769	-0.12381	-0.13964	-0.15516
0.4000	0.01824	0.00092	-0.01631	-0.03344	-0.05040	-0.06718	-0.08371
0.3000	0.31307	0.29535	0.27740	0.25925	0.24094	0.22250	0.20397
0.2000	0.70512	0.69050	0.67532	0.65959	0.64335	0.62662	0.60944
0.1000	1.33665	1.33330	1.32900	1.32376	1.31760	1.31054	1.30259
0.0500	1.93836	1.95083	1.96213	1.97227	1.98124	1.98906	1.99573
0.0400	2.12768	2.14591	2.16293	2.17873	2.19332	2.20670	2.21888
0.0250	2.52102	2.55222	2.58214	2.61076	2.63810	2.66413	2.68888
0.0200	2.70556	2.74325	2.77964	2.81472	2.84848	2.88091	2.91202
0.0100	3.27134	3.33035	3.38804	3.44438	3.49935	3.55295	3.60517
0.0050	3.82798	3.90973	3.99016	4.06926	4.14700	4.22336	4.29832
0.0020	4.55304	4.66651	4.77875	4.88971	4.99937	5.10768	5.21461
0.0010	5.09505	5.23353	5.37087	5.50701	5.64190	5.77549	5.90776
0.0005	5.63252	5.79673	5.95990	6.12196	6.28285	6.44251	6.60090
0.0001	6.86661	7.09277	7.31818	7.54272	7.76632	7.98888	8.21034

P	G =2.1	G =2.2	G =2.3	G =2.4	G =2.5	G =2.6	G =2.7
0.9999	-0.95234	-0.90908	-0.86956	-0.83333	-0.80000	-0.76923	-0.74074
0.9995	-0.95215	-0.90899	-0.86952	-0.83331	-0.79999	-0.76923	-0.74074
0.9990	-0.95188	-0.90885	-0.86945	-0.83328	-0.79998	-0.76922	-0.74074
0.9980	-0.95131	-0.90854	-0.86929	-0.83320	-0.79994	-0.76920	-0.74073
0.9950	-0.94945	-0.90742	-0.86863	-0.83283	-0.79973	-0.76909	-0.74067
0.9900	-0.94607	-0.90521	-0.86723	-0.83196	-0.79921	-0.76878	-0.74049
0.9800	-0.93878	-0.90009	-0.86371	-0.82959	-0.79765	-0.76779	-0.73987
0.9750	-0.93495	-0.89728	-0.86169	-0.82817	-0.79667	-0.76712	-0.73943
0.9600	-0.92295	-0.88814	-0.85486	-0.82315	-0.79306	-0.76456	-0.73765
0.9500	-0.91458	-0.88156	-0.84976	-0.81927	-0.79015	-0.76242	-0.73610
0.9000	-0.86938	-0.84422	-0.81929	-0.79472	-0.77062	-0.74709	-0.72422
0.8000	-0.76482	-0.75211	-0.73880	-0.72495	-0.71067	-0.69602	-0.68111
0.7000	-0.64125	-0.63833	-0.63456	-0.62999	-0.62463	-0.61854	-0.61176
0.6000	-0.49494	-0.49991	-0.50409	-0.50744	-0.50999	-0.51171	-0.51263
0.5704	-0.44628	-0.45329	-0.45953	-0.46499	-0.46966	-0.47353	-0.47660
0.5000	-0.31872	-0.32999	-0.34063	-0.35062	-0.35992	-0.36852	-0.37640
0.4296	-0.17030	-0.18504	-0.19933	-0.21313	-0.22642	-0.23915	-0.25129
0.4000	-0.09997	-0.11590	-0.13148	-0.14665	-0.16138	-0.17564	-0.18939
0.3000	0.18540	0.16682	0.14827	0.12979	0.11143	0.09323	0.07523
0.2000	0.59183	0.57383	0.55549	0.53683	0.51789	0.49872	0.47934
0.1000	1.29377	1.28412	1.27365	1.26240	1.25039	1.23766	1.22422
0.0500	2.00128	2.00570	2.00903	2.01128	2.01247	2.01263	2.01177
0.0400	2.22986	2.23967	2.24831	2.25581	2.26217	2.26743	2.27160
0.0250	2.71234	2.73451	2.75541	2.77506	2.79345	2.81062	2.82658
0.0200	2.94181	2.97028	2.99744	3.02330	3.04787	3.07116	3.09320
0.0100	3.65600	3.70543	3.75347	3.80013	3.84540	3.88930	3.93183
0.0050	4.37186	4.44398	4.51467	4.58393	4.65176	4.71815	4.78313
0.0020	5.32014	5.42426	5.52694	5.62818	5.72796	5.82629	5.92316
0.0010	6.03865	6.16816	6.29626	6.42292	6.54814	6.67191	6.79421
0.0005	6.75798	6.91370	7.06804	7.22098	7.37250	7.52258	7.67121
0.0001	8.43064	8.64971	8.86753	9.08403	9.29920	9.51301	9.72543

TABLE A-8 *(cont.)*

P	G =-0.0	G =-0.1	G =-0.2	G =-0.3	G =-0.4	G =-0.5	G =-0.6
0.9999	-3.71902	-3.93453	-4.15301	-4.37394	-4.59687	-4.82141	-5.04718
0.9995	-3.29053	-3.45513	-3.62113	-3.78820	-3.95605	-4.12443	-4.29311
0.9990	-3.09023	-3.23322	-3.37703	-3.52139	-3.66608	-3.81090	-3.95567
0.9980	-2.87816	-2.99978	-3.12169	-3.24371	-3.36566	-3.48737	-3.60872
0.9950	-2.57583	-2.66965	-2.76321	-2.85636	-2.94900	-3.04102	-3.13232
0.9900	-2.32635	-2.39961	-2.47226	-2.54421	-2.61539	-2.68572	-2.75514
0.9800	-2.05375	-2.10697	-2.15935	-2.21081	-2.26133	-2.31084	-2.35931
0.9750	-1.95996	-2.00688	-2.05290	-2.09795	-2.14202	-2.18505	-2.22702
0.9600	-1.75069	-1.78462	-1.81756	-1.84949	-1.88039	-1.91022	-1.93896
0.9500	-1.64485	-1.67279	-1.69971	-1.72562	-1.75048	-1.77428	-1.79701
0.9000	-1.28155	-1.29178	-1.30105	-1.30936	-1.31671	-1.32309	-1.32850
0.8000	-0.84162	-0.83639	-0.83044	-0.82377	-0.81638	-0.80829	-0.79950
0.7000	-0.52440	-0.51207	-0.49927	-0.48600	-0.47228	-0.45812	-0.44352
0.6000	-0.25335	-0.23763	-0.22169	-0.20552	-0.18916	-0.17261	-0.15589
0.5704	-0.17733	-0.16111	-0.14472	-0.12820	-0.11154	-0.09478	-0.07791
0.5000	0.0	0.01662	0.03325	0.04993	0.06651	0.08302	0.09945
0.4296	0.17733	0.19339	0.20925	0.22492	0.24037	0.25558	0.27047
0.4000	0.25335	0.26882	0.28403	0.29897	0.31362	0.32796	0.34198
0.3000	0.52440	0.53624	0.54757	0.55839	0.56867	0.57840	0.58757
0.2000	0.84162	0.84611	0.84986	0.85285	0.85508	0.85653	0.85718
0.1000	1.28155	1.27037	1.25824	1.24516	1.23114	1.21618	1.20028
0.0500	1.64485	1.61594	1.58607	1.55527	1.52357	1.49101	1.45762
0.0400	1.75069	1.71580	1.67999	1.64329	1.60574	1.56740	1.52830
0.0250	1.95996	1.91219	1.86360	1.81427	1.76427	1.71366	1.66253
0.0200	2.05375	1.99973	1.94499	1.88959	1.83361	1.77716	1.72033
0.0100	2.32635	2.25258	2.17840	2.10394	2.02933	1.95472	1.88029
0.0050	2.57583	2.48187	2.38795	2.29423	2.20092	2.10825	2.01644
0.0020	2.87816	2.75706	2.63672	2.51741	2.39942	2.28311	2.16884
0.0010	3.09023	2.94834	2.80786	2.66915	2.53261	2.39867	2.26780
0.0005	3.29053	3.12767	2.96698	2.80889	2.65390	2.50257	2.35549
0.0001	3.71902	3.50703	3.29921	3.09631	2.89907	2.70836	2.52507

P	G1=-0.7	G1=-0.8	G1=-0.9	G1=-1.0	G1=-1.1	G1=-1.2	G1=-1.3
0.9999	-5.27389	-5.50124	-5.72899	-5.95691	-6.18480	-6.41249	-6.63980
0.9995	-4.46189	-4.63057	-4.79899	-4.96701	-5.13449	-5.30130	-5.46735
0.9990	-4.10022	-4.24439	-4.38807	-4.53112	-4.67344	-4.81492	-4.95549
0.9980	-3.72957	-3.84981	-3.96932	-4.08802	-4.20582	-4.32263	-4.43839
0.9950	-3.22281	-3.31243	-3.40109	-3.48874	-3.57530	-3.66073	-3.74497
0.9900	-2.82359	-2.89101	-2.95735	-3.02256	-3.08660	-3.14944	-3.21103
0.9800	-2.40670	-2.45298	-2.49811	-2.54206	-2.58480	-2.62631	-2.66657
0.9750	-2.26790	-2.30764	-2.34623	-2.38364	-2.41984	-2.45482	-2.48855
0.9600	-1.96660	-1.99311	-2.01848	-2.04269	-2.06573	-2.08758	-2.10823
0.9500	-1.81864	-1.83916	-1.85856	-1.87683	-1.89395	-1.90992	-1.92472
0.9000	-1.33294	-1.33640	-1.33889	-1.34039	-1.34092	-1.34047	-1.33904
0.8000	-0.79002	-0.77986	-0.76902	-0.75752	-0.74537	-0.73257	-0.71915
0.7000	-0.42851	-0.41309	-0.39729	-0.38111	-0.36458	-0.34772	-0.33054
0.6000	-0.13901	-0.12199	-0.10486	-0.08763	-0.07032	-0.05297	-0.03560
0.5704	-0.06097	-0.04397	-0.02693	-0.00987	0.00719	0.02421	0.04116
0.5000	0.11578	0.13199	0.14807	0.16397	0.17968	0.19517	0.21040
0.4296	0.28516	0.29961	0.31368	0.32740	0.34075	0.35370	0.36620
0.4000	0.35565	0.36889	0.38186	0.39434	0.40638	0.41794	0.42899
0.3000	0.59615	0.60412	0.61146	0.61815	0.62415	0.62944	0.63400
0.2000	0.85703	0.85607	0.85426	0.85161	0.84809	0.84369	0.83841
0.1000	1.18347	1.16574	1.14712	1.12762	1.10726	1.08608	1.06413
0.0500	1.42345	1.38855	1.35299	1.31684	1.28019	1.24313	1.20578
0.0400	1.48852	1.44813	1.40720	1.36584	1.32414	1.28225	1.24028
0.0250	1.61099	1.55914	1.50712	1.45507	1.40314	1.35153	1.30042
0.0200	1.66325	1.60604	1.54886	1.49188	1.43529	1.37929	1.32412
0.0100	1.80621	1.73271	1.66001	1.58838	1.51808	1.44942	1.38267
0.0050	1.92580	1.83660	1.74919	1.66390	1.58110	1.50114	1.42439
0.0020	2.05701	1.94806	1.84244	1.74062	1.64305	1.55016	1.46232
0.0010	2.14053	2.01739	1.89894	1.78572	1.67825	1.57695	1.48216
0.0005	2.21328	2.07661	1.94611	1.82241	1.70603	1.59738	1.49673
0.0001	2.35015	2.18448	2.02891	1.88410	1.75053	1.62838	1.51752

TABLE A-8 (*cont.*)

P	G =-1.4	G =-1.5	G =-1.6	G =-1.7	G =-1.8	G =-1.9	G =-2.0
0.9999	-6.86661	-7.09277	-7.31818	-7.54272	-7.76632	-7.98888	-8.21034
0.9995	-5.63252	-5.79673	-5.95990	-6.12196	-6.28285	-6.44251	-6.60090
0.9990	-5.09505	-5.23353	-5.37087	-5.50701	-5.64190	-5.77549	-5.90776
0.9980	-4.55304	-4.66651	-4.77875	-4.88971	-4.99937	-5.10768	-5.21461
0.9950	-3.82798	-3.90973	-3.99016	-4.06926	-4.14700	-4.22336	-4.29832
0.9900	-3.27134	-3.33035	-3.38804	-3.44438	-3.49935	-3.55295	-3.60517
0.9800	-2.70556	-2.74325	-2.77964	-2.81472	-2.84848	-2.88091	-2.91202
0.9750	-2.52102	-2.55222	-2.58214	-2.61076	-2.63810	-2.66413	-2.68888
0.9600	-2.12768	-2.14591	-2.16293	-2.17873	-2.19332	-2.20670	-2.21888
0.9500	-1.93836	-1.95083	-1.96213	-1.97227	-1.98124	-1.98906	-1.99573
0.9000	-1.33665	-1.33330	-1.32900	-1.32376	-1.31760	-1.31054	-1.30259
0.8000	-0.70512	-0.69050	-0.67532	-0.65959	-0.64335	-0.62662	-0.60944
0.7000	-0.31307	-0.29535	-0.27740	-0.25925	-0.24094	-0.22250	-0.20397
0.6000	-0.01824	-0.00092	0.01631	0.03344	0.05040	0.06718	0.08371
0.5704	0.05803	0.07476	0.09132	0.10769	0.12381	0.13964	0.15516
0.5000	0.22535	0.23996	0.25422	0.26808	0.28150	0.29443	0.30685
0.4296	0.37824	0.38977	0.40075	0.41116	0.42095	0.43008	0.43854
0.4000	0.43949	0.44942	0.45873	0.46739	0.47538	0.48265	0.48917
0.3000	0.63779	0.64080	0.64300	0.64436	0.64488	0.64453	0.64333
0.2000	0.83223	0.82516	0.81720	0.80837	0.79868	0.78816	0.77686
0.1000	1.04144	1.01810	0.99418	0.96977	0.94496	0.91988	0.89464
0.0500	1.16827	1.13075	1.09338	1.05631	1.01973	0.98381	0.94871
0.0400	1.19842	1.15682	1.11566	1.07513	1.03543	0.99672	0.95918
0.0250	1.25004	1.20059	1.15229	1.10537	1.06001	1.01640	0.97468
0.0200	1.26999	1.21716	1.16584	1.11628	1.06864	1.02311	0.97980
0.0100	1.31815	1.25611	1.19680	1.14042	1.08711	1.03695	0.98995
0.0050	1.35114	1.28167	1.21618	1.15477	1.09749	1.04427	0.99499
0.0020	1.37981	1.30279	1.23132	1.16534	1.10465	1.04898	0.99800
0.0010	1.39408	1.31275	1.23805	1.16974	1.10743	1.05068	0.99900
0.0005	1.40413	1.31944	1.24235	1.17240	1.10901	1.05159	0.99950
0.0001	1.41753	1.32774	1.24728	1.17520	1.11054	1.05239	0.99990

P	G =-2.1	G =-2.2	G =-2.3	G =-2.4	G =-2.5	G =-2.6	G1=-2.7
0.9999	-8.43064	-8.64971	-8.86753	-9.08403	-9.29920	-9.51301	-9.72543
0.9995	-6.75798	-6.91370	-7.06804	-7.22098	-7.37250	-7.52258	-7.67121
0.9990	-6.03865	-6.16816	-6.29626	-6.42292	-6.54814	-6.67191	-6.79421
0.9980	-5.32014	-5.42426	-5.52694	-5.62818	-5.72796	-5.82629	-5.92316
0.9950	-4.37186	-4.44398	-4.51467	-4.58393	-4.65176	-4.71815	-4.78313
0.9900	-3.65600	-3.70543	-3.75347	-3.80013	-3.84540	-3.88930	-3.93183
0.9800	-2.94181	-2.97028	-2.99744	-3.02330	-3.04787	-3.07116	-3.09320
0.9750	-2.71234	-2.73451	-2.75541	-2.77506	-2.79345	-2.81062	-2.82658
0.9600	-2.22986	-2.23967	-2.24831	-2.25581	-2.26217	-2.26743	-2.27160
0.9500	-2.00128	-2.00570	-2.00903	-2.01128	-2.01247	-2.01263	-2.01177
0.9000	-1.29377	-1.28412	-1.27365	-1.26240	-1.25039	-1.23766	-1.22422
0.8000	-0.59183	-0.57383	-0.55549	-0.53683	-0.51789	-0.49872	-0.47934
0.7000	-0.18540	-0.16682	-0.14827	-0.12979	-0.11143	-0.09323	-0.07523
0.6000	0.09997	0.11590	0.13148	0.14665	0.16138	0.17564	0.18939
0.5704	0.17030	0.18504	0.19933	0.21313	0.22642	0.23915	0.25129
0.5000	0.31872	0.32999	0.34063	0.35062	0.35992	0.36852	0.37640
0.4296	0.44628	0.45329	0.45953	0.46499	0.46966	0.47353	0.47660
0.4000	0.49494	0.49991	0.50409	0.50744	0.50999	0.51171	0.51263
0.3000	0.64125	0.63833	0.63456	0.62999	0.62463	0.61854	0.61176
0.2000	0.76482	0.75211	0.73880	0.72495	0.71067	0.69602	0.68111
0.1000	0.86938	0.84422	0.81929	0.79472	0.77062	0.74709	0.72422
0.0500	0.91458	0.88156	0.84976	0.81927	0.79015	0.76242	0.73610
0.0400	0.92295	0.88814	0.85486	0.82315	0.79306	0.76456	0.73765
0.0250	0.93495	0.89728	0.86169	0.82817	0.79667	0.76712	0.73943
0.0200	0.93878	0.90009	0.86371	0.82959	0.79765	0.76779	0.73987
0.0100	0.94607	0.90521	0.86723	0.83196	0.79921	0.76878	0.74049
0.0050	0.94945	0.90742	0.86863	0.83283	0.79973	0.76909	0.74067
0.0020	0.95131	0.90854	0.86929	0.83320	0.79994	0.76920	0.74073
0.0010	0.95188	0.90885	0.86945	0.83328	0.79998	0.76922	0.74074
0.0005	0.95215	0.90899	0.86952	0.83331	0.79999	0.76923	0.74074
0.0001	0.95234	0.90908	0.86956	0.83333	0.80000	0.76923	0.74074

TABLE B-1 Sediment Yield Data

X_1	X_2	X_3	X_4	Y
.135	4.000	40.000	.000	.140
.135	1.600	40.000	.000	.120
.101	1.900	22.000	.000	.210
.353	17.600	2.000	-17.000	.150
.353	20.600	1.000	-18.000	.076
.492	31.000	.000	-14.000	.660
.466	70.000	23.000	.000	2.670
.466	32.200	57.000	.000	.610
.833	41.400	4.000	-12.000	.690
.085	14.900	22.000	19.000	2.310
.085	17.200	2.000	13.000	2.650
.085	30.500	15.000	12.000	2.370
.193	6.300	1.000	.000	.510
.167	14.000	3.000	16.000	1.420
.235	2.400	1.000	44.000	1.650
.448	19.700	3.000	-16.000	.140
.329	5.600	27.000	.000	.070
.428	4.600	12.000	.000	.220
.133	9.100	58.000	.000	.200
.149	1.600	28.000	.000	.180
.266	1.200	15.000	-7.000	.020
.324	4.000	48.000	.000	.040
.133	19.100	44.000	.000	.990
.133	9.800	32.000	.000	.250
.133	18.300	19.000	8.000	2.200
.356	13.500	58.000	.000	.020
.536	24.700	62.000	.000	.030
.155	2.200	27.000	.000	.036
.168	3.500	24.000	.000	.160
.673	27.900	22.000	.000	.370
.725	31.200	10.000	-5.000	.210
.275	21.000	64.000	.000	.350
.150	14.700	37.000	.000	.640
.150	24.400	29.000	.000	1.550
1.140	15.200	11.000	-7.000	.090
1.428	14.300	1.000	-20.000	.170
1.126	24.700	44.000	.000	.170

Source: E. M. Flakman, "Predicting Sediment Yield in Western United States," *Journal of the Hydraulics Division*, *ASCE*, Vol. 98(HY12):2073–2085, December 1972.

Y: sediment yield, expressed in acre-feet per square mile per year

X_1: climate variable, equals the average annual precipitation (inches) divided by average annual temperature (°F).

X_2: watershed slope variable, the weighted average slope of the watershed, as a percentage

X_3: coarse soil particle variable, the percent of soil particles coarser than 1 mm in the surface 5.1 cm of the soil profile.

X_4: soil aggregation variable, an indication of the aggregation or dispersion characteristics of clay-size particles 2.0×10^{-6} m or finer in size in the surface 5.1 cm of the soil profile.

TABLE B-2 Evaporation Data*

X_1	X_2	X_3	X_4	Y
78.	22.6	.30719	.63141	.22
68.	69.8	.12195	.23260	.17
73.	41.3	.27752	.46191	.14
76.	33.0	.15359	.36684	.14
74.	38.8	.17271	.29313	.10
74.	20.9	.30719	.41038	.08
79.	13.3	.31444	.64255	.17
69.	93.4	.12327	.18393	.05
67.	39.0	.29268	.32365	.01
72.	11.1	.21226	.41527	.10
74.	51.6	.16744	.25963	.04
63.	23.6	.41266	.37947	.24
61.	48.3	.42123	.35358	.02
62.	52.2	.46671	.34422	.16
57.	30.3	.45353	.30201	.25
62.	21.7	.45616	.37198	.26
72.	53.7	.39815	.43877	.03
62.	39.1	.46803	.36643	.22
70.	40.3	.39420	.40222	.27
64.	66.3	.22347	.26776	.10
66.	48.7	.32169	.35716	.04
64.	30.4	.44693	.32131	.16
71.	43.7	.47396	.44679	.18
73.	43.5	.37113	.47812	.22
76.	40.7	.24192	.42947	.21
76.	83.8	.15227	.44737	.17
58.	32.6	.47792	.31294	.01
66.	48.6	.41002	.35078	.16
62.	38.9	.47594	.37753	.22
68.	16.8	.46803	.47203	.08
75.	44.6	.15821	.25948	.28
70.	69.0	.31839	.30715	.04
74.	20.2	.42254	.47738	.21
76.	23.9	.43243	.52789	.22
63.	23.8	.34606	.42710	.23
80.	29.5	.43441	.35735	.23
83.	30.7	.34606	.39338	.20
83.	32.2	.48715	.43834	.26
84.	47.8	.49374	.45284	.32
66.	66.2	.39156	.48333	.43
80.	41.5	.45089	.61260	.17
78.	25.6	.46737	.53574	.20
76.	45.5	.47264	.53654	.40
74.	40.0	.44232	.50250	.45
78.	28.6	.47001	.52618	.12
81.	31.0	.45082	.62205	.22
81.	36.9	.36388	.53771	.12
80.	36.9	.37179	.47987	.17
80.	50.6	.27684	.46966	.13
82.	35.8	.38958	.48965	.17
81.	53.3	.34542	.50608	.25
79.	23.6	.45946	.51404	.22
82.	25.8	.40409	.65313	.30
82.	25.6	.46669	.60959	.39
81.	41.3	.44100	.63260	.40
82.	39.9	.43573	.65313	.27
79.	40.9	.47660	.22737	.22
83.	20.5	.45746	.66561	.11
85.	26.5	.45485	.71986	.28
84.	23.7	.46144	.63862	.27
85.	47.9	.45221	.70786	.28
85.	35.2	.44296	.61168	.25
84.	26.1	.42716	.54573	.30
86.	28.9	.43837	.57008	.31
83.	40.2	.43441	.50578	.26
82.	41.4	.43573	.47896	.35
82.	17.3	.46671	.58762	.27
82.	24.3	.47792	.54427	.23
80.	30.8	.47792	.55134	.12
84.	13.6	.31246	.53412	.12
83.	26.6	.46473	.57321	.23
85.	23.7	.42254	.62388	.27
87.	28.3	.40079	.65265	.26
84.	17.8	.45089	.59217	.28
82.	26.7	.49967	.58782	.30
84.	28.4	.37574	.59217	.26
82.	27.7	.42386	.56605	.26
82.	24.7	.38893	.60959	.26
85.	21.5	.35267	.68387	.28
86.	13.8	.44693	.69401	.29
89.	41.1	.34937	.74976	.30
86.	34.7	.32630	.43375	.19
82.	23.6	.27291	.44631	.21
82.	19.5	.42650	.48985	.10
80.	27.5	.42716	.42882	.19
81.	21.1	.32498	.46391	.06
79.	32.6	.45682	.48439	.10
80.	21.3	.27554	.44924	.26
78.	16.2	.34606	.41137	.22
78.	24.8	.27027	.16264	.15
76.	36.0	.17139	.20579	.08
80.	25.4	.40541	.61260	.21
80.	16.4	.36058	.61260	.24
82.	19.5	.39222	.56605	.26
80.	20.4	.40079	.57176	.29
82.	34.9	.39288	.54427	.24
80.	24.8	.42716	.53092	.26
82.	18.0	.43771	.70756	.26
80.	22.8	.42320	.60239	.27
83.	22.8	.35662	.47206	.33
82.	33.1	.27620	.50073	.24
80.	39.4	.18721	.34714	.18
76.	32.0	.06856	.07158	.05
66.	56.7	.33156	.28701	.03
70.	30.4	.45616	.48266	.15
76.	29.3	.44693	.59947	.24
73.	39.0	.37310	.51053	.33
78.	40.0	.39615	.63141	.18
84.	22.8	.37883	.76634	.20
84.	25.6	.40475	.73151	.30
83.	29.5	.21688	.64065	.21
82.	27.2	.23270	.34834	.12
76.	45.8	.21490	.32210	.15
77.	37.4	.32894	.54599	.11
82.	32.6	.28279	.70756	.22
82.	33.1	.41066	.62047	.17
80.	33.1	.41061	.62281	.14
84.	39.1	.17469	.39478	.32
78.	28.9	.44891	.42094	.06
81.	19.5	.44891	.48499	.24
83.	39.1	.38695	.51702	.26
82.	44.6	.26632	.45719	.25
80.	29.2	.45155	.51050	.05
83.	13.9	.44891	.60693	.15
86.	17.0	.43507	.69401	.29
86.	31.4	.42584	.63204	.24
84.	26.3	.31641	.51089	.32
83.	27.1	.25906	.47206	.25
82.	38.6	.39618	.43542	.11
80.	26.2	.24786	.39819	.26
78.	20.5	.44430	.45921	.09
81.	20.5	.43309	.48499	.26
83.	20.5	.43111	.49454	.25
84.	20.5	.39750	.53412	.23
84.	21.0	.37574	.58056	.25
86.	21.0	.40736	.65683	.20
73.	78.0	.12591	.16638	.24
74.	48.8	.37970	.59463	.23
74.	28.5	.30389	.52763	.22
78.	59.9	.15227	.35397	.20
74.	50.5	.19446	.34338	.07
69.	29.1	.17864	.26174	.06
69.	20.4	.37113	.37493	.10
74.	11.7	.37442	.46063	.17
79.	22.0	.40606	.52393	.12
81.	22.8	.20435	.37956	.20
77.	24.0	.30521	.36091	.08
76.	10.8	.33949	.46316	.12
77.	18.0	.34476	.48122	.20
73.	26.4	.27225	.42949	.22
73.	13.7	.38299	.47812	.16
73.	19.7	.38431	.48622	.21
74.	35.1	.25972	.41875	.12
78.	29.7	.26038	.46878	.16
80.	32.9	.28807	.52071	.16
75.	45.2	.35399	.48437	.13
74.	39.1	.37245	.52763	.20
78.	37.2	.31707	.53574	.12
76.	21.5	.26566	.45631	.11
74.	32.1	.33421	.50250	.10
69.	24.9	.34476	.36797	.18
70.	17.4	.33223	.37500	.18
72.	17.4	.32301	.44008	.16
72.	13.5	.32367	.43099	.19
78.	34.3	.27027	.43783	.20
76.	54.4	.11800	.38910	.18
65.	41.9	.32564	.44626	.07
63.	33.7	.33949	.51641	.23
63.	33.7	.16809	.50226	.21
61.	11.4	.32696	.51923	.12
66.	24.4	.31312	.58764	.17
68.	22.2	.30323	.62682	.15
68.	63.9	.28873	.60271	.15
77.	15.3	.25379	.45345	.06
76.	29.6	.22413	.34000	.10
72.	43.9	.27620	.36042	.12
74.	45.0	.22215	.41875	.10
74.	44.9	.27818	.36013	.14
74.	45.7	.25840	.46900	.13
73.	34.8	.31905	.46191	.17
73.	54.3	.21292	.25121	.17
73.	76.4	.15293	.26742	.11
75.	78.7	.22610	.37192	.08
77.	58.7	.12063	.17563	.04
74.	35.2	.28675	.49413	.01
67.	64.3	.31114	.35667	.19
67.	33.3	.25313	.38309	.15
67.	27.1	.30455	.38309	.14
73.	19.7	.24720	.46191	.11
69.	12.4	.22742	.44567	.09
64.	41.6	.30521	.36891	.15
50.	60.1	.31048	.24453	.05
51.	42.5	.30389	.25396	.04
55.	34.5	.28411	.17263	.14
52.	58.1	.29927	.22095	.21
50.	38.2	.28609	.24453	.09
57.	13.9	.27291	.30201	.14
67.	15.0	.30191	.43593	.18
65.	40.4	.29400	.32615	.16
66.	47.3	.19117	.20409	.04
77.	11.9	.17996	.33315	.04
74.	14.5	.28279	.60300	.08
69.	46.3	.27620	.42445	.25
60.	92.7	.18128	.18596	.28
61.	81.1	.14041	.13929	.16
66.	47.7	.17732	.19771	.08
70.	31.7	.15293	.24864	.05
68.	30.5	.26434	.36258	.07
64.	69.9	.26566	.32131	.19
59.	86.5	.21885	.22448	.21
62.	64.0	.18655	.18321	.12
63.	16.3	.25906	.33347	.12
65.	14.2	.26104	.35076	.11
66.	22.6	.21226	.34441	.08
72.	50.9	.22215	.32124	.04
54.	16.3	.21226	.17924	.09
61.	21.9	.25445	.29465	.06
56.	42.7	.27225	.27787	.01
56.	50.4	.01714	.11203	.01
36.	69.2	.27027	.13499	.06
43.	28.8	.25906	.16796	.01
54.	11.8	.13787	.17092	.02
61.	48.5	.14766	.16214	.01
40.	48.3	.25643	.14797	.02
39.	30.6	.25247	.13436	.02
45.	17.7	.23663	.16455	.06
54.	18.0	.19183	.23343	.04
56.	9.0	.19578	.26438	.03
50.	20.0	.23665	.20857	.01
57.	37.6	.20633	.22767	.08
65.	66.6	.17996	.26461	.05
65.	97.3	.20040	.31999	.05
59.	38.0	.08635	.18956	.10
66.	58.0	.20699	.19134	.04
71.	30.3	.11800	.32562	.04
59.	14.5	.23863	.28933	.11
59.	33.1	.05603	.15464	.06
55.	29.6	.21094	.26327	.08
60.	16.7	.24456	.34092	.10
57.	12.9	.25840	.31130	.12
58.	17.9	.25643	.31294	.10
58.	14.9	.24852	.33219	.12
66.	17.0	.13448	.22960	.11
58.	23.4	.02900	.06740	.05
38.	63.2	.25906	.14613	.17
38.	23.1	.25313	.16668	.11
52.	35.9	.19578	.18219	.10
63.	61.9	.18062	.29898	.04
69.	57.5	.06328	.25467	.05
50.	27.3	.23995	.22655	.04
56.	29.0	.19446	.26438	.10
65.	15.6	.15162	.34461	.10
65.	10.8	.08701	.19692	.06
56.	14.1	.19314	.22853	.02
58.	14.4	.18985	.25998	.07
64.	28.1	.09888	.14875	.05
51.	20.8	.21424	.22035	.02
49.	32.5	.22347	.25239	.13
55.	69.7	.06394	.05611	.07
59.	23.5	.02109	.02494	.01
48.	85.2	.23270	.17953	.01
51.	44.7	.21819	.21288	.01
58.	23.8	.05933	.25516	.04
39.	54.6	.23138	.15198	.08
37.	21.2	.11997	.13608	.03
38.	16.9	.21885	.13928	.05
41.	36.5	.21490	.13315	.04
48.	49.5	.18721	.16956	.04
37.	16.8	.18392	.15144	.01
42.	10.1	.17007	.15965	.02
58.	66.8	.10877	.11555	.06
36.	94.6	.15953	.12866	.12
36.	48.6	.22347	.13499	.11
41.	31.1	.21819	.15620	.08
49.	43.7	.15557	.14867	.03
60.	95.7	.22083	.29960	.03
47.	79.8	.22808	.19222	.01
44.	27.8	.20962	.17509	.05
54.	65.7	.02637	.06253	.05
50.	47.7	.22281	.23374	.18
49.	42.4	.22544	.23165	.15
48.	14.7	.22281	.23605	.10
52.	57.7	.22544	.24033	.16
44.	52.9	.20171	.20092	.13
49.	56.8	.04680	.14867	.10
52.	47.6	.10547	.15893	.02
50.	41.1	.21753	.20857	.01
41.	39.2	.21688	.16132	.10
48.	29.0	.15162	.15626	.09
49.	42.0	.20897	.24548	.19
50.	23.3	.20369	.20857	.07
43.	19.7	.21160	.19349	.08
44.	17.2	.19051	.19805	.08
49.	11.8	.11866	.13138	.05
48.	28.9	.20237	.21610	.05
45.	37.3	.05867	.15478	.08
48.	88.7	.05208	.09641	.16

TABLE B-2 (*cont.*)

X_1	X_2	X_3	X_4	Y	X_1	X_2	X_3	X_4	Y	X_1	X_2	X_3	X_4	Y
36.	39.8	.20501	.11179	.13	55.	19.0	.17601	.20284	.07	48.	49.0	.07581	.19615	.03
50.	48.3	.21094	.21936	.18	60.	21.4	.18985	.25311	.06	54.	15.1	.07515	.04168	.03
48.	36.2	.06790	.11304	.24	65.	57.0	.13184	.16615	.11	48.	24.6	.15293	.14296	.03
55.	86.2	.22149	.25895	.10	66.	77.9	.18260	.32527	.01	50.	29.5	.04680	.08271	.06
62.	9.6	.17337	.31646	.03	68.	90.4	.09163	.47887	.04	57.	23.8	.12063	.27877	.30
63.	16.8	.18985	.36222	.05	46.	24.3	.07910	.19150	.12	52.	9.8	.23401	.19381	.12
63.	15.8	.19051	.37372	.10	40.	75.5	.21226	.17017	.03	64.	40.6	.20897	.28561	.03
69.	46.2	.17007	.53763	.13	34.	41.6	.21753	.14170	.13	58.	113.8	.27093	.31775	.09
56.	42.4	.22017	.33608	.11	42.	38.3	.14173	.17828	.11	61.	39.9	.23665	.24643	.03
57.	22.6	.22083	.35776	.12	50.	30.3	.02044	.14025	.05	67.	97.7	.16744	.30383	.04
45.	75.0	.21490	.21134	.20	33.	63.6	.11866	.11709	.02	43.	106.3	.28741	.19073	.04
47.	55.6	.17469	.18902	.11	38.	28.1	.22083	.15070	.03	40.	40.7	.25181	.18250	.12
50.	50.4	.18523	.23015	.09	41.	20.5	.16348	.17668	.02	52.	33.4	.00659	.15893	.09
64.	32.9	.17930	.35701	.07	45.	48.3	.21424	.13394	.03	51.	63.2	.26038	.22035	.01
64.	20.1	.16612	.34511	.05	54.	56.9	.13052	.20008	.01	50.	20.1	.17601	.18699	.06
64.	38.3	.07779	.20231	.08	49.	62.4	.10020	.08989	.02	48.	26.4	.28345	.22940	.10
56.	76.5	.14898	.23301	.05	60.	37.6	.22413	.26860	.01	45.	76.6	.31444	.21431	.01
45.	60.1	.21292	.18752	.14	38.	32.2	.22149	.12766	.03	59.	38.6	.05405	.04988	.04
50.	16.7	.20303	.21217	.06	40.	38.6	.18655	.17510	.39	51.	84.9	.08438	.07469	.01

Source: Weather station at Tifton, Georgia.

X_1: mean daily temperature (°F)

X_2: wind speed (mi/day)

X_3: radiation (equiv. in./day)

X_4: vapor pressure deficit (equiv. in./day)

Y: evaporation (in./day)

TABLE B-3 French Broad River Water Yield Data

X_1	X_2	X_3	X_4	X_5	X_6	X_7	X_8	X_9	X_{10}	X_{11}	X_{12}	Y
6.5	5.5	2.8	4.6	2.3	57.3	44.6	42.1	4.08	2.60	1.27	8.00	3.30
9.0	6.5	5.5	2.8	3.3	66.5	57.3	44.6	4.77	4.08	2.60	7.70	3.40
2.7	9.0	6.5	5.5	3.4	70.0	66.5	57.3	5.75	4.77	4.08	6.30	2.70
6.4	2.7	9.0	6.5	2.7	74.4	70.0	66.5	4.84	5.75	4.77	5.90	1.70
5.0	6.4	2.7	9.0	1.7	75.4	74.4	70.0	4.56	4.84	5.75	5.20	1.30
10.1	5.0	6.4	2.7	1.3	67.6	75.4	74.4	3.40	4.56	4.84	6.30	2.80
9.2	10.1	5.0	6.4	2.8	59.9	67.6	75.4	1.66	3.40	4.56	7.20	4.40
2.1	9.2	10.1	5.0	4.4	45.1	59.9	67.6	1.77	1.66	3.40	10.00	2.40
4.0	2.1	9.2	10.1	2.4	40.7	45.1	59.9	.69	1.77	1.66	8.70	2.40
5.2	4.0	2.1	9.2	2.4	39.0	40.7	45.1	.91	.69	1.77	1.00	2.80
7.2	5.2	4.0	2.1	2.8	36.6	39.0	40.7	1.06	.91	.69	10.40	5.00
6.3	7.2	5.2	4.0	5.0	33.5	36.6	39.0	1.65	1.06	.91	10.10	3.90
3.3	6.3	7.2	5.2	3.9	58.2	33.5	36.6	4.37	1.65	1.06	8.20	4.60
3.5	3.3	6.3	7.2	4.6	61.1	58.2	33.5	5.42	4.37	1.65	7.70	2.40
5.1	3.5	3.3	6.3	2.4	71.1	61.1	58.2	4.78	5.42	4.37	5.90	1.70
3.5	5.1	3.5	3.3	1.7	73.4	71.1	61.1	4.85	4.78	5.42	5.40	1.30
8.6	3.5	5.1	3.5	1.3	74.1	73.4	71.1	4.30	4.85	4.78	5.10	2.50
4.3	8.6	3.5	5.1	2.5	68.1	74.1	73.4	3.00	4.30	4.85	6.40	1.40
5.0	4.3	8.6	3.5	1.4	58.9	68.1	74.1	1.78	3.00	4.30	5.80	1.90
1.4	5.0	4.3	8.6	1.9	47.4	58.9	68.1	1.44	1.78	3.00	7.70	1.20
1.7	1.4	5.0	4.3	1.2	34.7	47.4	58.9	1.27	1.44	1.78	8.30	1.20
3.2	1.7	1.4	5.0	1.2	33.5	34.7	47.4	.91	1.27	1.44	9.20	1.60
7.7	3.2	1.7	1.4	1.6	44.1	33.5	34.7	.98	.91	1.27	8.90	3.20
4.4	7.7	3.2	1.7	3.2	50.1	44.1	33.5	2.38	.98	.91	9.80	3.00
5.0	4.4	7.7	3.2	3.0	50.4	50.1	44.1	3.60	2.38	.98	11.00	3.50
4.6	5.0	4.4	7.7	3.5	60.0	50.4	50.1	4.03	3.60	2.38	8.20	2.70
7.5	4.6	5.0	4.4	2.7	68.6	60.0	50.4	4.22	4.03	3.60	5.40	2.40
5.5	7.5	4.6	5.0	2.4	72.5	68.6	60.0	4.56	4.22	4.03	5.00	2.00
14.7	5.5	7.5	4.6	2.0	72.0	72.5	68.6	3.93	4.56	4.22	4.90	5.30
2.0	14.7	5.5	7.5	5.3	68.4	72.0	72.5	2.90	3.93	4.56	5.90	2.40
2.1	2.0	14.7	5.5	2.4	55.3	68.4	72.0	1.47	2.90	3.93	6.70	1.50
6.2	2.1	2.0	14.7	1.5	50.4	55.3	68.4	1.66	1.47	2.90	9.60	2.20
10.1	6.2	2.1	2.0	2.2	39.7	50.4	55.3	.92	1.66	1.47	8.90	5.30
6.2	10.1	6.2	2.1	5.3	37.4	39.7	50.4	.88	.92	1.66	8.40	4.00
4.9	6.2	10.1	6.2	4.0	45.8	37.4	39.7	1.26	.88	.92	9.60	3.30
5.2	4.9	6.2	10.1	3.3	43.4	45.8	37.4	2.05	1.26	.88	10.00	4.00
5.5	5.2	4.9	6.2	4.0	52.9	43.4	45.8	3.70	2.05	1.26	8.70	4.60
2.9	5.5	5.2	4.9	4.6	69.4	52.9	43.4	5.70	3.70	2.05	6.50	2.20
8.7	2.9	5.5	5.2	2.2	70.2	69.4	52.9	4.32	5.70	3.70	6.20	2.90
3.8	8.7	2.9	5.5	2.9	73.7	70.2	69.4	5.06	4.32	5.70	5.70	1.70
3.6	3.8	8.7	2.9	1.7	72.3	73.7	70.2	4.87	5.06	4.32	5.60	1.40
3.7	3.6	3.8	8.7	1.4	65.4	72.3	73.7	2.73	4.87	5.06	6.50	1.10
4.7	3.7	3.6	3.8	1.1	58.3	65.4	72.3	1.75	2.73	4.87	6.40	1.80
4.6	4.7	3.7	3.6	1.8	46.0	58.3	65.4	1.20	1.75	2.73	8.20	1.60
2.6	4.6	4.7	3.7	1.6	34.4	46.0	58.3	.59	1.20	1.75	8.40	1.50
2.8	2.6	4.6	4.7	1.5	34.4	34.4	46.0	.61	.59	1.20	8.40	1.90
2.3	2.8	2.6	4.6	1.9	34.2	34.4	34.4	.81	.61	.59	9.50	1.50
9.4	2.3	2.8	2.6	1.5	51.0	34.2	34.4	3.27	.81	.61	10.20	5.10
4.3	9.4	2.3	2.8	5.1	57.6	51.0	34.2	5.02	3.27	.81	8.60	2.00
2.6	4.3	9.4	2.3	2.0	63.6	57.6	51.0	5.17	5.02	3.27	7.90	2.00
5.2	2.6	4.3	9.4	2.0	69.2	63.6	57.6	4.19	5.17	5.02	6.90	1.50
5.4	5.2	2.6	4.3	1.5	71.4	69.2	63.6	4.67	4.19	5.17	6.10	1.60
2.5	5.4	5.2	2.6	1.6	72.0	71.4	69.2	4.73	4.67	4.19	5.90	1.00
4.1	2.5	5.4	5.2	1.0	66.2	72.0	71.4	2.64	4.73	4.67	5.90	.90
.1	4.1	2.5	5.4	.9	59.6	66.2	72.0	1.82	2.64	4.73	5.70	.70
5.8	.1	4.1	2.5	.7	47.8	59.6	66.2	1.20	1.82	2.64	9.30	1.10
2.3	5.8	.1	4.1	1.1	31.4	47.8	59.6	.95	1.20	1.82	8.50	1.50
6.5	2.3	5.8	.1	1.5	38.1	31.4	47.8	1.00	.95	1.20	8.80	3.00
4.7	6.5	2.3	5.8	3.0	34.7	38.1	31.4	.77	1.00	.95	8.90	2.40
7.2	4.7	6.5	2.3	2.4	46.3	34.7	38.1	2.61	.77	1.00	10.80	4.20
8.5	7.2	4.7	6.5	4.2	57.3	46.3	34.7	3.50	2.61	.77	8.80	4.60

TABLE B-3 (*cont.*)

X_1	X_2	X_3	X_4	X_5	X_6	X_7	X_8	X_9	X_{10}	X_{11}	X_{12}	Y
1.8	8.5	7.2	4.7	4.6	65.3	57.3	46.3	4.96	3.50	2.61	6.70	3.10
3.0	1.8	8.5	7.2	3.1	72.8	65.3	57.3	5.66	4.96	3.50	5.80	1.60
8.1	3.0	1.8	8.5	1.6	72.9	72.8	65.3	4.46	5.66	4.96	5.10	1.80
9.9	8.1	3.0	1.8	1.8	71.7	72.9	72.8	3.93	4.46	5.66	6.60	2.40
8.0	9.9	8.1	3.0	2.4	65.3	71.7	72.9	2.42	3.93	4.46	5.60	2.40
11.5	8.0	9.9	8.1	2.4	52.0	65.3	71.7	1.85	2.42	3.93	7.90	8.50
3.6	11.5	8.0	9.9	8.7	49.6	52.0	65.3	1.66	1.85	2.42	7.90	2.70
5.0	3.6	11.5	8.0	2.7	40.5	49.6	52.0	.92	1.66	1.85	9.20	3.30
2.8	5.0	3.6	11.5	3.3	37.0	40.5	49.6	.94	.92	1.66	3.10	2.70
6.1	2.8	5.0	3.6	2.7	37.9	37.0	40.5	1.16	.94	.92	3.60	7.60
6.8	6.1	2.8	5.0	7.6	42.4	37.9	37.0	2.10	1.16	.94	3.20	4.30
4.4	6.8	6.1	2.8	4.3	57.4	42.4	37.9	4.58	2.10	1.16	1.20	3.20
4.9	4.4	6.8	6.1	3.2	65.8	57.4	42.4	5.76	4.58	2.10	1.80	2.80
5.2	4.9	4.4	6.8	2.8	66.8	65.8	57.4	4.77	5.76	4.58	.50	2.50
5.6	5.2	4.9	4.4	2.5	72.2	66.8	65.8	4.80	4.37	5.76	.80	1.80
4.6	5.6	5.2	4.9	1.8	71.1	72.2	66.8	4.78	4.80	4.37	.20	1.60
4.3	4.6	5.6	5.2	1.6	66.8	71.1	72.2	2.98	4.78	4.80	2.20	1.30
6.0	4.3	4.6	5.6	1.3	53.8	66.8	71.1	2.34	2.98	4.78	3.40	2.50
2.0	6.0	4.3	4.6	2.5	46.4	53.8	66.8	1.65	2.34	2.98	2.60	1.30
.4	2.0	6.0	4.3	1.3	40.3	46.4	53.8	1.10	1.65	2.34	3.70	1.10
4.1	.4	2.0	6.0	1.1	30.1	40.3	46.4	.97	1.10	1.65	6.70	1.10
8.5	4.1	.4	2.0	1.1	36.2	30.1	40.3	1.04	.97	1.10	2.20	4.60
3.2	8.5	4.1	.4	4.6	43.5	36.2	30.1	3.06	1.04	.97	2.40	3.40
5.9	3.2	8.5	4.1	3.4	52.0	43.5	36.2	3.17	3.06	1.04	.50	1.80
4.6	5.9	3.2	8.5	1.8	60.1	52.0	43.5	4.11	3.17	3.06	1.80	3.00
2.7	4.6	5.9	3.2	3.0	66.1	60.1	52.0	4.66	4.11	3.17	1.30	1.80
4.2	2.7	4.6	5.9	1.8	71.1	66.1	60.1	4.61	4.66	4.11	2.90	1.20
6.6	4.2	2.7	4.6	1.2	69.9	71.1	66.1	4.60	4.61	4.66	.40	1.50
5.8	6.6	4.2	2.7	1.5	62.9	69.9	71.1	2.53	4.60	4.61	2.40	1.70
6.0	5.8	6.6	4.2	1.7	51.7	62.9	69.9	2.14	2.53	4.60	2.70	2.10
4.6	6.0	5.8	6.6	2.1	45.0	51.7	62.9	1.20	2.14	2.53	.70	3.00
3.4	4.6	6.0	5.8	3.0	37.6	45.0	51.7	.73	1.20	2.14	3.90	2.10
3.1	3.4	4.6	6.0	2.1	38.7	37.6	45.0	1.05	.73	1.20	3.50	2.30
3.1	3.1	3.4	4.6	2.3	35.0	38.7	37.6	1.12	1.05	.73	3.00	2.00
3.4	3.1	3.1	3.4	2.0	49.8	35.0	38.7	3.06	1.12	1.05	1.30	2.50
1.9	3.4	3.1	3.4	2.5	57.6	49.8	35.0	5.04	3.06	1.12	1.80	1.50
5.9	1.9	3.4	3.1	1.5	59.7	57.6	49.8	4.41	5.04	3.06	2.50	1.70
8.7	5.9	1.9	3.4	1.7	66.8	59.7	57.6	4.54	4.41	5.04	.70	4.00
7.8	8.7	5.9	1.9	4.0	68.7	66.8	59.7	4.13	4.54	4.41	1.90	3.00
10.6	7.8	8.7	5.9	3.0	68.5	68.7	66.8	3.27	4.13	4.54	.50	4.00
3.6	10.6	7.8	8.7	4.0	60.2	68.5	68.7	2.80	3.27	4.13	1.40	2.30
3.6	3.6	10.6	7.8	2.3	53.5	60.2	68.5	1.96	2.81	3.27	1.00	2.00
3.2	3.6	3.6	10.6	2.0	42.7	53.5	60.2	1.98	1.96	2.81	4.20	2.00
7.6	3.2	3.6	3.6	2.0	41.8	42.7	53.5	.80	1.98	1.96	1.80	3.80
4.1	7.6	3.2	3.6	3.8	34.3	41.8	42.7	.87	.80	1.98	2.90	3.50
.8	4.1	7.6	3.2	3.5	32.4	34.3	41.8	1.50	.87	.80	9.20	1.60
6.7	.8	4.1	7.6	1.6	46.6	32.4	34.3	2.76	1.50	.87	2.60	3.40
3.5	6.7	.8	4.1	3.4	54.8	46.6	32.4	3.96	1.76	1.50	1.50	2.40
4.5	3.5	6.7	.8	2.4	61.0	54.8	46.6	4.49	3.96	2.76	.90	2.00
5.9	4.5	3.5	6.7	2.0	69.5	61.0	54.8	4.81	4.49	3.96	2.00	2.50
5.0	5.9	4.5	3.5	2.4	73.1	69.5	61.0	4.77	4.81	4.49	.90	1.50

X_1: monthly precipitation (in.) month t (current month)
X_2: monthly precipitation (in.) month $t-1$ (previous month)
X_3: monthly precipitation (in.) month $t-2$
X_4: monthly precipitation (in.) month $t-3$
X_5: total water yield (in.) month $t-1$ (previous month)
X_6: mean monthly temperature (°F) month t
X_7: mean monthly temperature (°F) month $t-1$
X_8: mean monthly temperature (°F) month $t-2$
X_9: total monthly evaporation (in.) month t
X_{10}: total monthly evaporation (in.) month $t-1$
X_{11}: total monthly evaporation (in.) month $t-2$
X_{12}: average wind speed (mph) month t
Y: total water yield (in.) month t

TABLE B-4 Indiana Peak Discharge Data

X_1	X_2	X_3	X_4	X_5	X_6	X_7	Y
121.0	12.8	19.5	11.0	341.0	9.5	.80	13300.
38.2	26.0	16.6	13.0	449.0	11.0	.90	12600.
188.0	5.5	25.1	15.0	265.0	10.5	.90	14100.
129.0	6.3	33.2	16.0	438.0	8.7	.80	13100.
41.9	15.4	14.0	17.5	435.0	12.0	.80	5500.
85.6	4.7	15.6	9.5	182.0	5.8	.50	3200.
133.0	4.6	20.1	10.0	150.0	7.5	.80	7700.
113.0	3.6	22.7	7.0	95.0	3.4	.30	600.
35.0	5.5	9.6	8.5	77.0	2.2	.30	460.
152.0	2.0	19.1	8.0	82.0	2.3	.70	2350.
146.0	3.3	24.1	9.0	100.0	2.8	.80	4100.
162.0	3.3	29.6	9.0	111.0	3.0	.80	4800.
24.7	4.5	12.7	9.5	72.0	4.5	.70	740.
132.0	7.2	35.9	12.0	252.0	5.4	.70	13100.
133.0	11.4	29.1	13.0	344.0	5.4	.70	16800.
35.5	10.2	12.5	10.5	165.0	5.7	.70	1580.
131.0	4.0	27.1	10.0	121.0	5.4	.70	3540.
40.4	6.2	15.0	9.5	110.0	4.6	.70	2400.
18.5	18.7	6.4	9.5	110.0	6.6	.70	3750.
169.0	7.2	31.8	11.5	273.0	5.3	.70	5800.
42.4	6.7	19.3	10.0	127.0	4.5	.70	1600.
103.0	15.2	17.4	10.5	153.0	7.5	.70	9000.
174.0	6.8	35.1	10.5	264.0	7.9	.70	11200.
23.9	18.8	9.2	11.0	189.0	4.5	.70	1820.
28.8	10.6	11.5	11.0	131.0	6.0	.70	3900.
14.6	19.8	7.6	13.0	291.0	11.5	.80	4150.
59.0	12.6	18.3	12.5	283.0	7.0	.70	9600.
184.0	5.8	30.8	12.0	251.0	5.2	.70	7800.
107.0	4.3	30.1	12.0	195.0	7.0	.70	8200.
91.4	10.3	33.7	12.5	401.0	7.4	.70	8200.
155.0	8.9	42.3	13.5	410.0	9.8	.90	14200.
77.2	9.4	32.6	14.0	318.0	1.5	1.00	14000.
85.9	12.2	29.1	13.5	391.0	9.0	1.00	13400.
198.0	9.2	43.2	14.0	448.0	9.1	1.00	29000.
38.2	13.0	10.6	14.0	312.0	10.0	.80	5700.
76.1	11.6	17.7	13.5	400.0	8.0	.70	7400.
120.0	9.0	34.7	13.5	436.0	9.5	.70	11700.
48.8	19.1	13.7	14.5	383.0	7.5	.80	8400.
60.7	12.7	16.5	15.0	390.0	7.0	.80	6300.
171.0	2.6	52.8	15.5	466.0	13.7	.80	5600.
69.2	7.4	22.6	6.5	159.0	4.0	.50	2300.
125.0	3.6	29.8	7.0	131.0	3.2	.50	2700.
160.0	3.2	36.0	8.0	173.0	3.1	.40	2400.
62.9	6.2	14.8	12.0	262.0	8.0	.40	2080.
78.7	4.7	22.5	8.0	206.0	6.6	.40	1800.
80.5	5.2	22.9	4.5	120.0	3.2	.30	530.
142.0	3.9	26.6	5.5	130.0	3.5	.30	650.
87.3	8.0	20.1	4.5	193.0	5.1	.50	1270.
116.0	1.2	23.2	5.5	100.0	3.0	.30	640.
132.0	5.0	12.9	7.0	65.0	2.5	.50	1420.
123.0	3.2	22.2	7.5	137.0	2.5	.40	1150.
54.7	2.3	21.1	7.0	102.0	6.5	.40	1600.
35.6	2.5	8.9	7.5	44.0	2.5	.40	340.
144.0	2.9	15.7	8.0	58.0	3.5	.40	1220.
203.0	2.5	18.6	8.0	63.0	4.5	.40	1670.
21.8	6.4	10.4	8.5	82.0	3.4	.70	750.
83.7	2.2	13.2	8.5	95.0	2.5	.70	2100.
44.8	6.4	21.4	8.5	193.0	3.0	.70	1970.

Source: L. G. Davis, *Floods in Indiana: Technical Manual for Estimating Their Magnitude and Frequency*, Geological Survey Circular 710, USGS, Reston, Va., 1974.

X_1: drainage area (mi^2)

X_2: channel slope (ft/mi), the difference in elevation at points 10% and 85% of the distance along the channel from a gaging station to the watershed divide, divided by the distance between the two points

X_3: channel length (mi), distance along a stream from a point of discharge to the watershed divide

X_4: precipitation index (in.), mean annual precipitation minus the sum of average annual evapotranspiration and mean annual snowfall (water equivalent)

X_5: watershed relief (ft), the difference in elevation between the highest point on the watershed perimeter and the stream at the point of discharge

X_6: drainage density (mi/mi^2), total length of streams in a watershed divided by drainage area

X_7: soil runoff coefficient, ratio of the volume of rainfall to the total volume of runoff occurring after the beginning of runoff

Y: instantaneous peak discharge (ft^3/sec) for a 10-year exceedence frequency

TABLE C-1 Chattooga River Watershed Data

Gage No. 2-1770, Chattooga River near Clayton, GA
DA = 207 sq. mi.

Year	Month	Precip. (in.)	Runoff (in.)	Temp. (°F)	Evap. (in.)	Year	Month	Precip. (in.)	Runoff (in.)	Temp. (°F)	Evap. (in.)
1953	1					1955	1	2.68	1.70	41.1	1.16
	2						2	8.97	3.54	43.3	2.36
	3						3	4.63	3.08	51.3	3.36
	4						4	6.70	4.48	61.2	5.95
	5						5	8.52	5.07	66.9	6.69
	6						6	5.26	2.88	67.4	6.98
	7						7	9.53	2.96	75.2	6.37
	8						8	3.39	2.38	75.7	6.24
	9						9	2.90	1.26	70.3	5.78
	10	3.51	1.00	59.7	3.94		10	1.44	1.19	58.1	3.95
	11	4.72	1.26	48.8	2.09		11	2.80	1.04	49.1	2.34
	12	10.33	4.00	40.6	2.79		12	1.95	1.02	39.9	1.40
1954	1	11.92	6.36	42.7	2.13	1956	1	2.23	0.86	39.9	1.61
	2	4.83	3.63	45.4	3.10		2	12.77	4.20	46.4	2.57
	3	7.40	4.28	48.3	3.33		3	5.30	3.45	49.5	4.30
	4	4.57	4.34	61.2	4.91		4	9.69	4.73	56.0	5.80
	5	2.62	3.10	60.0	5.19		5	2.90	2.80	66.5	6.33
	6	4.24	2.30	71.4	7.47		6	2.70	1.75	70.4	7.50
	7	3.68	1.40	77.0	8.22		7	8.11	1.97	73.3	6.84
	8	4.23	1.13	76.0	6.36		8	2.72	1.07	74.0	6.51
	9	0.44	0.65	71.9	6.13		9	6.00	0.94	66.3	5.40
	10	0.31	0.55	61.3	4.58		10	3.04	1.15	60.3	3.27
	11	3.54	0.84	46.5	1.88		11	2.55	1.27	47.8	1.86
	12	8.41	2.05	40.3	1.29		12	7.22	2.78	50.7	1.48

1957	1	10.59	3.10	43.2	1.68	1959	1	5.96	2.64	40.1	1.29
	2	7.55	5.40	48.9	1.81		2	5.92	2.51	44.7	1.57
	3	3.47	4.40	48.4	3.17		3	7.29	3.07	47.4	3.58
	4	8.34	6.37	60.4	5.33		4	7.27	4.28	54.9	4.46
	5	5.69	3.07	65.7	6.32		5	11.88	5.25	66.5	5.94
	6	7.32	3.65	71.9	6.08		6	1.19	5.10	70.9	5.80
	7	1.92	2.38	74.0	7.56		7	9.50	3.24	73.1	7.30
	8	2.17	1.43	72.3	7.08		8	3.23	1.99	75.0	5.90
	9	8.31	1.76	69.1	4.19		9	8.20	2.99	67.4	3.93
	10	6.58	2.97	54.7	2.76		10	13.93	5.83	60.8	3.20
	11	11.82	5.01	50.1	1.87		11	2.91	3.45	46.9	1.71
	12	6.79	5.19	43.4	1.33		12	5.58	3.83	42.0	1.61
1958	1	4.92	4.51	36.0	1.46	1960	1	7.75	4.50	40.4	1.30
	2	6.33	4.21	34.6	2.42		2	8.15	6.59	39.8	2.34
	3	6.19	4.56	45.4	2.78		3	8.42	5.74	36.8	3.93
	4	9.10	5.87	57.3	4.40		4	3.27	6.05	57.8	5.54
	5	5.04	6.08	65.6	5.44		5	2.88	3.46	58.6	6.27
	6	2.40	2.87	71.4	7.25		6	4.37	2.67	69.4	6.60
	7	12.76	4.65	73.7	4.20		7	4.98	2.39	72.5	7.84
	8	5.69	2.61	73.5	5.92		8	8.40	3.09	74.3	5.51
	9	1.23	1.52	67.4	5.38		9	8.87	2.42	68.9	4.43
	10	0.41	1.36	59.1	3.32		10	5.92	3.37	60.6	2.68
	11	3.36	1.14	54.5	2.18		11	1.40	2.15	48.7	1.89
	12	3.08	1.36	40.0	1.05		12	3.21	1.97	36.9	2.72

TABLE C-1 *(cont.)*

Gage No. 2-1770, Chattooga River near Clayton, GA
DA = 207 sq. mi.

Year	Month	Precip. (in.)	Runoff (in.)	Temp. (°F)	Evap. (in.)	Year	Month	Precip. (in.)	Runoff (in.)	Temp. (°F)	Evap. (in.)
1961	1	3.54	2.31	36.5	2.00	1963	1	5.12	2.65	36.8	1.53
	2	10.97	4.91	45.7	1.60		2	3.10	2.44	37.1	2.07
	3	6.42	5.42	50.7	3.10		3	11.00	7.07	51.5	4.51
	4	5.60	5.17	52.1	4.90		4	7.02	3.56	60.0	5.87
	5	3.23	3.38	61.3	4.80		5	4.39	3.48	64.3	6.34
	6	9.45	3.53	69.1	5.73		6	10.05	2.97	69.1	6.21
	7	6.32	2.94	72.0	6.35		7	7.13	3.92	72.0	6.14
	8	8.98	3.40	72.5	5.20		8	2.73	2.39	72.7	6.23
	9	2.65	2.57	69.8	4.99		9	3.13	1.75	67.8	5.06
	10	1.59	1.71	56.9	4.08		10	0.33	1.36	58.9	4.24
	11	5.81	2.25	53.0	2.70		11	6.67	1.96	48.4	2.49
	12	14.85	7.57	42.5	2.27		12	4.42	2.50	34.9	2.49
1962	1	7.88	5.81	40.5	1.96	1964	1	10.44	5.45	40.2	3.33
	2	6.90	4.86	48.8	2.65		2	5.98	4.10	37.2	1.32
	3	7.89	5.90	46.5	3.48		3	12.46	7.38	47.9	5.28
	4	6.07	7.16	53.3	4.19		4	11.65	8.80	57.5	4.95
	5	2.01	3.52	70.0	7.01		5	3.33	5.55	65.4	5.92
	6	6.79	3.50	71.4	5.77		6	4.54	2.84	72.1	7.48
	7	4.88	1.89	74.4	6.48		7	6.83	2.42	72.4	5.66
	8	4.54	1.51	73.2	6.66		8	8.19	2.57	72.0	5.46
	9	5.55	1.29	67.4	4.66		9	6.16	3.14	67.3	4.76
	10	4.42	2.06	59.3	3.45		10	10.69	8.49	54.1	3.60
	11	4.81	1.78	47.9	2.03		11	4.25	3.25	52.2	2.57
	12	4.13	1.84	38.5	2.07		12	7.00	5.09	43.4	2.42

1965	1	4.00	4.62	40.8	3.35
	2	7.16	5.57	41.5	4.62
	3	8.23	6.27	46.9	3.64
	4	4.91	5.36	60.4	4.94
	5	4.66	4.41	67.7	7.04
	6	6.52	4.37	69.7	5.61
	7	3.48	3.20	74.3	5.83
	8	3.41	2.61	73.3	5.18
	9	6.16	2.13	69.2	4.60
	10	4.52	4.13	56.7	3.32
	11	3.10	1.79	50.6	2.36
	12	0.94	1.55	43.4	1.95
1966	1	6.83	2.54	37.4	2.16
	2	10.61	7.03	41.2	2.30
	3	4.28	6.14	47.8	4.15
	4	9.33	4.25	55.6	4.50
	5	8.42	5.49	64.0	4.99
	6	3.14	3.28	69.6	6.43
	7	3.32	2.34	74.9	6.48
	8	7.84	2.55	73.6	5.34
	9	4.67	1.86	66.5	4.17
	10	6.14	2.61	56.8	3.35
	11	5.61	4.71	49.5	2.62
	12	4.35	3.12	41.6	1.94
1967	1	4.20	4.11	41.7	2.02
	2	4.86	3.43	39.4	2.26
	3	3.98	3.96	53.6	4.09
	4	3.74	2.64	61.0	5.89
	5	6.77	2.93	63.6	6.01
	6	7.60	6.41	69.3	5.58
	7	9.19	4.87	71.0	5.47
	8	16.01	5.25	70.9	5.02
	9	3.52	4.29	63.4	4.25
	10	5.80	3.30	56.4	3.63
	11	6.01	3.59	45.0	2.88
	12	11.01	6.66	45.2	2.92
1968	1	5.34	5.89	38.4	2.41
	2	1.40	3.24	35.9	2.50
	3	7.82	4.73	48.8	5.35
	4	4.59	4.24	57.5	4.00
	5	4.81	3.22	62.2	6.06
	6	3.90	3.44	69.5	5.53
	7	4.71	2.05	72.0	5.77
	8	1.42	1.54	72.9	6.63
	9	5.73	1.53	62.9	4.30
	10	4.38	1.71	56.2	3.21
	11	6.09	2.09	46.8	1.86
	12	6.65	2.98	37.0	2.21

TABLE C-1 (*cont.*)

Gage No. 2-1770, Chattooga River near Clayton, GA
DA = 207 sq. mi.

Year	Month	Precip. (in.)	Runoff (in.)	Temp. (°F)	Evap. (in.)	Year	Month	Precip. (in.)	Runoff (in.)	Temp. (°F)	Evap. (in.)
1969	1	5.78	3.47	38.2	2.71	1971	1	7.93	3.38	40.8	1.90
	2	5.77	4.90	41.1	2.12		2	8.15	4.87	41.3	3.41
	3	4.75	4.02	43.3	3.22		3	7.23	5.33	44.9	3.52
	4	6.97	5.24	58.2	5.22		4	3.27	3.68	55.6	5.48
	5	5.72	4.43	63.3	5.77		5	4.09	3.24	62.2	5.29
	6	6.68	5.24	72.3	6.33		6	6.84	2.21	72.0	6.14
	7	4.83	2.52	75.2	7.05		7	8.66	2.72	71.5	4.56
	8	8.44	4.16	70.4	4.87		8	3.91	4.36	71.9	5.25
	9	8.86	3.91	65.4	4.30		9	6.84	2.20	68.1	3.80
	10	3.12	3.07	59.0	3.17		10				
	11	5.99	3.63	49.1	1.82		11				
	12	5.24	3.41	38.5	2.63		12				
1970	1	3.93	3.38	33.9	1.86						
	2	3.07	3.05	41.9	3.06						
	3	5.14	3.44	49.2	4.14						
	4	4.97	3.53	59.4	4.18						
	5	3.84	2.43	65.1	5.23						
	6	7.95	3.00	69.8	5.27						
	7	7.72	2.01	73.8	6.42						
	8	5.83	2.15	73.0	5.22						
	9	2.88	1.32	71.2	5.29						
	10	9.85	2.43	60.1	3.14						
	11	4.79	2.89	43.1	1.67						
	12	3.51	2.03	41.6	2.21						

TABLE C-2 Monthly Streamflow for the Chestuee Creek Watershed, 1944–1961

Month	Streamflow
1	1.56
2	3.2[illegible]
3	5.16
4	3.47
5	1.70
6	.85
7	.56
8	.51
9	.3[illegible]
10	.71
11	.47
12	.3[illegible]
13	1.47
14	3.39
15	2.31
16	1.24
17	1.92
18	1.03
19	.55
20	.49
21	.30
22	.4[illegible]
23	.90
24	2.16
25	5.94
26	4.86
27	2.78
28	1.77
29	2.65
30	1.14
31	.91
32	.84
33	1.31
34	.8[illegible]
35	1.35
36	1.51
37	3.96
38	2.82
39	2.48
40	1.55
41	1.03
42	.8[illegible]
43	.67
44	.51
45	.3[illegible]
46	.39
47	.54
48	.74
49	1.35
50	5.72
51	3.52
52	2.27
53	.87
54	.5[illegible]
55	.5[illegible]
56	.51
57	.3[illegible]
58	.8[illegible]
59	3.8[illegible]
60	3.94
61	6.34
62	3.9[illegible]
63	2.86
64	2.65
65	1.7[illegible]
66	.29
67	2.7[illegible]
68	.7[illegible]
69	.6[illegible]
70	.7[illegible]
71	1.3[illegible]
72	1.6[illegible]
73	4.7[illegible]
74	3.9[illegible]
75	4.95
76	1.7[illegible]
77	1.86
78	2.54
79	1.3[illegible]
80	.9[illegible]
81	1.52
82	.9[illegible]
83	.7[illegible]
84	1.[illegible]
85	1.3[illegible]
86	4.35
87	5.7[illegible]
88	4.[illegible]
89	1.[illegible]
90	[illegible]
91	.9[illegible]
92	.59
93	.60
94	.48
95	2.9[illegible]
96	4.[illegible]1
97	4.[illegible]5
98	2.[illegible]1
99	4.55
100	1.6[illegible]
101	.[illegible]2
102	.7[illegible]
103	.4[illegible]
104	[illegible]
105	[illegible]
106	[illegible]
107	.4[illegible]
108	.4[illegible]
109	2.[illegible]7
110	[illegible]
111	.24
112	1.2[illegible]
113	[illegible]
114	1.[illegible]
115	1.5[illegible]
116	.54
117	.40
118	[illegible]
119	[illegible]
120	.5[illegible]
121	7.[illegible]
122	1.55
123	[illegible].66
124	[illegible].1[illegible]
125	1.[illegible]
126	[illegible]
127	[illegible]
128	.54
129	[illegible]
130	[illegible]
131	[illegible]
132	1.[illegible]
133	1.[illegible]
134	[illegible].7[illegible]
135	3.[illegible]8
136	[illegible].71
137	1.[illegible]
138	[illegible]
139	[illegible]
140	.4[illegible]
141	[illegible]
142	.41
143	[illegible]
144	.65
145	.6[illegible]
146	3.6[illegible]
147	3.2[illegible]
148	3.5[illegible]
149	1.55
150	[illegible]
151	[illegible]
152	.56
153	.49
154	.4[illegible]
155	.5[illegible]
156	1.46
157	1.91
158	[illegible].49
159	[illegible].93
160	[illegible].36
161	.97
162	.72
163	.52
164	.5[illegible]
165	.67
166	.8[illegible]
167	4.63
168	3.74
169	2.21
170	2.52
171	2.5[illegible]
172	2.56
173	[illegible].3[illegible]
174	.7[illegible]
175	.92
176	[illegible]
177	[illegible]
178	[illegible]
179	.3[illegible]
180	.3[illegible]
181	1.[illegible]
182	1.[illegible]
183	[illegible]
184	[illegible]
185	1.[illegible]
186	.7[illegible]
187	.5[illegible]
188	.3[illegible]
189	.4[illegible]
190	.4[illegible]
191	1.[illegible]
192	[illegible]
193	[illegible]
194	[illegible]
195	[illegible]
196	1.7[illegible]
197	.[illegible]1
198	.[illegible]4
199	.44
200	.[illegible]1
201	.5[illegible]
202	.51
203	.47
204	[illegible]
205	1.[illegible]
206	4.[illegible]
207	4.[illegible]5
208	[illegible].59
209	1.51
210	1.[illegible]
211	1.[illegible]
212	.7[illegible]
213	.4[illegible]
214	.3[illegible]
215	.37
216	7.32

TABLE C-3 Wildcat Creek Watershed Data

Gage No. 2.2050, Wildcat Creek near Lawrenceville, GA
DA = 1.59 sq. mi.

Class A

Year	Month	Precip. (in.)	Runoff (in.)	Temp. °F	Evap. (in.)	Year	Month	Precip. (in.)	Runoff (in.)	Temp. °F	Evap. (in.)
1953	1					1955	1	5.01	1.20	43.6	1.66
	2						2	5.91	1.92	47.6	2.95
	3						3	2.50	0.96	56.1	4.46
	4						4	4.27	0.90	65.9	6.99
	5						5	2.55	0.53	72.7	7.71
	6						6	2.92	0.34	72.9	7.42
	7						7	3.61	0.32	79.1	7.74
	8						8	2.14	0.12	80.1	6.85
	9						9	1.05	0.05	75.4	4.47
	10	1.25	0.34	64.0	4.38		10	1.29	0.09	61.9	4.51
	11	1.74	0.45	52.2	2.36		11	3.92	0.20	50.7	3.48
	12	8.17	1.59	43.8	2.44		12	1.29	0.23	43.4	2.07
1954	1	6.43	2.99	45.2	2.07	1956	1	2.13	0.32	41.5	1.61
	2	2.22	1.20	50.1	3.84		2	6.09	1.09	51.2	2.57
	3	4.22	1.35	53.1	4.82		3	7.50	2.52	53.9	4.30
	4	3.06	1.02	65.5	5.92		4	4.29	1.43	61.0	5.80
	5	2.50	0.62	64.2	6.46		5	3.51	2.32	72.4	6.33
	6	4.27	0.38	78.0	8.98		6	2.00	0.63	76.8	7.50
	7	2.63	0.15	82.1	8.30		7	4.83	2.47	79.3	6.84
	8	1.62	0.09	82.5	8.36		8	3.19	0.49	80.4	6.51
	9	0.51	0.09	77.8	7.44		9	7.11	0.84	72.0	5.40
	10	0.44	0.05	64.5	5.14		10	3.61	0.45	64.8	3.27
	11	3.23	0.24	49.5	2.70		11	2.75	0.68	50.5	1.86
	12	3.54	0.38	43.7	2.16		12	3.26	0.89	54.3	1.48

1957	1	4.49	1.48	45.9	1.68	1959	1	3.99	0.85	42.2	2.51
	2	2.82	1.14	54.7	3.03		2	3.35	1.22	47.7	3.54
	3	3.82	1.30	51.5	5.19		3	5.93	1.37	51.2	4.85
	4	5.47	2.93	64.3	6.96		4	3.84	1.01	63.0	6.37
	5	3.70	1.02	70.8	7.36		5	5.38	0.73	72.2	8.65
	6	3.34	0.45	77.8	7.79		6	2.89	0.94	76.6	9.24
	7	2.55	0.14	79.0	8.45		7	3.89	0.24	79.1	7.94
	8	0.92	0.06	79.5	8.22		8	2.02	0.07	80.6	8.32
	9	4.10	0.13	73.3	4.95		9	1.85	0.14	73.3	5.76
	10	2.29	0.35	58.4	3.41		10	5.02	0.31	64.8	5.52
	11	7.14	1.40	53.7	3.02		11	1.48	0.30	51.8	4.10
	12	3.46	1.36	46.7	2.47		12	2.93	0.54	46.3	2.33
1958	1	3.59	1.15	38.9	2.24	1960	1	8.62	2.21	44.2	2.50
	2	4.90	1.72	37.8	2.42		2	4.36	2.40	43.8	3.28
	3	6.03	2.21	49.8	4.17		3	4.61	1.81	41.8	4.76
	4	4.68	2.42	62.4	6.48		4	3.19	1.75	63.7	7.00
	5	2.55	0.94	70.7	7.33		5	1.93	0.73	68.5	8.71
	6	2.32	0.36	77.0	8.01		6	2.79	0.68	76.9	9.66
	7	5.93	0.59	78.8	8.39		7	1.02	0.28	80.3	9.24
	8	2.23	0.26	78.9	7.22		8	4.40	0.17	78.9	6.91
	9	2.25	0.13	74.0	7.24		9	5.95	0.28	74.0	5.77
	10	3.24	0.24	62.0	4.69		10	3.33	0.38	65.7	4.10
	11	2.37	0.28	56.5	4.05		11	1.66	0.33	52.4	3.22
	12	2.30	0.44	42.7	2.50		12	1.95	0.46	40.7	3.19

TABLE C-3 (*cont.*)

Gage No. 2-2050, Wildcat Creek near Lawrenceville, GA
DA = 1.59 sq. mi.

					Class A						
Year	Month	Precip. (in.)	Runoff (in.)	Temp. °F	Evap. (in.)	Year	Month	Precip. (in.)	Runoff (in.)	Temp. °F	Evap. (in.)
1961	1	2.55	0.54	38.9	2.42	1963	1	5.26	1.92	37.3	2.45
	2	11.18	4.02	48.8	2.81		2	2.91	1.44	39.1	3.15
	3	5.58	1.88	54.4	4.40		3	7.79	2.90	54.7	6.14
	4	5.79	2.03	55.3	6.89		4	7.55	2.51	61.2	7.14
	5	3.44	1.12	64.7	6.37		5	3.90	1.22	66.8	7.90
	6	5.52	1.32	72.2	7.41		6	9.86	2.08	73.9	7.40
	7	4.34	0.74	76.0	8.79		7	3.84	0.83	76.2	8.03
	8	5.16	1.11	74.1	6.15		8	0.50	0.23	78.0	8.09
	9	0.85	0.40	72.3	5.58		9	4.45	0.33	72.0	6.46
	10	0.00	0.33	60.1	5.16		10	0.00	0.29	64.7	5.41
	11	2.83	0.42	54.7	3.27		11	5.43	0.59	52.5	3.56
	12	10.82	2.85	41.7	2.73		12	4.51	1.40	37.5	2.49
1962	1	4.56	1.87	40.6	2.80	1964	1	8.51	2.49	41.9	2.71
	2	5.26	2.35	49.3	3.34		2	5.22	2.14	40.6	2.64
	3	4.98	2.06	47.5	4.53		3	10.20	4.60	50.8	5.28
	4	7.46	2.54	56.4	6.42		4	10.92	4.90	59.7	6.48
	5	0.60	0.71	74.0	10.31		5	2.87	2.46	70.3	8.04
	6	4.09	0.52	74.6	6.71		6	3.90	0.60	77.8	9.17
	7	7.13	0.38	78.7	8.05		7	6.95	1.02	77.2	6.91
	8	3.46	0.29	75.7	7.82		8	3.44	0.46	76.1	6.64
	9	3.83	0.23	70.4	6.65		9	1.89	0.26	72.1	7.42
	10	4.83	0.38	63.1	4.60		10	5.49	1.01	58.4	4.40
	11	6.80	1.37	49.2	2.87		11	3.08	0.81	54.8	3.51
	12	3.22	0.99	39.9	2.07		12	6.71	1.98	46.5	2.34

1965	1	3.93	1.50	45.5	3.61
	2	4.48	1.84	45.5	4.62
	3	8.51	2.93	50.4	4.08
	4	4.45	1.58	64.2	6.39
	5	3.84	0.97	71.6	8.58
	6	6.85	1.57	73.3	6.10
	7	2.11	0.62	77.5	7.50
	8	2.64	0.24	77.7	7.82
	9	4.39	0.30	72.7	5.35
	10	3.71	0.76	60.3	4.17
	11	3.16	0.52	55.4	3.17
	12	1.18	0.45	46.4	2.93
1966	1	6.20	1.50	39.3	2.16
	2	8.62	3.10	44.9	2.30
	3	5.27	2.55	50.6	5.18
	4	6.48	2.01	60.1	6.18
	5	6.26	2.66	67.6	6.38
	6	2.33	1.09	72.6	7.79
	7	2.05	0.60	79.3	8.16
	8	3.02	0.63	76.6	6.59
	9	1.95	0.32	70.9	5.92
	10	6.13	1.35	59.7	3.93
	11	4.47	1.30	53.1	2.88
	12	4.82	1.31	45.0	2.45
1967	1	5.19	2.45	45.6	2.47
	2	4.85	1.91	42.5	2.26
	3	3.32	1.85	56.9	5.66
	4	5.21	1.70	64.3	7.67
	5	5.41	1.32	66.5	7.78
	6	4.16	1.17	72.4	6.98
	7	10.03	2.55	74.4	6.16
	8	6.47	1.11	74.6	6.47
	9	2.15	0.60	67.8	4.84
	10	3.10	0.75	60.1	4.41
	11	7.90	1.86	48.9	3.35
	12	6.38	2.22	48.4	2.92
1968	1	6.55	2.61	40.5	2.41
	2	1.14	1.07	39.7	3.77
	3	4.72	2.12	52.3	5.35
	4	6.04	1.77	61.8	4.50
	5	5.03	1.14	66.7	7.08
	6	2.10	0.65	75.3	10.32
	7	5.14	0.74	77.5	7.44
	8	1.20	0.28	79.6	7.75
	9	3.64	0.31	70.4	5.72
	10	1.92	0.27	60.9	4.52
	11	6.83	0.77	50.6	3.62
	12	5.97	1.36	40.7	2.35

TABLE C-3 (*cont.*)

Gage No. 2.2050, Wildcat Creek near Lawrenceville, GA
DA = 1.59 sq. mi.

Class A

Year	Month	Precip. (in.)	Runoff (in.)	Temp. °F	Evap. (in.)	Year	Month	Precip. (in.)	Runoff (in.)	Temp. °F	Evap. (in.)
1969	1	5.19	1.76	41.6	3.46	1971	1	5.43	0.98	42.9	3.26
	2	3.18	1.23	43.5	4.55		2	4.04	1.68	44.3	5.23
	3	5.10	1.71	47.4	5.38		3	6.84	3.66	47.5	6.34
	4	6.97	3.18	62.7	6.03		4	3.82	1.21	60.8	6.89
	5	7.31	1.52	69.0	7.39		5	2.02	0.85	66.7	8.16
	6	0.85	0.60	78.4	8.23		6	2.93	0.35	77.2	8.96
	7	4.27	0.75	82.6	9.19		7	5.67	1.29	76.3	7.80
	8	3.94	0.91	76.2	6.70		8	5.69	1.12	76.7	6.56
	9	6.21	1.05	70.7	5.92		9	6.19	0.50	73.7	5.14
	10	1.01	0.76	63.1	4.64		10				
	11	3.24	1.09	50.0	3.06		11				
	12	4.64	1.21	43.2	3.44		12				
1970	1	4.29	1.33	35.9	1.86						
	2	2.13	1.24	43.9	3.55						
	3	5.59	2.30	52.9	4.70						
	4	2.53	1.10	64.4	6.67						
	5	2.82	0.86	70.1	9.19						
	6	5.00	0.51	74.6	8.15						
	7	5.19	0.41	78.7	9.24						
	8	0.50	0.63	79.1	7.31						
	9	2.20	0.16	77.0	7.16						
	10	4.92	0.22	65.2	4.65						
	11	0.61	0.25	49.5	3.09						
	12	2.45	0.50	47.4	3.10						

TABLE C-4 Conejos River Watershed Data for 1979

CONEJOS RIVER NEAR MAGOTE,CO -- DRAINAGE AREA 282 SQ. MI.

DAILY CLIMATOLOGICAL DATA FOR ZONE 1

DAY	APRIL			MAY			JUNE			JULY			AUGUST			SEPTEMBER		
	DEGREE DAYS	PREC. IN.	SNOW COVER	DEGREE DAYS	PREC. IN.	SNOW COVER	DEGREE DAYS	PREC. IN.	SNOW COVER	DEGREE DAYS	PREC. IN.	SNOW COVER	DEGREE DAYS	PREC. IN.	SNOW COVER	DEGREE DAYS	PREC. IN.	SNOW COVER
1	0.00	0.00	1.000	8.70	0.02	0.430	12.75	0.00	0.000	25.45	0.00	0.000	24.49	0.00	0.000	25.66	0.00	0.000
2	0.00	0.00	1.000	7.12	0.00	0.411	15.37	0.11	0.000	24.57	0.00	0.000	27.16	0.00	0.000	23.13	0.00	0.000
3	0.00	0.00	1.000	5.04	0.00	0.399	17.51	0.00	0.000	24.95	0.00	0.000	25.95	0.00	0.000	25.07	0.00	0.000
4	0.00	0.00	0.989	9.27	0.09	0.380	19.84	0.00	0.000	23.57	0.00	0.000	26.72	0.00	0.000	27.07	0.00	0.000
5	4.49	0.00	0.975	14.68	0.00	0.360	22.04	0.00	0.000	24.49	0.00	0.000	28.13	0.00	0.000	25.45	0.00	0.000
6	10.43	0.00	0.960	18.39	0.00	0.343	23.54	0.00	0.000	24.99	0.10	0.000	29.30	0.00	0.000	27.16	0.00	0.000
7	9.16	0.00	0.950	14.98	1.12	0.329	20.22	0.00	0.000	24.87	0.00	0.000	27.92	0.00	0.000	27.37	0.00	0.000
8	9.63	0.00	0.935	8.89	0.29	0.310	13.18	0.10	0.000	27.34	0.00	0.000	26.98	0.00	0.000	26.95	0.00	0.000
9	4.44	0.19	0.920	0.00	0.26	0.295	8.37	0.00	0.000	27.57	0.00	0.000	24.60	0.08	0.000	25.54	0.00	0.000
10	0.00	0.54	0.900	0.00	0.16	0.280	12.78	0.00	0.000	29.57	0.00	0.000	24.22	0.22	0.000	28.57	0.00	0.000
11	0.00	0.49	0.889	0.00	0.00	0.263	20.57	0.00	0.000	29.40	0.00	0.000	23.34	0.00	0.000	26.57	0.01	0.000
12	0.00	0.36	0.870	4.66	0.00	0.250	25.25	0.00	0.000	30.69	0.00	0.000	24.98	0.02	0.000	25.49	0.02	0.000
13	0.00	0.00	0.852	10.94	0.00	0.235	27.04	0.00	0.000	29.84	0.00	0.000	21.95	0.03	0.000	18.40	0.00	0.000
14	8.51	0.00	0.832	13.62	0.00	0.215	25.34	0.00	0.000	29.25	0.02	0.000	22.60	0.51	0.000	15.59	0.31	0.000
15	13.01	0.00	0.811	12.70	0.00	0.200	24.45	0.00	0.000	28.13	0.00	0.000	19.37	0.44	0.000	13.11	0.00	0.000
16	18.61	0.00	0.799	10.36	0.00	0.190	23.25	0.00	0.000	25.83	0.00	0.000	20.10	0.03	0.000	17.31	0.00	0.000
17	16.63	0.00	0.775	14.84	0.00	0.173	18.83	0.00	0.000	27.95	0.15	0.000	20.28	0.01	0.000	19.69	0.00	0.000
18	13.37	0.12	0.751	16.92	0.00	0.160	16.91	0.00	0.000	25.14	0.00	0.000	17.90	0.08	0.000	20.81	0.00	0.000
19	6.22	0.08	0.730	18.72	0.22	0.145	15.78	0.00	0.000	24.93	0.00	0.000	17.28	0.00	0.000	19.11	0.00	0.000
20	5.95	0.00	0.708	16.69	0.34	0.130	21.40	0.00	0.000	25.19	0.00	0.000	17.04	0.00	0.000	22.92	0.00	0.000
21	10.10	0.00	0.680	15.11	0.20	0.120	23.99	0.00	0.000	27.07	0.10	0.000	17.45	0.00	0.000	18.40	0.13	0.000
22	11.77	0.00	0.655	18.28	0.00	0.105	26.25	0.00	0.000	26.16	0.04	0.000	22.28	0.00	0.000	18.11	0.00	0.000
23	15.75	0.00	0.620	18.84	0.05	0.095	23.95	0.00	0.000	26.78	0.00	0.000	23.49	0.00	0.000	20.87	0.00	0.000
24	12.92	0.00	0.595	16.66	0.13	0.081	24.66	0.30	0.000	29.57	0.20	0.000	23.40	0.00	0.000	21.16	0.00	0.000
25	13.01	0.00	0.565	17.04	0.31	0.069	25.52	0.00	0.000	27.52	0.20	0.000	21.28	0.00	0.000	22.66	0.00	0.000
26	10.30	0.06	0.540	18.13	0.65	0.059	26.23	0.02	0.000	27.61	0.00	0.000	22.25	0.23	0.000	21.19	0.00	0.000
27	10.72	0.00	0.519	19.72	0.04	0.048	29.28	0.00	0.000	28.23	0.00	0.000	19.80	0.00	0.000	19.99	0.00	0.000
28	11.86	0.00	0.495	18.72	0.00	0.035	28.99	0.00	0.000	27.81	0.10	0.000	23.63	0.00	0.000	20.14	0.00	0.000
29	8.72	0.00	0.473	19.80	0.00	0.022	27.81	0.00	0.000	27.93	0.17	0.000	22.13	0.00	0.000	20.19	0.00	0.000
30	6.77	0.00	0.450	14.92	0.19	0.011	25.75	0.00	0.000	27.99	0.00	0.000	21.72	0.10	0.000	22.23	0.00	0.000
31				15.25	0.00	0.000				24.28	0.00	0.000	23.28	0.00	0.000			

TABLE C-4 *(cont.)*

CONEJOS RIVER NEAR MAGOTE,CO -- DRAINAGE AREA 282 SQ. MI.

DAILY CLIMATOLOGICAL DATA FOR ZONE 2

DAY	APRIL			MAY			JUNE			JULY			AUGUST			SEPTEMBER		
	DEGREE DAYS	PREC. IN.	SNOW COVER	DEGREE DAYS	PREC. IN.	SNOW COVER	DEGREE DAYS	PREC. IN.	SNOW COVER	DEGREE DAYS	PREC. IN.	SNOW COVER	DEGREE DAYS	PREC. IN.	SNOW COVER	DEGREE DAYS	PREC. IN.	SNOW COVER
1	0.00	0.00	1.000	0.00	0.00	1.000	7.43	0.00	0.460	20.36	0.00	0.000	20.55	0.00	0.000	20.80	0.00	0.000
2	0.00	0.00	1.000	0.00	0.00	1.000	10.74	0.11	0.420	20.18	0.00	0.000	22.30	0.00	0.000	17.11	0.00	0.000
3	0.00	0.00	1.000	0.00	0.00	1.000	10.80	0.00	0.385	19.86	0.00	0.000	20.86	0.00	0.000	20.68	0.00	0.000
4	0.00	0.00	1.000	1.17	0.02	1.000	14.05	0.00	0.350	19.18	0.00	0.000	20.24	0.00	0.000	22.68	0.00	0.000
5	0.55	0.00	1.000	7.05	0.00	1.000	16.49	0.00	0.320	20.55	0.00	0.000	22.11	0.00	0.000	20.36	0.00	0.000
6	8.12	0.00	1.000	10.98	0.00	1.000	17.99	0.00	0.299	21.05	0.10	0.000	22.36	0.00	0.000	22.30	0.00	0.000
7	4.30	0.00	1.000	7.11	2.38	1.000	13.74	0.00	0.270	20.24	0.00	0.000	21.67	0.00	0.000	22.74	0.00	0.000
8	3.61	0.00	1.000	1.48	0.57	1.000	5.55	0.10	0.245	21.55	0.00	0.000	19.11	0.00	0.000	21.86	0.00	0.000
9	0.00	0.39	1.000	0.00	0.55	0.990	3.74	0.00	0.220	23.18	0.00	0.000	17.42	0.08	0.000	19.99	0.00	0.000
10	0.00	1.14	1.000	0.00	0.25	0.980	8.62	0.00	0.200	25.18	0.00	0.000	17.74	0.22	0.000	24.18	0.00	0.000
11	0.00	1.04	1.000	0.00	0.00	0.971	16.18	0.00	0.182	25.93	0.00	0.000	17.55	0.00	0.000	22.18	0.01	0.000
12	0.00	0.76	1.000	0.00	0.00	0.960	19.93	0.00	0.165	26.99	0.00	0.000	17.11	0.02	0.000	21.55	0.02	0.000
13	0.00	0.00	1.000	1.92	0.00	0.950	21.49	0.00	0.149	24.05	0.00	0.000	16.86	0.03	0.000	14.93	0.00	0.000
14	1.80	0.00	1.000	3.67	0.00	0.940	19.55	0.00	0.130	23.93	0.02	0.000	15.42	0.51	0.000	15.12	0.31	0.000
15	6.30	0.00	1.000	2.29	0.00	0.929	19.36	0.00	0.115	22.11	0.00	0.000	14.74	0.44	0.000	9.87	0.00	0.000
16	15.37	0.00	1.000	1.79	0.00	0.912	17.93	0.00	0.103	20.05	0.00	0.000	12.92	0.03	0.000	14.31	0.00	0.000
17	10.61	0.00	1.000	9.05	0.00	0.900	13.05	0.00	0.090	22.86	0.15	0.000	16.12	0.01	0.000	15.99	0.00	0.000
18	8.74	0.25	1.000	10.67	0.00	0.886	6.73	0.00	0.079	23.06	0.00	0.000	14.43	0.08	0.000	17.81	0.00	0.000
19	0.00	0.17	1.000	12.24	0.45	0.870	11.62	0.00	0.065	22.62	0.00	0.000	13.12	0.00	0.000	15.87	0.00	0.000
20	0.86	0.00	1.000	12.99	0.30	0.851	17.93	0.00	0.055	21.49	0.00	0.000	11.49	0.00	0.000	16.67	0.00	0.000
21	2.92	0.00	1.000	11.87	0.32	0.838	20.05	0.00	0.045	22.68	0.10	0.000	12.36	0.00	0.000	14.93	0.13	0.000
22	3.67	0.00	1.000	14.12	0.00	0.818	20.93	0.00	0.032	21.30	0.04	0.000	18.12	0.00	0.000	14.87	0.00	0.000
23	10.43	0.00	1.000	13.05	0.07	0.798	18.86	0.00	0.022	22.62	0.00	0.000	19.55	0.00	0.000	16.24	0.00	0.000
24	6.67	0.00	1.000	11.80	0.23	0.775	19.80	0.30	0.015	25.18	0.20	0.000	19.93	0.00	0.000	16.30	0.00	0.000
25	6.30	0.00	1.000	11.49	0.67	0.750	22.74	0.00	0.008	24.74	0.20	0.000	17.12	0.00	0.000	17.80	0.00	0.000
26	3.36	0.13	1.000	12.11	0.26	0.720	23.68	0.02	0.000	24.37	0.00	0.000	16.93	0.23	0.000	17.49	0.00	0.000
27	4.24	0.00	1.000	13.24	0.09	0.690	25.12	0.00	0.000	25.68	0.00	0.000	12.86	0.00	0.000	16.05	0.00	0.000
28	3.29	0.00	1.000	12.24	0.00	0.660	25.05	0.00	0.000	24.81	0.10	0.000	17.61	0.00	0.000	18.06	0.00	0.000
29	2.24	0.00	1.000	12.86	0.00	0.620	24.81	0.00	0.000	25.62	0.17	0.000	16.11	0.00	0.000	16.49	0.00	0.000
30	0.00	0.00	1.000	8.67	0.39	0.580	20.43	0.00	0.000	24.05	0.00	0.000	15.24	0.10	0.000	19.68	0.00	0.000
31				9.93	0.00	0.520				20.12	0.00	0.000	19.12	0.00	0.000			

CONEJOS RIVER NEAR MAGOTE,CO -- DRAINAGE AREA 282 SQ. MI.

DAILY CLIMATOLOGICAL DATA FOR ZONE 3

DAY	APRIL DEGREE DAYS	APRIL PREC. IN.	APRIL SNOW COVER	MAY DEGREE DAYS	MAY PREC. IN.	MAY SNOW COVER	JUNE DEGREE DAYS	JUNE PREC. IN.	JUNE SNOW COVER
1	0.00	0.00	1.000	0.00	0.00	1.000	2.17	0.00	0.970
2	0.00	0.00	1.000	0.00	0.00	1.000	6.17	0.41	0.960
3	0.00	0.00	1.000	0.00	0.00	1.000	4.17	0.00	0.950
4	0.00	0.00	1.000	0.00	0.00	1.000	8.34	0.30	0.938
5	0.00	0.00	1.000	0.00	0.00	1.000	11.01	0.00	0.925
6	4.84	0.00	1.000	3.68	0.00	1.000	12.51	0.00	0.915
7	0.00	0.00	1.000	0.00	3.63	1.000	7.34	0.32	0.900
8	0.00	0.00	1.000	0.00	0.84	1.000	0.00	0.84	0.890
9	0.00	0.58	1.000	0.00	0.84	1.000	0.00	0.00	0.879
10	0.00	1.73	1.000	0.00	0.35	1.000	4.50	0.00	0.863
11	0.00	1.59	1.000	0.00	0.00	1.000	11.84	0.00	0.850
12	0.00	1.16	1.000	0.00	0.00	1.000	14.67	0.00	0.839
13	0.00	0.00	1.000	0.00	0.00	1.000	16.01	0.00	0.820
14	0.00	0.00	1.000	0.00	0.00	1.000	13.84	0.11	0.809
15	0.00	0.00	1.000	0.00	0.00	1.000	14.34	0.00	0.794
16	11.01	0.00	1.000	0.00	0.00	1.000	12.67	0.45	0.780
17	4.67	0.00	1.000	3.34	0.00	1.000	7.34	0.00	0.760
18	4.17	0.37	1.000	4.51	0.00	1.000	0.00	0.00	0.749
19	0.00	0.25	1.000	5.84	0.67	1.000	7.50	0.00	0.730
20	0.00	0.00	1.000	9.34	0.29	1.000	14.50	0.00	0.710
21	0.00	0.00	1.000	8.67	0.43	1.000	16.17	0.00	0.695
22	0.00	0.00	1.000	10.00	0.00	1.000	15.67	0.00	0.675
23	5.17	0.00	1.000	7.34	0.08	1.000	13.84	0.00	0.658
24	0.51	0.00	1.000	7.01	0.32	1.000	15.01	0.00	0.640
25	0.00	0.00	1.000	6.01	1.01	1.000	20.00	0.00	0.620
26	0.00	0.20	1.000	6.17	0.16	1.000	21.17	0.00	0.600
27	0.00	0.00	1.000	6.84	0.13	1.000	21.00	0.00	0.580
28	0.00	0.00	1.000	5.84	0.00	1.000	21.17	0.00	0.560
29	0.00	0.00	1.000	6.01	0.00	1.000	21.84	0.00	0.540
30	0.00	0.00	1.000	2.51	0.60	0.990	15.17	0.00	0.520
31				4.67	0.00	0.980			

DAY	JULY DEGREE DAYS	JULY PREC. IN.	JULY SNOW COVER	AUGUST DEGREE DAYS	AUGUST PREC. IN.	AUGUST SNOW COVER	SEPTEMBER DEGREE DAYS	SEPTEMBER PREC. IN.	SEPTEMBER SNOW COVER
1	15.34	0.00	0.500	16.67	0.00	0.013	16.01	0.00	0.000
2	15.84	0.00	0.478	17.51	0.00	0.005	11.17	0.00	0.000
3	14.84	0.05	0.455	15.84	0.00	0.000	16.34	0.00	0.000
4	14.84	0.00	0.435	13.84	0.00	0.000	18.34	0.00	0.000
5	16.67	0.00	0.415	16.17	0.00	0.000	15.34	0.00	0.000
6	17.17	0.00	0.395	15.51	0.00	0.000	17.51	0.00	0.000
7	15.67	0.00	0.375	15.51	0.00	0.000	18.17	0.00	0.000
8	15.84	0.00	0.355	11.34	0.33	0.000	16.84	0.00	0.000
9	18.84	0.00	0.335	10.34	0.11	0.000	14.51	0.00	0.000
10	20.84	0.00	0.312	11.34	0.17	0.000	19.84	0.00	0.000
11	22.50	0.00	0.299	11.84	1.33	0.000	17.84	0.00	0.000
12	23.34	0.05	0.280	9.34	0.78	0.000	17.67	0.00	0.000
13	18.34	0.03	0.260	11.84	0.32	0.000	11.50	0.00	0.000
14	18.67	0.06	0.245	8.34	0.44	0.000	14.50	0.42	0.000
15	16.17	0.07	0.220	10.17	0.60	0.000	6.67	0.00	0.000
16	14.34	0.00	0.211	5.84	0.11	0.000	11.34	0.00	0.000
17	17.84	0.00	0.200	12.00	0.00	0.000	12.34	0.00	0.000
18	21.00	0.00	0.183	11.00	0.00	0.000	14.84	0.00	0.000
19	20.34	0.00	0.169	9.00	0.00	0.000	12.67	0.00	0.000
20	17.84	0.01	0.159	6.01	0.00	0.000	10.51	1.18	0.000
21	18.34	0.07	0.140	7.34	0.00	0.000	11.50	0.00	0.000
22	16.51	0.00	0.129	14.00	0.00	0.000	11.67	0.00	0.000
23	18.50	0.00	0.115	15.67	0.00	0.000	11.67	0.00	0.000
24	20.84	0.00	0.100	16.50	0.00	0.000	11.51	0.00	0.000
25	22.00	0.00	0.090	13.00	0.10	0.000	13.01	0.00	0.000
26	21.17	0.00	0.079	11.67	0.00	0.000	13.84	0.00	0.000
27	23.17	0.00	0.069	6.01	0.00	0.000	12.17	0.00	0.000
28	21.84	3.10	0.055	11.67	0.00	0.000	16.00	0.00	0.000
29	23.34	0.02	0.045	10.17	0.00	0.000	12.84	0.00	0.000
30	20.17	0.01	0.033	8.84	0.00	0.000	17.17	0.00	0.000
31	16.00	0.00	0.025	15.00	0.00	0.000			

TABLE C-4 *(cont.)*

CONEJOS RIVER NEAR MAGOTE,CO -- DRAINAGE AREA 282 SQ. MI.

DAY	STREAMFLOW FOR APRIL		STREAMFLOW FOR MAY		STREAMFLOW FOR JUNE		STREAMFLOW FOR JULY		STREAMFLOW FOR AUGUST		STREAMFLOW FOR SEPTEMBER	
	COMPUTED CFS	ACTUAL CFS	COMPUTED CFS	ACTUAL CFS	COMPUTED CFS	ACTUAL CFS	COMPUTED CFS	ACTUAL CFS	COMPUTED CFS	ACTUAL CFS	COMPUTED CFS	ACTUAL CFS
1	63.	60.	425.	590.	1973.	2210.	2057.	2235.	411.	324.	92.	101.
2	58.	55.	390.	675.	1893.	1890.	1934.	2200.	364.	288.	84.	98.
3	54.	51.	356.	520.	2013.	2095.	1827.	2090.	321.	278.	76.	93.
4	50.	53.	328.	640.	2037.	2240.	1723.	1735.	284.	288.	70.	88.
5	48.	57.	342.	620.	2271.	2230.	1626.	1640.	252.	271.	64.	84.
6	69.	60.	446.	730.	2592.	2485.	1555.	1530.	224.	257.	59.	75.
7	145.	79.	626.	820.	2878.	2875.	1499.	1560.	200.	229.	54.	68.
8	187.	98.	734.	780.	2805.	2940.	1420.	1530.	182.	218.	50.	68.
9	216.	128.	685.	640.	2396.	2395.	1350.	1490.	180.	257.	47.	62.
10	205.	116.	594.	506.	2078.	1690.	1306.	1310.	179.	368.	43.	60.
11	183.	101.	518.	448.	2091.	2540.	1273.	1050.	199.	336.	40.	60.
12	164.	91.	456.	416.	2400.	1785.	1250.	1120.	245.	328.	38.	55.
13	148.	82.	425.	392.	2777.	2205.	1225.	1070.	260.	288.	36.	55.
14	140.	88.	421.	469.	3118.	2650.	1167.	990.	263.	268.	39.	60.
15	172.	109.	432.	980.	3310.	2845.	1116.	960.	316.	299.	61.	77.
16	262.	174.	422.	810.	3420.	3080.	1041.	1060.	353.	320.	57.	72.
17	446.	250.	434.	920.	3232.	2765.	934.	985.	321.	344.	52.	66.
18	546.	360.	600.	980.	2856.	2440.	863.	795.	286.	292.	48.	64.
19	594.	402.	808.	1405.	2430.	2070.	788.	720.	261.	212.	45.	60.
20	535.	378.	1074.	1665.	2241.	1630.	719.	616.	232.	177.	49.	59.
21	492.	378.	1380.	1630.	2207.	1670.	657.	566.	207.	177.	84.	64.
22	478.	418.	1624.	1605.	2196.	1860.	614.	561.	185.	158.	83.	70.
23	489.	486.	1859.	1680.	2161.	2080.	562.	690.	166.	151.	76.	70.
24	570.	506.	1988.	1965.	2104.	2175.	517.	530.	149.	142.	70.	64.
25	579.	534.	2079.	1925.	2110.	2055.	498.	489.	135.	131.	64.	57.
26	576.	530.	2158.	2010.	2144.	2090.	477.	480.	130.	122.	59.	48.
27	551.	488.	2220.	2200.	2178.	2185.	434.	476.	136.	151.	54.	48.
28	532.	583.	2249.	2325.	2190.	2255.	432.	468.	123.	125.	50.	48.
29	508.	619.	2225.	2485.	2192.	2210.	567.	468.	111.	109.	46.	46.
30	472.	588.	2188.	2670.	2172.	2260.	524.	460.	102.	109.	43.	45.
31			2076.	2360.			465.	384.	101.	109.		

TABLE D-1 Algorithm for Rainfall–Runoff Model of Chapter 8

```
  1*  C      RAINFALL-RUNOFF MODEL
  2*         DIMENSION IQDAYS(12),QSTOR1(10),QSTOR2(10)
  3*         DIMENSION PREC(366),OBSRO(366),TOTFLO(366),SRO(4),SROS(800)
  4*         DATA BLANK,DOT,X1,X2,X3,X4/' ','.','X','+','O','P'/
  5*         READ 12,GWS,SUBFLO,PINF,SROP,GWSM,BFPM,SFP,PET,SAT
  6*  C       CHECK LIMITS ON PARAMETERS
  7*         IF(PINF.GT.1.0) PINF=1.0
  8*         IF (PINF.LT.0.0) PINF=0.0
  9*         IF(SFP.GT.1.0)SFP=1.0
 10*         IF(SFP.LT.0.0)SFP=0.0
 11*  C      COMPUTE COEFFICIENTS FOR ROUTING SURFACE RO UNIT HYDROGRAPH
 12*         SUM=0.0
 13*         SRO(1)=1.0
 14*         SRO(2)=SROP
 15*         SRO(3)=SROP**2
 16*         SRO(4)=SROP**3
 17*         DO 14 I=1,4
 18*     14     SUM=SUM+SRO(I)
 19*         DO 16 I=1,4
 20*     16     SRO(I)=SRO(I)/SUM
 21*         PRINT 18,GWS,PET,SAT,SFP,      SUBFLO,GWSM,PINF,BFPM,SROP,(SRO(I),I=1,4)
 22*        I=1,4)
 23*     18  FORMAT(5X,'INITIAL PARAMETER ESTIMATES'/5X,'GWS =',F10.4/
 24*        1    5X,'PET = ',F10.4/ 5X,'SAT =',F10.4/5X,'SFP =',F10.4/
 25*        1                                   5X,'SUBFLO=',F10.5/5X,'GWSM='
 26*        1   ,F10.4/
 27*        15X,'PINF =',F10.4/5X,'BFPM=',F10.4/5X,'SROP =',F10.4///5X,
 28*        2'UNIT HYDROGRAPH ORDINATES',4F10.5/)
 29*      1  CONTINUE
 30*         READ 6,(IQDAYS(I),I=1,11)
 31*      6  FORMAT(10I5)
 32*         IDDD=1
 33*         READ 10,(PREC(I),I=1,365)
 34*         READ 10,(OBSRO(I),I=1,365)
 35*  C*******************************************************************************
 36*  C      THE FOLLOWING SHOULD BE USED ONLY WHEN CONVERTING CFS TO INCHES/DAY
 37*         DO 13 I=1,365
 38*     13  OBSRO(I)=0.00010953*OBSRO(I)
 39*  C*******************************************************************************
 40*     10  FORMAT(16F5.3)
 41*     12  FORMAT(9F5.3)
 42*  C      INITIALIZE ANNUAL SUMS
 43*         SUMOBS=0.0
 44*         SUMFLO=0.0
 45*         SUMPRE=0.0
 46*         SUMSFL=0.0
 47*         SUMEV=0.0
 48*         SUMBF=0.0
 49*         SUMRO=0.0
 50*         S2=0.0
 51*         S3=0.0
 52*         S4=0.0
 53*         DO 30 L=1,365
 54*            SWAT=PINF*PREC(L)
 55*            SFOS(L)=(1.0-PINF)*PREC(L)
 56*            GWP=0.0
```

TABLE D-1 (*cont.*)

```
 57*                IF(GWS.LT.GWSM) GWP=1.0-(GWS/GWSM)**2
 58*                GWS=GWS+GWAT*GWP
 59*                SUBFLO=SUBFLO+(1.0-GWP)*GWAT
 60*                ETCO=1.0
 61*                IF(SUBFLO.GT.SAT) GO TO 21
 62*                XX=SUBFLO/SAT
 63*                ETCO=0.25+1.5*(XX-1.0/3.0)
 64*                IF(XX.LT.0.3333) ETCO=0.75*XX
 65*                IF(XX.GT.0.6667) ETCO=0.75+0.75*(XX-2.0/3.0)
 66*       21       ET=ETCO*PET
 67*                IF(ET.GT.SUBFLO)ET=SUBFLO
 68*                EV=ET
 69*                SUBFLO=SUBFLO-ET
 70*                EXCESS=0.0
 71*                IF(GWS.GT.GWSM)EXCESS=GWS-GWSM
 72*                GWS=GWS-EXCESS
 73*                XX=GWS/GWSM
 74*                BFP=0.25+13.0*(XX-0.5)/7.0
 75*                IF(XX.LT.0.5)BFP=0.5*XX
 76*                IF(XX.GT.0.85)BFP=0.9+(XX-0.85)/1.5
 77*                BF=BFP*BFPM
 78*                IF(BF.GT.GWS)BF=GWS
 79*                GWS=GWS-BF
 80*                BF=BF+EXCESS
 81*                SFLO=SFP*SUBFLO
 82*                SUBFLO=SUBFLO-SFLO
 83*    C       COMPUTE SURFACE RUNOFF
 84*               IF (L.EQ.1) RO=SRO(1)*SROS(1)
 85*               IF (L.EQ.2) RO=SRO(1)*SROS(2)+SRO(2)*SROS(1)
 86*               IF (L.EQ.3) RO=SRO(1)*SROS(3)+SRO(2)*SROS(2)+SRO(3)*SROS(1)
 87*               IF (L.LT.4) GO TO 24
 88*               RO=0.0
 89*               DO 22 I=1,4
 90*       22      RO=RO+SRO(I)*SROS(L-I+1)
 91*       24   CONTINUE
 92*             IF(RO.LT.0.0)RO=0.0
 93*                TOTFLO(L)=RO+BF+SFLO
 94*                SUMSFL=SUMSFL+SFLO
 95*            IF(L.NE.IQDAYS(IDDD)) GO TO 20
 96*            QSTOR1(IDDD)=TOTFLO(L)
 97*            QSTOR2(IDDD)=OBSRO(L)
 98*            IDDD=IDDD+1
 99*            IF(IDDD.GT.10) IDDD=10
100*       20   CONTINUE
101*            SUMFLO=SUMFLO+TOTFLO(L)
102*            SUMPRE=SUMPRE+PREC(L)
103*            SUMEV=SUMEV+EV
104*            SUMBF=SUMBF+BF
105*            SUMRO=SUMRO+RO
106*            SUMOBS=SUMOBS+OBSRO(L)
107*            S2=S2+TOTFLO(L)**2
108*            S3=S3+OBSRO(L)**2
```

TABLE D-1 (*cont.*)

```
109*            S4=S4+TOTFLO(L)*OBSRO(L)
110*         31 CONTINUE
111*      C     COMPUTE SUMMARY STATISTICS
112*            SDO=((S3-SUMOBS**2/365.0)/365.0)**0.5
113*            SDP=((S2-SUMFLO**2/365.0)/365.0)**0.5
114*            R=(S4-SUMFLO*SUMOBS/365.0)/(365.0*SDO*SDP)
115*            RM=SDO*R/SDP
116*            IF(SDO.GT.SDP)RM=SDP*R/SDO
117*            PRINT 32,SUMPRE,SUMOBS,SUMFLO,SUMRO,SUMBF,SUMSFL,SUMEV,R,RM
118*         32 FORMAT(/5X,'ANNUAL TOTALS'/F10.4,2X,'ANNUAL PRECIP'/F10.4,2X,
119*           1'OBSERVED RUNOFF'/F10.4,2X,'PREDICTED RUNOFF'/F10.4,2X,
120*           2'SURFACE RUNOFF'/F10.4,2X,'BASEFLOW'/,F10.4,2X,'SUBSURFACE FLOW'/
121*           3     F10.4,2X,'EVAPORATION'
122*           3/F10.5,2X,'CORRELATION COEFFICIENT'
123*           1/F10.5,2X,'MODIFIED CORRELATOIN COEFF')
124*            PRINT 34,GWS,SUBFLO,(SROS(I),I=362,365)
125*         34 FORMAT(///'  ND OF YAR STORAG S'/5X,'GWS=',F10.4/5X,'SUBFLO=',F10.4
126*           14/5X,'SURFAC  RUNOFF STORAGR=',4F10.4/)
127*            BAL=SUMPRE-SUMEV-SUMFLO
128*            PRINT 35,BAL
129*         35 FORMAT(/5X,'ANNUAL WATER BALANCE=',F10.5/)
130*            IF(IQDAYS(1).EQ.0) GO TO 41
131*            PRINT 36
132*         36 FORMAT(///)
133*            DO 39 I=1,IDDD
134*         39 PRINT 40,IQDAYS(I),QSTOR1(I),QSTOR2(I)
135*         40 FORMAT(I5,2F15.2)
136*         41 CONTINUE
137*            CALL PLOT4(PREC,PREC,OBSRO,TOTFLO,365,365,365,365,400,400,400,
138*           1    400,80.,40.)
139*            CALL PLOT4(PREC,PREC,OBSRO,TOTFLO,365,365,365,365,400,400,400,
140*           1    400,80.,40.)
141*            GO TO 1
142*            END
```

TABLE D-2a Precipitation (in.) for Washington, D.C., 1962–1963

DAY	MONTH											
	OCT	NOV	DEC	JAN	FEB	MAR	APR	MAY	JUN	JUL	AUG	SEP
1	.00	.00	.00	.00	.00	.36	.34	.00	.00	.00	.05	.00
2	.00	.00	.00	.00	.00	.00	.01	.00	.26	.00	.00	.00
3	.00	.00	.00	.00	.00	.00	.00	.00	1.94	.00	.07	.10
4	2.11	.00	.00	.16	.00	.05	.00	.00	.00	.00	.00	.00
5	.00	1.62	.00	.00	.00	.25	.00	.06	.31	.00	.00	.00
6	.00	.00	.95	.00	.00	1.25	.00	.00	.00	.00	.00	.00
7	.00	.00	.00	.00	.00	.00	.00	.00	.76	.00	.00	.00
8	.00	.00	.00	.00	.00	.00	.03	.00	.00	.00	.00	.00
9	.00	1.35	.00	.00	.00	.00	.07	.00	.00	.00	.00	.00
10	.00	.43	.00	.00	.00	.23	.00	.04	.07	.00	.00	.00
11	.00	.00	.00	.24	.21	.00	.00	.00	.00	.00	.00	.00
12	.00	.00	.00	.42	.73	1.55	.00	.00	.00	.00	.00	.00
13	.00	.00	.00	.32	.00	.00	.00	.00	.00	.00	[illegible].10	.00
14	.00	.00	.00	.20	.00	.00	.00	.21	.27	.68	.01	.00
15	.00	.00	.00	.00	.00	.00	.13	.00	.00	.00	.00	.70
16	.00	.00	.00	.00	.00	.43	.00	.00	.00	.00	.00	.70
17	.00	.00	.00	.00	.00	.27	.10	.50	.00	.00	.00	.00
18	.00	.75	.00	.00	.00	.30	.00	.07	.00	.00	.00	.00
19	.00	.00	.00	.27	.70	.74	.00	.00	.00	.00	.00	.00
20	.00	.15	.00	.11	.00	.02	.25	.00	.00	.00	.00	.00
21	.00	.92	.00	.03	.00	.00	.00	.23	.35	.00	.00	.10
22	.00	.40	.55	.00	.00	.00	.14	.00	.00	.06	[illegible].50	.00
23	.00	.00	.00	.00	.00	.00	.33	.00	.00	.00	.00	.00
24	.00	.00	.00	.00	.17	.00	.00	.00	.00	.00	.00	.00
25	.14	.00	.76	.00	.00	.00	.00	.00	.00	.00	.00	.00
26	.01	.00	.00	.00	.00	.00	.00	.00	.00	.00	.00	.00
27	.00	.00	.00	.45	.00	.07	.00	.00	.00	.00	.00	.00
28	.00	.00	.00	.00	.00	.00	.00	.03	.00	.00	.00	.00
29	.00	.00	.35	.00		.00	.29	.40	1.01	.06	.74	.20
30	.00	.00	.00	.32		.00	.15	.00	.32	.00	.00	.00
31	.00		.00	.07		.00		.00		.00		

TABLE D-2b Precipitation (in.) for Washington, D.C., 1963–1964

	MONTH											
DAY	OCT	NOV	DEC	JAN	FEB	MAR	APR	MAY	JUN	JUL	AUG	SEP
1	.00	1.10	.00	.70	1.50	.00	.00	.00	.00	.00	.00	.00
2	.00	.10	.00	.00	.00	.00	.40	.00	.00	.00	.00	.00
3	.00	.00	.10	.30	.00	.00	.00	.00	.00	.00	.00	.00
4	.00	.00	.00	.00	.00	.00	.00	.00	.00	.00	.00	.00
5	.00	.00	.00	.00	.00	.00	.00	.00	.00	.00	.00	.00
6	.00	2.50	.00	.00	1.00	.00	.60	.00	.40	.00	.00	.00
7	.00	.30	.00	.30	.10	.00	.40	.00	.00	.00	.00	.00
8	.00	.00	.70	.00	.00	.00	.00	.00	.00	.00	.20	.00
9	.00	.00	.00	1.50	.00	.00	.00	.00	.00	.00	.00	.00
10	.00	.00	.00	.00	.00	.00	.00	.00	.00	.00	.00	.00
11	.00	.00	.00	.00	.20	.20	.00	.00	.00	.00	.00	.00
12	.00	.00	.20	.10	.00	.00	.00	.00	.00	.00	.00	.00
13	.00	.00	.00	.00	.10	.00	.00	.20	.00	.00	.00	.40
14	.00	.00	.30	.00	.00	.00	.50	.00	.50	.00	.00	.00
15	.00	.00	.00	.00	.40	.00	.10	.00	.20	.00	.00	.00
16	.00	.00	.00	.00	.30	.00	.00	.00	.00	.00	.00	.00
17	.00	.00	.00	.00	.00	.00	.00	.00	.00	.00	.00	.00
18	.00	.00	.10	.00	.70	.00	.00	.00	.00	.00	.00	.00
19	.00	.00	.00	.00	.50	.00	.40	.00	.10	.00	.00	1.50
20	.00	.00	.00	.00	.00	.00	.40	.00	.20	.00	.00	.00
21	.00	.00	.00	.00	.00	1.10	.00	.00	.00	.00	.00	.00
22	.00	.00	.00	.00	.00	.30	.10	.00	.00	.00	.00	.00
23	.40	.00	.30	.00	.00	.50	.10	.00	.00	.00	.00	.00
24	.00	.00	.00	.00	.00	.10	.00	.00	.00	.00	.00	.00
25	.00	.00	.00	.20	.00	.00	.00	.00	.00	.00	.00	.00
26	.00	.00	.00	.00	.00	.00	.00	.00	.00	.00	.00	.00
27	.00	.00	.00	.00	.00	.00	.00	.00	.00	.00	.00	.00
28	.00	.00	.00	.00	.00	.00	.10	.00	.00	.00	.00	.70
29	.00	1.40	.00	.00		.00	.00	.00	.00	.20	.00	.80
30	.00	.30	.00	.00		.20	.00	.00	.00	.00	.00	.70
31	.00		.00	.00		.10		.00		.00		

TABLE D-2c Streamflow (cfs) for Anacostia River at Colesville, 1962–1963

DAY	OCT	NOV	DEC	JAN	FEB	MAR	APR	MAY	JUN	JUL	AUG	SEP
	MONTH											
1	7.	3.	8.	7.	9.	13.	13.	14.	7.	12.	3.	3.
2	3.	5.	9.	3.	12.	15.	13.	11.	19.	6.	3.	3.
3	3.	12.	8.	9.	11.	14.	16.	11.	160.	5.	3.	3.
4	21.	10.	9.	9.	11.	14.	14.	9.	45.	4.	3.	3.
5	11.	8.	8.	8.	24.	73.	12.	9.	193.	4.	3.	4.
6	5.	7.	45.	9.	27.	253.	13.	10.	95.	4.	1.	4.
7	4.	6.	22.	9.	15.	50.	13.	9.	36.	4.	1.	3.
8	4.	5.	14.	8.	10.	23.	12.	8.	13.	3.	2.	3.
9	4.	8.	12.	9.	10.	19.	13.	3.	15.	3.	1.	3.
10	4.	111.	9.	10.	11.	19.	13.	3.	13.	3.	1.	3.
11	4.	15.	9.	13.	15.	13.	13.	3.	11.	3.	1.	3.
12	4.	9.	7.	107.	108.	134.	11.	7.	9.	3.	1.	3.
13	4.	9.	7.	37.	25.	53.	11.	7.	8.	3.	19.	3.
14	4.	9.	7.	35.	23.	32.	11.	3.	8.	12.	11.	3.
15	4.	8.	7.	13.	15.	23.	11.	8.	10.	7.	3.	3.
16	4.	7.	8.	11.	13.	47.	11.	7.	8.	4.	3.	11.
17	4.	5.	8.	11.	13.	72.	11.	3.	8.	4.	3.	5.
18	4.	24.	8.	13.	11.	31.	13.	12.	7.	3.	3.	4.
19	4.	10.	8.	14.	13.	47.	11.	8.	6.	3.	6.	3.
20	4.	11.	8.	46.	43.	73.	10.	9.	11.	3.	130.	3.
21	4.	23.	7.	29.	30.	30.	10.	9.	12.	3.	54.	3.
22	4.	112.	10.	15.	12.	23.	13.	7.	7.	3.	9.	4.
23	4.	19.	8.	16.	9.	15.	14.	7.	6.	3.	6.	3.
24	4.	17.	8.	13.	9.	12.	11.	7.	6.	3.	5.	3.
25	4.	11.	7.	10.	9.	12.	13.	7.	5.	3.	4.	3.
26	4.	13.	9.	9.	9.	50.	13.	6.	5.	2.	4.	3.
27	4.	9.	8.	9.	9.	30.	13.	7.	5.	2.	4.	3.
28	4.	9.	8.	9.	10.	13.	9.	7.	5.	3.	3.	3.
29	4.	9.	7.	7.		17.	10.	9.	4.	3.	4.	66.
30	4.	9.	7.	9.		16.	16.	16.	17.	2.	5.	9.
31	5.		7.	9.		16.		7.		2.		

TABLE D-2d Streamflow (cfs) for Anacostia River at Colesville, 1963–1964

DAY	OCT	NOV	DEC	JAN	FEB	MAR	APR	MAY	JUN	JUL	AUG	SEP
1	[illegible]	7.	24.	15.	24.	27.	22.	30.	9.	4.	2.	2.
2	5.	13.	17.	24.	32.	46.	26.	29.	9.	4.	4.	1.
3	5.	6.	15.	36.	17.	100.	37.	24.	9.	4.	18.	1.
4	4.	5.	17.	31.	15.	54.	24.	22.	9.	5.	9.	1.
5	4.	5.	12.	36.	16.	43.	21.	21.	8.	4.	4.	1.
6	4.	50.	12.	42.	75.	26.	37.	19.	17.	4.	4.	1.
7	4.	263.	10.	230.	70.	23.	47.	19.	13.	4.	5.	1.
8	[illegible]	23.	24.	37.	43.	25.	91.	18.	12.	13.	6.	1.
9	[illegible]	14.	30.	295.	35.	34.	44.	17.	9.	7.	4.	1.
10	4.	11.	13.	47.	22.	25.	27.	15.	9.	5.	3.	1.
11	4.	9.	14.	[illegible]	15.	40.	24.	15.	7.	5.	4.	1.
12	4.	9.	17.	[illegible]	17.	20.	22.	15.	7.	5.	4.	1.
13	7.	9.	17.	[illegible]	17.	19.	22.	22.	7.	19.	3.	1.
14	7.	8.	12.	17.	21.	25.	45.	20.	11.	6.	7.	1.
15	7.	7.	14.	15.	24.	29.	30.	15.	11.	5.	3.	1.
16	7.	7.	13.	[illegible]	51.	22.	27.	15.	12.	4.	3.	1.
17	3.	7.	13.	14.	44.	21.	23.	14.	7.	4.	4.	1.
18	[illegible]	7.	10.	14.	39.	18.	[illegible]	17.	7.	4.	4.	1.
19	7.	7.	11.	[illegible]	45.	17.	22.	13.	9.	4.	3.	5.
20	3.	7.	10.	[illegible]	51.	17.	31.	12.	16.	4.	3.	6.
21	4.	7.	9.	[illegible]	35.	38.	45.	11.	7.	4.	2.	2.
22	4.	7.	9.	47.	24.	115.	29.	13.	7.	4.	3.	3.
23	4.	22.	9.	32.	20.	57.	24.	11.	7.	4.	3.	3.
24	4.	21.	[illegible]	[illegible]	13.	31.	22.	11.	12.	3.	2.	2.
25	4.	11.	11.	131.	13.	26.	[illegible]	10.	9.	4.	3.	1.
26	4.	8.	17.	42.	34.	25.	13.	9.	6.	4.	2.	1.
27	4.	5.	17.	34.	32.	22.	13.	9.	5.	4.	1.	1.
28	4.	9.	13.	23.	25.	23.	24.	9.	5.	3.	1.	4.
29	4.	118.	15.	[illegible]		53.	24.	9.	5.	4.	3.	10.
30	5.	120.	9.	13.		22.	114.	9.	5.	3.	2.	24.
31	4.		[illegible]	13.		22.		9.		2.		

TABLE D-3a Precipitation (in.) for Anacostia River at Riverdale, 1963–1964

DAY	MONTH											
	OCT	NOV	DEC	JAN	FEB	MAR	APR	MAY	JUN	JUL	AUG	SEP
1	.00	.00	.00	.00	.00	.00	.00	.00	.00	.00	1.00	.10
2	.40	.00	.20	.10	.00	.00	.00	.00	.00	.00	.00	.00
3	.00	.00	.30	.00	.00	.00	.00	.00	.00	.00	.00	.00
4	.00	.00	.10	.00	.00	.70	.00	.00	.00	.00	.30	.00
5	.00	.00	.50	.00	.00	2.50	.00	.00	.00	.30	.00	.00
6	.00	.00	.10	.00	.00	.00	.00	.00	.00	.00	.00	.00
7	.00	.00	.00	.10	1.30	.00	.00	.00	.00	.00	.00	.00
8	.00	.00	.00	.00	.00	.10	.10	.00	.00	.00	.00	.00
9	.00	.00	.00	.40	.10	.00	.10	.00	.00	.00	.00	.00
10	.00	.00	.00	.70	.10	.00	.00	.00	.00	.40	.00	.00
11	.00	.20	.00	.00	.00	.00	.00	.00	.00	.90	.00	.00
12	.00	.00	1.20	.00	.10	.00	.00	.00	.00	.00	.00	.50
13	.00	.00	.00	.00	.00	.00	.00	.00	.00	.00	.00	1.10
14	.10	.00	.00	.00	.10	.00	.00	.00	.00	.00	.00	.00
15	.00	.00	.00	.00	.00	.00	.40	.00	.00	.00	.00	.00
16	.60	.00	.00	.10	.20	.00	.00	.30	.00	.00	.00	.00
17	.20	.00	.10	.00	.00	.60	.00	.00	.10	.00	.00	.00
18	.00	.00	.00	.00	.00	.10	.00	.00	.00	.30	.20	.00
19	.00	.30	.00	.00	.00	.00	.00	.00	.00	.00	.10	.00
20	.00	.00	.40	.00	.00	.00	.00	.00	.00	.00	.00	.00
21	.00	.00	.00	.00	.00	.00	.00	.00	.00	.00	.40	.00
22	.00	.00	.00	.00	.00	.00	.00	.00	.00	.00	.00	.00
23	.00	.00	.00	.40	.40	.00	.00	.00	.10	.00	.00	.00
24	.00	.00	.00	.00	.50	.00	.10	.20	.00	.00	.00	.00
25	.00	1.50	.00	.00	.70	.70	.10	.00	.00	.00	.00	.00
26	.60	.00	.60	.00	.00	.30	.00	.00	.00	.00	1.80	.00
27	.00	.00	.20	.00	.00	.00	.10	.00	.00	.00	.00	.00
28	.00	.00	.00	.00	.00	.00	.00	.00	.00	.00	.00	.00
29	.00	.00	.00	.00		.40	.00	.00	.00	.00	.00	.00
30	.00	.00	.00	.10		.00	.00	.00	.00	.00	.00	.00
31	.00		.00	.00		.00		.00		.00		

TABLE D-3b Streamflow (cfs) for Anacostia River at Riverdale, 1963–1964

DAY	MONTH											
	OCT	NOV	DEC	JAN	FEB	MAR	APR	MAY	JUN	JUL	AUG	SEP
1	35.	15.	29.	44.	47.	47.	73.	45.	31.	10.	39.	14.
2	151.	13.	26.	42.	44.	45.	73.	43.	32.	9.	27.	15.
3	61.	16.	44.	41.	42.	43.	73.	53.	73.	10.	15.	12.
4	35.	22.	45.	37.	39.	43.	64.	35.	38.	11.	14.	14.
5	24.	27.	37.	33.	37.	1520.	57.	32.	26.	25.	26.	13.
6	20.	23.	121.	31.	35.	803.	65.	32.	27.	17.	15.	12.
7	16.	21.	55.	31.	388.	180.	84.	46.	19.	10.	12.	11.
8	15.	22.	42.	113.	437.	136.	76.	44.	19.	12.	21.	11.
9	14.	18.	35.	105.	173.	83.	94.	38.	20.	10.	32.	11.
10	14.	17.	31.	120.	123.	74.	71.	34.	20.	18.	17.	9.
11	13.	19.	30.	130.	102.	65.	64.	30.	16.	739.	11.	11.
12	17.	20.	301.	124.	84.	59.	57.	26.	15.	110.	9.	27.
13	14.	20.	105.	77.	71.	55.	53.	25.	15.	47.	9.	187.
14	14.	21.	69.	66.	61.	53.	49.	24.	17.	26.	9.	50.
15	14.	21.	45.	56.	57.	51.	71.	23.	17.	19.	9.	27.
16	38.	21.	36.	45.	57.	50.	61.	39.	26.	17.	8.	21.
17	35.	22.	55.	45.	55.	64.	62.	45.	20.	15.	9.	16.
18	32.	20.	34.	45.	51.	107.	55.	29.	20.	98.	57.	17.
19	23.	32.	30.	42.	47.	149.	67.	24.	18.	25.	26.	15.
20	18.	54.	35.	42.	32.	121.	67.	21.	15.	14.	19.	14.
21	17.	34.	45.	38.	37.	98.	55.	21.	17.	10.	22.	13.
22	16.	27.	45.	47.	35.	86.	47.	22.	13.	10.	62.	17.
23	16.	23.	40.	65.	30.	76.	46.	21.	18.	10.	22.	12.
24	16.	27.	36.	394.	31.	69.	57.	31.	21.	11.	14.	26.
25	13.	186.	36.	230.	143.	60.	54.	24.	19.	10.	12.	25.
26	16.	159.	45.	127.	79.	392.	54.	23.	19.	9.	279.	14.
27	15.	65.	743.	79.	67.	202.	52.	19.	17.	9.	143.	11.
28	15.	43.	283.	77.	55.	106.	53.	28.	11.	9.	37.	11.
29	15.	35.	111.	61.		146.	46.	125.	9.	8.	19.	11.
30	14.	32.	67.	52.		119.	47.	45.	11.	8.	14.	10.
31	14.		51.	52.		67.		26.		9.		

Index

A

Absolute sensitivity, 406-7
Acceleration, 285
Accuracy, 69
Alternative hypothesis, 42–48, 220
Analysis, 8–10
Analysis of variance, 77, 79–80
Analytical optimization, 71–73, 305–6
Anisotropy, 160–62
Autocorrelation, 363, 366, 373–78
Autocovariance, 374
Autoregression, 382–97
Auxiliary function, 164, 166

B

Best linear unbiased estimation, 184
Bias, 93, 307
Binomial distribution, 33–34

C

Central tendency, 102
Characteristic equation, 449
Chi-square distribution, 32, 45, 83
Circular variates, 245–48
Coefficient of kurtosis, 30
Coefficient of skewness, 30
Coefficient of variation, 30, 225
Components analysis, 443, 456–62
Component sensitivity, 405–6
Components regression, 459–62
Confidence intervals, 34–35, 63, 171, 215–17
Constraints, 74–76
Continuous variable, 25
Convolution, 296
Correlation coefficient, 77, 204–5, 219, 220–21, 223, 374, 407
Correlogram, 374–78
Cross-correlation, 378–80
Cross-regression, 382–97

Cumulative distribution, 20, 55
Cyclical component, 365

D

Data transformation, 130–34
Deceleration, 285–86
Degrees of freedom, 32, 33
Deregularization, 163
Determinant, 449
Deterministic, 4, 201, 380
deWijsian semivariogram, 158
Digamma function, 107
Dimensionless variables, 30
Directional semivariogram, 160–61
Discrete variable, 24–25
Distribution of residuals, 201

E

Efficiency, 105, 205
Eigenstructure, 445
Eigenvalue, 449
Eigenvector, 449
Episodic component, 365
Erosivity, 291
Error of estimate, 384
Error variance, 73
Estimation, 151, 169–83
Exponential density function, 23
Exponential semivariogram, 158
Extreme-value distribution, 110, 124

F

F distribution, 32–33
F test, 77, 80, 220–21
Factor analysis, 477–92
Factorial, 33, 107
Filter, 365
Flowcharts, 61–62
Frequency distributions, 21

G

Gamma distribution, 106, 117, 291, 297, 410
Goodness of fit, 77–78
Graphical analysis, 102, 108

H

Histograms, 20–21
Hypothesis, 42
Hypothesis test, 42–49, 79

I

Incomplete gamma function, 118
Intercorrelation, 211–12, 376, 442

J

Jacobian, 133

K

Kolmogorov–Smirnov test, 47–49, 88–89
Kriging, 151, 170, 184–91, 293, 363
Kurtosis, 28, 30

L

Lag, 374, 376, 379
Lagrangian optimization, 74–77, 185
Least-squares, 73–74, 81–82, 201–5, 262
Level of significance, 42
Linear models, 199, 442, 492
Linear semivariogram, 157
Location parameter, 102
Logarithmic transforms, 243, 299

Logistic function, 291

M

Matrix inversion, 209
Maximum likelihood, 105–8
Mean, 25, 26, 106
Mean, confidence interval, 34–35
Mean, estimation, 152–53
Method of moments, 103–5, 296
Mode, 119
Model, 3
Model formulation, 5
Modeling, 68, 73, 99
Models, semivariogram, 157–59
Moment ratio analysis, 30, 61, 126–29, 380
Moments, 25
Moving average, 365–73
Multiple regression, 203–4, 218
Multivariate, 442–43, 448

N

Nash model, 298, 410
Nonlinear analysis, 492–503
Nonlinear least squares, 264, 271, 304, 492
Nonsystematic, 364, 365, 380
Normal distribution, 31
Normal equations, 74, 186, 203
Normal paper, 109
Nugget effect, 159
Null hypothesis, 42, 80, 220
Numerical optimization, 71, 262–303, 304

O

Objective function, 70, 71, 76, 304, 420
Optimization, analytical, 71–73, 305–6
Optimization, components of, 70
Optimization, numerical, 71, 262–303
Optimization, subjective, 71, 304–61
Orthogonality, 479

P

Parameter estimation, 101–8
Parametric, 4
Partial divided finite differences, 272
Partial regression coefficients, 77, 80, 224
Pattern search, 285
Pearson type III distribution, 119–21
Periodic component, 364
Persistence, 373
Phase angle, 364
Poisson distribution, 114–15
Population, 19, 24, 63, 98
Power semivariogram, 158
Prediction, 10, 100–101, 212–14
Principal components analysis, 456–77
Probabilistic analysis, 432–38
Probability, 19
Probability density function, 20, 98, 380
Probability paper, 108–10

R

Radius of influence, 152–53, 156
Rainfall–runoff model, 308
Random component, 201, 365
Random number generation, 55, 267, 432
Random variable, 25, 101
Rationality, 77, 223
Regression, 69, 73, 201–4, 382, 443

Regression, stepwise, 218–33
Regularization, 163
Relative sensitivity, 406–7
Reliability, 77
Reproducibility, 417
Residuals, 81–83, 201, 223, 225
Response surface, 264, 286–90
Return period, 110
Risk, 110
Rosenbrock method, 286
Rotation of axes, 455, 481

S

Sample, 19, 24, 63, 98
Sampling interval, 374, 376
Scale parameter, 103
Secular component, 364
Semivariogram, 154, 157, 363
Sensitivity, absolute, 406–7
Sensitivity, component, 405–6, 423
Sensitivity, initial value, 414–16
Sensitivity, parametric, 405
Sensitivity, relative, 406–7
Sensitivity analysis, 403–41
Sensitivity equation, 404
Separation of variation, 78
Serial correlation, 373
Shape parameter, 103
Significance level, 42
Sill, 156, 158
Simulation, 9, 49–60
Skewness, 27, 30
Smoothing, 366
Spectral analysis, 363
Spectral matrix, 454
Spherical semivariogram, 158
Standard deviation, 27, 29, 78
Standard error of estimate, 78, 82, 169
Standardized partial regression coefficients, 80–81, 223–24
Standardized variate, 31
Standard normal distribution, 31
Stationarity, 154, 160–61
Stationary points, 72
Stepwise regression, 218–33
Stochastic, 4, 100
Streamflow recession, 280–84
Student *t* distribution, 32, 43
Subjective optimization, 71, 304–61
Synthesis, 9
Systematic component, 364, 365–73, 380
Systematic search, 267, 287

T

t distribution, 32, 43
Taylor series, 273, 288, 404
Test statistic, 42
Time series, 362
Tolerance limit, 35
Trigonometric transform, 243–44
Type II error, 229

U

Unit hydrograph, 287, 296
Univariate analysis, 68, 150, 198

V

Variance, 27, 29, 78
Varimax, 481
Variogram, 155

W

Weighted mean, 152, 366